A Guide to Molecular Mechanics and Quantum Chemical Calculations

Warren J. Hehre

WAVEFUNCTION

Wavefunction, Inc.
18401 Von Karman Ave., Suite 370
Irvine, CA 92612

ISBN 1-890661-18-X

QD461.B754

Printed in the United States of America

Acknowledgements

This book derives from materials and experience accumulated at Wavefunction and Q-Chem over the past several years. Philip Klunzinger and Jurgen Schnitker at Wavefunction and Martin Head-Gordon and Peter Gill at Q-Chem warrant special mention, but the book owes much to members of both companies, both past and present. Special thanks goes to Pamela Ohsan and Philip Keck for turning a "sloppy manuscript" into a finished book.

To the memory of

Edward James Hehre

1912-2002

mentor and loving father.

Preface

Over the span of two decades, molecular modeling has emerged as a viable and powerful approach to chemistry. Molecular mechanics calculations coupled with computer graphics are now widely used in lieu of "tactile models" to visualize molecular shape and quantify steric demands. Quantum chemical calculations, once a mere novelty, continue to play an ever increasing role in chemical research and teaching. They offer the real promise of being able to complement experiment as a means to uncover and explore new chemistry.

There are fundamental reasons behind the increased use of calculations, in particular quantum chemical calculations, among chemists. Most important, the theories underlying calculations have now evolved to a stage where a variety of important quantities, among them molecular equilibrium geometry and reaction energetics, may be obtained with sufficient accuracy to actually be of use. Closely related are the spectacular advances in computer hardware over the past decade. Taken together, this means that "good theories" may now be routinely applied to "real systems". Also, computer software has now reached a point where it can be easily used by chemists with little if any special training. Finally, molecular modeling has become a legitimate and indispensable part of the core chemistry curriculum. Just like NMR spectroscopy several decades ago, this will facilitate if not guarantee its widespread use among future generations of chemists.

There are, however, significant obstacles in the way of continued progress. For one, the chemist is confronted with "too many choices" to make, and "too few guidelines" on which to base these choices. The fundamental problem is, of course, that the mathematical equations which arise from the application of quantum mechanics to chemistry and which ultimately govern molecular structure and properties cannot be solved. Approximations need to be made in order to realize equations that can actually be solved. "Severe" approximations may lead to methods which can be widely applied but may not yield accurate information. Less severe approximations may lead to methods which are more accurate but which are too costly to be routinely applied. In short, no one method of calculation is likely to be ideal for all applications, and the ultimate choice of specific methods rests on a balance between accuracy and cost.

This guide attempts to help chemists find that proper balance. It focuses on the underpinnings of molecular mechanics and quantum chemical methods, their relationship with "chemical observables", their performance in reproducing known quantities and on the application of practical models to the investigation of molecular structure and stability and chemical reactivity and selectivity.

Chapter 1 introduces *Potential Energy Surfaces* as the connection between structure and energetics, and shows how molecular equilibrium and transition-state geometry as well as thermodynamic and kinetic information follow from interpretation of potential energy surfaces. Following this, the guide is divided into four sections:

Section I. Theoretical Models (Chapters 2 to 4)

Chapters 2 and **3** introduce *Quantum Chemical Models* and *Molecular Mechanics Models* as a means of evaluating energy as a function of geometry. Specific models are defined. The discussion is to some extent "superficial", insofar as it lacks both mathematical rigor and algorithmic details, although it does provide the essential framework on which practical models are constructed.

Graphical Models are introduced and illustrated in **Chapter 4**. Among other quantities, these include models for presentation and interpretation of electron distributions and electrostatic potentials as well as for the molecular orbitals themselves. Property maps, which typically combine the electron density (representing overall molecular size and shape) with the electrostatic potential, the local ionization potential, the spin density, or with the value of a particular molecular orbital (representing a property or a reactivity index where it can be accessed) are introduced and illustrated.

Section II. Choosing a Model (Chapters 5 to 11)

This is the longest section of the guide. Individual chapters focus on the performance of theoretical models to account for observable quantities: *Equilibrium Geometries* (**Chapter 5**), *Reaction Energies* (**Chapter 6**), *Vibrational Frequencies and Thermodynamic Quantities* (**Chapter 7**), *Equilibrium Conformations* (**Chapter 8**), *Transition-State Geometries and Activation Energies* (**Chapter 9**) and *Dipole Moments* (**Chapter 10**). Specific examples illustrate each topic, performance statistics and graphical summaries provided and, based on all these, recommendations given. The number of examples provided in the individual chapters is actually fairly small (so as not to completely overwhelm the reader), but additional data are provided as **Appendix A** to this guide.

Concluding this section, *Overview of Performance and Cost* (**Chapter 11**), is material which estimates computation times for a number of "practical models" applied to "real molecules", and provides broad recommendations for model selection.

Section III. Doing Calculations (Chapters 12 to 16)

Because each model has its individual strengths and weaknesses, as well as its limitations, the best "strategies" for approaching "real problems" may involve not a single molecular mechanics or quantum chemical model, but rather a combination of models. For example, simpler (less costly) models may be able to provide equilibrium conformations and geometries for later energy and property calculations using higher-level (more costly) models, without seriously affecting the overall quality of results. Practical aspects or "strategies" are described in this section: *Obtaining and Using Equilibrium Geometries* (**Chapter 12**), *Using Energies for Thermochemical and Kinetic Comparisons* (**Chapter 13**), *Dealing with Flexible Molecules* (**Chapter 14**), *Obtaining and Using Transition-State Geometries* (**Chapter 15**) and *Obtaining and Interpreting Atomic Charges* (**Chapter 16**).

Section IV. Case Studies (Chapters 17 to 19)

The best way to illustrate how molecular modeling may actually be of value in the investigation of chemistry is by way of "real" examples. The first two chapters in this section illustrate situations where "numerical data" from calculations may be of value. Specific examples included have been drawn exclusively from organic chemistry, and have been divided broadly according to category: *Stabilizing "Unstable" Molecules* (**Chapter 17**), and *Kinetically-Controlled Reactions* (**Chapter 18**). Concluding this section is *Applications of Graphical Models* (**Chapter 19**). This illustrates the use of graphical models, in particular, property maps, to characterize molecular properties and chemical reactivities.

In addition to **Appendix A** providing *Supplementary Data* in support of several chapters in **Section II**, **Appendix B** provides a glossary of *Common Terms and Acronyms* associated with molecular mechanics and quantum chemical models.

At first glance, this guide might appear to be a sequel to an earlier book "*Ab Initio* Molecular Orbital Theory"[*], written in collaboration with Leo Radom, Paul Schleyer and John Pople nearly 20 years ago. While there are similarities, there are also major differences. Specifically, the present guide is much broader in its coverage, focusing on an entire range of computational models and not, as in the previous book, almost exclusively on Hartree-Fock models. In a sense, this simply reflects the progress which has been made in developing and assessing new computational methods. It is also a consequence of the fact that more and more "mainstream chemists" have now embraced computation. With this has come an increasing diversity of problems and increased realization that no single method is ideal, or even applicable, to all problems.

The coverage is also more broad in terms of "chemistry". For the most part, "*Ab Initio* Molecular Orbital Theory" focused on the structures and properties of organic molecules, accessible at that time

---

[*] W.J. Hehre, L. Radom, P.v.R. Schleyer and J.A. Pople, *Ab Initio Molecular Orbital Theory*, Wiley, New York, 1985.

using Hartree-Fock models. The present guide, while also strongly embracing organic molecules, also focuses on inorganic and organometallic compounds. This is, of course, a direct consequence of recent developments of methods to properly handle transition metals, in particular, semi-empirical models and density functional models.

Finally, the present guide is much less "academic" and much more "practical" than "*Ab Initio* Molecular Orbital Theory". Focus is not on the underlying elements of the theory or in the details of how the theory is actually implemented, but rather on providing an overview of how different theoretical models fit into the overall scheme. Mathematics has been kept to a minimum and for the most part, references are to monographs and "reviews" rather than to the primary literature.

This pragmatic attitude is also strongly reflected in the last section of the guide. Here, the examples are not so much intended to "show off" interesting chemistry, but rather to illustrate in some detail how computation can assist in elaborating chemistry.

This guide contains a very large quantity of numerical data derived from molecular mechanics and quantum chemical calculations using Spartan, and it is inconceivable that there are not numerous errors. The author alone takes full responsibility.

Finally, although the material presented in this guide is not exclusive to a particular molecular modeling program, it has been written with capabilities (and limitations) of the Spartan program in mind. The CD-ROM which accompanies the guide contains files readable by the Windows version of Spartan, in particular, relating to graphical models and to the example applications presented in the last section. These have been marked in text by the icon ⊗, x indicating the chapter number and y the number of the Spartan file in that chapter.

Table of Contents

Chapter 1

Potential Energy Surfaces

This chapter introduces potential energy surfaces as the connection between molecular structure and energetics.

Introduction

Every chemist has encountered a plot depicting the change in energy of ethane as a function of the angle of torsion about the carbon-carbon bond.

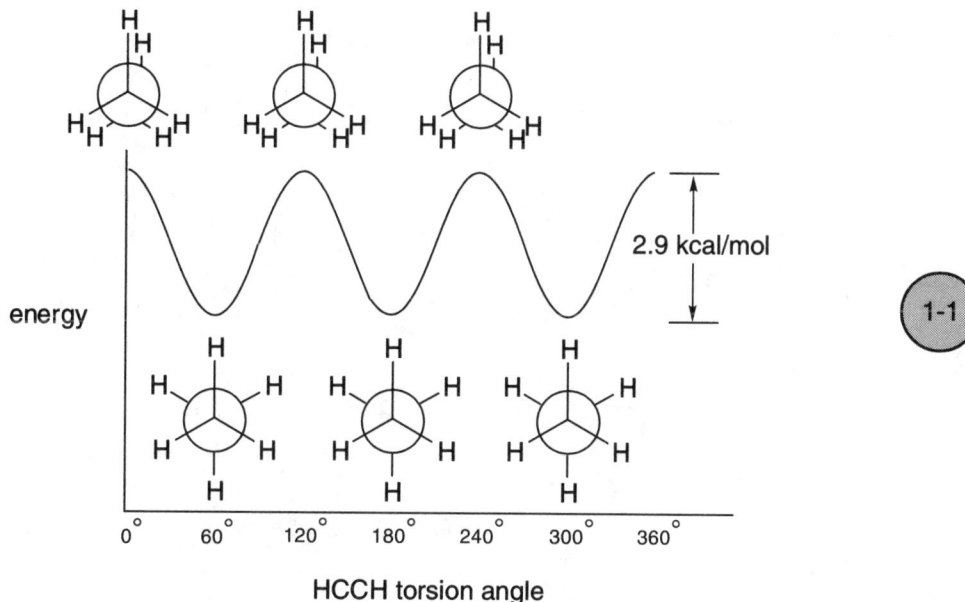

HCCH torsion angle

Full 360° rotation leads to three identical "staggered" structures which are energy minima, and three identical "eclipsed" structures which are energy maxima. The difference in energy between eclipsed and staggered structures of ethane, termed the barrier to rotation, is known experimentally to be 2.9 kcal/mol (12 kJ/mol). Note, that any physical measurements on ethane pertain only to its staggered structure, or

1

more precisely the set of three identical staggered structures. That is to say, eclipsed ethane does not exist in the sense that it is not possible to isolate it or to perform physical measurements on it. Rather, eclipsed ethane can only be "imagined" as a structure in between equivalent staggered forms.

Somewhat more complicated but also familiar is a plot of energy vs. the torsion angle involving the central carbon-carbon bond in *n*-butane.

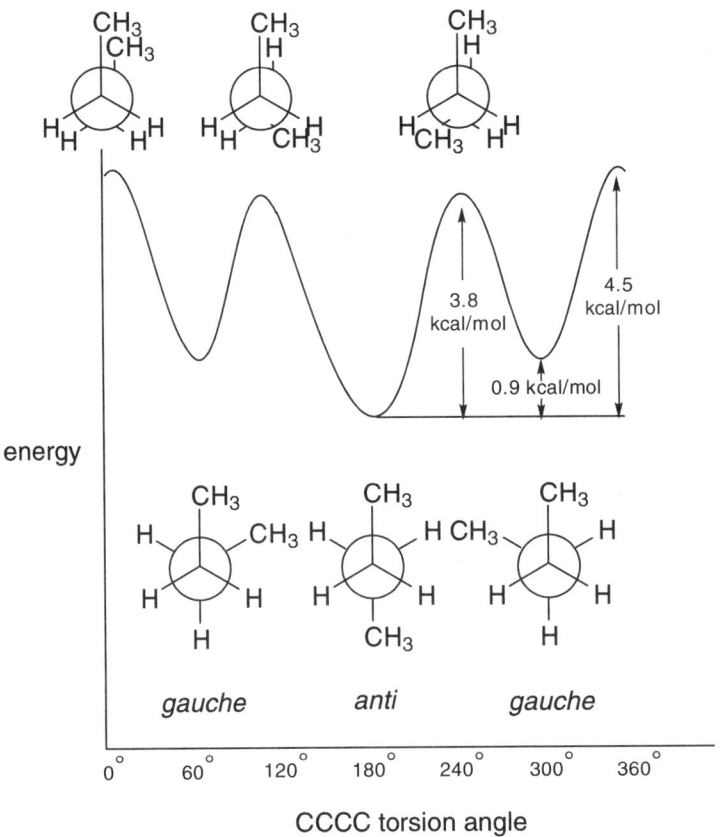

This plot also reveals three energy minima, corresponding to staggered structures, and three energy maxima, corresponding to eclipsed structures. In the case of *n*-butane, however, the three structures in each set are not identical. Rather, one of the minima, corresponding to a torsion angle of 180° (the *anti* structure), is lower in energy and distinct from the other two minima with torsion angles of approximately 60° and 300° (*gauche* structures), which are identical. Similarly, one of the energy maxima corresponding to a torsion angle

2

of 0°, is distinct from the other two maxima with torsion angles of approximately 120° and 240°, which are identical.

As in the case of ethane, eclipsed forms of *n*-butane do not exist, and correspond only to hypothetical structures in between *anti* and *gauche* minima. Unlike ethane, which is a single pure compound, any sample of *n*-butane is made up of two distinct compounds, *anti n*-butane and *gauche n*-butane. The relative abundance of the two compounds as a function of temperature is given by the Boltzmann equation (see discussion following).

The "important" geometrical coordinate in both of the above examples may clearly be identified as a torsion involving one particular carbon-carbon bond. Actually this is an oversimplification as other geometrical changes no doubt also occur during rotation around the carbon-carbon bond, for example, changes in bond lengths and angles. However, these are likely to be small and be safely ignored. However, it will not always be possible to identify a single "simple" geometrical coordinate. A good example of this is provided by the potential energy surface for "ring inversion" in cyclohexane.

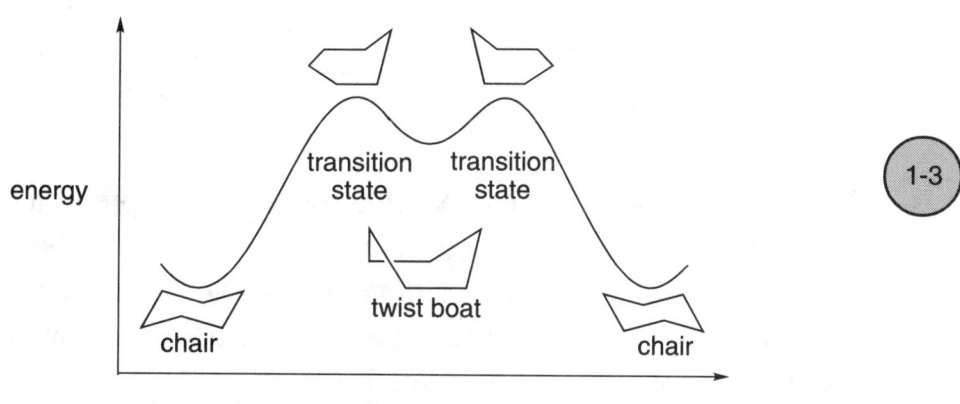

In this case, the geometrical coordinate connecting stable forms is not specified in detail (as in the previous two examples), but is referred to simply as the "reaction coordinate". Also the energy maxima have been designated as "transition states" as an indication that their structures may not be simply described (as the energy maxima for rotation in ethane and *n*-butane).

The energy surface for ring inversion in cyclohexane, like that for *n*-butane, contains three distinct energy minima, two of lower energy identified as "chairs", and one of higher energy identified as a "twist boat". In fact, the energy difference between the chair and twist-boat structures is sufficiently large (5.5 kcal/mol or 23 kJ/mol) that only the former can be observed at normal temperatures.[*]

All six carbons in the chair form of cyclohexane are equivalent, but the hydrogens divide into two sets of six equivalent *equatorial* hydrogens and six equivalent *axial* hydrogens.

However, only one kind of hydrogen can normally be observed, meaning that *equatorial* and *axial* positions interconvert via a low-energy process. This is the ring inversion process just described, in which one side of the ring bends upward while the other side bends downward.

According to the potential energy diagram on the previous page, the overall process actually occurs in two steps, with a twist-boat structure as a midway point (an "intermediate"). The two (equivalent) transition states leading to this intermediate adopt structures in which five of the ring carbons lie (approximately) in one plane.

The energy profile for ring inversion in cyclohexane may be rationalized given what has already been said about single-bond rotation in *n*-butane. Basically, the interconversion of chair cyclohexane into the twist-boat intermediate via the transition state can be viewed as a "restricted rotation" about one of the ring bonds.

---

[*] At room temperature, this would correspond to an equilibrium ratio of chair to twist-boat structures of >99:1.

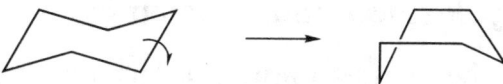

Correspondingly, the interconversion of the twist-boat intermediate into the other chair form can be viewed as rotation about the opposite ring bond. Overall, two independent "bond rotations", pausing at the high-energy (but stable) twist-boat intermediate, effect conversion of one chair structure into another equivalent chair, and at the same time switch *axial* and *equatorial* hydrogens.

Ethane, *n*-butane and cyclohexane all provide examples of the types of motions which molecules may undergo. Their potential energy surfaces are special cases of a general type of plot in which the energy is given as a function of reaction coordinate.

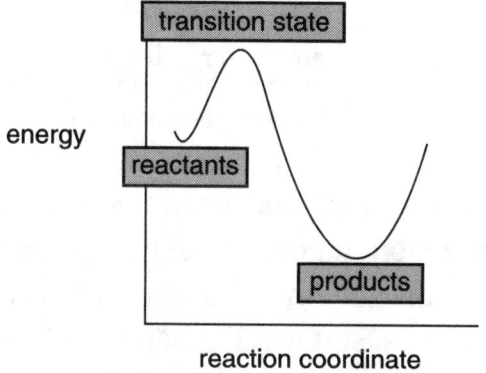

reaction coordinate

Diagrams like this ("reaction coordinate" diagrams) provide essential connections between important chemical observables - structure, stability, reactivity and selectivity - and energy. These connections are explored in the following sections.

Potential Energy Surfaces and Geometry

The positions of the energy minima along the reaction coordinate give the equilibrium structures of the reactants and products. Similarly, the position of the energy maximum gives the structure of the transition state.

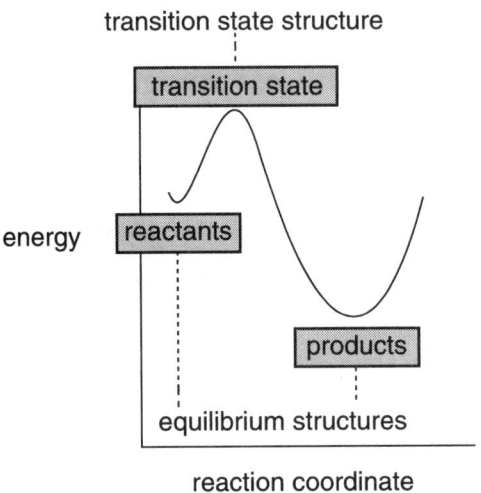

For example, where the "reaction" is rotation about the carbon-carbon bond in ethane, the reaction coordinate may be thought of as simply the HCCH torsion angle, and the structure may be thought of in terms of this angle alone. Thus, staggered ethane (both the reactant and the product) is a molecule for which this angle is 60° and eclipsed ethane is a molecule for which this angle is 0°.

60°   0°   60°

staggered ethane   eclipsed ethane   staggered ethane
"reactant"   "transition state"   "product"

A similar description applies to "reaction" of *gauche* n-butane leading to the more stable *anti* conformer. Again, the reaction coordinate may be thought of as a torsion about the central carbon-carbon bond, and the individual reactant, transition-state and product structures in terms of this coordinate.

| gauche *n*-butane | "transition state" | *anti n*-butane |
| "reactant" | | "product" |

Equilibrium structure (geometry) may be determined from experiment, given that the molecule can be prepared and is sufficiently long-lived to be subject to measurement.* On the other hand, the geometry of a transition state may not be established from measurement. This is simply because "it does not exist" in terms of a population of molecules on which measurements may be performed.

Both equilibrium and transition-state structure may be determined from calculation. The former requires a search for an energy minimum on a potential energy surface while the latter requires a search for an energy maximum. Lifetime or even existence is not a requirement.

---

* Note that where two or more structures coexist, e.g., *anti* and *gauche n*-butane, an experimental measurement can either lead to a single "average" structure or a "composite" of structures.

Potential Energy Surfaces and Thermodynamics

The relative stability of reactants and products is indicated on the potential surface by their relative heights. This gives the thermodynamics of reaction.[*]

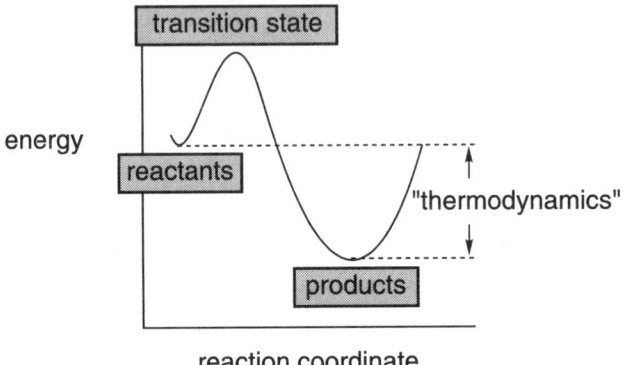

reaction coordinate

In the case of bond rotation in ethane, the "reactants" and "products" are the same and the reaction is said to be "thermoneutral". This is also the case for the overall ring-inversion motion in cyclohexane.

The more common case is, as depicted in the above diagram, where the energy of the products is lower than that of the reactants. This kind of reaction is said to be *exothermic*, and the difference in stabilities of reactant and product is simply the difference in their energies. For example, the "reaction" of *gauche n*-butane to *anti n*-butane is *exothermic*, and the difference in stabilities of the two conformers is simply the difference in their energies (0.9 kcal/mol or 3.8 kJ/mol).

Thermodynamics tells us that "if we wait long enough" the amount of products in an *exothermic* reaction will be greater than the amount

---

[*]   This is not strictly true. Thermodynamics depends on the relative free energies of reactants and products. Free energy is given by the enthalpy, ΔH, minus the product of the entropy, ΔS, and the (absolute) temperature.

$$\Delta G = \Delta H - T\Delta S$$

The difference between enthalpy and energy,

$$\Delta H = \Delta E + \Delta(PV)$$

may safely be ignored under normal conditions. The entropy contribution to the free energy cannot be ignored, although it is typically very small (compared to the enthalpy contribution) for many important types of chemical reactions. Its calculation will be discussed in **Chapter 7**. For the purpose of the present discussion, free energy and energy can be treated equivalently.

of reactants (starting material). The actual ratio of products to reactants also depends on the temperature and is given by the Boltzmann equation.

$$\frac{[products]}{[reactants]} = \exp\left[-(E_{products} - E_{reactants})/kT\right] \tag{1}$$

Here, $E_{products}$ and $E_{reactants}$ are the energies of products and reactants on the potential energy diagram, T is the temperature (in Kelvin) and k is the Boltzmann constant. The Boltzmann equation tells us exactly the relative amounts of products and reactants, [products]/[reactants], at infinite time.

Even small energy differences between major and minor products lead to large product ratios.

| energy difference | | product ratio |
kcal/mol	kJ/mol	major : minor
0.5	2	80 : 20
1	4	90 : 10
2	8	95 : 5
3	12	99 : 1

Chemical reactions can also be *endothermic*, which give rise to a reaction profile.

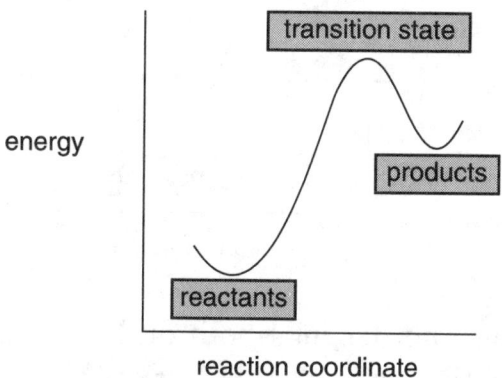

In this case, there would eventually be more reactants than products.

Where two or more different products may form in a reaction, thermodynamics tells us that if "we wait long enough", the product formed in greatest abundance will be that with the lowest energy irrespective of pathway.

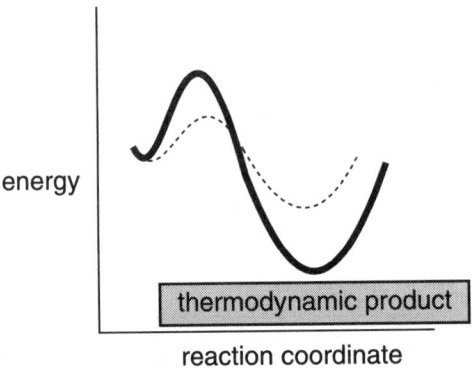

reaction coordinate

In this case, the product is referred to as the "thermodynamic product" and the reaction is said to be "thermodynamically controlled".

Potential Energy Surfaces and Kinetics

A potential energy surface also reveals information about the speed or rate at which a reaction will occur. This is the kinetics of reaction.

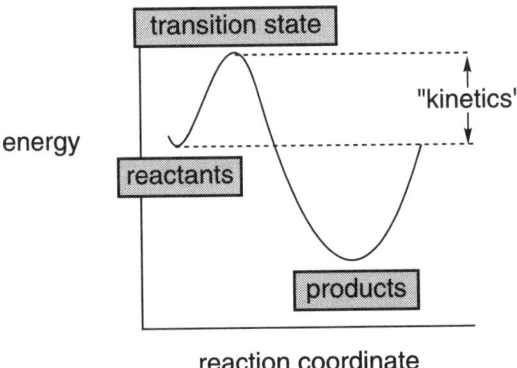

reaction coordinate

Absolute reaction rate depends both on the concentrations of the reactants, $[A]^a$, $[B]^b$..., where a, b... are typically integers or half integers, and a quantity termed the rate constant.

$$\text{rate} = \text{rate constant} \, [A]^a \, [B]^b \, [C]^c... \qquad (2)$$

The rate constant is given by the Arrhenius equation which depends on the temperature[*].

$$\text{rate constant} = A \exp [-(E_{\text{transition state}} - E_{\text{reactants}})/RT] \tag{3}$$

Here, $E_{\text{transition state}}$ and $E_{\text{reactants}}$ are the energies of the transition state and the reactants, respectively, T is the temperature and R is the gas constant. Note, that the rate constant (as well as the overall rate) does not depend on the relative energies of reactants and products ("thermodynamics") but only on the difference in energies between reactants and transition state. This difference is commonly referred to as the activation energy or the energy barrier, and is usually given the symbol $\Delta E^{\ddagger}$. Other factors such as the likelihood of encounters between molecules and the effectiveness of these encounters in promoting reaction are taken into account by way of the "A factor" multiplying the exponential. This is generally assumed to be constant for reactions involving a single set of reactants going to different products, or for reactions involving closely-related reactants.

In general, the lower the activation energy the faster the reaction. In the limit of a "zero barrier", reaction rate will be limited entirely by how rapidly molecules can move.[**] Such limiting reactions have come to be known as "diffusion controlled" reactions.

The product formed in greatest amount in a kinetically-controlled reaction (the kinetic product) is that proceeding via the lowest-energy transition state, irrespective of whatever or not this is lowest-energy product (the thermodynamic product).

---

[*] In addition to temperature, the "rate constant" also depends on pressure, but this dependence is usually ignored.

[**] In fact, "reactions without barriers" are fairly common. Further discussion is provided in **Chapter 15**.

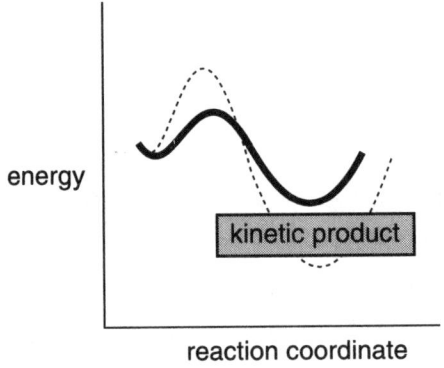

energy

reaction coordinate

Kinetic product ratios show dependence with activation energy differences which are identical to thermodynamic product ratios with difference in reactant and product energies (see box on page 9).

Thermodynamic vs. Kinetic Control of Chemical Reactions

The fact that there are two different and independent mechanisms controlling product distributions - thermodynamic and kinetic - is why some chemical reactions yield one distribution of products under one set of conditions and an entirely different distribution of products under a different set of conditions. It also provides a rationale for why organic chemists allow some reactions to "cook" for hours while they rush to quench others seconds after they have begun.

Consider a process starting from a single reactant (or single set of reactants) and leading to two different products (or two different sets of products) in terms of a reaction coordinate diagram.

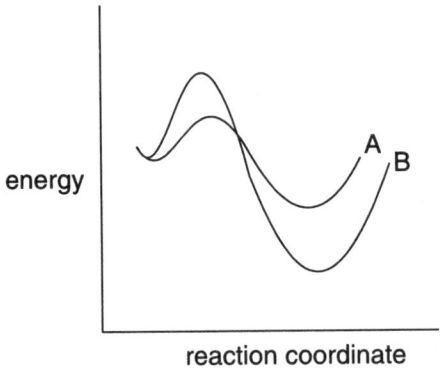

energy

reaction coordinate

According to this diagram, pathway A leads through the lower energy-transition state, but results in the higher-energy products. It is the kinetically-favored pathway leading to the kinetic product. Pathway B proceeds through the higher-energy transition state, but leads to the lower-energy products. It is the thermodynamically-favored pathway leading to the thermodynamic product. By varying conditions (temperature, reaction time, solvent) chemists can affect the product distribution.

Of course, the reaction coordinate diagram might be such that kinetic and thermodynamic products are the same, e.g., pathway B would be both the kinetic and thermodynamic pathway, and its product would be both the kinetic and thermodynamic product.

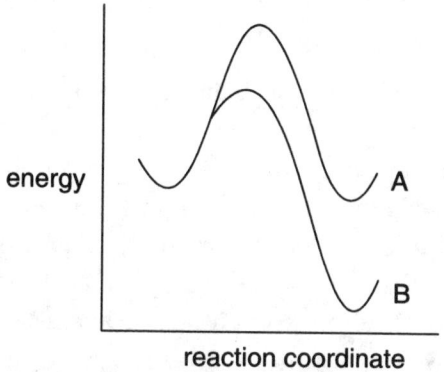

reaction coordinate

Here too, varying reaction conditions will affect product distribution, because the difference in activation energies will not be the same as the difference in product energies.

The exact distribution of products for any given chemical reaction depends on the reaction conditions; "continued cooking", i.e., long reaction times, yields the thermodynamic product distribution, while "rapid quenching" produces instead the kinetic distribution.

Radical cyclization reactions provide a good example of the situation where kinetic and thermodynamic products appear to differ. Cyclization of hex-5-enyl radical can either yield cyclopentylmethyl radical or cyclohexyl radical.

While cyclohexyl radical would be expected to be thermodynamically more stable than cyclopentylmethyl radical (six-membered rings are "less strained" than five-membered rings and 2° radicals are favored over 1° radicals), products formed from the latter dominate, e.g.

We will see in **Chapter 19** that calculations show cyclohexyl radical to be about 8 kcal/mol more stable than cyclopentylmethyl radical. Were the reaction under strict thermodynamic control, products derived from cyclopentylmethyl radical should not be observed at all. However, the transition state corresponding to radical attack on the "internal" double bond carbon (leading to cyclopentylmethyl radical) is about 3 kcal/mol lower in energy than that corresponding to radical attract on the "external" double bond carbon (leading to cyclohexyl radical). This translates into roughly a 99:1 ratio of major:minor products (favoring products derived from cyclopentylmethyl radical) in accord to what is actually observed. The reaction is apparently under kinetic control.

Potential Energy Surfaces and Mechanism

"Real" chemical reactions need not occur in a single step, but rather may involve several distinct steps and one or more "intermediates". The overall sequence of steps is termed a mechanism and may be represented by a reaction coordinate diagram.

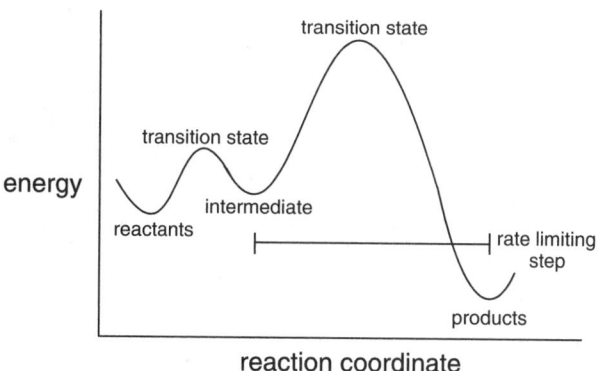

The thermodynamics of reaction is exactly as before, that is, related to the difference in energies between reactants and products. The intermediates play no role whatsoever. However, proper account of the kinetics of reaction does require consideration of all steps (and all transition states). Where one transition state is much higher in energy than any of the others (as in the diagram above) the overall kinetics may safely be assumed to depend only on this "rate limiting step".

In principle, mechanism may be established from computation, simply by first elucidating all possible sequences from reactants to products, and then identifying that particular sequence with the "fastest" rate-limiting step, that is, with the lowest-energy rate-limiting transition state. This is not yet common practice, but it likely will become so. If and when it does, calculations will provide a powerful supplement to experiment in elucidating reaction mechanisms.

Section I

Theoretical Models

As pointed out in the preface, a wide variety of different procedures or models have been developed to calculate molecular structure and energetics. These have generally been broken down into two categories, quantum chemical models and molecular mechanics models.

Quantum chemical models all ultimately stem from the Schrödinger equation first brought to light in the late 1920's. It treats molecules as collections of nuclei and electrons, without any reference whatsoever to "chemical bonds". The solution to the Schrödinger equation is in terms of the motions of electrons, which in turn leads directly to molecular structure and energy among other observables, as well as to information about bonding. However, the Schrödinger equation cannot actually be solved for any but a one-electron system (the hydrogen atom), and approximations need to be made. Quantum chemical models differ in the nature of these approximations, and span a wide range, both in terms of their capability and reliability and their "cost".

Although the origins of quantum chemical models will be detailed in the following chapter, it is instructive to "stand back" for an overall view. The place to start is the Hartree-Fock approximation, which when applied to the many-electron Schrödinger equation, not only leads directly to an important class of quantum chemical models (so-called Hartree-Fock molecular orbital models, or simply, molecular orbital models), but also provides the foundation for both simpler and more complex models. In effect, the Hartree-Fock approximation replaces the "correct" description of electron motions by a picture in which the electrons behave essentially as independent particles. Hartree-Fock models were first "put to the test" in the 1950's, soon after the first digital computers became available, and there is now a great deal of experience with their successes and failures. Except where transition metals are involved, Hartree-Fock models provide good descriptions of equilibrium geometries and conformations, and also perform well for many kinds of thermochemical comparisons. However, Hartree-Fock models fare poorly in accounting for the thermochemistry of reactions involving explicit bond making or bond breaking. Discussion is provided in **Section II**.

The failures of Hartree-Fock models can be traced to an incomplete description of "electron correlation" or, simply stated, the way in which the motion of one electron affects the motions of all the other electrons. Two fundamentally different approaches for improvement of Hartree-Fock models have emerged.

One approach is to construct a more flexible description of electron motions in terms of a combination of Hartree-Fock descriptions for ground and excited states. Configuration interaction (CI) and Møller-Plesset (MP) models are two of the most commonly used models of this type. The so-called second-order Møller-Plesset model (MP2) is the most practical and widely employed. It generally provides excellent descriptions of equilibrium geometries and conformations, as well as thermochemistry, including the thermochemistry of reactions where bonds are broken and formed. Discussion is provided in **Section II**.

An alternative approach to improve upon Hartree-Fock models involves including an explicit term to account for the way in which electron motions affect each other. In practice, this account is based on an "exact" solution for an idealized system, and is introduced using empirical parameters. As a class, the resulting models are referred to as density functional models. Density functional models have proven to be successful for determination of equilibrium geometries and conformations, and are (nearly) as successful as MP2 models for establishing the thermochemistry of reactions where bonds are broken or formed. Discussion is provided in **Section II**.

The Hartree-Fock approximation also provided the basis for what are now commonly referred to as semi-empirical models. These introduce additional approximations as well as empirical parameters to greatly simplify the calculations, with minimal adverse effect on the results. While this goal has yet to be fully realized, several useful schemes have resulted, including the popular AM1 and PM3 models. Semi-empirical models have proven to be successful for the calculation of equilibrium geometries, including the geometries of transition-metal compounds. They are, however, not satisfactory for thermochemical calculations or for conformational assignments. Discussion is provided in **Section II**.

The alternative to quantum chemical models are so-called molecular mechanics models. These do not start from an "exact-theory" (the Schrödinger equation), but rather from a simple but "chemically reasonable" picture of molecular structure. In this picture, molecules are made up of atoms and bonds (as opposed to nuclei and electrons), and atom positions are adjusted to best match known structural data (bond lengths and angles), as well as to accommodate non-bonded interactions. This is obviously much simpler than solving the Schrödinger equation for electron motions, but requires an explicit description of "chemical bonding", as well as a large amount of information about the structures of molecules[*]. It is in the use and extent of this information which distinguishes different molecular mechanics models.

The opening chapter in this section outlines a number of different classes of *Quantum Chemical Models* and provides details for a few specific models. It anticipates issues relating to "cost" and "capability" (to be addressed in detail in **Section II**). Similar treatment of *Molecular Mechanics Models* is provided in the second chapter in this section.

"Important" quantities which come out of molecular mechanics and quantum chemical models are typically related in terms of "numbers", e.g., the heat of a chemical reaction, or in terms of simple diagrams, e.g., an equilibrium structure. Other quantities, in particular those arising from quantum chemical models, may not be best expressed in this way, e.g., the distribution of electrons in molecules. Here computer graphics provides a vessel. This is addressed in the concluding chapter in this section, *Graphical Models*.

---

[*] In a sense, molecular mechanics is not a theory, but rather an elaborate interpolation scheme.

Chapter 2

Quantum Chemical Models

This chapter reviews models based on quantum mechanics starting from the Schrödinger equation. Hartree-Fock models are addressed first, followed by models which account for electron correlation, with focus on density functional models, configuration interaction models and Møller-Plesset models. All-electron basis sets and pseudopotentials for use with Hartree-Fock and correlated models are described. Semi-empirical models are introduced next, followed by a discussion of models for solvation.

Theoretical Models and Theoretical Model Chemistry

While it is not possible to solve the Schrödinger equation for a many-electron system, it may be assumed that were it possible the resulting molecular properties would exactly reproduce the corresponding experimental quantities. On the other hand, molecular properties resulting from solution of approximate Schrödinger equations would not be expected to be identical to experimentally-determined quantities. In fact, different approximations will lead to different results. We shall refer to a specific set of approximations to the Schrödinger equation as defining a theoretical model, and to the collective results of a particular theoretical model as a theoretical model chemistry.* It might be anticipated that the less severe the approximations which make up a particular theoretical model, the closer will be its results to experiment.

To the extent that it is possible, any theoretical model should satisfy a number of conditions. Most important is that it should yield a unique energy, among other molecular properties, given only the kinds and positions of the nuclei, the total number of electrons and the number

---

* These terms were introduced by John Pople, who in 1998 received the Nobel Prize in Chemistry for his work in bringing quantum chemical models into widespread use.

of unpaired electrons. A model should not appeal in any way to "chemical intuition". Also important, is that if at all possible, the magnitude of the error of the calculated energy should increase roughly in proportion to molecular size, that is, the model should be "size consistent". Only then is it reasonable to anticipate that reaction energies can be properly described. Somewhat less important, but highly desirable, is that the model energy should represent a bound to the exact (Schrödinger) energy, that is, the model should be "variational". Finally, a model needs to be "practical", that is, able to be applied not only to very simple or idealized systems, but also to problems which are actually of interest. Were this not an issue, then it would not be necessary to move beyond the Schrödinger equation itself.

Schrödinger Equation

Quantum mechanics describes molecules in terms of interactions among nuclei and electrons, and molecular geometry in terms of minimum energy arrangements of nuclei.[1] All quantum mechanical methods ultimately trace back to the Schrödinger equation, which for the special case of hydrogen atom (a single particle in three dimensions) may be solved exactly.[*]

$$\left[-\frac{1}{2} \nabla^2 - \frac{Z}{r} \right] \psi(\mathbf{r}) = E\psi(\mathbf{r}) \tag{1}$$

Here, the quantity in square brackets represents the kinetic and potential energy of an electron at a distance r from a nucleus of charge Z (1 for hydrogen). E is the electronic energy in atomic units and ψ, a function of the electron coordinates, $\mathbf{r}$, is a wavefunction describing the motion of the electron as fully as possible. Wavefunctions for the hydrogen atom are the familiar s, p, d... atomic orbitals. The square of the wavefunction times a small volume gives the probability of finding the electron inside this volume. This is termed the total electron density (or more simply the electron density), and corresponds to the electron density measured in an X-ray diffraction experiment.

---

[*]   This equation as well as multi-particle Schrödinger equation and all approximate equations which follow are given in so-called atomic units. This allows fundamental constants as well as the mass of the electron to be folded in.

Graphical representations of the electron density will be provided in **Chapter 4**, and connections drawn between electron density and both chemical bonding and overall molecular size and shape.

It is straightforward to generalize the Schrödinger equation to a multinuclear, multielectron system.

$$\hat{H}\Psi = E\Psi \qquad (2)$$

Here, Ψ is a many-electron wavefunction and $\hat{H}$ is the so-called Hamiltonian operator (or more simply the Hamiltonian), which in atomic units is given by.

$$\hat{H} = -\frac{1}{2}\sum_i^{\text{electrons}}\nabla_i^2 - \frac{1}{2}\sum_A^{\text{nuclei}}\frac{1}{M_A}\nabla_A^2 - \sum_i^{\text{electrons}}\sum_A^{\text{nuclei}}\frac{Z_A}{r_{iA}} + \sum\sum_{i<j}^{\text{electrons}}\frac{1}{r_{ij}} + \sum\sum_{A<B}^{\text{nuclei}}\frac{Z_A Z_B}{R_{AB}} \qquad (3)$$

Z is the nuclear charge, M_A is the ratio of mass of nucleus A to the mass of an electron, R_{AB} is the distance between nuclei A and B, r_{ij} is the distance between electrons i and j and r_{iA} is the distance between electron i and nucleus A.

The many-electron Schrödinger equation cannot be solved exactly (or at least has not been solved) even for a simple two-electron system such as helium atom or hydrogen molecule. Approximations need to be introduced to provide practical methods.

Born-Oppenheimer Approximation

One way to simplify the Schrödinger equation for molecular systems is to assume that the nuclei do not move. Of course, nuclei do move, but their motion is "slow" compared to the speed at which electrons move (the speed of light). This is called the Born-Oppenheimer approximation, and leads to an "electronic" Schrödinger equation.

$$\hat{H}^{el}\Psi^{el} = E^{el}\Psi^{el} \qquad (4)$$

$$\hat{H}^{el} = -\frac{1}{2}\sum_i^{\text{electrons}}\nabla_i^2 - \sum_i^{\text{electrons}}\sum_A^{\text{nuclei}}\frac{Z_A}{r_{iA}} + \sum\sum_{i<j}^{\text{electrons}}\frac{1}{r_{ij}} \qquad (5)$$

The term in equation 3 describing the nuclear kinetic energy is missing in equation 5 (it is zero), and the nuclear-nuclear Coulomb term in equation 3 is a constant. The latter needs to be added to the electronic energy, E^{el}, to yield the total energy, E, for the system.

$$E = E^{el} + \sum_{A < B}^{nuclei} \sum \frac{Z_A Z_B}{R_{AB}} \qquad (6)$$

Note that nuclear mass does not appear in the electronic Schrödinger equation. To the extent that the Born-Oppenheimer approximation is valid*, this means that mass effects (isotope effects) on molecular properties and chemical reactivities are of different origin.

Hartree-Fock Approximation

The electronic Schrödinger equation** is still intractable and further approximations are required. The most obvious is to insist that electrons move independently of each other. In practice, individual electrons are confined to functions termed molecular orbitals, each of which is determined by assuming that the electron is moving within an average field of all the other electrons. The total wavefunction is written in the form of a single determinant (a so-called Slater determinant). This means that it is antisymmetric upon interchange of electron coordinates.***

$$\Psi = \frac{1}{\sqrt{N!}} \begin{vmatrix} \chi_1(1) & \chi_2(1) \text{----} \chi_n(1) \\ \chi_1(2) & \chi_2(2) \text{----} \chi_n(2) \\ \vdots & \vdots \qquad \vdots \\ \chi_1(N) & \chi_2(N) \text{----} \chi_n(N) \end{vmatrix} \qquad (7)$$

---

*   All evidence points to the validity of the Born-Oppenheimer approximation with regard to the calculation of molecular structure and relative energetics among other important "chemical observables".

**   From this point on, we will use the terms "electronic Schrödinger equation" and "Schrödinger equation" interchangeably.

*** Antisymmetry is a requirement of acceptable solutions to the Schrödinger equation. The fact that the determinant form satisfies this requirement follows from the fact that different electrons correspond to different rows in the determinant. Interchanging the coordinates of two electrons is, therefore, equivalent to interchanging two rows in the determinant which, according to the properties of determinants, multiplies the value of the determinant by -1.

Here, χ_i is termed a spin orbital and is the product of a spatial function or molecular orbital, ψ_i, and a spin function, α or β.[*]

The set of molecular orbitals leading to the lowest energy are obtained by a process referred to as a "self-consistent-field" or SCF procedure. The archetypal SCF procedure is the Hartree-Fock procedure, but SCF methods also include density functional procedures. All SCF procedures lead to equations of the form.

$$f(i)\, \chi(\mathbf{x}_i) = \varepsilon \chi(\mathbf{x}_i) \tag{8}$$

Here, the Fock operator $f(i)$ can be written.

$$f(i) = -\frac{1}{2}\nabla_i^2 + \upsilon^{eff}(i) \tag{9}$$

$\mathbf{x}_i$ are spin and spatial coordinates of the electron i, χ are the spin orbitals and υ^{eff} is the effective potential "seen" by the electron i, which depends on the spin orbitals of the other electrons. The nature of the effective potential υ^{eff} depends on the SCF methodology.

LCAO Approximation

The Hartree-Fock approximation leads to a set of coupled differential equations (the Hartree-Fock equations), each involving the coordinates of a single electron. While they may be solved numerically, it is advantageous to introduce an additional approximation in order to transform the Hartree-Fock equations into a set of algebraic equations.

It is reasonable to expect that the one-electron solutions for many-electron molecules will closely resemble the (one-electron) solutions for the hydrogen atom. Afterall, molecules are made up of atoms, so why shouldn't molecular solutions be made up of atomic solutions? In practice, the molecular orbitals are expressed as linear combinations

---

[*] The fact that there are only two kinds of spin function (α and β), leads to the conclusion that two electrons at most may occupy a given molecular orbital. Were a third electron to occupy the orbital, two different rows in the determinant would be the same which, according to the properties of determinants, would cause it to vanish (the value of the determinant would be zero). Thus, the notion that electrons are "paired" is really an artifact of the Hartree-Fock approximation.

of a finite set (a basis set) of prescribed functions known as basis functions, ϕ.

$$\underset{\mu}{\overset{\text{basis functions}}{\psi_i = \sum c_{\mu i} \phi_\mu}} \tag{10}$$

c are the (unknown) molecular orbital coefficients, often referred to simply (and incorrectly) as the molecular orbitals. Because the ϕ are usually centered at the nuclear positions (although they do not need to be[*]), they are referred to as atomic orbitals, and equation 10 is termed the Linear Combination of Atomic Orbitals or LCAO approximation.

Roothaan-Hall Equations

The Hartree-Fock and LCAO approximations, taken together and applied to the electronic Schrödinger equation, lead to the Roothaan-Hall equations.[2]

$$\mathbf{Fc} = \varepsilon \mathbf{Sc} \tag{11}$$

Here, ε are orbital energies, $\mathbf{S}$ is the overlap matrix (a measure of the extent to which basis functions "see each other"), and $\mathbf{F}$ is the Fock matrix, which is analogous to the Hamiltonian in the Schrödinger equation. Its elements are given by.

$$F_{\mu\nu} = H_{\mu\nu}^{\text{core}} + J_{\mu\nu} - K_{\mu\nu} \tag{12}$$

H^{core} is the so-called core Hamiltonian, the elements of which are given by.

$$H_{\mu\nu}^{\text{core}} = \int \phi_\mu(\mathbf{r}) \left[-\frac{1}{2}\nabla^2 - \sum_A^{\text{nuclei}} \frac{Z_A}{r} \right] \phi_\nu(\mathbf{r}) \, d\mathbf{r} \tag{13}$$

Coulomb and exchange elements are given by.

$$J_{\mu\nu} = \overset{\text{basis functions}}{\sum_\lambda \sum_\sigma} P_{\lambda\sigma} (\mu\nu \mid \lambda\sigma) \tag{14}$$

$$K_{\mu\nu} = \frac{1}{2} \overset{\text{basis functions}}{\sum_\lambda \sum_\sigma} P_{\lambda\sigma} (\mu\lambda \mid \nu\sigma) \tag{15}$$

---

[*]   Insisting that the basis functions be nuclear centered eliminates the problem of having to specify their locations.

P is the so-called density matrix, the elements of which involve a product of two molecular orbital coefficients summed over all occupied molecular orbitals.[*]

$$P_{\lambda\sigma} = 2 \sum_i^{\substack{\text{occupied} \\ \text{molecular orbitals}}} c_{\lambda i} c_{\sigma i} \tag{16}$$

The product of an element of the density matrix and its associated atomic orbitals summed over all orbitals leads to the electron density. Further discussion is provided in **Chapter 4**.

$(\mu\nu \mid \lambda\sigma)$ are two-electron integrals, the number of which increases as the fourth power of the number of basis functions.

$$(\mu\nu \mid \lambda\sigma) = \iint \phi_\mu(\mathbf{r}_1)\phi_\nu(\mathbf{r}_1)\left[\frac{1}{r_{12}}\right]\phi_\lambda(\mathbf{r}_2)\phi_\sigma(\mathbf{r}_2)d\mathbf{r}_1 d\mathbf{r}_2 \tag{17}$$

Because they are so numerous, the evaluation and processing of two-electron integrals constitute the major time consuming steps.

Methods resulting from solution of the Roothaan-Hall equations are termed Hartree-Fock models. The corresponding energy for an infinite (complete) basis set is termed the Hartree-Fock energy. The term *Ab Initio* ("from the beginning") models is also commonly used to describe Hartree-Fock models, although this should be applied more generally to all models arising from "non-empirical" attempts to solve the Schrödinger equation.

Hartree-Fock models are well defined and yield unique properties. They are both size consistent and variational. Not only may energies and wavefunctions be evaluated from purely analytical (as opposed to numerical) methods, but so too may first and second energy derivatives. This makes such important tasks as geometry optimization (which requires first derivatives) and determination of vibrational frequencies (which requires second derivatives) routine. Hartree-Fock models and are presently applicable to molecules comprising upwards of 50 to 100 atoms.

---

[*] This will generally be the lowest-energy $\frac{1}{2}$ Ne molecular orbitals, where Ne is the total number of electrons.

Correlated Models

Hartree-Fock models treat the motions individual electrons as independent of one another. To do this, they replace "instantaneous interactions" between individual electrons by interactions between a particular electron and the average field created by all the other electrons. Because of this, electrons "get in each others way" to a greater extent than they should. This leads to overestimation of the electron-electron repulsion energy and to too high a total energy.[*] Electron correlation, as it is termed, accounts for coupling or "correlation" of electron motions, and leads to a lessening of the electron-electron repulsion energy (and to a lowering of the total energy). The correlation energy is defined as the difference between the Hartree-Fock energy and the experimental energy.

At this point, it is instructive to introduce a two-dimensional diagram onto which all possible theoretical models can be placed.[**]

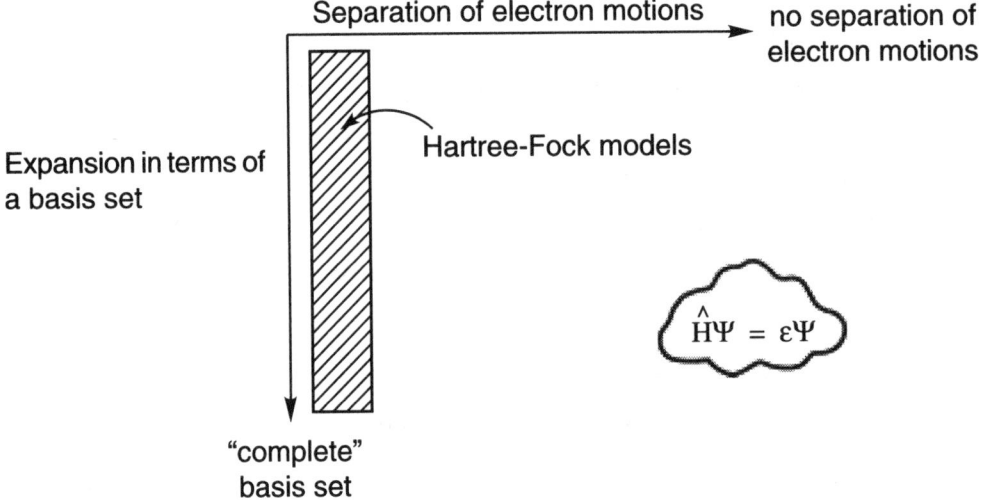

The horizontal axis relates the extent to which the motions of electrons in a many-electron system are independent of each other (uncorrelated). At the extreme left are found Hartree-Fock models,

---

[*] This is consistent with the fact that Hartree-Fock models are "variational", meaning that the Hartree-Fock energy is necessarily above the energy which would result upon solution of the Schrödinger equation.

[**] More precisely, this diagram allows all possible models *within the framework of the Born-Oppenheimer approximation*.

while fully-correlated models are found at the extreme right. Practical correlated models are located somewhere in between.

The vertical axis designates the basis set. At the top is a so-called minimal basis set, which involves the fewest possible functions (see discussion later in this chapter), while at the very bottom is a "complete" basis set. The "bottom" of the column of Hartree-Fock models (at the far left) is termed the Hartree-Fock limit. Note, that this limit is not the same as the exact solution of the Schrödinger equation (or experiment).

Proceeding all the way to the right (fully correlated) and all the way to the bottom (complete basis set) is functionally equivalent to solving exactly the exact Schrödinger equation. It cannot be realized. Note, however, if having occupied some position on the diagram, that is, some level of electron correlation and some basis set, significant motion down and to the right produces no change in a particular property of interest, then it can reasonably be concluded that further motion would also not result in change in this property. In effect, the "exact" solution has been achieved.

Although many different correlated models have been introduced, only three classes will be discussed here.[3] Density functional models introduce an "approximate" correlation term in an explicit manner. They offer the advantage of not being significantly more costly than Hartree-Fock models. The "quality" of density functional models obviously depends on the choice of this term, although it is not apparent how to improve on a particular choice. Configuration interaction models and Møller-Plesset models extend the flexibility of Hartree-Fock models by mixing ground-state and excited-state wavefunctions. They are significantly more costly than Hartree-Fock models. In the limit of "complete mixing" both configuration interaction and Møller-Plesset models lead to the "exact result", although in practice this limit cannot be reached.

Kohn-Sham Equations and Density Functional Models

One approach to the treatment of electron correlation is referred to as density functional theory. Density functional models have "at their heart" the electron density, $\rho(\mathbf{r})$, as opposed to the many-electron wavefunction, $\Psi(\mathbf{r}_1, \mathbf{r}_2,...)$. There are both distinct similarities and distinct differences between traditional wavefunction-based approaches (see following two sections) and electron-density-based methodologies. First, the essential building blocks of a many-electron wavefunction are single-electron (molecular) orbitals, which are directly analogous to the orbitals used in density functional methodologies. Second, both the electron density and the many-electron wavefunction are constructed from an SCF approach which requires nearly identical matrix elements.

The density functional theory of Hohenberg, Kohn and Sham[4] is based on the fact that the sum of the exchange and correlation energies of a uniform electron gas can be calculated exactly knowing only its density.[*] In the Kohn-Sham formalism, the ground-state electronic energy, E, is written as a sum of the kinetic energy, E_T, the electron-nuclear interaction energy, E_V, the Coulomb energy, E_J, and the exchange/correlation energy, E_{xc}.

$$E = E_T + E_V + E_J + E_{XC} \tag{18}$$

Except for E_T, all components depend on the total electron density, $\rho(\mathbf{r})$.

$$\rho(\mathbf{r}) = 2 \sum_i^{\text{orbitals}} |\psi_i(\mathbf{r})|^2 \tag{19}$$

Here, ψ_i are the so-called Kohn-Sham orbitals and the summation is carried out over pairs of electrons. Within a finite basis set (analogous to the LCAO approximation for Hartree-Fock models), the energy components may be written as follows.

$$E_T = \sum_\mu^{\text{basis functions}} \sum_\nu \int \phi_\mu(\mathbf{r}) \left[-\frac{1}{2} \nabla^2 \right] \phi_\nu(\mathbf{r}) \, d\mathbf{r} \tag{20}$$

---

[*]   For his discovery, leading up to the development of practical density functional models, Walter Kohn was awarded the Nobel Prize in Chemistry in 1998.

$$E_V = \sum_\mu^{\text{basis functions}} \sum_\nu P_{\mu\nu} \sum_A^{\text{nuclei}} \int \phi_\mu(\mathbf{r}) \left| \frac{Z_A}{|\mathbf{r} - \mathbf{R}_A|} \right| \phi_\nu(\mathbf{r}) d\mathbf{r} \tag{21}$$

$$E_J = \frac{1}{2} \sum_\mu^{\text{basis functions}} \sum_\nu \sum_\lambda \sum_\sigma P_{\mu\nu} P_{\lambda\sigma} (\mu\nu \mid \lambda\sigma) \tag{22}$$

$$E_{XC} = \int f(\rho(\mathbf{r}), \nabla\rho(\mathbf{r}), \dots) \, d\mathbf{r} \tag{23}$$

Z is the nuclear charge, R-r is the distance between the nucleus and the electron, P is the density matrix (equation 16) and $(\mu\nu|\lambda\sigma)$ are two-electron integrals (equation 17). f is an exchange/correlation functional, which depends on the electron density and perhaps as well the gradient of the density. Minimizing E with respect to the unknown orbital coefficients yields a set of matrix equations, the "Kohn-Sham equations", analogous to the Roothaan-Hall equations (equation 11).

$$\mathbf{Fc} = \varepsilon \mathbf{Sc} \tag{24}$$

Here the elements of the Fock matrix are given by.

$$F_{\mu\nu} = H_{\mu\nu}^{\text{core}} + J_{\mu\nu} - F_{\mu\nu}^{XC} \tag{25}$$

$H_{\mu\nu}^{\text{core}}$ and $J_{\mu\nu}$ are defined analogously to equations 13 and 14, respectively and $F_{\mu\nu}^{XC}$ is the exchange/correlation part, the form of which depends on the particular exchange/correlation functional employed. Note, that substitution of the Hartree-Fock exchange, $K_{\mu\nu}$, for $F_{\mu\nu}^{XC}$ yields the Roothaan-Hall equations.

Three types of exchange/correlation functionals are presently in use: (i) functionals based on the local spin density approximation, (ii) functionals based on the generalized gradient approximation, and (iii) functionals which employ the "exact" Hartree-Fock exchange as a component. The first of these are referred to as local density models, while the second two are collectively referred to as non-local models or alternatively as gradient-corrected models.

Density functional models are well-defined and yield unique results. They are neither size consistent nor variational. It should be noted that were the exact exchange/correlation functional known, then the density functional approach would be "exact". While "better" forms of such

31

functionals are constantly being developed, there is (at present) no systematic way to improve the functional to achieve an arbitrary level of accuracy. Density functional models, like Hartree-Fock models are applicable to molecules of moderate size (50-100 atoms).

Most modern implementations of density functional theory divide the problem into two parts. The first part, which involves everything except the exchange/correlation functional is done using the same analytical procedures employed in Hartree-Fock models. So-called "pure" density functional methods, including the local density model and non-local models such as the BP, BLYP and EDF1 models, require only the Hartree-Fock Coulomb terms ($J_{\mu\nu}$ from equation 14) and not the Hartree-Fock exchange terms ($K_{\mu\nu}$ from equation 15), and special algorithms based on multipole expansions have been developed as alternatives to conventional algorithms. These become competitive and ultimately superior to conventional algorithms for very large molecules, where "pure" density functional procedures will actually be significantly faster than Hartree-Fock models. So-called "hybrid" density functional models, such as the popular B3LYP model, make use of Hartree-Fock exchange terms. These do not benefit from multipole Coulomb methods and can never surpass Hartree-Fock models in computation speed.

The second part of the calculation involves dealing with the exchange/correlation functional. Analytical procedures have as yet to be developed to evaluate the required integrals, and numerical integration over a pre-specified grid is needed. The larger the number of grid points, the more precise will be the results of numerical integration and the more costly will be the calculation. Grid specification is an important part in the development of practical density functional methodology, and is an active and ongoing area of research.

Despite the fact that numerical integration is involved, "pseudoanalytical" procedures have been developed for calculation of first and second energy derivatives. This means that density functional models, like Hartree-Fock models are routinely applicable to determination of equilibrium and transition-state geometries and of vibrational frequencies.

Configuration Interaction Models[5]

In principle, density functional models are able to capture the full correlation energy. In practice, present generation methods exhibit a number of serious deficiencies in particular with regard to reaction energetics (see discussion in **Section II** of this guide), and wavefunction-based approaches for calculating the correlation energy are still required. These generally involve mixing the ground-state (Hartree-Fock) wavefunction with "excited-state" wavefunctions. Operationally, this entails implicit or explicit promotion of electrons from molecular orbitals which are occupied in the Hartree-Fock wavefunction to molecular orbitals which are unoccupied.

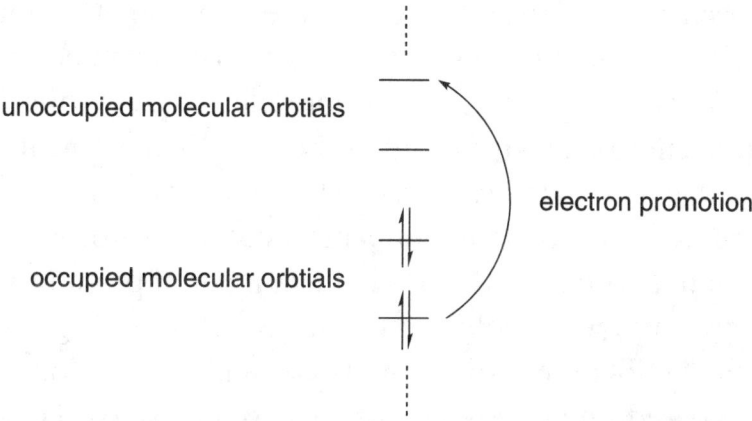

Conceptually, the most straightforward approach is the so-called full configuration interaction model. Here, the wavefunction is written as a sum, the leading term of which, Ψ_o, is the Hartree-Fock wavefunction, and remaining terms, Ψ_s, are wavefunctions derived from the Hartree-Fock wavefunction by electron promotions.

$$\Psi = a_o\Psi_o + \sum_{s>o} a_s\Psi_s \tag{26}$$

The unknown linear coefficients, a_s, are determined by solving equation 27.

$$\sum_s (H_{st} - E_i \delta_{st})a_{si} = 0 \qquad t = 0, 1, 2, \ldots \tag{27}$$

where the matrix elements are given by equation 28.

$$H_{st} = \int ... \int \; \Psi_s \hat{H} \, \Psi_t \, d\tau_1 \, d\tau_2 \, ... d\tau_n \qquad (28)$$

The lowest-energy from solution of equation 27 corresponds to the energy of the electronic ground state. The difference between this energy and the Hartree-Fock energy with a given basis set is the correlation energy for that basis set. As the basis set becomes more complete, the result of a full configuration interaction treatment will approach the exact solution of the Schrödinger equation. The full CI method is well-defined, size consistent and variational. It is, however, not practical except for very small systems, because of the very large number of terms in equation 26.

It is necessary to severely limit the number of electron promotions. One approach, referred to as the frozen-core approximation, eliminates any promotions from molecular orbitals which correspond essentially to (combinations of) inner-shell or core electrons.[*] While the total correlation energy arising from inner-shell promotions is not insignificant, experience suggests that this energy remains essentially unchanged in moving from one molecule to another. A more substantial approximation is to limit the number of promotions based on the total number of electrons involved, i.e., single-electron promotions, double-electron promotions, and so on. Configuration interaction based on single-electron promotions only, the so-called CIS method, leads to no improvement of the (Hartree-Fock) energy or wavefunction. The simplest procedure to actually lead to improvement over Hartree-Fock is the so-called CID method, which is restricted to double-electron promotions.

$$\Psi_{CID} = a_o \Psi_o + \overset{\overset{\text{molecular orbitals}}{\overset{\text{occ}\qquad\text{unocc}}{}}}{\underset{i<j}{\sum}\underset{}{\sum}\underset{a<b}{\sum}\sum} a_{ij}^{ab} \Psi_{ij}^{ab} \qquad (29)$$

A somewhat less restricted recipe, termed CISD, considers both single and double-electron promotions.[**]

---

[*]   In practice, one molecular orbital may be eliminated for each first-row element and four molecular orbitals may be eliminated for each second-row element.

[**]   While single-electron promotions do not themselves contribute, matrix elements involving both single and double-electron promotions do contribute if only weakly.

$$\Psi_{CISD} = a_0\Psi_0 + \overset{\substack{\text{molecular orbitals}\\\text{occ}\quad\text{unocc}}}{\underset{i\quad a}{\sum\sum}} a_i^a \Psi_i^a + \overset{\substack{\text{molecular orbitals}\\\text{occ}\quad\text{unocc}}}{\underset{i<j\quad a<b}{\sum\sum\sum\sum}} a_{ij}^{ab} \Psi_{ij}^{ab} \tag{30}$$

Solution of equation 27 for either CID or CISD methods is practical for reasonably large systems (with reasonable basis sets). First and second derivatives may be evaluated analytically, meaning that geometry optimizations and frequency calculations are routine. The methods are obviously well defined and they are variational. However, neither method (or any limited configuration interaction method) is size consistent. This can easily be seen by considering the CID description of a two-electron system, e.g., a helium atom, using just two basis functions. Here, there will be one occupied molecular orbital and one unoccupied molecular orbital, and the CID description is "exact" (within the confines of this basis set), meaning that all possible electron promotions have been considered. Next, consider the CID description of two helium atoms at infinite separation. It is not "exact" in that all possible electron promotions have not been considered. Thus, the energies of two helium atoms treated separately and two helium atoms at infinite separation will be different.

Møller-Plesset Models[6]

Another practical correlation energy scheme is the second-order Møller-Plesset model, or MP2. This is the simplest member of the class of so-called Møller-Plesset models, the basis of which is the recognition that, while the Hartree Fock wavefunction Ψ_0 and ground-state energy E_0 are approximate solutions to the Schrödinger equation, they are exact solutions to an analogous problem involving the Hartree-Fock Hamiltonian, $\hat{H}_0$, in place of the "exact" Hamiltonian, $\hat{H}$. Assuming that the Hartree-Fock wavefunction Ψ and energy are, in fact, very close to the exact wavefunction and ground-state energy E, the exact Hamiltonian can then be written in the following form.

$$\hat{H} = \hat{H}_0 + \lambda\hat{V} \tag{31}$$

Here, $\hat{V}$ is a small perturbation and λ is a dimensionless parameter. Expanding the exact wavefunction and energy in terms of the Hartree-Fock wavefunction and energy yields.

35

$$E = E^{(0)} + \lambda E^{(1)} + \lambda^2 E^{(2)} + \lambda^3 E^{(3)} + \ldots \tag{32}$$

$$\Psi = \Psi_0 + \lambda \Psi^{(1)} + \lambda^2 \Psi^{(2)} + \lambda^3 \Psi^{(3)} + \ldots \tag{33}$$

Substituting the expansions 31 to 33 into the Schrödinger equation and gathering terms in λ^n yields.

$$\hat{H}_0 \Psi_0 = E^{(0)} \Psi_0 \tag{34a}$$

$$\hat{H}_0 \Psi^{(1)} + \hat{V} \Psi_0 = E^{(0)} \Psi^{(1)} + E^{(1)} \Psi_0 \tag{34b}$$

$$\hat{H}_0 \Psi^{(2)} + \hat{V} \Psi^{(1)} = E^{(0)} \Psi^{(2)} + E^{(1)} \Psi^{(1)} + E^{(2)} \Psi_0 \tag{34c}$$

$$\vdots$$

Multiplying each of the equations 34 by Ψ_0 and integrating over all space yields the following expression for the nth order (MPn) energy.

$$E^{(0)} = \int \ldots \int \Psi_0 \hat{H}_0 \Psi_0 \, d\tau_1 \, d\tau_2 \ldots d\tau_n \tag{35a}$$

$$E^{(1)} = \int \ldots \int \Psi_0 \hat{V} \Psi_0 \, d\tau_1 \, d\tau_2 \ldots d\tau_n \tag{35b}$$

$$E^{(2)} = \int \ldots \int \Psi_0 \hat{V} \Psi^{(1)} \, d\tau_1 \, d\tau_2 \ldots d\tau_n \tag{35c}$$

$$\vdots$$

In this framework, the Hartree-Fock energy is the sum of the zero and first-order Møller-Plesset energies.

$$E^{(0)} = \int \ldots \int \Psi_0 (\hat{H}_0 + \hat{V}) \, \Psi_0 \, d\tau_1 \, d\tau_2 \ldots d\tau_n \quad E^{(0)} + E^{(1)} \tag{36}$$

The correlation energy can then be written.

$$E_{corr} = E_0^{(2)} + E_0^{(3)} + E_0^{(4)} + \ldots \tag{37}$$

The first term in equation 37 may be expanded as follows.

$$E^{(2)} = \overset{\substack{\text{molecular orbitals} \\ \text{occ} \qquad \text{unocc}}}{\underset{i<j}{\sum\sum} \underset{a<b}{\sum\sum}} (\varepsilon_a + \varepsilon_b + \varepsilon_i + \varepsilon_j)^{-1} [(ij \parallel ab)]^2 \tag{38}$$

ε_i, and ε_j are energies of occupied molecular orbitals, ε_a, and ε_b energies of unoccupied molecular orbitals, and integrals $(ij \parallel ab)$ over filled (i and j) and empty (a and b) molecular orbitals, account for changes in electron-electron interactions as a result of electron promotion,

$$(ij \parallel ab) = (ia \mid jb) - (ib \mid ja) \tag{39}$$

where the integrals $(ia \mid jb)$ involve molecular orbitals and not basis functions.

$$(ia \mid jb) = \int \psi_i(\mathbf{r_1})\psi_a(\mathbf{r_1})\left[\frac{1}{r_{12}}\right]\psi_j(\mathbf{r_2})\psi_b(\mathbf{r_2})d\mathbf{r_1}d\mathbf{r_2} \,. \tag{40}$$

The two are related by a simple transformation,

$$(ia \mid jb) = \overset{\text{basis functions}}{\sum_{\mu}\sum_{\nu}\sum_{\lambda}\sum_{\sigma}} c_{\mu i}\, c_{\nu j}\, c_{\lambda a}\, c_{\sigma b}\, (\mu\nu \mid \lambda\sigma) \tag{41}$$

where $(\mu\nu \mid \lambda\sigma)$ are given by equation 17.

Møller-Plesset theory terminated to second-order, or MP2, is perhaps the simplest model based on electron promotion which offers improvement over Hartree-Fock theory. It is well-defined and leads to unique results. MP2 is size consistent but it is not variational.[*] Analytical first energy derivatives are available making geometry (and transition-state geometry) optimization routine. Frequency evaluation typically needs to be performed by numerical differentiation of (analytical) first energy derivatives, but is still practical for molecules of moderate size. Higher-order Møller-Plesset models (MP3, MP4, etc.) have been formulated, but in practice are limited to very small systems. Also, analytical derivatives are not commonly available for these higher-order Møller-Plesset models, meaning that geometry optimization needs to be done numerically.

A number of different localized MP2 procedures ("LMP2") have been developed. The idea is to localize the Hartree-Fock orbitals prior to their use in the MP2 procedure. For sufficiently large molecules, this significantly reduces the number of integrals $(ij \parallel ab)$ which need to be calculated and processed and leads to reduction in both computational effort and overall memory and disk requirements. Localized MP3 and MP4 models are not presently available.

---

[*] Size consistency is a more important attribute than variational, and because of this, Møller-Plesset models are generally preferred over configuration interaction models.

Models for Open-Shell Molecules

While the vast majority of molecules may be described in terms of closed-shell electron configurations, that is, all electrons being paired, there are several important classes of molecules with one or more unpaired electrons. So-called free radicals are certainly the most recognizable. One way to treat open-shell molecules is by strict analogy with the treatment of closed-shell molecules, that is, to insist that electrons are either paired or are unpaired.

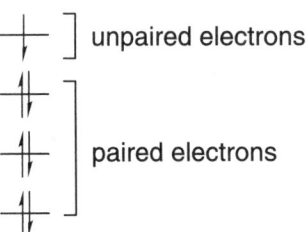

Such a procedure is referred to as "restricted" and individual models as restricted models, for example, restricted Hartree-Fock (or RHF) models.[7]

While the restricted procedure seems completely reasonable, it should be noted that it does not necessarily yield the lowest possible energy. An alternative procedure, termed "unrestricted" provides greater flexibility and may lead to a lower energy.[8] Here, electron pairing is not forced. Rather two different sets of electrons (corresponding to "spin up" and "spin down") are treated completely independently.

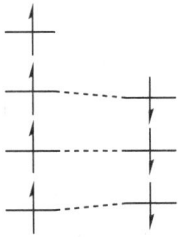

Unrestricted models, for example, the unrestricted Hartree-Fock (or UHF) model, are actually simpler and generally less costly than the corresponding restricted models, and because of this are much more widely used. Results for open-shell molecules provided in this book will make use of unrestricted models.

Models for Electronic Excited States[9]

Except for configuration interaction models, all of the procedures described thus far are strictly applicable only to the lowest-energy electronic state (the so-called ground state[*]). More precisely, they are applicable to the lowest-energy electronic state of given spin multiplicity, for example, the lowest-energy singlet state of methylene in addition to the lowest-energy triplet state. In principle, it is possible to apply theoretical models to higher-energy electronic states (so-called "excited states"), although the experience in doing so is much less than the experience with ground states. In part, this reflects the primary focus of experimental work on ground-state properties and reactivities and, because of this focus, a relative paucity of experimental data on excited-state species. Excited-state chemistry is, however, both relevant and interesting, and it is only a matter of time before application of theoretical models becomes commonplace.

The one common ground-state method which is directly applicable to excited states is the configuration interaction method. All that is required is to "focus" not on the lowest-energy combination of solutions resulting from promotion of electrons from a "reference" wavefunction, but instead on one of the higher-energy combinations. Whereas the lowest-energy combination necessarily pertains to the electronic ground state, different higher-energy combinations pertain to different excited states.

The simplest and most widely-employed method is the so-called configuration interaction singles or CIS method. This involves single-electron promotions only (from occupied molecular orbitals in the reference wavefunction to unoccupied molecular orbitals). Because there are relatively few of these, CIS is in fact practical for molecules of moderate complexity. As noted previously, single-electron promotions do not lead to improvement in either the ground-state wavefunction or energy over the corresponding Hartree-Fock

---

* Organic chemists, in particular, have the unfortunate habit of referring to a transition state as if it was not in fact a ground state, that is, implying that a transition state is an excited state. While a transition state corresponds to a molecule which is unstable with respect to motion along a single geometrical coordinate (the "reaction coordinate"), it also corresponds to the lowest-energy species for this structure. It is a ground state.

quantities. However, CIS does provide a simple means with which to describe non-ground-state wavefunctions and energies, and as such is applicable to the description of excited states. Also, as with Hartree-Fock, density functional and MP2 models, CIS readily lends itself to the analytical evaluation of first and second derivatives of the energy, making it a useful method for geometry (and transition-state geometry) calculation, as well as frequency evaluation.

Gaussian Basis Sets[10]

Basis sets for use in practical Hartree-Fock, density functional, Møller-Plesset and configuration interaction calculations make use of Gaussian-type functions. Gaussian functions are closely related to exponential functions, which are of the form of exact solutions to the one-electron hydrogen atom*, and comprise a polynomial in the Cartesian coordinates (x, y, z) followed by an exponential in r^2. Several series of Gaussian basis sets now have received widespread use and are thoroughly documented. A summary of "all electron" basis sets available in Spartan is provided in **Table 3-1**. Except for STO-3G and 3-21G, any of these basis sets can be supplemented with additional polarization functions and/or with diffuse functions. It should be noted that minimal (STO-3G) and split-valence (3-21G) basis sets, which lack polarization functions, are unsuitable for use with correlated models, in particular density functional, configuration interaction and Møller-Plesset models. Discussion is provided in **Section II**.

STO-3G Minimal Basis Set

The simplest possible atomic orbital representation is termed a minimal basis set. This comprises only those functions required to accommodate all of the electrons of the atom, while still maintaining its overall spherical symmetry. In practice, this involves a single (1s) function for hydrogen and helium, a set of five functions (1s, 2s, $2p_x$, $2p_y$, $2p_z$) for lithium to neon and a set of nine functions (1s, 2s, $2p_x$,

---

* The reason that Gaussian functions are used in place of exponential functions is that integrals, in particular electron repulsion integrals, which arise in approximate treatments are very difficult to evaluate using exponential functions but relatively easy to evaluate using Gaussian functions.

Table 3-1: All-Electron Gaussian Basis Sets Available in Spartan

basis set	available elements
STO-3G	H-Xe
3-21G, 3-21G(*)	H-Xe
6-31G*, 6-31G** 6-31+G*, 6-31+G** 6-31++G*, 6-31++G**	H-Kr
6-311G*, 6-311G** 6-311+G*, 6-311+G** 6-311++G*, 6-311++G**	H-Ar
cc-pVDZ	H-Ar, Ga-Kr
cc-pVTZ	H-Ar, Ga-Kr
cc-pVQZ	H-Ar, Ga-Kr

$2p_y$, $2p_z$, 3s, $3p_x$, $3p_y$, $3p_z$) for sodium to argon[*]. Third and fourth-row, main-group elements are treated in a similar manner. The STO-3G basis set for first-row transition metals comprises nine core functions (an "argon core") and nine functions describing the valence ($3d_{x^2-y^2}$, $3d_{z^2}$, $3d_{xy}$, $3d_{xz}$, $3d_{yz}$, 4s, $4p_x$, $4p_y$, $4p_z$). Second-row metals are treated in a similar manner. Each of the basis functions in the STO-3G representation is expanded in terms of three Gaussian functions, where the values of the Gaussian exponents and the linear coefficients have been determined by least squares as best fits to Slater-type (exponential) functions.

The STO-3G basis set has two obvious shortcomings: The first is that all basis functions are either themselves spherical or come in sets which, taken together, describe a sphere. This means that atoms with "spherical molecular environments" or "nearly spherical molecular environments" will be better described than atoms with "aspherical molecular environments". This suggests that comparisons among different molecules will be biased in favor of those incorporating the "most spherical" atoms. The second shortcoming follows from the fact that basis functions are atom centered. This restricts their flexibility to describe electron distributions between nuclei ("bonds").

Split-valence basis sets and polarization basis sets, respectively, have been formulated to address the two shortcomings. These are discussed in the following sections.

3-21G, 6-31G and 6-311G Split-Valence Basis Sets

The first shortcoming of a minimal basis set . . . bias toward atoms with "spherical" environments . . . may be addressed by providing two sets of valence basis functions ("inner" and "outer" functions). For example, proper linear combinations determined in the solution of the Roothaan-Hall equations allow for the fact that the p orbitals which make up a "tight" σ bond need to be more contracted than the p orbitals which make up a "looser" π bond.

---

[*] Note that, while 2p functions are not occupied in the lithium and beryllium atoms, they are required to provide proper descriptions where these are bonded to atoms with lone pairs. This is to allow "back bonding" from the lone pairs into empty orbitals. Note also, that 3p functions are required for sodium and magnesium, and so on.

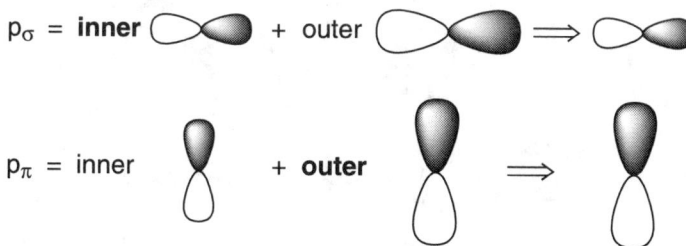

p_σ = **inner** + outer $\Longrightarrow$

p_π = inner + **outer** $\Longrightarrow$

A split-valence basis set represents core atomic orbitals by one set of functions and valence atomic orbitals by two sets of functions. Hydrogen is provided by two s-type functions, and main-group elements are provided two sets of valence s and p-type functions.

Among the simplest split-valence basis sets are 3-21G and 6-31G. Each core atomic orbital in the 3-21G basis set is expanded in terms of three Gaussians, while basis functions representing inner and outer components of valence atomic orbitals are expanded in terms of two and one Gaussians, respectively. 6-31G basis sets are similarly constructed, with core orbitals represented in terms of six Gaussians and valence orbitals split into three and one Gaussian components. Additional valence-shell splitting should lead to even greater flexibility. 6-311G basis sets split the valence functions into three parts instead of two, these being written in terms of three, one and one Gaussians, respectively.

Expansion coefficients and Gaussian exponents for 3-21G and 6-31G representations have been determined by Hartree-Fock energy minimization on atomic ground states. In the case of 6-311G representations, minimizations have been carried out at the MP2 level rather than at the Hartree-Fock level.

6-31G*, 6-31G**, 6-311G* and 6-311G** Polarization Basis Sets

The second shortcoming of a minimal (or split-valence) basis set . . . functions being centered only on atoms . . . may be addressed by providing d-type functions on main-group elements (where the valence orbitals are of s and p type), and (optionally) p-type functions on hydrogen (where the valence orbital is of s type). This allows displacement of electron distributions away from the nuclear positions.

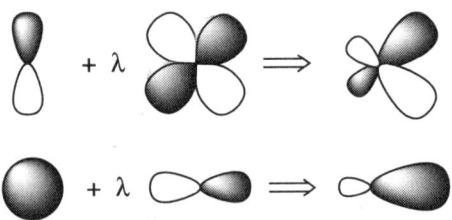

This can be thought about either in terms of "hybrid orbitals", e.g., pd and sp hybrids as shown above, or alternatively in terms of a Taylor series expansion of a function (d functions are the first derivatives of p functions, p functions are the first derivatives of s functions). While the first way of thinking is quite familiar to chemists (Pauling hybrids), the second offers the advantage of knowing what steps might be taken next to effect further improvement, i.e., adding second, third, . . . derivatives.

Among the simplest polarization basis sets are 6-31G* and 6-311G*, constructed from 6-31G and 6-311G, respectively, by adding a set of d-type polarization functions written in terms of a single Gaussian for each heavy (non-hydrogen) atom. A set of six second-order Gaussians is added in the case of 6-31G* while a set of five pure d-type Gaussians is added in the case of 6-311G*. Gaussian exponents for polarization functions have been chosen to give the lowest energies for representative molecules. Polarization of the s orbitals on hydrogen atoms is necessary for an accurate description of the bonding in many systems (particularly those in which hydrogen is a bridging atom). The 6-31G** basis set is identical to 6-31G*, except that it provides p-type polarization functions for hydrogen. Similarly, 6-311G** is identical to 6-311G* except for its description of hydrogen.

3-21G⁽*⁾ Basis Set

Experience suggests that d-type functions are required on second-row and heavier main-group elements even though they are not occupied in the free atoms (discussion is provided in **Section II**). This situation is very much like that found for alkali and alkaline-earth elements where p-type functions, while not occupied in the ground-state atoms, are required for proper description of bonding in molecules. Here, the absence of p functions leads to descriptions which are "too ionic" (bonds are too long). Similarly, the absence of d-functions of second-row and heavier main-group elements leads to bonds which are too long. These concerns apply not only to molecules with expanded valence octets (so-called "hypervalent molecules"), but also to normal-valent systems. One basis set which has proven to be quite successful for molecules incorporating heavy main-group elements is 3-21G$^{(*)}$, constructed from 3-21G basis sets by the addition of a set of d-type functions on second-row and heavier main-group elements only. Reference to 3-21G in this guide implies use of 3-21G$^{(*)}$ for second-row and heavier main-group elements.

cc-pVDZ, cc-pVTZ and cc-pVQZ Basis Sets

Most of the basis sets commonly used with correlated models, including density functional models, MP2 models and configuration interaction models, are in fact based on Hartree-Fock calculations.[*] Typically, Gaussian exponents and linear expansion coefficients are first determined to minimize the Hartree-Fock energy of the ground-state atom, and are then uniformly scaled to reflect the "tighter" nature of atoms in molecules. cc-pVDZ, cc-pVTZ and cc-pVQZ, "correlation consistent-polarized Valence Double (Triple, Quadruple) Zeta", basis sets are instead formulated to yield the lowest possible CISD ground-state atom energies. They should be better suited than basis sets as 6-31G* to capture most of the correlation energy (at least in the free atoms.)

cc-pVDZ basis sets for first-row atoms are made up of nine s-type Gaussians, four sets of p-type Gaussians and one set of d-type Gaussians, contracted to three s, two p and one d functions. The corresponding cc-pVTZ basis sets comprise ten s-type, five p-type, two d-type and one f-type Gaussians, contracted to four s, three p, two d and one f functions, and cc-pVQZ basis sets comprise twelve s-type, six p-type and three d-type, two f-type and one g-type Gaussians, contracted to five s, four p, three d, two f and one g functions. "Pure" d, f and g functions are employed. Unlike any of the basis sets previously discussed, these representations do not

---

[*] The exception is the 6-311G basis set which has been formulated using MP2 calculations on atoms rather than Hartree-Fock calculations.

attempt to share Gaussian exponents among functions of different angular quantum number, in particular s and p-type functions. Therefore, these basis sets are not as efficient computationally.

Basis Sets Incorporating Diffuse Functions

Calculations involving anions, e.g., absolute acidity calculations, often pose special problems. This is because the "extra electrons" may only be loosely associated with specific atoms (or pairs of atoms). In these situations, basis sets may need to be supplemented by diffuse s and p-type functions on heavy (non-hydrogen) atoms (designated by "+" as in 6-311+G**). It is also possible to add diffuse functions to hydrogens (designated by "++" as in 6-311++G**).

A similar situation may arise in calculations on excited states, although there is far too little experience at this point to generalize.

In practice, diffuse functions are often problematic in that they may lead to linear dependencies among basis functions. Steps need to be taken at the outset of the calculations to detect and eliminate such dependencies.

Pseudopotentials[11]

Calculations involving heavy elements, in particular, transition metals, can be simplified by explicitly considering only the valence, while replacing the core by some form of potential. This involves so-called "pseudopotentials". A summary of pseudopotentials available in Spartan is provided in **Table 3-2**. These are intended to be utilized only for heavy elements, and to be associated with specific all-electron basis sets for light elements. These associations are also indicated in the table.

In practice, use of pseudopotentials typically does not lead to significant improvement in computational speed. This is because most integrals involving core basis functions are vanishingly small and can be eliminated prior to actual calculation. What pseudopotentials accomplish, however, is extension of the range of methods for elements for which all-electron basis sets are not available.

Table 3-2: **Pseudopotentials and Associated All-Electron Basis Sets Available in Spartan**

basis set	available elements	basis set for lighter elements
LAV3P*, LAV3P**, LAV3P+*, LAV3P+**	Na-La, Hf-Bi	6-31G*, 6-31G**, 6-31+G*, 6-31+G** (H-Ne)
LACVP*, LACVP**, LACVP+*, LACVP+**	K-Cu, Rb-Ag, Cs-La, Hf-Au	6-31G*, 6-31G**, 6-31+G*, 6-31+G** (H-Ar), LAV3P*, LAV3P**, LAV3P+*, LAV3P+**(Zn-Kr, Cd-Xe, Hg-Bi)
LACV3P*, LACV3P**, LACV3P+* LACV3P+**	K-Cu, Rb-Ag, Cs-La, Hf-Au	6-311G*, 6-311G**, 6-311+G*, 6-311+G** (H-Ar), LAV3P*, LAV3P**, LAV3P+*, LAV3P+** (Zn-Kr, Cd-Xe, Hg-Bi)

Semi-Empirical Models[12]

Semi-empirical models follow directly from Hartree-Fock models. First, the size of the problem is reduced by restricting treatment to valence electrons only (electrons associated with the core are ignored). Next, the basis set is restricted to a minimal valence only representation. For main-group elements, this comprises a single s-type function and a set of p-type functions, e.g., $2s$, $2p_x$, $2p_y$, $2p_z$ for a first-row element, and for transition metals, a set of d-type functions, an s function and a set of p functions, e.g., $3d_{x^2-y^2}$, $3d_{z^2}$, $3d_{xy}$, $3d_{xz}$, $3d_{yz}$, $4s$, $4p_x$, $4p_y$, $4p_z$ for first-row transition metals. Hydrogen is represented by a single (1s) function. The only exception to this is the MNDO/d method for second-row (and heavier) main-group elements, used in conjunction with MNDO for hydrogen and first-row elements. This incorporates a set of d-type functions, in direct analogy to 3-21G$^{(*)}$ used in conjunction with 3-21G.

The central approximation, in terms of reducing overall computation, is to insist that atomic orbitals residing on different atomic centers do not overlap.

$$\int \phi_\mu \phi_\nu d\tau = 0 \quad \phi_\mu \text{ and } \phi_\nu \text{ not on the same atom} \qquad (42)$$

This is referred to as the Neglect of Diatomic Differential Overlap or NDDO approximation. It reduces the number of electron-electron interaction terms from $O(N^4)$ in the Roothaan-Hall equations to $O(N^2)$, where N is the total number of basis functions.

Additional approximations are introduced in order to further simplify the overall calculation, and more importantly to provide a framework for the introduction of empirical parameters. Except for models for transition metals, parameterizations are based on reproducing a wide variety of experimental data, including equilibrium geometries, heats of formation, dipole moments and ionization potentials. Parameters for PM3 for transition metals are based only on reproducing equilibrium geometries. The AM1 and PM3 models incorporate essentially the same approximations but differ in their parameterization.

Molecules in Solution

Much chemistry, perhaps most chemistry, is carried out not in the gas phase, but in solution. A wide variety of solvents are available to chemists. At one end of the spectrum is water which is both highly polar and highly structured. Water is unique among common solvents in that it is capable of forming hydrogen bonds to both (proton) donors and acceptors. At the other end of the spectrum are hydrocarbons such as decane, and relatively non-polar molecules such as methylene chloride. In the middle are a whole range of solvents such as tetrahydrofuran which differ both in their polarity and in their ability to act either as hydrogen-bond donors or acceptors.

Of course, quantum chemical calculations refer strictly to isolated "gas-phase" molecules. One can either hope that solvent effects will be small, as they apparently are in dealing with quantities such as molecular geometry, or that they will "cancel" where comparisons are made among similar systems. Where they are not small or where they cannot be made to cancel, some account needs to be made.

There are two general approaches to the treatment of molecules in solution:

i) Explicit approach. Here, one would perform a Monte-Carlo or molecular dynamics simulation on a molecule (the solute) "immersed" in a box containing a large but finite number of solvent molecules. This clearly has the advantage of being able to account for specific solvent-solute (and solvent-solvent) interactions, as well as for the effect of bulk solvent. The principle disadvantage of such an approach is computational cost. At the present time, only molecular mechanics models are practical, although "mixed" models, in which the solute is treated using quantum mechanics while the solvent is treated using molecular mechanics, should be in routine use within a few years.

ii) Implicit approach. This idea here is to replace specific solvent-solute interactions by an average field. In practice, this enters into the calculation as a term much like the core Hamiltonian in Hartree-Fock methodology (see equation 13), and does not add significantly to overall computation cost. Of course, because implicit models "do not know" about such things as hydrogen bonds, they are not likely to provide a wholly satisfactory account where specific interactions are of major importance.

Cramer/Truhlar Models for Aqueous Solvation[13]

Parameterized models proposed by Cramer, Truhlar and their coworkers are based on semi-empirical wavefunctions. The resulting "solvation energies" can then be added to "gas-phase" energies obtained at any level of calculation.

According to Cramer and Truhlar, the total solvation energy may be written as a sum of two terms, the first (E^{cavity}) accounting for the energy required to create a "cavity" in the solvent.

$$E^{cavity} = \sum_{A}^{solute\ atoms} \sigma_A S_A \tag{43}$$

The summation is over solute atoms. σ are empirical constants (parameters) specific to atom types and S are solvent-accessible surface areas.

The second term ($E^{electrostatic}$) accounts for electrostatic interactions between solvent and solute (once the solute is "placed" in the cavity).

$$E^{electrostatic} = -(1 - \frac{1}{\epsilon}) \left[\sum_{A}^{\substack{solute \\ atoms}} \sum_{B}^{\substack{images\ of \\ solute\ atoms}} Z_A Q_B \Gamma_{AB} \right.$$
$$\left. - \sum_{A}^{\substack{solute \\ atoms}} \sum_{A}^{\substack{basis\ functions \\ on\ A}} \sum_{B}^{\substack{images\ of \\ solute\ atoms}} P_{\mu\mu} Q_B \Gamma_{AB} \right] \tag{44}$$

Summations A are over solute atoms and summations B are over the images of the solute atoms. ϵ is the dielectric constant of the solvent, Z are solute atomic numbers, Q are "mirrors" of solute charges.

$$Q_B = -(1 - \frac{1}{\epsilon}) \; Z_A - \sum_{A}^{\substack{\text{basis functions} \\ \text{on A}}} P_{\mu\mu} \tag{45}$$

Γ are integrals representing the average Coulomb interaction between a solute atom and its image in the solvent and P are elements of the density matrix of the solute.

Nomenclature

Molecular modeling, like all other technical disciplines, has its own jargon. Much of this is described in **Appendix B** (**Common Terms and Acronyms**), and only one aspect will be addressed here. This concerns specification of theoretical model used for property calculation together with theoretical model used for equilibrium (or transition-state) geometry calculation.

Specification of theoretical model normally comprises two parts, separated by "/", i.e., type of calculation/basis set

type of calculation = not specified or HF (Hartree-Fock)

BP (Becke-Perdew density functional)

MP2 (second-order Møller-Plesset)

CIS (configuration interaction singles)

AM1 (AM1 semi-empirical)

MMFF (Merck molecular mechanics)

basis set = STO-3G (minimal)
3-21G (split valence)
6-31G* (polarization)

Specification of basis set alone implies a Hartree Fock calculation with that basis set, and no basis set is specified for semi-empirical and molecular mechanics calculation types.

If this is the complete designation, then this means that full optimization of equilibrium geometry (or transition-state geometry) is to be performed using the same type of calculation and basis set. If, however, the designation is followed by a "//" and then specification of a second "type of calculation/basis set", then this means that the calculation is to be preceded by equilibrium geometry (or transition-state geometry) optimization using this (second) type of calculation and basis set, i.e.

type of calculation/basis set//type of calculation/basis set

energy or property   geometry

6-31G*	Hartree-Fock 6-31G* calculation of energy and geometry
6-31G*//AM1	Hartree-Fock 6-31G* calculation of energy preceded by AM1 calculation of geometry
B3LYP/6-31G*//3-21G	B3LYP 6-31G* density functional calculation of energy preceded by Hartree-Fock 3-21G calculation of geometry
LMP2/6-311+G**//BP/6-31G*	localized MP2 calculation of energy with 6-311+G** basis set preceded by BP 6-31G* density functional calculation of geometry

References

1. Among the many excellent accounts of quantum mechanics applied to chemical systems are the following: (a) P.W. Aktins and R.S. Friedman, *Molecular Quantum Mechanics*, 3rd Ed., Oxford, 1997; (b) I.N. Levine, *(Quantum Chemistry*, 5th Ed., Prentice Hall, Upper Saddle River, NJ, 2000; (c) D.A. McQuarrie, *Quantum Chemistry*, University Science Books, Sausalito, CA, 1983.

2. Original papers: (a) C.C.J. Roothaan, *Rev. Mod. Phys.*, **23**, 69 (1951); (b) G.G. Hall, *Proc. Roy. Soc. (London)*, **A205**, 541 (1951). For reviews, see refs. 1 and: (c) W.J. Hehre, L. Radom, P.v.R. Schleyer and J.A. Pople, *Ab Initio Molecular Orbital Theory*, Wiley, New York, 1985; (d) A.R. Leach, *Molecular Modeling*, 2nd Ed., Prentice Hall, Upper Saddle River, NJ, 2001.

3. For a recent thorough treatment of correlated models, see: T. Helgaker, P. Jorgensen and J. Olsen, *Molecular Electronic Structure Theory*, Wiley, New York, 2000.

4. Original papers: (a) P. Hohenberg and W. Kohn, *Phys. Rev.*, **B136**, 864 (1964); (b) W. Kohn and L.J. Sham, *Phys. Rev.*, **A140**, 1133 (1965). For reviews see: refs. 1, 2d and: (c) R.G. Parr and W. Yang, *Density Functional Theory of Atoms and Molecules*, Oxford University Press, Oxford, 1988; (d) J.K. Labanowski and J.W. Andzelm, eds., *Density Functional Methods in Chemistry*, Springer-Verlag, New York, 1991; (e) J.M. Seminario and P. Politzer, eds., *Modern Density Functional Theory: A Tool for Chemistry*, Elsevier, Amsterdam, 1995.

5. For reviews see: refs. 1, 2(c) and 2(d).

6. Original paper: C. Møller and M.S. Plesset, *Phys. Rev.*, **46**, 618 (1934). For reviews see refs. 1, 2(c) and 2(d).

7. Refs. 2(a) and 2(b) and: (a) C.C. J. Roothaan, *Rev. Mod. Phys.*, **32**, 179 (1960); (b) J.S. Binkley, J.A. Pople and P.A. Dobosh, *Mol. Phys.*, **28**, 1423 (1974).

8. J.A. Pople and R.K. Nesbet, *J. Chem. Phys.*, **22**, 571 (1954).

9. For a review see: J.F. Stanton, J. Gauss, N. Ishikawa and M. Head-Gordon, *J. Chem. Phys.*, **103**, 4160 (1995).

10. For references to original papers, see ref. 2(c). Also: T.H. Dunning, *J. Chem. Phys.*, **90**, 1007 (1989), and following papers.

11. P.J. Hay and W.R. Wadt, *J. Chem. Phys.*, **82**, 270 (1985), and following papers.

12. Original papers: MNDO (a) M.J.S. Dewar and W.J. Thiel, *J. Amer. Chem. Soc.*, **99**, 4899 (1977); MNDO/d. (b) W. Thiel and A. Voityuk, *Theor. Chim.*

Acta, **81**, 391 (1992); (c) W. Thiel and A. Voityuk, Int. *J. Quantum Chem.*, **44**, 807 (1992). AM1. (d) M.J.S. Dewar, E.G. Zoebisch, E.F. Healy and J.J.P. Stewart, *J. Amer. Chem. Soc.*, **107**, 3908 (1985). PM3. (e) J.J.P. Stewart, *J. Computational Chem.*, **10**, 209 (1989). For reviews see refs. 1, 2(d) and 2 (f). T. Clark, *A Handbook of Computational Chemistry*, Wiley, New York, 1986. For a review of the original semi-empirical methods, see: (g) J.A. Pople and D.A. Beveridge, *Approximate Molecular Orbital Theory*, McGraw-Hill, New York, 1970.

13. (a) C.C. Chambers, G.D. Hawkins, C.J. Cramer and D.G. Truhlar, *J. Chem. Phys.*, **100**, 16385 (1996). For a review see: (b) C.J. Cramer and D.G. Truhlar, eds., *Structure and Reactivity in Aqueous Solution*, ACS Symposium Series, no. 568, American Chemical Society, Washington, D.C., 1994.

Chapter 3
Molecular Mechanics Models

This chapter describes the basis of molecular mechanics models and introduces the SYBYL and MMFF force fields. It also compares and contrasts molecular mechanics and quantum chemical models.

Introduction

Molecular mechanics describes molecules in terms of "bonded atoms", which have been distorted from some idealized geometry due to non-bonded van der Waals and Coulombic interactions.[1] This is fundamentally different from quantum chemical models, which make no reference whatsoever to chemical bonding. The success of molecular mechanics models depends on a high degree of transferability of geometrical parameters from one molecule to another, as well as predictable dependence of the parameters on atomic hybridization. For example, carbon-carbon single bond lengths generally fall in the small range from 1.45 to 1.55Å, and increase in length with increasing "p character" of the carbon hybrids. Thus, it is possible to provide a fairly accurate "guess" at molecular geometry in terms of bond lengths, bond angles and torsion angles, provided that the molecule has already been represented in terms of a particular valence structure. The majority of organic molecules fall into this category.

The molecular mechanics "energy" of a molecule is described in terms of a sum of contributions arising from distortions from "ideal" bond distances ("stretch contributions"), bond angles ("bend contributions") and torsion angles ("torsion contributions"), together with contributions due to "non-bonded" (van der Waals and Coulombic) interactions. It is commonly referred to as a "strain energy", meaning that it reflects the "strain" inherent to a "real" molecule relative to some idealized form.

$$E^{\text{strain}} = \overset{\text{bonds}}{\underset{A}{\sum}} E_A^{\text{stretch}} + \overset{\text{bond angles}}{\underset{A}{\sum}} E_A^{\text{bend}} + \overset{\text{torsion angles}}{\underset{A}{\sum}} E_A^{\text{torsion}} + \overset{\text{non-bonded atoms}}{\underset{A}{\sum}\underset{B}{\sum}} E_{AB}^{\text{non-bonded}} \quad (1)$$

The first three summations in equation 1 are over all "bonds", all "bond angles" and all "torsion angles", respectively. Thus, information about bonding is "part of the input" to a molecular mechanics calculation, in contrast to a quantum chemical calculation where it is "part of the output". The last summation in equation 1 is over all pairs of atoms which are not bonded.

Stretch and bend terms are most simply given in terms of quadratic ("Hook's law") forms.

$$E^{\text{stretch}}(r) = \frac{1}{2} k^{\text{stretch}} (r - r^{\text{eq}})^2 \quad (2)$$

$$E^{\text{bend}}(\alpha) = \frac{1}{2} k^{\text{bend}} (\alpha - \alpha^{\text{eq}})^2 \quad (3)$$

r and α are the bond distance and angle, respectively, r^{eq} and α^{eq} are the "ideal" (equilibrium) bond length and bond angle, respectively, taken either from experiment or from accurate quantum chemical calculations, and k^{stretch} and k^{bend}, so-called stretch and bend "force constants", respectively, are parameters. Molecular mechanics models may also include cubic and higher-order contributions, as well as "cross terms" to account for correlations between stretch and bend components. The degree of complexity depends on the availability of data on which to base parameters.

Proper description of the torsional potential requires a form that reflects its inherent periodicity. For example, the three-fold periodicity of rotation about the carbon-carbon bond in ethane may be described by the simple functional form.

$$E^{\text{torsion}}(\omega) = k^{\text{torsion3}} [1 - \cos 3(\omega - \omega^{\text{eq}})] \quad (4)$$

ω is the torsion angle, ω^{eq} is the ideal torsion angle and k^{torsion3} is treated as a parameter. Bond torsion contributions to the overall energy may also need to include terms which are one-fold and two-fold periodic.

$$E^{torsion}(\omega) = k^{torsion1}[1 - \cos(\omega - \omega^{eq})] + k^{torsion2}[1 - \cos 2(\omega - \omega^{eq})] + k^{torsion3}[1 - \cos 3(\omega - \omega^{eq})] \tag{5}$$

$k^{torsion1}$ and $k^{torsion2}$ are additional parameters. Equation 5 is a truncated Fourier series. The one-fold term accounts for the difference in energy between *cis* (0°) and *trans* (180°) conformers, and the two-fold term accounts for the difference in energy between planar (0°, 180°) and perpendicular (90°, 270°) conformers. (Further discussion is provided in **Chapter 14**.) Molecular mechanics models may also include higher-order terms and cross terms, as well as terms to account for asymmetrical environments. As with stretch and bend components, the degree of complexity depends on the availability of data on which to base parameters.

Non-bonded interactions typically involve a sum of van der Waals (VDW) interactions and Coulombic interactions.

$$E^{non\text{-}bonded}(r) = E^{VDW}(r) + E^{Coulombic}(r) \tag{6}$$

Additional non-bonded terms may be included to account explicitly for such interactions as hydrogen bonding.

Most commonly, van der Waals interactions are represented as a sum of a repulsive and attractive terms.

$$E^{VDW}(r) = \varepsilon\left[\left(\frac{r^o}{r}\right)^{12} - 2\left(\frac{r^o}{r}\right)^{6}\right] \tag{7}$$

r is the non-bonded distance, and ε and r^o are parameters. This functional form provides a very steep energy barrier inside the sum of van der Waals radii for the two atoms involved, and a shallow energy well at larger separations, and as such accounts both for the inherent size requirements of atoms, as well as for weak attractive forces between separated atoms.

The Coulombic term takes account of the interaction of charges.

$$E^{Coulombic}(r) = \frac{qq'}{r} \tag{8}$$

r is the non-bonded distance, and the atomic charges, q, may either be treated as parameters or be taken from quantum chemical calculations. The sum of atomic charges needs to be equal the total molecular charge, 0 in the case of a neutral molecule.

SYBYL and MMFF Force Fields

Molecular mechanics models differ both in the number and specific nature of the terms which they incorporate, as well as in the details of their parameterization. Taken together, functional form and parameterization, constitute what is termed a force field. Very simple force fields such as SYBYL[2], developed by Tripos, Inc., may easily be extended to diverse systems but would not be expected to yield quantitatively accurate results. On the other hand, a more complex force field such as MMFF94[3] (or more simply MMFF), developed at Merck Pharmaceuticals, while limited in scope to common organic systems and biopolymers, is better able to provide quantitative accounts of molecular geometry and conformation. Both SYBYL and MMFF are incorporated into Spartan.

Limitations of Molecular Mechanics Models

The primary advantage of molecular mechanics models (over any of the quantum chemical models described in the previous chapter) is their simplicity. Except for very small systems, computation cost is completely dominated by evaluation of non-bonded van der Waals and Coulombic terms, the number of which is given by the square of the number of atoms. However, the magnitude of these terms falls off rapidly with increasing interatomic distance and, in practice, computation cost scales linearly with molecular size for sufficiently large molecules. Molecular mechanics calculations may easily be performed on molecules comprising several thousand atoms. Additionally, molecular mechanics calculations are sufficiently rapid to permit extensive conformational searching on molecules containing upwards of 100-200 atoms. Conformational analysis is perhaps the single most important application of molecular mechanics.

The fact that molecular mechanics models are parameterized may also be seen as providing an advantage over quantum chemical models. It is possible, at least in principle, to construct molecular mechanics models which will accurately reproduce known experimental data, and hopefully will anticipate (unknown) data on closely-related systems.

There are important limitations of molecular mechanics models. First, they are limited to the description of equilibrium geometries and equilibrium conformations. Because the mechanics "strain energy" is specific to a given molecule (as a measure of how far this molecule deviates from its "ideal arrangement"), strain energies cannot be used in thermochemical calculations. Two important exceptions are calculations involving isomers with exactly the same bonding, e.g., comparison of *cis* and *trans*-2-butene, and conformational energy comparisons, where different conformers necessarily have exactly the same bonding.

Second, molecular mechanics calculations reveal nothing about bonding or, more generally, about electron distributions in molecules. As will become evident later, information about electron distributions is key to modeling chemical reactivity and selectivity. There are, however, important situations where purely steric effects are responsible for trends in reactivity and selectivity, and here molecular mechanics would be expected to be of some value.

Third, currently available force fields have not been parameterized to handle non-equilibrium forms, in particular, reaction transition states. Note, however, that there is no fundamental reason why this could not be done (using results from quantum chemical calculations rather than experiment as a basis for parameterization).

Finally, it needs to be noted that molecular mechanics is essentially an interpolation scheme, the success of which depends not only on good parameters, but also on systematics among related molecules. Molecular mechanics models would not be expected to be highly successful in describing the structures and conformations of "new" (unfamiliar) molecules outside the range of parameterization.

References

1. Reviews: (a) U. Burkert and N.L. Allinger, *Molecular Mechanics*, ACS Monograph no. 177, American Chemical Society, Washington D.C., 1982; (b) A.K. Rappe´ and C.J. Casewit, *Molecular Mechanics Across Chemistry*, University Science Books, Sausalito, CA, 1997.

2. M. Clark, R.D. Cramer III and N. van Opdensch, *J. Computational Chem.*, **10**, 982 (1989).

3. T.A. Halgren, *J. Computational Chem.*, **17**, 490 (1996), and following papers in this issue. Two different variations of MMFF have been published. The default choice in Spartan is the one in which nitrogen attached to an unsaturated carbon, e.g., the nitrogen in aniline, is allowed to pucker. The other variation assumes that the nitrogen attached to unsaturated carbon is planar.

Chapter 4

Graphical Models

This chapter introduces a number of "useful" graphical models, including molecular orbitals, electron densities, spin densities, electrostatic potentials and local ionization potentials, and relates these models both to molecular size and shape and molecular charge distributions. The chapter also introduces and illustrates "property maps" which simultaneously depict molecular size and shape in addition to a molecular property. Properties include the electrostatic potential, the value of the LUMO, the local ionization potential and the spin density.

Introduction

Among the quantities which have proven of value as graphical models are the molecular orbitals, the electron density, the spin density (for radicals and other molecules with unpaired electrons), the electrostatic potential and the local ionization potential. These may all be expressed as three-dimensional functions of the coordinates. One way to display them on a two-dimensional video screen (or on a printed page) is to define a surface of constant value, a so-called isovalue surface or, more simply, isosurface[*].

$$f(x,y,z) = \text{constant} \tag{1}$$

The value of the constant may be chosen to reflect a particular physical observable of interest, e.g., the "size" of a molecule in the case of display of electron density.

Graphical models need not be restricted to portraying a single quantity. Additional information may be presented in terms of a property map on top of an isosurface, where different colors may be used to portray

---

[*] Another common display involves a two-dimensional plane or "slice" which cuts into the overall three-dimensional function, and to demark equal value lines (contours) onto this slice.

different property values. Most common are maps on electron density surfaces. Here the surface may be used to designate overall molecular size and shape, and the colors to represent the value of some property at various locations on the surface. For example, the value of the electrostatic potential (the energy of interaction of a positive point charge with the nuclei and electrons of a molecule) mapped onto an electron density isosurface may be employed to distinguish regions on the surface which are electron rich ("basic" or subject to electrophilic attack) from those which are electron poor ("acidic" or subject to nucleophilic attack).

This chapter introduces and illustrates isosurface displays of molecular orbitals, electron and spin densities, electrostatic potentials and local ionization potentials, as well as maps of the lowest-unoccupied molecular orbital, the electrostatic and local ionization potentials and the spin density (on top of electron density surfaces). Applications of these models to the description of molecular properties and chemical reactivity and selectivity are provided in **Chapter 19** of this guide.

Because the images in this chapter are reproduced in black and white, some of the information they are intended to portray has been lost. This is especially true for property maps, where a spectrum of colors is used to convey the value of a particular property. All images in this chapter have been provided as Spartan files on an accompanying CD-ROM. These are marked by an icon (4-y) where 4 is the chapter number and y is the number of the Spartan file.

Molecular Orbitals

Chemists are familiar with the molecular orbitals of simple molecules. They recognize the σ and π orbitals of acetylene, and readily associate these with the molecule's σ and π bonds.

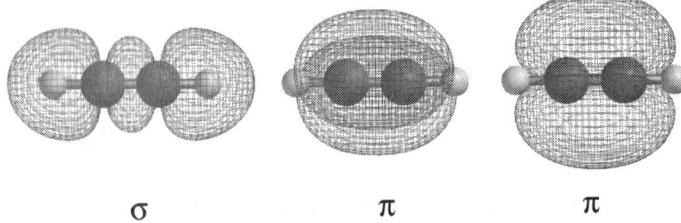

σ $\qquad$ π $\qquad$ π

Note, however, that even in such a simple case as this, molecular orbitals do not correspond one-to-one with bonds. For example, the highest-energy σ orbital in acetylene is clearly made up of both CC and CH bonding components. The reason, as pointed out in **Chapter 2**, is that molecular orbitals are written as linear combinations of nuclear-centered basis functions, and will generally be completely delocalized over the entire nuclear skeleton.

$$\psi_i = \overset{\text{basis functions}}{\underset{\mu}{\sum}} c_{\mu i}\phi_\mu \tag{2}$$

Molecular orbitals do not need to be directly involved in bonding to be informative. For example, the highest-occupied molecular orbital (HOMO) of sulfur tetrafluoride clearly reveals that the molecule incorporates a lone pair on sulfur pointing in the direction of the "missing *equatorial* bond".[*]

4-2

A simple example where the shape of the HOMO "foretells" of chemistry is found in cyanide anion.

4-3

Of course, cyanide acts as a nucleophile in S_N2 reactions, e.g.

$$:N\equiv C:^- \quad CH_3-I \longrightarrow :N\equiv C-CH_3 + I^-$$

---

[*] This is why SF_4 adopts a trigonal bipyramidal as opposed to a tetrahedral equilibrium geometry. Electrons, like atoms, take up space, which is the basis of such models as VSEPR theory.

The HOMO in cyanide is more concentrated on carbon (on the right) than on nitrogen suggesting, as is observed, that it will act as a carbon nucleophile. While at first glance, this might seem to be at odds with the fact that nitrogen is more electronegative than carbon, and thus more likely to hold the "negative charge", more careful consideration reveals that "all is as it should be". Because nitrogen is more electronegative than carbon, it "holds on to its electrons" better than does carbon, meaning that it will be the poorer nucleophile.

Molecular orbitals do not even need to be occupied to be informative. For example, the lowest-unoccupied molecular orbital (LUMO) of planar (top) and perpendicular (bottom) benzyl cation anticipate the difference in charge delocalization of the two systems.

It is into the LUMO, the energetically most accessible unfilled molecular orbital, that any further electrons will go. Hence, it may be thought of as demarking the location of positive charge in a molecule. The LUMO in planar benzyl cation is delocalized away from the formal cation center and onto the *ortho* and *para* ring carbons, in accord with classical resonance structures. On the other hand, the LUMO in perpendicular benzyl cation remains primarily localized on the benzylic carbon. Resonance theory suggests that delocalization ⌐f the positive charge leads to stabilization. Thus, planar benzyl cation ⌐re stable than perpendicular benzyl cation.

Examination of the LUMO of methyl iodide helps to "rationalize" why iodide leaves following attack by cyanide.

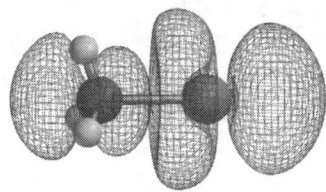

This orbital is antibonding between carbon and iodine, meaning that donation of the electron pair from cyanide will cause the CI bond to weaken and eventually break.

Molecular orbital descriptions offer a number of significant advantages over conventional resonance structures. For one, they often provide "more compact" descriptions, e.g., the LUMO in planar benzyl cation conveys the same information as four resonance structures. Second, orbital descriptions are quantitative, compared to resonance structures which are strictly qualitative. Finally, molecular orbital descriptions may be applied much more widely than resonance descriptions. Of course, molecular orbital descriptions cannot be generated "using a pencil" as can resonance structures, but rather require a computer. It can be argued that this does not constitute a disadvantage, but rather merely reflects a natural evolution of the tools available to chemists.

It was Woodward and Hoffmann[1] who first introduced organic chemists to the idea that so-called "frontier orbitals" (the HOMO and LUMO) often provide the key to understanding why some chemical reactions proceed easily whereas others do not.[*] For example, the fact that the HOMO in *cis*-1,3-butadiene is able to interact favorably with the LUMO in ethylene, suggests that the two molecules should readily combine in a concerted manner to form cyclohexene, i.e., Diels-Alder cycloaddition.

_____

* Fukui had earlier advanced similar ideas but not connected them as clearly to chemical reactivity. For their work, Hoffmann and Fukui shared the Nobel Prize in Chemistry in 1981.

4-6

LUMO

HOMO

On the other hand, interaction between the HOMO on one ethylene and the LUMO on another ethylene is not favorable, and concerted addition to form cyclobutane would not be expected.

4-7

LUMO

HOMO

"Orbital symmetry arguments" or the Woodward-Hoffmann rules, as they are now commonly referred to are, however, not easily extended beyond planar π systems. In great part, this is due to the difficulty of constructing and "sketching" by hand and visualizing molecular orbitals of three-dimensional systems, a situation which modern computer graphics has now completely altered.

Electron Density

The total electron density, or more simply, the electron density, $\rho(\mathbf{r})$, is a function of the coordinates $\mathbf{r}$, defined such that $\rho(\mathbf{r})d\mathbf{r}$ is the number of electrons inside a small volume $d\mathbf{r}$. This is what is measured in an X-ray diffraction experiment. For a (closed-shell) molecule, $\rho(\mathbf{r})$ is written in terms of a sum of products of basis functions, ϕ.

$$\rho(\mathbf{r}) = \overset{\text{basis functions}}{\sum_{\mu}\sum_{\nu}} P_{\mu\nu}\phi_{\mu}(\mathbf{r})\phi_{\nu}(\mathbf{r}) \tag{3}$$

P is the density matrix (equation 16 in **Chapter 2**). The electron density may be portrayed in terms of an isosurface (an isodensity surface) with the size and shape of the surface being given by the value of the density, for example, in cyclohexanone.

←——————— large density value

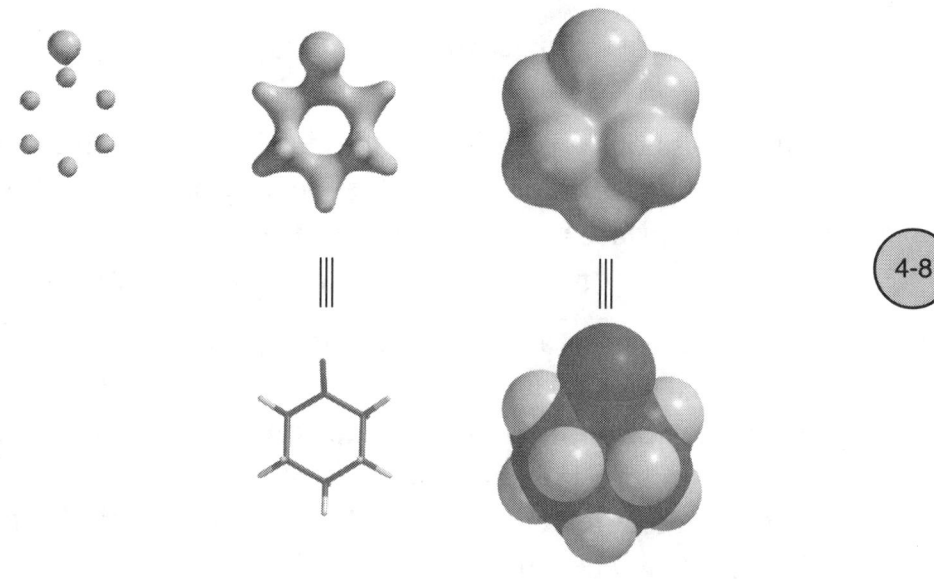

small density value ——→

Depending on this value, isodensity surfaces may either serve to locate atoms, delineate chemical bonds, or to indicate overall molecular size and shape. The regions of highest electron density surround the heavy (non-hydrogen) atoms in a molecule. Thus, the X-ray diffraction experiment locates atoms by identifying regions of high electron density. Also interesting, are regions of slightly lower electron density. For example, a 0.1 electrons/au^3 isodensity surface for cyclohexanone conveys essentially the same information as a conventional skeletal structure model, that is, it depicts the locations of bonds.

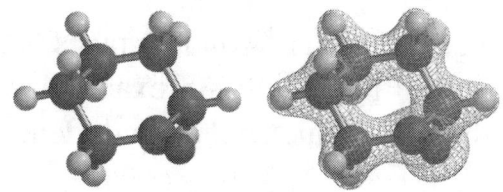

A different density surface (0.002 electrons/au³) serves to portray overall molecular size and shape. This is, of course, the same information portrayed by a conventional space-filling (CPK) model.*

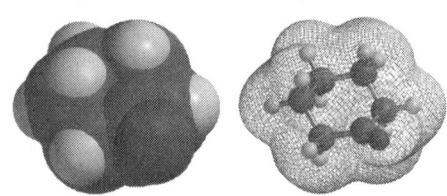

In the discussions which follow in this chapter and in the remainder of this guide, a 0.1 electrons/au³ isodensity surface will be referred to as a bond surface and a 0.002 electrons/au³ isodensity surface either as a size surface or more simply as a density surface.

Bond and size surfaces offer some significant advantages over conventional skeletal and space-filling models. Most important, bond surfaces may be applied to elucidate bonding and not only to portray "known" bonding. For example, the bond surface for diborane clearly shows a molecule with very little electron density concentrated between the two borons.

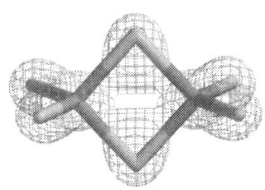

This suggests that the appropriate Lewis structure is the one which lacks a boron-boron bond, rather than the one which shows the two borons directly bonded.

Another important application of bond surfaces is to the description of the bonding in transition states. An example is the pyrolysis of ethyl formate, leading to formic acid and ethylene.

---

* This is not accidental. The radii used to define CPK models were originally chosen to reflect the space which molecules take up when they pack in solids.

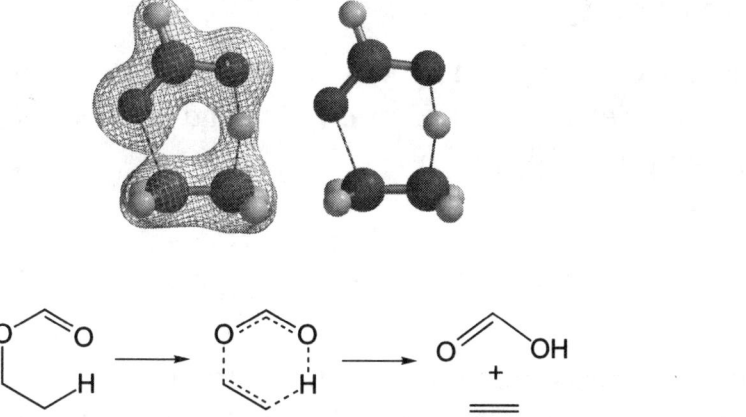

4-12

The bond surface offers clear evidence of a "late transition state". The CO bond is nearly fully cleaved and the migrating hydrogen is more tightly bound to oxygen (as in the product) than to carbon (as in the reactant). Further information may be obtained by replacing the static picture by a "movie", i.e., animation along the reaction coordinate (see discussion later in this chapter).

Electron density surfaces can also be used to uncover trends and build qualitative descriptions. For example, size surfaces for the isoelectronic molecules, methyl anion, ammonia and hydronium cation show a marked decrease in overall size.

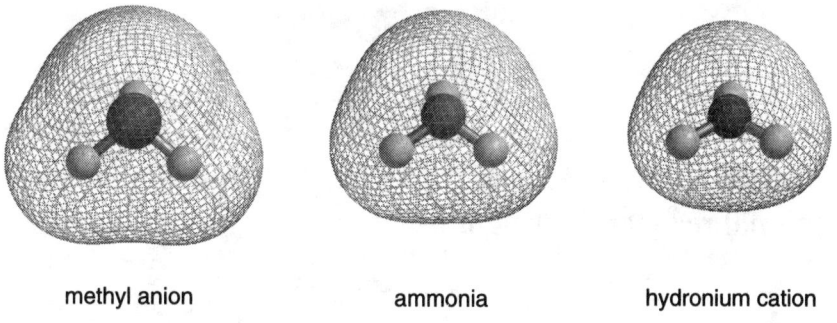

methyl anion   ammonia   hydronium cation

4-13

The reason for the trend is obvious; the greater the nuclear charge (for a given number of electrons) the more tightly will the electrons be held. This supports the notion that Coulombic attraction between nuclei and electrons is what holds molecules together.

Spin Density

The spin density, $\rho^{spin}(\mathbf{r})$, is defined as the difference in electron density formed by electrons of α spin, $\rho^{\alpha}(\mathbf{r})$, and the electron density formed by electrons of β spin, $\rho^{\beta}(\mathbf{r})$.[*]

$$\rho^{spin}(\mathbf{r}) = \rho^{\alpha}(\mathbf{r}) - \rho^{\beta}(\mathbf{r}) = \overset{\text{basis functions}}{\sum_{\mu}\sum_{\nu}} (P_{\mu\nu}^{\alpha} - P_{\mu\nu}^{\beta})\ \phi_{\mu}(\mathbf{r})\phi_{\nu}(\mathbf{r}) \tag{4}$$

The ϕ are basis functions, and the P are density matrices (analogous to equation 16 in **Chapter 2**, but with single-electron occupancy).

For closed-shell molecules (in which all electrons are paired), the spin density is zero everywhere. For open-shell molecules (in which one or more electrons are unpaired), the spin density indicates the distribution of unpaired electrons. Spin density is an obvious indicator of reactivity of radicals (in which there is a single unpaired electron). Bonds will be made to centers for which the spin density is greatest. For example, the spin density isosurface for allyl radical suggests that reaction will occur on one of the terminal carbons and not on the central carbon.

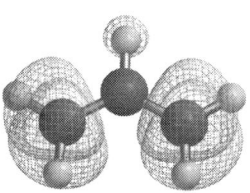

This is what is observed and, of course, is also anticipated using conventional resonance structures.

Closely related is the radical formed by removal of the hydroxyl hydrogen in vitamin E.

---

[*]   This assumes use of unrestricted SCF procedures for molecules with unpaired electrons as opposed to restricted open-shell SCF procedures (see discussion in **Chapter 2**). Spin densities constructed using restricted SCF procedures would only need to consider singly-occupied orbitals.

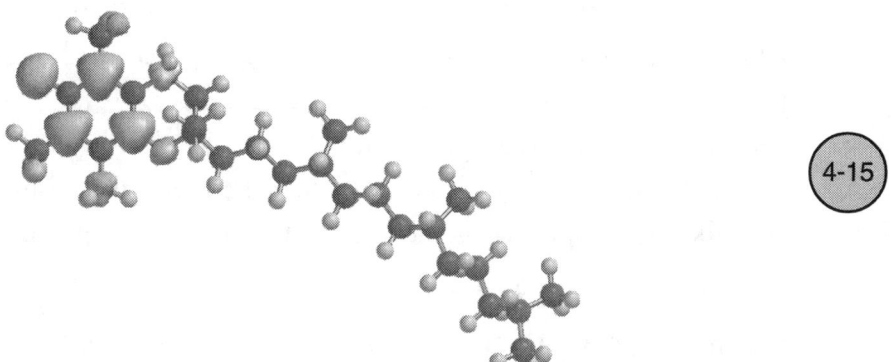

The spin density surface shows that the unpaired electron is not localized on oxygen, but is delocalized onto the benzene ring.

This suggests that vitamin E should be able to rapidly react with oxidizing agents (radicals) to give stable products that can be safely excreted.

Spin density surfaces offer significant advantages over resonance structures insofar as anticipating structure and reactivity. For one, while resonance structures are relatively easy to construct for simple "planar" systems (such as allyl radical), and their interpretation relatively straightforward, there is much less experience in applying resonance arguments to larger ("three-dimensional") systems. Additionally, resonance arguments are qualitative and completely inadequate for describing of subtle differences which are often critical in dictating structure, stability and reactivity, for example, differences caused by remote substituents or by changes in stereochemistry. In these situations, spin density surfaces are able to provide quantitative accounts.

Electrostatic Potential

The electrostatic potential, ε_p, is defined as the energy of interaction of a positive point charge located at p with the nuclei and electrons of a molecule.[2]

$$\varepsilon_p = \sum_A^{\text{nuclei}} \frac{Z_A}{R_{Ap}} - \sum_\mu^{\text{basis functions}} \sum_\nu P_{\mu\nu} \int \frac{\phi_\mu^*(\mathbf{r})\phi_\nu(\mathbf{r})}{r_p} \, d\mathbf{r} \qquad (5)$$

The first summation is over nuclei A. Z are atomic numbers and R_{AP} are distances between the nuclei and the point charge. The second pair of summations is over basis functions, ϕ. P is the density matrix (equation 16 in **Chapter 2**), and the integrals reflect Coulombic interactions between the electrons and the point charge, where r_p is the distance separating the electron and the point charge.

A surface for which the electrostatic potential is negative (a negative potential surface) delineates regions in a molecule which are subject to electrophilic attack, for example, above and below the plane of the ring in benzene, and in the ring plane above the nitrogen in pyridine.

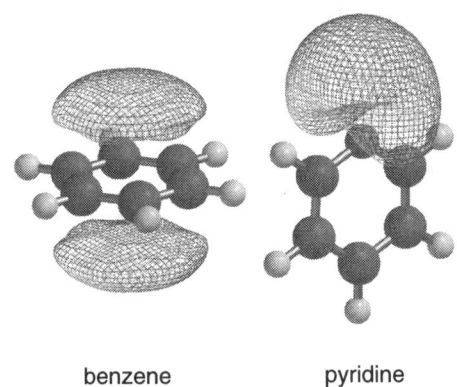

benzene   pyridine

While these two molecules are structurally very similar, potential surfaces make clear that this similarity does not carry over into their electrophilic reactivities.

More generally, negative potential surfaces serve to "outline" the location of the highest-energy electrons. For example, negative potential surfaces for trimethylamine, dimethyl ether and methyl fluoride are an artifact of the non-bonded "lone pairs" of electrons.

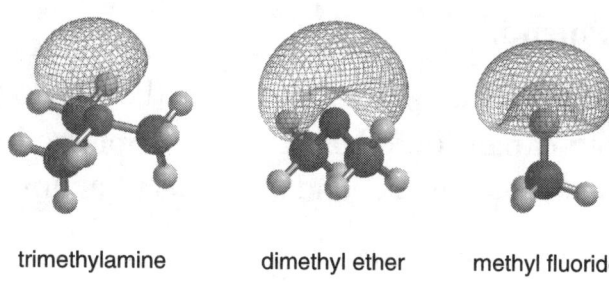

trimethylamine   dimethyl ether   methyl fluoride

The electrostatic potential surface for trimethylamine results from a single non-bonded valence molecular orbital (the HOMO), while the electrostatic potential surfaces for dimethyl ether and methyl fluoride result from a combination of two and three high-lying non-bonded molecular orbitals, respectively, i.e.

Visual comparison of the electrostatic potential surfaces for these three molecules uncovers a serious problem, and points out the need for caution in their interpretation. Note, that the "size" of the (attractive) potential for dimethyl ether is roughly the same as that for methyl fluoride and significantly larger than that for trimethylamine. Given that all three surfaces correspond to the same (negative) value of the potential, this suggests that dimethyl ether and methyl fluoride will be more likely to attract a proton than trimethylamine. The reality is different; the proton affinity of trimethylamine (in the gas phase) is much larger than the proton affinity of dimethyl ether, which in turn is much larger than the proton affinity of methyl fluoride.

One source of the "failure" has to do with the fact that the electrostatic potential does not take the energy of electron redistribution (the "polarization" energy) into account. This is likely to be more significant for trimethylamine (most polarizable) than for dimethyl ether, than for methyl fluoride (least polarizable). The problem can be addressed by explicitly taking the polarization energy into account.

Polarization Potential

The polarization potential, ε_p', is the next term (beyond the electrostatic potential) in the expansion of the energy of interaction of a point positive charge with the nuclei and electrons of a molecule.[3]

$$\varepsilon_p'(\mathbf{r}) = \overbrace{\sum_i^{occ} \sum_a^{unocc}}^{\substack{\text{molecular orbitals}}} \frac{1}{\epsilon_i - \epsilon_a} \left[\overbrace{\sum_\mu \sum_\nu}^{\substack{\text{basis functions}}} c_{\mu i} c_{\nu a} \int \frac{\phi_\mu(\mathbf{r}) \phi_\nu(\mathbf{r})}{r_p} \, d\mathbf{r} \right]^2 \tag{6}$$

ϵ_i and ϵ_a are energies of occupied and unoccupied molecular orbitals, respectively, and the c are the coefficients of the molecular orbitals. The outer two summations are over molecular orbitals, and the inner two summations are over basis functions.

The polarization potential provides the energy due to electronic reorganization of the molecule as a result of its interaction with a point positive charge.* The sum of the electrostatic and polarization potentials provides a better account of the energy of interaction of a point positive charge than available from the electrostatic potential alone. It properly orders the proton affinities of trimethylamine, dimethyl ether and fluoromethane.

Local Ionization Potential

Another quantity of some utility is the so-called local ionization potential, $I(\mathbf{r})$.[4] This is defined as the sum over orbital electron densities, $\rho_i(\mathbf{r})$ times absolute orbital energies, $|\epsilon_i|$, and divided by the total electron density, $\rho(\mathbf{r})$.

$$I(\mathbf{r}) = \overbrace{\sum_i}^{\substack{\text{occupied} \\ \text{molecular orbitals}}} \rho_i(\mathbf{r}) |\epsilon_i| \Big/ \rho(\mathbf{r}) \tag{7}$$

The local ionization potential is intended to reflect the relative ease of electron removal ("ionization") at any location around a molecule. For example, a surface of "low" local ionization potential for sulfur tetrafluoride demarks the areas which are most easily ionized.

---

*   Of course, the polarization potential does not account for electron transfer as would occur were a "real" electrophile to replace the point positive charge.

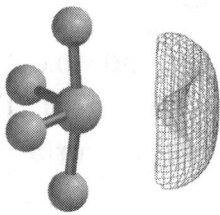

It is clearly recognizable as a lone pair on sulfur.

A more important application of the local ionization potential is as an alternative to the electrostatic potential as a graphical indicator of electrophilic reactivity. This is in terms of a property map rather than as an isosurface.[*] Further discussion is provided later in this chapter.

Property Maps

Additional information (a "property") may be added to any isosurface by using color to represent the value of the property. Colors at one end of the visible spectrum could represent "small" property values and at the other end, "large" property values. This gives rise to a model which actually conveys four dimensions of information.

3	dimensions conveying structure
+ 1	dimension conveying property value
4	dimensions

Most commonly, a property is mapped onto the size surface. This depicts overall molecular size and shape and, therefore, demarks surface regions "visible" to an incoming reagent.[**] There are situations where a property is more clearly visible when mapped instead onto a bond surface.

---

[*]   The local ionization potential does not fall to zero with increasing distance from the molecule. This makes its use as an isosurface problematic.

[**] This is not necessarily the same as depicting regions which are accessible to an incoming reagent.

Electrostatic Potential Map

The most commonly employed and (to date) most important property map is the electrostatic potential map. This gives the electrostatic potential at locations on a particular surface, most commonly a surface of electron density corresponding to overall molecular size (a size surface).

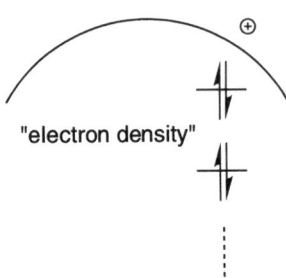

To see how an electrostatic potential map (and by implication any property map) is constructed, first consider both a size surface and a particular (negative) potential surface for benzene.

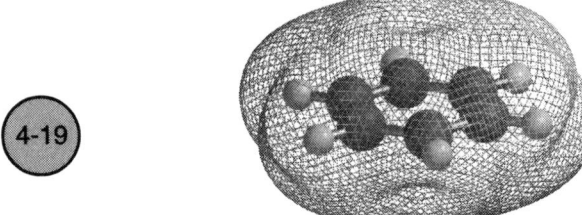 

size surface   negative potential surface

Both of these surfaces convey structure. The size surface reveals the size and shape of benzene, while the negative potential surface delineates in which regions surrounding benzene a particular (negative) electrostatic potential will be "felt".

Next, consider making a "map" of the value of the electrostatic potential on the size surface (an electrostatic potential map), using colors to designate values of the potential. This leaves the size surface unchanged (insofar as representing the size and shape of benzene), but replaces the grayscale image (conveying only structural information) by a color image (conveying the value of the electrostatic potential *in addition to* structure).

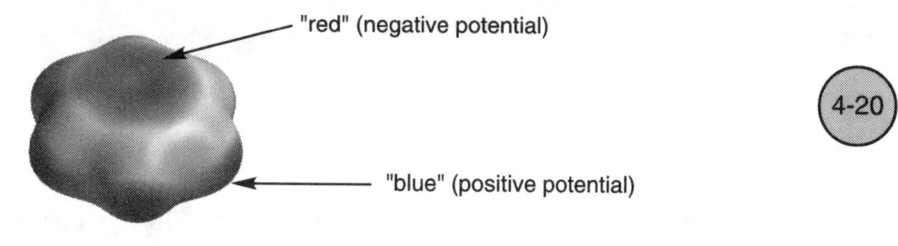

electrostatic potential map

In this example, colors near red represent large negative values of the potential, while colors near blue represent large positive values (orange, yellow and green represent intermediate values of the potential).[*] Note that the π system is "red", consistent with the (negative) potential surface previously shown.

Electrostatic potential maps have found a myriad of uses. Among them, they serve to quickly characterize various regions in a molecule as electron rich or electron poor, or neither rich or poor. For example, an electrostatic potential map of the zwitterionic form of β-alanine ($^+H_3NCH_2CH_2CO_2^-$) shows, as expected, positive charge (blue color) in the vicinity of the "protonated amine", negative charge (red color) in the vicinity of the "deprotonated carboxylic acid" and a central region which is "neutral" (green color).

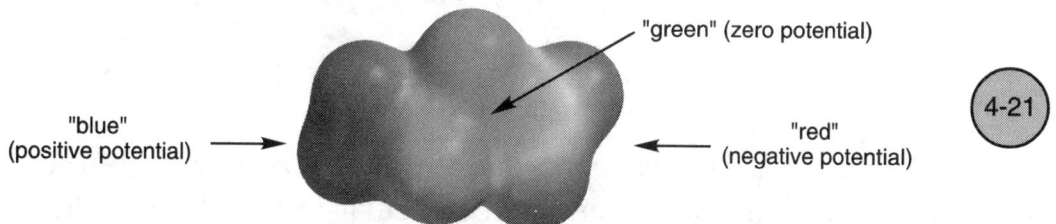

The presence of large "neutral" (green) regions in the electrostatic potential map for vitamin E suggests that the molecule will be soluble in lipids (as it must be in order to function as a trap for radicals).

---

*   This is the default color scheme used for all images in this guide and in Spartan.

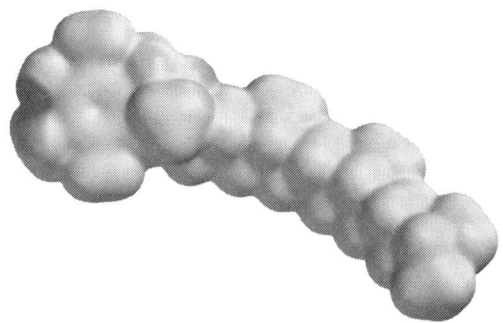

Another use of electrostatic potential maps is to distinguish between molecules in which charge is localized from those where it is delocalized. A good example involves comparison of electrostatic potential maps for planar (top) and perpendicular (bottom) structures of benzyl cation. (They have been drawn on the same (color) scale.)

"blue"
(positive potential)

The latter reveals heavy concentration of positive charge (blue color) on the benzylic carbon and perpendicular to the plane of the ring. This is consistent both with the notion that only a single resonance structure may be drawn, as well as with the fact that the LUMO is localized almost entirely on the benzylic carbon (see discussion earlier in this chapter). On the other hand, planar benzyl cation shows no such buildup of positive charge on the benzylic carbon, but rather delocalization onto *ortho* and *para* ring carbons, exactly as suggested by resonance theory, and by the LUMO which is delocalized over four centers.

> An organic chemist would "know" that benzyl cation is planar and not perpendicular because "four resonance structures are better than one". A physical chemist would reach the same conclusion based on Coulomb's law "separation of charge requires that energy be expended".

Electrostatic potential maps may also be employed to characterize transition states in chemical reactions. A good example is pyrolysis of ethyl formate (leading to formic acid and ethylene).

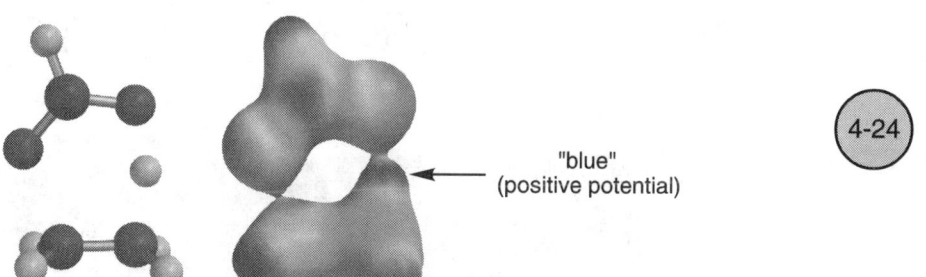

Here, the electrostatic potential map clearly shows that the hydrogen being transferred (from carbon to oxygen) is positively charged (it is an electrophile).*

"blue"
(positive potential)

(4-24)

S_N2 reactions provide an interesting example of the utility of electrostatic potential maps in rationalizing an experimental result, while challenging "conventional wisdom". It is well established that a nucleophile such as bromide reacts much faster with methyl bromide than it does with *tert*-butyl bromide. The reason normally cited is that while the transition state for the S_N2 reaction with methyl bromide is "uncrowded", that for the corresponding reaction with *tert*-butyl bromide is "sterically crowded". However, this interpretation does

---

*   In this example, a bond surface has been used in lieu of a size surface to better show the position of the migrating hydrogen.

not stand up to quantitative calculations which show that neither transition state is particularly crowded.

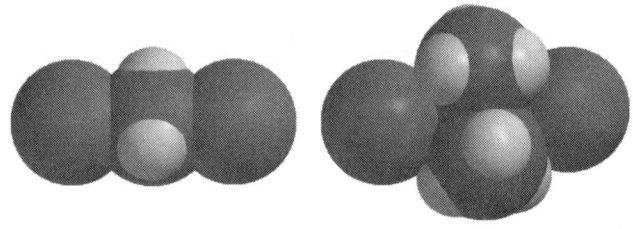

bromide + methyl bromide   bromide + *tert*-butyl bromide

What is going on? While a *tert*-butyl group presents the incoming nucleophile with a sterically-crowded environment, crowding in the transition state has been avoided simply by moving further apart from the central carbon (2.9Å vs. 2.5Å according to Hartree-Fock 3-21G calculations). There is, however, a price to pay. Increased atom separation – to minimize unfavorable steric interactions – leads to increased charge-charge separation, and to an increase in the electrostatic energy. This is clearly evident in comparison of electrostatic potential maps for the two transition states.

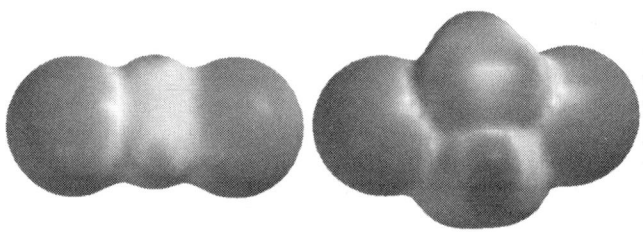

bromide + methyl bromide   bromide + *tert*-butyl bromide

Coulombs law, not sterics, is the reason behind the decrease in rate.[*]

---

[*]   There is an issue of symantics here. The atoms moved further apart to avoid steric crowding, and this is what caused the increase in electrostatic energy.

LUMO Map

Maps of "key" molecular orbitals may also lead to informative models. The most popular and (to date) most important of these is the so-called "LUMO map", in which the (absolute value) of the lowest-unoccupied molecular orbital (the LUMO) is mapped onto a size surface.

LUMO ——

non-bonded electron pair

"electron density"

A good example is provided by the LUMO map for cyclohexenone.

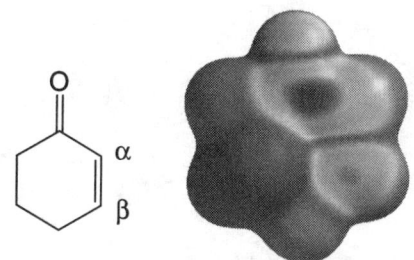

4-27

Recall that the LUMO shows which regions of a molecule are most electron deficient, and hence most subject to nucleophilic attack. One such region is over the carbonyl carbon, consistent with the observation that carbonyl compounds undergo nucleophilic addition at the carbonyl carbon. Another region is over the β carbon, again consistent with the known chemistry of α,β-unsaturated carbonyl compounds, in this case conjugate or Michael addition.

HO   CH$_3$

$\xleftarrow[\text{carbonyl addition}]{\text{CH}_3\text{Li}}$

$\xrightarrow[\text{Michael addition}]{(\text{CH}_3)_2\text{CuLi}}$

CH$_3$

Of course, the buildup of positive charge on the β carbon leading to possibility of Michael addition could have been easily anticipated from resonance arguments.

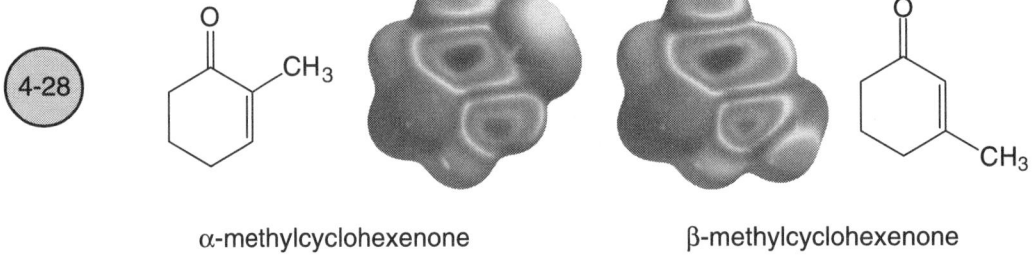

However, the LUMO map, like an experiment, has the advantage of "showing" the result ("you don't have to ask"). Also, resonance arguments could not readily account for changes in nucleophilic reactivity as a result of substitution on the ring, for example, methyl substitution. Here the LUMO maps suggest that substitution on the α carbon has little overall effect, whereas analogous β substitution significantly enhances reactivity at the carbonyl carbon.

α-methylcyclohexenone   β-methylcyclohexenone

Local Ionization Potential Map

Mapping the local ionization potential onto a size surface reveals those regions from which electrons are most easily ionized.

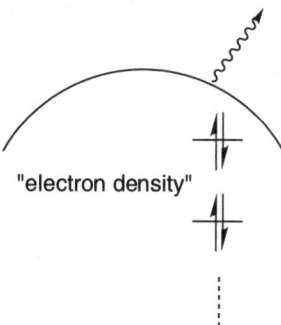

Such a representation is referred to as a local ionization potential map. Local ionization potential maps provide an alternative to electrostatic potential maps for revealing sites which may be particularly susceptible to electrophilic attack.[*] For example, local ionization potential maps show both the positional selectivity in electrophilic aromatic substitution (NH_2 directs *ortho/para*, and NO_2 directs *meta*), and the fact that π-donor groups (NH_2) activate benzene while electron-withdrawing groups (NO_2) deactivate benzene.

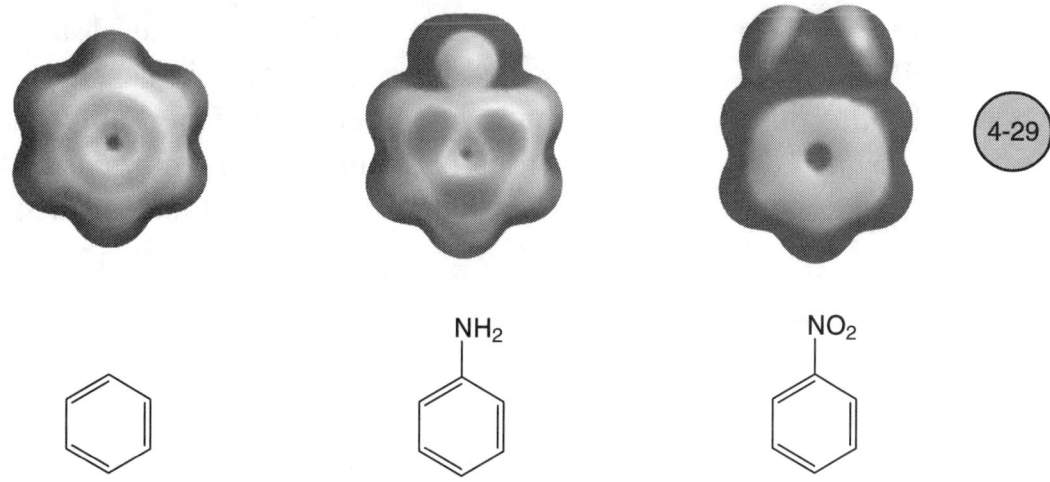

4-29

---

[*]   The author wishes to thank Dr. Denton Hoyer at Pfizer for pointing out the utility of local ionization potential maps for this purpose.

The success of graphical models, like the local ionization potential model for electrophilic reactions and the LUMO model for nucleophilic reactions, in accounting for chemical reactivity and selectivity requires that the transition state occurs early along the reaction coordinate. This being the case, it is reasonable to anticipate that the properties of the reactant, on which modeling is actually carried out, will closely resemble those of the transition state. The usual notion (the Hammond Postulate) is that *exothermic* reactions will have early transition states, and that the more *exothermic* the reaction the earlier will be the transition state. In practice, most reactions which are actually carried out in a laboratory need to be (and are) *exothermic*, or at least are not significantly *endothermic*. Therefore, graphical models would be expected to yield valid results in the majority of cases. Extreme cases are reactions which do not have transition states at all, but rather proceed without barrier from reactant to product (see discussion in **Chapter 15**). Here, the graphical models would also be expected to meet with success.

Spin Density Map

As discussed earlier in this chapter, the spin density of a radical indicates where its unpaired electron resides. This in turn allows qualitative assessment of radical stability. A radical in which the unpaired electron is localized onto a single center is likely to be more labile than a radical in which the unpaired electron is delocalized over several centers. An even more useful indicator of radical stability and radical reactivity is provided by a so-called spin density map. Like the other property maps considered in this chapter, this "measures" the value of the property (in this case the spin density) on an electron density surface corresponding to overall molecular size.

Animations

Chemists routinely "manipulate" physical models in an attempt to ascertain "what actually occurs" during a conformational change. A successful example of this is in showing first-time students of organic chemistry that interconversion between *anti* and *gauche* conformers of *n*-butane involves a simple rotation about the central carbon-carbon bond (see discussion in **Chapter 1**). Much less satisfactory is the attempt to show the interconversion of chair forms of cyclohexane. Here, computer animations provide a better alternative.

There is, however, much more to "computer animation" than merely looking at changes in structure. In principle, any graphical model can be animated and the change to its size and shape in addition to other properties can be monitored. A good example of this concerns the change in electrostatic potential map portraying S_N2 reaction of cyanide and methyl iodide in proceeding from reactants to products.

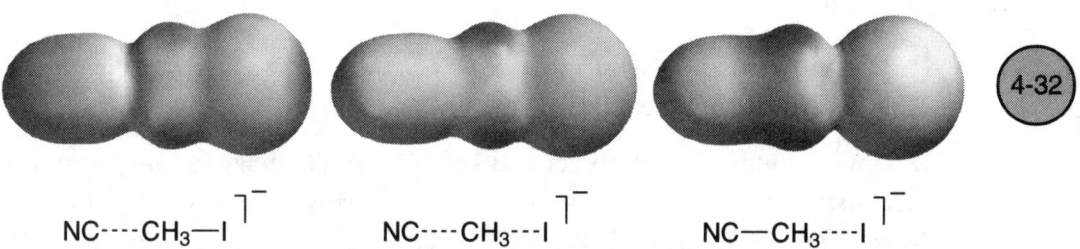

Here the buildup of (negative) charge on the reactants and on the products is significantly greater than the buildup at the transition state. Further, the buildup on the reactants is greater than the buildup on the products, indicating that iodide is the better "leaving group".

Additional animations show the positive nature of the hydrogen being transferred during pyrolysis of ethyl formate and the fact that the two new carbon-carbon bonds are formed at different "rates" during Diels-Alder cycloaddition of cyclopentadiene and acrylonitrile.

Choice of Quantum Chemical Model

Which of the quantum chemical models described in **Chapter 2** are likely to provide a suitable basis for the graphical descriptions outlined in this chapter? Experience suggests that even semi-empirical models generally provide reasonable qualitative descriptions of the sizes and shapes of molecular orbitals, electron densities, spin densities and electrostatic potentials. Semi-empirical models are, however, not as successful in describing subtle changes in size and shape in response to subtle changes in chemical structure or molecular environment. On the other hand, Hartree-Fock models have proven to be generally satisfactory, although split-valence or polarization basis sets are required. Correlated models, including density functional models and Møller-Plesset models also provide a satisfactory basis for graphical analyses, although there is little evidence to suggest that they provide better descriptions than Hartree-Fock models.

References

1. R.B. Woodward and R.Hoffmann, *The Conservation of Orbital Symmetry*, Verlag-Chemie, Weinhein, 1970.

2. Reviews include: (a) P. Politzer and J.S. Murray, *Reviews in Computational Chemistry*, vol. 2, K.B. Lipkowitz and D.B. Boyd, eds., VCH Publishers, New York, 1991; (b) G. Náray-Szabó and G.G. Ferenczy, *Chem. Rev.*, **95**, 829 (1995).

3. M.M. Francl, *J. Chem. Phys.*, **89**, 428 (1985).

4. (a) P. Sjoberg, J.S. Murray, T. Brinck and P. Politzer, *Can. J. Chem.*, **68**, 1440 (1990). For a review, see: (b) J.S. Murray and P. Politzer, in *Theoretical Organic Chemistry. Theoretical and Computational Chemistry*, vol. 5, C. Párkányi, ed., Elsevier, Amsterdam, p. 189, 1998; (c) P. Politzer, J.S. Murray and M.C. Concha, *Int. J. Quantum Chem.*, **88**, 19 (2002).

Section II

Choosing a Model

No single method of calculation is likely to be ideal for all applications. A great deal of effort has been expended defining the limits of different molecular mechanics and quantum chemical models, and judging the degree of success of different models. Most simply, success depends on the ability of a model to consistently reproduce known (experimental) data. This assumes that reliable experimental data are available, or at least, that errors in the data have been quantified. Molecular mechanics models are restricted to determination of geometries and conformations of stable molecules, for which high-quality experimental data are plentiful. Quantum chemical models also provide energy data, which may in turn be directly compared with experimental thermochemical data, and properties such as dipole moments which, may be compared directly with the corresponding experimental quantities. Quantum chemical models may also be applied to transition states. Here there are (and can be) no experimental structures with which to compare, although experimental kinetic data may be interpreted to provide information about activation energies. In this case, comparisons need to be made with the results of high-level quantum chemical calculations.

"Success" is not an absolute. Different properties and certainly different problems may require different levels of confidence to actually be of value. Neither is success sufficient. A model also needs to be "practical" for the task at hand. The nature and size of the system needs to be taken into account, as do the available computational resources and the experience and "patience" of the practitioner. Practical models usually do share one feature in common, in that they are not likely to be the "best possible" treatments which have been formulated. Compromise is almost always an essential component of model selection.

The material in this section seeks to answer two questions: "How well do the models reproduce what is known?" and "How costly are specific models in comparison with alternative models?" Concerns related to the practical application of molecular mechanics and quantum chemical models are deferred until **Section III**, while actual applications to chemical problems are put off until **Section IV**.

The response to the first question is spread over several chapters: *Equilibrium Geometries*, *Reaction Energies*, *Vibrational Frequencies and Thermodynamic Quantities*, *Equilibrium Conformations*, *Transition-State Geometries and Activation Energies* and *Dipole Moments*. Each considers a series of different models: SYBYL and MMFF94 molecular mechanics models[*]; MNDO, AM1 and PM3 semi-empirical molecular orbital models; Hartree-Fock models with STO-3G, 3-21G, 6-31G* and 6-311+G** basis sets; the local density model, BP, BLYP, EDF1 and B3LYP density functional models, and the MP2 model, all with the 6-31G* and 6-311+G** basis sets. More limited coverage is afforded to Hartree-Fock, EDF1, B3LYP and MP2 models to assess the individual effects of polarization functions on hydrogen (6-31G**), of diffuse functions (6-31+G*), and of additional valence-shell splitting (6-311G*), that is, the steps taken in moving between the 6-31G* and 6-311+G** basis sets.

The response to the second question is in terms of relative computation times for energy calculations, geometry optimizations and frequency evaluations on different size molecules. This is addressed in the final chapter of this section, *Overview and Cost*.

---

[*]   As commented previously, molecular mechanics models are applicable only to investigation of equilibrium geometries and conformations.

Chapter 5
Equilibrium Geometries

This chapter assesses the performance of molecular mechanics and quantum chemical models with regard to the calculation of equilibrium geometry. Several different classes of molecules are considered: one and two-heavy-atom, main-group hydrides, hydrocarbons, molecules with heteroatoms, hypervalent molecules, molecules with heavy main-group elements, transition-metal organometallic and inorganic molecules, a variety of reactive intermediates (carbocations, anions, carbenes and radicals) and hydrogen-bonded complexes. The chapter concludes with brief discussions of geometries of molecules in excited states and molecules in solution.

Introduction

Equilibrium geometries for upwards of four thousand small molecules have been determined experimentally in the gas phase, primarily by microwave spectroscopy and electron diffraction.[1] In the best cases, the experimental techniques are able to provide bond lengths and angles to within a few thousandths of an Å and a few tenths of a degree, respectively.[*] For larger systems, lack of data usually prohibits complete structure determination,[**] and some geometrical variables may have been assumed in the reported structure.

Equilibrium geometries for nearly four hundred thousand molecules have been determined in the solid (crystalline) phase primarily by

---

[*]   All bond distances and bond distance errors will be reported in Å and all bond angles and bond angle errors in degrees.

[**] Microwave spectroscopy provides only three "data points" (the principal moments of inertia) for each distinct molecule. Sufficient data to assign the 3N-6 independent geometrical variables (for a molecule with N atoms and no symmetry) is obtained by carrying out different isotopic substitutions. It is common practice to assume values for certain geometrical parameters, e.g., CH bond lengths, in order to reduce the number of isotopic substitutions required.

X-ray diffraction (and to a much lesser extent by neutron diffraction).[2,3] These are typically subject to larger uncertainties than the gas-phase structures due primarily to thermal motions in the crystal. In the best cases, bond lengths and angles are reliable to within one or two hundredths of an Å and one or two degrees, respectively. The exception are geometrical parameters involving hydrogens, which are generally not well described. The problem is that the X-ray experiment actually locates areas of high electron density, which in the case of hydrogens are not precisely the same as the nuclear positions. As a result, bonds involving hydrogen are typically too short by one or two tenths of an Å.

It also needs to be recognized that the geometries of molecules in crystals are not necessarily the same as those of isolated (gas-phase) molecules. They will be influenced by intermolecular interactions ("crystal packing forces"). However, comparisons of bond lengths and angles between gas and solid-phase structures suggest that these influences are likely to be small. Differences in conformations (torsion angles) may be much greater, leading to the possibility that overall "molecular shape" may differ significantly between the gas and crystalline phases. This is an issue of considerable importance, in particular with regard to the shapes of proteins, the structures of which are almost entirely known from X-ray diffraction.[4]

Finally, it needs to be stated that calculated geometries are also subject to errors in precision (due to incomplete convergence of the geometry optimization procedure). However, these errors may be reduced as much as desired.

Overall, except for very small molecules the structures of which have been completely determined in the gas phase by microwave spectroscopy, comparisons between calculated and measured equilibrium geometries below the level of 0.01Å and 1° for bond lengths and angles, respectively, and 5° for torsion angles are seldom meaningful. Within these bounds, comparisons with experiment should function to judge the quality of the calculations, although there is the real possibility that it is the experimental structure which is in error.

The sections which follow assess the performance of both molecular mechanics and quantum chemical models for bond length and angle calculations. Torsion angle comparisons and related comparisons involving the structures of "flexible rings" ("molecular shapes") are dealt with separately in **Chapter 8**. Several classes of molecules have been considered: hydrides incorporating one and two main-group elements, hydrocarbons, molecules with heteroatoms, molecules with hypervalent atoms, molecules with heavy main-group elements, transition-metal inorganics, carbonyls and organometallics, as well as a variety of carbocations, anions, carbenes and free radicals ("reactive intermediates") and hydrogen-bonded complexes. Most of the "raw data" has been relegated to **Appendix A5**, and has been summarized in terms of error statistics and plots. A variety of molecular mechanics and quantum chemical models have been examined: SYBYL and MMFF molecular mechanics models, Hartree-Fock models with STO-3G, 3-21G, 6-31G* and 6-311+G** basis sets, the local density model, BP, BLYP, EDF1 and B3LYP density functional models and the MP2 model, all with 6-31G* and 6-311+G** basis sets, and MNDO, AM1 and PM3 semi-empirical models. Additional basis sets "on the way" between 6-31G* and 6-311+G** are examined for selected models in a few cases in order to establish the effect of increased valence-shell splitting, diffuse functions and hydrogen polarization functions on calculated geometry.

Main-Group Hydrides

Calculated equilibrium geometries for hydrogen and main-group hydrides containing one and two heavy (non-hydrogen) atoms are provided in **Appendix A5** (**Tables A5-1** and **A5-10** for molecular mechanics models, **A5-2** and **A5-11** for Hartree-Fock models, **A5-3** and **A5-12** for local density models, **A5-4** to **A5-7** and **A5-13** to **A5-16** for BP, BLYP, EDF1 and B3LYP density functional models, **A5-8** and **A5-17** for MP2 models and **A5-9** and **A5-18** for MNDO, AM1 and PM3 semi-empirical models). Mean absolute errors in bond lengths are provided in **Tables 5-1** and **5-2** for one and two-heavy-atom systems, respectively.

Table 5-1: **Mean Absolute Errors in Bond Distances for One-Heavy-Atom, Main-Group Hydrides**

SYBYL	0.262			
MMFF	0.026			
MNDO	0.060			
AM1	0.038			
PM3	0.031			
	STO-3G	**3-21G**	**6-31G***	**6-311+G****
Hartree-Fock	0.051	0.026	0.016	0.013
local density	-	-	0.022	0.015
BP	-	-	0.023	0.017
BLYP	-	-	0.024	0.015
EDF1	-	-	0.018	0.012
B3LYP	-	-	0.012	0.005
MP2	-	-	0.017	0.005

Table 5-2: **Mean Absolute Errors in Heavy-Atom Bond Distances for Two-Heavy-Atom, Main-Group Hydrides**

SYBYL	0.133			
MMFF	0.042			
MNDO	0.072			
AM1	0.062			
PM3	0.044			
	STO-3G	**3-21G**	**6-31G***	**6-311+G****
Hartree-Fock	0.067	0.035	0.035	0.039
local density	-	-	0.028	0.023
BP	-	-	0.022	0.024
BLYP	-	-	0.034	0.030
EDF1	-	-	0.018	0.022
B3LYP	-	-	0.021	0.017
MP2	-	-	0.019	0.018

Only first and second-row hydrides have been considered for molecular mechanics models. Calculations on molecules incorporating heavier elements can be performed, but in general "good" parameters are not available. SYBYL performs acceptably for most hydrides containing C, N and O only, but often yields poor results for other systems. It cannot be recommended. On the other hand, MMFF generally performs quite well. Mean absolute errors for bond lengths are in the same range as small-basis-set Hartree-Fock models and smaller than those for semi-empirical models, and individual errors are seldom greater than 0.03Å. Overall, MMFF appears to be a good "first step" to determining equilibrium structure.

STO-3G and 3-21G basis sets are available through fourth-row elements, the 6-31G* basis set through third-row elements and the 6-311+G** basis set through second-row elements. Hartree-Fock, local density, density functional and MP2 calculations for hydrides containing one main-group element have been performed to these limits. However, calculations on hydrides containing two main-group elements have been restricted to molecules incorporating first and second-row elements only.

The Hartree-Fock STO-3G model provides a generally reasonable account of equilibrium geometry in main-group hydrides. The worst results are for alkali metal compounds where, with the exception of NaH, calculated bond distances are significantly shorter than experimental values. Significant errors also appear for systems with two highly electronegative elements, e.g., for F_2, where calculated bond distances are shorter than experimental values.

Hartree-Fock 3-21G calculations generally provide a better account. Bond distances are typically shorter than experimental values, but errors are usually small. The most conspicuous flaw with calculations at this level is that bond angles involving nitrogen are too large. For example, the bond angle in ammonia is 6° larger than the experimental value. It is not surprising, therefore, that the inversion barrier in ammonia, and nitrogen inversion barriers in general, calculated using the 3-21G model are too small (see **Chapter 8**). With due attention to such shortcomings, it is fair to suggest that the 3-21G model seems a reasonable choice for equilibrium geometry calculations.

6-31G* and 6-311+G** models provide very similar accounts of the geometries of main-group hydrides. Both models now properly account for the bond angle in ammonia. This alone might justify the additional computation cost of 6-31G* (over 3-21G) for a "low-cost" standard for equilibrium geometry calculations. It is, however, more difficult to justify the use of the 6-311+G** basis set for Hartree-Fock geometry calculations on molecules incorporating main-group elements, taking into account the difference in cost between the two models.

Note that "limiting" (6-311+G** basis set) Hartree-Fock bond lengths are consistently shorter than experimental distances, except for hydrides involving highly electropositive elements. This trend is clearly shown in **Figure 5-1**, and points to the systematic (as opposed to random) nature of the error in equilibrium bond lengths from ("limiting") Hartree-Fock models. This behavior may easily be rationalized. Treatment of electron correlation (as for example in the MP2 model) involves "promotion of electrons" from occupied molecular orbitals (in the Hartree-Fock wavefunction) to unoccupied molecular orbitals. As occupied molecular orbitals are (generally) net bonding in character, and as unoccupied molecular orbitals are (generally) net antibonding in character, any promotions should result in bond weakening (lengthening). This in turn suggests that bond lengths from ("limiting") Hartree-Fock models are necessarily shorter than "exact" values.

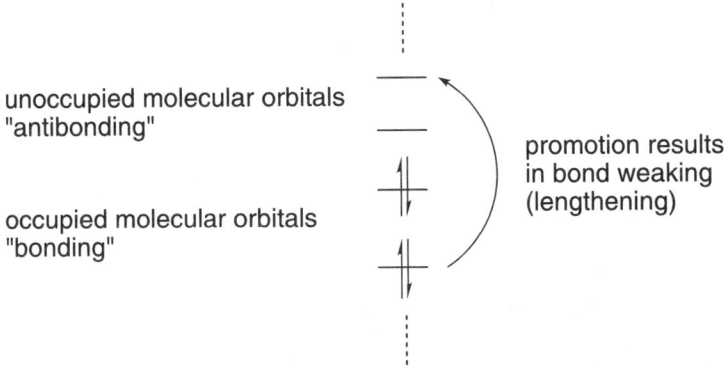

Figure 5-1: 6-311+G** vs. Experimental Heavy-Atom Bond Distances in Two-Heavy-Atom, Main-Group Hydrides

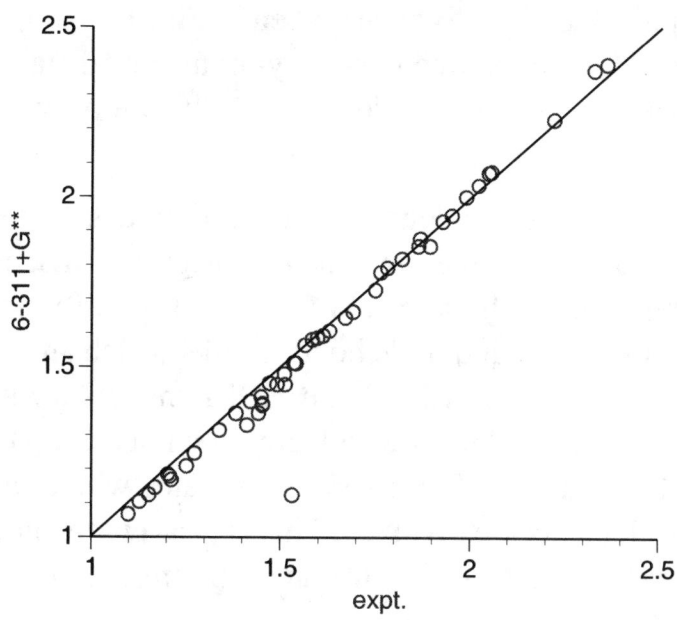

This simple picture is supported by the results of MP2 calculations, which show bond lengthening (over Hartree-Fock models). The resulting bond distances are generally (but not always) longer than experimental values. This is clearly seen in **Figure 5-2**, which relates MP2/6-311+G** to experimental heavy-atom bond distances. As with Hartree-Fock models, nearly identical results are provided with the smaller 6-31G* basis set.

Local density models and density functional models also give rise to systematic errors in heavy-atom bond distances. Local density models lead to single bond distances which are typically shorter than experimental values, but double bond lengths which are very close to experimental values. BP, BLYP and EDF1 models all lead to very similar bond lengths, which are typically (but not always) longer than experimental distances. The errors are greater where one (or two) second-row elements are involved. This is clearly seen in **Figure 5-3**, which compares EDF1/6-311+G** and experimental bond distances.*

B3LYP density functional models provide somewhat better bond length results than the other density functional models, generally very close to experimental distances and to those from MP2 calculations. As with the other density functional models, the errors are largest where one (or two) second-row elements are involved. This is apparent from **Figure 5-4**, which compares B3LYP/6-311+G** and experimental bond distances.

Given the similarity of results from B3LYP and MP2 calculations and the large difference in cost between the two models (which rapidly increases with increasing molecular size), the former would appear to be the better choice for accurate equilibrium geometry determinations. Also, given that results from B3LYP/6-31G* and B3LYP/6-311+G** models are nearly identical, the smaller basis set models would appear to be the better alternative for widespread application.

Semi-empirical models do not account for the geometries of main-group hydrides as well as any of the other quantum chemical models, with the notable exception of the STO-3G model. Overall, MNDO

---

* Bond lengths greater than 1.6Å necessarily involve one (or two) second-row elements.

Figure 5-2: MP2/6-311+G** vs. Experimental Heavy-Atom Bond Distances in Two-Heavy-Atom, Main-Group Hydrides

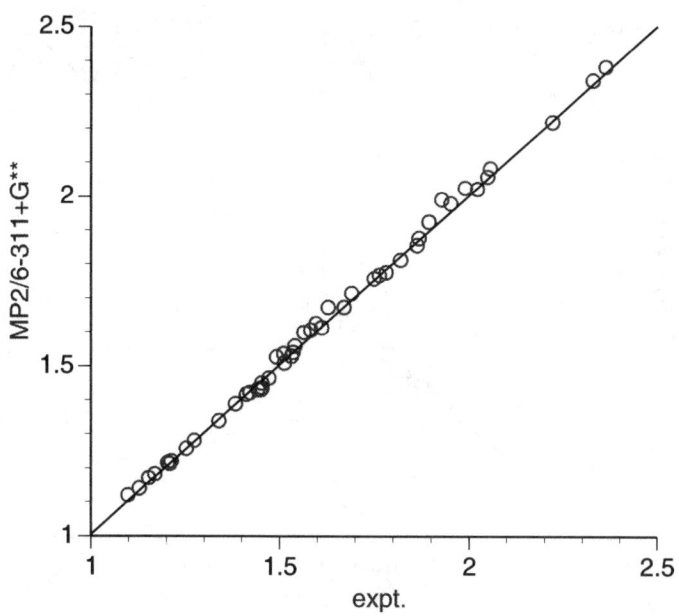

Figure 5-3: EDF1/6-311+G** vs. Experimental Heavy-Atom Bond Distances in Two-Heavy-Atom, Main-Group Hydrides

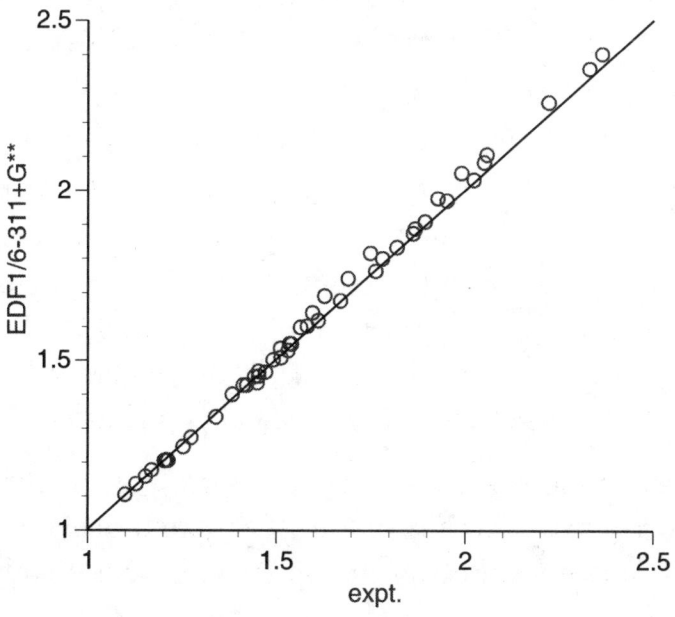

Figure 5-4: B3LYP/6-311+G** vs. Experimental Heavy-Atom Bond Distances in Two-Heavy-Atom, Main-Group Hydrides

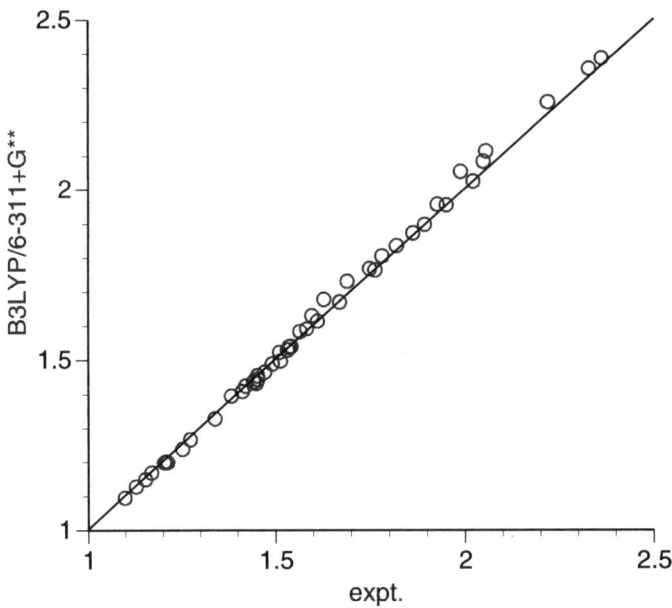

performs worse and PM3 best[*], although it should be noted that individual differences in bond lengths and angles among the three methods are quite small when they are restricted to molecules containing H, C N and O ("organic") molecules). Clearly, semi-empirical models have a role in structure determination if there are no practical alternatives available.

Hydrocarbons

Carbon-carbon bond lengths for a selection of hydrocarbons obtained from molecular mechanics calculations, Hartree-Fock calculations, local density calculations, density functional calculations, MP2 calculations and semi-empirical calculations are compared with experimental distances in **Table 5-3**. The same basis sets considered for main-group hydrides are utilized here. Mean absolute errors for each method have also been tabulated.

Overall, all models perform well in reproducing experimental bond lengths in hydrocarbons, as well as changes in bond lengths from one hydrocarbon to another. This includes both molecular mechanics models, the parameterizations for which have been heavily influenced by hydrocarbons. In fact, mean absolute errors for all models are very similar. Errors for MP2 and B3LYP density functional models are marginally lower than those for EDF1 density functional models, which in turn are marginally lower than those for the remaining models. The worst results are from Hartree-Fock STO-3G, local density and AM1 and PM3 semi-empirical models.

As was the case for main-group hydrides, "limiting" (6-311+G** basis set) Hartree-Fock models lead to bond distances which are generally shorter than experimental values, while hydrocarbon bond lengths from "limiting" density functional and MP2 models are both shorter and longer than measured distances. Note that bond lengths from Hartree-Fock, density functional and MP2 models using the 6-31G* basis set are nearly identical to those from the corresponding models with the 6-311+G** basis set. It appears that addition of

---

[*]   Note that PM3 has been parameterized for more elements than either MNDO or AM1 and
       comparisons are more numerous.

Table 5-3: Bond Distances in Hydrocarbons

| bond | hydrocarbon | molecular mechanics | | Hartree-Fock | | | | | expt. |
		SYBYL	MMFF	STO-3G	3-21G	6-31G*	6-311+G**	
C–C	but-1-yne-3-ene	1.441	1.419	1.459	1.432	1.439	1.438	**1.431**
	propyne	1.458	1.463	1.484	1.466	1.468	1.466	**1.459**
	1,3-butadiene	1.478	1.442	1.488	1.479	1.467	1.467	**1.483**
	propene	1.509	1.493	1.520	1.510	1.503	1.502	**1.501**
	cyclopropane	1.543	1.502	1.502	1.513	1.497	1.500	**1.510**
	propane	1.551	1.520	1.541	1.541	1.528	1.528	**1.526**
	cyclobutane	1.547	1.543	1.549	1.543	1.548	1.546	**1.548**
C=C	cyclopropene	1.317	1.302	1.277	1.282	1.276	1.276	**1.300**
	allene	1.305	1.297	1.288	1.292	1.296	1.295	**1.308**
	propene	1.339	1.339	1.308	1.316	1.318	1.320	**1.318**
	cyclobutene	1.327	1.345	1.314	1.326	1.322	1.323	**1.332**
	but-1-yne-3-ene	1.338	1.337	1.320	1.320	1.322	1.322	**1.341**
	1,3-butadiene	1.338	1.338	1.313	1.320	1.323	1.324	**1.345**
	cyclopentadiene	1.335	1.341	1.319	1.329	1.329	1.330	**1.345**
mean absolute error		0.011	0.010	0.018	0.011	0.011	0.010	–

Table 5-3: Bond Distances in Hydrocarbons (2)

bond	hydrocarbon	local density		BP		BLYP		EDF1		expt.
		6-31G*	6-311+G**	6-31G*	6-311+G**	6-31G*	6-311+G**	6-31G*	6-311+G**	
C–C	but-1-yne-3-ene	1.408	1.406	1.424	1.422	1.426	1.425	1.421	1.419	**1.431**
	propyne	1.441	1.437	1.461	1.458	1.466	1.463	1.458	1.454	**1.459**
	1,3-butadiene	1.439	1.439	1.458	1.457	1.462	1.461	1.455	1.453	**1.483**
	propene	1.482	1.481	1.505	1.503	1.511	1.510	1.502	1.499	**1.501**
	cyclopropane	1.496	1.496	1.516	1.516	1.521	1.510	1.510	1.510	**1.510**
	propane	1.511	1.512	1.536	1.536	1.543	1.543	1.533	1.532	**1.526**
	cyclobutane	1.535	1.537	1.559	1.559	1.566	1.567	1.554	1.554	**1.548**
C=C	cyclopropene	1.298	1.293	1.306	1.302	1.306	1.301	1.302	1.299	**1.300**
	allene	1.305	1.301	1.316	1.312	1.316	1.312	1.312	1.308	**1.308**
	propene	1.332	1.330	1.343	1.341	1.344	1.341	1.340	1.337	**1.318**
	cyclobutene	1.343	1.341	1.352	1.350	1.352	1.350	1.347	1.345	**1.332**
	but-1-yne-3-ene	1.344	1.338	1.353	1.350	1.354	1.350	1.349	1.346	**1.341**
	1,3-butadiene	1.340	1.338	1.352	1.349	1.353	1.350	1.348	1.346	**1.345**
	cyclopentadiene	1.351	1.350	1.361	1.359	1.361	1.359	1.356	1.355	**1.345**
mean absolute error		0.014	0.016	0.011	0.010	0.013	0.010	0.008	0.008	–

Table 5-3: Bond Distances in Hydrocarbons (3)

bond	hydrocarbon	B3LYP		MP2		semi-empirical			expt.
		6-31G*	6-311+G**	6-31G*	6-311+G**	MNDO	AM1	PM3	
C–C	but-1-yne-3-ene	1.424	1.423	1.429	1.430	1.417	1.405	1.414	**1.431**
	propyne	1.461	1.457	1.463	1.464	1.445	1.427	1.433	**1.459**
	1,3-butadiene	1.458	1.456	1.458	1.460	1.466	1.451	1.456	**1.483**
	propene	1.502	1.500	1.499	1.502	1.496	1.476	1.480	**1.501**
	cyclopropane	1.509	1.509	1.504	1.511	1.526	1.501	1.499	**1.510**
	propane	1.532	1.532	1.526	1.529	1.530	1.507	1.512	**1.526**
	cyclobutane	1.553	1.554	1.545	1.550	1.549	1.543	1.542	**1.548**
C=C	cyclopropene	1.295	1.291	1.303	1.305	1.328	1.318	1.314	**1.300**
	allene	1.307	1.304	1.313	1.314	1.306	1.298	1.297	**1.308**
	propene	1.333	1.331	1.338	1.341	1.340	1.331	1.328	**1.318**
	cyclobutene	1.341	1.339	1.347	1.352	1.355	1.354	1.349	**1.332**
	but-1-yne-3-ene	1.341	1.338	1.344	1.347	1.345	1.336	1.332	**1.341**
	1,3-butadiene	1.340	1.338	1.344	1.347	1.344	1.335	1.331	**1.345**
	cyclopentadiene	1.349	1.348	1.354	1.359	1.362	1.359	1.352	**1.345**
mean absolute error		0.006	0.008	0.007	0.006	0.012	0.017	0.015	–

polarization functions on hydrogen, addition of diffuse functions or further splitting of the valence shell is of little consequence to the geometries of hydrocarbons. As might be expected, the changes that do occur are smallest for Hartree-Fock models, where the 6-31G* basis set appears to provide a good approximation to the Hartree-Fock limit, at least insofar as hydrocarbon bond lengths.

It should be noted, however, that polarization functions on heavy (non-hydrogen) atoms have a much more significant (and beneficial) influence on hydrocarbon bond lengths calculated from density functional and MP2 models, than they do on bond lengths calculated using Hartree-Fock models. The reason is clear. Whereas Hartree-Fock models need only to describe molecular orbitals which are occupied (s and p-type orbitals in the case of hydrocarbons), correlated models also need proper descriptions of unoccupied orbitals which may involve d-type (polarization) functions. Documentation is provided in **Appendix A5**. **Table A5-19** compares hydrocarbon bond lengths from 6-31G basis set calculations (6-31G* minus polarization functions on carbon) and 6-31G* basis set calculations for EDF1, B3LYP and MP2 models. For all three models, the mean absolute error decreases significantly upon addition of polarization functions to the 6-31G basis set, and most individual bond lengths are in better agreement with their respective experimental values. This sensitivity to heavy-atom polarization functions is the primary reason that density functional and MP2 models with STO-3G and 3-21G basis sets cannot be recommended.

Molecules with Heteroatoms

Broadly similar conclusions follow from comparison of calculated and experimental C-N and C-O bond distances. A small selection of representative compounds is provided in **Table 5-4**.

Although all models considered provide a generally credible account of the experimental structural data, the variation among methods is somewhat greater than previously noted for hydrocarbons. Here, more so than in the previous comparison, it is evident that the "best performers" are B3LYP and MP2 models.

103

Table 5-4: Bond Distances in Molecules with Heteroatoms

bond	molecule	molecular mechanics		Hartree-Fock				expt.
		SYBYL	MMFF	STO-3G	3-21G	6-31G*	6-311+G**	
C-N	formamide	1.346	1.360	1.436	1.353	1.349	1.349	1.376
	methyl isocyanide	1.500	1.426	1.447	1.432	1.421	1.425	1.424
	trimethylamine	1.483	1.462	1.486	1.464	1.445	1.448	1.451
	aziridine	1.484	1.459	1.482	1.491	1.449	1.450	1.475
	nitromethane	1.458	1.488	1.531	1.493	1.479	1.484	1.489
C-O	formic acid	1.334	1.348	1.385	1.350	1.323	1.320	1.343
	furan	1.337	1.358	1.376	1.344	1.380	1.342	1.362
	dimethyl ether	1.437	1.421	1.433	1.433	1.391	1.391	1.410
	oxirane	1.395	1.433	1.433	1.470	1.402	1.400	1.436
	mean absolute error	0.026	0.008	0.028	0.019	0.018	0.018	–

Table 5-4: Bond Distances in Molecules with Heteroatoms (2)

bond	molecule	local density		BP		BLYP		EDF1		expt.
		6-31G*	6-311+G**	6-31G*	6-311+G**	6-31G*	6-311+G**	6-31G*	6-311+G**	
C-N	formamide	1.352	1.351	1.370	1.369	1.375	1.373	1.367	1.365	**1.376**
	methyl isocyanide	1.398	1.397	1.423	1.424	1.430	1.433	1.419	1.420	**1.424**
	trimethylamine	1.433	1.434	1.461	1.461	1.469	1.469	1.456	1.457	**1.451**
	aziridine	1.455	1.456	1.484	1.484	1.493	1.493	1.477	1.477	**1.475**
	nitromethane	1.477	1.472	1.510	1.513	1.521	1.525	1.507	1.511	**1.489**
C-O	formic acid	1.336	1.335	1.360	1.358	1.366	1.366	1.356	1.354	**1.343**
	furan	1.351	1.351	1.375	1.374	1.382	1.381	1.369	1.368	**1.362**
	dimethyl ether	1.390	1.393	1.418	1.422	1.426	1.431	1.414	1.417	**1.410**
	oxirane	1.415	1.417	1.442	1.443	1.450	1.454	1.435	1.436	**1.436**
mean absolute error		0.018	0.021	0.010	0.010	0.021	0.022	0.007	0.008	–

Table 5-4: Bond Distances in Molecules with Heteroatoms (3)

bond	molecule	B3LYP		MP2		semi-empirical			expt.
		6-31G*	6-311+G**	6-31G*	6-311+G**	MNDO	AM1	PM3	
C-N	formamide	1.362	1.358	1.362	1.364	1.389	1.367	1.392	**1.376**
	methyl isocyanide	1.420	1.422	1.426	1.426	1.424	1.395	1.433	**1.424**
	trimethylamine	1.455	1.455	1.455	1.455	1.464	1.445	1.480	**1.451**
	aziridine	1.473	1.472	1.474	1.478	1.479	1.455	1.484	**1.475**
	nitromethane	1.499	1.497	1.488	1.492	1.546	1.500	1.514	**1.489**
C-O	formic acid	1.347	1.346	1.351	1.348	1.354	1.357	1.344	**1.343**
	furan	1.364	1.363	1.367	1.361	1.367	1.395	1.378	**1.362**
	dimethyl ether	1.410	1.413	1.416	1.411	1.396	1.417	1.406	**1.410**
	oxirane	1.430	1.431	1.438	1.434	1.418	1.436	1.432	**1.436**
mean absolute error		0.005	0.005	0.005	0.006	0.015	0.014	0.013	–

The MMFF molecular mechanics model provides an excellent account of C-N and C-O bond distances. SYBYL also presents a credible account, although bond lengths for some systems are significantly in error. MMFF is clearly the better choice.

"Limiting" (6-311+G** basis set) Hartree-Fock bond lengths in the compounds examined are consistently shorter than experimental distances, the same result as seen for multiple bonds in hydrocarbons. Results obtained using the 6-31G* basis set are nearly identical, which suggests that the 6-31G* model closely reflects the Hartree-Fock limit insofar as bond length calculations for these types of systems.

As reflected by very similar error statistics, bond distances from local density calculations are generally close to those from corresponding Hartree-Fock calculations. For all cases considered, local density models underestimate experimental bond lengths. On the other hand, bond distances from BP, BLYP, EDF1 and B3LYP density functional models and from MP2 models are consistently longer than Hartree-Fock bond distances, and are generally (but not always) larger than experimental distances. Except for local density models and BLYP models, all density functional models and MP2 models lead to similar errors. (Errors from local density models and from BLYP models are significantly larger and neither is to be recommended.) In view of cost differences, density functional models, in particular EDF1 and B3LYP models, would appear to be more suitable than MP2 models for routine structure determinations for molecules with heteroatoms.

As was the case with hydrocarbons, 6-31G* and 6-311+G** basis sets lead to similar bond lengths for all density functional models as well as for the MP2 model. This is reflected in the mean absolute errors. It is difficult to justify use of the larger basis set models for routine structure determinations.

As with hydrocarbons, accurate descriptions of equilibrium structures for molecules with heteroatoms from density functional and MP2 models requires polarization basis sets. As shown in **Table A5-20** (**Appendix A5**), bond distances in these compounds obtained from (EDF1 and B3LYP) density functional models and from MP2 models with the 6-31G basis set (6-31G* minus polarization functions on heavy atoms) are significantly different (and much poorer) than those from corresponding 6-31G* basis set calculations. Note also that the effect of polarization functions on this set of molecules is much greater than that previously noted for hydrocarbons (**Table A5-19**).

Larger Molecules

Calculated heavy-atom bond distances in molecules with three or more first and/or second-row atoms are tabulated in **Appendix A5**: molecular mechanics models (**Table A5-21**), Hartree-Fock models (**Table A5-22**), local density models (**Table A5-23**), BP, BLYP, EDF1 and B3LYP density functional models (**Tables A5-24 to A5-27**), MP2 models (**Table A5-28**), and MNDO, AM1 and PM3 semi-empirical models (**Table A5-29**). Results for STO-3G, 3-21G, 6-31G* and 6-311+G** basis sets are provided for Hartree-Fock models, but as in previous comparisons, only 6-31G* and 6-311+G** basis sets are employed for local density, density functional and MP2 models.

A summary of mean absolute errors is provided in **Table 5-5**. Together with a series of plots of calculated vs. experimental bond distances, this further clarifies the performance of "practical" models: Hartree-Fock models with STO-3G, 3-21G and 6-31G* basis sets (**Figures 5-5 to 5-7**), the local density 6-31G* model (**Figure 5-8**), the BP/6-31G* model (**Figure 5-9**), the BLYP/6-31G* model (**Figure 5-10**), the EDF1/6-31G* model (**Figure 5-11**), the B3LYP/6-31G* model (**Figure 5-12**), the MP2/6-31G* model (**Figure 5-13**) and MNDO, AM1 and PM3 semi-empirical models (**Figures 5-14 to 5-16**).

Bond lengths from STO-3G calculations show considerable scatter over the full range of distances. The 3-21G model leads to much improved results, and the 6-31G* model to small additional improvements. Note that both 3-21G and 6-31G* calculations lead to bond distances which are generally shorter than experimental lengths where only first-row elements are involved (distances < 1.6Å), and to distances which are generally longer than experimental lengths where second-row elements are involved (distances > 1.6Å). This is the same

Table 5-5: Mean Absolute Errors in Heavy-Atom Bond Distances

SYBYL	0.051	
MMFF	0.034	
MNDO	0.048	
AM1	0.048	
PM3	0.037	

	STO-3G	3-21G	6-31G*	6-311+G**
Hartree-Fock	0.042	0.028	0.028	0.031
local density	-	-	0.018	0.019
BP	-	-	0.018	0.017
BLYP	-	-	0.029	0.027
EDF1	-	-	0.015	0.015
B3LYP	-	-	0.018	0.016
MP2	-	-	0.018	0.011

Figure 5-5: STO-3G vs. Experimental Heavy-Atom Bond Distances

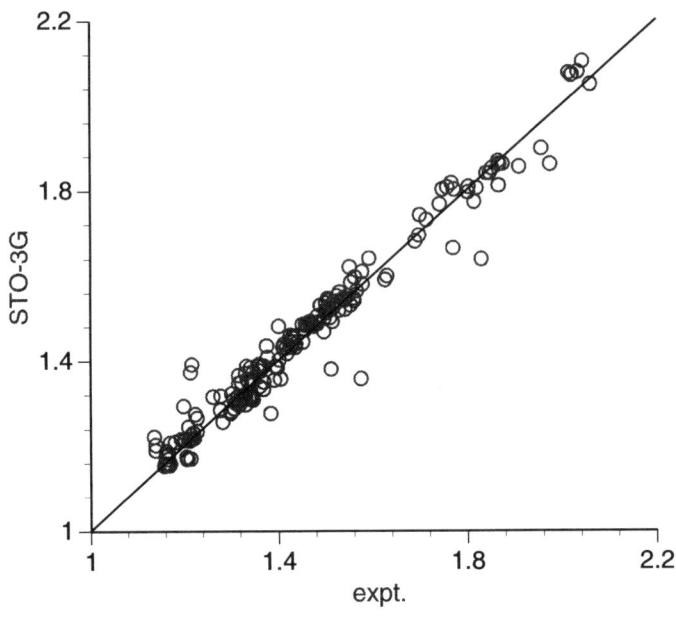

Figure 5-6: 3-21G vs. Experimental Heavy-Atom Bond Distances

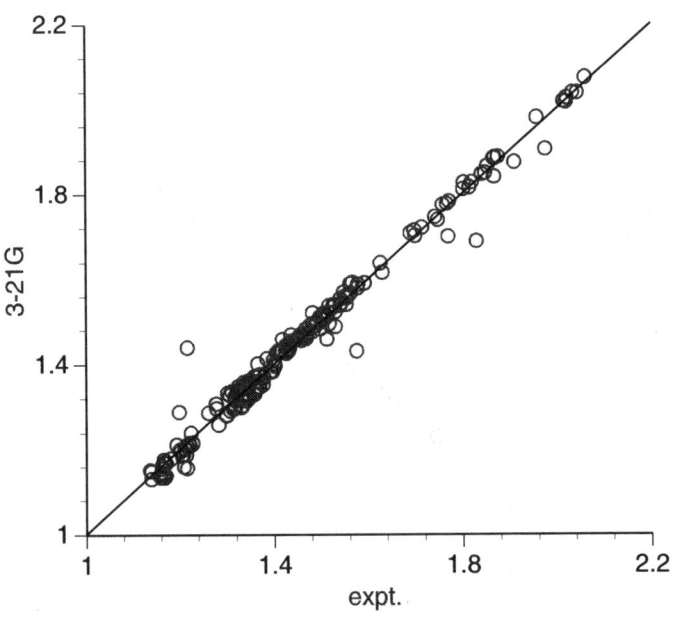

Figure 5-7: 6-31G* vs. Experimental Heavy-Atom Bond Distances

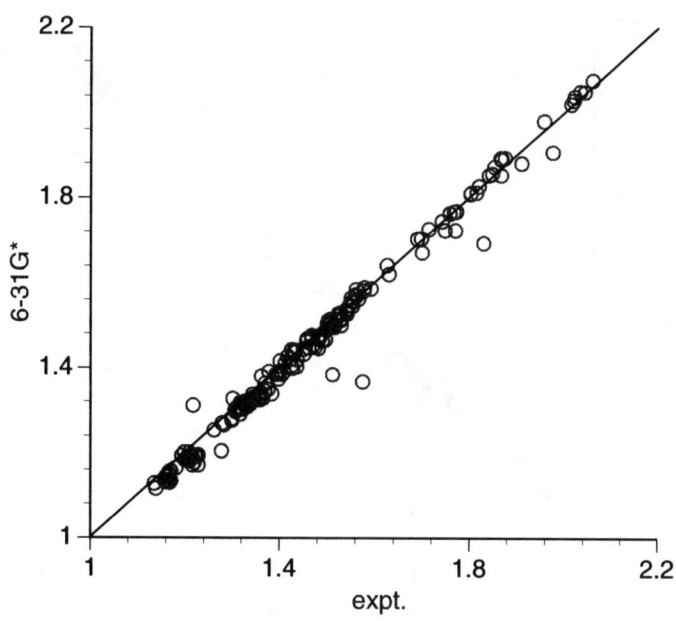

Figure 5-8: Local Density 6-31G* vs. Experimental Heavy-Atom Bond Distances

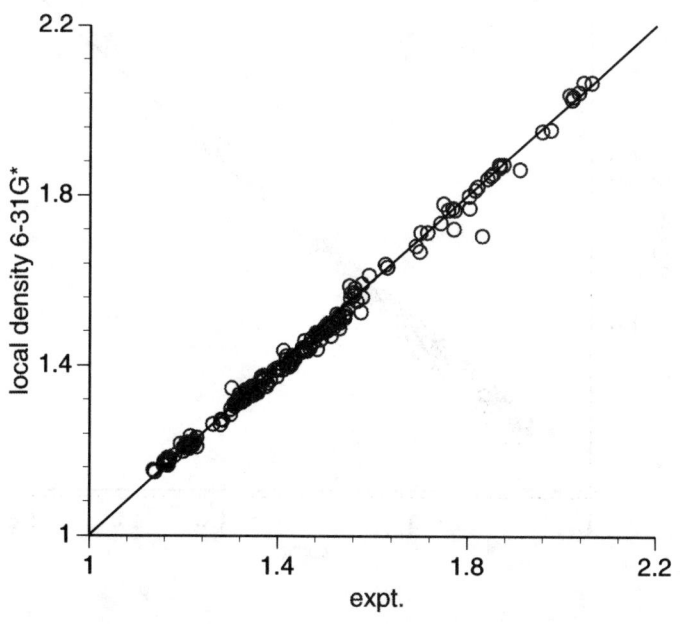

Figure 5-9: BP/6-31G* vs. Experimental Heavy-Atom Bond Distances

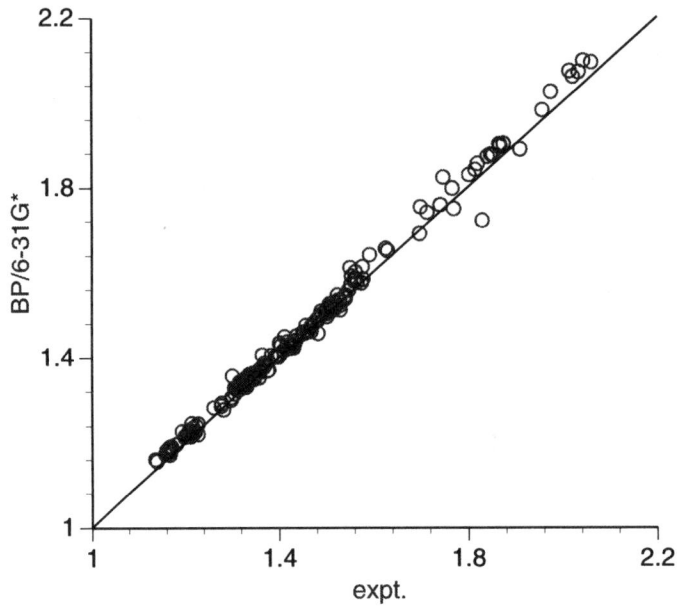

Figure 5-10:   BLYP/6-31G* vs. Experimental Heavy-Atom Bond Distances

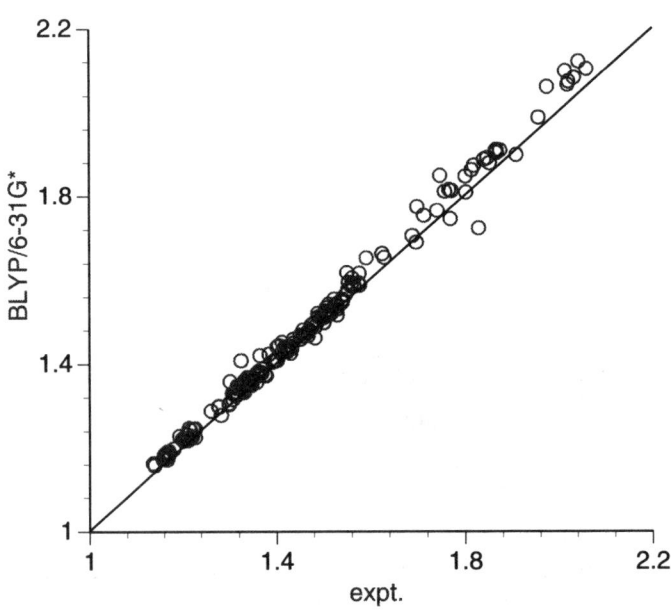

Figure 5-11:   EDF1/6-31G* vs. Experimental Heavy-Atom Bond Distances

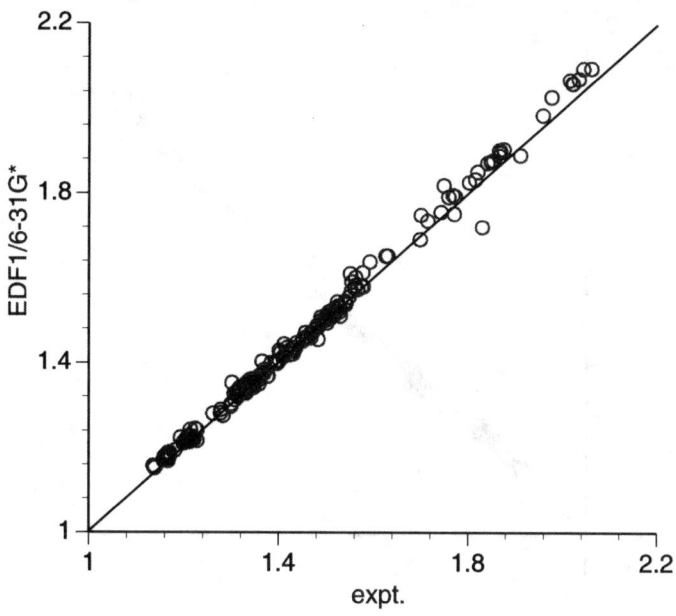

Figure 5-12:   B3LYP/6-31G* vs. Experimental Heavy-Atom Bond Distances

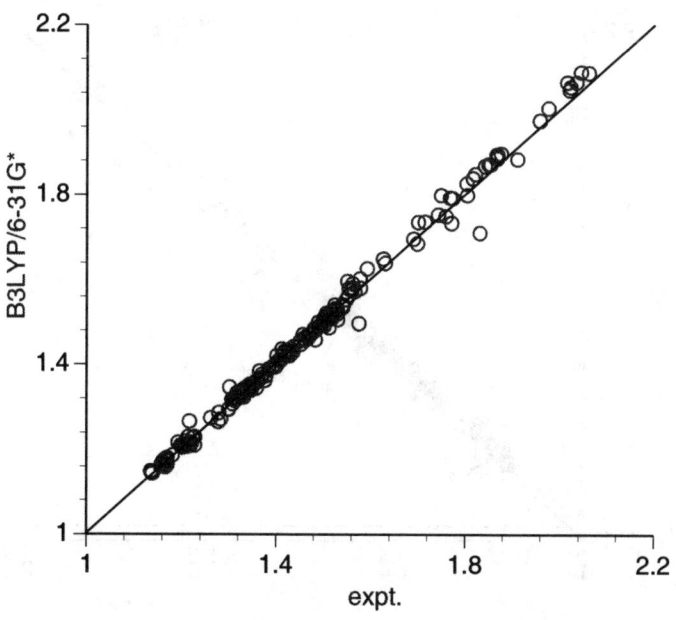

Figure 5-13: MP2/6-31G* vs. Experimental Heavy-Atom Bond Distances

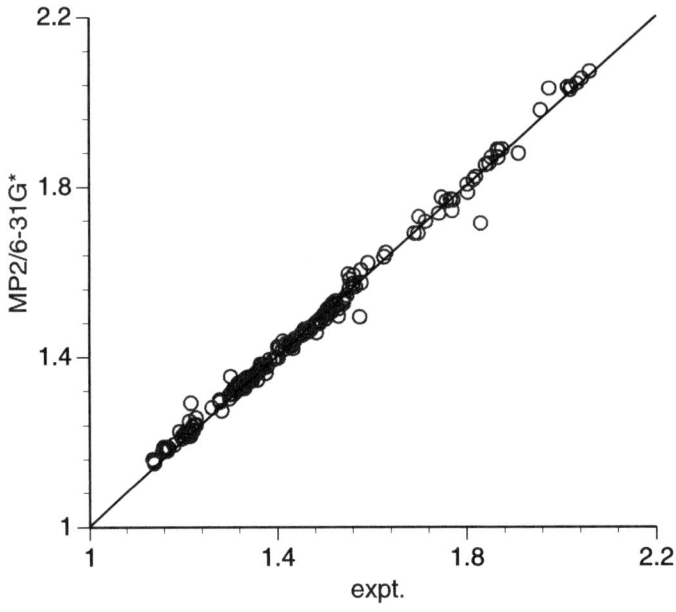

Figure 5-14: MNDO vs. Experimental Heavy-Atom Bond Distances

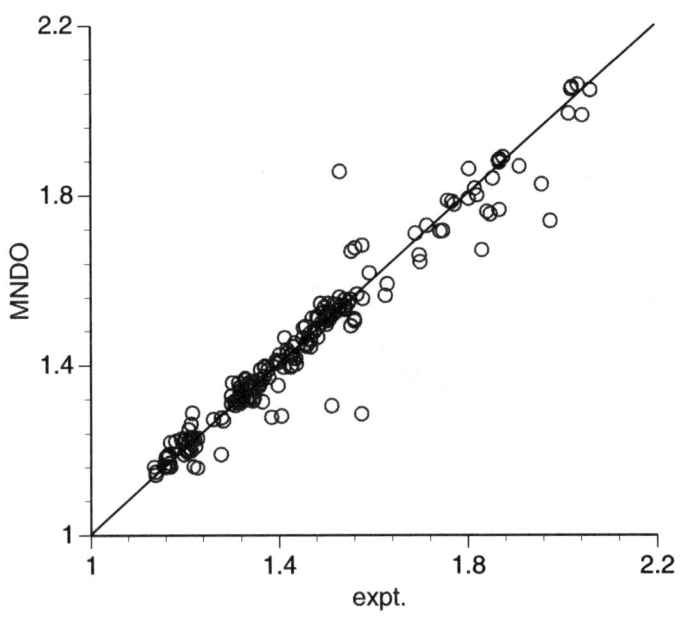

Figure 5-15: AM1 vs. Experimental Heavy-Atom Bond Distances

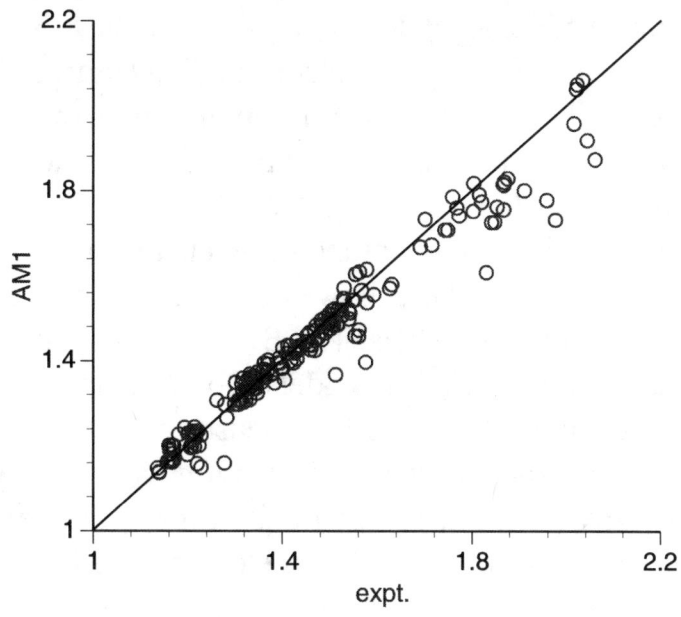

Figure 5-16: PM3 vs. Experimental Heavy-Atom Bond Distances

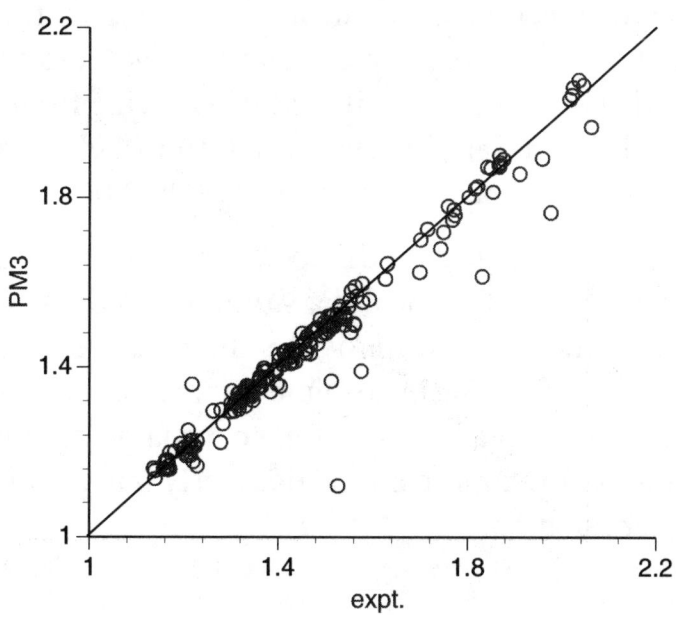

trend pointed out earlier for main-group hydrides, and has its origin in the effect of electron correlation (see preceding discussion).

The performance of the local density model with the 6-31G* basis set is somewhat better than that of the corresponding Hartree-Fock model. Most conspicuously, bond distances involving first-row elements only are generally shorter than experimental values. In terms of mean absolute errors, BP/6-31G*, EDF1/6-31G* and B3LYP/6-31G* models all show similar behavior (and similar to that of the local density 6-31G* model). Note that here, however, bond distances are typically longer than experimental values. The BLYP model stands out as the poorest of the density functional models. Bond distances from density functional calculations (except for local density calculations) are evenly distributed around the experimental lengths where only first-row elements are involved, but are generally longer than experimental values where one or more second-row elements are involved.

MP2/6-31G* bond distances are comparable to those from B3LYP/6-31G* calculations, although errors for bonds involving one or more second-row elements are generally somewhat smaller. As stated earlier, it is difficult to justify use of the more costly MP2/6-31G* model (over B3LYP/6-31G*) for geometry determinations.

On the basis of mean absolute errors, bond lengths from Hartree-Fock calculations, local density calculations and density functional calculations do not improve significantly in moving from the 6-31G* to the 6-311+G** basis set. However, results from the MP2/6-311+G** model are significantly better than those from the MP2/6-31G* model.

None of the semi-empirical models perform as well as Hartree-Fock models (except STO-3G), local density models, density functional models or MP2 models. PM3 provides the best overall description, although on the basis of mean absolute errors alone, all three models perform to an acceptable standard. Given the large difference in cost of application, semi-empirical models clearly have a role to play in structure determination.

Do structural variations anticipate interactions among substituents? Consider changes in CX bond lengths for disubstituted methanes, CH_2XY, from those in the corresponding monosubstituted systems, CH_3X.

Y \ X	Me	CMe₃	CN	OMe	F	SiMe₃
Me	-1	12	9	5	8	10
CMe₃	5	23	7	4	9	20
CN	8	14	4	-12	-21	34
OMe	-10	-3	21	-11	-2	17
F	-15	-9	10	-34	-27	20
SiMe₃	10	15	-4	14	24	6

Here, the change in bond length (in Å multiplied by 1000) is from Hartree-Fock 6-31G* calculations. Included are methyl and *tert*-butyl groups (both weak π donors), the cyano group (both a strong σ and π acceptor), the methoxy and fluoro groups (both σ acceptors and π donors) and the trimethylsilyl group (a σ donor and π acceptor).

Some of the changes could easily have been anticipated, for example, the large increase in bond length resulting from substitution by two bulky *tert*-butyl groups. Other changes are also recognizable, for example, the marked decrease in CO and CF bond lengths where two methoxy groups and/or fluorines are attached to the same carbon as due to the so-called "anomeric effect". The observed shortening of the CC bond in ethyl fluoride (relative to that in ethane), together with the lengthening of the CF bond (relative to that in methyl fluoride) is reminiscent of "hyperconjugation", i.e., participation of resonance structures of the form.

$$H^+$$
$$H_2C=CH_2$$
$$F^-$$

Similar effects are noted where CMe₃ replaces Me and/or where OMe replaces F. Perhaps the most interesting structural change, or lack of structural change, follows from substitution of two trimethylsilyl groups onto the same carbon. Note that all other substitutions involving this bulky group lead to significant lengthening of the CSi bond. The fact that there is only a slight increase in bond length, despite what must be significant steric repulsion between the two bulky trimethylsilyl groups, suggests interaction akin to the anomeric effect (see above). Further discussion is provided in **Chapter 6**, where the energetic effects of geminal interactions are considered.

5-1

117

Skeletal bond angles for the same series of molecules and from the same molecular mechanics and quantum chemical models are collected in **Appendix A5** (**Tables A5-30** to **A5-38**). A summary of mean absolute errors is provided in **Table 5-6**. Some of the data are also presented as plots: Hartree-Fock models with STO-3G, 3-21G and 6-31G* basis sets (**Figures 5-17** to **5-19**), the local density 6-31G* model (**Figure 5-20**), the BP/6-31G* model (**Figure 5-21**), the BLYP/6-31G* model (**Figure 5-22**), the EDF1/6-31G* model (**Figure 5-23**), the B3LYP/6-31G* model (**Figure 5-24**), the MP2/6-31G* model (**Figure 5-25**) and MNDO, AM1 and PM3 semi-empirical models (**Figures 5-26** to **5-28**).

As with calculated bond distances, Hartree-Fock bond angles show significant improvement in going from STO-3G to 3-21G basis sets, and lesser improvement in moving on to 6-31G*. Either of the latter two Hartree-Fock models appears to be suitable for bond angle calculations on moderate size organic molecules.

Results from local density models and BP, BLYP and EDF1 density functional models are, broadly speaking, comparable to those from 6-31G* models, consistent with similarity in mean absolute errors. As with bond length comparisons, BLYP models stand out as inferior to the other non-local models. Both B3LYP/6-31G* and MP2/6-31G* models provide superior results, and either would appear to be a suitable choice where improved quality is required.

Consistent with earlier remarks made for bond length comparisons, little if any improvement results in moving from the 6-31G* to the 6-311+G** basis set for Hartree-Fock, local density and density functional models, but significant improvement results for MP2 models.

All three semi-empirical models perform very poorly for skeletal bond angle calculation, and none can be recommended except where there is no alternative and where qualitative results may be sufficient.

Table 5-6: Mean Absolute Errors in Skeletal Bond Angles

	STO-3G	3-21G	6-31G*	6-311+G**
SYBYL	3.8			
MMFF	1.9			
MNDO	2.9			
AM1	2.9			
PM3	3.1			

	STO-3G	3-21G	6-31G*	6-311+G**
Hartree-Fock	1.7	1.7	1.4	1.3
local density	-	-	1.6	1.4
BP	-	-	0.9	0.9
BLYP	-	-	1.7	1.6
EDF1	-	-	1.3	1.3
B3LYP	-	-	1.4	1.4
MP2	-	-	1.5	0.7

Figure 5-17: STO-3G vs. Experimental Skeletal Bond Angles

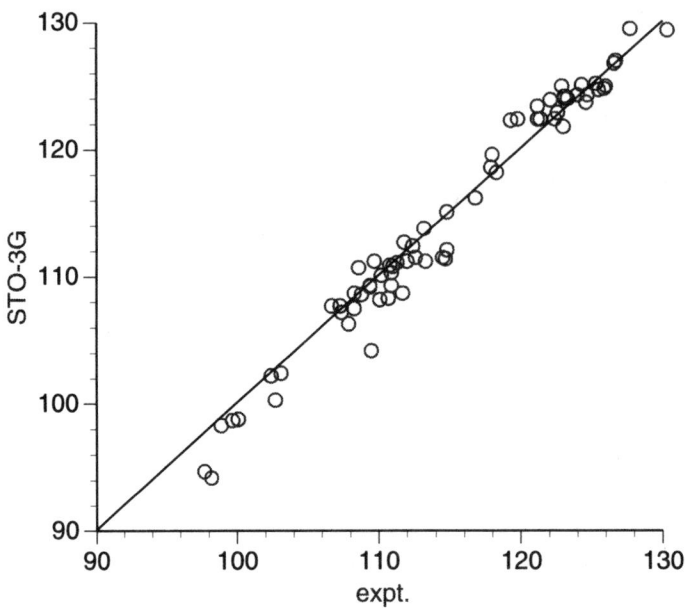

Figure 5-18: 3-21G vs. Experimental Skeletal Bond Angles

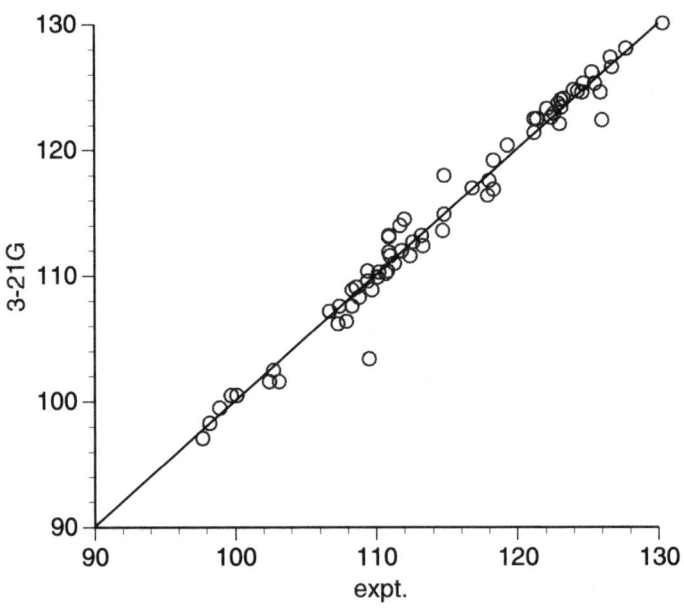

Figure 5-19: 6-31G* vs. Experimental Skeletal Bond Angles

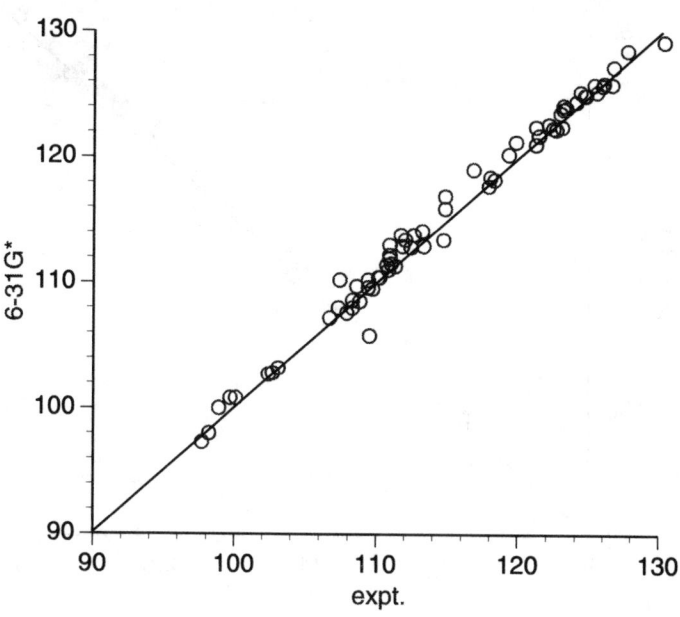

Figure 5-20: Local Density 6-31G* vs. Experimental Skeletal Bond Angles

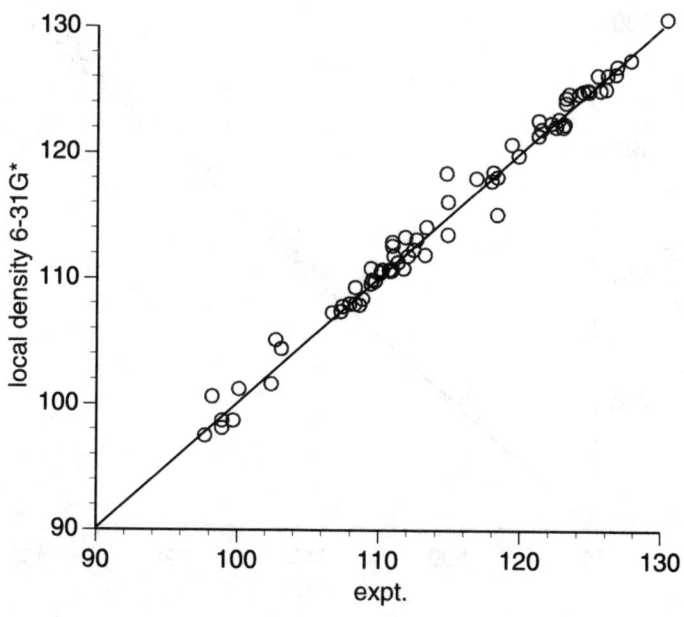

Figure 5-21: BP/6-31G* vs. Experimental Skeletal Bond Angles

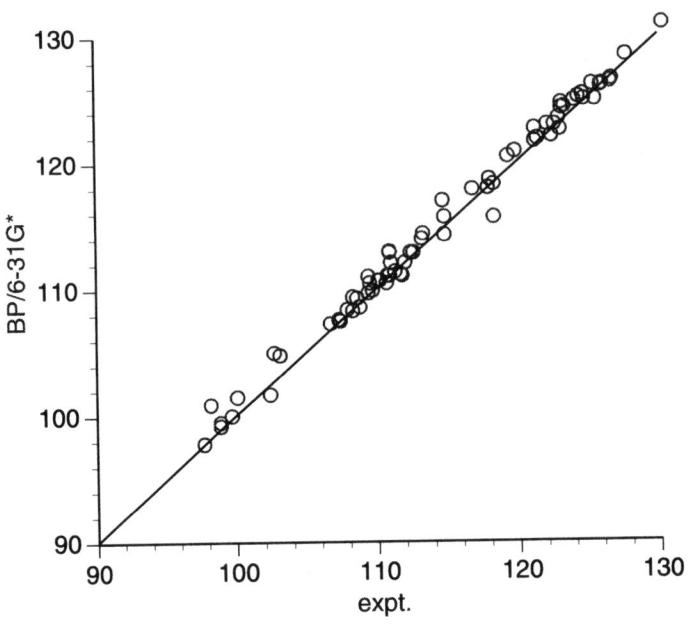

Figure 5-22: BLYP/6-31G* vs. Experimental Skeletal Bond Angles

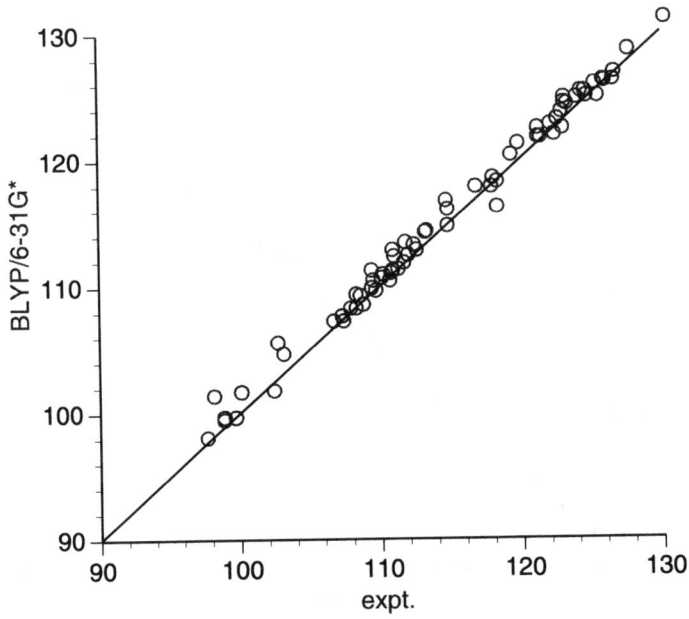

Figure 5-23:   EDF1/6-31G* vs. Experimental Skeletal Bond Angles

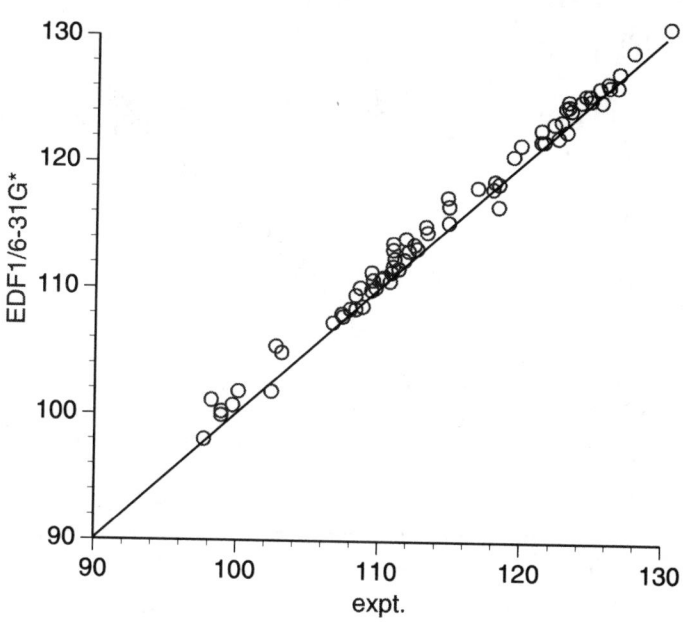

Figure 5-24:   B3LYP/6-31G* vs. Experimental Skeletal Bond Angles

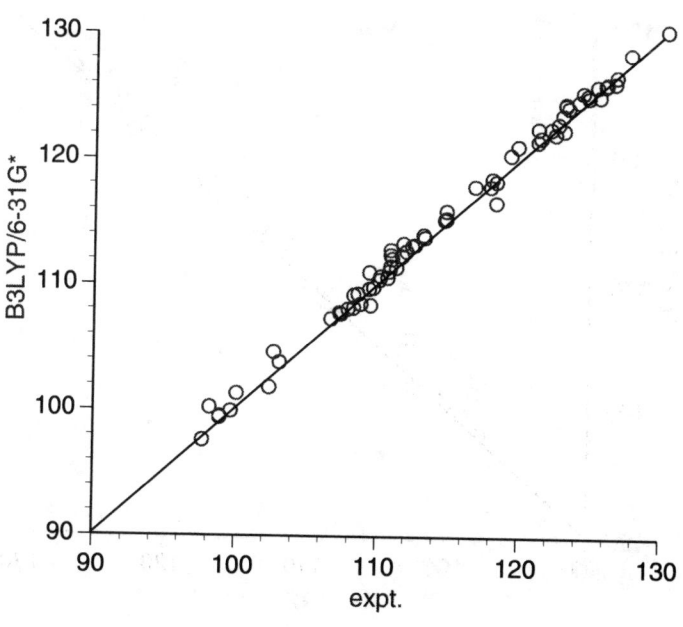

Figure 5-25: MP2/6-31G* vs. Experimental Skeletal Bond Angles

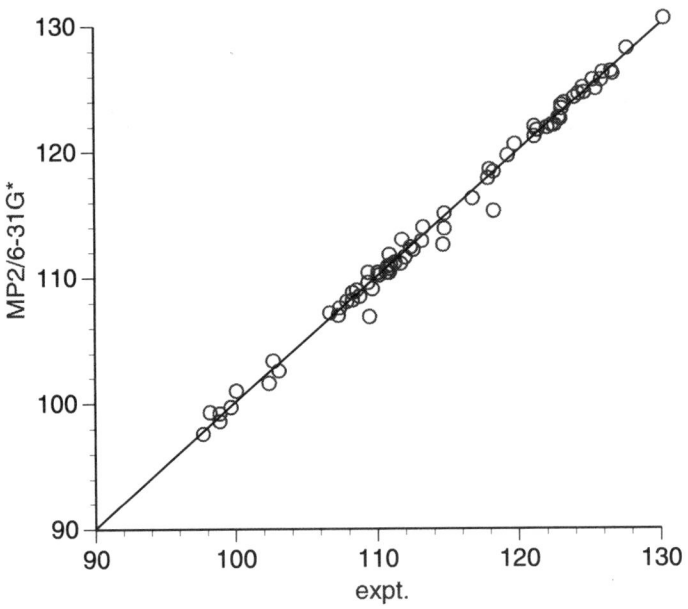

Figure 5-26: MNDO vs. Experimental Skeletal Bond Angles

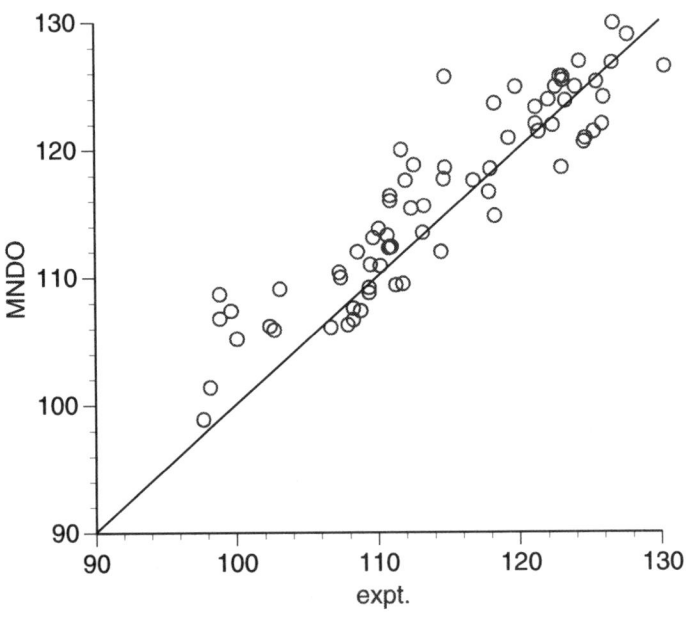

Figure 5-27: AM1 vs. Experimental Skeletal Bond Angles

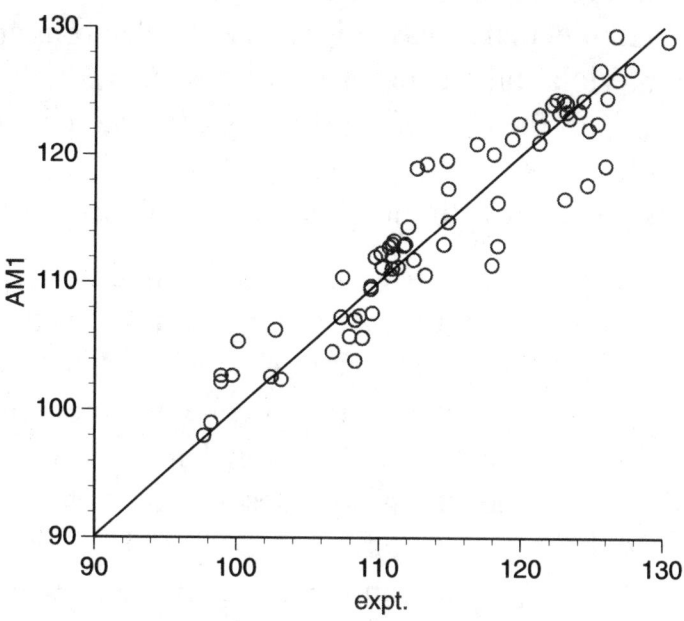

Figure 5-28: PM3 vs. Experimental Skeletal Bond Angles

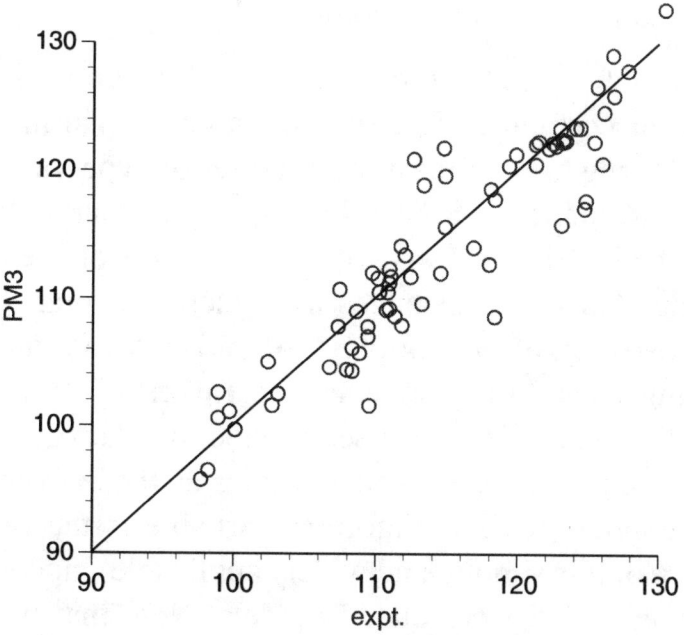

Hypervalent Molecules

Molecules comprising first-row elements (C-F) only nearly always satisfy the octet rule, that is, have eight valence electrons around each first-row atom. Violations occur for molecules incorporating elements on left-hand side of the *Periodic Table*, e.g., B, where there may not be sufficient electrons to satisfy the overall demand. Very seldom are molecules encountered with more than eight valence electrons.

On the other hand, second-row and heavier main-group elements frequently incorporate more than eight electrons in their valence shells, at least insofar as formal counting is involved. It may be argued whether the valence octet around the heavy element actually "expands" to involve functions not occupied in the ground-state atom (d-functions), or whether the proper description of such systems is largely ionic, with the heavy atom bearing significant positive charge, and "attached" atoms, significant negative charge (see discussion in **Chapter 16**). Whichever the case, so-called hypervalent molecules present a stringent test for quantum chemical models. Comparative data for heavy-atom bond lengths and skeletal bond angles are provided in **Table 5-7**. Molecular mechanics models have been excluded from the comparison, as they have not been specifically parameterized for hypervalent compounds.

Except for the STO-3G model, the performance of which is completely unacceptable, Hartree-Fock models perform remarkably well in describing the equilibrium structures of hypervalent systems. The poor performance of STO-3G is due in great part to the absence of d-type functions in the basis set for second-row elements, without which calculated bond distances are much too long. (Note, that STO-3G also performed poorly for normal-valent molecules incorporating second-row elements. See **Tables A5-11** and **A5-22** in **Appendix A5**.) The 3-21G basis set provides bond lengths and angles which are as close to experimental values as those from the larger basis set calculations. Ever more remarkable is the fact that the performance of this simple (and widely applicable) model is actually better than any of the correlated models. Note that bond lengths calculated with the 6-311+G** model are (nearly) always shorter

Table 5-7:   Heavy-Atom Bond Distances and Skeletal Bond Angles in Hypervalent Molecules

molecule	point group	geometrical parameter	Hartree-Fock				local density		expt.
			STO-3G	3-21G	6-31G*	6-311+G**	6-31G*	6-311+G**	
PF_5	D_{3h}	r (PF_{eq})	1.615	1.538	1.535	1.529	1.562	1.566	**1.534**
		r (PF_{ax})	1.614	1.566	1.568	1.567	1.586	1.596	**1.577**
$(CH_3)_3PO$	C_{3v}	r (PO)	1.620	1.478	1.474	1.471	1.503	1.500	**1.479**
		r (PC)	1.840	1.804	1.816	1.815	1.813	1.807	**1.813**
		< (CPC)	102.2	104.1	104.3	105.3	104.1	105.1	**106.0**
SO_2	C_{2v}	r (SO)	1.559	1.419	1.414	1.408	1.467	1.463	**1.431**
		< (OSO)	106.0	118.7	118.8	118.6	119.2	118.8	**119.3**
$(CH_3)_2SO$	C_s	r (SO)	1.821	1.490	1.485	1.486	1.504	1.508	**1.485**
		r (SC)	1.809	1.791	1.796	1.797	1.813	1.810	**1.799**
		< (CSO)	102.5	107.8	106.7	106.3	107.6	106.8	**106.7**
		< (CSC)	98.5	96.6	97.7	98.3	94.4	95.5	**96.6**
SF_4	C_{2v}	r (SF_{eq})	1.675	1.550	1.544	1.534	1.588	1.595	**1.545**
		r (SF_{ax})	1.675	1.617	1.632	1.643	1.655	1.691	**1.646**
		< ($F_{eq}SF_{eq}$)	138.3	101.7	142.7	102.4	101.6	100.9	**101.6**
		< ($F_{ax}SF_{ax}$)	138.3	169.8	189.9	170.8	170.8	172.7	**173.1**
SO_3	D_{3h}	r (SO)	1.599	1.411	1.405	1.397	1.455	1.450	**1.420**
$(CH_3)_2SO_2$	C_{2v}	r (SO)	1.851	1.438	1.437	1.431	1.431	1.465	**1.431**
		r (SC)	1.814	1.756	1.774	1.774	1.783	1.783	**1.777**
		< (OSO)	129.9	119.5	120.1	119.7	121.0	120.8	**121.0**
		< (CSC)	100.0	102.8	104.3	104.6	103.0	103.5	**103.3**
SF_6	O_h	r (SF)	1.652	1.550	1.554	1.548	1.587	1.598	**1.564**
ClF_3	C_{2v}	r (ClF_{ax})	1.776	1.676	1.672	1.690	1.708	1.752	**1.698**
		r (ClF_{eq})	1.795	1.601	1.579	1.577	1.641	1.662	**1.598**
		r ($F_{ax}ClF_{ax}$)	154.9	171.0	172.6	172.6	174.6	176.3	**175.0**
mean absolute error in bond distance			0.128	0.011	0.009	0.009	0.022	0.029	–
mean absolute error in bond angle			14.1	1.4	1.3	1.1	0.7	0.5	–

Table 5-7: Heavy-Atom Bond Distances and Skeletal Bond Angles in Hypervalent Molecules (2)

molecule	point group	geometrical parameter	BP 6-31G*	BP 6-311+G**	BLYP 6-31G*	BLYP 6-311+G**	EDF1 6-31G*	EDF1 6-311+G**	expt.
PF_5	D_{3h}	r (PF_{eq})	1.586	1.589	1.591	1.596	1.583	1.585	**1.534**
		r (PF_{ax})	1.611	1.620	1.618	1.629	1.609	1.617	**1.577**
$(CH_3)_3PO$	C_{3v}	r (PO)	1.516	1.514	1.518	1.517	1.511	1.509	**1.479**
		r (PC)	1.844	1.839	1.856	1.850	1.842	1.836	**1.813**
		< (CPC)	104.2	104.9	104.4	104.9	104.5	105.3	**106.0**
SO_2	C_{2v}	r (SO)	1.485	1.479	1.491	1.486	1.479	1.474	**1.431**
		< (OSO)	119.4	119.0	119.4	118.9	119.4	119.0	**119.3**
$(CH_3)_2SO$	C_s	r (SO)	1.523	1.526	1.530	1.533	1.517	1.519	**1.485**
		r (SC)	1.855	1.851	1.874	1.872	1.850	1.847	**1.799**
		< (CSO)	108.0	107.2	107.9	107.2	108.0	107.4	**106.7**
		< (CSC)	95.1	95.9	95.1	96.1	95.8	96.6	**96.6**
SF_4	C_{2v}	r (SF_{eq})	1.616	1.623	1.627	1.636	1.612	1.617	**1.545**
		r (SF_{ax})	1.690	1.726	1.701	1.744	1.687	1.722	**1.646**
		< ($F_{eq}SF_{eq}$)	101.5	101.2	102.0	101.1	101.4	101.2	**101.6**
		< ($F_{ax}SF_{ax}$)	172.5	174.0	172.5	174.7	172.9	174.5	**173.1**
SO_3	D_{3h}	r (SO)	1.473	1.467	1.480	1.474	1.468	1.461	**1.420**
$(CH_3)_2SO_2$	C_{2v}	r (SO)	1.485	1.482	1.490	1.488	1.479	1.475	**1.431**
		r (SC)	1.825	1.823	1.842	1.843	1.822	1.819	**1.777**
		< (OSO)	121.2	120.9	121.2	120.9	121.1	120.8	**121.0**
		< (CSC)	103.0	103.4	103.0	103.6	103.3	103.7	**103.3**
SF_6	O_h	r (SF)	1.619	1.630	1.630	1.644	1.615	1.625	**1.564**
ClF_3	C_{2v}	r (ClF_{ax})	1.750	1.794	1.772	1.818	1.747	1.791	**1.698**
		r (ClF_{eq})	1.681	1.702	1.700	1.726	1.675	1.695	**1.598**
		r ($F_{ax}ClF_{ax}$)	177.0	178.7	177.5	179.9	177.4	179.3	**175.0**
mean absolute error in bond distance			0.051	0.058	0.062	0.070	0.046	0.052	–
mean absolute error in bond angle			0.8	0.9	0.9	1.0	0.7	0.9	–

128

Table 5-7: Heavy-Atom Bond Distances and Skeletal Bond Angles in Hypervalent Molecules (3)

molecule	point group	geometrical parameter	B3LYP		MP2		semi-empirical			expt.
			6-31G*	6-311+G**	6-31G*	6-311+G**	MNDO	AM1	PM3	
PF_5	D_{3h}	$r(PF_{eq})$	1.569	1.571	1.569	1.562	1.573	1.535	1.528	1.534
		$r(PF_{ax})$	1.598	1.604	1.598	1.597	1.602	1.549	1.553	1.577
$(CH_3)_3PO$	C_{3v}	$r(PO)$	1.501	1.500	1.507	1.499	1.522	1.462	1.482	1.479
		$r(PC)$	1.836	1.830	1.820	1.816	1.798	1.649	1.824	1.813
		$<(CPC)$	104.3	105.0	104.2	104.5	107.5	104.4	103.5	106.0
SO_2	C_{2v}	$r(SO)$	1.464	1.458	1.478	1.469	1.492	1.429	1.442	1.431
		$<(OSO)$	119.2	118.7	119.8	119.2	116.9	108.0	106.1	119.3
$(CH_3)_2SO$	C_s	$r(SO)$	1.511	1.514	1.512	1.508	1.518	1.491	1.557	1.485
		$r(SC)$	1.838	1.835	1.809	1.806	1.819	1.739	1.818	1.799
		$<(CSO)$	107.5	106.9	107.4	106.7	109.5	105.7	104.5	106.7
		$<(CSC)$	95.7	96.7	95.8	95.9	101.2	99.7	99.4	96.6
SF_4	C_{2v}	$r(SF_{eq})$	1.595	1.596	1.588	1.578	1.642	1.545	1.596	1.545
		$r(SF_{ax})$	1.672	1.705	1.665	1.690	1.697	1.573	1.622	1.646
		$<(F_{eq}SF_{eq})$	102.1	101.3	102.2	101.9	107.8	103.7	121.2	101.6
		$<(F_{ax}SF_{ax})$	171.5	172.9	170.9	172.5	169.4	170.4	142.4	173.1
SO_3	D_{3h}	$r(SO)$	1.455	1.447	1.459	1.468	1.480	1.351	1.384	1.420
$(CH_3)_2SO_2$	C_{2v}	$r(SO)$	1.470	1.467	1.469	1.460	1.500	1.399	1.468	1.431
		$r(SC)$	1.810	1.809	1.785	1.784	1.809	1.690	1.793	1.777
		$<(OSO)$	120.8	120.6	121.1	121.0	118.1	117.6	118.3	121.0
		$<(CSC)$	103.3	103.9	103.6	103.4	100.5	99.4	100.0	103.3
SF_6	O_h	$r(SF)$	1.601	1.607	1.595	1.590	1.667	1.540	1.561	1.564
ClF_3	C_{2v}	$r(ClF_{ax})$	1.728	1.768	1.720	1.766	1.701	a	a	1.698
		$r(ClF_{eq})$	1.651	1.662	1.638	1.642	1.625			1.598
		$r(F_{ax}ClF_{ax})$	173.9	175.3	173.0	174.2	183.0			175.0
mean absolute error in bond distance			0.034	0.038	0.028	0.029	0.045	0.043	0.024	–
mean absolute error in bond angle			0.8	0.4	1.0	0.5	3.9	3.6	10.9	–

a) incorrect planar trigonal structure obtained

129

than experimental values consistent with previous experience about the behavior of "limiting" Hartree-Fock models.

None of the density functional models is as successful in accounting for geometries of hypervalent compounds as are Hartree-Fock models, irrespective of choice of basis set. Local density models actually perform best, consistent with the previous observation that structures at this level closely parallel those from Hartree-Fock models, but the bond length errors are still significantly greater than those for corresponding (same basis set) Hartree-Fock models. The performance of BP, BLYP and EDF1 models in describing the equilibrium geometries of hypervalent compounds is very poor. Bond distances are consistently longer than experimental values, sometimes significantly so. B3LYP models and MP2 models fare better, but are still inferior to Hartree-Fock models. The reason is unclear. The fact that the results do not improve with improvement in basis set suggests that this is not the underlying cause.

None of the semi-empirical models gives a good account of the geometries of hypervalent compounds. Bond length errors of a tenth of an Å or more are common, and bond angles often deviate from experiment by several degrees. Note also, that AM1 and PM3 models produce incorrect trigonal planar geometries for ClF_3, in contrast to the "T-shaped" structure known experimentally. While this is not a particularly important system, it does provide a warning flag; unless explicitly considered in the parameterization, semi-empirical models cannot be expected to properly account for "unusual" bonding situations. ClF_3 was included in the "training set" for the MNDO/d model (which is employed for the molecules here) and the calculations assign the correct geometry.

Molecules with Heavy Main-Group Elements

Both the quantity and quality of gas-phase experimental structural data rapidly diminish with incorporation of elements beyond the second row of the *Periodic Table*. Solid-phase structures abound,[2] but differences in detailed geometries from gas-phase structures due to crystal packing may be significant and preclude accurate comparisons with the calculations. There are, however, sufficient gas-phase data primarily on very small molecules to enable adequate assessment to be made.

Comparative data for heavy-atom bond lengths and skeletal bond angles for molecules incorporating one or more third or fourth-row, main-group elements are provided in **Appendix A5**: **Table A5-39** for Hartree-Fock models with STO-3G, 3-21G and 6-31G* basis sets, **Table A5-40** for the local density model, BP, BLYP, EDF1 and B3LYP density functional models and the MP2 model, all with the 6-31G* basis set, and in **Table A5-41** for MNDO, AM1 and PM3 semi-empirical models. 6-31G*, local density, density functional and MP2 calculations have been restricted to molecules with third-row elements only.[*] Also, molecular mechanics models have been excluded from the comparison. A summary of errors in bond distances is provided in **Table 5-8**.

The STO-3G model provides a generally unfavorable account of the geometries of this class of molecules. As seen in **Figure 5-29**, bond distances are almost always shorter than experimental lengths, sometimes significantly so. 3-21G and 6-31G* Hartree-Fock models provide similar results which are, for the most part, superior to those from STO-3G. As seen in **Figures 5-30** and **5-31**, calculated bond distances are now generally longer than experimental distances. While this might not seem to reflect the Hartree-Fock limit, which previous experience suggests leads to bond lengths which are too short, note that many of the molecules for which calculated bond lengths are larger than experimental distances involve electropositive elements,

---

[*] This is because the 6-31G* basis set has been defined though third-row elements only. Pseudopotentials could have been employed (and will be for molecules incorporating transition metals).

Table 5-8: **Mean Absolute Errors in Bond Distances for Molecules Incorporating Third and Fourth-Row, Main-Group Elements**

	STO-3G	3-21G	6-31G*
MNDO	0.046		
AM1	0.042		
PM3	0.080		

	STO-3G	3-21G	6-31G*
Hartree-Fock	0.061	0.035	0.023
local density	-	-	0.021
BP	-	-	0.039
BLYP	-	-	0.052
EDF1	-	-	0.036
B3LYP	-	-	0.032
MP2	-	-	0.034

Figure 5-29: **STO-3G vs. Experimental Bond Distances in Molecules Incorporating Third and Fourth-Row, Main-Group Elements**

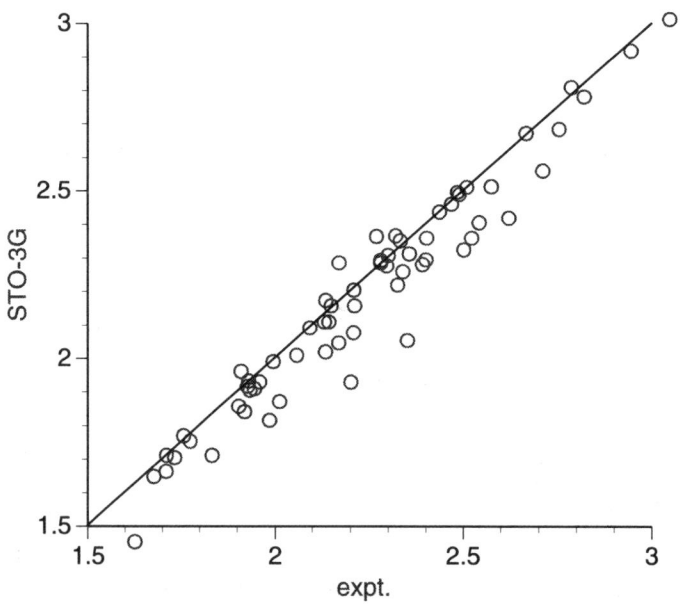

132

Figure 5-30: 3-21G vs. Experimental Bond Distances in Molecules Incorporating Third and Fourth-Row, Main-Group Elements

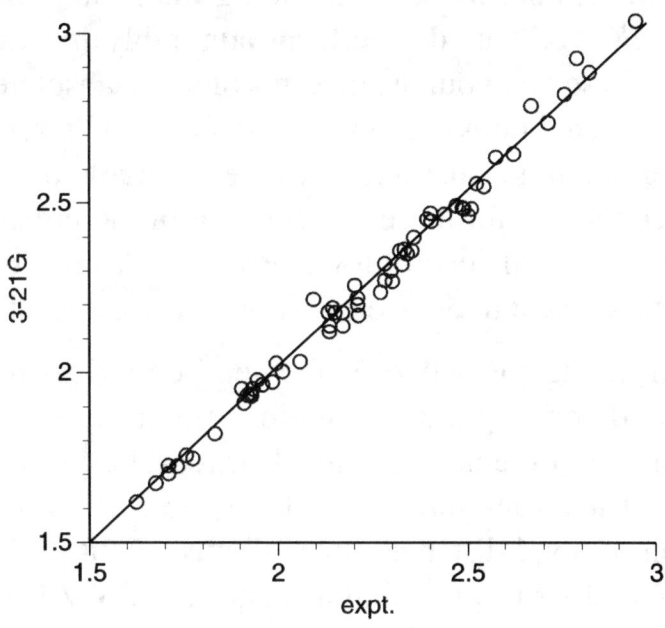

Figure 5-31: 6-31G* vs. Experimental Bond Distances in Molecules Incorporating Third-Row, Main-Group Elements

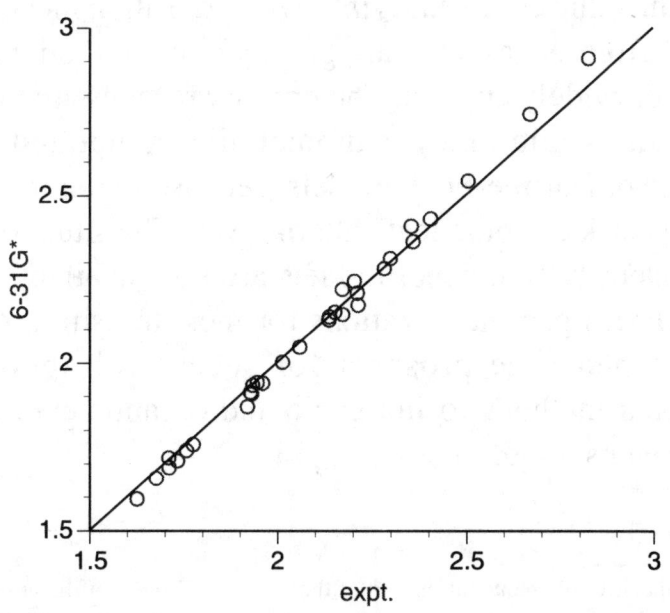

and not electronegative elements on which the previous generalizations were based.[*]

All density functional models (including the local density model) and the MP2/6-31G* model perform admirably in describing the structures of these compounds. In terms of mean absolute errors, the local density model fares best and the BLYP model fares worse. The former observation is consistent with the favorable performance of Hartree-Fock models for these systems and of the previously noted parallels in structural results for Hartree-Fock and local density models. **Figures 5-32** to **5-37** provide an overview.

It is not surprising that none of the semi-empirical models is as successful in describing the geometries of molecules incorporating heavy main-group elements as are Hartree-Fock models (except STO-3G), density functional models or the MP2/6-31G* model. Examination of bond distance plots (**Figures 5-38** to **5-40**), shows that MNDO and AM1 give similar results, and PM3 gives poorer results. This is consistent with the error statistics shown in **Table 5-8**.

Molecules with Transition Metals

Although the equilibrium geometries of more than 100,000 inorganic and organometallic compounds involving transition metals have been established from X-ray crystallography,[2,3] computational methods have not been widely applied. The primary reasons are the large size of typical inorganic and organometallic compounds, the poor performance of Hartree-Fock models (see discussion following) and the previous lack of "practical" alternatives. The situation is rapidly changing. Density functional models are now available, and semi-empirical (PM3) parameterizations for most transition metals have been developed. The prospect for successful applications of computational methods to inorganic and organometallic structural chemistry seems good.

---

[*]   For electropositive elements (at the left of the *Periodic Table*), unfilled molecular orbitals may be bonding in character and electron promotion will not necessarily lead to bond weakening (lengthening).

Figure 5-32: Local Density 6-31G* vs. Experimental Bond Distances in Molecules Incorporating Third-Row, Main-Group Elements

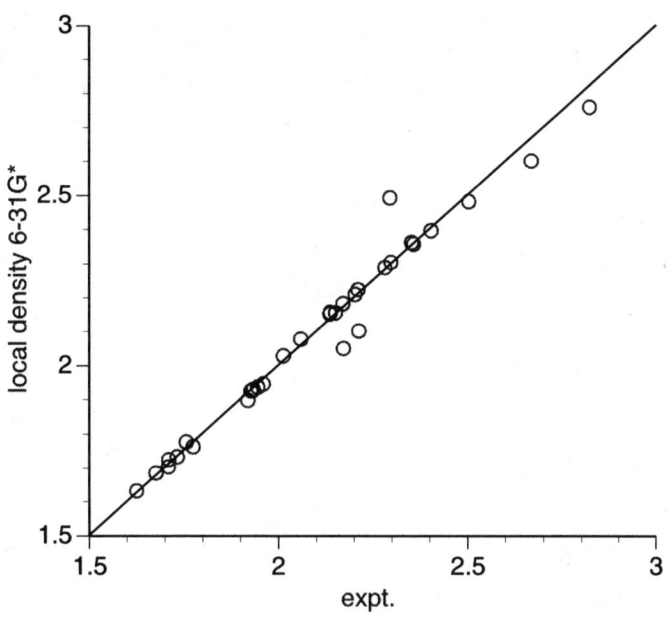

Figure 5-33: BP/6-31G* vs. Experimental Bond Distances in Molecules Incorporating Third-Row, Main-Group Elements

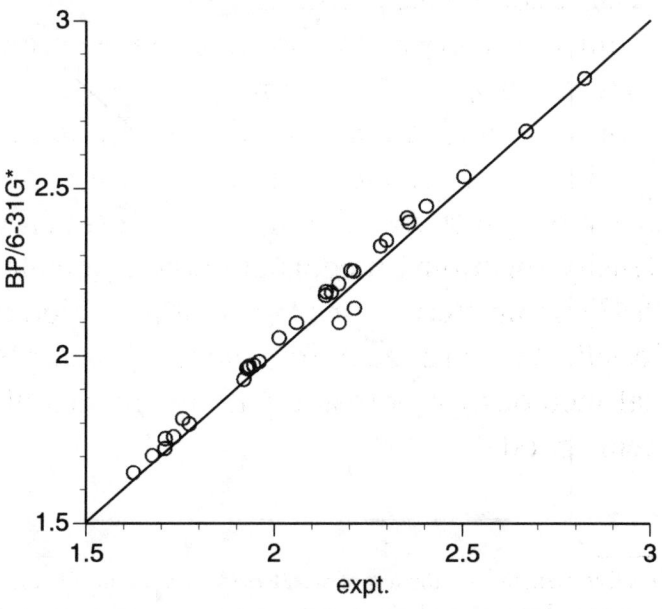

Figure 5-34: BLYP/6-31G* vs. Experimental Bond Distances in Molecules Incorporating Third-Row, Main-Group Elements

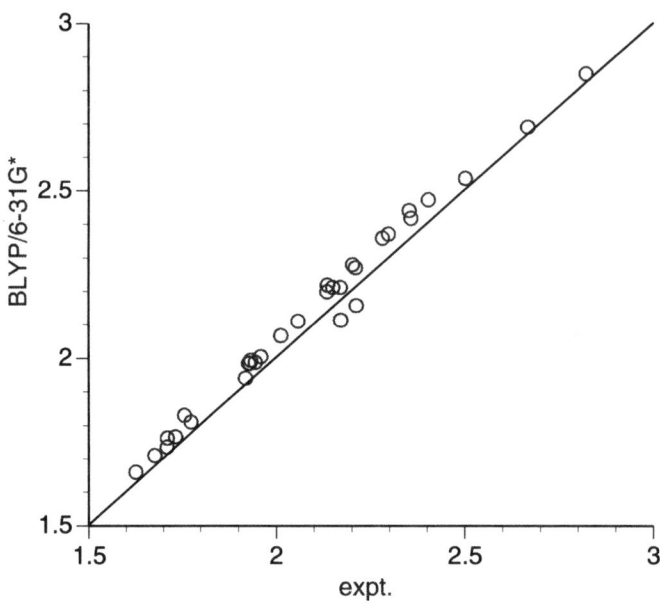

Figure 5-35: EDF1/6-31G* vs. Experimental Bond Distances in Molecules Incorporating Third-Row, Main-Group Elements

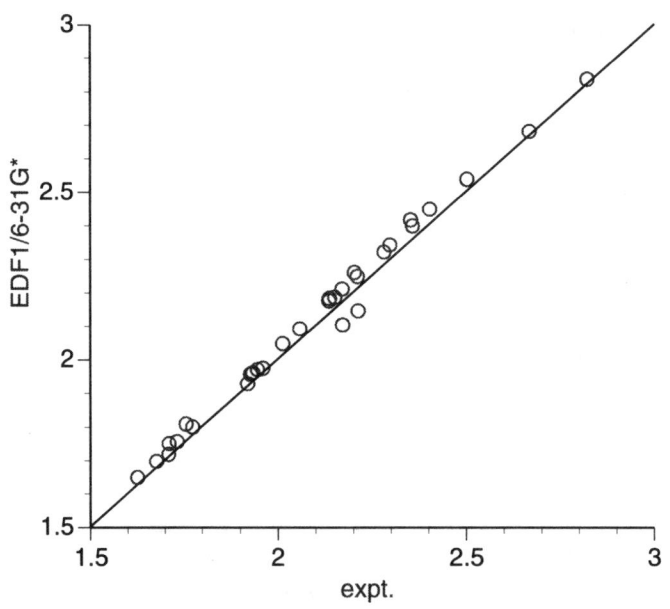

Figure 5-36: B3LYP/6-31G* vs. Experimental Bond Distances in Molecules Incorporating Third-Row, Main-Group Elements

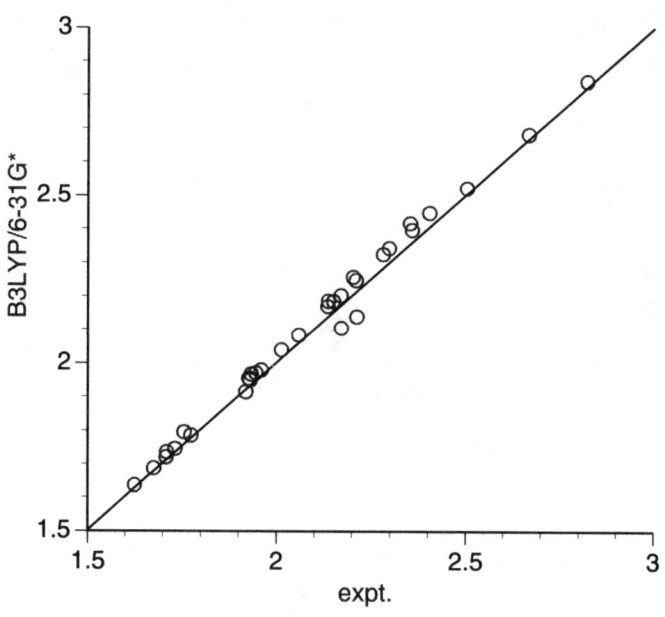

Figure 5-37: MP2/6-31G* vs. Experimental Bond Distances in Molecules Incorporating Third-Row, Main-Group Elements

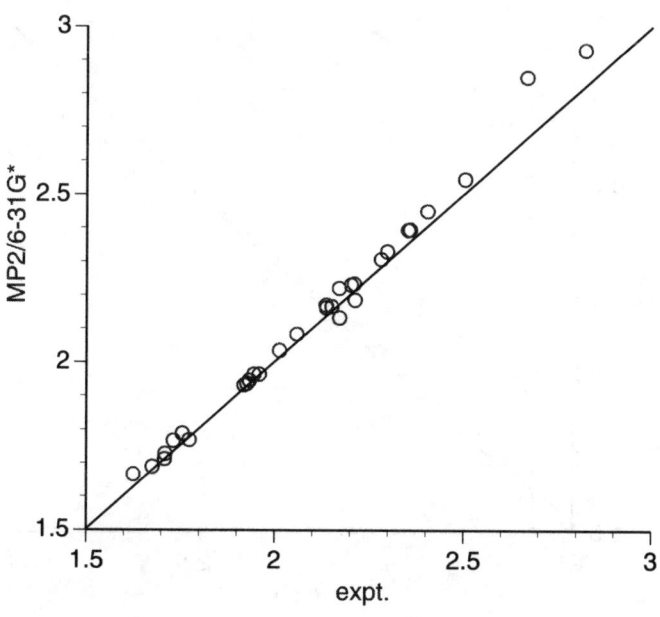

Figure 5-38: MNDO vs. Experimental Bond Distances in Molecules Incorporating Third and Fourth-Row, Main-Group Elements

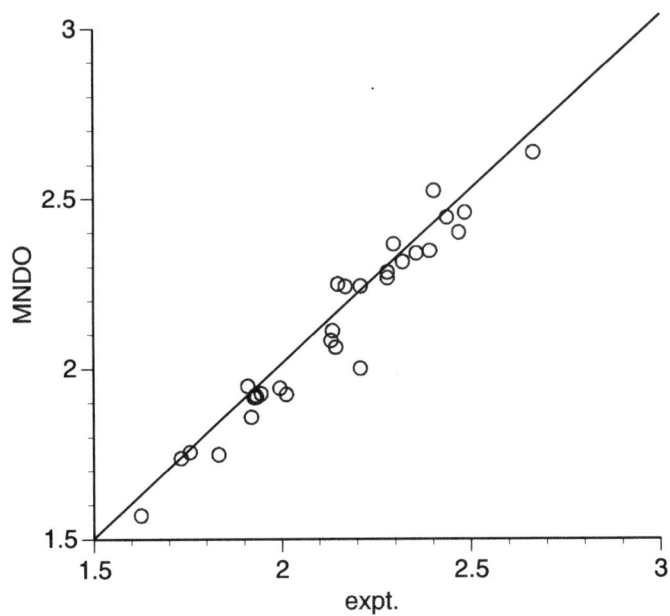

Figure 5-39: AM1 vs. Experimental Bond Distances in Molecules Incorporating Third and Fourth-Row, Main-Group Elements

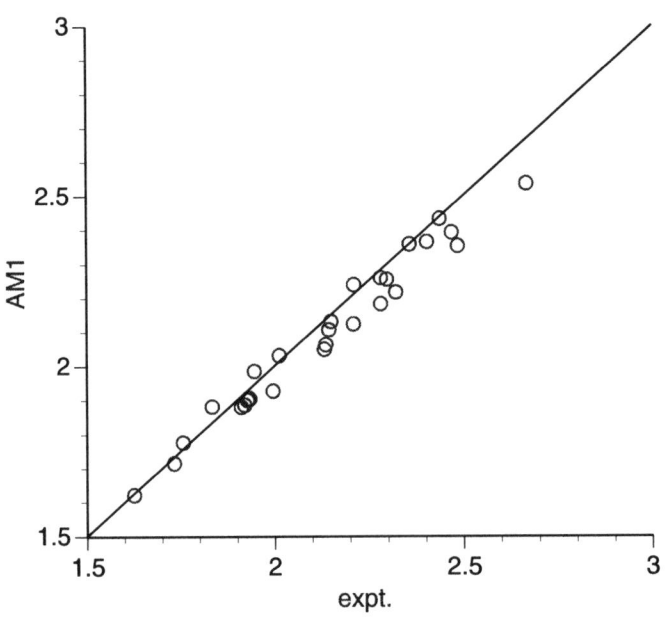

Figure 5-40: PM3 vs. Experimental Bond Distances in Molecules Incorporating Third and Fourth-Row, Main-Group Elements

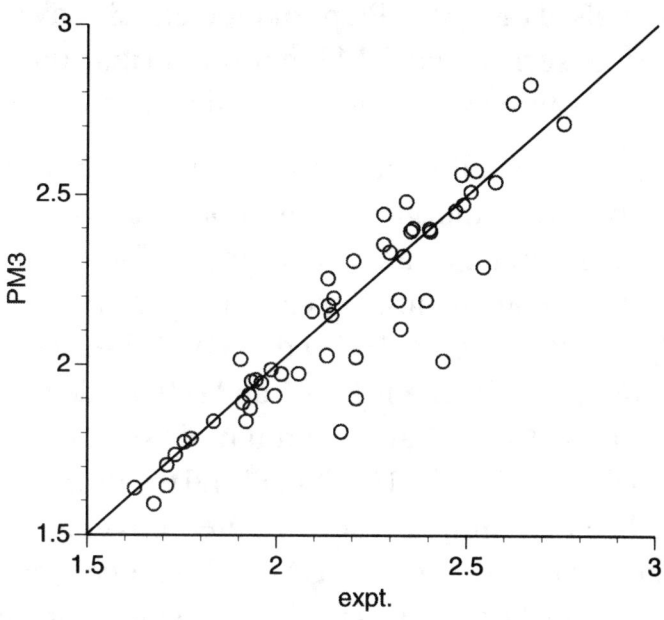

The 6-31G* basis set is presently available for first-row transition metals only (Sc-Zn). STO-3G and 3-21G basis sets are also available for second-row metals (Y-Cd), but are not recommended for use with correlated models. The LACVP* pseudopotential is available for all three transition series and PM3 parameterizations have been developed for most important metals in all three rows.

The present coverage is limited in the number of specific examples but extends to purely inorganic ("without carbon") compounds, coordination compounds, metal carbonyls (including bimetallic carbonyls) and organometallics. For each, the performance of Hartree-Fock models with STO-3G, 3-21G and 6-31G* all-electron basis sets, local density models, BP, BLYP, EDF1 and B3LYP density functional models all with the 6-31G* all-electron basis set and the LACVP* pseudopotential[*], the MP2/6-31G* model and the PM3 semi-empirical model have been examined. More limited coverage, PM3 semi-empirical and BP/6-31G* (or BP/LACVP*) models only, has been provided for skeletal bond angles in a series of larger organometallics, and divided according to class of molecule.

Transition-Metal Inorganic Compounds

Bond length comparisons in transition-metal oxides, halides and oxyhalides are provided in **Table 5-9**. On the basis of mean absolute errors, Hartree-Fock descriptions improve greatly from STO-3G to 3-21G models and less so moving on to the 6-31G* model. However, individual errors are often very large and Hartree-Fock models cannot be recommended. The performance of local density models is better, and BP, BLYP, EDF1 and B3LYP density functional models much better. None of the latter stand out as being any better (or worse) than the others. Mean absolute errors from density functional models with the all-electron 6-31G* basis set and with the LACVP* pseudopotential are essentially the same. With the conspicuous exception of CuF and CuCl, individual geometries from all-electron and pseudopotential calculations are also very similar.

---

[*] Note that the LACVP* pseudopotential reverts to the 6-31G* basis set for non-metals.

The MP2/6-31G* model does not perform as well as any of the density functional models. As for Hartree-Fock models, most individual systems are well described but some are very poorly described. This behavior is perhaps not unexpected, as MP2 models are based on the use of Hartree-Fock wavefunctions. This means that a single electronic configuration is assumed to be "better" than all other configurations, a situation that is probably unreasonable for this class of compounds.

The PM3 model qualitatively accounts for equilibrium geometries in these compounds, but does not afford the quantitative descriptions available from density functional models. None of the systems is particularly poorly described, but individual bond length errors are often significant (as reflected in the large mean absolute error). The PM3 model certainly has a role in surveying the geometries of transition-metal inorganic compounds, but it is not a replacement for better models.

Transition-Metal Coordination Compounds

Bond length comparisons involving a collection of coordination compounds are provided in **Table 5-10**. Hartree-Fock models and the local density model yield very poor results and cannot be recommended.[*] On the other hand, density functional models with the (all-electron) 6-31G* basis set are moderately successful in accounting for the structure of these compounds. None of the models stands out as being particularly better (or worse) than any of the others. The BP/LACVP* model also provides a solid account of bond lengths in these compounds, but oddly enough, the other density functional models with pseudopotentials do not. In particular, very large errors in bond lengths occur in copper complexes. In view of the success of the corresponding models with the all-electron 6-31G* basis set, this result points to a possible problem for the LACVP* pseudopotential for copper (see also comments in previous section).

---

[*] Interestingly, geometries from Hartree-Fock and local density models with the 6-31G* basis set are markedly different, in contrast to the similarity normally seen in dealing with main-group compounds.

Table 5-9: Bond Distances in Transition-Metal Oxides, Halides and Oxyhalides

molecule	point group	geometrical parameter	Hartree-Fock			local density		BP		BLYP		expt.
			STO-3G	3-21G	6-31G*	6-31G*	LACVP*	6-31G*	LACVP*	6-31G*	LACVP*	expt.
ScF_3	D_{3h}	r (ScF)	1.85	1.81	1.83	1.80	1.81	1.83	1.83	1.84	1.84	**1.91**
TiF_4	T_d	r (TiF)	1.70	1.72	1.73	1.73	1.73	1.75	1.75	1.76	1.76	**1.75**
$TiCl_4$	T_d	r (TiCl)	2.17	2.17	2.17	2.14	2.16	2.18	2.19	2.20	2.20	**2.17**
VF_5	D_{3h}	r (VF$_{ax}$)	1.65	1.70	1.70	1.72	1.72	1.75	1.75	1.76	1.75	**1.73**
		r (VF$_{eq}$)	1.62	1.67	1.66	1.69	1.69	1.72	1.72	1.73	1.73	**1.71**
VOF_3	C_{3v}	r (VO)	1.49	1.52	1.50	1.56	1.57	1.58	1.58	1.59	1.59	**1.57**
		r (VF)	1.62	1.69	1.70	1.70	1.70	1.73	1.72	1.74	1.73	**1.73**
$VOCl_3$	C_{3v}	r (VO)	1.47	1.52	1.48	1.55	1.56	1.57	1.58	1.58	1.58	**1.57**
		r (VCl)	2.11	2.12	2.14	2.12	2.12	2.15	2.16	2.17	2.18	**2.14**
CrO_2F_2	C_{2v}	r (CrO)	1.44	1.51	1.49	1.55	1.56	1.57	1.58	1.58	1.59	**1.58**
		r (CrF)	1.58	1.67	1.68	1.69	1.68	1.71	1.70	1.72	1.71	**1.72**
CrO_2Cl_2	C_{2v}	r (CrO)	1.43	1.51	1.49	1.55	1.56	1.57	1.57	1.58	1.58	**1.58**
		r (CrCl)	2.09	2.11	2.13	2.10	2.11	2.14	2.15	2.15	2.17	**2.13**
CrO_4^{2-}	T_d	r (CrO)	1.52	1.60	1.60	1.65	1.65	1.67	1.67	1.68	1.68	**1.66**
MnO_4^-	T_d	r (MnO)	1.46	1.55	1.55	1.59	1.59	1.62	1.61	1.63	1.62	**1.63**
CuF	$C_{\infty v}$	r (CuF)	1.58	1.61	1.73	1.63	1.70	1.66	1.74	1.67	1.75	**1.75**
$CuCl$	$C_{\infty v}$	r (CuCl)	2.20	1.96	2.11	1.95	2.08	1.99	2.13	2.01	2.14	**2.05**
mean absolute error			0.10	0.06	0.05	0.04	0.03	0.02	0.02	0.02	0.02	–

Table 5-9: Bond Distances in Transition-Metal Oxides, Halides and Oxyhalides (2)

molecule	point group	geometrical parameter	EDF1		B3LYP		MP2	semi-empirical	expt.
			6-31G*	LACVP*	6-31G*	LACVP*	6-31G*	PM3	
ScF_3	D_{3h}	r (ScF)	1.83	1.83	1.83	1.83	1.85	a	1.91
TiF_4	T_d	r (TiF)	1.75	1.75	1.74	1.74	1.75	1.84	1.75
$TiCl_4$	T_d	r (TiCl)	2.18	2.19	2.18	2.18	2.17	2.22	2.17
VF_5	D_{3h}	r (VF$_{ax}$)	1.75	1.75	1.73	1.73	1.74	1.76	1.73
		r (VF$_{eq}$)	1.72	1.72	1.70	1.70	1.72	1.72	1.71
VOF_3	C_{3v}	r (VO)	1.57	1.58	1.56	1.56	1.61	1.62	1.57
		r (VF)	1.73	1.72	1.72	1.71	1.76	1.72	1.73
$VOCl_3$	C_{3v}	r (VO)	1.56	1.57	1.55	1.56	2.63	1.63	1.57
		r (VCl)	2.16	2.16	2.15	2.15	2.17	2.04	2.14
CrO_2F_2	C_{2v}	r (CrO)	1.57	1.57	1.55	1.56	1.66	1.61	1.58
		r (CrF)	1.72	1.71	1.71	1.70	1.77	1.75	1.72
CrO_2Cl_2	C_{2v}	r (CrO)	1.57	1.57	1.55	1.56	–	1.62	1.58
		r (CrCl)	2.14	2.15	2.13	2.12	–	2.03	2.13
CrO_4^{2-}	T_d	r (CrO)	1.67	1.67	1.65	1.65	1.69	1.68	1.66
MnO_4^-	T_d	r (MnO)	1.62	1.61	1.60	1.60	1.58	1.66	1.63
CuF	$C_{\infty v}$	r (CuF)	1.67	1.75	1.67	1.74	1.65	1.82	1.75
$CuCl$	$C_{\infty v}$	r (CuCl)	2.00	2.14	2.01	2.14	1.99	2.12	2.05
mean absolute error			0.02	0.02	0.02	0.02	0.04	0.05	–

a) PM3 model has not been parameterized for scandium

Table 5-10: Metal-Ligand Bond Distances in Transition-Metal Coordination Compounds

coordination compound	coordination	bond to	Hartree-Fock			local density		BP		BLYP	
			STO-3G	3-21G	6-31G*	6-31G*	LACVP*	6-31G*	LACVP*	6-31G*	LACVP*
$FeCN_6^{4-}$	octahedral	CN	2.31	1.91	2.33	1.89	1.89	1.95	1.96	1.99	2.00
$CoCN_6^{3-}$	octahedral	CN	2.15	1.91	2.05	1.86	1.86	1.91	1.92	1.94	1.95
$Co(NH_3)_6^{3+}$	octahedral	NH_3	2.08	2.04	2.04	1.94	1.96	2.00	2.03	2.03	2.06
$NiCN_4^{2-}$	square planar	CN	2.02	2.00	2.03	1.80	1.83	1.85	1.88	1.87	1.91
$Ni(NH_3)_4^{2+}$	square planar	NH_3	2.01[a]	2.00	2.03	1.86	1.90	1.92	1.96	1.94	1.99
$CuCN_3^{2-}$	trigonal planar	CN	b	1.99	2.08	1.86	1.93	1.90	1.93	1.92	2.02
$CuCN_4^{3-}$	tetrahedral	CN	1.64	2.01	2.27	1.93	2.02	2.00	2.02	2.03	2.18
mean absolute error			0.20	0.06	0.16	0.08	0.07	0.04	0.04	0.04	0.09

coordination compound	coordination	bond to	EDF1		B3LYP		MP2	semi-empirical	expt.
			6-31G*	LACVP*	6-31G*	LACVP*	6-31G*	PM3	
$FeCN_6^{4-}$	octahedral	CN	1.95	1.96	2.01	2.04	1.85	1.91	**1.92**
$CoCN_6^{3-}$	octahedral	CN	1.91	1.92	1.94	1.95	1.86	1.92	**1.98**
$Co(NH_3)_6^{3+}$	octahedral	NH_3	2.01	2.04	2.01	2.03	1.97	1.98	**1.98**
$NiCN_4^{2-}$	square planar	CN	1.85	1.89	1.88	1.91	1.80	1.90	**1.86**
$Ni(NH_3)_4^{2+}$	square planar	NH_3	1.93	1.98	1.94	1.97	1.92	1.90	**2.04**
$CuCN_3^{2-}$	trigonal planar	CN	1.91	2.02	1.93	2.04	1.88	1.99	**1.93**
$CuCN_4^{3-}$	tetrahedral	CN	2.01	2.18	2.04	2.20	1.96	2.01	**2.00**
mean absolute error			0.04	0.07	0.05	0.09	0.07	0.05	–

a) distorts from square planar
b) calculation results in unreasonable geometry

144

Paralleling its behavior for inorganic compounds, the MP2/6-31G* model does not provide a completely satisfactory description of bond lengths in coordination compounds, certainly not of the quality provided by density functional models. It is difficult to recommend its use, especially in light of its high cost (relative to density functional models).

PM3 semi-empirical calculations furnish a solid account of the geometries of transition-metal coordination compounds. Most distances are within a few hundredths of an Å of their experimental values, but large errors appear for a few compounds. PM3 can be recommended (with due caution) for preliminary structure determinations of transition-metal coordination compounds.

Further examples of coordinate bonds are found in metal carbonyl complexes. Metal carbon (carbon monoxide) bond distances in a selection of (first-row) transition-metal carbonyls and transition-metal organometallics are examined in **Table 5-11**. As expected, Hartree-Fock models do not perform well. The 6-31G* model is clearly superior to the STO-3G and 3-21G models (both of which lead to completely unreasonable geometries for several compounds), but still exhibits unacceptable errors. For example, the model shows markedly different lengths for the *axial* and *equatorial* bonds in iron pentacarbonyl, in contrast to experiment where they are nearly the same. Hartree-Fock models cannot be recommended.

Density functional models provide a solid account of metal-carbon (carbon monoxide) bond lengths in these compounds. The local density model is inferior to the other density functional models, but still leads to acceptable results. Note that bond lengths from local density 6-31G* calculations differ greatly from those from the corresponding Hartree-Fock calculations. As noted previously, the overall similarity of structural results for the two models seen for main-group compounds does not appear to extend to transition-metal systems. Among the BP, BLYP, EDF1 and B3LYP models, none stands out as particularly better or particularly worse than the others. The LACVP* pseudopotential usually (but not always) leads to longer metal-carbon bond lengths than the all-electron 6-31G* basis set,

Table 5-11: Metal-Carbon (CO) Bond Distances in Transition-Metal Carbonyls and Organometallics

metal carbonyl	bond	Hartree-Fock			local density		BP		BLYP		expt.
		STO-3G	3-21G	6-31G*	6-31G*	LACVP*	6-31G*	LACVP*	6-31G*	LACVP*	
$CrCO_6$		1.79	1.97	1.99	1.86	1.86	1.90	1.90	1.93	1.92	1.91
Mn_2CO_{10}	ax	a	1.93	1.95	1.76	1.75	1.80	1.79	1.82	1.81	1.80
	eq		1.93	1.98	1.80	1.80	1.84	1.84	1.86	1.86	1.87
$FeCO_5$	ax	2.02	2.01	2.03	1.76	1.76	1.80	1.80	1.82	1.82	1.81
	eq	1.64	1.84	1.85	1.75	1.75	1.79	1.79	1.81	1.82	1.82
Fe_2CO_9	terminal	1.75	1.96	1.97	1.77	1.76	1.81	1.80	1.83	1.83	1.8
	bridge	2.07	2.10	2.08	1.94	1.95	1.99	2.00	2.02	2.03	1.9
Co_2CO_8	terminal	a	a	a	1.75	1.75	1.79	1.80	1.81	1.82	1.78
	terminal				1.76	1.76	1.80	1.81	1.82	1.83	1.80
	bridge				1.89	1.90	1.94	1.96	1.96	1.98	1.92
$NiCO_4$		1.58	1.83	1.86	1.76	1.78	1.80	1.83	1.82	1.85	1.82
CO_3Cr (benzene)		1.75	1.93	1.95	1.80	1.80	1.84	1.83	1.86	1.85	1.84
CO_4Cr (Dewar benzene)	ax	1.87	1.92	1.96	1.85	1.84	1.89	1.88	1.91	1.91	1.86
	eq	1.95	1.97	1.95	1.81	1.81	1.85	1.84	1.86	1.86	1.83
$CO_5Cr=C(Me)NH(Me)$	ax	1.95	1.92	1.94	1.84	1.84	1.88	1.87	1.90	1.90	1.86-1.88
	eq	1.96	1.91	1.96	1.85	1.85	1.89	1.89	1.91	1.91	1.88-1.91
CO_3Fe (cyclobutadiene)		a	1.96	1.98	1.73	1.73	1.77	1.77	1.79	1.79	1.79
CO_3Fe (butadiene)		a	2.00	2.04	1.73	1.73	1.77	1.77	1.79	1.79	1.76
CO_4Fe (acetylene)	ax	a	2.04	2.13	1.76	1.76	1.80	1.79	1.82	1.82	1.77
	eq		1.98	2.01	1.74	1.74	1.78	1.78	1.80	1.80	1.76
CO_4Fe (ethylene)	ax	a	2.02	2.07	1.76	1.75	1.79	1.80	1.81	1.82	1.78
	eq		1.96	1.97	1.74	1.74	1.78	1.78	1.78	1.80	1.81
CO_3Co (allyl)		a	1.92	1.98	1.75	1.74	1.78	1.78	1.79	1.80	1.77
mean absolute error		–	0.13	0.16	0.04	0.04	0.02	0.02	0.03	0.03	–

Table 5-11: Metal-Carbon (CO) Bond Distances in Transition-Metal Carbonyls and Organometallics (2)

metal carbonyl	bond	EDF1 6-31G*	EDF1 LACVP*	B3LYP 6-31G*	B3LYP LACVP*	MP2 6-31G*	semi-empirical PM3	expt.
$CrCO_6$		1.90	1.90	1.92	1.92	1.88	1.93	**1.91**
Mn_2CO_{10}	ax	1.79	1.78	1.80	1.80	-	1.79	**1.80**
	eq	1.84	1.84	1.85	1.86	-	1.85	**1.87**
$FeCO_5$	ax	1.80	1.80	1.81	1.82	1.69	1.79	**1.81**
	eq	1.79	1.79	1.80	1.81	1.77	1.74	**1.82**
Fe_2CO_9	terminal	1.81	1.80	1.82	1.82	-	1.77	**1.8**
	bridge	1.99	2.00	1.99	2.01	-	1.99	**1.9**
Co_2CO_8	terminal	1.79	1.80	1.80	1.81	-	1.81	**1.78**
	terminal	1.80	1.81	1.81	1.82	-	1.82	**1.80**
	bridge	1.94	1.96	1.94	1.96	-	1.96	**1.92**
$NiCO_4$		1.80	1.84	1.81	1.85	1.77	1.80	**1.82**
CO_3Cr (benzene)		1.83	1.83	1.85	1.85	1.79	1.90	**1.84**
CO_4Cr (Dewar benzene)	ax	1.89	1.88	1.90	1.90	1.90	1.96	**1.86**
	eq	1.84	1.84	1.86	1.86	1.79	1.92	**1.83**
$CO_5Cr=C(Me)NH(Me)$	ax	1.88	1.87	1.89	1.88	1.83	1.91	**1.86-1.88**
	eq	1.89	1.88	1.90	1.90	1.87	1.91	**1.88-1.91**
CO_3Fe (cyclobutadiene)		1.77	1.77	1.78	1.78	1.64	1.74	**1.79**
CO_3Fe (butadiene)		1.77	1.77	1.78	1.79	1.65	1.75	**1.76**
CO_4Fe (acetylene)	ax	1.80	1.80	1.82	1.83	1.69	1.82	**1.77**
	eq	1.78	1.78	1.79	1.79	1.76	1.75	**1.76**
CO_4Fe (ethylene)	ax	1.79	1.79	1.81	1.81	1.69	1.81	**1.78**
	eq	1.78	1.78	1.79	1.79	1.76	1.75	**1.81**
CO_3Co (allyl)		1.77	1.78	1.78	1.80	1.66	1.81	**1.77**
mean absolute error		0.02	0.02	0.02	0.02	0.06	0.04	-

a) calculation results in unreasonable geometry.

and usually (but not always) to improved agreement with the experimental data. However, the differences are not great.

The MP2/6-31G* model does not perform nearly as well as any of the density functional models, paralleling the behavior previously noted for other transition-metal compounds. Metal-carbon bond lengths are consistently shorter than experimental distances, sometimes by a tenth of an Å or more. It cannot be recommended.

The PM3 semi-empirical model turns in a surprisingly good account of metal-carbon (carbon monoxide) bond distances in these compounds. While PM3 is not as good as the best of the (density functional) models, individual bond lengths are typically within a few hundredths of an Å from their respective experimental values, and larger deviations are uncommon. In view of cost considerations, PM3 certainly has a role in transition-metal structural chemistry.

Transition-Metal Organometallics

Calculated and experimental geometries for organometallics involving first-row transition metals are compared in **Table 5-12**. The tabulated information is limited to metal-carbon bond distances to the organic ligands, e.g., the two unique distances involving the allyl ligand in $(CO)_3Co$ (allyl). (Metal-carbon monoxide ligand distances have previously been addressed.)

The 6-31G* model turns in a poor performance. With a single exception, all bonds are longer than the experimental distances, sometimes by as much as 0.1 to 0.2Å. STO-3G and 3-21G models do not exhibit such consistency, and calculated bond distances for both are often quite far from their respective experimental values. Hartree-Fock models cannot be trusted to account for the geometries of organometallic compounds.

Density functional models provide a much better account. The local density model does the poorest and BP and B3LYP models do the best, but the differences are not great. As with metal-carbon (carbon monoxide) lengths, bond distances from all-electron 6-31G* calculations are usually (but not always) shorter than those obtained using the LACVP* pseudopotential, typically by a few hundredths of an Å. Either provides a good account.

As with metal-carbon monoxide bonds, the MP2/6-31G* model does not lead to results of the same calibre as those from density functional models (except local density models). The model actually shows the opposite behavior as 6-31G*, in that bond lengths are consistently shorter than experimental values, sometimes significantly so. In view of its poor performance and the considerable cost of MP2 models (relative to density functional models), there seems little reason to employ them for structural investigations on organometallics.

The PM3 model gives a respectable account of metal-carbon bond lengths in transition-metal organometallics. This is not at all surprising. Parameter selection for PM3 was largely based on producing good geometries for organometallics. PM3 appears to be a reasonable low-cost alternative to density functional calculations for providing geometries of transition-metal organometallics. There will, no doubt, be cases where the performance of the semi-empirical model will be poor, and due caution needs to be exercised.

Bimetallic Carbonyls

Calculated metal-metal bond distances in three bimetallic carbonyl compounds are compared with experimental values in **Table 5-13**.* While the sample is much too small to draw meaningful generalizations, the excellent performance of density functional models (local density and BLYP models excluded) is noteworthy, as is the good showing of PM3. The latter is actually somewhat of a surprise as PM3 parameterizations were carried out using single metal compounds only. Hartree-Fock models turn in a very poor performance, with several of the calculated geometries being completely unreasonable. Calculations with the MP2/6-31G* model have not been performed due both to cost considerations and to the poor showing of the method for other transition-metal systems.

---

*   Metal-carbon monoxide ligand bond lengths in these same compounds have already been provided in **Table 5-11**.

Table 5-12: Metal-Carbon Bond Distances in Transition-Metal Organometallics

organometallic	bond	Hartree-Fock			local density		BP		BLYP		expt.
		STO-3G	3-21G	6-31G*	6-31G*	LACVP*	6-31G*	LACVP*	6-31G*	LACVP*	
Cp$_2$Ti (ethylene)	ethylene	2.05	2.19	2.28	2.15	2.15	2.18	2.19	2.20	2.21	**2.16**
Cr (benzene)$_2$	benzene	2.04	2.17	2.18	2.09	2.09	2.14	2.15	2.17	2.21	**2.22**
CO$_3$Cr (benzene)	benzene	2.10	2.24	2.27	2.15	2.17	2.21	2.23	2.25	2.30	**2.22**
CO$_4$Cr (Dewar benzene)	Dewar benzene C$_1$	1.99	2.25	2.51	2.23	2.24	2.31	2.34	2.38	2.43	**2.33**
CO$_5$Cr=C(Me)NH(Me)	=C	2.20	2.11	2.20	2.01	2.01	2.07	2.07	2.10	2.11	**2.08-2.12**
FeCp$_2$	cyclopentadienyl	2.09	2.20	2.20	1.99	2.00	2.03	2.05	2.06	2.09	**2.06**
CO$_3$Fe (cyclobutadiene)	cyclobutadiene	a	2.10	2.11	2.00	2.01	2.05	2.05	2.07	2.08	**2.06**
CO$_3$Fe (butadiene)	butadiene C$_1$	a	2.09	2.09	2.05	2.07	2.10	2.13	2.13	2.17	**2.14**
	butadiene C$_2$		2.19	2.19	2.01	2.02	2.06	2.07	2.09	2.11	**2.06**
CO$_4$Fe (acetylene)	acetylene	a	2.03	2.01	2.04	2.05	2.08	2.11	2.11	2.15	**2.08**
CO$_4$Fe (ethylene)	ethylene	a	2.03	2.03	2.06	2.07	2.12	2.15	2.15	2.20	**2.12**
CO$_3$Co (allyl)	allyl C$_1$	a	2.14	2.20	2.03	2.07	2.09	2.13	2.12	2.17	**2.10**
	allyl C$_2$		2.07	2.11	1.96	1.97	2.01	2.03	2.04	2.07	**1.99**
Ni (allyl)$_2$	allyl C$_1$	2.13	2.10	2.10	1.92	1.97	1.97	2.04	2.00	2.07	**2.02**
	allyl C$_2$	2.25	2.11	2.11	1.92	1.94	1.96	1.99	1.98	2.02	**1.98**
mean absolute error		-	0.07	0.10	0.07	0.05	0.02	0.02	0.03	0.05	-

Table 5-12: Metal-Carbon Bond Distances in Transition-Metal Organometallics (2)

organometallic	bond	EDF1		B3LYP		MP2	semi-empirical	expt.
		6-31G*	LACVP*	6-31G*	LACVP*	6-31G*	PM3	
Cp₂Ti (ethylene)	ethylene	2.18	2.18	2.18	2.19	2.16	2.15	**2.16**
Cr (benzene)₂	benzene	2.14	2.15	2.16	2.17	2.13	2.21	**2.22**
CO₃Cr (benzene)	benzene	2.22	2.26	2.23	2.27	2.18	2.23	**2.22**
CO₄Cr (Dewar benzene)	Dewar benzene C₁	2.32	2.36	2.36	2.41	2.28	2.20	**2.33**
CO₅Cr=C(Me)NH(Me)	=C	2.08	2.09	2.09	2.10	2.02	2.06	**2.08-2.12**
FeCp₂	cyclopentadienyl	2.03	2.05	2.05	2.08	1.94	2.09	**2.06**
CO₃Fe (cyclobutadiene)	cyclobutadiene	2.04	2.05	2.04	2.05	1.98	2.07	**2.06**
CO₃Fe (butadiene)	butadiene C₁	2.10	2.13	2.10	2.13	2.16	2.05	**2.14**
	butadiene C₂	2.06	2.07	2.07	2.08	1.99	2.13	**2.06**
CO₄Fe (acetylene)	acetylene	2.08	2.11	2.08	2.10	2.22	2.00	**2.08**
CO₄Fe (ethylene)	ethylene	2.12	2.15	2.12	2.16	2.21	2.01	**2.12**
CO₃Co (allyl)	allyl C₁	2.09	2.14	2.09	2.13	1.96	2.02	**2.10**
	allyl C₂	2.01	2.03	2.02	2.04	1.93	2.02	**1.99**
Ni (allyl)₂	allyl C₁	1.97	2.04	1.99	2.05	1.85	2.03	**2.02**
	allyl C₂	1.95	1.99	1.97	2.01	1.82	2.06	**1.98**
mean absolute error		0.02	0.03	0.02	0.03	0.09	0.05	–

a) calculation results in unreasonable geometry.

Table 5-13: Metal-Metal Bond Distances in Bimetallic Transition-Metal Carbonyls

metal carbonyl	Hartree-Fock			local density		BP		BLYP		expt.
	STO-3G	3-21G	6-31G*	6-31G*	LACVP*	6-31G*	LACVP*	6-31G*	LACVP*	
Mn_2CO_{10}	a	3.36	3.28	2.78	2.78	2.96	2.95	3.06	3.05	**2.98**
Fe_2CO_9	2.63	2.54	2.58	2.43	2.44	2.50	2.52	2.55	2.57	**2.52**
Co_2CO_8	a	a	a	2.46	2.47	2.53	2.56	2.58	2.61	**2.52**
mean absolute error	–	–	–	0.12	0.11	0.02	0.02	0.06	0.07	–

metal carbonyl	EDF1		B3LYP		semi-empirical	expt.
	6-31G*	LACVP*	6-31G*	LACVP*	PM3	
Mn_2CO_{10}	3.01	3.00	3.00	2.99	2.92	**2.98**
Fe_2CO_9	2.50	2.51	2.51	2.52	2.60	**2.52**
Co_2CO_8	2.54	2.56	2.55	2.57	2.56	**2.52**
mean absolute error	0.02	0.03	0.02	0.02	0.06	–

a) calculation resutls in unreasonable geometry

Organometallics with Second and Third-Row Transition Metals

Coverage of molecules incorporating second and third-row transition metals is more limited, in that only pseudopotential models and the PM3 model are available, and more targeted, in that it builds an experience gained with first-row metal systems. Only three models are assessed: BP and B3LYP density functional models with the LACVP* pseudopotential and the PM3 semi-empirical model. Hartree-Fock, local density and MP2 models have been excluded due to their unfavorable performance with first-row metals, and BLYP and EDF1 density functional models have been excluded because, based on experience with first-row compounds, their performance would be expected to be nearly identical to those from the corresponding BP and B3LYP models. Finally, only organometallic compounds have been considered.

Data are presented in **Table 5-14**. The experimental data derive from X-ray crystallography.[2] In most cases, tabulated bond distances correspond either to average values from "identical" parameters in the same structure, from different determinations on the same molecule or from parameters in several closely-related molecules. As such, they exhibit considerable uncertainty, and comparisons with calculated bond lengths below a threshold of 0.02Å are probably meaningless.

Both density functional models provide a reasonable account of metal-ligand bond lengths in all compounds. This includes a variety of π–bonded ligands, coordinate bonds to carbon monoxide and trimethylphosphine, and single and double bonds to carbon. Bonds to π ligands are consistently longer than experimental values, typically by 0.02 to 0.05Å, while bonds to CO ligands are almost always within the experimental range. Both models properly reproduce the difference in single and double bonds to carbon in the tantalum methylidene complex. In terms of mean absolute errors, the BP/LACVP* model is superior to B3LYP/LACVP*, but the difference is not great.

The PM3 model also provides a solid account. In terms of mean absolute error, it is as good as either of the more costly density functional models. It would appear to be an excellent choice for preliminary structure surveys of organometallics.

Table 5-14: Metal-Ligand Bond Distances in Second and Third-Row Transition-Metal Organometallics

organometallic	bond	BP LACVP*	B3LYP LACVP*	PM3	expt.
$Cp_2Zr(PMe_3)$(ethylene)	cyclopentadienyl	2.58	2.61	2.56	**2.50-2.52**
	ethylene	2.37	2.36	2.46	**2.33-2.35**
	PMe_3	2.75	2.80	2.78	**2.70**
$Cp_2Ta(CH_3)=CH_2$	cyclopentadienyl	2.48	2.50	2.44	**2.41-2.43**
	methyl	2.26	2.25	2.22	**2.27**
	$=CH_2$	2.00	1.99	2.00	**2.04**
CO_3Mo(benzene)	benzene	2.45	2.49	2.39	**2.36-2.38**
	CO	1.96	1.96	1.95	**1.96-1.97**
CO_2Mo (Cp)(allyl)	cyclopentadienyl	2.42	2.44	2.36	**2.31-2.35**
	allyl C_1	2.38	2.40	2.26	**2.32-2.36**
	allyl C_2	2.26	2.28	2.23	**2.21-2.22**
	CO	1.95	1.96	2.02	**1.94-1.96**
CO_3W(benzene)	benzene	2.42	2.44	2.34	**2.36-2.37**
	CO	1.96	1.97	2.02	**1.95-1.96**
$RuCp_2$	cyclopentadienyl	2.25	2.26	2.18	**2.18-2.19**
CpRh(cyclobutadiene)	cyclopentadienyl	2.30	2.33	2.19	**2.21-2.22**
	cyclobutadiene	2.16	2.16	2.13	**2.10-2.11**
Cl_2Pd(cyclooctatetraene)	cyclooctatetraene	2.35	2.30	2.15	**2.20-2.22**
	Cl	2.39	2.39	2.28	**2.27-2.29**
mean absolute error		0.04	0.06	0.03	–

Bond Angles Involving Transition-Metal Centers

Another key indicator of structure in organometallic compounds are bond angles involving the metal center. Trends in bond angles among closely-related systems may relate to steric and/or electronic requirements, and the ability of calculations to reproduce such trends should, therefore, reflect on their ability to account for such factors.

An example is provided by the structures of Group IV metallacycles, $LL'MCl_2$, where ligands L and L´ are cyclopentadienyl based and M is Ti or Zr. As a class, these compounds act as catalyst precursors in homogenous (Ziegler-Natta) polymerization of olefins, e.g.

$$Cp_2ZrCl_2 \longrightarrow \longrightarrow Cp_2Zr^+R \xrightarrow{\ H_2C=CH_2\ } Cp_2\overset{H_2C=CH_2}{Zr^+}R \longrightarrow Cp_2\overset{H_2C=CH_2}{Zr^+}{\cdot\cdot}R \longrightarrow Cp_2Zr^+CH_2CH_2R \longrightarrow$$

It might be expected that the bond angle involving the centroids of the cyclopentadienyl rings and the metal center would anticipate the amount of "space" available to an incoming olefin and, therefore, reasonably expected to correlate with reactivity and/or selectivity of Ziegler-Natta processes.

Figures 5-41 and **5-42** compare CpTiCp´ (centroid) bond angles in titanium cyclopentadienyl dichloride complexes from PM3 and BP/6-31G* calculations, respectively, with experimental values for these and other compounds dealt with in this section from X-ray crystallography.[2] Due to practical limitations, the data used for comparison with the density functional calculations are a subset of that used in comparison with PM3. Both models perform well in separating those systems where the cyclopentadienyl rings are spread far apart from those where they are closer together.

Similar comments can be made for related comparisons involving CpZrCp´ (centroid) bond angles in analogous zirconium complexes (**Figures 5-43** and **5-44** for PM3 and BP/LACVP* calculations, respectively), and involving InZrIn´ (centroid) bond angles in zirconium indenyl dichloride complexes (**Figures 5-45** and **5-46** for PM3 and BP/LACVP* calculations, respectively). Again, the basic trends are reproduced, and again errors resulting from the semi-empirical

Figure 5-41: PM3 vs. Experimental CpTiCp´ (Centroid) Bond Angles in Titanium Metallacycles, CpCp´TiCl$_2$

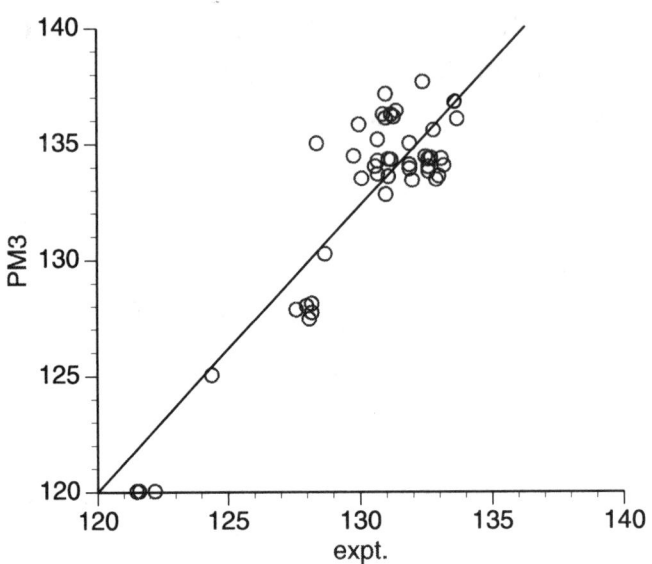

Figure 5-42: BP/6-31G* vs. Experimental CpTiCp´ (Centroid) Bond Angles in Titanium Metallacycles, CpCp´TiCl$_2$

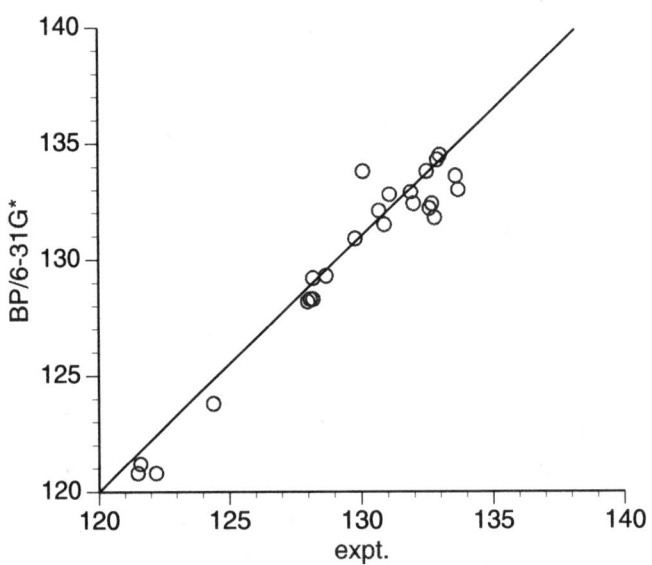

Figure 5-43: PM3 vs. Experimental CpZrCp´ (Centroid) Bond Angles in Zirconium Metallacycles, CpCp´ZrCl₂

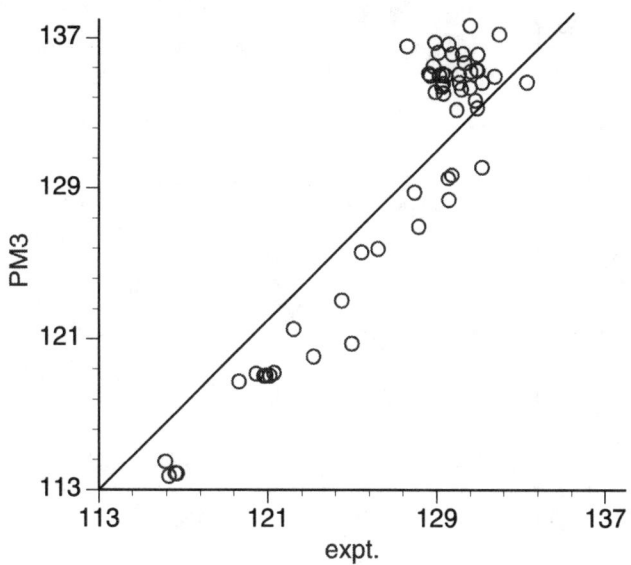

Figure 5-44: BP/LACVP* vs. Experimental CpZrCp´ (Centroid) Bond Angles in Zirconium Metallacycles, CpCp´ZrCl₂

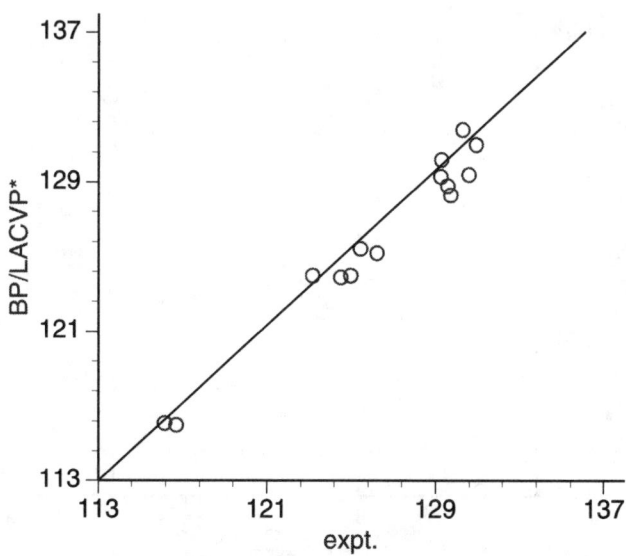

Figure 5-45: PM3 vs. Experimental InZrIn´ (Centroid) Bond Angles in Zirconium Metallacycles, InIn´ZrCl$_2$

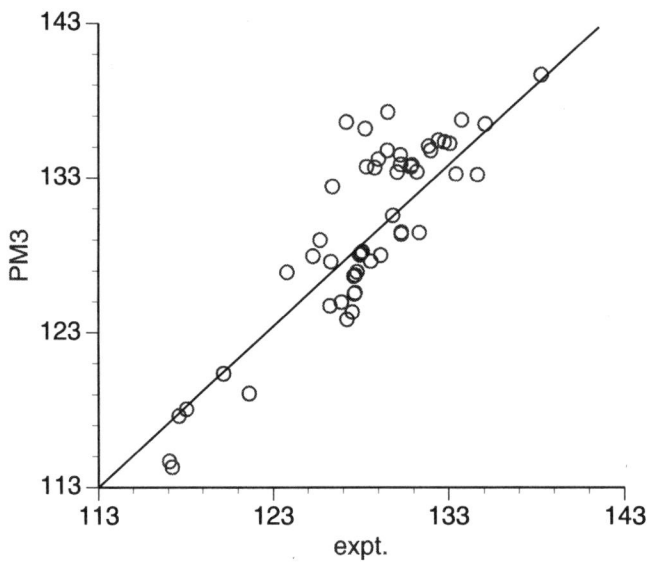

Figure 5-46: BP/LACVP* vs. Experimental InZrIn´ (Centroid) Bond Angles in Zirconium Metallacycles, InIn´ZrCl$_2$

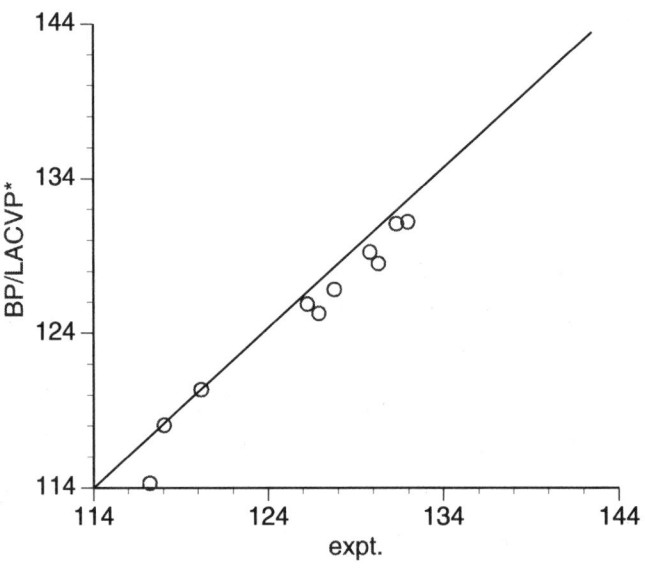

calculations are somewhat larger than those from the density functional calculations. (Note, however, that the data sets are also larger.)

In all three sets of comparisons, both levels of calculation provide an acceptable account of structural changes. BP/LACVP* calculations fare better than the PM3 calculations although, as mentioned previously, these constitute a subset of "smaller" systems.

An entirely different kind of bond angle comparison is provided in **Figures 5-47** and **5-48**, respectively for PM3 and BP/LACVP* calculations. This involves the Ta=CR bond angle involving the sp^2 ("carbene") carbon in a series of tantalum carbenes, R'R''R'''Ta=CHR. These complexes are all electron deficient, meaning that tantalum has fewer than its normal complement of 18 valence electrons. Ideally, this angle should be 120°, but in fact it is much larger (approaching 180°). One explanation for the structural change is that the CH bond contributes its pair of electrons to tantalum-carbon bonding, in the limit leading to a "metal carbyne".

This is a close analogy to what is commonly known by organic chemists as hyperconjugation, where an electron-deficient carbocation center "grabs electrons" from a neighboring CH bond.

The PM3 model does a good job in reproducing the experimental structural data over a very wide range (bond angles from a nearly "normal" 130° to a nearly linear 170°). There are significant deviations, but these need to be put in the context that the potential surface for angular distortion is likely to be very flat. The density functional calculations are also successful, although again there are some large individual errors.

Figure 5-47: PM3 vs. Experimental Ta=CR Bond Angles in Tantalum Carbenes R´R´´R´´Ta=CHR

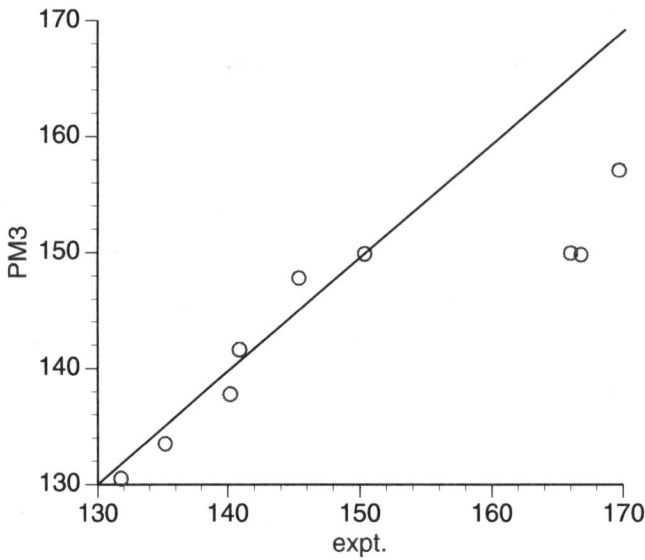

Figure 5-48: BP/LACVP* vs. Experimental Ta=CR Bond Angles in Tantalum Carbenes R´R´´R´´´Ta=CHR

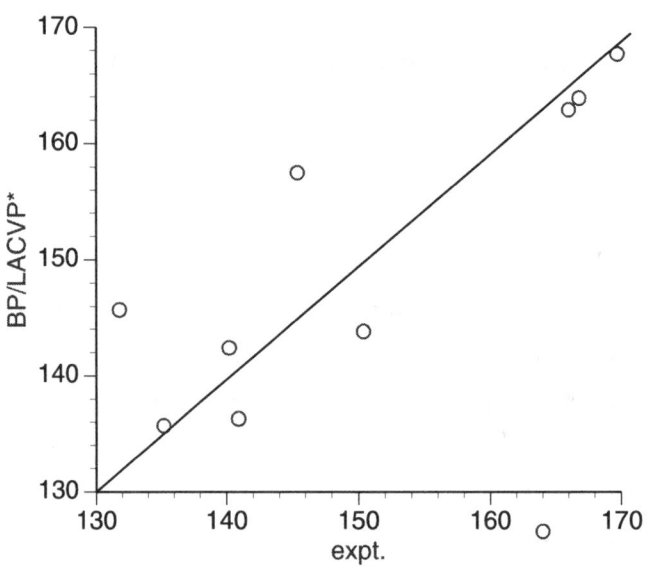

Reactive Intermediates

Reactive intermediates play an important role in many organic reactions, and are often invoked in mechanistic explanations. As their name "intermediates" has come to imply, such molecules are often difficult to isolate, let alone characterize, and high-quality experimental data relating to the structures, stabilities and other properties of intermediates are generally scarce.

The present coverage is divided according to the class of intermediate: carbocations, anions, carbenes (and related divalent compounds) and radicals. Within each class, the usual set of models is assessed, the exception being that molecular mechanics models have been excluded. These have not been explicitly parameterized for charged species or molecules with unpaired electrons, and cannot be expected to perform favorably. Note that in some, but not all cases, the quality of the experimental structural data is not up to the same standard as for other small molecules. This is particularly true for carbocations and anions, where differences among counterions may lead to large differences in structure.

Carbocations

A small number of carbocations have been crystallized and their geometries determined by X-ray diffraction.[2]* Although the influence of counterions is of some concern, in many cases these can be made sufficiently bulky to preclude their encroaching too close. Here, the measured crystal geometry should closely reflect that of the (isolated) charged species.

Comparative data for a few particularly interesting systems is provided in **Table 5-15**. STO-3G, 3-21G and 6-31G* Hartree-Fock models, local density models, BP, BLYP, EDF1 and B3LYP density functional models all with the 6-31G* basis set, the MP2/6-31G* model and MNDO, AM1 and PM3 semi-empirical models have been examined.

---

* Experimental gas-phase structures of charged species are virtually non-existent. It is impossible to establish sufficiently high concentrations for conventional spectroscopy. Also, microwave spectroscopy, the principal technique for accurate structure determinations, cannot be applied to charged species.

Table 5-15: Structures of Carbocations

carbocation	geometrical parameter	Hartree-Fock			local density	expt.
		STO-3G	3-21G*	6-31G*	6-31G*	
	$r(C^+C)$	1.51	1.48	1.48	1.45	**1.35-1.37**
	$r(C^+C_\alpha)$	1.41	1.40	1.41	1.41	**1.41**
	$r(C_\alpha C_\beta)$	1.43	1.42	1.42	1.42	**1.43**
	$r(C_\beta C_\gamma)$	1.37	1.37	1.37	1.38	**1.37**
	$r(C_\alpha C_\delta)$	1.40	1.39	1.39	1.40	**1.40**
	$r(C^+C_\alpha)$	1.52	1.49	1.48	1.48	**1.49**
	$r(C^+C_\beta)$	1.50	1.45	1.45	1.42	**1.41-1.44**
	$r(C^+C_\gamma)$	2.33	2.25	2.24	2.06	**2.09-2.11**
	$r(C_\beta C_\gamma)$	1.59	1.65	1.62	1.69	**1.71-1.74**
	$r(C^+C_\alpha)$	1.55	1.55	1.54	1.57	**1.51-1.55**
	$r(C^+C_\beta)$	1.76	1.87	1.72	1.68	**1.71-1.72**
	$r(C_\alpha C_\beta)$	1.51	1.52	1.48	1.48	**1.48**
	$r(C_\beta C_\gamma)$	1.41	1.39	1.40	1.42	**1.43**
	$r(C_\alpha C_\beta)$	1.53	1.51	1.50	1.48	**1.46-1.50**
	$r(C_\beta C_\gamma)$	1.37	1.36	1.36	1.38	**1.35-1.38**
	$r(C_\gamma C_\delta)$	1.44	1.43	1.42	1.42	**1.40-1.45**
	$r(C^+C_\alpha)$	1.49	1.46	1.46	1.44	**1.44**
	$r(C_\alpha C_\beta)$	1.59	1.62	1.60	1.60	**1.61**
	$<(C_\alpha C^+ C_\alpha')$	117	118	118	118	**118**
	$r(CC)$	1.38	1.38	1.36	1.38	**1.35-1.37**

Table 5-15: Structures of Carbocations (2)

carbocation	geometrical parameter	BP 6-31G*	BLYP 6-31G*	EDF1 6-31G*	B3LYP 6-31G*	MP2 6-31G*	expt.
	$r(C^+C)$	1.47	1.48	1.47	1.47	1.46	**1.35-1.37**
	$r(C^+C_\alpha)$	1.43	1.43	1.42	1.42	1.41	**1.41**
	$r(C_\alpha C_\beta)$	1.44	1.44	1.43	1.43	1.43	**1.43**
	$r(C_\beta C_\gamma)$	1.39	1.39	1.39	1.38	1.38	**1.37**
	$r(C_\alpha C_\delta)$	1.41	1.41	1.41	1.40	1.40	**1.40**
	$r(C^+C_\alpha)$	1.49	1.50	1.49	1.49	1.51	**1.49**
	$r(C^+C_\beta)$	1.44	1.44	1.44	1.43	1.41	**1.41-1.44**
	$r(C^+C_\gamma)$	2.16	2.22	2.17	2.18	1.85	**2.09-2.11**
	$r(C_\beta C_\gamma)$	1.72	1.74	1.70	1.71	1.89	**1.71-1.74**
	$r(C^+C_\alpha)$	1.58	1.57	1.57	1.56	1.56	**1.51-1.55**
	$r(C^+C_\beta)$	1.74	1.79	1.72	1.74	1.71	**1.71-1.72**
	$r(C_\alpha C_\beta)$	1.50	1.51	1.49	1.49	1.48	**1.48**
	$r(C_\beta C_\gamma)$	1.43	1.42	1.43	1.42	1.42	**1.43**
	$r(C_\alpha C_\beta)$	1.52	1.52	1.51	1.51	1.50	**1.46-1.50**
	$r(C_\beta C_\gamma)$	1.40	1.40	1.39	1.38	1.39	**1.35-1.38**
	$r(C_\gamma C_\delta)$	1.44	1.45	1.44	1.44	1.43	**1.40-1.45**
	$r(C^+C_\alpha)$	1.47	1.47	1.46	1.46	1.45	**1.44**
	$r(C_\alpha C_\beta)$	1.63	1.65	1.63	1.63	1.62	**1.61**
	$<(C_\alpha C^+C_\alpha')$	118	118	117	118	118	**118**
	$r(CC)$	1.39	1.40	1.39	1.38	1.39	**1.35-1.37**

163

Table 5-15: Structures of Carbocations (3)

carbocation	geometrical parameter	semi-empirical			expt.
		MNDO	AM1	PM3	
	$r(C^+C)$	1.50	1.46	1.45	**1.35-1.37**
	$r(C^+C_\alpha)$	1.42	1.41	1.41	**1.41**
	$r(C_\alpha C_\beta)$	1.45	1.43	1.43	**1.43**
	$r(C_\beta C_\gamma)$	1.39	1.38	1.38	**1.37**
	$r(C_\alpha C_\delta)$	1.42	1.40	1.40	**1.40**
	$r(C^+C_\alpha)$	1.51	1.48	1.46	**1.49**
	$r(C^+C_\beta)$	1.52	1.50	1.47	**1.41-1.44**
	$r(C^+C_\gamma)$	2.45	2.40	2.36	**2.09-2.11**
	$r(C_\beta C_\gamma)$	1.59	1.57	1.59	**1.71-1.74**
	$r(C^+C_\alpha)$	1.54	1.54	1.59	**1.51-1.55**
	$r(C^+C_\beta)$	2.04	2.25	2.44	**1.71-1.72**
	$r(C_\alpha C_\beta)$	1.56	1.55	1.55	**1.48**
	$r(C_\beta C_\gamma)$	1.41	1.37	1.38	**1.43**
	$r(C_\alpha C_\beta)$	1.54	1.49	1.50	**1.46-1.50**
	$r(C_\beta C_\gamma)$	1.40	1.38	1.38	**1.35-1.38**
	$r(C_\gamma C_\delta)$	1.45	1.42	1.42	**1.40-1.45**
	$r(C^+C_\alpha)$	1.50	1.47	1.46	**1.44**
	$r(C_\alpha C_\beta)$	1.59	1.57	1.58	**1.61**
	$<(C_\alpha C^+ C_\alpha{}')$	118	117	118	**118**
	$r(CC)$	1.42	1.40	1.39	**1.35-1.37**

Aside from *tert*-butyl cation, where the experimental X-ray structure is clearly suspect, and aside from semi-empirical methods applied to cations with multi-center bonding, all models perform quite well for all systems. Subtle (and not so subtle) effects such as the alternation in bond distances in benzyl and in heptamethylbenzenium cations, or the greatly elongated carbon-carbon bonds in the derivative of adamantyl cation are generally well reproduced. In fact, except for systems capable of multi-center bonding, the differences among the various methods are fairly modest. (Of course, this is due in part to "lack of precision" in the experimental data.)

The two systems capable of multi-center bonding in the collection, derivatives of bicyclohexenyl cation and of norbornyl cation, warrant additional comment. All three semi-empirical models show species with essentially localized cationic centers, clearly at odds with the experimental X-ray data, and with the results of the other models. Semi-empirical models are clearly not to be trusted for structure determinations in molecules where multi-center bonding is a strong possibility. On the other hand, the 6-31G* model and all density functional models (including the local density model) provide quite similar (and quite close to experimental) structures for these two cations. While the MP2/6-31G* geometry for the bicyclohexenyl cation is very similar to geometries from Hartree-Fock and density functional calculations (and to the experimental X-ray structure), the MP2/6-31G* geometry for the norbornyl cation derivative is significantly different from the other calculated geometries and from the experimental structure. Specifically, the MP2 calculations show a more symmetrical geometry with the bridging methylene group nearly equidistant from the two base ring carbons. It is likely that the potential energy surface is very shallow with regard to the position of the bridging methylene group.

Anions

While experimental gas-phase geometries of (isolated) anions are virtually non-existent, "anionic fragments" are common occurrences in solid-phase (crystal) structures.[2,3] The extent to which these fragments actually bear negative charge obviously depends on detailed environment, as should experimental "anion" geometries. Thus, experimental bond lengths and angles for anions need to be given in terms of a range of values rather than a single value, and comparisons with calculated geometries need to be treated accordingly.

A comparison of calculated and experimental anion geometries are provided in **Table 5-16**. Included are Hartree-Fock models with STO-3G, 3-21G, 6-31G* and 6-311+G** basis sets, local density models, BP, BLYP, EDF1 and B3LYP density functional models and MP2 models, all with 6-31G* and 6-311+G** basis sets, and MNDO, AM1 and PM3 semi-empirical models. Experimental bond lengths are given as ranges established from examination of distances in a selection of different systems, that is, different counterions, and mean absolute errors are relative to the "closest" experimental distance.

Given the uncertainties in the experimental data, it is not surprising that all models, including the simplest Hartree-Fock models and the three semi-empirical models, exhibit broadly similar performance. Errors exceeding a few hundredths of a Å (outside the range of experimental lengths) are uncommon. In the absence of true gas-phase experimental data, or (in the future) much higher-level calculations, one can only speculate that the performance of the various models here will roughly parallel their performance for neutral molecules where more precise comparisons with experiment can be made.

Experimental data aside, it is interesting to note that bond lengths change only slightly on moving from the 6-31G* to 6-311+G** basis set, independent of model type. This suggests that incorporation of diffuse ("+") functions into the basis set, while important with regard to energy calculations (see discussion following in **Chapter 6**), are relatively unimportant insofar as calculated geometry is concerned. This has considerable practical ramifications given the large increase in "cost" in going to the larger basis set.

Table 5-16: Heavy-Atom Bond Distances in Anions

anion	point group	geometrical parameter	Hartree-Fock				local density		BP		BLYP		expt.
			STO-3G	3-21G	6-31G*	6-311+G**	6-31G*	6-311+G**	6-31G*	6-311+G**	6-31G*	6-311+G**	
BF_4^-	T_d	r (BF)	1.39	1.40	1.39	1.40	1.40	1.41	1.42	1.43	1.43	1.43	1.40-1.43
BCl_4^-	T_d	r (BCl)	1.88	1.88	1.87	1.88	1.86	1.86	1.89	1.88	1.90	1.90	1.83-1.86
CN^-	$C_{\infty v}$	r (CN)	1.16	1.17	1.16	1.16	1.19	1.18	1.20	1.19	1.20	1.19	1.15
$CH_3CO_2^-$	C_s	r (CC)	1.63	1.58	1.55	1.55	1.56	1.55	1.59	1.57	1.60	1.58	1.50-1.54
		r (CO)	1.26	1.25	1.23	1.23	1.26	1.25	1.27	1.27	1.27	1.27	1.25-1.27
$C(CN)_3^-$	D_{3h}	r (CC)	1.42	1.40	1.41	1.41	1.40	1.40	1.42	1.41	1.42	1.42	1.38-1.42
		r (CN)	1.16	1.15	1.15	1.14	1.18	1.17	1.19	1.18	1.19	1.18	1.14-1.16
$C_5H_5^-$	D_{5h}	r (CC)	1.39	1.41	1.40	1.41	1.41	1.41	1.42	1.42	1.42	1.43	1.38-1.40
$C_7H_7^{-a}$	C_{2v}	r ($C-C_\alpha$)	1.35	1.37	1.37	1.38	1.38	1.38	1.39	1.40	1.39	1.40	1.39-1.42
		r ($C_\alpha C_\beta$)	1.46	1.45	1.45	1.44	1.44	1.44	1.46	1.45	1.46	1.46	1.42-1.44
		r ($C_\beta C_\gamma$)	1.36	1.37	1.37	1.37	1.38	1.38	1.39	1.39	1.39	1.39	1.37-1.40
		r ($C_\gamma C_\delta$)	1.40	1.40	1.40	1.40	1.40	1.40	1.42	1.42	1.42	1.42	1.37-1.40
N_3^-	$D_{\infty h}$	r (NN)	1.20	1.18	1.16	1.15	1.19	1.19	1.20	1.20	1.21	1.20	1.12-1.18
NO_2^-	C_{2v}	r (NO)	1.29	1.29	1.23	1.22	1.27	1.25	1.29	1.27	1.30	1.28	1.24
NO_3^-	D_{3h}	r (NO)	1.31	1.28	1.23	1.22	1.26	1.26	1.28	1.27	1.29	1.28	1.22-1.27
PF_6^-	O_h	r (PF)	1.64	1.60	1.61	1.60	1.63	1.64	1.65	1.66	1.66	1.67	1.56-1.60
NCN^{2-}	$D_{\infty v}$	r (CN)	1.23	1.24	1.23	1.23	1.26	1.25	1.27	1.26	1.27	1.26	1.25
CO_3^{2-}	D_{3h}	r (CO)	1.33	1.31	1.29	1.28	1.31	1.30	1.33	1.32	1.33	1.32	1.29
mean absolute error			0.02	0.01	0.01	0.01	0.01	0.01	0.02	0.02	0.03	0.02	–

Table 5-16: Heavy-Atom Bond Distances in Anions (2)

anion	point group	geometrical parameter	EDF1		B3LYP		MP2		semi-empirical			expt.
			6-31G*	6-311+G**	6-31G*	6-311+G**	6-31G*	6-311+G**	MNDO	AM1	PM3	
BF_4^-	T_d	r (BF)	1.42	1.43	1.41	1.42	1.41	1.42	1.39	1.36	1.36	1.40-1.43
BCl_4^-	T_d	r (BCl)	1.89	1.88	1.88	1.88	1.86	1.86	1.82	1.83	1.79	1.83-1.86
CN^-	$C_{\infty v}$	r (CN)	1.19	1.18	1.18	1.18	1.20	1.20	1.18	1.18	1.17	1.15
$CH_3CO_2^-$	C_s	r (CC)	1.58	1.57	1.58	1.57	1.56	1.56	1.55	1.53	1.54	1.50-1.54
		r (CO)	1.26	1.26	1.26	1.26	1.26	1.26	1.26	1.27	1.26	1.25-1.27
$C(CN)_3^-$	D_{3h}	r (CC)	1.41	1.41	1.41	1.41	1.41	1.41	1.40	1.40	1.39	1.38-1.42
		r (CN)	1.18	1.18	1.17	1.16	1.19	1.18	1.17	1.17	1.17	1.14-1.16
$C_5H_5^-$	D_{5h}	r (CC)	1.42	1.42	1.41	1.42	1.41	1.42	1.42	1.42	1.41	1.38-1.40
$C_7H_7^{-a}$	C_{2v}	r (C-C$_\alpha$)	1.39	1.40	1.38	1.39	1.38	1.40	1.37	1.36	1.36	1.39-1.42
		r (C$_\alpha$C$_\beta$)	1.46	1.45	1.45	1.45	1.45	1.45	1.46	1.44	1.44	1.42-1.44
		r (C$_\beta$C$_\gamma$)	1.39	1.39	1.38	1.38	1.39	1.39	1.39	1.38	1.37	1.37-1.40
		r (C$_\gamma$C$_\delta$)	1.41	1.41	1.41	1.41	1.41	1.41	1.41	1.40	1.40	1.37-1.40
N_3^-	$D_{\infty h}$	r (NN)	1.20	1.19	1.19	1.18	1.22	1.21	1.17	1.17	1.17	1.12-1.18
NO_2^-	C_{2v}	r (NO)	1.28	1.27	1.27	1.26	1.28	1.26	1.21	1.21	1.23	1.24
NO_3^-	D_{3h}	r (NO)	1.27	1.27	1.26	1.26	1.27	1.26	1.24	1.23	1.24	1.22-1.27
PF_6^-	O_h	r (PF)	1.65	1.66	1.64	1.65	1.64	1.64	1.70	1.57	1.59	1.56-1.60
NCN^{2-}	$D_{\infty v}$	r (CN)	1.27	1.26	1.26	1.25	1.27	1.27	1.23	1.24	1.25	1.25
CO_3^{2-}	D_{3h}	r (CO)	1.32	1.32	1.31	1.31	1.32	1.31	1.30	1.30	1.29	1.29
mean absolute error			0.01	0.02	0.01	0.01	0.02	0.01	0.02	0.01	0.01	–

a) benzyl anion

168

Carbenes and Related Compounds

Carbenes and related compounds are among those reactive intermediates for which gas-phase experimental data exist.[1] Some of those are compared to calculated geometries in **Table 5-17**, drawn from a larger collection provided in **Appendix A5** (**Tables A5-42** to **A5-49**). Except for methylene (CH_2), where both singlet and triplet states have been considered, only singlet-state molecules have been examined. The usual theoretical models have been assessed. Mean absolute errors in bond lengths and angles based on the full data set have also been provided.

Triplet methylene is known to be bent with a bond angle of approximately 136°. This is closely reproduced by all Hartree-Fock models (except for STO-3G which yields a bond angle approximately 10° too small), as well as local density models, BP, BLYP, EDF1 and B3LYP density functional models and MP2 models. Semi-empirical models also suggest a bent structure, but with an HCH angle which is much too large.

With the exception of MNDO and AM1, all models show a bond angle in singlet methylene which is significantly less than tetrahedral, in accord with experimental data. This is consistent with the notion that the σ lone pair "takes up more space" than a CH bond. This is, of course, one of the fundamental premises of VSEPR theory. Calculations and experiment also concur that the bond angle in difluorocarbene is several degrees less than tetrahedral. However, they disagree with regard to the bond angle in dichlorocarbene. Aside from semi-empirical models and STO-3G and 3-21G Hartree-Fock models, all models show a nearly tetrahedral value, which is approximately 2° less than the corresponding bond angle in methylene chloride. The experimental data, which suggests a bond angle of around 100° (but with a very large uncertainty), is clearly suspect.

Again dismissing semi-empirical and STO-3G models, calculations and experiment are in close accord with regard to bond angles in silylene (near to 90°) and its difluoro and dichloro analogues (near to 100°).

Table 5-17: Structures of Carbenes and Related Compounds

carbene	point group	geometrical parameter	Hartree-Fock				local density		BP		BLYP		expt.
			STO-3G	3-21G	6-31G*	6-311+G**	6-31G*	6-311+G**	6-31G*	6-311+G**	6-31G*	6-311+G**	
CH₂	C₂ᵥ	r (CH)	1.123	1.102	1.097	1.098	1.136	1.128	1.134	1.127	1.132	1.124	1.111
		<(HCH)	100.5	104.7	103.1	103.6	99.1	100.5	99.0	100.4	99.1	100.8	102.4
CH₂[a]	C₂ᵥ	r (CH)	1.082	1.071	1.071	1.072	1.093	1.091	1.090	1.088	1.089	1.086	1.078
		<(HCH)	125.6	131.3	130.7	131.9	134.8	137.6	133.6	135.8	133.6	136.0	136
CF₂	C₂ᵥ	r (CF)	1.323	1.321	1.283	1.276	1.308	1.301	1.330	1.326	1.335	1.331	1.304
		<(FCF)	102.7	104.0	104.5	105.0	103.8	104.4	103.8	104.2	103.9	104.3	104.8
CCl₂	C₂ᵥ	r (CCl)	1.803	1.737	1.712	1.710	1.741	1.725	1.773	1.761	1.788	1.775	1.76
		<(ClCCl)	106.8	100.0	110.4	110.3	109.0	109.6	109.1	109.5	109.1	109.6	100±9
SiH₂	C₂ᵥ	r (SiH)	1.458	1.506	1.509	1.510	1.547	1.543	1.546	1.542	1.545	1.540	1.516
		<(HSiH)	91.5	93.4	93.4	93.7	89.7	89.8	90.4	90.6	90.7	91.0	92.1
SiF₂	C₂ᵥ	r (SiF)	1.602	1.587	1.592	1.598	1.610	1.626	1.631	1.648	1.633	1.640	1.590
		<(FSiF)	93.2	99.2	99.6	99.2	100.9	100.0	101.5	100.8	101.8	101.3	100.8
SiCl₂	C₂ᵥ	r (SiCl)	2.109	2.071	2.093	2.083	2.092	2.089	2.119	2.119	2.135	2.136	2.083
		<(ClSiCl)	98.2	101.5	101.5	101.3	101.3	101.1	102.2	102.0	102.7	102.4	102.8
mean absolute error in bond distance			0.035	0.020	0.021	0.022	0.016	0.017	0.035	0.025	0.039	0.034	–
mean absolute error in bond angle			4.3	2.4	2.3	2.0	2.5	2.0	2.6	1.7	2.6	1.6	–

Table 5-17: Structures of Carbenes and Related Compounds (2)

carbene	point group	geometrical parameter	EDF1 6-31G*	EDF1 6-311+G**	B3LYP 6-31G*	B3LYP 6-311+G**	MP2 6-31G*	MP2 6-311+G**	semi-empirical MNDO	semi-empirical AM1	semi-empirical PM3	expt.
CH_2	C_{2v}	r (CH)	1.128	1.121	1.119	1.114	1.109	1.110	1.091	1.103	1.092	**1.111**
		< (HCH)	99.2	100.5	100.3	101.6	102.0	101.9	111.1	110.6	103.7	**102.4**
CH_2^a	C_{2v}	r (CH)	1.086	1.083	1.082	1.080	1.078	1.079	1.052	1.062	1.063	**1.078**
		< (HCH)	133.5	135.6	133.3	135.4	131.6	132.5	152.9	151.7	150.0	**136**
CF_2	C_{2v}	r (CF)	1.327	1.322	1.313	1.308	1.315	1.303	1.304	1.312	1.298	**1.304**
		< (FCF)	104.0	104.4	104.0	104.5	104.2	104.8	108.3	106.0	106.3	**104.8**
CCl_2	C_{2v}	r (CCl)	1.765	1.753	1.754	1.742	1.718	1.709	1.748	1.672	1.591	**1.76**
		< (ClCCl)	109.5	109.9	109.3	109.8	109.9	110.3	113.9	118.1	120.0	**100 ±9**
SiH_2	C_{2v}	r (SiH)	1.540	1.536	1.531	1.528	1.519	1.510	1.437	1.457	1.513	**1.516**
		< (HSiH)	90.7	91.0	91.1	91.5	92.5	92.5	99.5	101.0	94.8	**92.1**
SiF_2	C_{2v}	r (SiF)	1.629	1.646	1.616	1.633	1.617	1.625	1.681	1.612	1.575	**1.590**
		< (FSiF)	101.6	101.0	101.1	100.3	101.0	100.2	102.9	97.1	95.3	**100.8**
$SiCl_2$	C_{2v}	r (SiCl)	2.115	2.114	2.110	2.109	2.075	2.074	2.064	2.027	2.000	**2.083**
		< (ClSiCl)	102.6	102.5	102.0	101.8	101.7	101.6	106.2	104.4	101.9	**102.8**
mean absolute error in bond distance			0.026	0.022	0.022	0.019	0.016	0.016	0.032	0.042	0.092	–
mean absolute error in bond angle			2.6	1.6	2.4	1.4	2.1	1.6	5.7	6.1	4.8	–

a) triplet state

Radicals

Calculated geometries for a small number of diatomic and small polyatomic free radicals are compared with experimental structures in **Table 5-18**. These have been drawn from a somewhat larger collection provided in **Appendix A5 (Tables A5-50 to A5-57)**. Except for triplet oxygen, all radicals possess a single unpaired electron (they are doublets). The usual set of theoretical models has been examined. All calculations involve use of the unrestricted open-shell SCF approach, where electrons of different spin occupy different orbitals, as opposed to the restricted open-shell SCF approach, where "paired electrons" are confined to the same orbital (see **Chapter 2** for more detailed discussion).

The STO-3G model and all three semi-empirical models produce unsatisfactory results for several systems. For example, the STO-3G bond length in triplet oxygen is nearly two tenths of an Å longer than the experimental value and the HNH bond angle in amino radical is nearly 20° too large. These models are not to be trusted for equilibrium geometry calculations on systems with unpaired electrons. All other models perform much better, although there are some surprises. For example, while bond lengths from density functional calculations do not change significantly in moving from the 6-31G* to the 6-311+G** basis set, the corresponding results from both Hartree-Fock and MP2 calculations often show large changes. Where they do, the larger basis set calculations perform better. In fact, this is one of the few instances where the behavior of B3LYP and MP2 calculations do not parallel each other.

Triplet oxygen deserves special attention if for no other reason that it is the only "common" non-closed shell molecule. As expected, the "limiting" (6-311+G** basis set) Hartree-Fock bond distance is too short. The corresponding B3LYP model provides a bond length in nearly exact agreement with experiment, while all other density functional models and especially the MP2 model significantly overestimate the bond distance.

Table 5-18: Structures of Diatomic and Small Polyatomic Radicals

radical	point group	electronic state	geometrical parameter	Hartree-Fock				local density		BP		expt.
				STO-3G	3-21G	6-31G*	6-311+G**	6-31G*	6-311+G**	6-31G*	6-311+G**	
NH_2	C_{2v}	$^2B_1$	r (NH)	1.015	1.026	1.013	1.012	1.044	1.039	1.046	1.041	**1.024**
			< (HNH)	131.3	106.0	104.3	104.6	101.7	102.7	101.2	103.1	**103.13**
CN	$C_{\infty v}$	$^2\Sigma^+$	r (CN)	1.235	1.180	1.162	1.154	1.178	1.169	1.178	1.179	**1.172**
HCO	C_s	$^2A'$	r (CO)	1.253	1.180	1.159	1.152	1.184	1.174	1.184	1.186	**1.175**
			r (CH)	1.101	1.095	1.106	1.108	1.141	1.137	1.141	1.137	**1.125**
			< (HCO)	126.3	129.0	126.3	126.9	123.1	124.0	123.1	124.0	**124.9**
NO	$C_{\infty v}$	$^2\Pi$	r (NO)	1.186	1.201	1.127	1.118	1.160	1.148	1.173	1.162	**1.151**
HOO	C_s	$^2A''$	r (OO)	1.396	1.434	1.369	1.299	1.313	1.310	1.314	1.343	**1.335**
			r (OH)	1.002	0.973	0.955	0.948	0.998	0.991	0.998	0.990	**0.977**
			< (HOO)	100.1	103.3	102.9	106.5	105.6	106.7	106.9	105.5	**104.1**
O_2	$D_{\infty h}$	$^3\Sigma_g^-$	r (OO)	1.398	1.240	1.168	1.158	1.214	1.204	1.231	1.223	**1.208**

Table 5-18: Structures of Diatomic and Small Polyatomic Radicals (2)

radical	point group	electronic state	geometrical parameter	BLYP		EDF1		B3LYP		expt.
				6-31G*	6-311+G**	6-31G*	6-311+G**	6-31G*	6-311+G**	
NH_2	C_{2v}	$^2B_1$	r (NH)	1.047	1.040	1.040	1.035	1.034	1.030	**1.024**
			<(HNH)	101.1	102.3	101.2	102.1	102.1	103.1	**103.3**
CN	$C_{\infty v}$	$^2\Sigma^+$	r (CN)	1.187	1.179	1.183	1.176	1.174	1.166	**1.172**
HCO	C_s	$^2A'$	r (CO)	1.196	1.187	1.191	1.183	1.183	1.174	**1.175**
			r (CH)	1.141	1.135	1.137	1.133	1.129	1.125	**1.125**
			<(HCO)	123.0	123.8	123.0	123.9	123.6	124.4	**124.9**
NO	$C_{\infty v}$	$^2\Pi$	r (NO)	1.176	1.165	1.169	1.158	1.159	1.148	**1.151**
HOO	C_s	$^2A''$	r (OO)	1.358	1.358	1.338	1.336	1.332	1.328	**1.335**
			r (OH)	0.998	0.990	0.992	0.985	0.984	0.977	**0.977**
			<(HOO)	104.6	105.3	105.0	105.7	105.1	105.9	**104.1**
O_2	$D_{\infty h}$	$^3\Sigma_g^-$	r (OO)	1.239	1.232	1.227	1.218	1.215	1.206	**1.208**

Table 5-18: Structures of Diatomic and Small Polyatomic Radicals (3)

radical	point group	electronic state	geometrical parameter	MP2		semi-empirical			expt.
				6-31G*	6-311+G**	MNDO	AM1	PM3	
NH_2	C_{2v}	$^2B_1$	r (NH) <(HNH)	1.028 103.3	1.025 102.8	1.002 104.5	0.996 107.3	0.987 110.8	**1.024** **103.3**
CN	$C_{\infty v}$	$^2\Sigma^+$	r (CN)	1.136	1.130	1.155	1.149	1.157	**1.172**
HCO	C_s	$^2A'$	r (CO) r (CH) <(HCO)	1.192 1.123 123.3	1.183 1.122 124.3	1.185 1.075 143.1	1.192 1.083 141.1	1.166 1.089 136.5	**1.175** **1.125** **124.9**
NO	$C_{\infty v}$	$^2\Pi$	r (NO)	1.143	1.134	1.123	1.115	1.127	**1.151**
HOO	C_s	$^2A''$	r (OO) r (OH) <(HOO)	1.396 0.981 101.2	1.311 0.972 105.3	1.208 0.976 112.3	1.177 1.010 112.5	1.266 0.957 107.5	**1.335** **0.977** **104.1**
O_2	$D_{\infty h}$	$^3\Sigma_g^-$	r (OO)	1.247	1.224	1.134	1.085	1.169	**1.208**

175

Hydrogen-Bonded Complexes

Hydrogen bonding is certainly the most studied (and some would argue the most "important") of intermolecular interactions. There are literally thousands of experimental X-ray crystal structures which reveal hydrogen-bonding interactions.[2,3] The skeptic will be quick to point out that hydrogen-bond energies (typically a few kcal/mol) are of the same order of magnitude as crystal-packing energies, and that the experimental X-ray structures may not accurately reflect those of isolated systems. However, taken as a whole, the systematics in hydrogen-bond lengths[*] and directionalities revealed by X-ray crystallography present an overwhelming case.

Experimental gas-phase structures of hydrogen-bonded complexes are quite rare and quite limited in their "information content". Dimers and some mixed complexes involving such species as water, hydrogen fluoride, hydrogen chloride and hydrogen cyanide have been investigated using microwave spectroscopy[1], and "reasonable estimates" of intermolecular distances separating heavy atoms, e.g., the OO distance in water dimer, are available. It is primarily to this data which we compare the performance of quantum chemical models.

Representative examples are provided in **Table 5-19**. Only a single (intermolecular) distance is examined for each system, underlying the fact that the experimental structure data are incomplete. The usual quantum chemical models have been surveyed. Comparisons with molecular mechanics models have not been included even though force fields such as MMFF have been explicitly parameterized to reproduce known hydrogen-bond distances.

There is a very wide variation in the quality of results from the different models. MNDO and AM1 semi-empirical models, the STO-3G model and both local density models are completely unsatisfactory. The 3-21G model, all density functional models with the 6-31G* basis set and the PM3 model fare better, while 6-31G*

---

[*]   As previously mentioned, hydrogens are difficult to locate precisely in the X-ray diffraction experiment, and what is accurately known are the positions of the two heavy atoms involved in the hydrogen bond.

Table 5-19: Bond Distances in Hydrogen-Bonded Complexes

hydrogen-bonded complex	bond	Hartree-Fock				local density		BP		expt.
		STO-3G	3-21G	6-31G*	6-311+G**	6-31G*	6-311+G**	6-31G*	6-311+G**	
	OO	2.73	2.80	2.98	3.00	2.63	2.71	2.84	2.89	**2.98**
	OF	2.63	2.57	2.72	2.72	2.50	2.51	2.63	2.63	**2.69**
	FF	2.57	2.60	2.72	2.83	2.33[a]	2.56	2.46[a]	2.74	**2.79**
	ClF	2.79	3.13	3.36	3.45	2.96	3.05	3.11	3.27	**3.37**
	NF	3.03	2.81	2.92	2.90	2.67	2.60	2.83	2.75	**2.80**
	OO	2.60	2.69	2.73	2.80	2.50	2.47	2.65	2.63	**2.76**
	NO	2.62	2.82	3.00	3.01	2.72	2.68	2.86	2.85	–
mean absolute error		0.27	0.15	0.05	0.05	0.31	0.24	0.16	0.07	–

Table 5-19: Bond Distances in Hydrogen-Bonded Complexes (2)

hydrogen-bonded complex	bond	BLYP		EDF1		B3LYP		expt.
		6-31G*	6-311+G**	6-31G*	6-311+G**	6-31G*	6-311+G**	
	OO	2.87	2.93	2.91	3.01	2.85	2.91	**2.98**
	OF	2.65	2.67	2.67	2.68	2.65	2.64	**2.69**
	FF	2.49[a]	2.77	2.52[a]	2.86	2.48[a]	2.74	**2.79**
	ClF	3.15	3.32	3.20	3.48	3.15	3.31	**3.37**
	NF	2.85	2.79	2.89	2.82	2.83	2.77	**2.80**
	OO	2.70	2.70	2.70	2.69	2.70	2.68	**2.76**
	NO	2.90	2.91	2.92	2.93	2.89	2.89	–
mean absolute error		0.15	0.03	0.12	0.05	0.15	0.05	–

Table 5-19: Bond Distances in Hydrogen-Bonded Complexes (3)

hydrogen-bonded complex	bond	MP2		semi-empirical			expt.
		6-31G*	6-311+G**	MNDO	AM1	PM3	
	OO	2.92	2.91	3.89[b]	3.06[b]	2.77	**2.98**
	OF	2.69	2.66	3.95	2.41[b]	2.68	**2.69**
	FF	2.64	2.79	3.92	2.91	2.68	**2.79**
	ClF	3.24	3.36	c	3.39	3.00	**3.37**
	NF	2.88	2.81	3.98	3.43	2.76	**2.80**
	OO	2.75	2.69	4.05	3.07	2.74	**2.76**
	NO	2.93	2.90	4.33	3.07	2.81	–
mean absolute error		0.08	0.02	d	d	0.15	–

a) gives incorrect structure b) gives incorrect bifurcated structure c) dissociates d) insufficient data

179

and 6-311+G** Hartree-Fock models, all density functional models with the 6-311+G** basis set and the MP2/6-311+G** model perform admirably. Of these, the MP2 model is the best. Results for 6-31G**, 6-31+G* and 6-311G* basis sets, intermediate between 6-31G* and 6-311+G**, are provided for Hartree-Fock, EDF1, B3LYP and MP2 models in **Appendix A5** (**Tables A5-58** to **A5-61**). With the exception of Hartree-Fock models, these show that the majority of the improvements comes from the addition of diffuse functions. The significant (and beneficial) effect of diffuse functions is not unexpected, given that the distances involved in hydrogen bonding are much longer than "normal" (covalent) distances.

Geometries of Excited States

A number of methods have been proposed for calculations of the geometries of molecules in excited states. These include CIS (Configuration Interaction Singles) and variations on CIS to account for the effect of double substitutions, as well as so-called time dependent density functional models. Except for CIS (the simplest of the methods) there is very little practical experience. There is also very little solid experimental data on the geometries of excited-state molecules.

Results for CIS/6-31G* and CIS/6-311+G** calculations on the first ($n \rightarrow \pi^*$) excited state of formaldehyde are provided below.

	CIS/6-31G*	CIS/6-311+G**	expt.
r(CO)	1.258	1.248	1.321
r(CH)	1.092	1.095	1.097
<(HCH)	112.5	112.5	117.2
<(HHCO)	135.8	138.4	154.3

Two changes from the ground-state are apparent. For one, $n \rightarrow \pi^*$ excited state is puckered. Both CIS models are in accord. Second, the CO bond has elongated (from 1.208Å in the ground state). Elongation is seen, but its magnitude underestimated.

Structures of Molecules in Solution

To what extent does the equilibrium geometry of a molecule change in moving from the gas phase into solution? The question is of great importance because, whereas calculations refer strictly to isolated (gas-phase) species, experimental structural data follow from diverse sources: gas, liquid, solution and most commonly the solid state. In the absence of proven theoretical models to calculate equilibrium structure in real media, the only way to answer such a question is to compare gas-phase experimental structures with those obtained in solution or in the solid state. This is beyond the scope of the present treatment, and we limit ourselves to a few general remarks:

i) "Solvent effects" on the equilibrium bond lengths and angles in uncharged molecules appear to be fairly modest. This is consistent with the notion that the energy required for significant bond stretching or angle bending is quite large compared to typical (neutral) solute-solvent interactions. It is also consistent with available comparisons which have been made between gas[1] and solid-phase[2,3] equilibrium geometries.

ii) Larger "solvent effects" would be expected for charged species. Interaction energies may now be comparable to bond stretching/angle bending energies, and some changes in geometry are to be expected. In the limit of fully localized charge, it is possible that the "charged atom" will covalently bond to the "solvent", causing a great change in local geometry.

iii) The solvent may lead to a change in tautomeric form. The extreme cases are amino acids. For example, while in the gas phase, glycine is most stable as an "uncharged" molecule, i.e., $H_2NCH_2CO_2H$, in water it prefers a zwitterionic structure, i.e., $^+H_3NCH_2CO_2^-$. Many less dramatic examples exist, in particular, with heterocyclic compounds. Where multiple tautomers exist, it will at the very least be necessary to examine the role of solvent in altering relative gas-phase energies.

iv) Solvation is also known to effect changes in conformation. For example, polypeptides in the gas phase would be expected

181

to adopt "linear" (stretched out) structures as opposed to compact (globular) structures found in solution (and naturally for proteins).

Pitfalls

There are actually very few. Modern optimization techniques practically guarantee location of a minimum energy structure, and only where the initial geometry provided is "too symmetric" will this not be the outcome. With a few notable exceptions (Hartree-Fock models applied to molecules with transition metals), Hartree-Fock, density functional and MP2 models provide a remarkably good account of equilibrium structure. Semi-empirical quantum chemical models and molecular mechanics models, generally fare well where they have been explicitly parameterized. Only outside the bounds of their parameterization is extra caution warranted. Be on the alert for surprises. While the majority of molecules assume the structures expected of them, some will not. Treat "unexpected" results with skepticism, but be willing to alter preconceived beliefs.

References

1. G. Graner, E. Hirota, T. Tijima, K. Kuchitsu, D.A. Ramsay, J. Vogt and N. Vogt, *Structure Data of Free Polyatomic Molecules. Molecules Containing Three or Four Carbon Atoms*, Landolt-Börnstein New Series II, vol 25C, W. Martienssen, ed., 2000, and earlier volumes in this series.

2. (a) F.H. Allen, S. Bellard, M.D. Brice, T.W.A. Hummelink, B.G. Hummelink-Peters, O. Kennard, W.D.S. Motherwell, J.R. Rogers and D.G. Watson, *Acta Cryst.*, **B35**, 2331 (1979); (b) F.H. Allen, *ibid.*, **B58**, 380 (2002).

3. (a) G. Bergerfoff, R. Hundt, R. Sievers and I.D. Brown, *J. Chem. Inf. Comput. Sci.*, **23** 66 (1983); (b) A. Belsky, M. Hellenbrandt, V.L. Karen and P. Luksch, *Acta Cryst.*, **B58**, 364 (2002).

4. H.M. Berman, J. Westbrook, Z. Feng, G. Gilliland, T.N. Bhat, H. Weissig, I.N. Shindyalov and P.E. Bourne, *Nucleic Acids Res.*, **28**, 235 (2000), and references therein.

Chapter 6

Reaction Energies

This chapter assesses the performance of quantum chemical models with regard to the calculation of reaction energies. Several different reaction classes are considered: homolytic and heterolytic bond dissociation reactions, hydrogenation reactions, isomerization reactions and a variety of isodesmic reactions. The chapter concludes with a discussion of reaction energies in solution.

Introduction

Chemical reactions may be divided into one of several categories depending on the extent to which overall bonding is maintained (**Table 6-1**). Reactions which lead to a change in the total number of electron pairs (bonds and non-bonded lone pairs) are at one extreme. Homolytic bond dissociation reactions provide an example.

$$H{-}F \longrightarrow H^{\bullet} + F^{\bullet} \qquad\qquad \textit{homolytic bond dissociation}$$

Comparisons between reactants and transition states (required for calculation of absolute activation energies) also typify situations in which the total number of electron pairs is not likely to be conserved. These will be considered separately in **Chapter 9**.

Less disruptive are reactions in which the total number of electron pairs is maintained, but a chemical bond is converted to a non-bonded lone pair or vice versa. Heterolytic bond dissociation reactions provide an example.

$$H{-}F \longrightarrow H^{+} + F^{-} \qquad\qquad \textit{heterolytic bond dissociation}$$

Reactions defining absolute acidity, e.g., the reaction above for the acidity of HF, and absolute basicity are important special cases. Some comparisons between transition states and reactants will also likely fall into this category. These will be considered in **Chapter 9**.

Table 6-1: Reaction Types

characteristics	examples
no conservation of number of electron pairs	homolytic bond dissociation, comparison of different electronic states
conservation of number of electron pairs, but no conservation of number of bonds	heterolytic bond dissociation, absolute acidity and basicity comparisons
conservation of number of bonds and number of non-bonded lone pairs, but no conservation of number of each kind of bond or number of each kind of non-bonded lone pair	hydrogenation, structural isomerism
conservation of number of each kind of bond and number of each kind of non-bonded lone pair (*isodesmic* reactions)	bond separation, regio and stereo isomerization, relative acidity and basicity comparisons

Reactions in which both the total number of bonds and the total number of non-bonded lone pairs are conserved are even less disruptive. Two examples are given below.

$$H_2C{=}CH_2 + 2H_2 \longrightarrow 2CH_4 \qquad\qquad\qquad hydrogenation$$

$$\overline{CH_2CH_2CH_2} \longrightarrow CH_3CH{=}CH_2 \qquad structural\ isomerization$$

At the other extreme are reactions in which the number of each kind of formal chemical bond (and each kind of non-bonded lone pair) are conserved. These are *isodesmic* ("equal bond") reactions. Examples include the processes below.

$$H_3C{-}C{\equiv}CH + CH_4 \longrightarrow H_3C{-}CH_3 + HC{\equiv}CH \qquad bond\ separation$$

$$(CH_3)_3NH^+ + NH_3 \longrightarrow (CH_3)_3N + NH_4^+ \qquad proton\ transfer$$

In addition, all regio and stereochemical comparisons (including comparisons involving transition states instead of "normal" molecules) are *isodesmic* reactions, as are conformation changes. (*Isodesmic* comparisons involving transition states are discussed in **Chapter 9** and conformation changes are discussed in **Chapter 8**.) Thus, *isodesmic* processes constitute a large and important class of reactions.

It is likely that different quantum chemical models will perform differently in each of these situations. Processes which involve net loss or gain of an electron pair are likely to be problematic for Hartree-Fock models, which treat the electrons as essentially independent particles, but less so for density functional models and MP2 models, which attempt to account for electron correlation. Models should fare better for processes in which reactants and products are similar and benefit from cancellation of errors, than those where reactants and products are markedly different. The only exception might be for semi-empirical models, which have been explicitly parameterized to reproduce individual experimental heats of formation, and might not be expected to benefit from error cancellation.

This chapter assesses the performance of quantum chemical models with regard to reaction thermochemistry. Considered are Hartree-Fock models with STO-3G, 3-21G, 6-31G* and 6-311+G** basis sets, local density models and BP, BLYP, EDF1 and B3LYP density functional models with 6-31G* and 6-311+G** basis sets, MP2/6-31G* and MP2/6-311+G** models, and MNDO, AM1 and PM3 semi-empirical models. Molecular mechanics models are not applicable to the description of reaction thermochemistry. Several different classes of reactions are examined: homolytic bond dissociation reactions, absolute acidity and basicity comparisons, typifying heterolytic bond dissociation reactions, hydrogenation reactions, reactions relating multiple and single bonds, structural isomer comparisons, and a variety of *isodesmic* reactions, including bond separation reactions, relative acid and base strength comparisons, and comparisons between regio and stereoisomers.

Experimental thermochemical data are of widely varying quality.[1] While heats of formation for hydrocarbons and oxygen-containing compounds (their combustion products) are often quite accurately known (± 1 kcal/mol*), data for compounds with nitrogen, halogens and heavy main-group elements may be significantly in error. Comparisons with the results of calculations below the level of ± 2 kcal/mol are, in most cases, likely to be meaningless. There are exceptions. For example, relative gas-phase acidities and basicities, as well as the thermochemistry of related ion-molecule reactions, have typically been established to better than ± 1 kcal/mol[2].

Homolytic Bond Dissociation Reactions

Energies for a selection of homolytic bond dissociation reactions of two-heavy-atom hydrides are provided in **Table 6-2**. These have been drawn from a larger collection found in **Appendix A6** (**Tables A6-1** to **A6-8**). A summary of mean absolute deviations from G3 calculations[3] (based on the full collection) is provided in **Table 6-3**.

Hartree-Fock models provide very poor results, irrespective of basis set. Reaction energies are consistently smaller than experimental enthalpies, sometimes significantly so. This could easily have been anticipated. Hartree-Fock models provide an incomplete account of electron correlation, and will do better the fewer the total number of

---

*   All reaction energies and errors in reaction energies will be reported in kcal/mol.

Table 6-2:   Homolytic Bond Dissociation Energies

bond dissociation reaction	Hartree-Fock				local density		BP		BLYP	
	STO-3G	3-21G	6-31G*	6-311+G**	6-31G*	6-311+G**	6-31G*	6-311+G**	6-31G*	6-311+G**
$CH_3\text{-}CH_3 \rightarrow CH_3^{\bullet} + CH_3^{\bullet}$	96	68	70	66	123	118	99	95	96	91
$CH_3\text{-}NH_2 \rightarrow CH_3^{\bullet} + NH_2^{\bullet}$	73	59	58	57	119	117	93	91	89	86
$CH_3\text{-}OH \rightarrow CH_3^{\bullet} + OH^{\bullet}$	67	53	59	58	128	124	101	98	97	93
$CH_3\text{-}F \rightarrow CH_3^{\bullet} + F^{\bullet}$	66	59	69	69	147	143	119	116	117	113
$NH_2\text{-}NH_2 \rightarrow NH_2^{\bullet} + NH_2^{\bullet}$	44	37	34	33	104	103	75	74	70	69
$HO\text{-}OH \rightarrow OH^{\bullet} + OH^{\bullet}$	22	3	0	-2	93	86	65	59	61	55
$F\text{-}F \rightarrow F^{\bullet} + F^{\bullet}$	5	-29	-33	-39	85	73	58	47	56	44

bond dissociation reaction	EDF1		B3LYP		MP2		semi-empirical			G3	expt.
	6-31G*	6-311+G**	6-31G*	6-311+G**	6-31G*	6-311+G**	MNDO	AM1	PM3		
$CH_3\text{-}CH_3 \rightarrow CH_3^{\bullet} + CH_3^{\bullet}$	97	92	97	92	99	97	69	77	74	**96**	**97**
$CH_3\text{-}NH_2 \rightarrow CH_3^{\bullet} + NH_2^{\bullet}$	90	88	89	87	92	93	69	75	68	**91**	**93**
$CH_3\text{-}OH \rightarrow CH_3^{\bullet} + OH^{\bullet}$	98	95	96	93	98	98	82	88	83	**97**	**98**
$CH_3\text{-}F \rightarrow CH_3^{\bullet} + F^{\bullet}$	117	114	113	110	113	112	104	110	101	**–**	**114**
$NH_2\text{-}NH_2 \rightarrow NH_2^{\bullet} + NH_2^{\bullet}$	72	71	70	69	73	74	59	62	50	**72**	**73**
$HO\text{-}OH \rightarrow OH^{\bullet} + OH^{\bullet}$	61	55	54	49	55	52	39	57	46	**53**	**55**
$F\text{-}F \rightarrow F^{\bullet} + F^{\bullet}$	53	42	42	32	38	29	30	60	59	**38**	**38**

Table 6-3: **Mean Absolute Deviations from G3 Calculations in Energies of Homolytic Bond Dissociation Reactions**

MNDO	14			
AM1	16			
PM3	16			

	STO-3G	3-21G	6-31G*	6-311+G**
Hartree-Fock	24	39	38	40
local density	—	—	27	22
BP	—	—	5	3
BLYP	—	—	4	5
EDF1	—	—	4	4
B3LYP	—	—	3	5
MP2	—	—	3	4

electron pairs. The (radical) products of homolytic bond dissociation contain one fewer electron pair than the reactant and, relative to the reactant, will have too low an energy.*

Local density models also provide a poor account of homolytic bond dissociation energies. The direction of the errors is the opposite as noted for Hartree-Fock models (reaction energies are too large), but the magnitudes of the errors are comparable. In fact, the average of Hartree-Fock and local density homolytic bond dissociation energies is typically quite close to the experimental energy.**

Density functional models provide an altogether different picture. BP, BLYP, EDF1 and B3LYP models all provide a respectable (and broadly similar) account of homolytic bond dissociation energies where one or more carbons are involved, but differ substantially where both atoms are highly electronegative, e.g., F_2. Density functional calculations with the 6-311+G** basis set yield better results overall than the corresponding 6-31G* calculations, but except for bond dissociation in hydrogen peroxide and in F_2, i.e., bonds between two highly electronegative elements, the differences are not great.

The performance of MP2 models is also good, and quite similar to that of the corresponding (same basis set) B3LYP models. As with B3LYP models and other density functional models, bond dissociation energies from MP2/6-31G* and MP2/6-311+G** calculations are quite similar, except where two highly electronegative elements are involved in the bond. In view of the difference in cost between MP2 and B3LYP models, the latter seems the obvious choice for this purpose.

Calculations have been performed in order to dissect the observed changes in bond dissociation energies between models with 6-31G* and 6-311+G** basis sets. These are provided in **Appendix A6**, **Tables A6-9** to **A6-11** for EDF1, B3LYP and MP2 models, respectively. No significant changes in bond dissociation energies are noted as a result

---

* Consider bond dissociation in H_2. The product (H˙) contains only a single electron and its energy is given exactly by Hartree-Fock theory. The reactant (H_2) contains two electrons and its energy is too positive. Therefore, the bond dissociation energy is too small.

** It was this observation which gave rise to so-called hybrid density functional models, such as the B3LYP model. Here, the Hartree-Fock exchange energy is added to the exchange energy from a particular density functional model with one or more adjustable parameters.

of addition of polarization functions on hydrogen (6-31G** basis set), not unreasonable given that the hydrogens are not directly involved. However, addition of diffuse functions (6-31+G* basis set) and increased splitting of the valence shell (6-311G* basis set) both lead to significant changes.

Semi-empirical models do not provide an adequate description of bond dissociation energies and should not be used for this purpose. Errors are not systematic, as was the case for Hartree-Fock models (bond energies too small) and local density models (bond energies too large). Rather, significant errors in both directions are observed.

Singlet-Triplet Separation in Methylene

Another type of process which may lead to a net gain or loss in the number of electron pairs is that involving change in electronic state. For example, the ground state of methylene, CH_2, is a triplet with one electron in an in-plane molecular orbital and one electron in an out-of-plane molecular orbital.

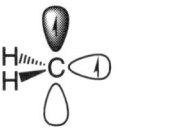
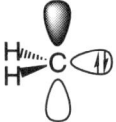

triplet methylene   singlet methylene

It possesses one fewer electron pair than the corresponding singlet state in which both electrons reside in an in-plane orbital (the out-of-plane orbital is empty).

It is to be expected that Hartree-Fock models will unduly favor the triplet state over the singlet simply because it contains one fewer electron pair. "Limiting" (6-311+G** basis set) Hartree-Fock calculations (**Table 6-4**) show this to be the case, yielding a singlet-triplet energy separation of 29 kcal/mol (in favor of the triplet), compared to an experimental estimate of approximately 10 kcal/mol (in favor of the triplet). Density functional calculations (including local density calculations) using the 6-311+G** basis set show energy separations in range of 11-15 kcal/mol (favoring the triplet), while

Table 6-4: **Singlet-Triplet Energy Separation in Methylene[a]**

	33			
MNDO	33			
AM1	34			
PM3	42			

	STO-3G	3-21G	6-31G*	6-311+G**
Hartree-Fock	40	36	31	29
local density	–	–	15	14
BP	–	–	16	15
BLYP	–	–	12	11
EDF1	–	–	15	13
B3LYP	–	–	14	12
MP2	–	–	21	17

a) triplet ground state; experimental energy separation is estimated to be 10 kcal/mol.

corresponding calculations using the 6-31G* basis set suggest slightly larger values. Overall, the EDF1 and B3LYP models appear to perform the best (and the BP model the worst), but the differences are not great. Interestingly, MP2 models do not perform as well. The MP2/6-311+G** model shows a singlet-triplet energy separation of 17 kcal/mol, somewhat greater than the experimental estimate.

All three semi-empirical models yield similar results, indicating that triplet methylene is much more stable than the corresponding singlet than is actually the case.

Heterolytic Bond Dissociation Reactions

Unlike homolytic bond dissociation, heterolytic bond dissociation occurs without loss of an electron pair. It might be expected that even models which take incomplete account of electron correlation would provide a reasonable description of the energetics of heterolytic bond dissociation. In the hypothetical limit of dissociation of a fully ionic bond, the electron pair in the undissociated molecule is already localized on a single atom (it is a lone pair) and is not even disrupted upon dissociation. In this case, the dissociation energy is given by Coulomb's law and corresponds to the energy of charge separation. Bond dissociation in a molecule like sodium chloride probably closely approaches this situation. More common is heterolytic dissociation of a polar covalent bond. Here, an electron pair which was previously (unequally) shared by two atoms is moved onto a single atom. A bond is destroyed but the electron pair maintains.

Perhaps the two most important heterolytic bond dissociation reactions are those used to define "absolute" acidity and basicity.

$$A{-}H \longrightarrow A{:}^- + H^+$$

$$B{-}H^+ \longrightarrow B{:} + H^+$$

Both reactions involve dissociation of a polar covalent bond to hydrogen and both lead to a "free" proton. While absolute acidities and basicities are rarely if ever (directly) measured experimentally, they provide a good opportunity to assess the performance of different models with regard to the energetics of heterolytic bond dissociation.

Absolute Basicities

Table 6-5 compares absolute basicities obtained from Hartree-Fock STO-3G, 3-21G, 6-31G* and 6-311+G** models, local density models, BP, BLYP, EDF1 and B3LYP density functional models and MP2 models, all with 6-31G* and 6-311+G** basis sets and MNDO, AM1 and PM3 semi-empirical models, with experimental enthalpies. All reactions are highly *endothermic* because they lead to a free proton.*

The STO-3G model provides a poor account of absolute basicities although, with the single exception of methane, yields the correct ordering of basicities. Results steadily improve in moving to the bigger basis sets. While the mean absolute deviation from experiment of the 6-311+G** model is only 5 kcal/mol (over a range of nearly 100 kcal/mol), it is difficult to identify systematic trends. The absolute basicities of some molecules are overestimated while those of others are underestimated.

Density functional models and MP2 models show more consistent behavior. With the 6-311+G** basis set, calculated basicities are generally very close to experimental values. The corresponding results with the 6-31G* basis set are generally not as good, although the differences are not that great. In terms of mean absolute errors, local density models perform the worst, and B3LYP/6-311+G** and MP2/6-311+G** models perform the best.

Semi-empirical models provide a completely unsatisfactory account of absolute basicities. They should not be used for this purpose.

Absolute Acidities

Comparisons between calculated and experimental absolute acidities are provided in **Table 6-6**. The reactions are even more *endothermic* than those for absolute basicity, due to the additional penalty arising from charge separation. With the exception of the local density 6-311+G** model, mean absolute errors here are significantly larger than those uncovered in absolute basicity comparisons, even though

---

* The situation is very different in water, where the energy of hydration of the proton is much larger than the hydration energies of the base or protonated base.

Table 6-5: Absolute Basicities[a]

base, B	Hartree-Fock				local density		BP		expt.
	STO-3G	3-21G	6-31G*	6-311+G**	6-31G*	6-311+G**	6-31G*	6-311+G**	
HF	118	132	122	120	128	117	129	120	**116**
N_2	141	126	118	120	120	119	123	123	**122**
CH_4	118	110	117	127	133	136	133	137	**130**
HCl	141	128	130	135	131	133	135	138	**144**
CO	176	143	143	144	146	143	150	148	**145**
C_2H_2	194	163	169	166	157	153	163	158	**156**
C_2H_4	210	170	175	175	166	162	171	168	**168**
H_2O	229	192	175	175	175	169	177	171	**179**
H_2S	223	170	172	175	169	169	174	175	**183**
PH_3	244	194	197	200	185	185	191	191	**197**
NH_3	259	227	217	216	215	208	217	212	**213**
mean absolute error	30	10	7	5	6	7	6	4	–

Table 6-5: Absolute Basicities (2)

base, B	BLYP		EDF1		B3LYP		MP2		semi-empirical			expt.
	6-31G*	6-311+G**	6-31G*	6-311+G**	6-31G*	6-311+G**	6-31G*	6-311+G**	MNDO	AM1	PM3	
HF	129	118	129	120	127	118	125	121	210	181	235	116
N_2	123	122	125	125	122	122	119	123	236	223	220	122
CH_4	131	134	133	138	129	133	124	132	265	235	242	130
HCl	135	137	137	139	134	136	132	140	223	220	205	144
CO	150	147	151	150	148	146	149	150	191	193	197	145
C_2H_2	164	157	164	160	163	158	155	156	207	207	213	156
C_2H_4	171	167	172	170	171	168	168	168	204	200	206	168
H_2O	176	169	177	172	176	171	175	173	195	203	213	179
H_2S	174	174	176	176	173	174	172	178	174	181	177	183
PH_3	191	191	193	194	192	193	194	198	176	187	117	197
NH_3	217	210	219	214	217	212	217	214	171	158	156	213
mean absolute error	6	4	6	4	5	3	5	3	59	52	65	–

a) energy (enthalpy) of reaction: $BH^+ \rightarrow B + H^+$

Table 6-6: Absolute Acidities[a]

acid, AH	Hartree-Fock				local density		BP		expt.
	STO-3G	3-21G	6-31G*	6-311+G**	6-31G*	6-311+G**	6-31G*	6-311+G**	
CH_4	560	462	457	434	450	415	454	422	**409**
NH_3	547	462	444	422	439	404	442	410	**397**
H_2O	565	450	429	406	429	389	432	394	**384**
C_2H_2	496	405	403	386	397	376	401	381	**370**
HF	602	432	409	381	417	367	420	372	**366**
SiH_4	510	390	388	385	371	364	380	374	**364**
PH_3	525	387	383	376	375	362	382	370	**363**
H_2S	506	364	360	356	357	346	363	353	**345**
HCN	463	379	370	355	371	349	375	355	**344**
HCl	411	337	335	333	336	327	342	333	**328**
mean absolute error	153	40	31	16	27	3	32	9	–

Table 6-6: Absolute Acidities (2)

acid, AH	BLYP		EDF1		B3LYP		MP2		semi-empirical			expt.
	6-31G*	6-311+G**	6-31G*	6-311+G**	6-31G*	6-311+G**	6-31G*	6-311+G**	MNDO	AM1	PM3	
CH_4	456	421	458	426	457	425	458	428	69	66	65	409
NH_3	443	408	445	414	444	413	444	414	54	60	41	397
H_2O	433	391	434	397	432	396	429	398	55	45	36	384
C_2H_2	403	381	404	385	404	383	403	385	47	34	26	370
HF	421	369	421	374	418	373	411	377	43	77	32	366
SiH_4	382	375	384	378	384	378	386	385	17	-6	-15	364
PH_3	384	370	385	374	384	372	383	378	12	-20	-15	363
H_2S	364	352	366	357	363	354	363	360	-6	-13	-15	345
HCN	378	354	378	358	375	355	374	357	20	13	-5	344
HCl	343	332	344	336	341	333	339	340	-43	-13	-31	328
mean absolute error	35	8	35	13	33	11	33	15	-	-	-	-

a) energy (enthalpy) of reaction $AH \rightarrow A^- + H^+$

both reactions effect heterolytic bond fracture. Note, that the effect of diffuse functions in the basis set is much larger than previously noted for absolute basicity comparisons. This is to be expected as comparisons are now being made between a neutral molecule and an anion. Finally, all calculated acidities are larger than experimental values. All of this points to the fact that anions (the products) are more difficult to describe than neutral molecules (the reactants), and that the level of error cancellation seen in absolute basicity comparisons will not be found here. It is highly likely that basis sets even larger than 6-311+G**, with multiple sets of diffuse functions will be needed.[*]

The excellent performance of the local density model with the 6-311+G** basis set is unexpected and not easily explained. The mean absolute error is only 3 kcal/mol, a factor of three lower than any other model examined, and the largest single deviation is only 7 kcal/mol (out of 400 kcal/mol).

Semi-empirical calculations provide a very poor account of absolute acidities. Even ignoring the large (~ 350 kcal/mol) systematic error, the calculations even fail to reproduce the ordering of acidities in these compounds. Semi-empirical models should not be employed for this purpose.

Absolute Lithium Cation Affinities

Absolute basicity comparisons represent just one example of an entire class of reactions in which an electrophile, E^+, is added to a neutral molecule.

$$B + E^+ \longrightarrow BE^+$$

Important other examples include electrophiles involved in aromatic alkylation, acylation and nitration. While in some instances quantitative gas-phase heats of formation are available, primarily indirectly from measurements of gas-phase proton affinities, there are insufficient

---

[*] As noted above, calculated acidities are all too large, meaning that some portion of the overall error is systematic. This anticipates the notion that improvement could be realized by choosing one of the compounds as a "standard", and relating the acidities of the other compounds to this standard. Discussion is provided later in this chapter.

data for these and other like cases to permit meaningful comparisons with the results of calculations to be made. However, there are ample high-quality gas-phase data for several other "electrophiles". One such class of electrophiles are alkali metal cations such as Li^+.

A comparison of calculated and experimentally measured lithium cation affinities, e.g., energies (enthalpies) where E^+ is Li^+, is provided in **Table 6-7**. The usual selection of models has been surveyed, except that AM1 has not been parameterized for lithium.

First note that the reaction energies are a factor of five smaller than the corresponding proton transfer energies (absolute basicities; see **Table 6-5**). Lithium cation is not as "unhappy" as a free-proton. Also note that the experimental data span a range of only 19 kcal/mol, much smaller than for the analogous protonation reaction. For example, while the enthalpies of protonation of ammonia and water are separated by 34 kcal/mol, the corresponding enthalpies for lithium addition differ by only 5 kcal/mol. Except for Hartree-Fock STO-3G and 3-21G models and the two semi-empirical models, all levels of calculation reproduce this range and generally provide a credible account of absolute lithium cation affinities. The best (in terms of smallest mean absolute error) are the Hartree-Fock model, BP, BLYP, EDF1 and B3LYP density functional models and the MP2 model with the 6-311+G** basis set. The performance of the corresponding local density model is slightly less satisfactory. The performance of all models with the 6-31G* basis set is much inferior. This should come as no surprise, as the products are likely to have significant "anion character", i.e.

$$Li-X \longleftrightarrow Li^+ \quad X^-$$

This means that diffuse basis functions will be required, just as they were for absolute acidity comparisons. Note, however, that in the case of absolute acidities, none of the models, except the local density model, gave satisfactory results even with the 6-311+G** basis set.

While neither semi-empirical model is really satisfactory, it should be noted that (at least in terms of mean absolute error) MNDO is much superior to PM3.

Table 6-7: Absolute Lithium Cation Affinities[a]

molecule	Hartree-Fock				local density		BP		expt.
	STO-3G	3-21G	6-31G*	6-311+G**	6-31G*	6-311+G**	6-31G*	6-311+G**	
MeCl	45	23	20	23	24	25	22	22	24
MeF	69	49	35	32	37	31	34	28	31
Me$_2$S	70	32	31	32	35	–	32	32	32
H$_2$O	79	57	40	36	45	39	40	34	35
H$_2$CO	74	53	41	40	40	38	36	34	36
HCN	62	46	38	37	39	38	36	35	36
MeOH	78	58	42	39	45	41	41	37	37
C$_6$H$_6$	85	44	41	38	38	43	34	36	38
NH$_3$	75	56	44	41	50	44	46	40	40
Me$_2$O	77	58	41	40	44	41	40	38	40
MeCHO	82	59	46	45	48	45	43	41	41
MeNH$_2$	75	56	45	42	50	45	45	41	42
Me$_2$NH	74	55	44	42	48	45	44	41	42
Me$_3$N	72	54	42	41	45	41	41	40	42
MeCN	71	54	45	45	49	48	44	43	43
mean absolute error	35	13	3	1	5	3	2	1	–

Table 6-7: Absolute Lithium Cation Affinities (2)

molecule	BLYP		EDF1		B3LYP		MP2		semi-empirical		expt.
	6-31G*	6-311+G**	6-31G*	6-311+G**	6-31G*	6-311+G**	6-31G*	6-311+G**	MNDO	PM3	
MeCl	25	24	22	22	24	24	24	24	10	28	**24**
MeF	38	31	34	28	37	32	36	30	50	13	**31**
Me$_2$S	34	–	32	32	34	33	33	32	45	20	**32**
H$_2$O	43	36	40	34	42	37	42	35	31	21	**35**
H$_2$CO	39	36	36	34	40	38	40	35	38	26	**36**
HCN	38	36	36	35	38	37	37	35	41	33	**36**
MeOH	44	39	41	37	44	40	43	38	31	19	**37**
C$_6$H$_6$	34	37	39	36	42	38	44	38	49	18	**38**
NH$_3$	48	41	45	39	48	42	47	41	35	29	**40**
Me$_2$O	43	40	40	37	43	41	43	39	31	18	**40**
MeCHO	46	43	43	41	46	44	44	40	39	29	**41**
MeNH$_2$	48	42	45	40	48	43	47	42	36	27	**42**
Me$_2$NH	47	42	44	40	46	43	47	42	36	24	**42**
Me$_3$N	44	41	42	39	45	40	45	41	35	22	**42**
MeCN	47	45	45	43	47	46	44	42	41	35	**43**
mean absolute error	4	1	2	1	4	2	4	1	7	14	–

a) energies of reactions B + Li$^+$ → BLi$^+$

Hydrogenation Reactions

Hydrogenation reactions typify processes in which the total number of bonds and total number of lone pairs are conserved, but the numbers of each kind of formal bond are not conserved. For example, hydrogenation of ethane leads to the destruction of one CC bond and one HH bond, but to the creation of two new CH bonds.

$$CH_3-CH_3 + H_2 \longrightarrow 2CH_4$$

It might be expected that methods that provide inadequate treatment of electron correlation, specifically Hartree-Fock models, would perform better here than they would in their description of reactions in which total bond count is not conserved. However, bonding changes associated with hydrogenation are great and Hartree-Fock models might still prove unacceptable.

The usual selection of theoretical models has been used to provide energies for complete hydrogenation of a small selection of two-heavy-atom, main-group hydrides. These are compared with experimental reaction enthalpies as well as reaction energies from G3 calculations[3] in **Table 6-8**. The data have been drawn from a larger collection of hydrogenation reactions found in **Appendix A6 (Tables A6-12** to **A6-19**). A summary of mean absolute deviations in hydrogenation energies from G3 values for all methods is provided in **Table 6-9**. These are based on the full set of data.

The STO-3G model provides a very poor account of hydrogenation energies. Hartree-Fock models with the 3-21G and 6-31G* basis sets offer better results, but large errors still exist for some systems (16 kcal/mol for hydrogenation of acetylene at 6-31G*). Clearly these levels are inadequate for quantitative descriptions. Except for hydrogenation of F_2, the 6-311+G** model performs well for the hydrogenation reactions in **Table 6-8**. However, the more extensive collection provided in **Table A6-12**, reveals large discrepancies between 6-311+G** and experimental hydrogenation energies. In some cases, however, it is the experimental data that are suspect.

Table 6-8: Energies of Hydrogenation Reactions

hydrogenation reaction	Hartree-Fock				local density		BP		BLYP	
	STO-3G	3-21G	6-31G*	6-311+G**	6-31G*	6-311+G**	6-31G*	6-311+G**	6-31G*	6-311+G**
$CH_3CH_3 + H_2 \rightarrow 2CH_4$	-19	-25	-22	-21	-17	-17	-18	-19	-19	-20
$CH_3NH_2 + H_2 \rightarrow CH_4 + NH_3$	-20	-30	-27	-28	-21	-25	-23	-26	-22	-27
$CH_3OH + H_2 \rightarrow CH_4 + H_2O$	-16	-28	-27	-31	-22	-29	-23	-29	-22	-29
$CH_3F + H_2 \rightarrow CH_4 + HF$	-8	-22	-23	-29	-19	-29	-18	-28	-16	-27
$NH_2NH_2 + H_2 \rightarrow 2NH_3$	-28	-47	-46	-50	-37	-45	-38	-46	-38	-46
$HOOH + H_2 \rightarrow 2H_2O$	-31	-67	-82	-93	-68	-85	-65	-81	-62	-80
$FF + H_2 \rightarrow 2HF$	-29	-98	-126	-149	-106	-137	-99	-129	-94	-126
$H_2C{=}CH_2 + 2H_2 \rightarrow 2CH_4$	-91	-71	-66	-61	-70	-66	-61	-58	-57	-55
$HC{\equiv}CH + 3H_2 \rightarrow 2CH_4$	-154	-124	-121	-112	-135	-126	-115	-108	-108	-101

hydrogenation reaction	EDF1		B3LYP		MP2		semi-empirical			G3	expt.
	6-31G*	6-311+G**	6-31G*	6-311+G**	6-31G*	6-311+G**	MNDO	AM1	PM3		
$CH_3CH_3 + H_2 \rightarrow 2CH_4$	-19	-20	-19	-20	-16	-17	-5	5	6	**-18**	**-19**
$CH_3NH_2 + H_2 \rightarrow CH_4 + NH_3$	-23	-27	-24	-27	-23	-26	-11	-3	3	**-26**	**-26**
$CH_3OH + H_2 \rightarrow CH_4 + H_2O$	-23	-30	-24	-30	-25	-31	-16	-6	1	**-30**	**-30**
$CH_3F + H_2 \rightarrow CH_4 + HF$	-18	-28	-18	-29	-21	-32	-11	-17	-9	–	**-29**
$NH_2NH_2 + H_2 \rightarrow 2NH_3$	-39	-46	-40	-47	-42	-48	-27	-23	-13	**-48**	**-48**
$HOOH + H_2 \rightarrow 2H_2O$	-66	-83	-69	-85	-75	-92	-84	-78	-53	**-87**	**-86**
$FF + H_2 \rightarrow 2HF$	-100	-130	-105	-134	-116	-145	-128	-121	-90	**-135**	**-133**
$H_2C{=}CH_2 + 2H_2 \rightarrow 2CH_4$	-58	-56	-62	-59	-58	-58	-41	-24	-16	**-57**	**-57**
$HC{\equiv}CH + 3H_2 \rightarrow 2CH_4$	-109	-103	-116	-108	-104	-104	-84	-57	-37	**-105**	**-105**

Table 6-9: **Mean Absolute Deviation from G3 Calculations in Energies of Hydrogenation Reactions**

		STO-3G	3-21G	6-31G*	6-311+G**
MNDO	16				
AM1	22				
PM3	38				
Hartree-Fock		25	10	7	7
local density		—	—	11	8
BP		—	—	10	3
BLYP		—	—	13	4
EDF1		—	—	11	3
B3LYP		—	—	9	3
MP2		—	—	9	3

Local density models also show high sensitivity to basis set. The local density 6-31G* model leads to large errors in hydrogenation energies, which (except for hydrogenation of ethylene and acetylene) are greatly diminished when the 6-311+G** basis set is employed. Similar basis set sensitivity for hydrogenation energies is seen for all other density functional models and for the MP2 model. The individual effects of polarization functions on hydrogen, diffuse functions and increased valence-shell splitting on hydrogenation energies calculated from 6-31G*, EDF1/6-31G*, B3LYP/6-31G* and MP2/6-31G* models are examined in **Appendix A6** (**Tables A6-20** to **A6-23**). In most cases, diffuse functions play the largest role, but the two other basis set extensions are also significant. BP/6-311+G**, BLYP/6-311+G**, EDF1/6-311+G** and B3LYP/6-311+G** models all provide solid and very similar accounts of the energetics of hydrogenation. The MP2/6-311+G** model is also successful, although a large error exists for hydrogenation of F_2.

Semi-empirical models provide a completely unacceptable account of hydrogenation energies.

Reactions Relating Multiple and Single Bonds

The fact that Hartree-Fock models do as well as they do in describing the energetics of hydrogenation reactions (see discussion in previous section) suggests that merely maintaining the total number of electron pairs may be sufficient to effect significant cancellation of electron correlation effects. The worst cases for "limiting" (6-311+G** basis set) Hartree-Fock models involved hydrogenation of molecules with bonds between two highly electronegative elements, e.g., hydrogenation of F_2, and hydrogenation of multiple bonds. To get around the latter, and perhaps to capitalize as much as possible on the simplicity of Hartree-Fock models, consider another series of reactions which relate the energies of molecules incorporating multiple bonds to those of molecules incorporating an equal number of analogous single bonds. For example, the energy of ethylene (incorporating a carbon-carbon double bond) would be related to the energy of two ethane molecules (each incorporating a CC single bond), while the energy of acetylene (incorporating a CC triple bond) would be related to the energy of three ethanes.

$$CH_2{=}CH_2 + 2CH_4 \longrightarrow 2CH_3{-}CH_3$$

$$HC{\equiv}CH + 4CH_4 \longrightarrow 3CH_3{-}CH_3$$

Data are provided in **Table 6-10**, with the same calculation models previously examined for hydrogenation reactions. As might be expected from the experience with hydrogenation reactions, Hartree-Fock models with 6-31G* and 6-311+G** basis sets perform relatively well. In fact, they turn in the lowest mean absolute errors of any of the models examined. The performance of density functional models (excluding local density models) and MP2 models with both 6-31G* and 6-311+G** basis sets is not much worse. On the other hand, local density models yield very poor results in all cases showing reactions which are too *exothermic*. The reason is unclear. Semi-empirical models yield completely unacceptable results, consistent with their performance for hydrogenation reactions.

Structural Isomerization

Which of several possible isomers is most stable, and what are the relative energies of any "reasonable" alternatives, are two of the most commonly asked thermochemical questions. The ability to pick out the lowest-energy structure and rank the energies of higher-energy isomers is essential to the success of any model. The general case, where isomers differ substantially in bonding, is dealt with in this section. An important special case involving regio and stereoisomers, i.e., molecules which have the same component bonds and differ only in detailed environment, is considered in **Chapter 12**.

Calculated relative energies for a small selection of structural isomers are compared with experimental values and with the results of G3 calculations[3] in **Table 6-11**. These have been drawn from a much more extensive set of comparisons found in **Appendix A6 (Tables A6-24** to **A6-31)**. Mean absolute errors from the full set of comparisons are collected in **Table 6-12**, and a series of graphical comparisons involving Hartree-Fock, EDF1, B3LYP and MP2 models

Table 6-10: Energies of Reactions Relating Multiple and Single Bonds

reaction	Hartree-Fock				local density		expt.
	STO-3G	3-21G	6-31G*	6-311+G**	6-31G*	6-311+G**	
$CH_2=CH_2 + 2CH_4 \rightarrow 2CH_3-CH_3$	-53	-21	-22	-18	-36	-32	**-20**
$CH_2=NH + CH_4 + NH_3 \rightarrow 2CH_3-NH_2$	-38	-12	-7	-5	-22	-20	**-12**
$CH_2=O + CH_4 + H_2O \rightarrow 2CH_3-OH$	-32	-7	1	3	-14	-10	**1**
$H_2C=S + CH_4 + H_2S \rightarrow 2CH_3-SH$	-52	-20	-20	-19	-32	-29	**-20**
$HN=NH + 2NH_3 \rightarrow 2NH_2-NH_2$	-18	-1	16	17	-3	-1	**28**
$HC{\equiv}CH + 4CH_4 \rightarrow 3CH_3-CH_3$	-97	-49	-55	-48	-84	-75	**-49**
$HC{\equiv}N + 2CH_4 + 2NH_3 \rightarrow 3CH_3-NH_2$	-37	5	3	7	-32	-25	**4**
$^-C{\equiv}O^+ + 2CH_4 + 2H_2O \rightarrow 3CH_3-OH$	-23	16	27	34	-16	-4	**27**
$^-C{\equiv}S^+ + 2CH_4 + 2H_2S \rightarrow 3CH_3-SH$	-88	-46	-45	-39	-77	-70	**-25**
$N{\equiv}N + 4NH_3 \rightarrow 3NH_2-NH_2$	49	88	109	114	58	65	**107**
mean absolute error	42	9	5	6	29	24	–

Table 6-10: Energies of Reactions Relating Multiple and Single Bonds (2)

reaction	BP		BLYP		EDF1		expt.
	6-31G*	6-311+G**	6-31G*	6-311+G**	6-31G*	6-311+G**	
$CH_2=CH_2 + 2CH_4 \rightarrow 2CH_3-CH_3$	-24	-20	-19	-15	-20	-16	-20
$CH_2=NH + CH_4 + NH_3 \rightarrow 2CH_3-NH_2$	-8	-6	-4	-2	-5	-2	-12
$CH_2=O + CH_4 + H_2O \rightarrow 2CH_3-OH$	-1	2	2	6	3	6	1
$H_2C=S + CH_4 + H_2S \rightarrow 2CH_3-SH$	-20	-18	-16	-13	-16	-14	-20
$HN=NH + 2NH_3 \rightarrow 2NH_2-NH_2$	12	14	16	18	16	11	28
$HC\equiv CH + 4CH_4 \rightarrow 3CH_3-CH_3$	-60	-52	-51	-41	-53	-44	-49
$HC\equiv N + 2CH_4 + 2NH_3 \rightarrow 3CH_3-NH_2$	-5	1	3	10	2	9	4
$^-C\equiv O^+ + 2CH_4 + 2H_2O \rightarrow 3CH_3-OH$	10	21	16	29	17	29	27
$^-C\equiv S^+ + 2CH_4 + 2H_2S \rightarrow 3CH_3-SH$	-51	-45	-42	-35	-43	-37	-25
$N\equiv N + 4NH_3 \rightarrow 3NH_2-NH_2$	87	94	95	103	96	103	107
mean absolute error	11	8	7	7	7	7	–

Table 6-10: Energies of Reactions Relating Multiple and Single Bonds (3)

reaction	B3LYP		MP2		semi-empirical			expt.
	6-31G*	6-311+G**	6-31G*	6-311+G**	MNDO	AM1	PM3	
$CH_2=CH_2 + 2CH_4 \rightarrow 2CH_3–CH_3$	-23	-19	-25	-24	-31	-34	-27	**-20**
$CH_2=NH + CH_4 + NH_3 \rightarrow 2CH_3–NH_2$	-8	-5	-7	-7	-17	-17	-15	**-12**
$CH_2=O + CH_4 + H_2O \rightarrow 2CH_3–OH$	-1	3	3	4	-9	-15	-3	**1**
$H_2C=S + CH_4 + H_2S \rightarrow 2CH_3–SH$	-19	-17	-18	-19	-31	-31	-35	**-20**
$HN=NH + 2NH_3 \rightarrow 2NH_2–NH_2$	13	15	17	17	9	10	10	**28**
$HC\equiv CH + 4CH_4 \rightarrow 3CH_3–CH_3$	-58	-48	-55	-52	-69	-72	-53	**-49**
$HC\equiv N + 2CH_4 + 2NH_3 \rightarrow 3CH_3–NH_2$	-3	4	8	7	-21	-21	-16	**4**
$^-C\equiv O^+ + 2CH_4 + 2H_2O \rightarrow 3CH_3–OH$	13	25	26	32	-20	-29	-3	**27**
$^-C\equiv S^+ + 2CH_4 + 2H_2S \rightarrow 3CH_3–SH$	-49	-42	-46	-42	-91	-93	-86	**-25**
$N\equiv N + 4NH_3 \rightarrow 3NH_2–NH_2$	92	99	112	114	60	59	57	**107**
mean absolute error	9	6	7	6	25	32	24	–

Table 6-11: Relative Energies of Structural Isomers

formula (reference)	isomer	Hartree-Fock				local density		G3	expt.
		STO-3G	3-21G	6-31G*	6-311+G**	6-31G*	6-311+G**		
C_2H_3N (acetonitrile)	methyl isocyanide	24	21	24	21	25	25	–	21
C_2H_4O (acetaldehyde)	oxirane	11	34	31	32	22	23	26	27
$C_2H_4O_2$ (acetic acid)	methyl formate	7	13	13	17	13	18	16	18
C_2H_6O (ethanol)	dimethyl ether	-1	6	7	11	6	12	12	12
C_3H_4 (propyne)	allene	17	3	2	2	-4	-3	1	1
	cyclopropene	30	40	26	28	15	18	24	22
C_3H_6 (propene)	cyclopropane	-4	14	8	10	1	3	8	7
C_4H_6 (1,3-butadiene)	2-butyne	-13	4	7	7	7	9	9	9
	cyclobutene	-13	18	13	15	3	6	13	11
	bicyclo [1.1.0] butane	12	46	30	33	13	16	28	26

Table 6-11: Relative Energies of Structural Isomers (2)

formula (reference)	isomer	BP		BLYP		EDF1		G3	expt.
		6-31G*	6-311+G**	6-31G*	6-311+G**	6-31G*	6-311+G**		
C_2H_3N (acetonitrile)	methyl isocyanide	25	25	25	25	25	25	–	21
C_2H_4O (acetaldehyde)	oxirane	25	27	28	30	26	27	26	27
$C_2H_4O_2$ (acetic acid)	methyl formate	12	17	10	15	12	17	16	18
C_2H_6O (ethanol)	dimethyl ether	6	12	4	11	6	12	12	12
C_3H_4 (propyne)	allene	-4	-3	-4	-3	-4	-3	1	1
	cyclopropene	19	21	30	25	18	21	24	22
C_3H_6 (propene)	cyclopropane	6	7	10	11	5	7	8	7
C_4H_6 (1,3-butadiene)	2-butyne	9	9	9	10	8	8	9	9
	cyclobutene	9	12	14	17	10	12	13	11
	bicyclo [1.1.0] butane	23	26	31	35	23	26	28	26

Table 6-11: Relative Energies of Structural Isomers (3)

formula (reference)	isomer	B3LYP		MP2		semi-empirical			G3	expt.
		6-31G*	6-311+G**	6-31G*	6-311+G**	MNDO	AM1	PM3		
C_2H_3N (acetonitrile)	methyl isocyanide	27	24	29	27	41	31	31	–	21
C_2H_4O (acetaldehyde)	oxirane	28	29	27	28	27	33	36	26	27
$C_2H_4O_2$ (acetic acid)	methyl formate	12	16	14	18	16	12	15	16	18
C_2H_6O (ethanol)	dimethyl ether	5	11	9	14	12	10	9	12	12
C_3H_4 (propyne)	allene	-3	-2	5	5	3	3	7	1	1
	cyclopropene	22	24	23	24	27	31	28	24	22
C_3H_6 (propene)	cyclopropane	8	9	4	5	6	11	10	8	7
C_4H_6 (1,3-butadiene)	2-butyne	8	9	4	5	-4	2	-1	9	9
	cyclobutene	12	15	8	9	2	16	7	13	11
	bicyclo [1.1.0] butane	28	31	21	22	35	48	38	28	26

Table 6-12: Mean Absolute Errors in Relative Energies of Structural Isomers

MNDO	8			
AM1	7			
PM3	6			

	STO-3G	3-21G	6-31G*	6-311+G**
Hartree-Fock	13	6	3	3
local density	—	—	5	3
BP	—	—	3	2
BLYP	—	—	4	3
EDF1	—	—	3	2
B3LYP	—	—	3	2
MP2	—	—	3	3

with both 6-31G* and 6-311+G** basis sets and MNDO, AM1 and PM3 models have been provided in **Figures 6-1** to **6-11**.

Hartree-Fock models with STO-3G and 3-21G basis sets are unsatisfactory. On the other hand, the corresponding 6-31G* and 6-311+G** models generally lead to reasonable results. The worst results involve comparisons between small rings and unsaturated acyclics, e.g., oxirane vs. acetaldehyde. Here, the "limiting" (6-311+G** basis set) Hartree-Fock model generally leads to the ring isomer being "too stable". The opposite behavior is seen for the corresponding local density model, where comparisons involving small rings generally tilt in favor of the unsaturated acyclic.

Better accounts of relative isomer energies are provided by density functional models and by MP2 models. With both 6-31G* and 6-311+G** basis sets, BP, EDF1 and MP2 models perform best and BLYP models perform worst, although the differences are not great. In terms of mean absolute errors, all models improve upon replacement of the 6-31G* by the 6-311+G** basis set. With some notable exceptions, individual errors also decrease in moving from the 6-31G* to 6-311+G** basis sets. (A further breakdown of basis set effects is provided in **Tables A6-32** to **A6-35** in **Appendix A6**.) The improvements are, however, not great in most cases, and it may be difficult to justify of the extra expense incurred in moving from 6-31G* to the larger basis set.

The comparison between propyne and allene warrants additional comment. Experimentally, propyne is the more stable by approximately 2 kcal/mol, an observation which is reproduced by Hartree-Fock models but is somewhat exaggerated by MP2 models. Note, however, that all density functional models (including local density models) show the reverse order of isomer stabilities with allene being more stable than propyne. This is another instance where the behavior of B3LYP and MP2 models do not mimic each other.

None of the semi-empirical models provides a satisfactory account of the relative energies of structural isomers. Individual errors >10 kcal/mol are commonplace and some systems show much larger errors. Semi-empirical models cannot be recommended.

Figure 6-1: 6-31G* vs. Experimental Relative Energies of Structural Isomers

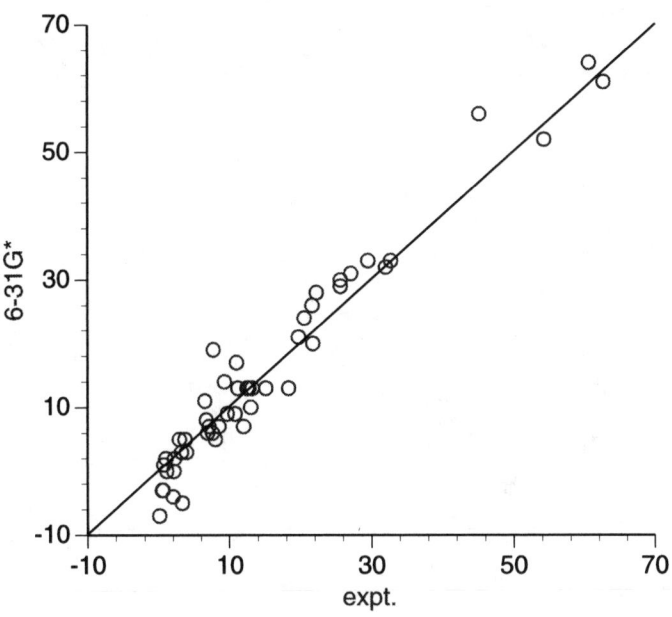

Figure 6-2: 6-311+G** vs. Experimental Relative Energies of Structural Isomers

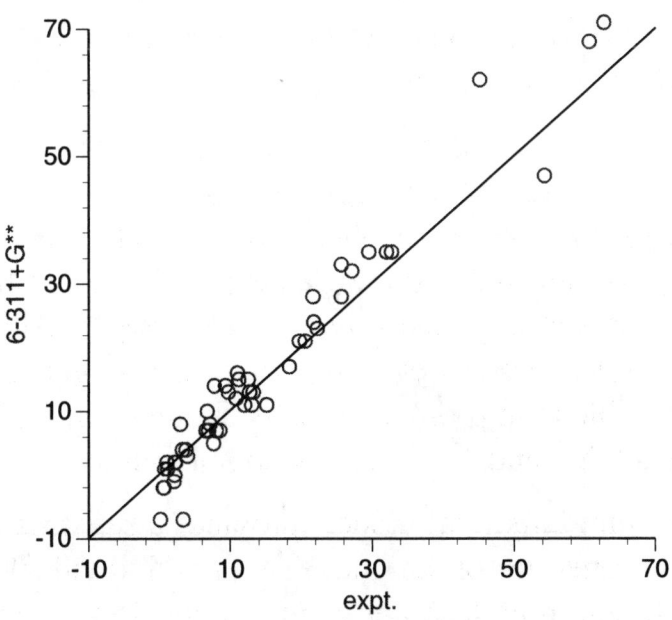

Figure 6-3: EDF1/6-31G* vs. Experimental Relative Energies of Structural Isomers

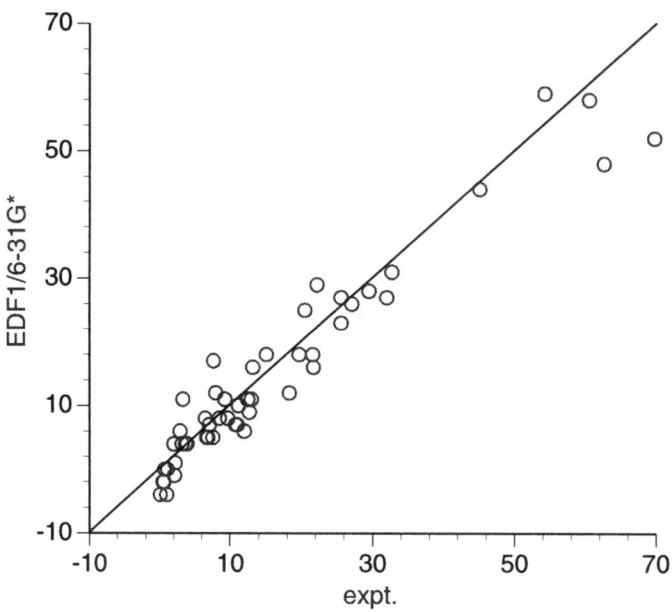

Figure 6-4: EDF1/6-311+G** vs. Experimental Relative Energies of Structural Isomers

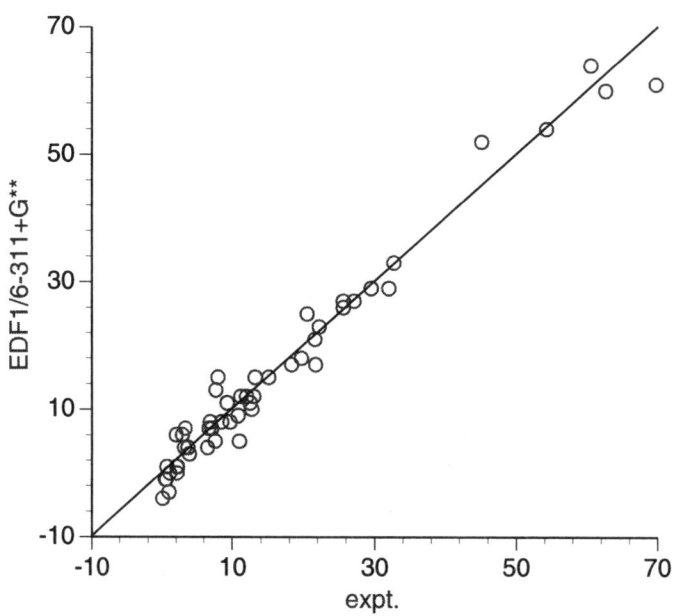

Figure 6-5: **B3LYP/6-31G* vs. Experimental Relative Energies of Structural Isomers**

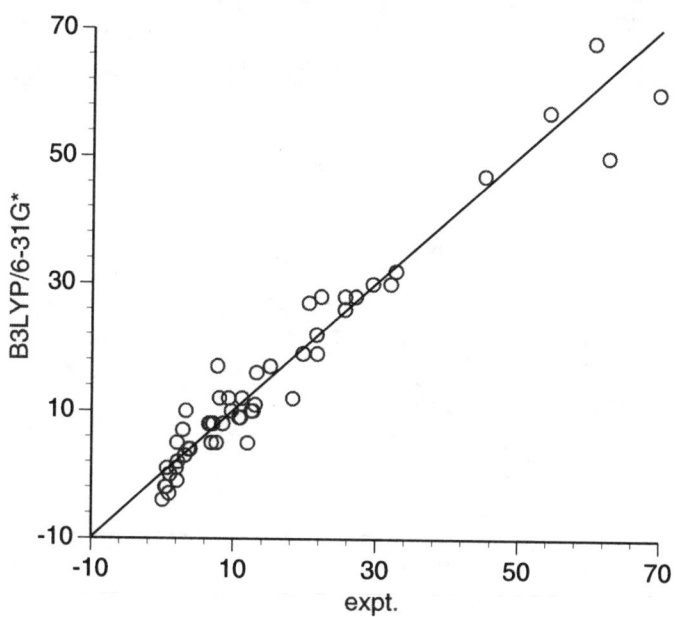

Figure 6-6: **B3LYP/6-311+G** vs. Experimental Relative Energies of Structural Isomers**

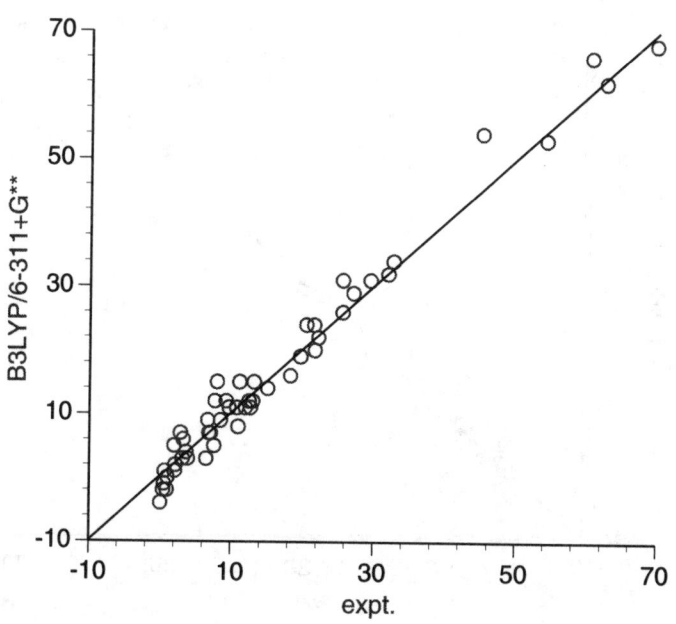

Figure 6-7: MP2/6-31G* vs. Experimental Relative Energies of Structural Isomers

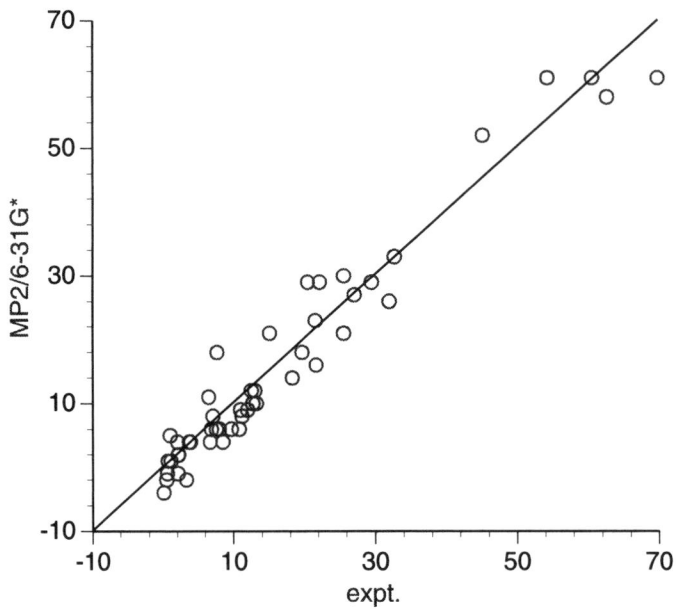

Figure 6-8: MP2/6-311+G** vs. Experimental Relative Energies of Structural Isomers

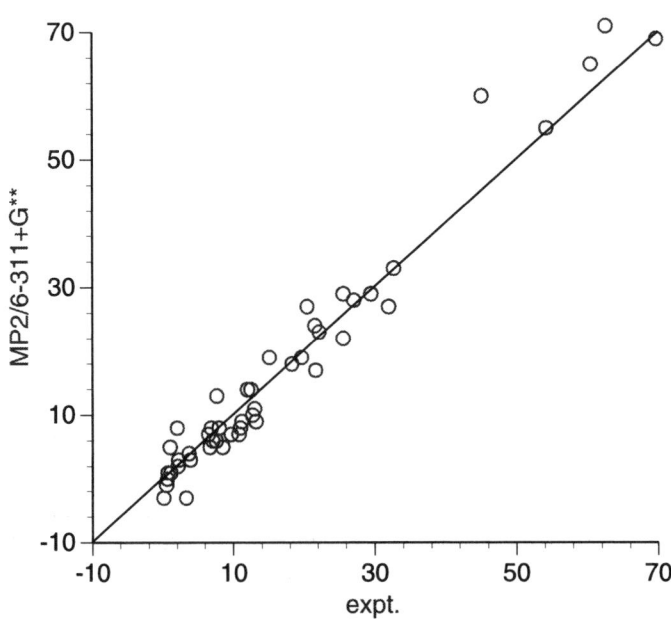

Figure 6-9: MNDO vs. Experimental Relative Energies of Structural Isomers

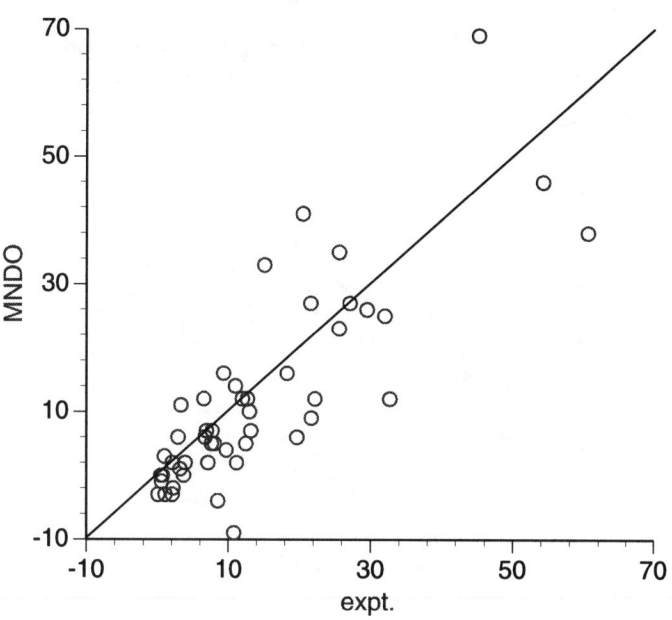

Figure 6-10:AM1 vs. Experimental Relative Energies of Structural Isomers

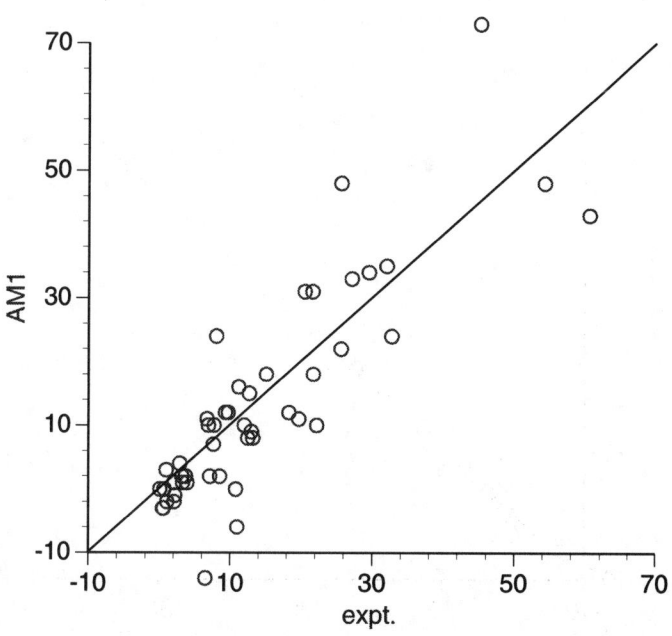

Figure 6-11: PM3 vs. Experimental Relative Energies of Structural Isomers

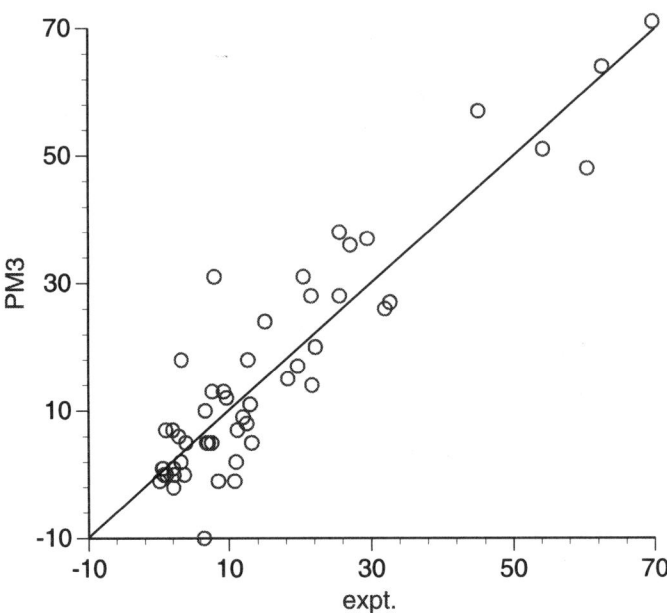

Isodesmic Reactions

The term *"isodesmic"* was coined to designate a process in which the numbers of each kind of chemical bond are conserved, and only detailed bonding environments differed between reactants and products.[4] The hope was that this would lead to significant cancellation of errors, and that even relatively simple models, in particular Hartree-Fock models, would provide an acceptable account of overall energetics. The comparisons which follow examine the extent to which such a conjecture is true.

At first glance it might appear that relatively few reactions are *isodesmic*. However, as pointed out in the introduction to this chapter, many thermochemical comparisons may be written in terms of *isodesmic* processes. Particularly important examples include comparisons of acid and base strengths relative to "standard" acids and bases, respectively, and comparisons among regio and stereoisomers. In addition, conformational energy comparisons may be thought of as *isodesmic* reactions. (These are discussed separately in **Chapter 8**.) All in all, *isodesmic* reactions constitute an important class of processes. Therefore, it would be highly desirable were practical calculation models able to yield accurate thermochemistry for *isodesmic* processes.

As will be discussed in **Chapter 13**, calculated energies of one particular class of *isodesmic* reactions, so-called bond separation reactions, may be combined with experimental or high-quality calculated thermochemical data in order to lead directly to "accurate" heats of formation. These in turn can be used in whatever types of thermochemical comparisons are of interest. We start our assessment of *isodesmic* processes with bond separation reactions. Following this, we consider description of bond dissociation energies, hydrogenation energies and acid and base strengths in terms of *isodesmic* processes, that is, not as absolute quantities but expressed relative to "standard" compounds.

Bond Separation Reactions

A bond separation reaction takes any molecule comprising three or more heavy (non-hydrogen) atoms into the set of simplest (two-heavy-atom) molecules containing the same bonds. The only requirement is that bonding must be defined in terms of a single valence (Lewis) structure or set of equivalent valence structures. This in turn guarantees that the bond separation energy is unique.[*]

Bond separation energies from Hartree-Fock models with STO-3G, 3-21G, 6-31G* and 6-311+G** basis sets, local density models, BP, BLYP, EDF1 and B3LYP density functional models and MP2 models all with 6-31G* and 6-311+G** basis sets and MNDO, AM1 and PM3 semi-empirical models are compared with values based on G3 energies[3] and on experimental thermochemical data in **Table 6-13**. These have been abstracted from a much larger collection found in **Appendix A** (**Tables A6-36** to **A6-43**). A summary of mean absolute deviations from G3 values in calculated bond separation energies (based on the full data set) is provided in **Table 6-14**.

First note that bond separation energies based on G3 energies and experimental heats of formation are nearly identical where the number of reactant molecules is the same as the number of product molecules. Bond separation energies from G3 calculations for reactions in which the number of reactant molecules is greater than the number of product molecules are too *exothermic* (or not sufficiently *endothermic*). This discrepancy is the result of comparing 0K data (G3 energies) with 298K data (measured heats of formation.) A temperature correction may be applied (see **Chapter 7**), which would bring all G3 bond separation energies into close accord with the corresponding experimental energies. For the present purpose (assessing the

---

[*] This requirement presents a serious problem only where two (or more) different reasonable Lewis structures may be drawn, for example for pyridazine.

In this case, it is necessary to (arbitrarily) choose one of the structures, leading to ambiguity in the definition of the bond separation reaction.

Table 6-13: Energies of Bond Separation Reactions

molecule	bond separation reaction	Hartree-Fock				local density		G3	expt.
		STO-3G	3-21G	6-31G*	6-311+G**	6-31G*	6-311+G**		
isobutane	$CH(CH_3)_3 + 2CH_4 \rightarrow 3CH_3CH_3$	1	4	2	2	6	7	6.4	7.5
trimethylamine	$(CH_3)_3N + 2NH_3 \rightarrow 3CH_3NH_2$	3	6	5	6	11	12	11.6	11.1
tetrafluoromethane	$CF_4 + 3CH_4 \rightarrow 4CH_3F$	54	62	58	41	75	54	–	52.8
tetrachloromethane	$CCl_4 + 3CH_4 \rightarrow 4CH_3Cl$	-21	-26	-29	-26	-5	-2	-2.4	-1.8
propene	$CH_3CHCH_2 + CH_4 \rightarrow CH_3CH_3 + CH_2CH_2$	5	4	4	4	7	6	4.9	5.4
propyne	$CH_3CCH + CH_4 \rightarrow CH_3CH_3 + HCCH$	8	8	8	8	12	10	7.8	7.6
allene	$CH_2CCH_2 + CH_4 \rightarrow 2CH_2CH_2$	0	-3	-4	-6	4	3	-3.2	-2.8
1,3-butadiene	$CH_2CHCHCH_2 + 2CH_4 \rightarrow CH_3CH_3 + 2CH_2CH_2$	13	11	11	10	18	17	13.2	14.2
benzene	⬡ $+ 6CH_4 \rightarrow 3CH_3CH_3 + 3CH_2CH_2$	70	60	58	55	73	68	61.0	64.2
cyclopropane	△ $+ 3CH_4 \rightarrow 3CH_3CH_3$	-45	-31	-26	-25	-30	-28	-23.5	-19.6
oxacyclopropane	O△ $+ 2CH_4 + H_2O \rightarrow CH_3CH_3 + 2CH_3OH$	-35	-31	-22	-19	-22	-20	-14.1	-10.5
thiophene	S⬠ $+ 4CH_4 + H_2S \rightarrow CH_3CH_3 + 2CH_2CH_2 + 2CH_3SH$	41	30	29	27	45	44	40.1	29.2
bicyclo[1.1.0]butane	⧖ $+ 6CH_4 \rightarrow 5CH_3CH_3$	-106	-76	-63	-60	-67	-63	-55.0	-45.5

Table 6-13: Energies of Bond Separation Reactions (2)

molecule	bond separation reactoin	BP		BLYP		EDF1		G3	expt.
		6-31G*	6-311+G**	6-31G*	6-311+G**	6-31G*	6-311+G**		
isobutane	$CH(CH_3)_3 + 2CH_4 \rightarrow 3CH_3CH_3$	4	3	3	3	1	1	6.4	7.5
trimethylamine	$(CH_3)_3N + 2NH_3 \rightarrow 3CH_3NH_2$	6	7	7	7	5	6	11.6	11.1
tetrafluoromethane	$CF_4 + 3CH_4 \rightarrow 4CH_3F$	64	41	60	38	62	41	–	52.8
tetrachloromethane	$CCl_4 + 3CH_4 \rightarrow 4CH_3Cl$	-11	-9	-13	-11	-14	-12	-2.4	-1.8
propene	$CH_3CHCH_2 + CH_4 \rightarrow CH_3CH_3 + CH_2CH_2$	6	5	5	5	5	4	4.9	5.4
propyne	$CH_3CCH + CH_4 \rightarrow CH_3CH_3 + HCCH$	11	9	10	9	11	9	7.8	7.6
allene	$CH_2CCH_2 + CH_4 \rightarrow 2CH_2CH_2$	2	1	2	0	2	0	-3.2	-2.8
1,3-butadiene	$CH_2CHCHCH_2 + 2CH_4 \rightarrow CH_3CH_3 + 2CH_2CH_2$	16	14	16	14	16	13	13.2	14.2
benzene	+ $6CH_4 \rightarrow 3CH_3CH_3 + 3CH_2CH_2$	71	65	67	62	71	64	61.0	64.2
cyclopropane	+ $3CH_4 \rightarrow 3CH_3CH_3$	-23	-22	-24	-21	-20	-20	-23.5	-19.6
oxacyclopropane	+ $2CH_4 + H_2O \rightarrow CH_3CH_3 + 2CH_3OH$	-14	-13	-15	-12	-11	-9	-14.1	-10.5
thiophene	+ $4CH_4 + H_2S \rightarrow CH_3CH_3 + 2CH_2CH_2 + 2CH_3SH$	42	40	38	37	43	41	40.1	29.2
bicyclo[1.1.0]butane	+ $6CH_4 \rightarrow 5CH_3CH_3$	-53	-52	-54	-50	-48	-46	-55.0	-45.5

Table 6-13: Energies of Bond Separation Reactions (3)

molecule	bond separation reaction	B3LYP		MP2		semi-empirical			G3	expt.
		6-31G*	6-311+G**	6-31G*	6-311+G**	MNDO	AM1	PM3		
isobutane	$CH(CH_3)_3 + 2CH_4 \rightarrow 3CH_3CH_3$	3	3	6	7	-9	-5	1	6.4	7.5
trimethylamine	$(CH_3)_3N + 2NH_3 \rightarrow 3CH_3NH_2$	7	7	11	11	-7	-6	1	11.6	11.1
tetrafluoromethane	$CF_4 + 3CH_4 \rightarrow 4CH_3F$	60	40	64	46	6	8	49	–	52.8
tetrachloromethane	$CCl_4 + 3CH_4 \rightarrow 4CH_3Cl$	-16	-14	-26	-3	-29	-21	6	-2.4	-1.8
propene	$CH_3CHCH_2 + CH_4 \rightarrow CH_3CH_3 + CH_2CH_2$	5	5	5	5	3	1	5	4.9	5.4
propyne	$CH_3CCH + CH_4 \rightarrow CH_3CH_3 + HCCH$	10	9	8	7	9	3	5	7.8	7.6
allene	$CH_2CCH_2 + CH_4 \rightarrow 2CH_2CH_2$	1	-1	-2	-3	-1	-4	-1	-3.2	-2.8
1,3-butadiene	$CH_2CHCHCH_2 + 2CH_4 \rightarrow CH_3CH_3 + 2CH_2CH_2$	15	13	14	14	6	3	10	13.2	14.2
benzene	$+ 6CH_4 \rightarrow 3CH_3CH_3 + 3CH_2CH_2$	67	62	71	68	37	28	50	61.0	64.2
cyclopropane	$+ 3CH_4 \rightarrow 3CH_3CH_3$	-25	-23	-24	-23	-35	-44	-32	-23.5	-19.6
oxacyclopropane	$+ 2CH_4 + H_2O \rightarrow CH_3CH_3 + 2CH_3OH$	-17	-15	-13	-14	-34	-46	-34	-14.1	-10.5
thiophene	$+ 4CH_4 + H_2S \rightarrow CH_3CH_3 + 2CH_2CH_2 + 2CH_3SH$	38	36	46	45	22	13	18	40.1	29.2
bicyclo[1.1.0]butane	$+ 6CH_4 \rightarrow 5CH_3CH_3$	-59	-55	-55	-55	-91	-113	-81	-55.0	-45.5

Table 6-14: Mean Absolute Deviations from G3 Calculations in Energies of Bond Separation Reactions

MNDO	16
AM1	20
PM3	11

	STO-3G	3-21G	6-31G*	6-311+G**
Hartree-Fock	10	8	6	7
local density	—	—	6	4
BP	—	—	6	4
BLYP	—	—	4	4
EDF1	—	—	4	4
B3LYP	—	—	4	4
MP2	—	—	4	3

performance of practical models for calculation of bond separation energies) the correction is ignored and comparisons are made directly to G3 energies (rather than to experimental energies).

STO-3G calculations provide an uneven account of bond separation energies. For example, the stability of benzene is overestimated (its bond separation reaction is too *endothermic*), while the stability of cyclopropane is underestimated (its bond separation reaction is too *exothermic*). The overall quality of Hartree-Fock bond separation energies improves with increasing size of the underlying basis set. 3-21G energies still show unacceptably large errors, but results from 6-31G* and 6-311+G** Hartree-Fock calculations are typically only a few kcal/mol removed from the G3 values. Some larger discrepancies exist, in particular for benzene and for small-ring systems, and this is reflected in high mean absolute deviations. With the exception of tetrafluoromethane, bond separation energies from 6-31G* and 6-311+G** models are nearly identical. Data provided in **Appendix A6** (see **Table A6-44**) shows that this difference is due almost entirely to the incorporation of diffuse functions rather than to the addition of polarization functions on hydrogen or of increased splitting of the valence shell.

Local density models yield bond separation energies of similar quality to those from corresponding (same basis set) Hartree-Fock models. Bond separation energies for isobutane and for trimethylamine, which were underestimated with Hartree-Fock models, are now well described. However, local density models do an even poorer job than Hartree-Fock models with benzene and with small-ring compounds.

BP, BLYP, EDF1 and B3LYP density functional models all lead to significant improvements over both Hartree-Fock and local density models, at least in terms of mean absolute deviations. While most reactions are better described, there are exceptions. Most notable among these is the bond separation reaction for tetrachloromethane. All four models show a highly *exothermic* reaction in contrast with both G3 and experimental results which show a nearly thermoneutral reaction. Similar, but somewhat smaller, effects are seen for isobutane and trimethylamine. As was the case with Hartree-Fock calculations, the only large energy change noted in going from the 6-31G* to the 6-311+G** basis set is for tetrafluoromethane. Results for intermediate basis sets, provided in **Appendix A6** (see **Tables A6-45** and **A6-46** for EDF1 and B3LYP models, respectively), show that this is primarily due to incorporation of diffuse functions (as was also the case for Hartree-Fock models).

A much more satisfactory account of bond separation energies is provided by the MP2/6-311+G** model. Individual energies are now typically only 1 or 2 kcal/mol from their respective G3 values, although larger discrepancies are seen for benzene and for thiophene (delocalized systems). The MP2/6-31G* model does not lead to satisfactory results for a number of systems, including tetrafluoromethane and tetrachloromethane. Results for intermediate basis sets (see **Table A6-47** in **Appendix A6**) show that while the different in tetrafluoromethane can be attributed mainly to the addition of diffuse functions, the differences noted in tetrachloromethane and other systems examined appear to be due to a combination of factors. This is unfortunate, as it would be useful from a practical standpoint to find a smaller (and less costly) alternative to 6-311+G** for use in MP2 calculations. However, the MP2/6-311+G** model would be routinely applicable to fairly large molecules if only energy (and not geometry) were required. Discussion is provided in **Chapter 12**.

Semi-empirical models are completely unsatisfactory in describing the energetics of bond separation reactions and should not be employed for this purpose. PM3 provides the best account of the three, but individual errors are often in excess of 10 kcal/mol.

The ability of quantum chemical calculations to routinely and reliably account for reaction energies provides a powerful tool for chemists to investigate interactions among substituents. For example, the energetics of *isodesmic* reactions,

$$X-CH_2-Y \ + \ CH_4 \ \longrightarrow \ CH_3X \ + \ CH_3Y$$

reveal whether geminal substituents X and Y interact destructively, constructively or not at all. Data for the same set of systems examined in **Chapter 5** for bond length changes are as follows:

Y \ X	Me	CMe₃	CN	OMe	F	SiMe₃
Me	1	–	–	–	–	–
CMe	-2	-11	–	–	–	–
CN	1	1	-10	–	–	–
OMe	4	4	-1	13	–	–
F	6	6	-4	15	14	–
SiMe₃	-2	-7	2	-2	-2	-1

Like the bond length changes, these are based on 6-31G* calculations. Positive numbers (in kcal/mol) reflect constructive interaction between X and Y substituents, while negative numbers reflect destructive interaction.

There is a strong parallel between reaction energies and previously noted bond length changes. Substituent combinations which lead to bond shortening (two electronegative groups as OMe or F, or one of these groups together with an alkyl substituent) lead to stabilization. On the contrary, combinations of groups which lead to bond lengthening lead to destabilization. Discussion for individual systems has already been provided in **Chapter 5**.

The near-zero energy uncovered for interaction of two trimethylsilyl groups attached to the same carbon warrants additional comment, in that severe steric repulsion might have been anticipated between the two bulky substituents. As mentioned previously, trimethylsilyl is both a σ–donor substituent and a π -acceptor substituent. Placed on a tetrahedral carbon center, this suggests that π donation from one substituent be followed by σ acceptance from the other substituent, and vice versa.

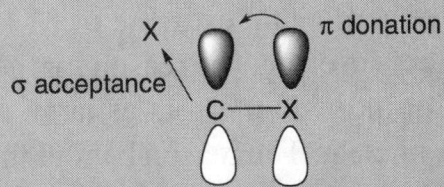

This is exactly the reverse situation found in dealing with attachment of two methoxy groups or two fluorines (σ acceptors and π donors) onto a tetrahedral carbon, also known as the anomeric effect, yet leads to exactly the same energetic consequences.

6-1

Relative Bond Dissociation Energies

It is clear that proper description of the energetics of homolytic bond dissociation requires models that account for electron correlation. Are correlated models also needed for accurate descriptions of relative homolytic bond dissociation energies where the relevant reactions are expressed as *isodesmic* processes? A single example suggests that they may not be. **Table 6-15** compares calculated and measured CH bond dissociation energies in hydrocarbons, R–H, relative to the CH bond energy in methane as a "standard", i.e.

$$R{-}H + CH_3^{\bullet} \longrightarrow R^{\bullet} + CH_4$$

The experimental data span a very wide range from the CH bond in acetylene (27 kcal/mol stronger than that in methane) to the CH bond in cycloheptatriene (31 kcal/mol weaker). This presumably reflects the stability (in instability) of the radical product more than it does the hydrocarbon reactant. The usual models have been surveyed.

With the exception of STO-3G and both MP2 models, all models (including semi-empirical models) provide a credible account of relative CH bond energies. In terms of mean absolute error, BP and B3LYP models with the 6-311+G** basis set are best and Hartree-Fock 3-21G and 6-31G* models, local density 6-31G* models and semi-empirical models are worst. More careful scrutiny turns up sizeable individual errors which may in part be due to the experimental data. For example, the "best" of the models appear to converge on a CH bond dissociation for cycloheptatriene which is 35-37 kcal/mol less than that in methane (the reference compound) compared with the experimental estimate of 31 kcal/mol. It is quite possible that the latter is in error. The reason for the poor performance of MP2 models, with individual errors as large as 16 kcal/mol (for cycloheptatriene) is unclear. The reason behind the unexpected good performance of all three semi-empirical models is also unclear.

With due attention to the noted failures, it is evident that relative homolytic bond dissociation energies, unlike absolute homolytic bond dissociation energies, can be reasonably well described with simple and practical models.

Table 6-15: CH Bond Dissociation Energies in Hydrocarbons Relative to Methane

hydrocarbon	Hartree-Fock				local density		BP		expt.
	STO-3G	3-21G	6-31G*	6-311+G**	6-31G*	6-311+G**	6-31G*	6-311+G**	
acetylene	-6	-19	-20	-23	-24	-28	-25	-28	-27
benzene	19	-1	-1	-4	-1	-3	-3	-5	-6
ethylene	7	-3	-3	-5	-1	-2	-3	-4	-4
cyclopropane	-2	-7	-5	-5	2	0	0	-1	4
ethane	4	3	3	3	7	7	5	5	6
propane[a]	8	4	5	5	12	11	10	9	9
propyne[b]	17	15	14	12	21	19	19	17	10
isobutane	12	6	8	7	16	14	14	12	12
propene	28	21	21	19	23	22	21	20	16
toluene	38	22	23	21	21	20	19	18	19
cyclopentadiene	34	22	23	22	29	26	26	24	23
cycloheptatriene	47	32	34	32	39	37	37	35	31
mean absolute error	11	5	4	3	5	3	3	2	–

231

Table 6-15: CH Bond Dissociation Energies in Hydrocarbons Relative to Methane (2)

hydrocarbon	BLYP		EDF1		B3LYP		MP2		semi-empirical			expt.
	6-31G*	6-311+G**	6-31G*	6-311+G**	6-31G*	6-311+G**	6-31G*	6-311+G**	MNDO	AM1	PM3	
acetylene	-25	-29	-24	-27	-26	-29	-37	-39	-55	-49	-47	**-27**
benzene	-3	-5	-3	-4	-5	-6	–	–	-11	-12	-7	**-6**
ethylene	-3	-4	-3	-4	-5	-4	-11	-11	-7	-5	0	**-4**
cyclopropane	0	-1	0	-1	-1	-2	-4	-5	-3	-3	3	**4**
ethane	5	5	5	5	5	5	3	3	6	6	8	**6**
propane[a]	10	9	10	10	9	8	6	5	13	11	15	**9**
propyne[b]	19	18	19	18	18	16	4	–	11	10	11	**10**
isobutane	13	12	14	13	12	11	8	6	20	16	21	**12**
propene	22	21	22	21	21	20	12	11	16	15	16	**16**
toluene	19	20	19	18	18	17	–	–	18	15	15	**19**
cyclopentadiene	27	25	27	25	26	24	17	13	18	15	17	**23**
cycloheptatriene	38	41	38	36	37	35	15	–	30	27	28	**31**
mean absolute error	3	3	3	3	3	2	7	7	5	5	5	–

a) formation of 2-propyl radical
b) formation of propargyl radical

232

Relative Hydrogenation Energies

Energies for hydrogenation reactions are often employed to probe for unusual stability or instability. For example, the observation that the first step in the hydrogenation of benzene (to 1,3-cyclohexadiene) is slightly *endothermic* while the remaining two steps (to cyclohexene and then to cyclohexane) are strongly *exothermic*, i.e.

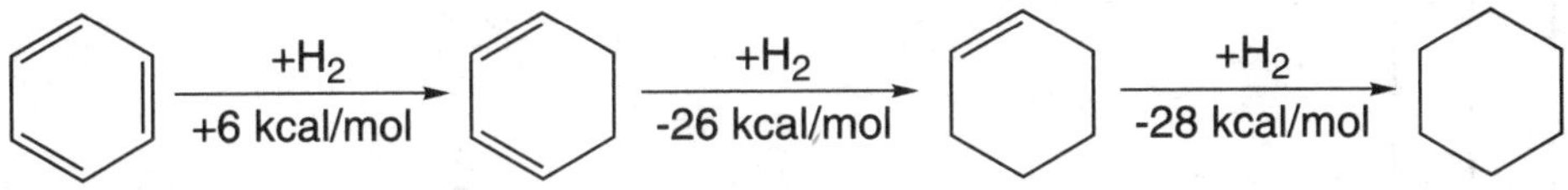

clearly reveals benzene's unusual "aromatic" stability. Indeed the difference in energies between the first step in the hydrogenation of benzene and that for a "typical" olefin (or example, cyclohexene) provides a quantitative measure of the aromatic stabilization.

On the other hand, the observation that hydrogenation of cyclopropene is much more *exothermic* than would be expected for a typical cycloalkene, reveals the effect of ring strain.

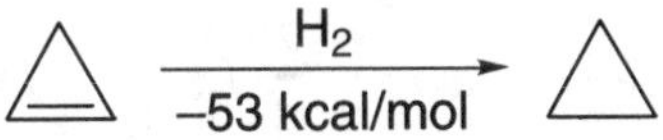

Both of the above reactions may be written as *isodesmic* reactions by relating them to an appropriate "standard" hydrogenation reaction, e.g., for cyclohexene.

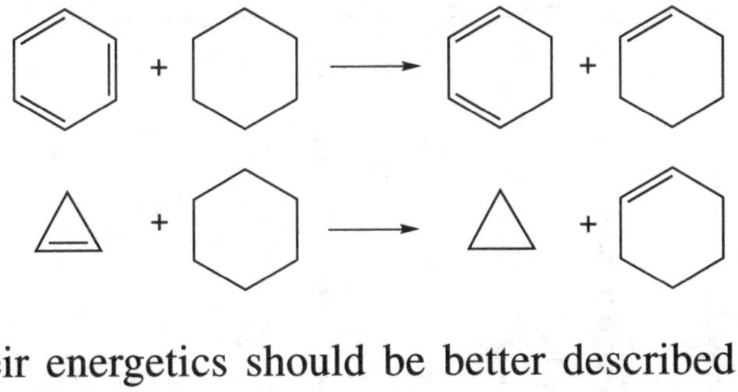

As such, their energetics should be better described by quantum chemical models which are simple enough for routine application than the energetics of hydrogenation in the absence of a reference reaction (see discussion earlier in this chapter).

Table 6-16 provides a comparison of calculated and experimental hydrogenation energies for alkenes and closely-related compounds

Table 6-16: Hydrogenation Energies of Alkenes and Related Compounds Relative to Cyclohexene

alkene	Hartree-Fock				local density		BP		expt.
	STO-3G	3-21G	6-31G*	6-311+G**	6-31G*	6-311+G**	6-31G*	6-311+G**	
cyclopropene	27	32	30	30	27	28	27	28	25.1
1,2-dimethylcyclopropene	17	25	20	21	17	19	16	18	14.7
allene	11	10	14	14	9	10	10	11	12.6
1-methylmethylenecyclopropene	7	9	10	11	7	8	7	8	9.7
bicyclo [2.2.1] heptadiene	10	12	13	13	11	11	11	10	6.4
bicyclo [2.2.1] heptene	7	6	7	7	4	5	5	5	4.5
cyclobutene	8	8	6	6	4	4	4	5	2.3
cyclopentene	-1	-1	-1	-1	-2	-2	-2	-1	-1.7
1,3-butadiene	-1	-3	-1	-1	-3	-2	-2	-2	-1.9
1,3-cyclohexadiene	-3	-1	0	0	-2	-2	-2	-2	-1.9
cycloheptene	-2	-2	-1	-1	-1	-1	-1	-1	-2.2
cyclopentadiene	-3	-2	-2	-2	-6	-6	-4	-3	-4.6
cis-cyclooctene	-5	-6	-4	-4	-4	-4	-4	-4	-5.6
cis-cyclodecene	-8	-9	-7	-7	-6	-6	-6	-6	-7.9
benzene	-43	-40	-37	-36	-39	-37	-38	-36	-34.0
mean absolute error	2	3	3	3	2	2	2	2	–

Table 6-16: Hydrogenation Energies of Alkenes and Related Compounds Relative to Cyclohexene (2)

alkene	BLYP		EDF1		B3LYP		MP2		semi-empirical			expt.
	6-31G*	6-311+G**	6-31G*	6-311+G**	6-31G*	6-311+G**	6-31G*	6-311+G**	MNDO	AM1	PM3	
cyclopropene	27	27	27	28	28	28	27	27	32	29	26	25.1
1,2-dimethylcyclopropene	16	18	15	17	17	19	18	20	17	22	20	14.7
allene	10	11	11	11	11	12	13	13	14	11	15	12.6
1-methylmethylenecyclopropene	7	8	6	7	7	8	9	10	4	5	5	9.7
bicyclo [2.2.1] heptadiene	11	11	11	11	11	11	9	9	13	13	11	6.4
bicyclo [2.2.1] heptene	5	5	5	5	5	6	4	3	11	12	10	4.5
cyclobutene	5	5	5	5	5	5	4	5	18	18	15	2.3
cyclopentene	-1	-1	-1	-1	-1	-1	-2	-2	5	3	1	-1.7
1,3-butadiene	-2	-2	-2	-1	-2	-2	-2	-1	4	1	3	-1.9
1,3-cyclohexadiene	-2	-2	-2	-2	-2	-2	-2	-2	0	-1	-1	-1.9
cycloheptene	-1	-1	-1	-1	-1	-1	-2	-2	-1	-1	0	-2.2
cyclopentadiene	-4	-4	-4	-4	-4	-4	-5	-5	8	6	3	-4.1
cis-cyclooctene	-4	-4	-4	-4	-4	-4	-5	–	-4	-3	-2	-5.6
cis-cyclodecene	-6	-6	-7	-6	-6	-6	-7	–	-6	-5	-1	-7.9
benzene	-38	-36	-38	-37	-39	-37	-41	-39	-18	-24	-23	-34.0
mean absolute error	2	2	2	2	2	2	1	2	6	6	5	–

relative to the corresponding hydrogenation energy of cyclohexene. In terms of mean absolute errors, there is little to distinguish among the various Hartree-Fock, density functional and MP2 models. Hartree-Fock models fare worst and MP2 models best, but the differences are not great and all models offer quantitative descriptions. In strong contrast to the situation with absolute hydrogenation energies, models with 6-31G* and 6-311+G** basis sets offer nearly identical results. There is little to justify use of the larger representation for this purpose.

Results from semi-empirical calculations are not as good as those from other models, but somewhat better than might have been anticipated on the basis of previous reaction energy comparisons (note, however, their favorable performance for relative CH bond dissociation energies).

To what extent is 1,6-methanocyclodeca-1,3,5,7,9-pentaene stabilized by aromaticity? The X-ray crystal structure suggests a fully delocalized π system. The ten carbons which make up the base are very nearly coplanar and all CC bonds are intermediate in length between "normal" single and double linkages, just as they are in naphthalene.

1.38 (1.37)
1.42 (1.42)
1.40 (1.41)

1.37 (1.36)
1.41 (1.42)
1.42 (1.42)

Bond lengths from 6-31G* Hartree-Fock calculations (shown in parentheses) are in close accord.

Hydrogenation energies provide additional evidence. Hydrogenation of naphthalene (to 1,2-dihydronaphthalene) is 25 kcal/mol more *exothermic* than hydrogenation of cyclohexene (the reference compound) according to 6-31G* calculations. This is smaller than the corresponding hydrogenation energy for benzene (-37 kcal/mol relative to cyclohexene), but still suggests considerable "aromatic" stabilization. On the other hand, hydrogenation of 1,6-methanocyclodeca-1,3,5,7,9-pentaene is predicted to be only 10 kcal/mol less than that of cyclohexene.

6-2

Relative Acidities and Basicities

Another important type of *isodesmic* reaction compares acid (or base) strength to that of a closely-related standard compound, for example, the basicity of trimethylamine relative to that of ammonia as a standard. This differs fundamentally from absolute acid (or base) strength comparisons, which are heterolytic bond dissociations and which significantly alter overall bonding.[*]

A comparison of calculated and measured proton affinities (basicities) of nitrogen bases relative to the proton affinity of ammonia as a standard is provided in **Table 6-17**. The calculations correspond to the usual theoretical models, and the experimental data derive from equilibrium measurements in the gas phase.[2] The data span a large range; the proton affinity of the strongest base examined, quinuclidine, is some 27 kcal/mol greater than that of the weakest base, ammonia.

With the exception of semi-empirical models, all models provide very good descriptions of relative nitrogen basicities. Even STO-3G performs acceptably; compounds are properly ordered and individual errors rarely exceed 1-2 kcal/mol. One unexpected result is that neither Hartree-Fock nor any of the density functional models improve on moving from the 6-31G* to the 6-311+G** basis set (local density models are an exception). Some individual comparisons improve, but mean absolute errors increase significantly. The reason is unclear. The best overall description is provided by MP2 models. Unlike bond separation energy comparisons (see **Table 6-11**), these show little sensitivity to underlying basis set and results from the MP2/6-31G* model are as good as those from the MP2/6-311+G** model.

Semi-empirical models are poor for relative base strength comparisons, paralleling their behavior for most other *isodesmic* reactions. They should not be used for this purpose.

Additional comparisons between calculated and experimental relative base strengths are found in **Appendix A6** (**Tables A6-48** and

---

[*] In practice, experimental determinations of acidity and basicity are rarely if ever "absolute" measurements, but rather measurements relative to given standards, although the standards may be structurally unrelated to the systems at hand.

Table 6-17: Proton Affinities of Nitrogen Bases Relative to Ammonia[a]

base, B	Hartree-Fock				local density		BP		expt.
	STO-3G	3-21G	6-31G*	6-311+G**	6-31G*	6-311+G**	6-31G*	6-311+G**	
aniline	9	1	7	6	0	2	4	5	6.7
methylamine	9	10	11	12	8	9	9	10	9.1
aziridine	9	16	14	16	8	11	9	12	11.2
ethylamine	13	13	14	15	11	12	12	14	11.8
dimethylamine	15	17	18	19	12	15	15	17	15.5
pyridine	18	14	18	19	13	17	15	18	16.0
tert-butylamine	19	18	19	20	17	18	18	19	16.1
cyclohexylamine	19	18	20	20	16	18	19	19	16.3
azetidine	20	22	22	23	15	17	18	20	18.0
pyrrolidine	22	23	23	25	18	20	21	23	19.8
trimethylamine	20	21	22	24	14	18	18	21	20.0
piperidine	23	22	24	25	19	20	21	23	21.1
diazabicyclooctane	23	26	28	30	19	23	22	26	23.5
N-methylpyrrolidine	25	25	26	28	19	22	22	25	24.3
N-methylpiperidine	27	26	28	30	20	24	24	27	25.7
quinuclidine	29	29	31	34	23	27	26	30	27.1
mean absolute error	2	2	3	4	3	1	1	2	–

Table 6-17: Proton Affinities of Nitrogen Bases Relative to Ammonia (2)

base, B	BLYP		EDF1		B3LYP		MP2		semi-empirical			expt.
	6-31G*	6-311+G**	6-31G*	6-311+G**	6-31G*	6-311+G**	6-31G*	6-311+G**	MNDO	AM1	PM3	
aniline	5	6	5	6	5	6	7	5	5	2	3	6.7
methylamine	10	11	10	11	10	11	10	11	2	2	-2	9.1
aziridine	10	13	10	12	11	13	9	11	5	3	-6	11.2
ethylamine	14	15	13	14	13	14	12	13	4	4	0	11.8
dimethylamine	16	18	15	18	16	18	16	18	2	3	-4	15.5
pyridine	16	19	16	19	16	19	13	15	12	6	0	16.0
tert-butylamine		20	19	20	19	21	17	17	7	11	6	16.1
cyclohexylamine	19	20	20	20	19	20	18	18	7	11	5	16.3
azetidine	19	21	18	20	19	21	18	19	8	9	0	18.0
pyrrolidine	22	24	21	23	22	24	21	22	6	8	0	19.8
trimethylamine	19	23	17	22	19	23	19	22	1	4	-6	20.0
piperidine	22	25	22	24	22	24	21	22	6	10	3	21.1
diazabicyclooctane	24	28	24	27	24	28	23	25	5	5	-5	23.5
N-methylpyrrolidine	24	29	23	26	24	26	23	25	4	9	-2	24.3
N-methylpiperidine	23	28	25	28	25	28	24	26	4	11	1	25.7
quinuclidine	28	31	29	31	28	32	27	29	11	10	1	27.1
mean absolute error	1	3	1	2	1	3	1	1	12	11	18	–

a) energy of reaction: $BH^+ + NH_3 \rightarrow B + NH_4^+$

A6-49 for oxygen bases relative to water and carbonyl compounds relative to formaldehyde, respectively).

It might be expected that the calculations will have greater difficulty in properly accounting for relative acidities than they did for relative basicities. In particular, models which make use of the 6-31G* basis set, which lacks diffuse ("+") functions, are likely to be unsatisfactory. While diffuse functions are relatively unimportant for cations (resulting from protonation of a base), they are quite important for anions (resulting from deprotonation of an acid). The data in **Table 6-18** test the extent to which such an expectation is realized. Here, acidities for a series of "CH acids" are referred to the acidity of methane. While the acids are all very similar[*], the resulting anions show considerable structural diversity and the acidities span a large range. As with the basicity comparisons presented previously, the experimental data follow from gas-phase equilibrium measurements[2] and are believed to be accurate to within one or two kcal/mol.

Basis set effects are similar for all models. Specifically, 6-31G* basis set models, which lack diffuse functions, clearly lead to unsatisfactory results, while the corresponding 6-311+G** models, which include diffuse functions, all perform well. STO-3G and 3-21G Hartree-Fock models also lead to poor results. Individual errors (for 6-311+G** models) are typically kept to 2-3 kcal/mol and are only rarely greater than 5 kcal/mol. The largest single error is 9 kcal/mol (the acidity of cyclopentadiene using the MP2/6-311+G** model). In short, there is very little to distinguish from among the different models with the 6-311+G** basis set.

Semi-empirical models are entirely unsatisfactory in describing relative acidities, just as they were in describing relative basicities.

An additional comparison between calculated and experimental relative acid strengths for substituted phenols relative to the parent compound is found in **Appendix A6 (Tables A6-50)**.

---

[*] Except for propyne, which is known to deprotonate from the sp hybridized carbon, all acids deprotonate from sp^3 carbon.

A special but important case of relative acidity/basicity comparisons involves reactions which differ only by remote (from the "reaction site") substitution. The example here relates to acidities of p-substituted benzoic acids to that of the parent compound. Data are provided in **Table 6-19**. Results from STO-3G, 3-21G and 6-31G* Hartree-Fock models, local density models, BP, BLYP, EDF1 and B3LYP density functional models and MP2 models all with the 6-31G* basis-set, and MNDO, AM1 and PM3 semi-empirical models are provided.

All models provide a reasonable account of the effects of remote substituents on the acidity of benzoic acid. The performance of STO-3G and 3-21G models is comparable to their performance for amine basicities. Also noteworthy is the fact that the 6-31G* basis set is adequate for these types of comparisons, that is, the effect of diffuse functions (in the 6-311+G** basis set) largely cancels. Also encouraging (and unexpected), is the excellent account provided by all three semi-empirical models.

A related comparison of remote substituent effects on acidities is provided by the data in **Table 6-20**. Here, the acidities of p-substituted benzoic acids complexed to chromium tricarbonyl are related to the acidity of the parent compound.

$$p\text{-}XC_6H_4CO_2^- + C_6H_5CO_2H \longrightarrow p\text{-}XC_6H_4CO_2H + C_6H_5CO_2^-$$
$$\underset{Cr(CO)_3}{|} \qquad \underset{Cr(CO)_3}{|} \qquad\qquad \underset{Cr(CO)_3}{|} \qquad \underset{Cr(CO)_3}{|}$$

Only local density, BP, BLYP, EDF1 and B3LYP density functional all with the 6-31G* basis set and PM3 models have been examined. All produce similar results which closely parallel the known relative acidities of the uncomplexed benzoic acids (see data in **Table 6-19**), although, the overall range of substituent effects is somewhat reduced. Of special note is the (apparently) favorable performance of PM3, paralleling its behavior for relative acidities in uncomplexed benzoic acids. While experimental data on the acidities of complexed benzoic acids are unavailable, the consistency of results among the various models lends credence to their validity.

Table 6-18: Acidities of Carbon Acids Relative to Methane[a]

acid, AH	Hartree-Fock				local density		BP		expt.
	STO-3G	3-21G	6-31G*	6-311+G**	6-31G*	6-311+G**	6-31G*	6-311+G**	
propene	44	38	32	26	44	32	40	29	**26.1**
tetramethylsilane	23	38	34	22	40	23	38	22	**26.9**
toluene	62	48	42	31	59	39	55	37	**34.9**
propyne[b]	62	53	50	41	49	33	50	35	**35.3**
acetonitrile	64	58	55	45	64	48	61	46	**43.4**
1,3-pentadiene	70	57	51	40	68	50	64	47	**44.2**
acetaldehyde	57	65	60	50	71	56	67	53	**49.6**
nitromethane	85	91	78	64	80	65	76	62	**58.9**
cyclopentadiene	104	81	73	63	86	67	81	64	**60.9**
mean absolute error	22	17	11	3	20	5	15	3	–

Table 6-18: Acidities of Carbon Acids Relative to Methane (2)

acid, AH	BLYP		EDF1		B3LYP		MP2		semi-empirical			expt.
	6-31G*	6-311+G**	6-31G*	6-311+G**	6-31G*	6-311+G**	6-31G*	6-311+G**	MNDO	AM1	PM3	
propene	40	29	41	30	39	29	37	29	48	46	46	26.1
tetramethylsilane	38	22	37	21	37	22	39	24	47	39	30	26.9
toluene	54	36	54	37	52	36	49	35	64	60	60	34.9
propyne[b]	50	34	50	35	50	36	51	38	29	35	41	35.3
acetonitrile	60	45	61	46	60	46	58	44	56	55	59	43.4
1,3-pentadiene	63	46	64	47	61	46	58	44	66	64	64	44.2
acetaldehyde	66	53	67	54	66	53	66	52	62	62	60	49.6
nitromethane	77	63	77	63	78	64	76	59	87	86	92	58.9
cyclopentadiene	80	62	81	64	80	64	85	70	82	78	80	60.9
mean absolute error	16	3	17	3	16	3	16	3	18	16	17	–

a) energy of reaction: $CH_4 + A^- \rightarrow CH_3^- + AH$
b) deprotonation occurs at alkyne carbon

243

Table 6-19: Acidites of *p*-Substituted Benzoic Acids Relative to Benzoic Acid[a]

p-substituent, X	Hartree-Fock			local density	BP	BLYP	EDF1	B3LYP	MP2	semi-empirical			expt.
	STO-3G	3-21G	6-31G*	6-31G*	6-31G*	6-31G*	6-31G*	6-31G*	6-31G*	MNDO	AM1	PM3	
NH$_2$	-2	-6	-4	-6	-5	-5	-5	-5	-3	-2	-2	-1	-3.4
Me	-1	-1	-1	-1	-1	-1	-1	-1	-1	1	0	0	-1.0
OMe	-1	-1	-2	-2	-2	-2	-2	-2	-1	3	0	0	-0.8
CMe$_3$	-1	-1	-1	0	0	0	-1	0	0	1	0	0	0.1
F	2	4	3	2	2	2	2	2	3	4	4	4	3.0
Cl	8	6	6	5	6	6	6	6	5	5	4	3	4.5
CHO	5	8	8	10	10	10	10	10	8	6	6	5	7.2
CF$_3$	7	10	8	8	8	8	8	8	8	10	9	8	7.7
NO$_2$	13	14	14	14	13	14	14	14	12	13	14	13	11.9
CN	11	11	12	12	12	12	12	12	11	8	8	8	12.2
mean absolute error	1	1	1	1	1	1	1	1	0	2	1	1	–

a) energy of reaction: p-XC$_6$H$_4$CO$_2^-$ + C$_6$H$_5$CO$_2$H $\rightarrow$ p-XC$_6$H$_4$CO$_2$H + C$_6$H$_5$CO$_2^-$

Table 6-20: Acidities of *p*-Substituted Benzoic Acid Chromium Tricarbonyl Organometallics Relative to Benzoic Acid Chromium Tricarbonyl[a]

p-substituent,X	local density 6-31G*	BP 6-31G*	BLYP 6-31G*	EDF1 6-31G*	B3LYP 6-31G*	semi-empirical PM3	expt.[b]
Me	-1	-1	-1	-1	-1	0	-1
F	2	2	2	2	2	3	3
Cl	3	3	3	3	3	2	5
CF$_3$	5	5	5	6	6	6	8
CN	8	7	7	7	7	5	12

a) energy of reaction: p-XC$_6$H$_4$CO$_2^-$ + C$_6$H$_5$CO$_2$H $\longrightarrow$ p-XC$_6$H$_4$CO$_2$H + C$_6$H$_5$CO$_2^-$
     Cr(CO)$_3$ Cr(CO)$_3$ Cr(CO)$_3$ Cr(CO)$_3$

b) experimental data relate to reaction energies for "free" benzoic acids

Reaction Energies in Solution

As discussed in **Chapter 2**, treatment of molecular structure, energetics and properties in solvent lags well behind treatment of analogous quantities in the gas phase. At the present time, there are only two viable options. The first is to completely ignore solvation, that is, to formulate energy comparisons in such a way that solvent effects will largely cancel. The second is to make use of so-called reaction field models, such as the Cramer/Truhlar SM5.4 model[5], which are practical for routine application. Note that while such types of models may, in principle, be applied to any solvent, adequate data for their parameterization is presently limited to water. The SM5.4 model cannot be expected to account for absolute quantities, in particular, absolute solvation energies, although it is reasonable to expect that it will be able to account for relative solvation energies among closely-related systems.

Three different sets of experimental aqueous-phase pKa's allow us to judge to what extent solvent effects can be ignored and, where they cannot be ignored, assess the performance of the SM5.4 model in accounting for solvation. The first involves a diverse set of carboxylic acids[*] and the second a diverse series of alcohols and phenols[**]. Calculated acidities (relative to acetic acid in the case of carboxylic acids and relative to ethanol in the case of alcohols and phenols) have been obtained from the Hartree-Fock 6-311+G** model. Previous comparisons with gas-phase acidities suggest that this should be as satisfactory as any other model for this purpose (see, for example, **Tables 6-18** and **A6-50**). 6-31G* geometries have been used in place of 6-311+G** geometries in order to save computation time. (See

---

[*] pKa values in parentheses: 2,2,3,3,3-pentafluoropropanoic acid (-0.41), trichloroacetic acid (0.51), trifluoroacetic acid (0.52), nitroacetic acid (1.48), propiolic acid (1.89), pyruvic acid (2.39), 1-naphthoic acid (3.60), formic acid (3.75), 2-naphthoic acid (4.14), benzoic acid (4.20), acrylic acid (4.25), acetic acid (4.76), propanoic acid (4.87), cyclohexanecarboxylic acid (4.9), cyclopentanecarboxylic acid (4.99).

[**] pKa values in parentheses: 1-naphthol (9.39), 2-naphthol (9.59), phenol (10.0), 2,2,2-trifluoroethanol (12.4), propargyl alcohol (13.6), 2-chloroethanol (14.3), 2-methoxyethanol (14.8), benzyl alcohol (15.4), allyl alcohol (15.5), methanol (15.5), ethanol (15.9), 1-butanol (16.1), 1-propanol (16.1), 2-propanol (17.1), 2-butanol (17.6), *tert*-butyl alcohol (19.2).

Chapter 12 for a discussion of the consequences of using approximate geometries for energy calculations.)

Figure 6-12 compares calculated acidities for carboxylic acids (relative to the acidity of acetic acid) to experimental aqueous-phase pKa's. The overall correlation is good, and suggests a high level of cancellation of solvent effects in these closely-related systems.

A similar comparison is provided in **Figure 6-13** for acidities of alcohols, phenols and closely-related compounds. Here, the correlation between the gas-phase data (relative to ethanol) and the aqueous-phase pKa's is not as good as noted above for carboxylic acids, perhaps reflecting greater differences among the compounds. In particular, "sterically encumbered" molecules such as *tert*-butyl alcohol are found (in the gas-phase calculations) to be significantly stronger acids than they actually are in solution. Still in all, the gas-phase calculations generally reproduce the known trends in aqueous-phase acidities of alcohols.

In view of the above comparisons, it might be expected that solvation energy corrections to gas-phase acidities would be of little importance in the case of carboxylic acids, but of greater importance in the case of alcohols. **Figures 6-14** and **6-15** show this to be essentially true. Here, solvation energies obtained from the Cramer/Truhlar SM5.4 model have been added to the gas-phase energies of neutral and anionic species prior to calculation of acidities. The plot for carboxylic acids (**Figure 6-14**) is essentially identical to the previous (gas-phase) comparison (**Figure 6-12**) in terms of overall "quality of fit". Note, however, that solvation has significantly reduced the overall range of acidities. The plot for alcohols (**Figure 6-15**), however, shows a significant improvement in quality over the corresponding gas-phase plot (**Figure 6-13**). No longer are sterically-encumbered molecules as *tert*-butyl alcohol outliers.

A third comparison points out that there will be exceptions to the notion that solvent effects will cancel in comparisons among what appears to be very-closely-related systems, and that "gas-phase" calculations will lead to acceptable results. Among the most quoted

Figure 6-12: 6-311+G** vs. Experimental Aqueous-Phase Relative Acidities of Carboxylic Acids

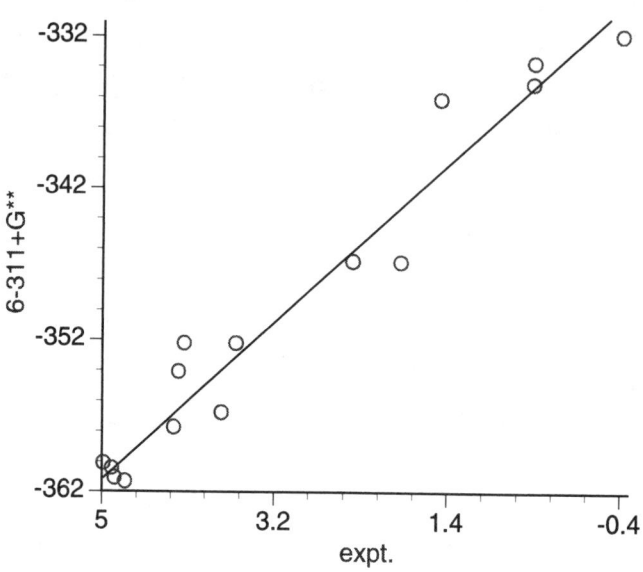

Figure 6-13: 6-311+G** vs. Experimental Aqueous-Phase Relative Acidities of Alcohols, Phenols and Related Compounds

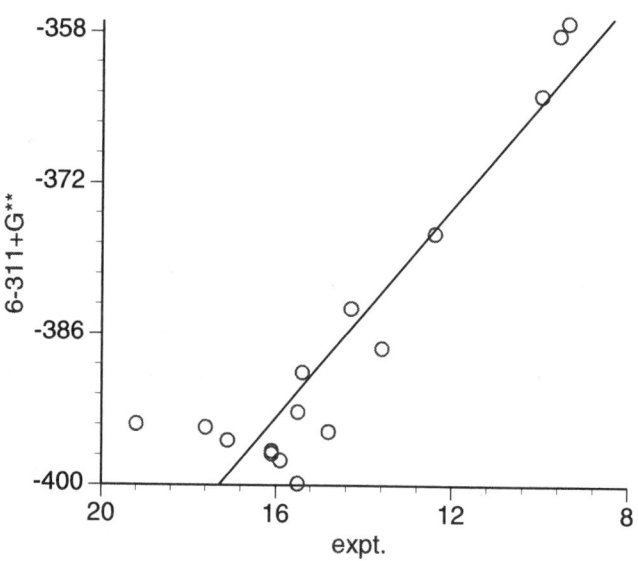

Figure 6-14: **Solvent-Corrected 6-311+G** vs. Experimental Aqueous-Phase Relative Acidities of Carboxylic Acids**

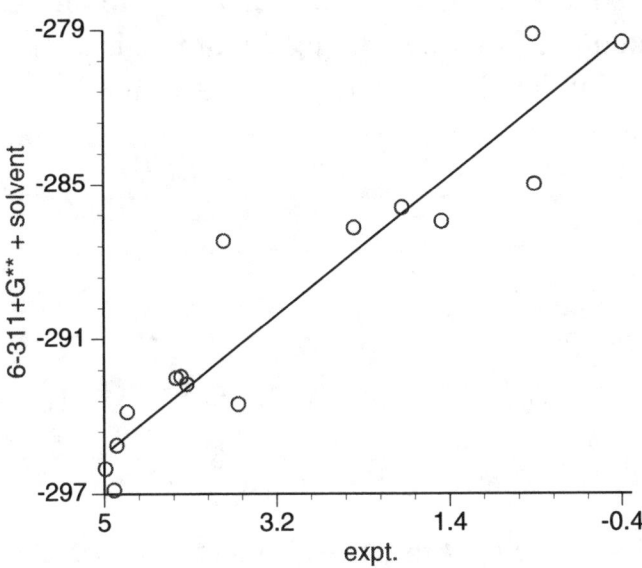

Figure 6-15: **Solvent-Corrected 6-311+G** vs. Experimental Aqueous-Phase Relative Acidities of Alcohols, Phenols and Related Compounds**

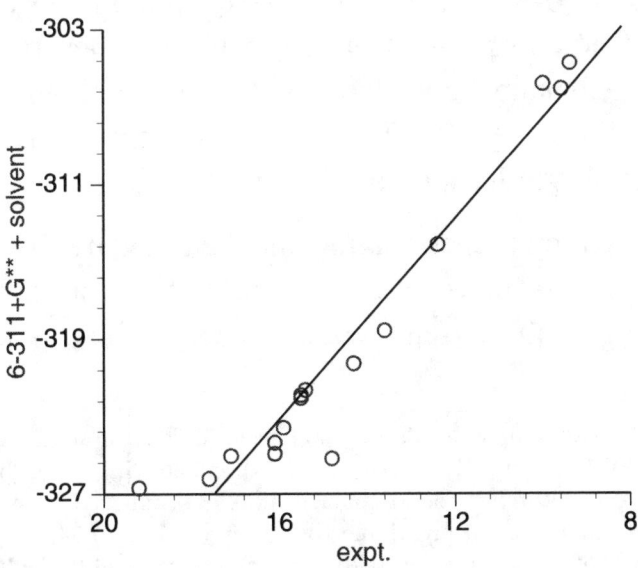

and simplest examples concerns the relative basicities of ammonia, methylamine, dimethylamine and trimethylamine. The experimental data clearly show that, while increased methyl substitution leads to a monotonic increase in gas-phase basicity of ammonia, the effect in water is markedly different. Here, methylamine is actually less basic than ammonia, while dimethylamine is only slightly more basic.

	ΔH for $B + NH_4^+ \rightarrow BH^+ + NH_3$	
	gas	aqueous
NH_3	0	0
$MeNH_2$	-9.1	0.7
Me_2NH	-15.5	-0.45
Me_3N	-20.0	-3.67

More generally, as seen in **Figure 6-16**, there is no correlation between calculated (Hartree-Fock 6-31G*) and experimental aqueous-phase basicities of amines.* (As shown earlier in this chapter, Hartree-Fock and other simple calculation models are quite successful in reproducing relative gas-phase basicities in amines. Therefore, a plot of measured gas-phase basicities vs. measured aqueous-phase basicities would be expected to show poor correlation.) On the other hand, calculated (6-31G*) relative basicities of amines corrected for the effects of aqueous solvation using the Cramer/Truhlar SM5.4 model shows reasonable correlation with the experimental (aqueous-phase) data (**Figure 6-17**). This further confirms that the simple solvation model is at least qualitatively correct.

Overall, the situation is encouraging, insofar as even simple treatments of solvation seem to correct, at least qualitatively, for the conspicuous limitations of gas-phase treatments.

---

* Experimental aqueous-phase enthalpies of protonation (relative to ammonia) are as follows: pyridine (-7.69), diazobicyclo [2.2.2] octane (-5.19), aniline (-5.10), trimethylamine (-3.67), triallylamine (-3.66), aziridine (-3.43), quinuclidine (-1.34), dimethylamine (-0.45), azetidine (0.10), diethylamine (0.25), piperidine (0.28), pyrrolidine (0.54), allylamine (0.58), diethylamine (0.68), methylamine (0.70), diisopropylamine (1.07), ethylamine (1.23), n-propylamine (1.36), isopropylamine (1.48), cyclohexylamine (1.82).

Figure 6-16: **6-31G* vs. Experimental Aqueous-Phase Relative Basicities of Amines**

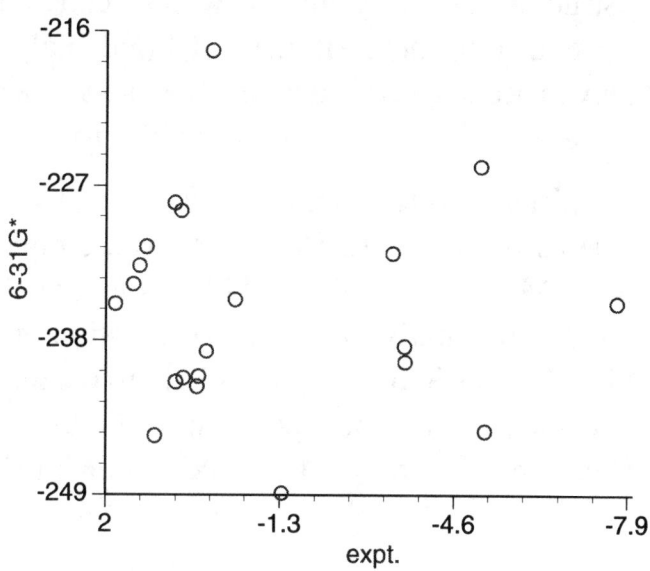

Figure 6-17: **Solvent-Corrected 6-31G* vs. Experimental Aqueous-Phase Relative Basicities of Amines**

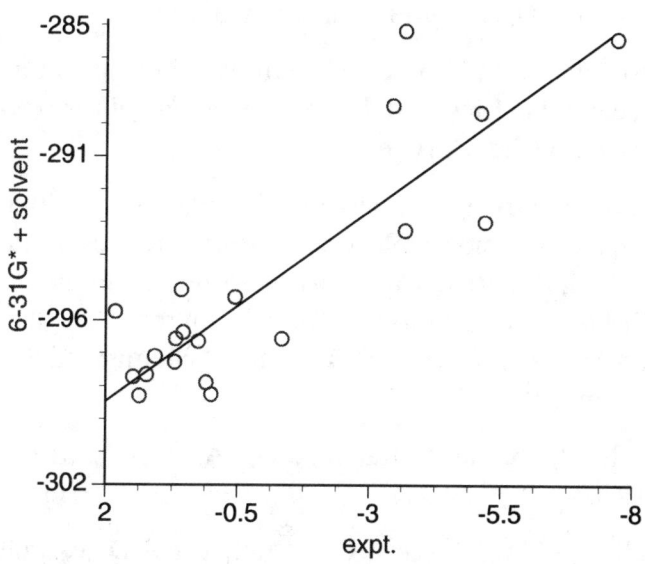

Pitfalls

There are many. Reaction energies are not as easy to calculate reliably as equilibrium structures. Clearly, models which account for electron correlation are required where significant bond making or bond breaking are likely to occur. Clearly, underlying basis sets which include diffuse functions are needed where anions are destroyed or created.

Clearly, semi-empirical models cannot be fully trusted for reaction energy comparisons of any kind. However, even simple correlated models such as the MP2/6-311+G** and B3LYP/6-311+G** models are able to provide quantitative accounts of reaction energies even where significant changes in bonding are likely to occur. Far simpler models, including the 6-31G* model perform admirably for *isodesmic* comparisons where a high degree of error cancellation is to be expected.

References

1. Compendia of experimental thermochemical data on neutral molecules: (a) S.W. Benson, F.R. Cruickshank, D.M. Golden, G.R. Haugen, H.E. O'Neal, A.S. Rogers, R. Shaw and R. Walsh, *Chem. Rev.*, **69**, 279 (1969); (b) J.D. Cox and G. Pilcher, ***Thermochemistry of Organic and Organometallic Compounds***, Academic Press, New York, 1970; (c) S.W. Benson, ***Thermochemical Kinetics***, 2nd Ed., Wiley, New York, 1976; (d) J.B. Pedley and J. Rylance, ***CATCH Tables***, University of Sussex, 1977.

2. For an up-to-date, on-line source of thermochemical data for ion-molecule reactions, see: S.G. Lias and J. Bartmess, ***Gas Phase Ion Chemistry*** "webbook.nist.gov/chemistry/ion".

3. G3 is a "recipe" involving a variety of different models with the purpose of providing accurate thermochemical data. Original reference : (a) L.A. Curtiss, K. Raghavachari, P.C. Redfern, V. Rassolov and J.A. Pople, *J. Chem. Phys.*, **109**, 7764 (1998). For an up-to-date, on-line source of G3 data see: (b) L.A. Curtiss, ***Computational Thermochemistry***, "chemistry.anl.gov/compmat/ comptherm.htm"

4. W.J. Hehre, R. Ditchfield, L. Radom and J.A. Pople, *J. Amer. Chem. Soc.*, **94**, 4796 (1970).

5. C.C. Chambers, G.D. Hawkins, C.J. Cramer and D.G. Truhlar, *J. Chem. Phys.*, **100**, 16385 (1996).

Chapter 7

Vibrational Frequencies and Thermodynamic Quantities

This chapter assesses the performance of quantum chemical models with regard to the calculation of vibrational frequencies, and describes the evaluation of thermodynamic quantities resulting from vibrational frequencies.

Introduction

As described in **Chapter 1**, potential energy surfaces provide a basis for understanding the relationship between molecular structure and stability. The one-dimensional potential energy surface, or "reaction coordinate" diagram as it is commonly known, is familiar to all chemists, with stable molecules corresponding to energy minima along the reaction coordinate and transition states corresponding to energy maxima. The problem is that such a diagram cannot be constructed (or at least cannot be visualized) for many-dimensional systems, i.e., beyond simple diatomic molecules. However, the underlying principle that stable molecules (energy minima) will be interconnected by smooth pathways passing through well-defined transition states remains the same. The only problem is to identify these "special points" (stable molecules and transition states) in the absence of a picture. Vibrational frequencies are key to this task.

The vibrational frequency of a diatomic molecule (a one-dimensional system) is proportional to the square root of force constant (the second derivative of the energy with respect to the interatomic distance) divided by the reduced mass (which depends on the masses of the two atoms).

$$\text{frequency} \propto \sqrt{\frac{\text{force constant}}{\text{reduced mass}}} \qquad (1)$$

This assumes that the molecule is at a stationary point[*] (the first derivative of the energy with respect to the interatomic distance is zero), which in the case of a diatomic molecule necessarily corresponds to the equilibrium structure.

The force constant gives the curvature of the potential energy surface in the vicinity of the stationary point. A small value means the surface is "shallow", while a large value means that it is "steep". Equally important is the sign of the force constant. A positive force constant corresponds to an energy minimum, and gives rise to a real frequency, while a negative force constant corresponds to an energy maximum and gives rise to an imaginary frequency[**] (which cannot be measured).

The many-dimensional case is complicated by the fact that each geometrical coordinate is associated, not with a single force constant, but rather with a set of force constants corresponding to all possible second energy derivatives involving this coordinate and all other coordinates. At first glance, it would appear that it is not possible to determine whether a particular stationary point corresponds to an energy minimum, an energy maximum or something in between. However, it is possible to replace the original geometrical coordinates (bond lengths, angles, etc.) by a new set of coordinates (normal coordinates) which guarantee that only the "diagonal" force constants involving these (new) coordinates will be non-zero. By examining the sign of the (diagonal) force constant associated with each normal coordinate (or the corresponding vibrational frequency), it is possible, therefore, to fully characterize a particular stationary point. Stable molecules will be characterized by real frequencies. It is postulated that reaction transition states will possess one (and only one) imaginary

---

* This follows from an expansion of the energy in terms of a Taylor series:

$$E = E^\circ + E' + E'' + \text{higher-order terms}$$

E° is a constant and E' is zero if the structure is at a stationary point (an energy minimum or a transition state). If this is not the case, interpretation of E'' in terms of a frequency will be meaningless. Higher-order terms are assumed to be unimportant, but in fact lead to systematic discrepancies between calculated and measured frequencies on the order of 5%. For small molecules, it is possible to factor out these higher-order terms leading to so-called "harmonic" frequencies, which in turn are directly comparable to calculated results.

** The reduced mass is necessarily a positive number and the sign of the quantity inside the square root depends on the sign of the force constant.

frequency. The geometrical coordinate corresponding to the imaginary frequency is the reaction coordinate. This is the assumption behind what is commonly referred to as transition-state theory.

In addition to their role in characterizing structures on a potential energy surface, vibrational frequencies, along with molecular geometry, are the essential ingredients for calculation of a number of thermodynamic quantities. These follow from straightforward application of statistical mechanics.[1] Entropy is certainly the most important of these, primarily for its contribution to the free energy. Also important are the zero-point energy and the change in enthalpy with temperature, quantities which are needed if the results of calculations, pertaining to stationary (non-vibrating) molecules at 0K, are to be related to real laboratory measurements.

Except for very low values ($< 600 \text{ cm}^{-1}$)[*], frequencies can normally be measured to high precision ($< 5 \text{ cm}^{-1}$) using infrared or Raman spectroscopy.[2] Similar or better precision is available for frequencies calculated analytically (Hartree-Fock, density functional and semi-empirical models), but somewhat lower precision results where numerical differentiation is required (MP2 models).

Diatomic Molecules

Calculated vibrational frequencies for diatomic molecules containing first and/or second-row elements only are compared with experimental values in **Table 7-1**. The usual theoretical models, excluding molecular mechanics models, have been examined. Where harmonic frequencies[**] are available, these have also been tabulated.

Aside from the STO-3G model, all Hartree-Fock models lead to similar results. Except where highly electropositive alkali metals are involved, calculated frequencies are consistently larger than measured values, typically by 10-12% (5-6% if comparisons are made instead to

---

[*] frequencies and errors in frequencies will be reported in cm^{-1}.

[**] As previously commented, harmonic frequencies are measured values which have been corrected to remove non-quadratic components. They directly correspond to calculated frequencies. The corrections require data on isotopically-substituted systems and are typically available only for small molecules.

255

Table 7-1: Vibrational Frequencies for Diatomic Molecules

molecule	Hartree-Fock				local density		BP		expt.	
	STO-3G	3-21G	6-31G*	6-311+G**	6-31G*	6-311+G**	6-31G*	6-311+G**	harmonic	measured
H_2	5481	4662	4649	4589	4230	4208	4335	4313	4401	4460
LiH	1866	1426	1450	1432	1365	1381	1356	1369	1406	1360
HF	4475	4065	4357	4490	3927	4010	3874	3976	4139	3962
NaH	2174	1168	1322	1180	1182	1172	1146	1135	–	1172
HCl	3373	3152	3185	3146	2888	2891	2867	2872	–	2890
Li_2	409	340	341	337	336	342	330	331	351	346
LiF	1298	1082	1031	927	1014	907	965	864	914	898
LiCl	806	625	629	643	637	651	615	623	–	641
CO	2643	2315	2240	2431	2175	2187	2115	2122	2170	2143
CS	1577	1418	1426	1432	1274	1296	1236	1254	–	1285
N_2	2670	2613	2758	2734	2408	2405	2349	2342	2360	2331
F_2	1677	1297	1245	1224	1090	995	1014	923	923	891
FCl	1033	884	914	873	805	750	746	698	–	784
NaF	916	680	593	536	572	535	536	505	–	536
NaCl	532	351	359	356	359	359	338	334	–	363
Cl_2	695	564	599	604	540	530	505	500	–	560
mean absolute error	438	131	159	147	51	40	44	37	36	–

Table 7-1: Vibrational Frequencies for Diatomic Molecules (2)

molecule	BLYP		EDF1		B3LYP		MP2		semi-empirical			expt.	
	6-31G*	6-311+G**	6-31G*	6-311+G**	6-31G*	6-311+G**	6-31G*	6-311+G**	MNDO	AM1	PM3	harmonic	measured
H_2	4369	4340	4404	4371	4448	4414	4533	4528	4293	4341	4478	4401	4460
LiH	1371	1382	1361	1372	1400	1409	1395	1437	1466	-	1476	1406	1360
HF	3810	3940	3902	4024	3978	4098	4041	4198	4594	4459	4352	4139	3962
NaH	1153	1142	1139	1126	1188	1177	1182	1186	-	-	1004	-	1172
HCl	2832	2840	2909	2911	2937	2934	3049	3087	3162	2658	2704	-	2890
Li_2	333	333	327	326	343	343	341	342	436	-	288	351	346
LiF	971	869	966	865	1003	894	999	879	928	-	984	914	898
LiCl	620	626	619	627	630	639	636	660	550	-	573	-	641
CO	2106	2113	2133	2137	2209	2212	2119	2124	2383	2268	2306	2170	2143
CS	1220	1239	1250	1267	1292	1308	1311	1324	1365	946	1138	-	1285
N_2	2337	2328	2372	2362	2458	2445	2175	2176	2739	2744	2640	2360	2331
F_2	987	893	1011	922	1064	984	1008	899	1694	1365	1323	923	891
FCl	717	670	750	700	780	733	801	733	1174	1082	825	-	784
NaF	539	504	531	498	564	520	581	521	-	-	381	-	536
NaCl	339	333	334	330	350	345	371	367	-	-	307	-	363
Cl_2	483	478	507	502	519	515	548	546	766	794	596	-	560
mean absolute error	51	41	38	37	44	45	55	52	270	302	146	36	-

257

harmonic frequencies). The reason for this behavior is directly related to the fact that bond distances from Hartree-Fock models are consistently shorter than experimental values (except where highly electropositive elements are involved). As pointed out in **Chapter 5**, electron promotion from occupied molecular orbitals, which inevitably are bonding in character, into unoccupied molecular orbitals, which inevitably are antibonding in character, as implicit in electron correlation schemes, necessarily leads to bond lengthening and to a decrease in force constant. Thus, frequencies from correlated models, which in their limit must approach experimental (harmonic) frequencies, must be smaller than frequencies from Hartree-Fock models, meaning that Hartree-Fock frequencies must be larger than experimental values. The situation is less clear where elements such as lithium and sodium are involved. Here, the lowest-energy unoccupied molecular orbitals are likely to be bonding in character, and electron promotion from occupied to unoccupied orbitals will not necessarily lead to decrease in force constant. Indeed, Hartree-Fock frequencies for alkali metal compounds are (in their limit) sometimes less (or only marginally larger) than experimental values.

On the basis of mean absolute errors, all density functional models (including local density models) and MP2 models are superior to Hartree-Fock models in reproducing experimental frequencies for diatomic molecules. All models exhibit markedly similar behavior (as judged from mean absolute errors), and yield individual frequencies which are both larger and smaller than experimental values. It is interesting to note that the magnitudes of the error between calculated and measured frequencies (ranging from 37 to 55 cm^{-1} depending on the model) are very similar to the difference between measured and harmonic frequencies (36 cm^{-1}).

Semi-empirical models generally turn in a poor account of vibrational frequencies in these systems, both in terms of mean absolute errors and for individual systems. In part this reflects their (poor) performance for the geometries of many diatomic molecules.

Main-Group Hydrides

Calculated vibrational frequencies for main-group hydrides containing one first or second-row element are provided in **Appendix A7 (Tables A7-1** to **A7-8)**, and compared both with experimentally measured values and, where available, with "harmonic" experimental frequencies. The same theoretical models considered for diatomic molecules are also examined here. A summary of mean absolute errors for symmetric stretching frequencies (only) is provided in **Table 7-2**.

The performance of Hartree-Fock models here closely parallels their performance for diatomic molecules. Frequencies are nearly always larger than experimental values, typically by 10-12%. This appears to apply not only to stretching frequencies, but also to frequencies associated with bending motions.

In terms of both mean absolute error (in symmetric stretching frequencies) and of individual frequencies, density functional models perform significantly better than Hartree-Fock models. As with diatomic molecules, local density models appear to provide the best overall account, but the performance of the other models (except for B3LYP models) is not much different. B3LYP models and MP2 models do not "appear" to fare as well in their descriptions of frequencies in one-heavy-atom hydrides[*], and the performance of each "appears" to worsen in moving from the 6-31G* to the 6-311+G** basis set.

As with diatomic molecules, the performance of semi-empirical models in dealing with frequencies in one-heavy-atom hydrides is very poor. These techniques are not to be trusted for this purpose.

---

[*] While experimental harmonic frequencies are limited, there are sufficient data to suggest that were they used instead of measured frequencies, local density models would fare worse and MP2 models would fare better. For example, mean absolute errors (based on limited data) for the local density 6-311+G** model is 121 cm^{-1} while that for the MP2/6-311+G** model is 49cm^{-1}.

Table 7-2: Mean Absolute Errors in Symmetric Stretching Frequencies for One-Heavy-Atom Hydrides

		STO-3G	3-21G	6-31G*	6-311+G**
MNDO	344				
AM1	246				
PM3	370				
Hartree-Fock		590	183	229	260
local density		—	—	22	28
BP		—	—	30	23
BLYP		—	—	53	28
EDF1		—	—	36	46
B3LYP		—	—	55	83
MP2		—	—	113	166

CH₃X Molecules

CH₃X molecules provide an excellent opportunity to assess the ability of the calculations both to reproduce gross trends in measured vibrational frequencies, for example, trends in CX stretching frequencies, as well as to account for what are presumed to be subtle differences associated with the methyl rotor with change in X. Data are provided in **Appendix A7 (Tables A7-9** to **A7-16)** for the usual collection of theoretical models. The reader can easily verify that the same comments made for diatomic molecules and for one-heavy-atom, main-group hydrides generally apply here as well.

Calculated CX stretching frequencies for these compounds (repeating the data in **Appendix A7**) are provided in **Table 7-3** and compared to measured values. As expected, "limiting" (6-311+G** basis set) Hartree-Fock frequencies are all larger than experimental values. In fact, with the sole exception of methyl chloride at the 3-21G level, Hartree-Fock frequencies are always larger than experimental frequencies, irrespective of choice of basis set.

Frequencies from "limiting" local density calculations are also consistently larger than experimental values, the only exception being for methanol where the calculated CO stretching frequency is slightly smaller than the measured value. The similar behavior of local density and Hartree-Fock calculations with regard to stretching frequencies is consistent with what has been previously noted for equilibrium structures (see discussion in **Chapter 5**). On the other hand, frequencies from "limiting" BP, BLYP and EDF1 density functional calculations and "limiting" MP2 calculations are always smaller than measured values, while "limiting" B3LYP density functional calculations, which turn in the best performance in terms of mean absolute error, yield frequencies which are both higher and lower than experimental values. All of these models produce a satisfactory account of changes in CX stretching frequencies.

MNDO and AM1 semi-empirical models are not successful in reproducing experimental CX stretching frequencies. The PM3 model provides a much better account, and is the method of choice if semi-empirical models need to be utilized.

Table 7-3: CX Stretching Frequencies for CH₃X Molecules

molecule	Hartree-Fock				local density		BP		BLYP	
	STO-3G	3-21G	6-31G*	6-311+G**	6-31G*	6-311+G**	6-31G*	6-311+G**	6-31G*	6-311+G**
CH₃CH₃	1194	1002	1060	1052	1042	1028	995	980	1011	959
CH₃NH₂	1260	1096	1149	1138	1101	1090	1045	1031	1023	1005
CH₃OH	1208	1090	1164	1148	1047	1022	1031	1004	1011	977
CH₃F	1355	1141	1187	1156	1138	1075	1063	1000	1037	965
CH₃SiH₃	870	738	725	723	704	705	675	676	664	665
CH₃SH	953	744	776	764	721	714	685	677	650	651
CH₃Cl	936	716	782	773	732	751	710	699	678	665
mean absolute error	216	43	83	70	32	21	12	28	31	55

molecule	EDF1		B3LYP		MP2		semi-empirical			expt.
	6-31G*	6-311+G**	6-31G*	6-311+G**	6-31G*	6-311+G**	MNDO	AM1	PM3	
CH₃CH₃	1001	989	1011	996	1050	1033	1217	1248	1138	**995**
CH₃NH₂	1056	1041	1071	1056	1101	1089	1331	1350	1103	**1044**
CH₃OH	1043	1018	1067	1042	1083	1076	1439	1362	1164	**1033**
CH₃F	1069	1007	1091	1033	1106	1076	1480	1402	1203	**1049**
CH₃SiH₃	677	678	688	689	716	717	815	677	647	**700**
CH₃SH	693	685	702	693	759	754	934	754	753	**708**
CH₃Cl	716	707	721	710	785	785	942	837	678	**732**
mean absolute error	15	19	21	14	49	38	277	202	91	–

As seen from the data in **Table 7-4** (abstracted from frequencies provided in **Appendix A7**) symmetric methyl group CH stretching frequencies change with substitution. The smallest value is for methylamine (chosen as the reference compound) and the largest is for methyl chloride. (Ethane has been excluded from this comparison as the symmetric stretch here involves all six hydrogens.)

Except for the STO-3G model, Hartree-Fock models provide a reasonable account insofar as properly ordering the symmetric stretching frequencies. Mean absolute errors (20 to 26 cm^{-1}) are, however, quite large in view of the small range of frequencies (experimentally 117 cm^{-1}). Density functional models lead to lower mean absolute errors but only if the 6-311+G** basis set is used instead of 6-31G*. Otherwise the stretching frequency for methylsilane in particular is much larger than is observed. MP2 models lead to significantly larger mean absolute errors, but actually provide a fairly uniform account of symmetric stretching frequencies in these compounds.

Semi-empirical models are completely unsatisfactory. The MNDO model performs worst and the PM3 model performs best (paralleling the behavior that was previously noted in other frequency comparisons), but none is successful in properly ordering the frequencies.

Characteristic Frequencies

One of the main "routine" uses of infrared spectroscopy is identification of specific functional groups present in an "unknown" molecule and, as a result, further characterization of the unknown. By far the most common example involves the carbonyl group. Location of a strong band in the infrared in the vicinity of 1730cm^{-1} is almost certain "proof" that carbonyl functionality is present. This confidence is based on the fact that the "characteristic frequency" (the CO stretch in this case) is "isolated," that is to say, it is sufficiently far removed from the other bands in the infrared spectrum to not be confused with them. It also assumes that carbonyl groups in different chemical environments will exhibit "similar" characteristic

Table 7-4: Symmetric Methyl Group CH Stretching Frequencies in CH_3X Molecules[a]

molecule	Hartree-Fock				local density		BP		BLYP	
	STO-3G	3-21G	6-31G*	6-311+G**	6-31G*	6-311+G**	6-31G*	6-311+G**	6-31G*	6-311+G**
CH_3NH_2	3540	3135	3158	3122	2889	2888	2878	2881	2876	2878
CH_3OH	-23	44	28	45	24	25	26	26	29	33
CH_3SiH_3	23	39	38	34	117	95	186	85	100	80
CH_3F	-29	94	75	71	46	66	68	70	71	77
CH_3SH	14	89	80	74	120	99	117	98	119	99
CH_3Cl	29	122	106	102	133	118	135	115	140	119
mean absolute error	85	21	20	24	22	15	35	13	19	12

molecule	EDF1		B3LYP		MP2		semi-empirical			expt.
	6-31G*	6-311+G**	6-31G*	6-311+G**	6-31G*	6-311+G**	MNDO	AM1	PM3	
CH_3NH_2	2917	2917	2971	2962	3060	3041	3316	3104	3140	**2820**
CH_3OH	25	27	26	26	16	14	-16	45	1	**24**
CH_3SiH_3	102	84	83	67	54	40	-17	67	91	**78**
CH_3F	66	69	66	70	55	52	-18	18	1	**110**
CH_3SH	114	97	104	88	73	57	4	63	65	**111**
CH_3Cl	132	113	123	108	90	75	6	46	54	**117**
mean absolute error	18	12	12	15	30	41	97	60	51	–

a) relative to methylamine (except for methylamine)

frequencies. In fact, subtle differences among frequencies can be employed to elucidate differences in chemical environments.

To what extent are calculations able to reproduce the uniformity of calculated frequencies such as the C=O stretching frequency? To what extent are they able to reproduce the subtle changes which follow changes in chemical environment? Two series of calculations provide limited assessment. The first involves C=C stretching frequencies (relative to the stretching frequency in ethylene), and the second involves C=O stretching frequencies (relative to the stretching frequency in acetone). Data for Hartree-Fock calculations, with the STO-3G, 3-21G, 6-31G* and 6-311+G** basis sets, local density calculations, BP, BLYP, EDF1 and B3LYP calculations, and MP2 calculations all with the 6-31G* and 6-311+G** basis sets and MNDO, AM1 and PM3 calculations are provided in **Appendix A7** (**Tables A7-17** to **A7-24** for C=C stretching frequencies and **Tables A7-25** to **A7-32** for C=O stretching frequencies). Summaries of mean absolute errors in stretching frequencies (referred to the appropriate standards) are provided in **Tables 7-5** and **7-6** for the two classes of compounds.

C=C stretching frequencies experimentally range from 1570 cm^{-1} in cyclobutene to 1872 cm^{-1} in tetrafluoroethylene (see appropriate tables in **Appendix A7**). All levels of calculation reproduce the basic trend in frequencies but, on the basis of mean absolute errors, show widely different performance (**Table 7-5**). Local density and MP2 models with the 6-311+G** basis set perform best and semi-empirical models and density functional models (except the B3LYP model) with the 6-31G* basis set perform worst. Hartree-Fock models with the 3-21G and larger basis sets also turn in good performance.

Similar comments apply to C=O stretching frequencies, except that here density functional models perform better than Hartree-Fock models (the reverse of the trend noted for C=C stretching frequencies). In terms of absolute values, errors are of comparable magnitude to those noted for C=C stretching frequencies. Note, however, that the overall range of values is much smaller (only 64 cm^{-1} compared to 302 cm^{-1} for C=C stretching frequencies), meaning that in terms of percentage differences, the errors are much larger.

Table 7-5: **Mean Absolute Errors in C=C Stretching Frequencies Relative to Ethylene**

MNDO	23			
AM1	36			
PM3	23			

	STO-3G	3-21G	6-31G*	6-311+G**
Hartree-Fock	31	17	15	20
local density	—	—	12	10
BP	—	—	25	18
BLYP	—	—	34	26
EDF1	—	—	32	18
B3LYP	—	—	20	13
MP2	—	—	14	10

Table 7-6: **Mean Absolute Errors in C=O Stretching Frequencies Relative to Acetone**

MNDO	36			
AM1	34			
PM3	34			

	STO-3G	3-21G	6-31G*	6-311+G**
Hartree-Fock	41	33	18	15
local density	—	—	12	12
BP	—	—	12	12
BLYP	—	—	14	14
EDF1	—	—	13	13
B3LYP	—	—	15	14
MP2	—	—	9	9

Infrared and Raman Intensities

In principle, infrared and Raman intensities are both measurable and calculable quantities, allowing the performance of the theoretical models to be assessed. In practice, however, the experimental data are qualitative at best and typically limited to such descriptors as "very weak", "strong", etc.. This makes comparison with the results of calculation difficult, and we defer any attempt to do this to a future time.

Thermodynamic Quantities

Calculated vibrational frequencies, along with calculated equilibrium geometries, may be employed to yield a variety of thermodynamic quantities. The most important of these from the present perspective are associated with bringing energetic data obtained from calculation into juxtaposition with that obtained in a real experiment. The former are energies of non-vibrating molecules at 0K, while the latter are free energies at some finite temperature. Standard thermodynamic relationships provide necessary connections:

$$\Delta G = \Delta H - T\Delta S \tag{2}$$

$$\Delta H = \Delta E + P\Delta V \tag{3}$$

G is the free energy, H is the enthalpy, S is the entropy, E is the energy, and T, P and V are the temperature, pressure and volume.

Entropy

The absolute entropy may be written as a sum of terms:

$$S = S_{tr} + S_{rot} + S_{vib} + S_{el} - nR \left[\ln (nN_0) - 1 \right] \tag{4}$$

$$S_{tr} = nR \left\{ \frac{3}{2} + \ln \left[\left(\frac{2\pi MkT}{2} \right)^{3/2} \left(\frac{nRT}{P} \right) \right] \right\} \tag{5}$$

$$S_{rot} = nR \left\{ \frac{3}{2} + \ln \left[\frac{(\pi V_A V_B V_C)^{1/2}}{s} \right] \right\} \tag{6}$$

$$S_{vib} = nR \sum_i \left\{ (u_i e^{u_i} - 1)^{-1} - \ln (1 - e^{-u_i}) \right\} \tag{7}$$

$$S_{el} = nR \ln \omega_{el} \qquad (8)$$

where

$$\nu_A = h^2/8\pi I_A kT, \ \nu_B = h^2/8\pi I_B kT, \ \nu_C = h^2/8\pi I_C kT \qquad (9)$$

$$\mu_i = h\nu_i/kT \qquad (10)$$

n is the number of moles, M is the molecular mass, I_A, I_B, I_C are the principal moments of inertia, s is the symmetry number, ν_i are the vibrational frequencies and ω_{el} is the degeneracy of the electronic ground state. R, k and h are the gas constant, Boltzmann's constant and Planck's constant, respectively and N_0 is Avogadro's number.

Molecular structure enters into the rotational entropy component, and vibrational frequencies into the vibrational entropy component. The translational entropy component cancels in a (mass) balanced reaction, and the electronic component is most commonly zero. Note that the vibrational contribution to the entropy goes to ∞ as ν goes to 0. This is a consequence of the linear harmonic oscillator approximation used to derive equation 7, and is inappropriate. Vibrational entropy contributions from frequencies below 300 cm^{-1} should be treated with caution.

Correction for Non-Zero Temperature

The change in enthalpy from 0K to a finite temperature (T) is given by:

$$\Delta H(T) = H_{trans}(T) + H_{rot}(T) + \Delta H_{vib}(T) + RT \qquad (11)$$

$$H_{trans}(T) = \frac{3}{2} RT \qquad (12)$$

$$H_{rot}(T) = \frac{3}{2} RT \ (RT \text{ for a linear molecule}) \qquad (13)$$

$$\Delta H_{vib}(T) = H_{vib}(T) - H_{vib}(0) = Nh \sum_i^{\text{normal modes}} \frac{\nu_i}{(e^{h\nu_i/kT} - 1)} \qquad (14)$$

This requires knowledge of the vibrational frequencies, the same information as required for evaluation of the vibrational contribution to the entropy.

Correction for Zero-Point Vibrational Energy

The residual vibrational energy of a molecule at 0K is given by:

$$H_{vib}(0) = \varepsilon_{zero\text{-}point} = \frac{1}{2} h \sum_{i}^{normal\ modes} v_i \qquad (15)$$

This requires knowledge of the vibrational frequencies, the same information that is needed for entropy calculation and correction of the enthalpy for finite temperature.

Pitfalls

Frequency calculation, while costly, is now routine. Except for very low frequencies (where experimental data are often uncertain) even procedures which require numerical differentiation are sufficiently precise. The only pitfall is to make certain that the structure is at a stationary point, that is, an energy minimum or a saddle point.

References

1. See for example: D.A. McQuarrie, *Statistical Mechanics*, Harper and Row, New York, 1976.

2. Experimental compendia of vibrational frequencies include: (a) T. Shinanouchi, *Tables of Molecular Vibrational Frequencies. Consolidated Volume I*, NSRDS-NBS 39, National Bureau of Standards, Washington, D.C., 1972, and following volumes in this series; (b) K.P. Huber and G Herzberg, *Molecular Spectra and Molecular Structure. IV. Constants for Diatomic Molecules*, Van Nostrand Reinhold, New York, 1979; (c) M.W. Chase, Jr., *NIST-JANAF Thermochemical Tables*, 4th Ed., National Institute of Standards and Techonology, Washington, D.C., 1998.

Chapter 8
Equilibrium Conformations

This chapter assesses the ability of molecular mechanics and quantum chemical models to properly assign preferred conformation, and to account quantitatively for differences in conformer energy as well as for barriers to rotation and inversion. The chapter ends with a discussion of ring inversion in cyclohexane.

Introduction

More than any other factors, single-bond conformation and ring conformation dictate overall molecular size and shape. Thus, proper assignment of ground-state conformation is a very important task for calculation.

> The importance of single-bond conformation is never more apparent than for polypeptides. Here, distinct local domains involving α-helices and β-sheets (among other structures) occur commonly, and these in turn dictate overall (tertiary) structure of proteins and ultimately protein function. Interestingly, proteins appear to exhibit well-defined shapes, that is, exist as a single conformer or a very few closely-related conformers. This is the reason that they can be crystallized and their structures determined, and is certainly a major factor behind the ability of proteins to direct specific chemical reactions.

This chapter assesses the ability of both molecular mechanics and quantum chemical models to correctly assign the lowest-energy conformational arrangements in flexible molecules as well as account for energy differences between alternative conformers. It also assesses the performance of different models with regard to the calculation of barriers to single-bond rotation and pyramidal inversion.

Experimental data on conformational energy differences derive mainly from abundance measurements on equilibrium mixtures (containing different conformers). The obvious difficulty is that a particular conformer must be sufficiently abundant in order to be detected, which in practice means that conformers need to be separated by no more than a few kcal/mol.[*] Another problem is that different conformers are likely to possess very similar (or nearly identical) measurable properties, and it may be difficult to distinguish one from another in an equilibrium mixture. All in all, "high-quality" experimental data exist for perhaps 100 molecules, most of them very simple organic molecules.[1] The overall uncertainty in these data (conformational energy differences) is on the order of a few tenths of a kcal/mol, although conformational energy differences in some individual systems are more accurately known.

Barriers to single-bond rotation and pyramidal inversion derive principally from microwave spectroscopy, from vibrational spectroscopy in the far infrared and (for the larger barriers) from NMR. Although the number of systems for which data are available is limited (and the systems themselves primarily limited to very small molecules), in some cases barriers are known to high accuracy (to within 0.1 kcal/mol).

While experimental data on conformational energy differences is limited, much more is known about the preferred conformations (shapes) of molecules. Information from gas-phase experiments (primarily microwave spectroscopy and electron diffraction) relates directly to isolated molecules, albeit small molecules. Much more abundant are data on crystalline solids from X-ray diffraction, including data on larger molecules.[2] However, these data need to be carefully interpreted. Conformational energy differences are of the same order of magnitude as crystal packing energies and molecular shape may change from the gas to the solid.

It might be anticipated that computational models would provide good accounts of conformational energy differences and rotation/inversion

---

[*]   All conformational energy differences and errors in conformational energy differences will be reported in kcal/mol.

barriers. Molecular mechanics models in general, and the MMFF molecular mechanics model, in particular, have been specifically parameterized to reproduce the known conformations of small organic molecules. Conformational changes are (extreme) examples of *isodesmic* reactions and, aside from semi-empirical models, quantum chemical models might be expected to provide accurate energy differences. However, it needs to be recognized that conformational energy differences are typically very small (on the order of 1-5 kcal/mol), and even small errors might lead to incorrect assignment of lowest-energy conformer.

Conformational Energy Differences in Acyclic Molecules

A comparison of calculated and experimentally measured conformational energy differences for a small selection of "single-rotor" acyclic systems is provided in **Table 8-1**. The experimental data for some systems are subject to large uncertainties, and too much weight should not be placed on quantitative comparisons.

SYBYL molecular mechanics is completely unsatisfactory for describing conformational energy differences in acyclic systems, and should not be employed for this purpose. On the other hand, the MMFF mechanics model provides a good account of all systems examined.* In fact, the performance of MMFF is significantly better than any of the semi-empirical models, and in the same league as the best of the Hartree-Fock, local density, density functional and MP2 models (see discussion following).

Except for systems where the difference in energy between the conformers is very small, even the STO-3G Hartree-Fock model properly assigns ground-state conformation. However, conformational energy differences from STO-3G calculations show large errors in some cases. Results from 3-21G calculations are generally even worse, and the simplest Hartree-Fock model to provide a reliable (and for the most part quantitative) account of conformational energy differences is the 6-31G* model. Except for formic acid and methyl

---

*   It should be noted that many of the molecules used in the conformation energy comparisons presented here have been drawn from the "training set" used to determine MMFF parameters.

273

Table 8-1: Conformational Energy Differences in Acyclic Molecules

molecule	low-energy/high-energy conformer	molecular mechanics		Hartree Fock				local density		expt.
		SYBYL	MMFF	STO-3G	3-21G	6-31G*	6-311+G**	6-31G*	6-311+G**	
n-butane	*trans/gauche*	0.6	0.8	0.8	0.8	1.0	1.0	0.5	0.4	**0.67**
1-butene	*skew/cis*	1.0	0.3	0.9	0.8	0.7	0.7	-0.6	-0.7	**0.22**
1,3-butadiene	*trans/gauche*	1.0	2.5	1.8	2.7	3.1	3.2	3.7	3.5	**2.89**
acrolein	*trans/cis*	0.0	2.0	0.4	0.0	1.7	2.0	1.7	2.2	**1.70**
N-methylformamide	*trans/cis*	-0.1	1.3	0.3	1.6	1.1	1.1	1.4	1.5	**1.4**
N-methylacetamide	*trans/cis*	-0.5	2.6	2.4	3.8	3.0	2.9	2.3	2.2	**2.3**
formic acid	*cis/trans*	-0.4	4.9	4.4	7.3	6.1	5.4	5.3	4.6	**3.90**
methyl formate	*cis/trans*	-0.3	5.3	3.9	7.1	6.2	6.0	5.6	5.6	**4.75**
methyl acetate	*cis/trans*	4.0	8.3	9.4	9.7	9.4	10.8	8.2	8.2	**8.5**
propanal	*eclipsed/anti*	-0.1	0.5	0.1	1.8	1.1	0.8	1.9	1.7	**0.67**
2-methylpropanal	*eclipsed/anti*	-0.2	0.6	0.0	1.5	0.8	0.4	1.4	0.9	**0.78**
1,2-difluoroethane	*gauche/anti*	0.0	0.6	0.3	-0.8	-0.5	0.2	1.3	1.8	**0.56**
1,2-dichloroethane	*anti/gauche*	0.0	1.2	1.4	1.8	1.9	1.9	1.2	0.9	**1.08**
ethanol	*anti/gauche*	0.0	0.2	-0.2	-0.1	0.1	0.3	-0.5	-0.2	**0.12**
methyl ethyl ether	*anti/gauche*	0.5	1.5	1.3	1.0	1.7	1.8	0.9	1.1	**1.5**
methyl vinyl ether	*cis/skew*	-3.1	2.2	1.0	3.5	2.0	1.8	3.8	3.4	**1.7**
mean absolute error		2.1	0.3	0.6	1.1	0.6	0.6	0.6	0.6	–

Table 8-1: Conformational Energy Differences in Acyclic Molecules (2)

molecule	low-energy/ high-energy conformer	BP		BLYP		EDF1		expt.
		6-31G*	6-311+G**	6-31G*	6-311+G**	6-31G*	6-311+G**	
n-butane	trans/gauche	0.8	0.8	0.8	0.9	0.8	1.0	0.67
1-butene	skew/cis	0.4	0.3	0.6	0.6	0.8	0.7	0.22
1,3-butadiene	trans/gauche	3.7	3.6	3.7	3.7	4.0	3.8	2.89
acrolein	trans/cis	1.8	2.2	1.7	2.2	1.8	2.3	1.70
N-methylformamide	trans/cis	0.8	1.0	0.8	1.0	0.7	0.8	1.4
N-methylacetamide	trans/cis	2.7	2.3	2.7	2.3	2.5	2.4	2.3
formic acid	cis/trans	5.0	4.3	4.9	4.2	4.9	4.2	3.90
methyl formate	cis/trans	5.0	5.0	4.9	4.9	4.9	4.7	4.75
methyl acetate	cis/trans	7.2	8.4	7.3	8.5	8.7	8.6	8.5
propanal	eclipsed/anti	1.2	1.0	1.1	0.8	1.0	0.8	0.67
2-methylpropanal	eclipsed/anti	0.8	0.4	0.7	0.3	0.6	0.2	0.78
1,2-difluoroethane	gauche/anti	0.6	1.1	0.6	0.9	0.5	1.0	0.56
1,2-dichloroethane	anti/gauche	1.7	1.4	1.8	1.7	1.7	1.6	1.08
ethanol	anti/gauche	-0.5	-0.1	-0.5	0.0	-0.6	-0.1	0.12
methyl ethyl ether	anti/gauche	1.2	1.4	1.3	1.5	1.5	1.6	1.5
methyl vinyl ether	cis/skew	2.5	2.3	2.1	1.8	1.9	1.6	1.7
mean absolute error		0.5	0.3	0.5	0.3	0.4	0.3	–

275

Table 8-1:   Conformational Energy Differences in Acyclic Molecules (3)

molecule	low-energy/ high-energy conformer	B3LYP		MP2		semi-empirical			expt.
		6-31G*	6-311+G**	6-31G*	6-311+G**	MNDO	AM1	PM3	
n-butane	*trans/gauche*	0.8	0.9	0.7	0.5	0.6	0.7	0.5	**0.67**
1-butene	*skew/cis*	0.4	0.4	0.5	0.5	1.3	0.6	1.0	**0.22**
1,3-butadiene	*trans/gauche*	3.6	3.5	2.6	2.5	0.3	0.8	1.7	**2.89**
acrolein	*trans/cis*	1.7	2.2	1.5	2.2	0.4	0.2	0.4	**1.70**
N-methylformamide	*trans/cis*	0.7	1.0	1.0	1.2	0.6	-0.5	-1.5	**1.4**
N-methylacetamide	*trans/cis*	2.7	2.4	2.8	2.4	-1.1	0.4	-0.5	**2.3**
formic acid	*cis/trans*	5.2	4.5	6.1	4.6	3.7	7.4	4.3	**3.90**
methyl formate	*cis/trans*	5.3	5.3	6.2	5.7	2.9	5.6	1.9	**4.75**
methyl acetate	*cis/trans*	7.8	9.1	11.0	9.8	5.9	5.3	1.2	**8.5**
propanal	*eclipsed/anti*	1.2	0.9	1.1	0.9	-0.5	-0.7	-0.7	**0.67**
2-methylpropanal	*eclipsed/anti*	0.8	0.4	1.0	0.4	-0.2	-0.4	-0.8	**0.78**
1,2-difluoroethane	*gauche/anti*	0.4	0.8	-0.5	0.8	0.3	-0.5	1.4	**0.56**
1,2-dichloroethane	*anti/gauche*	1.7	1.6	1.9	1.4	1.1	0.8	0.6	**1.08**
ethanol	*anti/gauche*	-0.3	0.0	0.1	0.0	0.4	-1.6	-1.9	**0.12**
methyl ethyl ether	*anti/gauche*	1.4	1.5	1.7	1.4	0.9	-0.4	-1.0	**1.5**
methyl vinyl ether	*cis/skew*	2.3	2.0	1.9	2.6	-0.6	2.1	1.7	**1.7**
mean absolute error		0.5	0.3	0.7	0.4	1.2	1.5	1.8	–

formate, individual conformer energy differences from 6-31G* calculations are within 1 kcal/mol of their respective experimental values and, except for 1,2-difluoroethane, the model properly assigns the lowest-energy conformer. The 6-311+G** model yields only marginal improvements over 6-31G*, and there seems little justification in its use (in lieu of 6-31G*).

All density functional models (including local density models) perform well, both in their assignment of ground-state conformer, and in reproducing experimental conformational energy differences. Except for local density models, use of the 6-311+G** basis set in lieu of 6-31G* generally leads to improvement. (Local density models, like Hartree-Fock models, are less sensitive to improvements in basis set.) Differences are not large (on the order of a few tenths of a kcal/mol at most), however, and it may be difficult to justify use of the larger (and computationally much more expensive) basis set. Note that the large individual errors seen for Hartree-Fock models for formic acid and methyl formate have now disappeared. Also, except for ethanol where the known (experimental) energy difference between *anti* and *gauche* conformers is minuscule, all density functional models lead to correct assignment of lowest-energy conformer.

Surprisingly, the MP2/6-31G* model is not as satisfactory as any of the density functional models, both insofar as mean absolute error and in terms of individual errors. Use of the 6-311+G** basis set in place of 6-31G* leads to marked improvement, and the results are now of comparable quality to those of the best density functional models. Given the large difference in cost between density functional and MP2 models, and given the apparent need for basis sets larger than 6-31G* for the latter, it seems difficult to recommend use of MP2 models for the purpose of conformational analysis involving acyclic systems.

MNDO, AM1 and PM3 models are unsatisfactory for assignment of ground-state conformer and for calculation of conformational energy differences in acyclic systems. While this could have been anticipated, given the poor performance of semi-empirical models for other *isodesmic* processes (see discussion in **Chapter 6**), it is nevertheless disappointing. In many cases, semi-empirical models either yield the wrong ground-state conformer or produce energy differences which are far smaller than experimental values. Semi-empirical models should not be employed for conformational assignments in acyclic molecules.

Conformational Energy Differences in Cyclic Molecules

Similar comments apply to cyclic systems (**Table 8-2**). SYBYL molecular mechanics is completely unsatisfactory for establishing relative conformer stabilities, while MMFF appears to be quite well suited for this purpose. The only unsatisfactory case for the MMFF model is 2-chlorotetrahydropyran, where the noted preference for an *axial* chlorine (usually attributed to the anomeric effect[3]) is not reproduced. Caution should be exercised in the application of MMFF to carbohydrates where the anomeric effect may lead to significant conformational preferences.

Hartree-Fock, local density, density functional and MP2 models properly assign ground-state conformation in these systems, the sole exception being an incorrect assignment for piperidine with the STO-3G model. In fact, all Hartree-Fock models, and local density and all density functional models with the 6-31G* basis set lead to very similar results. Overall, the B3LYP/6-31G* model performs best (in terms of lowest mean absolute error) and the EDF1/6-31G* model worst, but the differences are small. For Hartree-Fock and density functional models, substitution of the 6-311+G** basis set for 6-31G* leads neither to significant improvements nor significant degradations in performance. Note, however, that the sample is very small and generalizations should be avoided.

The MP2/6-31G* model performs better than any of the density functional models with the same basis set. This is due primarily to an improved result for 2-chlorotetrahydropyran. MP2/6-31G* and MP2/6-311+G** models give rise to nearly identical results.

As with acyclic systems, semi-empirical models provide a poor account of the ground-state conformation and conformational energy differences in cyclic systems. While all three models typically yield reasonable results for hydrocarbons, results for other systems are not acceptable. The performance of the PM3 model with regard to the

278

Table 8-2: Conformational Energy Differences in Cyclic Molecules

molecule	low-energy/high energy conformer	molecular mechanics		Hartree Fock				local density		expt.
		SYBYL	MMFF	STO-3G	3-21G	6-31G*	6-311+G**	6-31G*	6-311+G**	
methylcyclohexane	*equatorial/axial*	1.4	1.4	1.8	1.9	2.3	2.3	1.6	2.0	**1.75**
tert-butylcyclohexane	*equatorial/axial*	7.4	6.3	6.4	6.5	6.1	6.2	4.6	4.9	**5.4**
cis-1,3-dimethylcyclohexane	*equatorial/axial*	5.7	5.1	5.8	6.3	6.5	6.6	5.2	5.0	**5.5**
fluorocyclohexane	*equatorial/axial*	0.0	-0.4	0.2	-0.7	-0.3	0.1	-0.2	0.1	**0.16**
chlorocyclohexane	*equatorial/axial*	0.2	-0.3	0.5	0.7	1.0	0.9	0.1	-0.1	**0.5**
piperidine	*equatorial/axial*	-0.2	0.9	-0.4	0.3	0.8	0.9	0.3	0.8	**0.53**
N-methylpiperidine	*equatorial/axial*	0.5	3.3	2.0	2.0	3.6	3.9	3.5	4.0	**3.15**
2-chlorotetrahydropyran	*axial/equatorial*	0.1	-0.1	3.2	3.6	2.5	2.7	4.6	4.5	**1.8**
2-methylcyclohexanone	*equatorial/axial*	0.8	1.3	1.5	3.0	2.3	2.0	2.7	2.2	**2.10**
3-methylcyclohexanone	*equatorial/axial*	0.7	0.5	1.3	1.1	1.7	1.7	1.0	0.9	**1.36,1.55**
4-methylcyclohexanone	*equatorial/axial*	1.2	1.3	1.8	1.7	2.1	2.1	1.5	1.3	**1.75,2.10**
mean absolute error		1.5	0.7	0.9	0.9	0.5	0.7	0.9	0.9	–

Table 8-2: Conformational Energy Differences in Cyclic Molecules (2)

molecule	low-energy/ high energy conformer	BP		BLYP		EDF1		expt.
		6-31G*	6-311+G**	6-31G*	6-311+G**	6-31G*	6-311+G**	
methylcyclohexane	*equatorial/axial*	2.2	2.4	2.2	2.5	2.8	2.7	**1.75**
tert-butylcyclohexane	*equatorial/axial*	5.3	5.2	5.0	5.2	5.9	5.8	**5.4**
cis-1,3-dimethylcyclohexane	*equatorial/axial*	6.0	5.8	6.0	6.0	6.8	6.7	**5.5**
fluorocyclohexane	*equatorial/axial*	0.1	0.3	-0.2	0.2	0.1	0.4	**0.16**
chlorocyclohexane	*equatorial/axial*	0.8	0.7	1.0	0.9	1.3	1.1	**0.5**
piperidine	*equatorial/axial*	0.1	0.5	0.1	0.6	0.1	0.5	**0.53**
N-methylpiperidine	*equatorial/axial*	3.2	3.9	3.4	4.1	3.7	4.1	**3.15**
2-chlorotetrahydropyran	*axial/equatorial*	4.4	4.3	4.4	4.3	3.7	3.7	**1.8**
2-methylcyclohexanone	*equatorial/axial*	2.4	2.0	2.4	1.9	2.5	2.0	**2.10**
3-methylcyclohexanone	*equatorial/axial*	1.5	1.5	1.6	1.7	1.9	1.9	**1.36,1.55**
4-methylcyclohexanone	*equatorial/axial*	2.0	1.9	2.1	2.0	2.4	2.3	**1.75,2.10**
mean absolute error		0.7	0.8	0.8	0.9	0.9	0.8	–

Table 8-2: Conformational Energy Differences in Cyclic Molecules (3)

| molecule | low-energy/ high energy conformer | B3LYP | | MP2 | | semi-empirical | | | expt. |
		6-31G*	6-311+G**	6-31G*	6-311+G**	MNDO	AM1	PM3	
methylcyclohexane	*equatorial/axial*	2.1	2.4	1.9	1.7	1.0	1.4	1.1	**1.75**
tert-butylcyclohexane	*equatorial/axial*	5.3	5.3	5.6	5.1	3.9	5.1	1.1	**5.4**
cis-1,3-dimethylcyclohexane	*equatorial/axial*	6.0	6.0	5.7	5.3	3.1	4.2	2.2	**5.5**
fluorocyclohexane	*equatorial/axial*	-0.2	0.2	-0.7	0.1	0.3	1.3	0.3	**0.16**
chlorocyclohexane	*equatorial/axial*	0.9	0.7	0.7	0.5	0.7	0.8	0.4	**0.5**
piperidine	*equatorial/axial*	0.3	0.7	0.6	0.8	0.5	-2.7	-2.2	**0.53**
N-methylpiperidine	*equatorial/axial*	3.4	4.0	3.6	3.7	1.3	-1.4	-1.3	**3.15**
2-chlorotetrahydropyran	*axial/equatorial*	3.7	3.7	2.8	2.9	1.8	3.6	3.1	**1.8**
2-methylcyclohexanone	*equatorial/axial*	2.4	2.0	2.2	1.5	-0.2	0.6	-0.1	**2.10**
3-methylcyclohexanone	*equatorial/axial*	1.6	1.6	0.9	1.6	0.8	1.3	1.1	**1.36,1.55**
4-methylcyclohexanone	*equatorial/axial*	2.0	2.0	1.5	1.3	1.0	1.4	1.0	**1.75,2.10**
mean absolute error		0.5	0.7	0.4	0.5	0.8	2.1	2.7	–

equatorial/axial energy difference in *tert*-butylcyclohexane warrants additional comment. PM3 assigns the correct (*equatorial*) ground-state conformation, but yields an energy difference to the *axial* form of only about 1 kcal/mol. This is far smaller than the experimental *equatorial/axial* separation of 5.5 kcal/mol, and of the same order of magnitude as the calculated PM3 difference in methylcyclohexane. A similar problem exists for *cis*-1,3-dimethylcyclohexane. Both can be traced to a tendency of non-bonded hydrogens to attract each other in the PM3 model.*

Barriers to Rotation and Inversion

Closely related to conformational energy differences are barriers to single-bond rotation and to pyramidal inversion. Here the experimental data are restricted to very small systems and derive primarily from microwave spectroscopy, from vibrational spectroscopy in the far infrared and from NMR, but are generally of high quality. Comparisons with calculated quantities are provided in **Table 8-3** for single-bond rotation barriers and **Table 8-4** for inversion barriers. The same models considered for conformational energy differences have been surveyed here.

As with conformational energy differences, SYBYL and MMFF molecular mechanics show marked differences in performance for rotation/inversion barriers. MMFF provides a good account of single-bond rotation barriers. Except for hydrogen peroxide and hydrogen disulfide, all barriers are well within 1 kcal/mol of their respective experimental values. Inversion barriers are more problematic**. While the inversion barrier in ammonia is close to the experimental value, barriers in trimethylamine and in aziridine are much too large, and inversion barriers in phosphine and (presumably) trimethylphosphine are smaller than their respective experimental quantities. Overall,

---

*   This problem has been addressed in a modification of the PM3 model, available in Spartan, which introduces a hydrogen-hydrogen repulsive term. With this modification, the *equatorial/axial* energy differences in *tert*-butylcyclohexane increases to 4.8 kcal/mol, and the *equatorial/axial* energy difference in *cis*-1,3-dimethylcyclohexane to 4.3 kcal/mol.

**   Because molecular mechanics has not been explicitly parameterized for reaction transition states, optimization algorithms are presently limited to finding minima (and not saddle points). Only where transition states have higher symmetry than reactants may they be located.

MMFF mechanics is not well suited to this problem. SYBYL mechanics is unsuitable for both single-bond rotation and pyramidal inversion barriers.

The performance of Hartree-Fock models for rotation/inversion barrier calculations gradually improves with increasing size of basis set. Both STO-3G and 3-21G models lead to reasonable accounts of single-bond rotation barriers, although problems occur for some individual systems. For example, the 3-21G model incorrectly assigns the lowest-energy conformation of hydrogen peroxide (as *trans* planar instead of twisted). Serious problems also occur for inversion barriers. The STO-3G model leads to inversion barriers in amines which are too high, consistent with the previously noted trend to overestimate pyramidalization at nitrogen (see **Chapter 5**). Calculated barriers in phosphine and (presumably) trimethylphosphine are also too large. On the other hand, the 3-21G model leads to nitrogen inversion barriers which are too small, paralleling the trend of that method to underestimate pyramidalization in amines. In fact, 3-21G structures for both cyanamide and aniline are planar, at odds with their (slightly) puckered experimental geometries. In terms of mean absolute errors, rotation barriers from 6-31G* and 6-311+G** models are as good as those from corresponding (same basis set) local density, density functional and MP2 models (see below). Aside from those in hydrogen peroxide and hydrogen disulfide, all rotation barriers from either model are within a few tenths of a kcal/mol from their respective experimental values. Inversion barriers show greater errors, but the overall quality (as measured by mean absolute errors) for 6-31G* and 6-311+G** models is not much inferior to that of the corresponding local density, density functional or MP2 models (see below).

All density functional models (including local density models) yield similar mean absolute errors for single-bond rotation barriers. Mean absolute errors either stay the same or improve slightly upon replacement of the 6-31G* basis set by 6-311+G**. The largest errors (and the largest variations among the different models) occur for hydrogen disulfide. Greater sensitivity to basis set is seen for inversion barriers. In nearly all cases, replacement of 6-31G* by 6-311+G**

Table 8-3: Barriers to Rotation

molecule	molecular mechanics		Hartree-Fock				local density		BP		
	SYBYL	MMFF	STO-3G	3-21G	6-31G*	6-311+G**	6-31G*	6-311+G**	6-31G*	6-311+G**	expt.
CH_3-CH_3	3.8	3.2	2.9	2.7	3.0	3.1	2.9	2.8	2.7	3.8	2.9
CH_3-NH_2	1.4	2.4	2.8	2.0	2.4	2.1	2.3	1.9	2.5	2.0	2.0
CH_3-OH	3.7	1.2	2.0	1.5	1.4	1.1	1.7	1.1	1.5	1.1	1.1
CH_3-SiH_3	0.8	1.8	1.3	1.4	1.4	1.5	1.5	1.4	1.4	2.1	1.7
CH_3-PH_2	2.5	2.3	1.9	2.0	2.0	2.0	2.0	1.9	1.8	1.6	2.0
CH_3-SH	1.4	1.8	1.5	1.4	1.4	1.3	1.4	1.3	1.3	1.2	1.3
CH_3-CHO	0.0	2.0	1.1	1.1	1.0	1.2	1.2	1.2	1.1	1.1	1.2
HO-OH (*cis* barrier)	0.3	8.7	9.1	11.7	9.2	8.6	9.6	8.3	8.8	8.3	7.0
(*trans* barrier)	0.0	0.3	0.1	0.0[a]	0.9	0.9	0.9	0.9	0.7	0.7	1.1
HS-SH (*cis* barrier)	4.4	8.8	6.1	8.8	8.5	7.8	9.2	9.1	8.5	7.6	6.8
(*trans* barrier)	-0.2	8.1	2.9	6.2	6.1	5.4	9.0	5.4	5.6	4.8	6.8
mean absolute error	2.2	0.8	1.0	0.8	0.6	0.5	0.7	0.5	0.6	0.6	–

Table 8-3: Barriers to Rotation (2)

molecule	BLYP		EDF1		B3LYP		MP2		semi-empirical			expt.
	6-31G*	6-311+G**	6-31G*	6-311+G**	6-31G*	6-311+G**	6-31G*	6-311+G**	MNDO	AM1	PM3	
CH₃-CH₃	2.7	2.6	2.9	2.6	2.8	2.7	3.1	3.1	1.0	1.3	1.3	**2.9**
CH₃-NH₂	2.4	1.9	2.5	2.0	2.4	2.0	2.6	2.2	1.1	1.3	1.2	**2.0**
CH₃-OH	1.5	1.0	1.5	1.1	1.4	1.0	1.5	1.2	0.7	1.0	0.9	**1.1**
CH₃-SiH₃	1.4	1.4	0.7	1.4	1.4	1.4	1.5	1.5	0.4	0.4	0.6	**1.7**
CH₃-PH₂	1.8	1.7	1.8	1.7	1.9	1.8	2.1	2.0	0.7	0.8	0.6	**2.0**
CH₃-SH	1.3	1.1	1.3	1.2	1.3	1.2	1.5	1.3	0.5	0.7	0.7	**1.3**
CH₃-CHO	1.1	1.1	1.1	1.1	1.1	1.1	1.0	1.0	0.2	0.4	0.6	**1.2**
HO-OH(*cis* barrier)	8.5	8.2	8.8	8.4	8.9	8.5	9.4	8.9	6.9	7.0	7.3	**7.0**
(*trans* barrier)	0.6	0.6	0.7	0.8	0.7	0.7	0.6	1.0	0.0	0.1	0.0	**1.1**
HS-SH (*cis* barrier)	7.9	7.1	8.5	7.7	8.2	7.4	8.7	7.9	5.0	6.1	6.3	**6.8**
(*trans* barrier)	5.2	4.7	5.7	5.1	5.5	4.9	5.9	5.3	1.9	2.1	0.2	**6.8**
mean absolute error	0.6	0.5	0.7	0.5	0.6	0.6	0.7	0.5	1.6	1.5	1.5	–

a) calculated equilibrium structure is *trans* planar

Table 8-4: Barriers to Inversion

molecule	molecular mechanics		Hartree-Fock				local density		BP		expt.
	SYBYL	MMFF	STO-3G	3-21G	6-31G*	6-311+G**	6-31G*	6-311+G**	6-31G*	6-311+G**	
cyanamide	0.0	0.3	3.3	0.0[a]	1.1	0.9	0.4	0.1	1.0	0.6	**1.9,2.0**
aniline	–	1.8	4.2	0.0[a]	1.6	1.3	0.5	0.3	1.3	0.8	**~2**
trimethylamine	2.5	12.0	10.9	6.8	9.0	8.9	9.0	8.6	8.8	8.3	–
dimethylamine	–	–	10.4	4.8	5.6	4.9	4.5	3.3	5.5	4.3	–
methylamine	–	–	10.5	2.5	6.0	4.9	4.8	3.2	6.2	4.4	**4.8**
ammonia	0.0	6.6	11.1	1.6	6.5	4.6	5.2	3.1	6.7	4.3	**5.8**
piperidine	–	–	9.9	3.0	6.1	5.5	4.8	3.7	5.7	4.6	**6.1**
N-methylpiperidine	–	–	10.7	5.9	8.8	9.1	8.2	8.4	8.1	8.2	**8.7**
aziridine	54.5	25.9	27.3	13.0	19.4	18.4	15.0	13.5	17.7	16.3	**>11.6,>12**
trimethylphosphine	7.2	27.7	65.6	51.6	48.0	48.2	39.9	40.2	39.1	39.9	–
phosphine	3.3	16.5	61.3	38.8	37.8	36.0	33.3	31.4	34.6	32.9	**31.5**
mean absolute error											–

286

Table 8-4: Barriers to Inversion (2)

molecule	BLYP		EDF1		B3LYP		MP2		semi-empirical			expt.
	6-31G*	6-311+G**	6-31G*	6-311+G**	6-31G*	6-311+G**	6-31G*	6-311+G**	MNDO	AM1	PM3	
cyanamide	1.2	0.6	1.1	0.6	0.9	0.5	1.6	1.6	5.7	1.7	6.0	1.9,2.0
aniline	1.2	0.8	1.2	0.8	1.2	–	2.1	–	4.2	1.0	4.5	~2
trimethylamine	8.6	8.3	8.0	7.6	8.5	8.3	11.1	11.4	1.7	4.6	8.4	–
dimethylamine	5.6	4.3	5.3	4.2	5.3	4.2	6.5	5.8	4.0	4.1	8.4	–
methylamine	6.4	4.3	6.2	4.5	5.9	4.2	6.6	5.4	7.6	4.2	9.1	4.8
ammonia	6.9	4.2	6.8	4.5	6.4	4.0	6.6	5.0	11.6	4.2	10.0	5.8
piperidine	5.7	4.5	5.5	4.6	5.6	4.5	6.9	5.8	4.8	4.0	7.7	6.1
N-methylpiperidine	7.8	8.0	7.5	7.6	8.0	8.2	10.3	10.7	2.3	4.4	7.7	8.7
aziridine	17.9	16.4	17.6	16.2	17.7	16.3	20.2	18.8	20.6	15.7	19.5	>11.6,>12
trimethylphosphine	40.3	40.8	39.3	39.2	41.1	42.1	44.5	45.4	19.6	37.3	36.7	–
phosphine	35.2	33.6	34.8	33.2	35.5	33.6	35.4	35.2	30.0	32.8	22.6	31.5
mean absolute error	0.6	0.5	0.7	0.5	0.6	0.6	0.7	0.5	1.6	1.5	1.5	–

a) calculated equilibrium structure is planar

leads to lowering of the barrier, usually (but not always) bringing it into better accord with experiment. The change in barrier parallels the previously noted change in bond angles about nitrogen and phosphorous (see **Tables A5-34** to **A5-38**).

MP2 models provide broadly similar results to the best of the density functional models for both rotation and inversion barriers. For rotation barriers, the MP2/6-311+G** model provides improvement over MP2/6-31G*. On the other hand, the two models yield very similar inversion barriers, perhaps reflecting the fact that bond angles involving nitrogen and phosphorous change only slightly between the two.

Semi-empirical models are markedly inferior to all other models dealt with (except the SYBYL molecular mechanics model) for barrier calculations. Major trends in rotation barriers are often not reproduced, for example, the nearly uniform decrement in rotation barrier from ethane to methylamine to methanol. None of the semi-empirical models is better than the others in this regard. One the other hand, AM1 is clearly superior to MNDO and PM3 in accounting for nitrogen inversion barriers. All in all, semi-empirical models are not recommended for barrier calculations.

In addition to rotation and inversion, there are other mechanisms by which conformational interconversion may occur. One of these, "pseudorotation", is most easily seen in a molecule like phosphorous pentafluoride which adopts a trigonal bipyramidal equilibrium geometry with distinct *equatorial* and *axial* fluorines.

However, *equatorial* and *axial* fluorines rapidly interchange. The two out-of-plane *equatorial* fluorines bend outward at the same time as the top *axial* fluorine bends downward and the bottom *axial* fluorine bends upward.

8-1

Ring Inversion in Cyclohexane

Another important conformational process, "ring inversion", is best typified by cyclohexane. This molecule undergoes motion in which *axial* and *equatorial* ring positions interconvert.

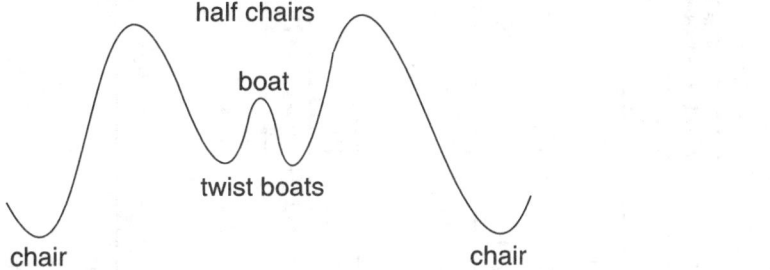

This is now known to involve two distinct transition states ("half-chair" and "boat") and an intermediate ("twist boat") in between.

half chairs

boat

8-2

twist boats

chair   chair

The overall barrier to ring inversion is well established experimentally, and is believed to correspond to the energy difference between chair and half-chair structures. Less certain are the relative energies of chair and twist-boat conformers and the energy of the boat transition structure, although a small range of values for each of these quantities has been established experimentally. Comparison of the experimental data with the results of calculations is provided in **Table 8-5**. The

Table 8-5: Energy Profile for Cyclohexane Inversion[a]

structure	Hartree-Fock				local density		BP		BLYP	
	STO-3G	3-21G	6-31G*	6-311+G**	6-31G*	6-311+G**	6-31G*	6-311+G**	6-31G*	6-311+G**
half-chair (1st transition state)	11.6	12.9	12.3	12.4	12.9	12.5	11.4	11.0	11.1	10.6
twist-boat (intermediate)	6.1	6.5	6.8	6.9	6.8	6.5	6.4	6.2	6.4	6.2
boat (2nd transition state)	7.0	7.6	7.8	8.0	8.0	7.7	7.2	7.0	7.1	6.9

structure	EDF1		B3LYP		MP2		semi-empirical			expt.
	6-31G*	6-311+G**	6-31G*	6-311+G**	6-31G*	6-311+G**	MNDO	AM1	PM3	
half-chair (1st transition state)	11.1	10.8	11.5	11.2	12.8	12.3	3.6	6.5	7.5	**10.7-11.5**
twist-boat (intermediate)	6.4	6.2	6.5	6.4	6.6	6.3	2.4	3.2	4.1	**4.7-6.2**
boat (2nd transition state)	7.1	6.9	7.3	7.2	8.1	7.9	2.5	3.5	4.4	**5.7-7.7**

a) energies relative to chair structure

same models as used in previous conformational energy difference and barrier comparisons are examined here.

Except for semi-empirical models, all calculation models yield basically similar results both for the two "barrier heights" and for the relative energies of the two stable conformers. All density functional models (excluding local density models) yield overall inversion barriers which are inside the experimental range (10.7 to 11.5 kcal/mol), but show that the twist-boat intermediate is somewhat higher in energy than the experiments suggest. The density functional calculations concur with experiment that the boat transition state is only slightly higher in energy than the twist-boat intermediate.

Hartree-Fock, local density and MP2 models all yield barrier heights which are slightly larger than those from density functional models, and are outside the experimental range. Additionally, the energies of the twist-boat intermediate and boat transition state (relative to the chair conformer) are also slightly higher.

Semi-empirical models do not provide good descriptions of the energy barrier to ring inversion in cyclohexane. The MNDO model underestimates the barrier by a factor of three, and the AM1 and PM3 models by almost a factor of two. This behavior is consistent with previous experience in dealing with single-bond rotation barriers.

Pitfalls

Determination of equilibrium conformation is beset with two serious and closely-related problems. The first is that, except for molecules with only a few degrees of conformational freedom, it will only rarely be possible (or at least practical) to thoroughly explore conformation space. While search algorithms have improved greatly in recent years (and will no doubt continue to improve), they still do not "guarantee" that the best structure located is actually the lowest-energy form of the molecule. It is possible that a completely different and unexplored region of the overall conformational energy surface contains a better (lower energy) structure. Further discussion is provided in **Chapter 14**.

The second problem also reflects the exceptional difficulty of exploring complex conformational energy surfaces. Quite simply, only the "lowest-cost" methods are applicable to anything but molecules with only a few degrees of conformational freedom. In practice and at the present time, this translates to molecular mechanics models. (Semi-empirical quantum chemical models might also represent practical alternatives, except for the fact that they perform poorly in this role.) Whereas molecular mechanics models such as MMFF seem to perform quite well, the fact of the matter is, outside the range of their explicit parameterization, their performance is uncertain at best.

References

1. Compilations of experimental data relating to conformational energy differences and rotation/inversion barriers may be found in: (a) T.A. Halgren and R.B. Nachbar, *J. Computational Chem.*, **17**, 587 (1996); (b) T.A. Halgren, *ibid.*, **20**, 730 (1999).

2. (a) F.H. Allen, S. Bellard, M.D. Brice, T.W.A. Hummelink, B.G. Hummelink-Peters, O. Kennard, W.D.S. Motherwell, J.R. Rogers and D.G. Watson, *Acta Cryst.*, **B35**, 2331 (1979); (b) F.H. Allen, *ibid.*, **B58**, 380 (2002).

Chapter 9

Transition-State Geometries and Activation Energies

This chapter assesses the performance of quantum chemical models with regard to the calculation of both absolute and relative activation energies. It also attempts to judge the ability of different models to properly describe the geometries of transition states using structures calculated from high-level models as a standard.

Introduction

Quantum chemical calculations need not be limited to the description of the structures and properties of stable molecules, that is, molecules which can actually be observed and characterized experimentally. They may as easily be applied to molecules which are highly reactive ("reactive intermediates") and, even more interesting, to molecules which are not minima on the overall potential energy surface, but rather correspond to species which connect energy minima ("transition states" or "transition structures")*. In the latter case, there are (and there can be) no experimental structure data. Transition states do not exist in the sense that they can be observed let alone characterized. However, the energies of transition states, relative to energies of reactants, may be inferred from experimental reaction rates, and qualitative information about transition-state geometries may be inferred from such quantities as activation entropies and activation volumes as well as kinetic isotope effects.

---

* Special cases of these involving "transition states" for rotation about single bonds, inversion of pyramidal nitrogen and phosphorus centers and ring inversion in cyclohexane, have been discussed in the previous chapter. The only difference is that these conformational processes are typically well described in terms of a simple motion, e.g., rotation about a single bond, whereas the motion involved in a chemical reaction is likely to be more complex.

293

Transition-State Geometries

The complete absence of experimental data on transition-state geometries complicates assessment of the performance of different models, and is a primary reason why molecular mechanics calculations are presently limited to the description of stable molecules. In the absence of experimental data, it is still possible to assess the performance of different models by assuming that some particular (high-level) model yields "reasonable" geometries, and then to compare the results of the other models with this "standard". Unfortunately, determination and verification of transition-state geometries is still difficult (see discussion in **Chapter 15**), and the "best" models may not yet be practical for any but the simplest systems. Here, the MP2/6-311+G** model has been selected as a "standard". While this model is not simple enough for widespread application, it does appear to reproduce the limited experimental data, i.e., absolute activation energies (see discussion following). Of course, there is really no way of knowing whether or not it is equally successful in describing the geometries of transition states.

"Key" bond distances in transition states for the Claisen rearrangement of allyl vinyl ether, the ene reaction involving 1-pentene and CO_2 extrusion from a cyclic ester obtained from STO-3G, 3-21G, 6-31G* and 6-311+G** Hartree-Fock models, local density models, BP, BLYP, EDF1 and B3LYP density functional models and MP2 models, all with the 6-31G* and 6-311+G** basis sets, and MNDO, AM1 and PM3 semi-empirical models are provided in **Table 9-1**.[*] These have been abstracted from a larger collection of transition-state geometries given in **Appendix A9** (**Tables A9-1** to **A9-8**). A listing of mean absolute deviations in "key" transition-state bond distances from MP2/6-311+G** bond lengths (the "standard") deriving from the full collection of transition states is provided in **Table 9-2**. The latter should only be used with caution to judge the performance of different models. Not only is the quality of the "standard" in question, but so too is the selection of transition states.

---

[*]   All bond distances and bond distance deviations will be reported in Å.

The most conspicuous difference between the data presented in **Table 9-1** and previous comparisons involving equilibrium bond distances (see **Chapter 5**) are the much larger variations among different models. This should not come as a surprise. Transition states represent a "compromise situation" where some bonds are being broken while others are being formed, and the potential energy surface around the transition state would be expected to be "flat."

In terms of mean absolute deviations from the "standard", all four Hartree-Fock models yield similar results. Inspection of individual systems (see also **Table A9-1**) shows that structures from 6-31G* and 6-311+G** models are nearly identical, but often differ significantly from those from STO-3G and 3-21G models. While it is difficult to justify use of the 6-311+G** basis set in place of 6-31G* for transition-state determinations using Hartree-Fock models, a case can be made for use of 6-31G* over STO-3G or 3-21G. As is reasonable, the largest deviations among different models correspond to making and breaking single bonds. In such situations the potential energy surface would be expected to be quite "flat" and large changes in geometry would be expected to lead only to small changes in energy.

In terms of mean absolute deviations from the "standard", local density models, all density functional models, excluding BLYP models, and the MP2/6-31G* model perform better than any of the Hartree-Fock models in providing transition-state geometries. Note, however, that local density models fail to find a reasonable transition state for the SO2 elimination reaction (see **Table A9-2**). As with Hartree-Fock models, significant differences in individual bond distances exist. For the most part, geometries obtained from a given density functional model with the 6-31G* basis set are very similar to geometries from the same model with the larger 6-311+G** basis set. However, differences in individual places on the order of 0.1Å occur in some instances. A case for use of the 6-311+G** basis set in transition-state geometry optimizations using density functional models can perhaps be made, but cost considerations may mitigate against this.

In terms of mean absolute deviation from the "standard", two of the three semi-empirical models are as successful as any of the Hartree-

Table 9-1:   Key Bond Distances in Transition States for Organic Reactions

reaction	transition state	bond length	Hartree-Fock				local density		BP		BLYP	
			STO-3G	3-21G	6-31G*	6-311+G**	6-31G*	6-311+G**	6-31G*	6-311+G**	6-31G*	6-311+G**
		a	1.63	1.88	1.92	1.96	1.71	1.77	1.87	1.93	1.96	2.02
		b	1.32	1.29	1.26	1.26	1.30	1.29	1.30	1.30	1.30	1.30
		c	1.40	1.37	1.37	1.37	1.38	1.38	1.39	1.39	1.39	1.39
		d	1.90	2.14	2.27	2.32	2.10	2.17	2.31	2.39	2.42	2.51
		e	1.40	1.38	1.38	1.38	1.39	1.38	1.39	1.39	1.39	1.39
		f	1.43	1.39	1.39	1.39	1.41	1.40	1.41	1.41	1.41	1.41
		a	1.43	1.40	1.40	1.40	1.43	1.42	1.43	1.42	1.43	1.42
		b	1.37	1.37	1.38	1.38	1.37	1.37	1.39	1.39	1.40	1.39
		c	2.00	2.11	2.12	2.13	2.12	2.12	2.12	2.14	2.15	2.21
		d	1.39	1.40	1.40	1.40	1.39	1.39	1.41	1.41	1.42	1.42
		e	1.53	1.45	1.45	1.43	1.59	1.55	1.53	1.49	1.48	1.42
		f	1.22	1.35	1.36	1.37	1.24	1.26	1.29	1.31	1.34	1.37
		a	1.41	1.39	1.38	1.38	1.40	1.40	1.41	1.40	1.40	1.40
		b	1.37	1.37	1.37	1.37	1.37	1.37	1.39	1.39	1.39	1.39
		c	1.99	2.12	2.26	2.28	2.24	2.21	2.20	2.19	2.18	2.17
		d	1.24	1.23	1.22	1.22	1.23	1.23	1.26	1.25	1.26	1.26
		e	2.04	1.88	1.74	1.72	1.79	1.77	1.78	1.76	1.78	1.76
		f	1.38	1.40	1.43	1.43	1.40	1.40	1.43	1.43	1.44	1.44

296

Table 9-1: Key Bond Distances in Transition States for Organic Reactions (2)

reaction	transition state	bond length	EDF1 6-31G*	EDF1 6-311+G**	B3LYP 6-31G*	B3LYP 6-311+G**	MP2 6-31G*	MP2 6-311+G**	MNDO	AM1	PM3
		a	1.89	1.95	1.90	1.96	1.80	**1.80**	1.46	1.58	1.68
		b	1.29	1.29	1.29	1.28	1.31	**1.30**	1.34	1.32	1.30
		c	1.39	1.38	1.38	1.38	1.38	**1.39**	1.41	1.41	1.40
		d	2.34	2.42	2.31	2.38	2.20	**2.22**	1.88	1.84	1.94
		e	1.39	1.38	1.38	1.38	1.39	**1.39**	1.40	1.40	1.39
		f	1.41	1.40	1.40	1.40	1.41	**1.41**	1.48	1.43	1.42
		a	1.43	1.42	1.42	1.41	1.43	**1.43**	1.45	1.41	1.41
		b	1.39	1.39	1.39	1.39	1.39	**1.39**	1.44	1.39	1.39
		c	2.11	2.14	2.11	2.14	2.02	**2.07**	1.65	2.02	1.97
		d	1.41	1.41	1.41	1.41	1.41	**1.41**	1.44	1.40	1.40
		e	1.51	1.47	1.48	1.44	1.55	**1.53**	1.85	1.44	1.51
		f	1.29	1.30	1.32	1.34	1.25	**1.25**	1.18	1.33	1.29
		a	1.40	1.40	1.40	1.39	1.40	**1.40**	1.40	1.40	1.39
		b	1.39	1.39	1.38	1.38	1.38	**1.38**	1.39	1.37	1.38
		c	2.20	2.19	2.18	2.18	2.08	**2.06**	2.37	2.25	2.02
		d	1.25	1.25	1.24	1.24	1.25	**1.24**	1.26	1.25	1.24
		e	1.76	1.74	1.78	1.76	1.83	**1.83**	1.65	1.71	1.93
		f	1.43	1.43	1.42	1.42	1.41	**1.41**	1.46	1.41	1.40

Table 9-2: Mean Absolute Deviations from MP2/6-311+G** of Key Bond Distances in Transition States for Organic Reactions

	STO-3G	3-21G	6-31G*	6-311+G**
MNDO	0.11			
AM1	0.05			
PM3	0.05			

	STO-3G	3-21G	6-31G*	6-311+G**
Hartree-Fock	0.05	0.05	0.05	0.05
local density	-	-	0.04	0.03
BP	-	-	0.03	0.04
BLYP	-	-	0.05	0.06
EDF1	-	-	0.04	0.04
B3LYP	-	-	0.03	0.04
MP2	-	-	0.01	-

Fock models in describing transition-state geometries. The MNDO model is much inferior according to this measure. Note, however, that all three semi-empirical models lead to very poor results in a number of specific situations (see **Table A9-8**). For example, all fail to provide a "reasonable" transition state for the Cope rearrangement of 1,5-hexadiene. Additionally, the PM3 model fails to find a "reasonable" transition state for sigmatropic rearrangement of 1,3-pentadiene. No doubt, there are many other reactions where similar failings will appear. While semi-empirical models certainly provide a useful function of exploring reactions, caution must be urged in their application.

Absolute Activation Energies

In order to extract an "experimental" activation energy from a measured reaction rate, it is first necessary to postulate a rate law. This generally takes the form,

$$\text{rate} = k\,[A]^a\,[B]^b\,[C]^c \text{ ---} \tag{1}$$

where k (the rate constant) is assumed to be independent of reagent concentrations, [A], [B], [C],..., and a, b, c are most commonly integers or half integers. The experimental rate law may be used to imply a mechanism, and mechanisms which fail to obey the rate law may be discarded. However, it will usually be the case that more than one mechanism can be found to satisfy the rate law.

The rate constant is not really a constant but depends on temperature. This is typically expressed using the Arrenhius equation,

$$k = Ae^{-\varepsilon_A/RT} \tag{2}$$

where the pre-exponential A and the activation energy ε_A are "parameters", R is the gas constant and T is the temperature (in K). A and ε_A are determined by measuring reaction rates over a (small) temperature range, and fitting the data to equation 2.

Given a mechanism which is consistent with the experimental rate law, an "experimental" activation energy may be calculated, and interpreted as the difference in energies between reactants and transition state. Association of this energy difference with ε_A in

299

equation 2, requires the further assumption that all reactants pass through the transition state. In effect, this implies that all reactants have the "same energy", or that none has energy in excess of that needed to reach the transition state. This is the essence of transition-state theory. While this might be reasonable in the condensed phase, it is probably less applicable to the gas phase. Liken the former to a crowded highway where all vehicles necessarily travel at nearly the same speed, and the latter to an "empty" highway where different vehicles may travel at widely different speeds.

It goes without saying that direct comparison of calculated (absolute) activation energies with experimental ε_A parameters is likely to prove problematic in some situations. For this reason, it is perhaps better to judge the performance of individual models by comparison with activation energies calculated from a standard reference. This standard has been chosen as MP2/6-311+G**, the same level used as a standard to judge transition-state geometries.

Absolute activation energies for a small series of organic reactions are provided in **Table 9-3**.* Results from Hartree-Fock models with STO-3G, 3-21G, 6-31G* and 6-311+G** basis sets, local density models, BP, BLYP, EDF1 and B3LYP density functional models with 6-31G* and 6-311+G** basis sets, the MP2/6-31G* model and MNDO, AM1 and PM3 semi-empirical models are compared with those from the "standard" (MP2/6-311+G**). Experimental activation energies, where available, are also provided.

Overall, the performance of Hartree-Fock models is very poor. In most cases, activation energies are overestimated by large amounts. This is not surprising in view of previous comparisons involving homolytic bond dissociation energies (see **Table 6-2**), which were too small.** In terms of mean absolute deviation from the "standard" (MP2/6-311+G**) calculations, STO-3G yields the poorest results and 3-21G the best results. 6-31G* and 6-311+G** models provide nearly identical activation energies (just as they did for transition-

---

*   All activation energies and activation energy deviations will be reported in kcal/mol.

**   The argument that might be given here is that a transition state is typically more tightly bound than the reactants, meaning that correlation effects will be greater.

state geometries). The bottom line is that Hartree-Fock models are unsatisfactory for absolute activation energy calculations.

On the basis of mean absolute deviations from "standard" absolute activation energies alone, all density functional models, including local density models, exhibit very similar behavior. Oddly enough, 6-31G* basis set models consistently turn in better results than the corresponding 6-311+G** models. Examination of individual reactions reveals significant differences among the models. Local density models consistently underestimate activation energies, but to varying degrees. For both Diels-Alder and dipolar cycloaddition reactions, they lead to nearly zero activation energies (approximately 10 kcal/mol below the "standard" values), but give activation energies for the closely-related Cope and Claisen rearrangements which are within a few kcal/mol of their respective "standard" values. In one case, SO_2 elimination, they fail to find a "reasonable" transition state. In short, it is difficult to anticipate the behavior of local density models, and they cannot be recommended for this purpose.

Individual activation energies from BP, BLYP, EDF1 and B3LYP density functional models are similar (and different from those of Hartree-Fock and local density models). They are both smaller and larger than "standard" values, but typically deviate by only a few kcal/mol. The most conspicuous exception is for Diels-Alder cycloaddition of cyclopentadiene and ethylene. Density functional models show activation energies around 20 kcal/mol, consistent with the experimental estimate for the reaction but significantly larger than the 9 kcal/mol value obtained from MP2/6-311+G** calculations. Overall, density functional models appear to provide an acceptable account of activation energies, and are recommended for use. Results from 6-31G* and 6-311+G** basis sets are very similar, and it is difficult to justify use of the latter.

The MP2/6-31G* model provides very similar activation energies to standard (MP2/6-311+G**) values, consistent with the insensitivity to basis set shown by Hartree-Fock and density functional models. Differences of 2-3 kcal/mol are common, and the largest difference is

Table 9-3:   Absolute Activation Energies for Organic Reactions

reaction	Hartree-Fock				local density		BP		BLYP		MP2
	STO-3G	3-21G	6-31G*	6-311+G**	6-31G*	6-311+G**	6-31G*	6-311+G**	6-31G*	6-311+G**	6-311+G**
$CH_3NC \longrightarrow CH_3CN$	56	57	46	45	41	41	39	38	39	39	41
$HCO_2CH_2CH_3 \longrightarrow HCO_2H + C_2H_4$	96	62	70	67	46	42	46	43	46	42	56
(Diels–Alder, benzene)	56	46	57	58	20	22	27	28	30	31	26
(Diels–Alder, with O)	50	42	49	50	22	23	23	23	23	23	26
(cyclopentadiene + ethylene)	35	30	40	43	0	1	13	16	19	23	9
$+ C_2H_4$	106	75	85	84	48	46	50	49	51	50	55
$HCNO + C_2H_2 \longrightarrow$ (isoxazole)	21	25	35	37	1	1	7	10	8	12	9
(Cope, hexadiene)	62	55	59	58	26	25	32	32	36	36	34
(cyclobutane → butadiene)	80	42	47	45	36	34	33	31	31	34	34
$+ CO_2$	108	59	60	58	41	37	33	31	31	28	41
$+ SO_2$	66	49	49	48	a	a	17	15	12	10	22
mean absolute deviation from MP2/6-311+G**	35	17	24	22	6	6	4	5	5	7	–

302

Table 9-3: Absolute Activation Energies for Organic Reactions (2)

reaction	EDF1		B3LYP		MP2		semi-empirical			expt.
	6-31G*	6-311+G**	6-31G*	6-311+G**	6-31G*	6-311+G**	MNDO	AM1	PM3	
CH$_3$NC — CH$_3$CN	40	39	41	40	43	41	68	83	58	**38**
HCO$_2$CH$_2$CH$_3$ — HCO$_2$H + C$_2$H$_4$	48	45	53	49	60	**56**	79	64	60	**40,44**
	31	32	34	35	28	**26**	a	a	a	**36**
	26	26	29	29	26	**26**	39	32	35	**31**
	21	24	20	23	12	**9**	56	28	32	**20**
+ C$_2$H$_4$	53	52	58	57	60	**55**	80	67	61	–
HCNO + C$_2$H$_2$	12	16	12	16	8	**9**	27	22	102	–
	35	35	39	39	38	**34**	82	83	a	–
	34	33	36	33	37	**34**	50	35	40	–
+ CO$_2$	34	32	40	37	44	**41**	85	62	66	–
+ SO$_2$	18	16	22	20	25	**22**	78	43	56	–
mean absolute deviation from MP2/6-311+G**	4	6	4	5	3	–	32	19	24	–

a)   reasonable transition state cannot be found

5 kcal/mol (for the ene reaction). Where MP2 models are appropriate, the 6-31G* basis set appears to provide a satisfactory account.

Semi-empirical models provide a wholly unsuitable account of absolute activation energies. MNDO turns in the poorest performance and AM1 the best, but all are unsatisfactory, and none should be used for this purpose. This is a similar situation to that revealed previously for thermochemical comparisons (see **Chapter 6**).

Relative Activation Energies

While knowledge of absolute activation energies is no doubt important in some situations, there are numerous other situations where it is not. For example, a proper account of remote substituent effects or changes in regio and/or stereochemistry on kinetic product distributions does not require knowledge about absolute activation energies, but only about relative activation energies. Like the thermochemical comparisons discussed in **Chapter 6**, it might be anticipated that activation energy comparisons formulated in such a way as to benefit from error cancellation, in particular, *isodesmic* comparisons, would be better described by simple quantum chemical models than comparisons which do not try to benefit from cancellation of errors.

A good example for which experimental data are available, involves activation energies for Diels-Alder cycloadditions of different cyanoethylenes as dienophiles with cyclopentadiene, relative to the addition of acrylonitrile with cyclopentadiene as a standard.

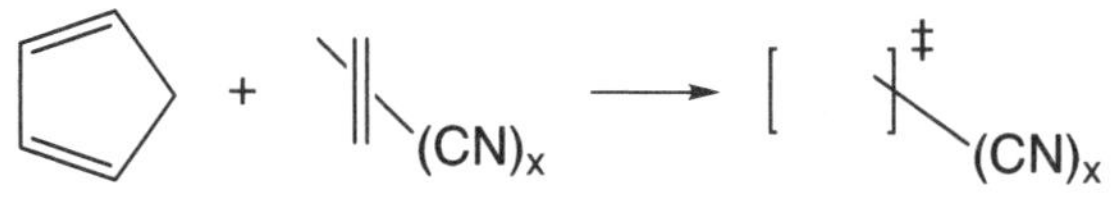

relative to

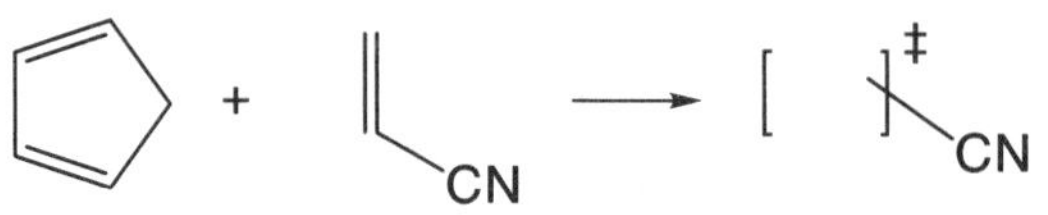

or

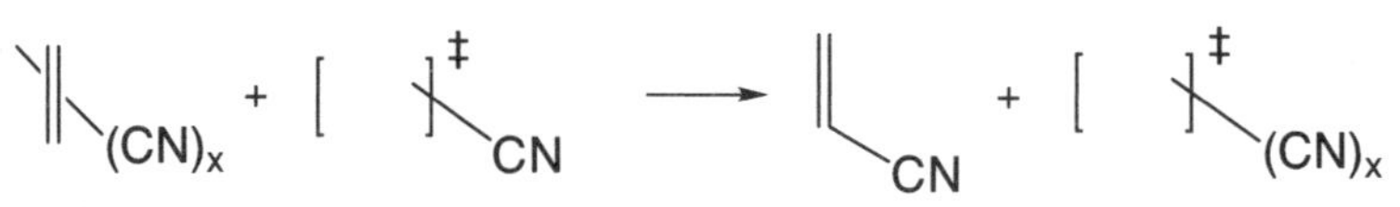

304

[ ]‡ denotes the transition state for the reaction. This is an *isodesmic* reaction, and previous experience with reaction thermochemistry suggests that its energy should be well described with Hartree-Fock models (but probably not with semi-empirical models; see **Table 6-10**). Density functional models would also be expected to yield acceptable results, but no better than those obtained from Hartree-Fock models.

Data are provided in **Table 9-4**. Hartree-Fock calculations have been limited to STO-3G, 3-21G and 6-31G* basis sets, and local density, density functional and MP2 calculations have been limited to the 6-31G* basis set. Results from AM1 and PM3 semi-empirical calculations have also been provided. MNDO calculations failed to locate reasonable transition states.

Both 3-21G and 6-31G* models provide an excellent account of relative activation energies in these systems, paralleling their performance in accounting for the thermochemistry of *isodesmic* reactions (see **Chapter 6**). The STO-3G model is not successful, again consistent with its performance in *isodesmic* thermochemical comparisons.

The local density 6-31G* model provides nearly identical results to the corresponding Hartree-Fock model. However, BP, BLYP, EDF1 and density functional models fail to provide a satisfactory account of relative activation energies in these systems. Except for cycloaddition involving 1,1-dicyanoethylene, all of these models underestimate the effect of increased nitrile substitution (on the dienophile) in decreasing the activation barrier. The B3LYP model gives the best account of the three, and the BLYP and EDF1 models the worst account but the differences are not great. The reason for the shortcoming in the performance of density functional models is not apparent.[*]

The MP2/6-31G* model provides a good account of relative activation energies in those systems (although it is no better than that provided by the corresponding Hartree-Fock model). This is another instance where the behavior of MP2 and B3LYP models diverge.

---

[*] Limited calculations have been performed with the 6-311+G** basis set and show similar problems. Therefore, the failure of density functional models of this instance does not appear to be related to limitations in the basis set.

Table 9-4: Relative Activation Energies of Diels-Alder Cycloadditions of Cyclopentadiene and Electron-Deficient Dienophiles[a]

dienophile	Hartree-Fock			local density	BP	BLYP
	STO-3G	3-21G	6-31G*	6-31G*	6-31G*	6-31G*
trans-1,2-dicyanoethylene	-1	-4	-3	-3	-1	-1
cis-1,2-dicyanoethylene	0	-3	-3	-2	-1	0
1,1-dicyanoethylene	-5	-7	-8	-7	-7	-7
tricyanoethylene	-4	-8	-9	-8	-5	-5
tetracyanoethylene	-4	-11	-11	-10	-5	-3
mean absolute error	4	1	0	1	3	4

dienophile	EDF1	B3LYP	MP2	semi-empirical		expt.
	6-31G*	6-31G*	6-31G*	AM1	PM3	
trans-1,2-dicyanoethylene	0	-2	-5	1	1	-2.6
cis-1,2-dicyanoethylene	0	-1	-4	2	1	-3.8
1,1-dicyanoethylene	-7	-7	-7	-1	0	-7.2
tricyanoethylene	-4	-6	-10	1	1	-9.2
tetracyanoethylene	-2	-6	-15	2	3	-11.2
mean absolute error	4	2	1	8	8	–

a) energy of reaction ⬠ + ‖(CN)x ⟶ (bicyclic CNx) + ‖CN relative to: ⬠ + ‖CN ⟶ (bicyclic CN)

Both AM1 and PM3 semi-empirical models provide a poor account of relative activation energies. With one exception, they indicate that increasing the electron deficiency of the dienophile leads to an increase in barrier height and not the decrease which is known to occur.

The overall conclusion is that relative activation energies among closely-related systems can be accurately described using quantum chemical models. The surprise is that such comparisons (like relative energy comparisons) are perhaps best carried out with Hartree-Fock models (or local density models) instead of with density functional or MP2 models.

Regio and stereochemical preferences in kinetically-controlled reactions may also be expressed as *isodesmic* processes. For example, the regioselectivity of (*endo*) addition of 2-methylcyclopentadiene with acrylonitrile comes down to the difference in energy the transition states leading to *meta* and *para* products, respectively.

or

Similarly, the difference in energy between *syn* and *anti* transition states for (*endo*) addition of 5-methylcyclopentadiene with acrylonitrile accounts for the stereochemistry of this reaction.

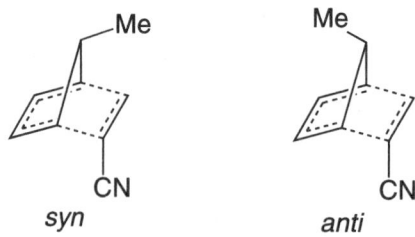

syn   anti

The experimental regio and stereochemical data as a function of substitution on cyclopentadiene may be summarized as follows:

i)   Electron-donor substituents in the 1 position lead preferentially to *ortho* products, and in the 2 position to *para* products.

ii) Strong electron-donor substituents such as OMe lead to greater regioselectivity than weak donor substituents such as Me.

iii) Electron-donor substituents in the 1 position lead to greater regioselectivity (for *ortho* products) than the same substituents in the 2 position (for *para* products).

iv) Alkyl substituents in the 5 position lead preferentially to *anti* products, while alkoxy substituents lead to *syn* products.

The results of Hartree-Fock calculations with STO-3G, 3-21G and 6-31G* basis sets, EDF1/6-31G* and B3LYP/6-31G* density functional calculations, MP2/6-31G* calculations and PM3 semi-empirical calculations are provided in **Table 9-5**.

All levels of calculation (including semi-empirical calculations) provide a qualitatively correct account of the experimental regio and stereochemical preferences. The only (apparent) exceptions are that both B3LYP/6-31G* and MP2/6-31G* models show modest preferences for *meta* products in cycloaddition of 2-methylcyclopentadiene and acrylonitrile. Note, in particular, the success of the calculations in properly assigning the more crowded *syn* product for the cycloaddition of 5-methoxycyclopentadiene and acrylonitrile. Also note the large magnitude for the preference. Clearly factors other than sterics are at work.

Table 9-5: Regio and Stereoselectivity in Diels-Alder Cycloadditions of Substituted Cyclopentadienes with Acrylonitrile[a]

position and substitutent on cyclopentadiene	Hartree-Fock			EDF1	B3LYP	MP2	semi-empirical		expt.
	STO-3G	3-21G	6-31G*	6-31G*	6-31G*	6-31G*	AM1	PM3	
regioselection									
1-Me	ortho (1.0)	ortho (0.9)	ortho (1.4)	ortho (2.2)	ortho (1.6)	ortho (0.7)	ortho (0.7)	ortho (0.4)	**ortho**
1-OMe	ortho (3.0)	ortho (3.3)	ortho (4.2)	ortho (5.4)	ortho (4.6)	ortho (1.9)	ortho (1.8)	ortho (1.0)	**ortho**
2-Me	para (0.5)	para (0.1)	para (0.6)	para (0.3)	meta (0.1)	meta (0.7)	para (0.9)	para (0.9)	**para**
2-OMe	para (1.6)	para (2.4)	para (2.8)	para (2.4)	para (2.2)	–	para (2.4)	para (1.8)	**para**
stereoselection									
5-Me	anti (3.1)	anti (2.5)	anti (1.0)	anti (1.6)	anti (0.9)	anti (1.0)	anti (2.0)	anti (0.8)	**anti**
5-OMe	syn (3.4)	syn (7.0)	syn (6.6)	syn (4.9)	syn (5.7)	syn (6.4)	syn (1.8)	syn (0.5)	**syn**

a)

309

Solvent Effects on Activation Energies

The S_N2 reaction is perhaps the single most familiar mechanism in all of organic chemistry. It involves approach of a nucleophile (Nu) to a tetrahedral carbon opposite to some leaving group (X). This geometry allows transfer of lone pair electrons on the nucleophile into an unoccupied σ^* orbital localized on the CX bond. Substitution occurs with the inversion of configuration at carbon and via a trigonal bipyramidal transition state in which bonds to both the incoming nucleophile and outgoing X group are greatly elongated over normal single-bond values.

The prevailing view is that S_N2 displacement occurs in one step, that is, without formation of any intermediates.

While this may be the case in solution where the solvent affords significant stabilization to both the incoming nucleophile and to the leaving group, it is certainly not the correct mechanism for the S_N2 reaction in the gas phase, at least where the incoming nucleophile is (negatively) charged. Here the overall reaction profile involves two stable ion-molecule complexes (intermediates), one involving the incoming nucleophile and the reactant, and the other involving the leaving group and the product.

310

In fact, approach of a charged nucleophile to a neutral reagent is energetically downhill in the gas phase, and any barrier to S_N2 displacement in solution is a "solvent effect". The underlying reason is that charges on both the nucleophile and the leaving group are much more highly localized than the charge on the transition state, and are much better stabilized by solvent.

While the S_N2 reaction represents an extreme case, it is clear that the solvent is capable of selectively stabilizing (or destabilizing) one product over another in a thermodynamically-controlled reaction, or one transition state over another in a kinetically-controlled reaction. Differentiation might be effected by steric and/or electronic considerations.

Consider, for example, *endo/exo* selectivity in the Diels-Alder cycloaddition of cyclopentadiene and 2-butanone. In cyclopentadiene as a solvent, the observed *endo/exo* product ratio is 80:20 (*endo* preferred), corresponding to a transition state energy difference on the order of 0.5 kcal/mol. With water as the solvent, this ratio increases to 95:5, corresponding to an energy difference on the order of 2 kcal/mol. Hartree-Fock 6-31G* calculations on the respective *endo* and *exo* transition states are largely in accord. Uncorrected for solvent, they show a very slight (0.3 kcal/mol) preference for *endo* in accord with the data in (non-polar) cyclopentadiene. This preference increases to 1.5 kcal/mol when the "solvent" is added (according to the Cramer/

Truhlar model[1]). Electrostatics may be responsible. The dipole moment for the *endo* transition state is slightly larger than that for the *exo* transition state (3.4 vs. 3.2 debyes), meaning that a polar medium will stabilize it to greater extent.

One should not place too much faith in what appears to be excellent agreement between calculations and experiment in a particular case (or be discouraged too much by what might appear to be complete disagreement in another). The important point to make is that computational tools have a role to play in "engineering" solvent control over reactions, and that these tools will become more and more adept in fulfilling this role.

Pitfalls

Characterization of transition-state geometries and energetics and ultimately reaction mechanisms remains a challenge for quantum-chemical models. The complete absence of experimental structural data and the need to interpret experimental reaction rates in terms of transition-state theory greatly complicates assessment of the theory, but it also increases its value as an exploratory tool. Nowhere is the problem more acute than in dealing with reactions in solution.

References

1. C.C. Chambers, G.D. Hawkins, C.J. Cramer and D.G. Truhlar, *J. Chem. Phys.*, **100**, 16385 (1996).

Chapter 10

Dipole Moments

This chapter assesses the performance of quantum chemical models with regard to the calculation of dipole moments. Several different classes of molecules, including diatomic and small polyatomic molecules, hydrocarbons, molecules with heteroatoms and hypervalent molecules are considered. The chapter concludes with assessment of the ability of quantum chemical models to calculate what are often subtle differences in dipole moments for different conformers.

Introduction

Chemists commonly include "formal charges" as part of structural formulas. While such a practice might be viewed simply as "chemical bookkeeping", it serves as well to anticipate molecular properties and chemical reactivity. However useful charges may be, the fact is that they may neither be determined from any experiment nor calculated in a unique manner. (Discussion of why this is so is provided in **Chapter 16**.) The closest one can get is a measure of overall molecular polarity as contained in the dipole moment. This is a measurable and calculable quantity[*].

This chapter assesses the performance of quantum chemical models with regard to the calculation of the magnitudes of dipole moments.[1] (Too little experimental information is available about the sign and/ or direction of dipole moments or about higher moments to make comparisons of these quantities with the results of calculations of value.) Coverage is divided according to type of molecule: diatomics and small polyatomics, hydrocarbons, molecules with heteroatoms and hypervalent molecules. Models examined include Hartree-Fock

---

[*]   Note, however, that the dipole moment for a charged molecule depends on choice of origin, and therefore is not unique. Dipole moments for ions cannot be measured and calculated dipole moments for charged molecules are meaningless.

models with STO-3G, 3-21G, 6-31G* and 6-311+G** basis sets, local density models, BP, BLYP, EDF1 and B3LYP density functional models, all with 6-31G* and 6-311+G** basis sets, MP2/6-31G* and MP2/6-311+G** models and MNDO, AM1 and PM3 semi-empirical models. Most data on individual molecules has been relegated to **Appendix A10**. However, mean absolute errors for each of the different models for each class of molecule have been summarized in text, and overall comparisons for selected models presented as graphs.*

Diatomic and Small Polyatomic Molecules

Dipole moments for a selection of diatomic and small polyatomic molecules from Hartree-Fock models are provided in **Table A10-1**, local density models in **Table A10-2**, BP, BLYP, EDF1 and B3LYP density functional models in **Tables A10-3** to **A10-6**, respectively, MP2 models in **Table A10-7** and MNDO, AM1 and PM3 semi-empirical models in **Table A10-8**. Included in the comparisons are weakly polar molecules like carbon monoxide, polar molecules like ammonia and "ionic" molecules like sodium chloride. Plots of calculated vs. experimental dipole moments are provided for all four Hartree-Fock models (**Figures 10-1** to **10-4**), the two EDF1 models (**Figures 10-5** and **10-6**), the two B3LYP models (**Figures 10-7** and **10-8**), the two MP2 models (**Figures 10-9** and **10-10**) and the PM3 model (**Figure 10-11**). A summary of mean absolute errors is given in **Table 10-1**. BP and BLYP density functional models would yield similar plots to that from the EDF1 models, and MNDO and AM1 semi-empirical models similar plots to that of the PM3 model. Square markers (□) designate diatomic and small polyatomic molecules.

The STO-3G model provides a very non-uniform account of dipole moments in these compounds (see **Figure 10-1**). Calculated dipole moments for extremely polar ("ionic") molecules like lithium chloride are almost always much smaller than experimental values, while dipole moments for moderately polar molecules such as silyl chloride are often larger, and dipole moments for other molecules like carbon

---

* All dipole moments and dipole moment errors will be reported in debyes.

Figure 10-1: STO-3G vs. Experimental Dipole Moments

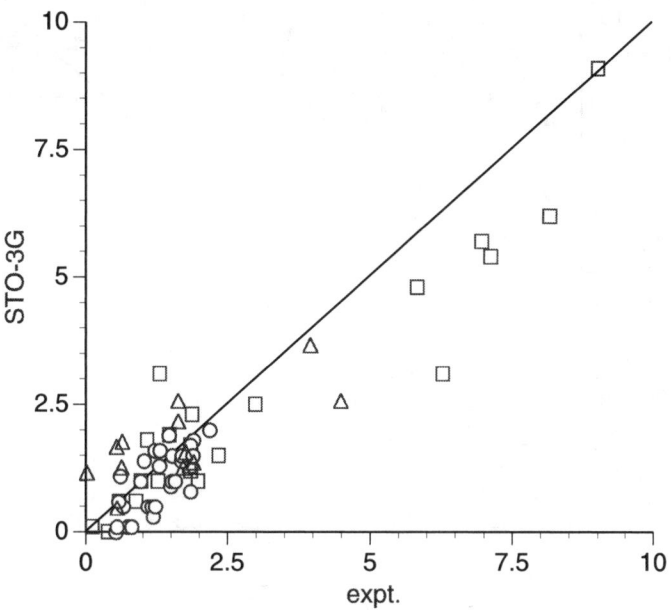

Figure 10-2: 3-21G vs. Experimental Dipole Moments

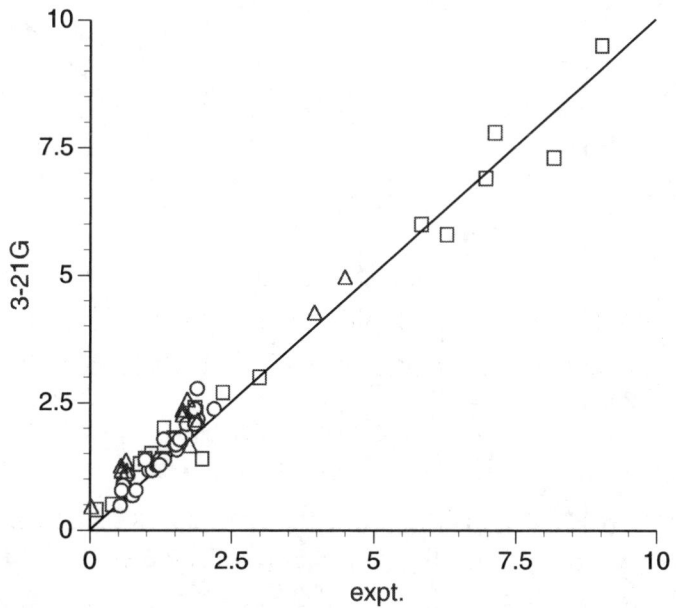

☐ Diatomic and small polyatomic molecules
○ Molecules with heteroatoms
△ Hypervalent molecules

Figure 10-3: 6-31G* vs. Experimental Dipole Moments

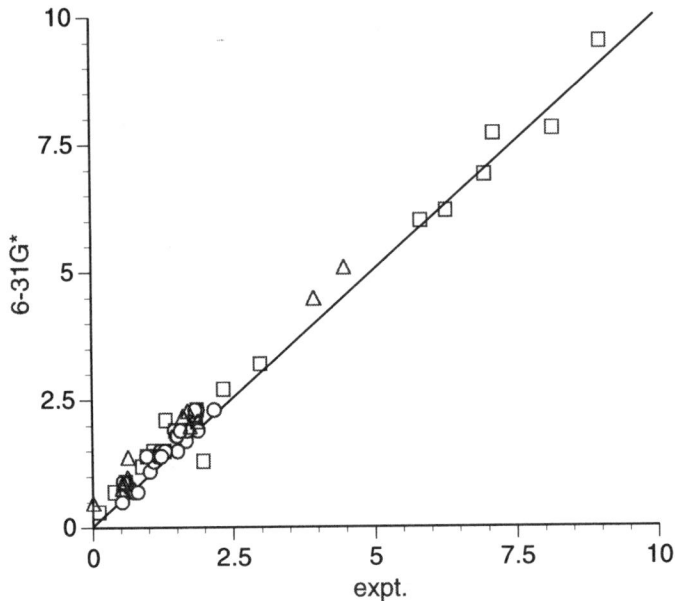

Figure 10-4: 6-311+G** vs. Experimental Dipole Moments

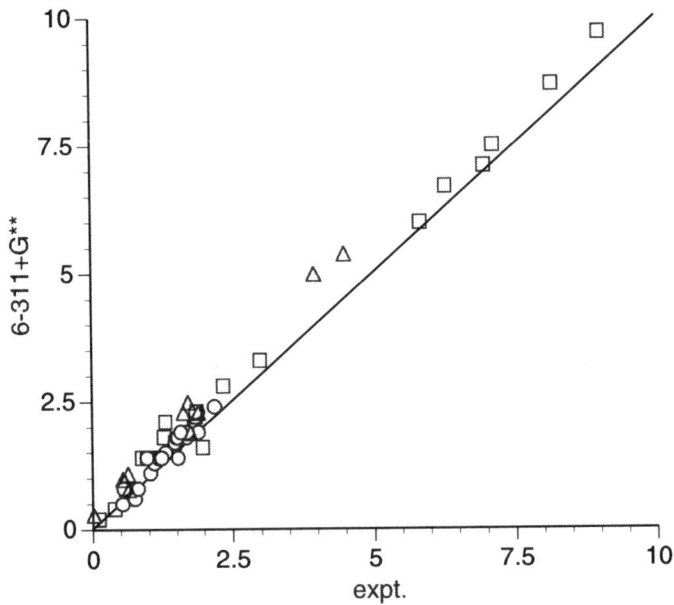

☐ Diatomic and small polyatomic molecules
○ Molecules with heteroatoms
△ Hypervalent molecules

Figure 10-5: EDF1/6-31G* vs. Experimental Dipole Moments

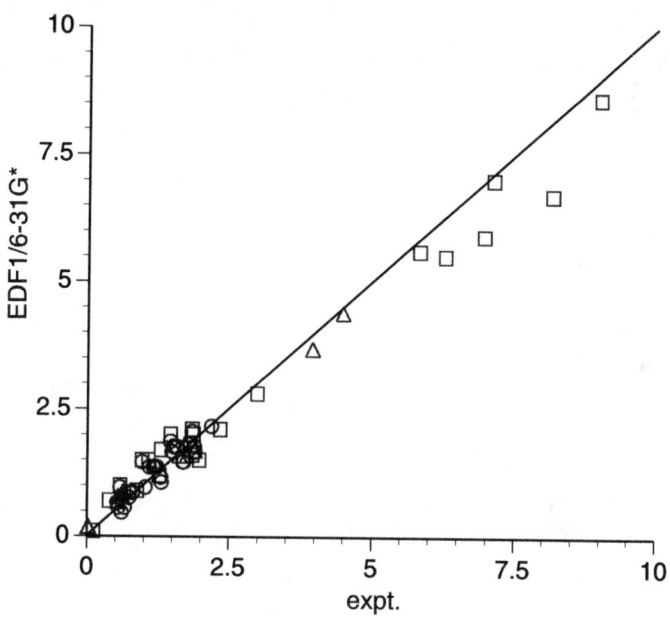

Figure 10-6: EDF1/6-311+G** vs. Experimental Dipole Moments

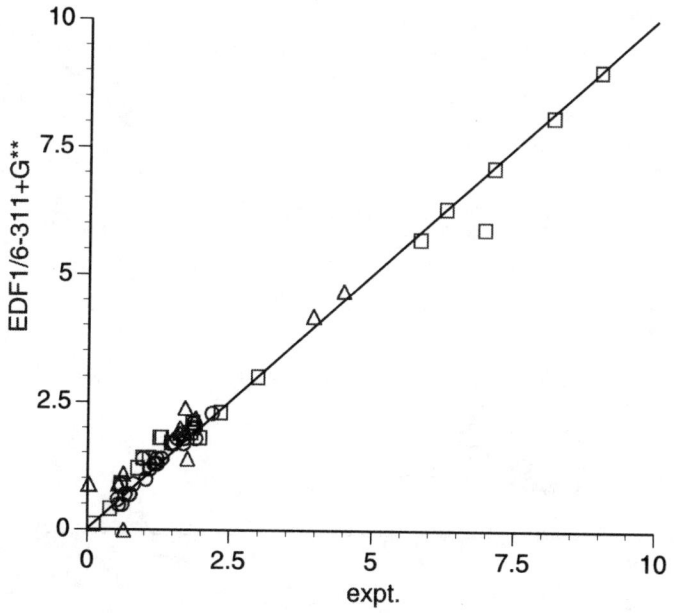

☐ Diatomic and small polyatomic molecules
○ Molecules with heteroatoms
△ Hypervalent molecules

Figure 10-7:   B3LYP/6-31G* vs. Experimental Dipole Moments

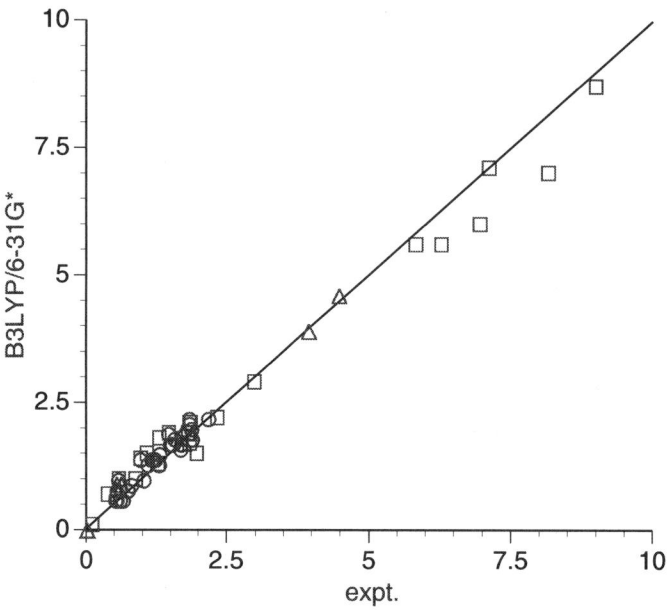

Figure 10-8:   B3LYP/6-311+G** vs. Experimental Dipole Moments

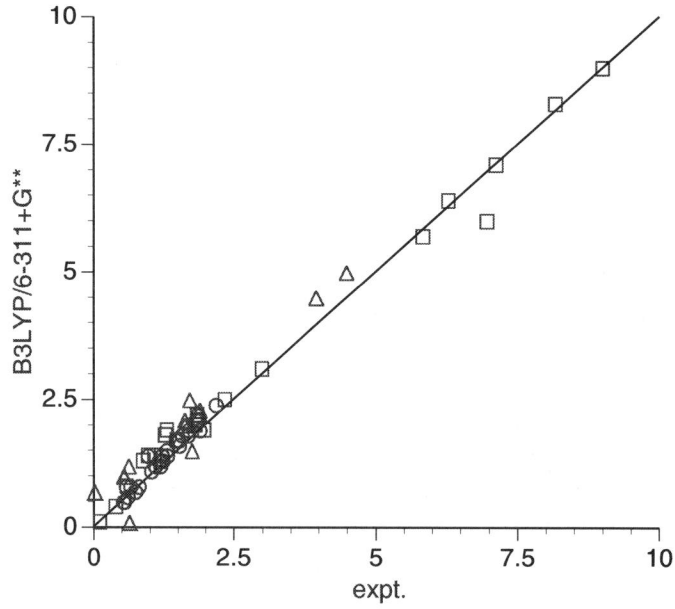

☐ Diatomic and small polyatomic molecules
○ Molecules with heteroatoms
△ Hypervalent molecules

Figure 10-9: MP2/6-31G* vs. Experimental Dipole Moments

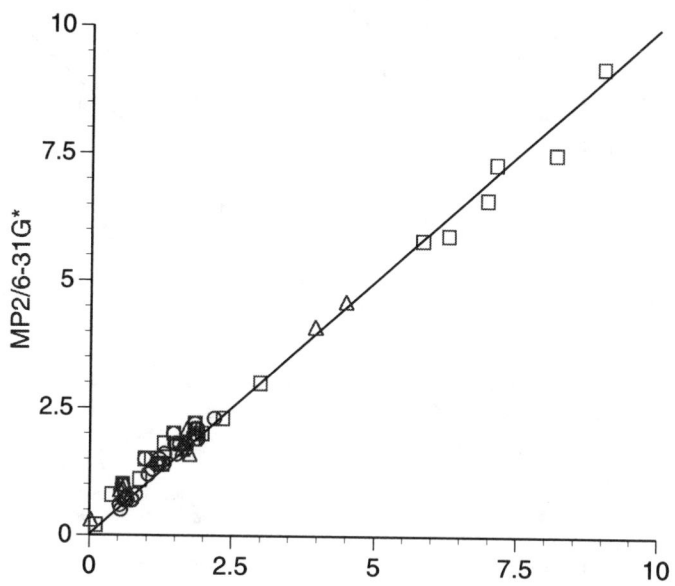

Figure 10-10: MP2/6-311+G** vs. Experimental Dipole Moments

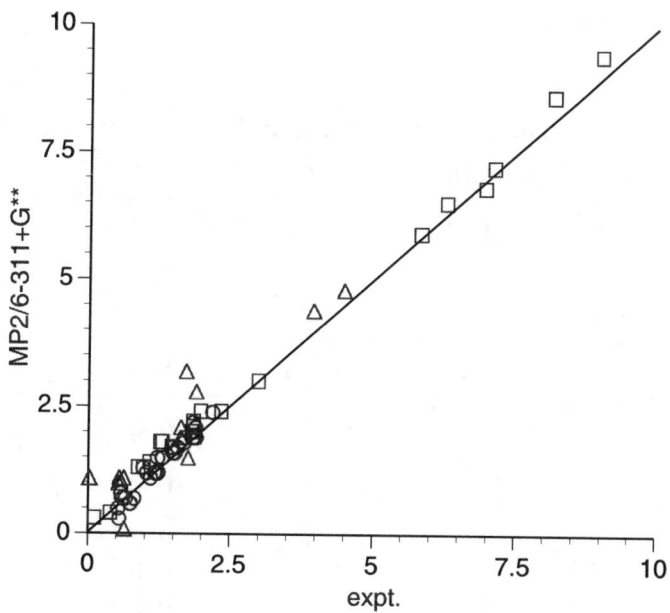

□　Diatomic and small polyatomic molecules
○　Molecules with heteroatoms
△　Hypervalent molecules

Figure 10-11: PM3 vs. Experimental Dipole Moments

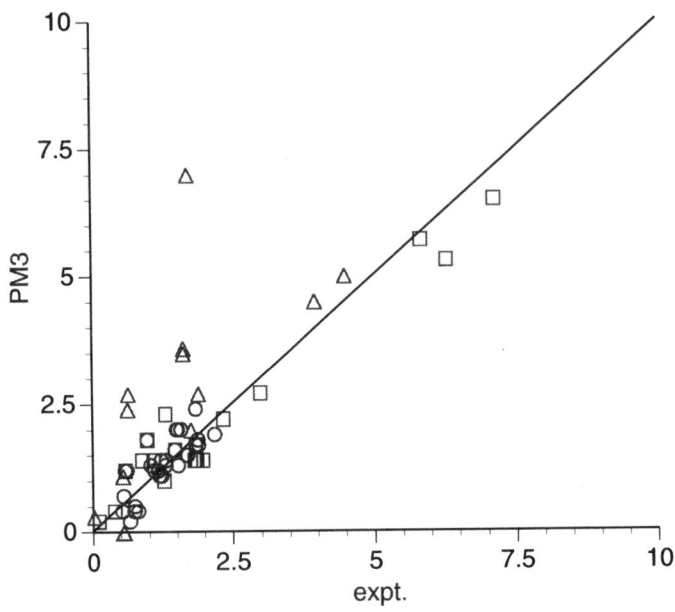

☐ Diatomic and small polyatomic molecules
○ Molecules with heteroatoms
△ Hypervalent molecules

Table 10-1: Mean Absolute Errors in Dipole Moments for Diatomic and Small Polyatomic Molecules

MNDO	0.4			
AM1	0.4			
PM3	0.4			

	STO-3G	3-21G	6-31G*	6-311+G**
Hartree-Fock	0.8	0.4	0.3	0.4
local density	—	—	0.4	0.3
BP	—	—	0.4	0.2
BLYP	—	—	0.4	0.2
EDF1	—	—	0.4	0.2
B3LYP	—	—	0.3	0.2
MP2	—	—	0.2	0.2

monoxide and hydrogen sulfide are in nearly perfect agreement with their respective experimental values. All in all, it is not apparent how to anticipate errors in the calculations.

On the other hand, Hartree-Fock models using larger basis sets provide a much more consistent account. As clearly apparent from **Figures 10-2** to **10-4**, calculated dipole moments are nearly always larger than experimental values. This can be rationalized using similar arguments to those earlier employed to account for systematic errors in "limiting" Hartree-Fock bond lengths (too short; see **Chapter 5**) and frequencies primarily associated with bond stretching (too large; see **Chapter 7**). Molecular orbitals which are occupied in the Hartree-Fock description will tend to be concentrated on the more electronegative atoms, while unoccupied molecular orbitals will tend to be concentrated on the less electronegative elements. Therefore, electron promotion from occupied to unoccupied molecular orbitals, as implicit in electron correlation schemes such as MP2, will have the effect of moving electrons from "where they are" in the Hartree-Fock description (on the more electronegative atoms) to "where they are not" (on the less electronegative atoms). In effect, electron correlation will act to reduce overall separation of charge and, therefore, lower the dipole moment. This implies that dipole moments from ("limiting") Hartree-Fock models will be too large, which is exactly what is observed.

As seen from comparison of data in **Tables A10-1** and **A10-2** , local density models parallel the behavior of the corresponding Hartree-Fock models. Except for highly polar ("ionic") lithium and sodium compounds, dipole moments are generally larger than experimental values. Recall that local density models typically (but not always) exhibit the same systematic errors in bond lengths (too short) and stretching frequencies (too large) as Hartree-Fock models.

All density functional models exhibit similar behavior with regard to dipole moments in diatomic and small polyatomic molecules. **Figures 10-6** (EDF1) and **10-8** (B3LYP) show clearly that, except for highly polar (ionic) molecules, "limiting" (6-311+G** basis set) dipole moments are usually (but not always) larger than experimental values.

Individual errors are typically quite small (on the order of a few tenths of a debye at most), and even highly polar and ionic molecules are reasonably well described. Comparison of results from 6-31G* and 6-311+G** density functional models (**Figure 10-5** vs. **10-6** for the EDF1 model and **Figure 10-7** vs. **10-8** for the B3LYP model) clearly reveals that the smaller basis set is not as effective, in particular with regard to dipole moments in highly polar and ionic molecules. Here, the models underestimate the experimental dipole moments, sometimes by 1 debye or more.

MP2 models also provide a good account of dipole moments for these compounds. The largest individual errors are for highly polar and ionic compounds, where the MP2/6-31G* model generally leads to dipole moments which are too small, while the MP2/6-311+G** model generally leads to dipole moments which are too large. Comparison of **Figures 10-9** and **10-10** reveals that differences between dipole moments calculated using the two MP2 models are not as great as those previously noted for the corresponding density functional models.

Semi-empirical models generally turn in a respectable account of dipole moments in these compounds. None of the models stand out as being particularly better (or particularly worse) than the others. While there are a few very bad cases (for example, the AM1 dipole moment in phosphine is four times larger than the experimental value), most of the calculated moments fall within a few tenths of a debye of their respective experimental values. Comparison of **Figure 10-11** (for the PM3 model) with the other figures clearly shows, however, that semi-empirical models are not as successful as the other models in accounting for dipole moments in these compounds.

Overall, the best descriptions are from density functional and MP2 models with the 6-311+G** basis set. Hartree-Fock models (except STO-3G), local density models and semi-empirical models generally perform adequately, although some systems (in particular highly-polar and ionic molecules) exhibit large errors.

Hydrocarbons

Dipole moments for hydrocarbons are small (typically less than 1 debye), and provide a good test of different models to reproduce subtle effects. A small selection of data is provided in **Table 10-2**, for the same models used previously for diatomic and small polyatomic molecules.[*]

Hartree-Fock, local density, density functional and MP2 models provide a credible account of dipole moments in hydrocarbons. Even STO-3G and 3-21G (Hartree-Fock) models appear to be suitable. Not only is the mean absolute error very low (0.1 debye or less), but all models properly account for a variety of subtle trends in the experimental data, for example, the increase in dipole moment in cyclopropene in response to methyl substitution on the double bond. Finally, note that there is very little difference in the performance of any of the models with 6-31G* and 6-311+G** basis sets.

None of the semi-empirical models are as successful. Mean absolute errors are two to three times larger than for other models, and individual molecules often show errors larger than the dipole moments themselves. In addition, a number of subtle effects are not properly reproduced. Overall, semi-empirical models do not offer a good choice for dipole moment calculations in hydrocarbons (and presumably as well in other molecules of low polarity).

Molecules with Heteroatoms

Data on molecules containing nitrogen, oxygen, silicon, phosphorous and sulfur not only provide additional examples, but also allow assessment of the ability of the different models to reproduce known changes in dipole moments accompanying structural variations. For example, dipole moments in methylamines are known experimentally to decrease with increasing methyl substitution,

$$NH_3 > MeNH_2 > MeNH_2 > Me_3N$$

---

[*]   Data for hydrocarbons has been excluded from the plots of calculated vs. experimental dipole moments, simply because of their small magnitudes.

Table 10-2: Dipole Moments in Hydrocarbons

formula	hydrocarbon	Hartree-Fock				local density		BP		expt.
		STO-3G	3-21G	6-31G*	6-311+G**	6-31G*	6-311+G**	6-31G*	6-311+G**	
C$_3$H$_4$	propyne	0.5	0.7	0.6	0.8	0.8	0.9	0.7	0.9	0.75
	cyclopropene	0.6	0.5	0.6	0.5	0.5	0.5	0.5	0.5	0.45
C$_3$H$_6$	propene	0.3	0.3	0.3	0.4	0.4	0.5	0.4	0.5	0.36
C$_3$H$_8$	propane	0.0	0.0	0.1	0.1	0.1	0.1	0.1	0.1	0.08
C$_4$H$_4$	but-1-yne-3-ene	0.4	0.5	0.5	0.5	0.4	0.4	0.3	0.4	0.4
C$_4$H$_6$	cyclobutene	0.1	0.1	0.0	0.1	0.2	0.3	0.2	0.2	0.13
	1,2-butadiene	0.3	0.4	0.4	0.5	0.4	0.4	0.4	0.5	0.40
	1-butyne	0.5	0.7	0.7	0.8	0.7	0.9	0.7	0.9	0.80
	methylenecyclopropane	0.2	0.3	0.4	0.5	0.5	0.5	0.4	0.5	0.40
	bicyclo[1.1.0]butane	0.6	0.8	0.7	0.7	0.8	0.8	0.8	0.7	0.68
	1-methylcyclopropene	0.8	0.9	0.9	0.9	1.0	1.0	0.9	0.9	0.84
C$_4$H$_8$	isobutene	0.4	0.5	0.5	0.6	0.6	0.7	0.5	0.6	0.50
	cis-2-butene	0.1	0.2	0.1	0.2	0.2	0.3	0.2	0.3	0.26
	cis-1-butene	0.3	0.4	0.4	0.5	0.5	0.6	0.5	0.5	0.44
	methylcyclopropane	0.1	0.1	0.1	0.1	0.1	0.2	0.1	0.1	0.14
C$_4$H$_{10}$	isobutane	0.0	0.1	0.1	0.1	0.1	0.2	0.1	0.1	0.13
C$_5$H$_6$	cyclopentadiene	0.3	0.4	0.3	0.4	0.6	0.6	0.5	0.6	0.42
mean absolute error		0.1	0.1	0.1	0.0	0.1	0.1	0.1	0.1	–

Table 10-2: Dipole Moments in Hydrocarbons (2)

formula	hydrocarbon	BLYP		EDF1		B3LYP		expt.
		6-31G*	6-311+G**	6-31G*	6-311+G**	6-31G*	6-311+G**	
C_3H_4	propyne	0.7	0.9	0.7	0.9	0.7	0.9	**0.75**
	cyclopropene	0.5	0.4	0.5	0.5	0.5	0.5	**0.45**
C_3H_6	propene	0.4	0.4	0.4	0.4	0.4	0.4	**0.36**
C_3H_8	propane	0.1	0.1	0.1	0.1	0.1	0.1	**0.08**
C_4H_4	but-1-yne-3-ene	0.3	0.4	0.3	0.4	0.3	0.4	**0.4**
C_4H_6	cyclobutene	0.1	0.2	0.2	0.2	0.1	0.2	**0.13**
	1,2-butadiene	0.4	0.4	0.4	0.5	0.4	0.4	**0.40**
	1-butyne	0.7	0.9	0.7	0.9	0.7	0.9	**0.80**
	methylenecyclopropane	0.4	0.5	0.4	0.5	0.4	0.5	**0.40**
	bicyclo[1.1.0]butane	0.8	0.7	0.8	0.7	0.8	0.7	**0.68**
	1-methylcyclopropene	0.9	0.9	0.9	0.9	0.9	0.9	**0.84**
C_4H_8	isobutene	0.5	0.6	0.5	0.6	0.5	0.6	**0.50**
	cis-2-butene	0.2	0.3	0.2	0.3	0.2	0.3	**0.26**
	cis-1-butene	0.4	0.5	0.4	0.5	0.4	0.5	**0.44**
	methylcyclopropane	0.1	0.1	0.1	0.1	0.1	0.1	**0.14**
C_4H_{10}	isobutane	0.1	0.1	0.1	0.1	0.1	0.1	**0.13**
C_5H_6	cyclopentadiene	0.5	0.5	0.5	0.5	0.4	0.5	**0.42**
mean absolute error		0.0	0.0	0.1	0.1	0.0	0.0	–

Table 10-2: Dipole Moments in Hydrocarbons (3)

formula	hydrocarbon	MP2		semi-empirical			expt.
		6-31G*	6-311+G**	MNDO	AM1	PM3	
C_3H_4	propyne	0.6	0.7	0.1	0.4	0.4	**0.75**
	cyclopropene	0.5	0.4	0.5	0.4	0.4	**0.45**
C_3H_6	propene	0.3	0.3	0.0	0.2	0.2	**0.36**
C_3H_8	propane	0.1	0.1	0.0	0.0	0.0	**0.08**
C_4H_4	but-1-yne-3-ene	0.3	0.3	0.1	0.2	0.2	**0.4**
C_4H_6	cyclobutene	0.1	0.2	0.1	0.2	0.2	**0.13**
	1,2-butadiene	0.3	0.4	0.0	0.2	0.2	**0.40**
	1-butyne	0.6	0.7	0.1	0.4	0.3	**0.80**
	methylenecyclopropane	0.3	0.3	0.0	0.1	0.2	**0.40**
	bicyclo[1.1.0]butane	0.8	0.8	0.4	0.4	0.4	**0.68**
	1-methylcyclopropene	0.8	0.8	0.4	0.6	0.6	**0.84**
C_4H_8	isobutene	0.4	0.5	0.1	0.4	0.4	**0.50**
	cis-2-butene	0.2	0.2	0.0	0.2	0.2	**0.26**
	cis-1-butene	0.3	0.4	0.0	0.2	0.3	**0.44**
	methylcyclopropane	0.1	0.2	0.1	0.1	0.1	**0.14**
C_4H_{10}	isobutane	0.1	0.1	0.0	0.0	0.0	**0.13**
C_5H_6	cyclopentadiene	0.4	0.5	0.2	0.5	0.5	**0.42**
mean absolute error		0.1	0.1	0.3	0.2	0.2	–

while dipole moments in methylphosphines show a completely different ordering.

$$PH_3 < MePH_2 < Me_3P < Me_2PH$$

Dipole moments in amines obtained from the same set of models used in previous comparisons are compared with experimental values in **Table 10-3**. These and further data (on oxygen, silicon, phosphorous and sulfur compounds) are given in **Appendix A10** (**Tables A10-9** to **A10-16**), and mean absolute errors corresponding to the full set are summarized in **Table 10-4**. In addition, graphical comparisons of calculated and experimental dipole moments have been provided for selected models (**Figures 10-1** to **10-11**). Circular markers (○) designate molecules with heteroatoms.

Comparison of **Figure 10-1** (STO-3G model) and **Figure 10-2** (3-21G model) clearly reveals significant improvement in dipole moments (and presumably more "realistic" descriptions of overall charge distributions). Differences seen in moving to the HF/6-31G* model (**Figure 10-3**) are more subtle. In fact, in terms of mean absolute errors, 3-21G and 6-31G* models, local density models and all density functional models with the 6-31G* basis set are indistinguishable. All also reproduce the above mentioned trend in amine dipole moments. However, other criteria reveal differences. For example, while 6-31G* and 6-311+G** Hartree-Fock models fail to reproduce the increase in dipole moment observed upon substituting the methyl group in methylamine by a phenyl ring, local density models, density functional models and MP2 models properly account for the noted change in dipole moment.

Dipole moments from 6-31G* and 6-311+G** calculations are nearly identical both for the amines (**Table 10-3**) and for the full set of molecules found in **Appendix A10**. Comparison of **Figures 10-3** and **10-4** further drives home the point. It is difficult to justify use of the larger basis set for this purpose. On the other hand, dipole moments resulting from all density functional models (including local density models) and from MP2 models, display some sensitivity to basis set. In most (but not all) cases, agreement with experiment improves, as generally reflected by the mean absolute errors in **Table 10-4**, and by

Table 10-3: Dipole Moments in Amines

molecule	Hartree-Fock				local density		BP		BLYP	
	STO-3G	3-21G	6-31G*	6-311+G**	6-31G*	6-311+G**	6-31G*	6-311+G**	6-31G*	6-311+G**
trimethylamine	1.1	0.9	0.7	0.8	0.5	0.4	0.5	0.5	0.5	0.5
dimethylamine	1.4	1.2	1.1	1.1	1.0	1.0	1.0	1.0	1.0	1.0
ethylamine	1.6	1.4	1.5	1.5	1.5	1.3	1.5	1.4	1.4	1.4
methylamine	1.6	1.4	1.5	1.5	1.4	1.3	1.5	1.4	1.5	1.4
ammonia	1.9	1.8	1.9	1.7	2.0	1.7	2.0	1.7	1.9	1.7
aniline	1.5	1.6	1.5	1.4	2.0	1.9	1.8	1.7	1.8	1.6
aziridine	1.8	2.2	1.9	1.9	1.7	1.8	1.7	1.8	1.7	1.9
pyridine	2.0	2.4	2.3	2.4	2.2	2.4	2.2	2.3	2.1	2.3

molecule	EDF1	B3LYP		MP2		semi-empirical			expt.
	6-311+G**	6-31G*	6-311+G**	6-31G*	6-311+G**	MNDO	AM1	PM3	
trimethylamine	0.5	0.6	0.6	0.7	0.7	0.8	1.0	1.2	**0.61**
dimethylamine	1.0	1.0	1.1	1.2	1.2	1.2	1.2	1.3	**1.03**
ethylamine	1.4	1.4	1.4	1.5	1.5	1.5	1.6	1.4	**1.22**
methylamine	1.1	1.5	1.4	1.6	1.5	1.5	1.5	1.4	**1.31**
ammonia	1.9	1.9	1.7	2.0	1.7	1.8	1.9	1.6	**1.47**
aniline	1.8	1.7	1.6	1.6	1.6	1.5	1.5	1.3	**1.53**
aziridine	1.7	1.8	1.9	1.9	1.9	1.8	1.8	1.7	**1.90**
pyridine	2.2	2.2	2.4	2.3	2.4	2.0	2.0	1.9	**2.19**

Table 10-4: Mean Absolute Errors in Dipole Moments for Molecules with Heteroatoms

MNDO	0.3			
AM1	0.3			
PM3	0.3			

	STO-3G	3-21G	6-31G*	6-311+G**
Hartree-Fock	0.4	0.2	0.2	0.2
local density	—	—	0.2	0.2
BP	—	—	0.2	0.1
BLYP	—	—	0.2	0.1
EDF1	—	—	0.2	0.1
B3LYP	—	—	0.2	0.1
MP2	—	—	0.2	0.1

graphical comparisons in **Figures 10-5** to **10-8** for EDF1 and B3LYP models and **Figures 10-9** and **10-10** for MP2 models. Indeed, the performance of density functional models (except the local density model) with the 6-311+G** basis set and the MP2/6-311+G** model is excellent for these compounds, with errors seldom exceeding a few tenths of a debye. They are clearly the methods of choice for accurate dipole moment descriptions.

The performance of semi-empirical models in accounting for dipole moments both in amines, and more generally for molecules incorporating heteroatoms, is not acceptable. Not only are individual errors often very large, but observed trends in dipole moments with differing substitution are not always reproduced.

Somewhat smaller errors result from comparison of dipole moments among closely-related molecules. The data in **Table 10-5** provide an example for dipole moments in carbonyl compounds relative to the dipole moment in acetone. Except for semi-empirical models, all of the models considered reproduce the observed trend in dipole moments in these systems, mainly that dipole moments in aldehydes are consistently smaller than that in acetone, while dipole moments in cyclic ketones are consistently larger. In terms of mean absolute errors (and discounting semi-empirical models) B3LYP and MP2 models perform best and Hartree-Fock and local density models worst, but the differences are not great. Much larger differences in performance are however, seen in individual systems. For example, in its "limit" (6-311+G** basis set) the local density model greatly overestimates the difference in dipole moment between formaldehyde and acetone (the "standard"). Other density functional models also overestimate this difference but to lesser extent, while "limiting" Hartree-Fock and MP2 models provide excellent accounts.

Cyclopentanone warrants special attention. While the measured dipole moment is 0.32 debyes larger than that for acetone, and while all models (except semi-empirical models) show an increase in dipole moment over acetone, none come close to "known" difference. Most disturbing is the MP2/6-311+G** result, which shows the dipole moment in cyclopentanone to be only 0.11 debyes larger than that in acetone.

Table 10-5: Dipole Moments in Carbonyl Compounds Relative to Acetone

molecule	Hartree-Fock				local density		expt.
	STO-3G	3-21G	6-31G*	6-311+G**	6-31G*	6-311+G**	
formaldehyde	-0.48	-0.48	-0.45	-0.53	-0.80	-0.85	-0.57
cis-propanal	-0.16	-0.28	-0.24	-0.28	-0.34	-0.38	-0.41
acetaldehyde	-0.14	-0.16	-0.14	-0.16	-0.21	-0.24	-0.20
eclipsed 2-methylpropanal	-0.13	-0.20	-0.17	-0.21	-0.29	-0.34	-0.24
pivaldehyde	-0.12	-0.15	-0.16	-0.21	-0.27	-0.35	-0.27
pinacolone	-0.04	-0.11	-0.14	-0.18	-0.21	-0.28	-0.18
anti-propanal	-0.10	-0.06	-0.05	-0.05	-0.14	-0.15	-0.07
anti-2-methylpropanal	-0.09	-0.02	-0.04	-0.05	-0.12	-0.16	-0.07
cyclobutanone	-0.02	-0.08	0.04	0.04	-0.09	-0.11	-0.04
2-methylcyclohexanone	0.08	0.09	0.12	0.12	0.04	0.06	0.05
3-methylcyclohexanone	0.13	0.25	0.26	0.30	0.22	0.27	0.13
4-methylcyclohexanone	0.15	0.29	0.30	0.34	0.25	0.31	0.14
3-methylcyclopentanone	0.09	0.15	0.20	0.26	0.07	0.13	0.21
cyclohexanone	0.13	0.24	0.27	0.31	0.22	0.29	0.32
cyclopentanone	0.05	0.07	0.16	0.20	0.02	0.08	0.32
mean absolute error	0.11	0.09	0.09	0.08	0.09	0.10	–

Table 10-5: Dipole Moments in Carbonyl Compounds Relative to Acetone (2)

molecule	BP		BLYP		EDF1		expt.
	6-31G*	6-311+G**	6-31G*	6-311+G**	6-31G*	6-311+G**	
formaldehyde	-0.70	-0.73	-0.68	-0.70	-0.69	-0.72	**-0.57**
cis-propanal	-0.28	-0.30	-0.37	-0.29	-0.25	-0.28	**-0.41**
acetaldehyde	-0.18	-0.19	-0.18	-0.20	-0.18	-0.20	**-0.20**
eclipsed 2-methylpropanal	-0.21	-0.23	-0.19	-0.22	-0.18	-0.21	**-0.24**
pivaldehyde	-0.18	-0.24	-0.16	-0.22	-0.15	-0.21	**-0.27**
pinacolone	-0.16	-0.22	-0.14	-0.21	-0.15	-0.22	**-0.18**
anti-propanal	-0.08	-0.07	-0.07	-0.06	-0.04	-0.07	**-0.07**
anti-2-methylpropanal	-0.05	-0.07	-0.04	-0.06	-0.04	-0.06	**-0.07**
cyclobutanone	-0.06	-0.07	-0.06	-0.06	-0.05	-0.05	**-0.04**
2-methylcyclohexanone	0.10	0.13	0.12	0.15	0.12	0.14	**0.05**
3-methylcyclohexanone	0.26	0.32	0.27	0.34	0.27	0.33	**0.13**
4-methylcyclohexanone	0.30	0.36	0.31	0.38	0.31	0.37	**0.14**
3-methylcyclopentanone	0.10	0.16	0.11	0.18	0.11	0.17	**0.21**
cyclohexanone	0.26	0.33	0.27	0.35	0.27	0.34	**0.32**
cyclopentanone	0.05	0.12	0.06	0.13	0.06	0.12	**0.32**
mean absolute error	0.08	0.08	0.08	0.08	0.09	0.08	–

Table 10-5: Dipole Moments in Carbonyl Compounds Relative to Acetone (3)

molecule	B3LYP		MP2		semi-empirical			expt.
	6-31G*	6-311+G**	6-31G*	6-311+G**	MNDO	AM1	PM3	
formaldehyde	-0.63	-0.66	-0.49	-0.55	-0.35	-0.60	-0.62	-0.57
cis-propanal	-0.28	-0.30	-0.26	-0.31	-0.19	-0.33	-0.31	-0.41
acetaldehyde	-0.18	-0.19	-0.15	-0.18	-0.13	-0.23	-0.24	-0.20
eclipsed 2-methylpropanal	-0.20	-0.23	-0.20	-0.24	-0.26	-0.36	-0.33	-0.24
pivaldehyde	-0.18	-0.23	-0.18	-0.25	-0.18	-0.33	-0.37	-0.27
pinacolone	-0.15	-0.20	-0.14	-0.20	-0.10	-0.15	-0.10	-0.18
anti-propanal	-0.08	-0.07	-0.07	-0.08	-0.15	-0.24	-0.26	-0.07
anti-2-methylpropanal	-0.06	-0.07	-0.05	-0.08	-0.13	-0.25	-0.31	-0.07
cyclobutanone	-0.04	-0.04	-0.03	-0.09	-0.11	-0.25	-0.27	-0.04
2-methylcyclohexanone	0.10	0.13	0.08	0.09	-0.01	-0.06	-0.06	0.05
3-methylcyclohexanone	0.25	0.32	0.23	0.26	0.08	0.08	0.05	0.13
4-methylcyclohexanone	0.29	0.36	0.27	0.31	0.08	0.08	0.05	0.14
3-methylcyclopentanone	0.12	0.19	0.13	0.15	-0.02	-0.04	-0.01	0.21
cyclohexanone	0.26	0.33	0.24	0.28	0.06	0.05	0.03	0.32
cyclopentanone	0.08	0.14	0.08	0.11	-0.05	-0.11	-0.07	0.32
mean absolute error	0.07	0.07	0.08	0.06	0.13	0.14	0.15	–

Hypervalent Molecules

The "octet rule" is one of the cornerstones of chemical bonding theory. While the vast majority of molecules conform, "apparent" exceptions occur for molecules incorporating second-row (and heavier) main-group elements. "Apparent" refers to the fact that molecules such as dimethylsulfoxide and dimethylsulfone may either be represented in terms of structures with ten and twelve valence electrons, respectively, surrounding sulfur, or as zwitterions with the normal complement of eight valence electrons (see also discussions in **Chapters 5** and **16**).

$$
\begin{array}{ccccc}
\mathrm{CH_3}\!\!\diagdown & & \mathrm{CH_3}\!\!\diagdown & & \\
 & \mathrm{S=O} & \text{vs.} & \mathrm{S^+\!-O^-} & \mu = 4.0 \text{ debyes} \\
\mathrm{CH_3}\!\!\diagup & & \mathrm{CH_3}\!\!\diagup & &
\end{array}
$$

$$
\begin{array}{ccccc}
\mathrm{CH_3}\!\!\diagdown\quad\diagup\mathrm{O} & & \mathrm{CH_3}\!\!\diagdown\quad\diagup\mathrm{O^-} & & \\
\quad\ \mathrm{S} & \text{vs.} & \quad\ \overset{2+}{\mathrm{S}} & & \mu = 4.5 \text{ debyes} \\
\mathrm{CH_3}\!\!\diagup\quad\diagdown\mathrm{O} & & \mathrm{CH_3}\!\!\diagup\quad\diagdown\mathrm{O^-} & &
\end{array}
$$

While hypervalent molecules may not be very important as a class[*], they do "push the limits" of conventional bonding theory, and provide a fairly stringent test of quantum chemical models. Comparisons with experimental data for Hartree-Fock models are provided in **Tables A10-17**, for local density models in **Table A10-18**, for BP, BLYP, EDF1 and B3LYP density functional models in **Tables A10-19** to **A10-22**, for MP2 models in **Tables A10-23** and for MNDO, AM1 and PM3 semi-empirical models in **Table A10-24**. A summary of mean absolute errors is given in **Table 10-6**, and graphical comparisons of calculated vs. experimental dipole moments for selected models are provided in **Figures 10-1** to **10-11**. Triangular markers (△) designate hypervalent molecules.

"Limiting" (6-311+G** basis set) Hartree-Fock models consistently overestimate the magnitudes of dipole moments in hypervalent compounds. The largest error is 1 debye (for dimethylsulfoxide), but errors for several other compounds are 0.5 debye or greater. The 6-31G* model provides comparable (and generally slightly better) dipole moments, whereas results from STO-3G and 3-21G models

---

[*]   Exceptions to this generalization include molecules with sulfoxide and sulfone groups.

Table 10-6: **Mean Absolute Errors in Dipole Moments for Hypervalent Molecules**

	STO-3G	3-21G	6-31G*	6-311+G**
MNDO	1.0			
AM1	1.4			
PM3	1.4			
Hartree-Fock	0.8	0.6	0.4	0.5
local density	—	—	0.1	0.4
BP	—	—	0.2	0.4
BLYP	—	—	0.3	0.5
EDF1	—	—	0.2	0.4
B3LYP	—	—	0.2	0.5
MP2	—	—	0.3	0.6

are poorer. The poor performance of the Hartree-Fock 3-21G and 6-31G* models with regard to dipole moments is perhaps unexpected in view of their favorable account of the geometries of hypervalent compounds (see **Chapter 5**), and is certainly disappointing. Graphical comparisons (**Figures 10-1** to **10-4**) provide a clear overview of the performance of Hartree-Fock models.

In terms of mean absolute errors, local density models with the 6-31G* basis set perform better than the corresponding Hartree-Fock model, as well as (and generally better than) any of the density functional models, and better than MP2 models. This parallels previously noted behavior for equilibrium geometries of hypervalent compounds (see **Table 5-8**).

Mean absolute errors indicate that there is actually a significant degradation of performance for density functional models and MP2 models in moving from the 6-31G* to the 6-311+G** basis set. (Mean errors in bond lengths also increase; see **Table 5-8**.) Significantly larger errors are also seen for individual systems (1.5 debye for $FClO_3$ at the MP2/6-311+G** level) and errors approaching 1 debye are common. This behavior is also seen in the plots of calculated vs. experimental dipole moments (**Figures 10-5** vs. **10-6** for EDF1 models, **Figure 10-7** vs. **10-8** for B3LYP models and **Figure 10-9** vs. **10-10** for MP2 models). The reason behind the behavior is unclear, just as is the reason for the poor performance of density functional models and the MP2 model for equilibrium structures in these compounds unclear (see **Chapter 5**). In almost all cases, dipole moments from density functional and MP2 models are larger than experimental values, suggesting bonding that is too ionic. Such an interpretation is consistent with the fact that calculated bond distances involving the "hypervalent atom" are always longer than experimental distances (see **Table 5-8**).

Dipole moments for hypervalent molecules calculated from semi-empirical models are generally larger than experimental values (sometimes by a factor of two or more), suggesting descriptions which are too ionic. **Figure 10-11** provides an overview for the PM3 model. Semi-empirical models should not be used.

Dipole Moments for Flexible Molecules

Only rarely will dipole moments be known from experiment for specific conformers of a flexible molecule. In most cases, these follow from microwave spectroscopy and will generally be limited to quite small molecules. More commonly, measured dipole moments for flexible molecules will represent a Boltzmann average over all accessible conformers. As such, they will depend on the temperature, low temperatures biasing in favor of the low-energy conformers and higher temperatures taking more account of higher-energy conformers.

Proper treatment of dipole moments (or other properties) in flexible molecules thus involves evaluating Boltzmann populations (see discussion in **Chapter 14**). This is beyond our present scope and coverage will be limited to assessing the performance of different models with regard to their ability to reproduce the change in dipole moment with change in conformer. Data are presented in **Table 10-7**, with the usual series of theoretical models surveyed.

With some notable exceptions, all models perform reasonably well in accounting for change in dipole moment with change in conformer. Note in particular, the very large change in dipole moment in going from the *cis* to the *trans* conformer of formic acid. This closely parallels the large difference in energy between the two conformers and presumably is responsible in great part for the strong preference for the *cis* conformer.[*]

expt. rel. E	O	3.9
expt. μ	1.4	3.8

All models (except STO-3G and semi-empirical models) reproduce the significant reduction in dipole moment observed in going from *equatorial* to *axial* fluorocyclohexane, but all underestimate the even

---

[*] Except for fluorocyclohexane and chlorocyclohexane, the lower-energy conformer for all molecules examined has the smaller dipole moment. This is in accord with Coulomb's law (increased separation of charge leads to increased energy).

Table 10-7: Conformational Dependence of Dipole Moments[a]

molecule	conformer	Hartree-Fock				local density		expt.
		STO-3G	3-21G	6-31G*	6-311+G**	6-31G*	6-311+G**	
formic acid	cis	0.63	1.40	1.60	1.68	1.45	1.52	1.42
	trans	2.99	4.55	4.37	4.46	3.83	4.05	3.79
	Δ	2.36	3.15	2.77	2.78	2.38	2.53	2.37
ethyl formate	gauche	0.90	1.79	2.01	2.14	1.78	1.94	1.81
	cis	1.03	1.99	2.18	2.30	2.03	2.18	1.98
	Δ	0.13	0.20	0.17	0.16	0.25	0.24	0.17
propanal	eclipsed	1.76	2.86	2.88	3.06	2.51	2.82	2.52
	anti	1.82	3.08	3.07	3.29	2.71	3.05	2.86
	Δ	0.06	0.22	0.19	0.23	0.20	0.23	0.34
2-methylpropanal	eclipsed	1.79	2.94	2.95	3.13	2.56	2.86	2.69
	anti	1.83	3.12	3.08	3.29	2.73	3.04	2.86
	Δ	0.04	0.18	0.13	0.16	0.17	0.18	0.17
1-fluoropropane	gauche	1.19	2.16	1.90	2.14	1.48	1.94	1.90
	trans	1.24	2.33	2.00	2.25	1.63	2.09	2.05
	Δ	0.05	0.17	0.10	0.11	0.15	0.15	0.15
ethyl phosphine	:PCC gauche	0.55	1.28	1.35	1.38	1.53	1.37	1.22
	:PCC trans	0.50	1.32	1.37	1.40	1.53	1.38	1.23
	Δ	-0.05	0.04	0.05	0.02	0.00	0.01	0.01
ethane thiol	CCSH trans	0.99	1.80	1.86	1.89	1.83	1.76	1.58
	CCSH gauche	0.99	1.85	1.89	1.90	1.88	1.78	1.61
	Δ	0.00	0.05	0.03	0.01	0.05	0.02	0.03
fluorocyclohexane	equatorial	1.34	2.40	2.10	2.42	1.74	2.29	2.11
	axial	1.18	1.96	1.77	2.06	1.38	1.87	1.81
	Δ	-0.16	-0.44	-0.33	-0.36	-0.36	-0.42	-0.30
chlorocyclohexane	equatorial	2.92	2.88	2.77	2.84	2.49	2.53	2.44
	axial	2.67	2.55	2.49	2.55	2.14	2.20	1.91
	Δ	-0.25	-0.33	-0.28	-0.29	-0.35	-0.33	-0.53
mean absolute error in Δ		0.12	0.15	0.11	0.10	0.06	0.08	–

338

Table 10-7: Conformational Dependence of Dipole Moments (2)

molecule	conformer	BP		BLYP		EDF1		expt.
		6-31G*	6-311+G**	6-31G*	6-311+G**	6-31G*	6-311+G**	
formic acid	cis	1.40	1.46	1.37	1.43	1.41	1.47	1.42
	trans	3.74	3.94	3.67	3.93	3.73	3.92	3.79
	Δ	2.34	2.48	2.30	2.50	2.32	2.45	2.37
ethyl formate	gauche	1.88	2.04	1.88	2.08	1.94	2.13	1.81
	cis	2.08	2.22	2.08	2.26	2.12	2.27	1.98
	Δ	0.20	0.18	0.20	0.18	0.18	0.14	0.17
propanal	eclipsed	2.47	2.77	2.43	2.78	2.49	2.78	2.52
	anti	2.67	3.00	2.63	3.00	2.67	2.99	2.86
	Δ	0.20	0.23	0.20	0.22	0.18	0.21	0.34
2-methylpropanal	eclipsed	2.54	2.84	2.51	2.85	2.56	2.85	2.69
	anti	2.70	3.00	2.66	3.01	2.70	3.00	2.86
	Δ	0.16	0.16	0.15	0.16	0.14	0.15	0.17
1-fluoropropane	gauche	1.56	2.03	1.57	2.11	1.58	2.04	1.90
	trans	1.69	2.17	1.70	2.25	1.70	2.17	2.05
	Δ	0.13	0.14	0.13	0.14	0.12	0.13	0.15
ethyl phosphine	:PCC gauche	1.44	1.31	1.37	1.25	1.43	1.32	1.22
	:PCC trans	1.45	1.32	1.36	1.25	1.43	1.33	1.23
	Δ	0.01	0.01	-0.01	0.00	0.00	0.01	0.01
ethane thiol	CCSH trans	1.82	1.76	1.79	1.76	1.80	1.77	1.58
	CCSH gauche	1.85	1.77	1.81	1.75	1.83	1.77	1.61
	Δ	0.03	0.01	0.02	-0.01	0.03	0.00	0.03
fluorocyclohexane	equatorial	1.81	2.37	1.83	2.47	1.83	2.38	2.11
	axial	1.47	1.98	1.49	2.07	1.51	2.01	1.81
	Δ	-0.34	-0.39	-0.34	-0.40	-0.32	-0.37	-0.30
chlorocyclohexane	equatorial	2.64	2.67	2.69	2.76	2.63	2.66	2.44
	axial	2.29	2.34	2.34	2.44	2.31	2.35	1.94
	Δ	-0.35	-0.43	-0.35	-0.32	-0.32	-0.31	-0.53
mean absolute error in Δ		0.05	0.05	0.06	0.06	0.06	0.07	–

Table 10-7: Conformational Dependence of Dipole Moments (3)

molecule	conformer	B3LYP		MP2		semi-empirical			expt.
		6-31G*	6-311+G**	6-31G*	6-311+G**	MNDO	AM1	PM3	
formic acid	cis	1.43	1.50	1.30	1.32	1.49	1.48	1.51	**1.42**
	trans	3.87	4.02	3.95	3.98	3.92	4.02	3.93	**3.79**
	Δ	2.44	2.52	2.65	2.66	2.43	2.54	2.44	**2.37**
ethyl formate	gauche	1.89	2.07	1.74	1.81	1.62	1.47	1.55	**1.81**
	cis	2.09	2.24	1.93	1.98	1.76	1.67	1.72	**1.98**
	Δ	0.20	0.17	0.19	0.17	0.14	0.20	0.17	**0.17**
propanal	eclipsed	2.54	2.84	2.57	2.64	2.33	2.59	2.47	**2.52**
	anti	2.74	3.07	2.71	2.86	2.37	2.68	2.52	**2.86**
	Δ	0.20	0.21	0.14	0.22	0.04	0.09	0.05	**0.34**
2-methylpropanal	eclipsed	2.62	2.91	2.57	2.70	2.26	2.56	2.45	**2.69**
	anti	2.76	3.07	2.72	2.87	2.38	2.67	2.47	**2.86**
	Δ	0.14	0.16	0.15	0.17	0.14	0.11	0.02	**0.17**
1-fluoropropane	gauche	1.64	2.10	1.76	2.01	1.82	1.64	1.54	**1.90**
	trans	1.77	2.23	1.90	2.15	1.89	1.73	1.60	**2.05**
	Δ	0.13	0.13	0.14	0.14	0.07	0.09	0.06	**0.15**
ethyl phosphine	:PCC gauche	1.38	1.29	1.38	1.22	0.83	1.95	1.14	**1.22**
	:PCC trans	1.38	1.30	1.40	1.23	0.84	2.14	1.17	**1.23**
	Δ	0.00	0.01	0.02	0.01	0.01	0.19	0.03	**0.01**
ethane thiol	CCSH trans	1.80	1.78	1.78	1.66	1.36	1.79	1.98	**1.58**
	CCSH gauche	1.83	1.78	1.82	1.66	1.42	1.89	2.06	**1.61**
	Δ	0.03	0.00	0.04	0.00	0.06	0.10	0.08	**0.03**
fluorocyclohexane	equatorial	1.89	2.43	1.99	2.31	2.01	1.83	1.75	**2.11**
	axial	1.56	2.04	1.62	1.92	1.87	1.64	1.59	**1.81**
	Δ	-0.33	-0.39	-0.37	-0.39	-0.14	-0.19	-0.16	**-0.30**
chlorocyclohexane	equatorial	2.65	2.71	2.45	2.35	2.15	1.91	1.71	**2.44**
	axial	2.32	2.39	2.14	2.04	2.01	1.74	1.58	**1.91**
	Δ	-0.33	-0.32	-0.31	-0.31	-0.14	-0.17	-0.13	**-0.53**
mean absolute error in Δ		0.06	0.07	0.09	0.08	0.12	0.14	0.13	–

a) dipole moment for lower-energy conformer first, followed by that for the higher-energy conformer, followed by the difference (Δ).

larger reduction (in the same direction) seen in chlorocyclohexane. All models account for the modest but noticeable increases in dipole moments from the more stable to the less stable conformers of ethyl formate, propanal and 2-methylpropanal and 1-fluoropropane, and all models show little change in the dipole moments of ethyl phosphine and ethane thiol, in line with the experimental data.

In terms of mean absolute errors (associated with the change in dipole moment from one conformer to another), density functional models perform best and Hartree-Fock and semi-empirical models perform worst. The performance of MP2 models is intermediate.

References

1. Experimental data has been taken from the following compendia: R.D. Nelson, D.R. Lide and A.A. Maryott, *Selected Values of Electric Dipole Moments for Molecules in the Gas Phase*, NSRDA-NBS 10, U.S. Government Printing Office, Washington, D.C., 1967; (b) A.L. McClellan, *Tables of Experimental Dipole Moments*, W.H. Freeman, San Francisco, 1963; (c) *ibid.*, vol. 2, Rahara Enterprises, El Cerritos, CA, 1974; (d) K.P. Huber and G. Herzberg, *Molecular Spectra and Molecular Structure. IV. Constants of Diatomic Molecules*, Van Nostrand Reinhold, New York, 1979.

Chapter 11

Overview of Performance and Cost

This chapter addresses the relative "cost" of molecular mechanics and quantum chemical models for energy and equilibrium geometry calculations as well as for frequency evaluations. Taken together with performance issues addressed in previous chapters, this allows broad recommendations to be made regarding selection of an appropriate model.

Introduction

Beyond its ability to account for what is known, the second important consideration in the selection of an appropriate molecular mechanics or quantum chemical model is its "cost". It is really not possible to estimate precisely how much computer time a particular calculation will require, as many factors remain uncertain. In addition to the size of the system at hand and the choice of model (both of which can be precisely defined), there are issues the "quality" of the guess (which in turn relates to the "experience" of the user) and the "inherent difficulty" of the problem (some things are easier than others). It is possible, however, to provide representative examples to help distinguish applications which are practical from those which are clearly not.

Computation Times

Relative times for the MMFF molecular mechanics model, the AM1 semi-empirical model, Hartree-Fock models with 3-21G, 6-31G* and 6-311+G** basis sets, EDF1 and B3LYP density functional models and the MP2 model with 6-31G* and 6-311+G** basis sets, for energy calculations, geometry optimizations and frequency evaluations, and the localized MP2 (LMP2) model with 6-31G* and

6-311+G** basis sets (energies only), for camphor ($C_{10}H_{16}O$, a typical "small" organic molecule), for morphine ($C_{17}H_{19}NO_2$, a typical "medium size" organic molecule) and (energies and geometry optimizations only) for triacetyldynemicin A ($C_{36}H_{25}NO_{12}$, a typical "large" organic molecule) are given in **Table 11-1**. These are relative to times for Hartree-Fock 3-21G energy calculations on camphor, morphine and triacetyldynemicin A, respectively. MNDO and PM3 semi-empirical models will yield similar times to AM1, and local density and BP and BLYP density functional models will yield similar times to EDF1 models.

Molecular mechanics calculations do not "show up" on the chart. They are at least an order of magnitude less costly than the simplest (semi-empirical) quantum chemical calculations, and the ratio between the two increases rapidly with increasing molecular size. Molecular mechanics is really the only viable alternative at present for molecules comprising more than a few hundred atoms. It is also likely to be the only practical alternative for conformational searching on molecules with more than a few degrees of freedom.

The cost of evaluating the energy using the Hartree-Fock 3-21G model is two orders of magnitude greater than that for obtaining an equilibrium geometry using the AM1 semi-empirical model. This ratio should maintain with increasing size, as both semi-empirical and Hartree-Fock models scale as the cube of number of basis functions.[*] Geometry optimization using 3-21G is approximately an order of magnitude more costly than energy calculation. This ratio should increase with increasing molecule size, due to an increase in the number of geometrical variables and a corresponding increase in the number of steps required for optimization[**]. The cost difference for both energy evaluation and

---

[*] Both Hartree-Fock and density functional models actually formally scale as the fourth power of the number of basis functions. In practice, however, both scale as the cube or even lower power. Semi-empirical models appear to maintain a cubic dependence. "Pure" density functional models (excluding hybrid models such as B3LYP which require the Hartree-Fock exchange) can be formulated to scale linearly for sufficiently large systems. MP2 models scale formally as the fifth power of the number of basis functions, and this dependence does not diminish significantly with increasing number of basis functions.

[**] The number of steps required for geometry optimization formally scales with the number of independent variables. In practice, optimization typically requires on the order of a quarter to a half the number of steps as the number of independent variables. Transition-state optimization typically requires two to three times the number of steps as geometry optimization.

Table 11-1: Relative Computation Times

model	camphor[a]			morphine[b]			triacetyldynemicin A[c]	
	energy	geometry[d]	frequency	energy	geometry[e]	frequency	energy	geometry[f]
MMFF								
AM1	too small to measure							0.1
3-21G	1	5	27	1[g]	12	38	1[h]	60
6-31G*	7	30	150	8	110	–	7	540
6-311+G**	42	180	–	40	–	–	–	–
EDF1/6-31G*	12	57	710	10	140	–	5	240
EDF1/6-311+G**	60	290	–	50	–	–	–	–
B3LYP/6-31G*	13	65	720	12	160	–	10	610
B3LYP/6-311+G**	85	370	–	76	–	–	–	–
MP2/6-31G*	27	270	–	80	2000	–	320	–
MP2/6-311+G**	260	–	–	650	–	–	–	–
LMP2/6-31G*	25	–	–	50	–	–	90	–
LMP2/6-311+G**	140	–	–	270	–	–	–	–

a) $C_{10}H_{16}O$; 131 basis functions for 3-21G, 197 basis functions for 6-31G* and 338 basis functions for 6-311+G**
b) $C_{17}H_{19}NO_3$; 227 basis functions for 3-21G, 353 basis functions for 6-31G* and 576 basis functions for 6-311+G**
c) $C_{36}H_{25}NO_{12}$; 491 basis functions for 3-21G and 785 basis functions for 6-31G*
d) assumes 4 optimization steps
e) assumes 12 optimization steps
f) assumes 36 optimization steps
g) approximately 4 relative to 3-21G energy calculation on camphor
h) approximately 27 relative to 3-21G energy calculation on camphor

geometry optimization between (Hartree-Fock) 3-21G and 6-31G* calculations is on the order of five or ten times.

EDF1 density functional calculations are only slightly more costly than Hartree-Fock calculations with the same basis set for small and medium size molecules, and actually less costly for large molecules. B3LYP calculations are roughly 50% more costly than Hartree-Fock calculations. This applies both to energy calculations and to geometry optimization. Note, however, that density functional calculations (and MP2 calculations) require basis sets which are larger than those needed for Hartree-Fock calculations. In particular, while the 3-21G basis set is often sufficient for Hartree-Fock geometry optimizations, it is not acceptable as a basis for density functional or MP2 geometry optimizations (see discussion in **Chapter 5**).

MP2 calculations are much more costly than comparable (same basis set) Hartree-Fock and density functional calculations. In practice, their application is much more limited than either of these models. Localized MP2 (LMP2) energy calculations are similar in cost to MP2 calculations for small molecules, but the cost differential rapidly increases with increasing molecular size. Still they are close to an order of magnitude more costly than Hartree-Fock or density functional calculations.

Summary

Taking both quality of results and "cost" into account, is it possible to say with certainty which model is the "best" for a particular application? Probably not, although rough guidelines can certainly be set. **Table 11-2** provides an overview of the classes of methods discussed in this guide with regard to the calculation of equilibrium and transition-state geometries, conformations and reaction energetics. Equilibrium geometry "assessment" has been subdivided depending on whether or not transition metals are involved, and thermochemical "assessment" has been divided between non-*isodesmic* and *isodesmic* processes. For each task, the methods are "graded": **G** (good), **F** (fair) and **P** (poor). While the grading is very rough, it allows the obvious trends to be exposed:

Table 11-2: Performance and Cost of Models

task	molecular mechanics	semi-empirical	Hartree-Fock	density functional		LMP2	MP2
				EDF1	B3LYP		
geometry (organic)	F→G	G	G	G	G	N/A	G
geometry (transition metals)	P	G	P	G	G	N/A	F
transition-state geometry	N/A	F→G	G	G	G	N/A	G
conformation	F→G	P	F→G	G	G	N/A	G
thermochemistry (non *isodesmic*)	N/A	P	F→G	G	G	G	G
thermochemistry (*isodesmic*)	N/A	P	G	G	G	G	G
cost	very low	low	moderate	moderate	moderate	high	high

G = good   F= fair   P = poor   N/A = not applicable

i) All models provide a good account of equilibrium geometries for organic molecules and all quantum chemical models of transition-state geometries for organic reactions. This is not to suggest that all models are "equal", but rather that all meet "minimum standards". In particular, low-cost molecular mechanics and semi-empirical models often provide very good equilibrium geometries, and only rarely yield very poor geometries.

Hartree-Fock models with basis sets larger than 6-31G* do not provide significantly improved descriptions of either equilibrium or transition-state geometries over the 6-31G* model and, in most cases, the 3-21G model. Note, that MP2 and density functional models require basis sets which incorporate polarization functions to yield acceptable geometries. There is very little difference in geometries obtained from MP2 and density functional models with 6-31G* and 6-311+G** basis sets.

Hartree-Fock models are not reliable for geometry calculations on compounds incorporating transition metals, but the PM3 semi-empirical model and density functional models provide good accounts. While MP2 models provide reasonable geometries for many systems, structures for some transition-metal compounds are significantly in error.

ii) Hartree-Fock models (6-31G* and larger basis sets), MP2 models and density functional models all generally provide good descriptions of conformational energy differences in organic compounds. Semi-empirical models and the SYBYL molecular mechanics model do not provide acceptable results, but the MMFF molecular mechanics model appears to do an excellent job.

iii) Hartree-Fock, density functional and MP2 models all generally provide good accounts of the energetics of *isodesmic* reactions. MP2 models with the 6-311+G** basis set are, however, required to provide a uniformly excellent account of bond separation energies. MP2 and density functional models are more reliable than Hartree-Fock models for describing the energetics of non-*isodesmic* reactions although, except for reactions which involve bond making or breaking, 6-31G* and larger-basis-set Hartree-Fock models generally also yield acceptable results.

Density functional and MP2 models are needed to accurately account for the energetics of reactions where bonds are broken or formed and to describe absolute activation energies. Hartree-Fock models are unsatisfactory, but properly account for relative activation energies expressed in terms of *isodesmic* processes. MP2 models are also satisfactory here but density functional models sometimes lead to problems.

Semi-empirical models are unsatisfactory in describing the energetics of all types of reactions, *isodesmic* processes included.

Recommendations

Molecular mechanics models are restricted to the description of molecular equilibrium geometry and conformation. They are the method of choice for conformational searching on complex systems.

Semi-empirical models are particularly attractive for:

i) Equilibrium structure determinations for large molecules, where the cost of *Hartree-Fock*, *density functional* and *MP2 models* may be prohibitive.

ii) Transition-state geometry optimizations, where the cost of *Hartree-Fock*, *density functional* and *MP2 models* may be prohibitive.

iii) Equilibrium and transition-state geometry optimizations involving transition metals, where *Hartree-Fock models* are known to produce poor results, and where the cost of *density functional* and *MP2 models* may be prohibitive.

Semi-empirical models are unsuitable for :

i) Calculation of reaction energies, even the energies of *isodesmic* processes.

ii) Calculation of conformational energy differences.

Hartree-Fock models are particularly attractive for:

i)   Equilibrium and transition-state geometry optimizations on medium-size molecules where transition metals are not present, where increased accuracy over that available from *semi-empirical models* is required, and where the cost of *MP2 models* may be prohibitive.

ii) Calculation of reaction energies (except reactions involving net bond making or breaking), where *semi-empirical models* yield unacceptable results, and where the cost of *MP2 models* may be prohibitive.

Hartree-Fock models are unsuitable for:

i)   Equilibrium and transition-state geometry optimizations on transition-metal inorganic and organometallic molecules.

ii) Calculation of reaction energies which involve net bond making or breaking and calculation of absolute activation energies.

Density functional models and *MP2 models* are needed for accurate descriptions of the thermochemistry of reactions which involve net bond making or breaking, and for calculation of absolute activation energies. In practice, *MP2 models* may only be applied to relatively small molecules, whereas *density functional models* are roughly comparable in cost to *Hartree-Fock models* for molecules of moderate size.

Density functional models are particularly attractive for:

i)   Accurate equilibrium (and transition-state) geometry calculations where the cost of *MP2 models* may be prohibitive.

ii) Calculations on inorganic and organometallic systems where *Hartree-Fock models* are not suitable, and where the cost of *MP2 models* may be prohibitive.

iii) Thermochemical calculations, in particular, those which involve net bond making or breaking, and absolute activation energy calculations.

Section III
Doing Calculations

Models available for calculation of molecular structures, relative stabilities, and other properties differ in "cost" of application by several orders of magnitude (see **Table 11-1**). Molecular mechanics and semi-empirical models are the least costly and correlated methods such as the MP2 model are the most costly. In between are Hartree-Fock models and density functional models. While in general, the most costly models provide the best overall descriptions, comparisons provided in the previous section suggest that even semi-empirical and small-basis-set Hartree-Fock models perform quite well in certain tasks. For example, equilibrium and transition-state geometries appear to be well described even with semi-empirical models and certainly with small-basis-set Hartree-Fock models. Might these structures replace geometries from higher-level calculations for relative energy and property calculations? Also, while semi-empirical models provide a poor account of reaction energetics, Hartree-Fock models, even with small basis sets, often perform well, except in situations where there is net bond making or breaking. Is it possible to formulate energetic comparisons which avoid this pitfall? Finally, for those comparisons which necessarily involve bond making or breaking, are LMP2 models as reliable as MP2 models, or might density functional models replace much more costly MP2 models?

These and related issues are the subjects of chapters in this section. A pair of chapters address issues associated with *Obtaining and Using Equilibrium Geometries* and *Obtaining and Using Transition-State Geometries*, respectively. Other chapters discuss practical issues associated with *Using Energies for Thermochemical and Kinetic Comparisons* and which arise in *Dealing with Flexible Molecules*. The section concludes with an outline of methods available for *Obtaining and Interpreting Atomic Charges*.

351

Chapter 12

Obtaining and Using Equilibrium Geometries

This chapter addresses a number of practical issues associated with establishing, verifying and using equilibrium geometries. It provides guidelines on how best to calculate geometry, and criteria for establishing whether or not a calculated geometry actually corresponds to an energy minimum. The bulk of the chapter focuses on "choice of geometry", and considers under what conditions a geometry from one model might be used for energy and property evaluation with another "better" model.

Introduction

The energy of a molecule as well as other properties depend on its geometry. Even small changes in geometry can lead to significant changes in total energy and/or other properties. Proper choice of molecular geometry is therefore quite important in carrying out computational studies. What geometry is best? Experimental geometries would seem to be the obvious choice, given that they are available and are accurate. The trouble is, of course, that accurate experimental geometries are often not available. Accurate gas-phase structure determinations using such techniques as microwave spectroscopy are very tedious and have generally been restricted to very small molecules. Even then they are often not "complete", certain geometrical parameters having been "assumed". X-ray structure determinations on solid samples are routine, but one must be concerned about the role of the crystalline environment in altering geometry and particularly conformation. Ions present special problems. Gas-phase data are unavailable, and different counterions present in crystals lead to different geometries, sometimes significantly so. Finally, only a few experimental geometries exist for reactive "short-lived" molecules, let alone for molecular complexes including hydrogen-bonded complexes. All in all, use of experimental geometries in computational studies is not usually a viable alternative.

Another approach is to employ "idealized" geometries. This should be reasonable given the very high degree of systematics exhibited by a large range of geometries, in particular, those of organic molecules. However, as mentioned above, energies and other properties may be sensitive to subtle changes in geometry. For example, a major reason that the dipole moment of trimethylamine is smaller than the dipole moment of ammonia is the change in the local geometry about nitrogen. In order to avoid (or lessen) steric interactions, the CNC bond angle in trimethylamine increases beyond its "ideal" value (the HNH bond angle in ammonia where there is no crowding). As a consequence, the nitrogen lone pair in trimethylamine is significantly less directed, and the dipole moment is reduced. (In the limit of a "planar" nitrogen center, the dipole moment would be zero.) Were both ammonia and trimethylamine constrained to have the same bond angle about nitrogen, then the relative magnitudes of the two dipole moments would not be properly reproduced. Another problem with the use of "idealized" geometries is that the geometries of many of the most interesting molecules may differ greatly from the norm. All in all, "idealized" geometries also do not offer a good solution.

In the final analysis, there is usually little choice but to obtain geometries directly from calculation. This is not as difficult a chore as it might appear. For one, and as shown in the previous section, equilibrium geometries can be reliably calculated even with relatively simple models. Second, equilibrium geometry calculation is fully automated in modern computer programs such as Spartan, and therefore requires no more human effort than use of an experimental or standard geometry. Geometry calculation is, however, significantly more costly than energy calculation, and there are a number of issues which need to be raised. These issues constitute the subject of the present chapter.

Obtaining Equilibrium Geometries

As documented in **Chapter 5**, small-basis-set Hartree-Fock models generally provide a reasonable account of molecular equilibrium geometries, at least for organic molecules. Semi-empirical models also usually provide a good account of geometries of organic molecules, and the PM3 model also performs well for organometallic compounds. However, semi-empirical models are likely to be less satisfactory for classes of molecules for which they have not been explicitly parameterized, e.g., charged species and radicals. (In practice, situations where geometries obtained from semi-empirical models are outlandish are actually quite rare.) Given their overall success, however, it is nearly always advantageous to utilize low-cost semi-empirical models or small-basis-set Hartree-Fock models to provide a guess at equilibrium geometry. In practice, modern computer programs such as Spartan allow such a "two-step optimization" to be carried out automatically.

Verifying Calculated Equilibrium Geometries

Geometry optimization is an iterative process. The energy and its first derivatives with respect to all geometrical coordinates are calculated for the guess geometry, and this information is then used to project a new geometry. This process needs to continue until the lowest-energy or "optimized" geometry is reached. Three criteria must be satisfied before a geometry is accepted as optimized. First, successive geometry changes must not lower the energy by more than a specified (small) value. Second, the energy gradient (first derivative of the energy with respect to geometrical distortions) must closely approach zero. Third, successive iterations must not change any geometrical parameter by more than a specified (small) value.

In principle, geometry optimization carried out in the absence of symmetry, i.e., in C_1 symmetry, must result in a local minimum. On the other hand, imposition of symmetry may result in a geometry which is not a local minimum. For example, optimization of ammonia constrained to a planar trigonal geometry (D_{3h} symmetry) will result in a geometry which is an energy maximum in one dimension. Indeed,

355

this geometry corresponds to the transition state for pyramidal inversion in ammonia[*], The most conservative tactic is always to optimize geometry in the absence of symmetry. If this is not practical, and if there is any doubt whatsoever that the "symmetrical structure" actually corresponds to an energy minimum, then it is always possible to verify that the geometry located indeed corresponds to a local minimum by calculating vibrational frequencies for the final (optimized) geometry. These should all be real numbers. The presence of an imaginary frequency indicates that the corresponding coordinate is not an energy minimum.

The default criteria in Spartan have been chosen in an attempt to assure that calculated bond distances, bond angles and dihedral angles are within 0.005Å, 0.5° and 1° of their respective "exact" values. This will normally be sufficient, but these criteria can be "tightened" to more closely approach the "exact" equilibrium geometry. Alternatively, convergence criteria can be "loosened" to save computation time.

Geometry optimization does not guarantee that the final geometry has a lower energy than any other geometry of the same molecular formula. All that is guarantees is that the geometry corresponds to a local minimum, that is, a geometry the energy of which is lower than that of any "similar geometry". However, the resulting structure may still not be the lowest-energy structure possible for the molecule. There may be other local minima which are accessible via low-energy rotations about single bonds or puckering of rings, and which are actually lower in energy[**]. The full collection of local minima are referred to as conformers. Finding the lowest-energy conformer or global minimum requires repeated geometry optimization starting with different initial geometries (see discussion in **Chapter 14**).

---

[*] Transition states for a number of simple "reactions" can be located simply by geometry optimization subject to an overall symmetry constraint.

[**] Of course, entirely different structures or *isomers* are also possible, although these are not normally thought of as interconnected by low-energy processes. For the purpose of the present discussion, isomers are considered as independent molecules which are not accessible to each other.

Using "Approximate" Equilibrium Geometries to Calculate Thermochemistry

Is it always necessary to utilize "exact" equilibrium geometries in carrying out thermochemical comparisons, or are there situations where "approximate" geometries will suffice?

This is a question of considerable practical importance, given that optimization of equilibrium geometry can easily require one or two orders of magnitude more computation than an energy (property) calculation at a single geometry (see **Chapter 11**). Rephrased, the question might read:

Is the substantial added effort required to produce a proper optimized geometry effort well spent?

This issue is addressed by reference to some of the same types of reactions previously discussed in **Chapter 6**, in particular, isomerization reactions, bond separation reactions and reactions relating acid and base strength to those of standards. For each reaction class, comparisons are made using four different models: 6-31G*, EDF1/6-31G*, B3LYP/6-31G* and MP2/6-31G*, and several different choices of geometry: AM1 or PM3, 3-21G and/or 6-31G* and "exact". The choice of these models is deliberate and reflects the practical focus of this section. All four models have generally been shown to provide reasonable energies for the classes of reactions considered (see **Chapter 6**), yet are simple enough for widespread application. It is to be expected that other (more costly) models will give rise to similar conclusions. The choice of models for geometry calculation is also deliberate. All have been shown to produce reliable results (see **Chapter 5**) and all may be extended to systems of considerable size.

Although the purpose of this section is not to assess the performance of the various models with regard to reaction thermochemistry (this has already been addressed at length in **Chapter 6**), experimental data and G3 data[1] have been provided where available. This allows comparison of the magnitude of errors brought about from the use of approximate geometries, relative to errors inherent to use of a particular theoretical model in describing a particular type of reaction.

The effect of use of approximate geometries on the relative energies of structural isomers is examined in **Table 12-1** for the 6-31G* model, **Tables 12-2** and **12-3** for EDF1/6-31G* and B3LYP/6-31G* density functional models and **Table 12-4** for the MP2/6-31G* model. AM1 and 3-21G geometries, in addition to "exact" geometries, have been considered for the Hartree-Fock model, and AM1, 3-21G and 6-31G* geometries (in addition to "exact" geometries) have been considered for density functional and MP2 models. Only small changes in relative energies in response to changes in equilibrium geometries are noted for all cases. In fact, mean absolute errors between calculated and experimental isomer energies are either unchanged or actually reduced through use of approximate geometries.

Bond separation reactions (discussed in **Chapter 6**) are of considerable practical importance in that they may be used in conjunction with limited experimental data (or high-quality calculated data) to estimate heats of formation, which in turn may be used for whatever thermochemical comparisons may be desired. (Discussion is provided in **Chapter 13**.) In the interest of cost savings, it would be highly desirable for bond separation energies to be based on approximate geometries. A few examples are provided in **Tables 12-5** to **12-8**. Here, bond separation energies from "full" 6-31G*, EDF1/6-31G*, B3LYP/6-31G* and MP2/6-31G* calculations, respectively, are compared with calculations based on "approximate" reactant and product geometries.

As with structural isomer comparisons, detailed choice of reactant and product geometries has little overall effect on the quality of results for all four levels of calculation. 3-21G geometries generally provide closer agreement to "exact" results than do AM1 geometries and 6-31G* geometries (for EDF1, B3LYP and MP2 models) closer still, but the differences are not great. As previously noted in **Chapter 6**, MP2 models, in particular, lead to better (more consistent) overall results than either Hartree-Fock or density functional models, even though all models provide a respectable account. (Small differences in quality may, however, be relevant in providing heats of formation; see **Chapter 13** for a discussion.) It may be advisable to spend available resources on obtaining a "better" energy rather than on obtaining a "better" equilibrium geometry.

Table 12-1: Effect of Choice of Geometry on Relative Energies of Structural Isomers. 6-31G* Model

formula (reference)	isomer	geometry			
		AM1	3-21G	6-31G*	expt.
C_2H_3N (acetonitrile)	methyl isocyanide	21	21	24	**21**
C_2H_4O (acetaldehyde)	oxirane	30	33	31	**26**
$C_2H_4O_2$ (acetic acid)	methyl formate	15	14	13	**18**
C_2H_6O (ethanol)	dimethyl ether	8	7	7	**12**
C_3H_4 (propyne)	allene	1	2	2	**2**
	cyclopropene	26	27	26	**22**
C_3H_6 (propene)	cyclopropane	8	8	8	**7**
C_4H_6 (1,3-butadiene)	2-butyne	10	7	7	**7**
	cyclobutene	13	13	13	**9**
	bicyclo [1.1.0] butane	32	31	30	**23**
mean absolute error		3	3	3	–

Table 12-2: Effect of Choice of Geometry on Relative Energies of Structural Isomers. EDF1/6-31G* Model

formula (reference)	isomer	geometry				
		AM1	3-21G	6-31G*	EDF1/ 6-31G*	expt.
C_2H_3N (acetonitrile)	methyl isocyanide	25	25	25	25	**21**
C_2H_4O (acetaldehyde)	oxirane	25	26	26	26	**26**
$C_2H_4O_2$ (acetic acid)	methyl formate	12	12	12	12	**18**
C_2H_6O (ethanol)	dimethyl ether	6	6	6	6	**12**
C_3H_4 (propyne)	allene	-5	-4	-4	-4	**2**
	cyclopropene	18	18	18	18	**22**
C_3H_6 (propene)	cyclopropane	5	5	6	5	**7**
C_4H_6 (1,3-butadiene)	2-butyne	10	8	8	8	**7**
	cyclobutene	10	10	9	10	**9**
	bicyclo [1.1.0] butane	25	23	23	23	**23**
mean absolute error		4	3	3	3	–

Table 12-3: **Effect of Choice of Geometry on Relative Energies of Structural Isomers. B3LYP/6-31G* Model**

| formula (reference) | isomer | geometry | | | | |
		AM1	3-21G	6-31G*	B3LYP/ 6-31G*	expt.
C_2H_3N (acetonitrile)	methyl isocyanide	24	24	24	27	**21**
C_2H_4O (acetaldehyde)	oxirane	27	29	28	28	**26**
$C_2H_4O_2$ (acetic acid)	methyl formate	12	12	12	12	**18**
C_2H_6O (ethanol)	dimethyl ether	6	6	5	5	**12**
C_3H_4 (propyne)	allene	-4	-3	-3	-3	**2**
	cyclopropene	21	22	21	22	**22**
C_3H_6 (propene)	cyclopropane	7	7	8	8	**7**
C_4H_6 (1,3-butadiene)	2-butyne	10	8	8	8	**7**
	cyclobutene	12	12	12	12	**9**
	bicyclo [1.1.0] butane	30	28	28	28	**23**
mean absolute error		4	3	3	4	–

Table 12-4: **Effect of Choice of Geometry on Relative Energies of Structural Isomers. MP2/6-31G* Model**

| formula (reference) | isomer | geometry | | | | |
		AM1	3-21G	6-31G*	MP2/ 6-31G*	expt.
C_2H_3N (acetonitrile)	methyl isocyanide	29	27	27	29	**21**
C_2H_4O (acetaldehyde)	oxirane	27	28	27	27	**26**
$C_2H_4O_2$ (acetic acid)	methyl formate	14	14	14	14	**18**
C_2H_6O (ethanol)	dimethyl ether	9	9	9	9	**12**
C_3H_4 (propyne)	allene	3	4	4	5	**2**
	cyclopropene	22	23	22	23	**22**
C_3H_6 (propene)	cyclopropane	4	4	4	4	**7**
C_4H_6 (1,3-butadiene)	2-butyne	7	5	5	4	**7**
	cyclobutene	8	8	8	8	**9**
	bicyclo [1.1.0] butane	23	21	20	21	**23**
mean absolute error		2	3	3	3	–

Table 12-5: Effect of Choice Geometry on Energies of Bond Separation Reactions. 6-31G* Model

molecule	bond separation reaction	geometry			G3	expt.
		AM1	3-21G	6-31G*		
propane	$CH_3CH_2CH_3 + CH_4 \rightarrow 2CH_3CH_3$	1	1	1	2.3	2.6
dimethyl ether	$CH_3OCH_3 + H_2O \rightarrow 2CH_3OH$	3	3	3	5.1	5.4
propene	$CH_3CHCH_2 + CH_4 \rightarrow CH_3CH_3 + CH_2CH_2$	4	4	4	4.9	5.4
acetaldehyde	$CH_3CHO + CH_4 \rightarrow CH_3CH_3 + H_2CO$	10	10	10	10.6	11.4
propyne	$CH_3CCH + CH_4 \rightarrow CH_3CH_3 + HCCH$	7	8	8	7.8	7.6
formamide	$NH_2CHO + CH_4 \rightarrow CH_3NH_2 + H_2CO$	31	32	31	–	30.8
benzene	⬡ + $6CH_4 \rightarrow 3CH_3CH_3 + 3CH_2CH_2$	60	57	58	61.0	64.2
cyclopropane	△ + $3CH_4 \rightarrow 3CH_3CH_3$	-24	-26	-26	-23.5	-19.6
oxacyclopropane	△(O) + $2CH_4 + H_2O \rightarrow CH_3CH_3 + 2CH_3OH$	-19	-21	-22	-14.1	-10.5
bicyclo[1.1.0]butane	▷ + $6CH_4 \rightarrow 5CH_3CH_3$	-62	-64	-63	-55.0	-45.5
mean absolute deviation from G3		4	5	5	–	–

Table 12-6: Effect of Choice Geometry on Energies of Bond Separation Reactions. EDF1/6-31G* Model

molecule	bond separation reaction	geometry				G3	expt.
		AM1	3-21G	6-31G*	EDF1/6-31G*		
propane	$CH_3CH_2CH_3 + CH_4 \rightarrow 2CH_3CH_3$	1	1	1	1	2.3	2.6
dimethyl ether	$CH_3OCH_3 + H_2O \rightarrow 2CH_3OH$	3	3	3	3	5.1	5.4
propene	$CH_3CHCH_2 + CH_4 \rightarrow CH_3CH_3 + CH_2CH_2$	5	5	5	5	4.9	5.4
acetaldehyde	$CH_3CHO + CH_4 \rightarrow CH_3CH_3 + H_2CO$	11	12	11	11	10.6	11.4
propyne	$CH_3CCH + CH_4 \rightarrow CH_3CH_3 + HCCH$	10	10	10	11	7.8	7.6
formamide	$NH_2CHO + CH_4 \rightarrow CH_3NH_2 + H_2CO$	35	36	35	35	–	30.8
benzene	$+ 6CH_4 \rightarrow 3CH_3CH_3 + 3CH_2CH_2$	71	70	70	71	61.0	64.2
cyclopropane	$+ 3CH_4 \rightarrow 3CH_3CH_3$	-19	-21	-21	-20	-23.5	-19.6
oxacyclopropane	$+ 2CH_4 + H_2O \rightarrow CH_3CH_3 + 2CH_3OH$	-10	-12	-11	-11	-14.1	-10.5
bicyclo[1.1.0]butane	$+ 6CH_4 \rightarrow 5CH_3CH_3$	-47	-48	-49	-48	-55.0	-45.5
mean absolute error deviation from G3		2	2	2	2	–	–

Table 12-7: Effect of Choice Geometry on Energies of Bond Separation Reactions. B3LYP/6-31G* Model

molecule	bond separation reaction	geometry				G3	expt.
		AM1	3-21G	6-31G*	B3LYP/6-31G*		
propane	$CH_3CH_2CH_3 + CH_4 \rightarrow 2CH_3CH_3$	2	1	1	1	2.3	2.6
dimethyl ether	$CH_3OCH_3 + H_2O \rightarrow 2CH_3OH$	4	3	3	3	5.1	5.4
propene	$CH_3CHCH_2 + CH_4 \rightarrow CH_3CH_3 + CH_2CH_2$	5	5	5	5	4.9	5.4
acetaldehyde	$CH_3CHO + CH_4 \rightarrow CH_3CH_3 + H_2CO$	11	10	11	11	10.6	11.4
propyne	$CH_3CCH + CH_4 \rightarrow CH_3CH_3 + HCCH$	10	11	10	10	7.8	7.6
formamide	$NH_2CHO + CH_4 \rightarrow CH_3NH_2 + H_2CO$	33	35	34	34	–	30.8
benzene	$\bigcirc + 6CH_4 \rightarrow 3CH_3CH_3 + 3CH_2CH_2$	68	67	67	67	61.0	64.2
cyclopropane	$\triangle + 3CH_4 \rightarrow 3CH_3CH_3$	-24	-25	-26	-23	-23.5	-19.6
oxacyclopropane	$\triangle_O + 2CH_4 + H_2O \rightarrow CH_3CH_3 + 2CH_3OH$	-16	-18	-17	-17	-14.1	-10.5
bicyclo[1.1.0]butane	$\bowtie + 6CH_4 \rightarrow 5CH_3CH_3$	-57	-59	-59	-59	-55.0	-45.5
mean absolute error deviation from G3		3	4	4	3	–	–

Table 12-8: Effect of Choice Geometry on Energies of Bond Separation Reactions. MP2/6-31G* Model

molecule	bond separation reaction	geometry				G3	expt.
		AM1	3-21G	6-31G*	MP2/6-31G*		
propane	$CH_3CH_2CH_3 + CH_4 \rightarrow 2CH_3CH_3$	3	2	2	2	**2.3**	**2.6**
dimethyl ether	$CH_3OCH_3 + H_2O \rightarrow 2CH_3OH$	5	5	5	5	**5.1**	**5.4**
propene	$CH_3CHCH_2 + CH_4 \rightarrow CH_3CH_3 + CH_2CH_2$	5	5	5	5	**4.9**	**5.4**
acetaldehyde	$CH_3CHO + CH_4 \rightarrow CH_3CH_3 + H_2CO$	11	11	11	11	**10.6**	**11.4**
propyne	$CH_3CCH + CH_4 \rightarrow CH_3CH_3 + HCCH$	7	8	8	8	**7.8**	**7.6**
formamide	$NH_2CHO + CH_4 \rightarrow CH_3NH_2 + H_2CO$	33	34	33	33	**–**	**30.8**
benzene	$+ 6CH_4 \rightarrow 3CH_3CH_3 + 3CH_2CH_2$	73	72	72	71	**61.0**	**64.2**
cyclopropane	$+ 3CH_4 \rightarrow 3CH_3CH_3$	-22	-24	-24	-24	**-23.5**	**-19.6**
oxacyclopropane	$+ 2CH_4 + H_2O \rightarrow CH_3CH_3 + 2CH_3OH$	-12	-14	-14	-13	**-14.1**	**-10.5**
bicyclo[1.1.0]butane	$+ 6CH_4 \rightarrow 5CH_3CH_3$	-55	-56	-56	-55	**-55.0**	**-45.5**
mean absolute error deviation from G3		2	3	3	2	–	–

The effect of choice of geometry on relative acid and base strengths is considered next. As shown in **Chapter 6** for the case of nitrogen bases, Hartree-Fock, density functional and MP2 models are all quite successful in accounting for relative base strengths. The present data (**Tables 12-9** to **12-12** for 6-31G*, EDF1/6-31G*, B3LYP/6-31G* and MP2/6-31G* models, respectively) show that the overall quality of results is not greatly affected by detailed choice of geometry. Use of AM1 geometries in place of "exact" structures actually improves the quality of the results from the 6-31G* model (using mean absolute error as the sole criterion). However, it has a (small) detrimental effect on the quality of the calculated results from the two density functional models and from the MP2 model. In these three cases, use of 6-31G* geometries in place of "exact" geometries does not change the overall mean absolute error.

Comparisons involving relative strengths of *p*-substituted benzoic acids are provided in **Tables 12-13** to **12-16**. As with relative base strength comparisons, results from Hartree-Fock, EDF1, B3LYP and MP2 models with the 6-31G* basis set are examined as a function of underlying geometry. Here, PM3, 3-21G, 6-31G* and "exact" geometries are employed. The noted sensitivity to structure is even less than noted in previous *isodesmic* comparisons, reflecting significant cancellation of errors among very similar systems. Certainly, effort expended to provide "exact" equilibrium geometries in situations such as this is difficult to justify.

The final set of comparisons, provided in **Tables 12-17** to **12-20**, examines the effect of structure on energy differences between different regio and stereochemical products of Diels-Alder cycloadditions of substituted cyclopentadienes with acrylonitrile. Because the energy differences between products are typically very small (often just a few tenths of a kcal/mol), this case provides a more stringent test of the strategy of using approximate geometries than previous examples. The same four calculation levels are employed, with AM1, 3-21G and 6-31G* "approximate" geometries.

Table 12-9: **Effect of Choice Geometry on Proton Affinities of Nitrogen Bases Relative to Ammonia.**[a] **6-31G* Model**

base, B	geometry		expt.
	AM1	6-31G*	
aniline	5	8	**6.7**
methylamine	9	11	**9.1**
aziridine	12	14	**11.2**
ethylamine	12	14	**11.8**
dimethylamine	15	18	**15.5**
pyridine	16	18	**16.0**
tert-butylamine	18	19	**16.1**
cyclohexylamine	18	20	**16.3**
azetidine	21	22	**18.0**
pyrrolidine	20	23	**19.8**
trimethylamine	19	22	**20.0**
piperidine	22	24	**21.1**
diazabicyclooctane	24	28	**23.5**
N-methylpyrrolidine	25	26	**24.3**
N-methylpiperidine	28	28	**25.7**
quinuclidine	28	31	**27.1**
mean absolute error	1	3	–

a) energy of reaction: $BH^+ + NH_3 \rightarrow B + NH_4^+$

Table 12-10: Effect of Choice Geometry on Proton Affinities of Nitrogen Bases Relative to Ammonia.[a] EDF1/6-31G* Model

base, B	geometry			expt.
	AM1	6-31G*	EDF1/6-31G*	
aniline	3	4	6	**6.7**
methylamine	8	10	10	**9.1**
aziridine	8	10	10	**11.2**
ethylamine	11	13	13	**11.8**
dimethylamine	13	15	15	**15.5**
pyridine	13	16	16	**16.0**
tert-butylamine	18	19	19	**16.1**
cyclohexylamine	18	19	20	**16.3**
azetidine	17	18	18	**18.0**
pyrrolidine	18	21	21	**19.8**
trimethylamine	16	18	17	**20.0**
piperidine	20	22	22	**21.1**
diazabicyclooctane	19	23	24	**23.5**
N-methylpyrrolidine	25	23	23	**24.3**
N-methylpiperidine	22	24	25	**25.7**
quinuclidine	24	27	29	**27.1**
mean absolute error	3	1	1	–

a) energy of reaction: $BH^+ + NH_3 \rightarrow B + NH_4^+$

Table 12-11: **Effect of Choice Geometry on Proton Affinities of Nitrogen Bases Relative to Ammonia.[a] B3LYP/6-31G* Model**

base, B	geometry			expt.
	AM1	6-31G*	B3LYP/6-31G*	
aniline	3	5	5	**6.7**
methylamine	9	10	10	**9.1**
aziridine	9	11	11	**11.2**
ethylamine	12	13	13	**11.8**
dimethylamine	13	16	16	**15.5**
pyridine	14	16	16	**16.0**
tert-butylamine	17	19	19	**16.1**
cyclohexylamine	18	19	19	**16.3**
azetidine	18	19	19	**18.0**
pyrrolidine	18	22	22	**19.8**
trimethylamine	16	19	19	**20.0**
piperidine	21	22	22	**21.1**
diazabicyclooctane	20	24	24	**23.5**
N-methylpyrrolidine	25	24	24	**24.3**
N-methylpiperidine	22	25	25	**25.7**
quinuclidine	24	28	28	**27.1**
mean absolute error	2	1	1	–

a) energy of reaction: $BH^+ + NH_3 \rightarrow B + NH_4^+$

Table 12-12: **Effect of Choice Geometry on Proton Affinities of Nitrogen Bases Relative to Ammonia.[a] MP2/6-31G* Model**

base, B	geometry			expt.
	AM1	6-31G*	MP2/6-31G*	
aniline	3	5	7	**6.7**
methylamine	9	10	10	**9.1**
aziridine	8	10	9	**11.2**
ethylamine	11	12	12	**11.8**
dimethylamine	14	16	16	**15.5**
pyridine	11	13	13	**16.0**
tert-butylamine	16	17	17	**16.1**
cyclohexylamine	17	18	18	**16.3**
azetidine	18	18	18	**18.0**
pyrrolidine	17	21	21	**19.8**
trimethylamine	17	20	19	**20.0**
piperidine	20	21	21	**21.1**
diazabicyclooctane	19	23	23	**23.5**
N-methylpyrrolidine	26	23	23	**24.3**
N-methylpiperidine	22	24	24	**25.7**
quinuclidine	23	27	27	**27.1**
mean absolute error	2	1	1	–

a) energy of reaction: $BH^+ + NH_3 \rightarrow B + NH_4^+$

Table 12-13: **Effect of Choice of Geometry on Acidities of *p*-Substituted Benzoic Acids Relative to Benzoic Acid.[a] 6-31G* Model**

p-substituent, X	geometry			expt.
	PM3	3-21G	6-31G*	
NH$_2$	-3	-4	-4	**-3.4**
Me	-1	-1	-1	**-1.0**
OMe	-1	-2	-2	**-0.8**
CMe$_3$	-1	-1	0	**0.1**
F	3	3	3	**3.0**
Cl	6	6	6	**4.5**
CHO	8	8	8	**7.2**
CF$_3$	9	8	8	**7.7**
NO$_2$	14	14	14	**11.9**
CN	12	12	12	**12.2**
mean absolute error	1	1	1	–

a) energy of reaction: p-XC$_6$H$_4$CO$_2$H + C$_6$H$_5$CO$_2^-$ → p-XC$_6$H$_4$CO$_2^-$ + C$_6$H$_5$CO$_2$H

Table 12-14: **Effect of Choice of Geometry on Acidities of *p*-Substituted Benzoic Acids Relative to Benzoic Acid.[a] EDF1/6-31G* Model**

p-substituent, X	geometry				expt.
	PM3	3-21G	6-31G*	EDF1/6-31G*	
NH$_2$	-5	-6	-6	-5	**-3.4**
Me	-1	-1	-1	-1	**-1.0**
OMe	-2	-2	3	-2	**-0.8**
CMe$_3$	-1	-1	-2	0	**0.1**
F	2	2	2	2	**3.0**
Cl	5	6	5	6	**4.5**
CHO	9	9	8	10	**7.2**
CF$_3$	8	8	8	8	**7.7**
NO$_2$	14	14	13	14	**11.9**
CN	12	12	12	12	**12.2**
mean absolute error	1	1	1	1	–

a) energy of reaction: p-XC$_6$H$_4$CO$_2$H + C$_6$H$_5$CO$_2^-$ → p-XC$_6$H$_4$CO$_2^-$ + C$_6$H$_5$CO$_2$H

Table 12-15: **Effect of Choice of Geometry on Acidities of *p*-Substituted Benzoic Acids Relative to Benzoic Acid.[a] B3LYP/6-31G* Model**

p-substituent, X	geometry				expt.
	PM3	3-21G	6-31G*	B3LYP/6-31G*	
NH_2	-4	-6	-6	-5	**-3.4**
Me	-1	-1	-1	-1	**-1.0**
OMe	-2	-2	3	-2	**-0.8**
CMe_3	0	0	-2	0	**0.1**
F	3	3	3	2	**3.0**
Cl	5	6	6	6	**4.5**
CHO	9	9	7	10	**7.2**
CF_3	8	8	8	8	**7.7**
NO_2	14	14	14	14	**11.9**
CN	12	12	12	12	**12.2**
mean absolute error	1	1	1	1	–

a) energy of reaction: p-$XC_6H_4CO_2H$ + $C_6H_5CO_2^-$ → p-$XC_6H_4CO_2^-$ + $C_6H_5CO_2H$

Table 12-16: **Effect of Choice of Geometry on Acidities of *p*-Substituted Benzoic Acids Relative to Benzoic Acid.[a] MP2/6-31G* Model**

p-substituent, X	geometry				expt.
	PM3	3-21G	6-31G*	MP2/6-31G*	
NH_2	-3	-4	-5	-3	**-3.4**
Me	-1	-1	-1	-1	**-1.0**
OMe	-1	-1	4	-1	**-0.8**
CMe_3	0	0	-1	0	**0.1**
F	3	3	3	3	**3.0**
Cl	5	5	5	5	**4.5**
CHO	8	8	6	8	**7.2**
CF_3	8	8	8	8	**7.7**
NO_2	12	12	12	12	**11.9**
CN	11	11	11	11	**12.2**
mean absolute error	0	0	1	0	–

a) energy of reaction: p-$XC_6H_4CO_2H$ + $C_6H_5CO_2^-$ → p-$XC_6H_4CO_2^-$ + $C_6H_5CO_2H$

Table 12-17: Effect of Choice of Geometry on Relative Energies of Regio and Stereochemical Products of Diels-Alder Cycloadditions of Substituted Cyclopentadienes with Acrylonitrile.[a] 6-31G* Model

position and substituent	product geometry		
on cyclopentadiene	AM1	3-21G	6-31G*
regioselection			
1-Me	*ortho* (0.1)	*ortho* (0.2)	*ortho* (0.2)
1-OMe	*meta* (0.6)	*meta* (0.4)	*meta* (0.5)
2-Me	*para* (0.1)	*para* (0.4)	*para* (0.5)
2-OMe	*para* (1.3)	*para* (1.2)	*para* (1.2)
stereoselection			
5-Me	*anti* (0.6)	*anti* (0.5)	*anti* (0.6)
5-OMe	*syn* (4.6)	*syn* (4.3)	*syn* (4.5)

a)

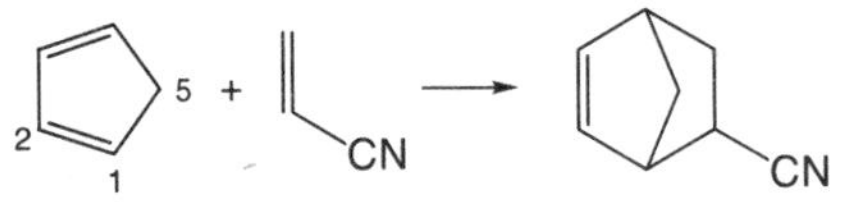

Table 12-18: Effect of Choice of Geometry on Relative Energies of Regio and Stereochemical Products of Diels-Alder Cycloadditions of Substituted Cyclopentadienes with Acrylonitrile.[a] EDF1/6-31G* Model

position and substituent	product geometry			
on cyclopentadiene	AM1	3-21G	6-31G*	EDF1/6-31G*
regioselection				
1-Me	*ortho* (0.1)	none	*ortho* (0.1)	*ortho* (0.1)
1-OMe	*meta* (0.2)	*meta* (0.1)	*meta* (0.1)	none
2-Me	*para* (0.1)	*para* (0.3)	*para* (0.1)	*para* (0.8)
2-OMe	*para* (1.1)	*para* (1.0)	*para* (1.0)	*para* (0.8)
stereoselection				
5-Me	*anti* (0.7)	*anti* (0.5)	*anti* (0.6)	*anti* (0.6)
5-OMe	*syn* (4.0)	*syn* (3.6)	*syn* (3.9)	*syn* (4.0)

a)

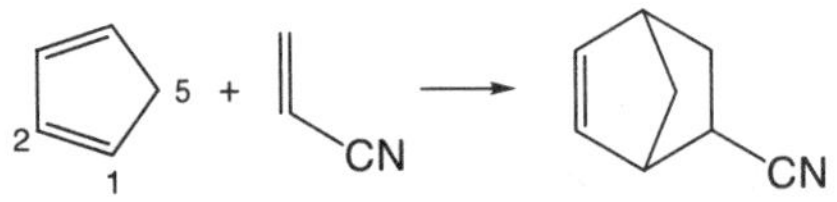

Table 12-19: **Effect of Choice of Geometry on Relative Energies of Regio and Stereochemical Products of Diels-Alder Cycloadditions of Substituted Cyclopentadienes with Acrylonitrile.[a] B3LYP/6-31G* Model**

position and substituent on cyclopentadiene	product geometry			
	AM1	3-21G	6-31G*	B3LYP/6-31G*
regioselection				
1-Me	*ortho* (0.2)	*ortho* (0.2)	*ortho* (0.2)	*ortho* (0.2)
1-OMe	*meta* (0.2)	none	none	*ortho* (0.1)
2-Me	*meta* (0.2)	*meta* (0.1)	*meta* (0.2)	*meta* (0.1)
2-OMe	*para* (1.0)	*para* (0.9)	*para* (0.9)	*para* (0.8)
stereoselection				
5-Me	*anti* (0.4)	*anti* (0.3)	*anti* (0.4)	*anti* (0.3)
5-OMe	*syn* (4.3)	*syn* (4.0)	*syn* (4.2)	*syn* (4.2)

a)

Table 12-20: **Effect of Choice of Geometry on Relative Energies of Regio and Stereochemical Products of Diels-Alder Cycloadditions of Substituted Cyclopentadienes with Acrylonitrile.[a] MP2/6-31G* Model**

position and substituent on cyclopentadiene	product geometry			
	AM1	3-21G	6-31G*	MP2/6-31G*
regioselection				
1-Me	*ortho* (0.6)	*ortho* (0.6)	*ortho* (0.6)	*ortho* (0.7)
1-OMe	none	*ortho* (0.4)	*ortho* (0.4)	*ortho* (0.2)
2-Me	*meta* (0.3)	*meta* (0.4)	*meta* (0.3)	*meta* (0.4)
2-OMe	*para* (0.8)	*para* (0.6)	*para* (0.8)	*para* (0.6)
stereoselection				
5-Me	*anti* (0.3)	*anti* (0.1)	*anti* (0.1)	*anti* (0.1)
5-OMe	*syn* (4.7)	*syn* (4.9)	*syn* (4.9)	*syn* (4.9)

a)

While experimental thermochemical data on these reactions are unavailable[*], in all cases the calculated data show little sensitivity to choice of equilibrium geometry. Use of approximate geometries for such types of comparisons offers the promise of large cost savings with negligible detrimental effect.

The overall recommendation from the examples provided in this section is very clear: make use of either semi-empirical or small-basis-set Hartree-Fock equilibrium geometries in constructing thermochemical comparisons based on higher-level models. While some caution is clearly needed in dealing with systems where specific calculation models are known to produce "poor" geometries, e.g., 3-21G calculations on amines, in general the errors resulting from the use of approximate geometries are very small.

One final word of caution. It has already been pointed out (**Chapter 5**) that "limiting" Hartree-Fock geometries sometimes differ significantly from structures obtained from MP2 or density functional models using the same basis set. For example, bonds involving two highly-electronegative elements, e.g., the OO bond in hydrogen peroxide, are poorly described at the Hartree-Fock limit but well described using correlated techniques. On the other hand, the structures of many hypervalent compounds are poorly reproduced using the MP2 model and density functional models, whereas Hartree-Fock schemes provide good descriptions. In both of these situations, use of Hartree-Fock models to provide geometries for energy calculations at correlated levels may lead to unacceptable errors.

While these types of cases are uncommon, caution needs to be exercised. The best (and only) advice which can be given in operating in areas where there is little or no prior experience, is to test any proposed strategy on related systems for which experimental data are available or, where experimental data are unavailable, to perform higher-level calculations.

---

[*] Experimental data on kinetic regio and stereoselectivity of these same reactions are available, and has previously been discussed in **Chapter 9**.

Using Localized MP2 Models to Calculate Thermochemistry

Chapter 6 assessed the performance of Hartree-Fock, density functional and MP2 models for a variety of thermochemical calculations. It was concluded that density functional models, in particular, EDF1/6-311+G** and B3LYP/6-311+G** models, and the MP2/6-311+G** model provided the best overall description, although Hartree-Fock models also provided solid accounts in many situations. One important area where there are problems with both Hartree-Fock and density functional models is for bond separation reactions. While both Hartree-Fock models and density functional models generally lead to bond separation energies in good agreement with experiment (and with the results of G3 calculations[1]), significant problems occur for a number of systems. Even the MP2/6-31G* model fails to provide an entirely satisfactory account of bond separation energies, and, only the MP2/6-311+G** model provides results which are sufficiently accurate to be used for determination of heats of formation (see **Chapter 13**).

However, MP2/6-311+G** calculations rapidly become prohibitive with increasing molecular size, both because of time required and memory and disk usage. Any savings translate directly into their increased range of application. One such savings is to make use of geometries from simpler calculation models, in particular, Hartree-Fock models. This completely eliminates the need for costly geometry optimizations using MP2 models, without seriously affecting the overall quality of results. Discussion has been provided earlier in this chapter.

Further savings might be realized by basing the MP2 calculation on Hartree-Fock orbitals which have been localized. In practice, this does not greatly reduce calculation times (until the molecule becomes spatially quite large), but does significantly reduce both memory and disk demands. Data provided in **Tables 12-21** and **12-22** make a strong case for use of localized MP2 (LMP2) models. Calculated bond separation energies (**Table 12-21**) and calculated relative proton affinities for nitrogen bases (**Table 12-22**) from LMP2/6-311+G**// 6-31G* and MP2/6-311+G**//6-31G* models are virtually identical.

Table 12-21: Performance of Localized MP2 Models on Energies of Bond Separation Reactions

molecule	bond separation reaction	LMP2/6-311+G**// 6-31G*	MP2/6-311+G**// 6-31G*	G3	expt.
isobutane	$CH(CH_3)_3 + 2CH_4 \rightarrow 3CH_3CH_3$	6	7	6.4	7.5
trimethylamine	$(CH_3)_3N + 2NH_3 \rightarrow 3CH_3NH_2$	11	11	11.6	11.1
tetrafluoromethane	$CF_4 + 3CH_4 \rightarrow 4CH_3F$	46	46	–	52.8
tetrachloromethane	$CCl_4 + 3CH_4 \rightarrow 4CH_3Cl$	-3	-3	-2.4	-1.8
propene	$CH_3CHCH_2 + CH_4 \rightarrow CH_3CH_3 + CH_2CH_2$	5	5	4.9	5.4
propyne	$CH_3CCH + CH_4 \rightarrow CH_3CH_3 + HCCH$	7	7	7.8	7.6
allene	$CH_2CCH_2 + CH_4 \rightarrow 2CH_2CH_2$	-3	-3	-3.2	-2.8
1,3-butadiene	$CH_2CHCHCH_2 + 2CH_4 \rightarrow CH_3CH_3 + 2CH_2CH_2$	13	14	13.2	14.2
benzene	$+ 6CH_4 \rightarrow 3CH_3CH_3 + 3CH_2CH_2$	68	68	61.0	64.2
cyclopropane	$+ 3CH_4 \rightarrow 3CH_3CH_3$	-24	-24	-23.5	-19.6
oxacyclopropane	$+ 2CH_4 + H_2O \rightarrow CH_3CH_3 + 2CH_3OH$	-14	-14	-14.1	-10.5
thiophene	$+ 4CH_4 + H_2S \rightarrow CH_3CH_3 + 2CH_3SH + 2CH_2CH_2$	44	45	40.1	29.2
bicyclo[1.1.0]butane	$+ 6CH_4 \rightarrow 5CH_3CH_3$	-56	-56	-55.0	-45.5
mean absolute deviation from G3		1	2	–	–

Table 12-22: **Performance of Localized MP2 Models on Proton Affinities of Nitrogen Bases Relative to Ammonia[a]**

base, B	LMP2/6-311+G**// 6-31G*	MP2/6-311+G**// 6-31G*	expt.
aniline	5	5	**6.7**
methylamine	11	10	**9.1**
aziridine	12	12	**11.2**
ethylamine	13	13	**11.8**
dimethylamine	18	18	**15.5**
pyridine	15	15	**16.0**
tert-butylamine	17	17	**16.1**
cyclohexylamine	18	17	**16.3**
azetidine	19	19	**18.0**
pyrrolidine	22	22	**19.8**
trimethylamine	22	22	**20.0**
piperidine	23	22	**21.1**
diazabicyclooctane	25	25	**23.5**
N-methylpyrrolidine	25	25	**24.3**
N-methylpiperidine	27	26	**25.7**
quinuclidine	29	29	**27.1**
mean absolute error	2	1	–

a) energy of reaction: $BH^+ + NH_3 \rightarrow B + NH_4^+$

Using "Approximate" Equilibrium Geometries to Calculate Molecular Properties

The fact that energies of chemical reactions generally do not show large variations with choice of equilibrium structure is a consequence of the "shallowness" of potential energy surfaces in the vicinity of minima. Because other properties will not be at minima at these "special" points, it is to be expected that they could be more sensitive to choice of geometry. For example, it has already been pointed out that the dipole moments of amines are very sensitive to the local geometry about nitrogen; the further from planarity the larger the dipole moment. For example, the dipole moment in trimethylamine, where non-bonded interactions among methyl groups force the angle about nitrogen to open up, is much smaller than the dipole moment in ammonia, where such interactions are absent. Any method that provides a poor description of geometry about nitrogen, for example, the 3-21G model which leads to bond angles which are typically several degrees too large, will likely provide a poor basis for dipole moment calculations.

Tables 12-23 to **12-26** examine the effect of geometry on dipole moments in a small collection of hydrocarbons and amines. Single-point 6-31G*, EDF1/6-31G*, B3LYP/6-31G* and MP2/6-31G* dipole moment calculations have been carried out using MMFF, AM1 and (except for the Hartree-Fock calculations) 6-31G* geometries, and compared with dipole moments obtained from "exact" structures. While subtle differences exist, for the most part they are very small. In fact, using mean absolute error as a criterion, there is little to differentiate dipole moments obtained from use of approximate geometries from those calculated using "exact" geometries.

The same advice already provided in the previous sections applies here. In the absence of prior experience, perform sufficient calculations to judge the sensitivity of property of choice of geometry. Only then can confidence be established in a particular choice of model.

Table 12-23: **Effect of Choice of Geometry on Dipole Moments in Hydrocarbons and in Amines. 6-31G* Model**

molecule	geometry			expt.
	MMFF	AM1	6-31G*	
propyne	0.6	0.6	0.6	**0.75**
cyclopropene	0.5	0.5	0.6	**0.45**
propene	0.3	0.3	0.3	**0.36**
cyclobutene	0.0	0.0	0.0	**0.13**
bicyclo [1.1.0] butane	0.8	0.6	0.7	**0.68**
trimethylamine	0.9	0.6	0.7	**0.61**
dimethylamine	1.2	1.0	1.1	**1.03**
methylamine	1.6	1.4	1.5	**1.31**
ammonia	2.0	1.8	1.9	**1.47**
aziridine	1.9	1.8	1.9	**1.90**
mean absolute error	0.2	0.1	0.1	–

Table 12-24: **Effect of Choice of Geometry on Dipole Moments in Hydrocarbons and in Amines. EDF1/6-31G* Model**

molecule	geometry				expt.
	MMFF	AM1	6-31G*	EDF1/6-31G*	
propyne	0.7	0.7	0.7	0.7	**0.75**
cyclopropene	0.5	0.5	0.5	0.5	**0.45**
propene	0.4	0.4	0.4	0.4	**0.36**
cyclobutene	0.2	0.1	0.2	0.2	**0.13**
bicyclo [1.1.0] butane	0.8	0.6	0.7	0.8	**0.68**
trimethylamine	0.6	0.4	0.5	0.5	**0.61**
dimethylamine	1.0	0.9	1.0	1.0	**1.03**
methylamine	1.5	1.3	1.4	1.1	**1.31**
ammonia	1.8	1.7	1.8	1.9	**1.47**
aziridine	1.7	1.6	1.7	1.7	**1.90**
mean absolute error	0.1	0.1	0.1	0.1	–

Table 12-25: **Effect of Choice of Geometry on Dipole Moments in Hydrocarbons and in Amines. B3LYP/6-31G* Model**

molecule	geometry				expt.
	MMFF	AM1	6-31G*	B3LYP/6-31G*	
propyne	0.7	0.6	0.7	0.7	**0.75**
cyclopropene	0.5	0.5	0.5	0.5	**0.45**
propene	0.4	0.3	0.4	0.4	**0.36**
cyclobutene	0.1	0.1	0.1	0.1	**0.13**
bicyclo [1.1.0] butane	0.8	0.6	0.7	0.8	**0.68**
trimethylamine	0.7	0.4	0.5	0.6	**0.61**
dimethylamine	1.1	0.9	1.0	1.0	**1.03**
methylamine	1.5	1.3	1.4	1.5	**1.31**
ammonia	1.9	1.7	1.8	1.9	**1.47**
aziridine	1.7	1.6	1.7	1.8	**1.90**
mean absolute error	0.1	0.1	0.1	0.1	–

Table 12-26: **Effect of Choice of Geometry on Dipole Moments in Hydrocarbons and in Amines. MP2/6-31G* Model**

molecule	geometry				expt.
	MMFF	AM1	6-31G*	MP2/6-31G*	
propyne	0.6	0.6	0.6	0.6	**0.75**
cyclopropene	0.5	0.5	0.5	0.5	**0.45**
propene	0.3	0.3	0.3	0.3	**0.36**
cyclobutene	0.1	0.1	0.1	0.1	**0.13**
bicyclo [1.1.0] butane	0.9	0.6	0.8	0.8	**0.68**
trimethylamine	0.8	0.5	0.7	0.7	**0.61**
dimethylamine	1.2	1.0	1.1	1.2	**1.03**
methylamine	1.6	1.4	1.5	1.6	**1.31**
ammonia	2.0	1.8	1.9	2.0	**1.47**
aziridine	1.8	1.7	1.8	1.9	**1.90**
mean absolute error	0.2	0.1	0.1	0.2	–

There are situations where "exact" equilibrium structures must be used. The most conspicuous is for the calculation of vibrational frequencies, as well as thermodynamic properties such as entropies obtained from calculated frequencies. As already discussed in **Chapter 7**, this is because the frequencies derive from the second derivative term, E'', in a Taylor series expansion of the total energy.

$$E = E^0 + E' + E'' + ... \qquad (1)$$

Here, $E^°$ is a constant , and E' (the first derivative term) is assumed to be rigorously zero. (Higher-order terms are generally ignored). If E' is not zero (meaning that the structure is not an energy minimum), then the interpretation given to E'' is no longer correct.

References

1. Original reference : (a) L.A. Curtiss, K. Raghavachari, P.C. Redfern, V. Rassolov and J.A. Pople, *J. Chem. Phys.*, **109**, 7764 (1998). For an up-to-date, on-line source of G3 data see: (b) L.A. Curtiss, *Computational Thermochemistry*, "chemistry.anl.gov/compmat/comptherm.htm"

Chapter 13

Using Energies for Thermochemical and Kinetic Comparisons

This chapter outlines strategies for using calculated energies to obtain accurate estimates for heats of formation. These in turn may be employed for diverse thermochemical and kinetic comparisons.

Introduction

The discussion in **Chapter 6** centered around the use of quantum chemical models to calculate reaction thermochemistry. A number of important conclusions were reached:

i)   Correlated models (in particular, MP2 and density functional models) with moderate to large basis sets, including one or more sets of polarization functions as well as diffuse functions, are required for the accurate description of processes such as homolytic bond dissociation reactions in which the total number of electron pairs is not conserved. Hartree-Fock models yield bond dissociation energies which are consistently too small while local density models yield energies which are too large.

Closely related is the need for correlated models to account for absolute activation energies. Here too, bonds are being made or broken and the number of electron pairs may not be conserved.

ii)   The energetics of reactions in which the total number of electron pairs is conserved, including heterolytic bond dissociation reactions in which a bond is exchanged for a non-bonded lone pair, and comparisons among structural isomers, where reactants and products differ in the kinds of bonds, are generally well described using Hartree-Fock models. Moderate to large basis sets including polarization functions and, in the case of heterolytic bond dissociation reactions in which anions are produced, diffuse functions are required. Correlated models (including density functional and MP2 models) also perform well for these classes of reactions.

iii) The energetics of *isodesmic* reactions are generally well described using both Hartree-Fock and correlated models, including density functional and MP2 models. Small to moderate basis sets usually give acceptable results for Hartree-Fock models, although larger basis sets are required for use with correlated models.

Relative activation energies, when written in terms of *isodesmic* reactions, are also well described using Hartree-Fock models. MP2 models also provide satisfactory results but density functional models are problematic.

iv) Semi-empirical models are unsatisfactory (or at best unreliable) for the description of the energetics of all classes of reactions. Even most *isodesmic* reactions are poorly represented.

The primary recommendation to follow from these generalizations (aside from needing to exercise caution in the use of semi-empirical models for energetics comparisons of any kind) is to make use of *isodesmic* reactions wherever possible. Where this is not possible, the recommendation is to write reactions in which the total number of chemical bonds is conserved.

While it is straightforward to obtain "theoretical" heats of formation from processes which greatly disrupt bonding, e.g., the G3 recipe[1], it is also possible to make use of *isodesmic* reactions together with limited experimental data, or alternatively data from high-level quantum chemical calculations, to estimate heats of formation. Once in hand, these can be used for whatever thermochemical comparisons are desired. The key is to find an *isodesmic* reaction which is both uniquely defined, and which leads to products with known heats of formation. This is the subject of the present chapter.

Calculating Heats of Formation from Bond Separation Reactions

As previously described in **Chapter 6**, a bond separation reaction "breaks down" any molecule comprising three or more heavy (non-hydrogen) atoms, and which can be represented in terms of a classical valence structure, into the simplest set of two-heavy-atom molecules containing the same component bonds. For example, the bond separation reaction for methylhydrazine breaks the molecule into methylamine and hydrazine, the simplest molecules incorporating CN and NN single bonds, respectively.

$$CH_3NHNH_2 + NH_3 \longrightarrow CH_3NH_2 + NH_2NH_2$$

A molecule of ammonia needs to be added to the left to achieve stoichiometric balance.

A bond separation reaction is uniquely defined[*]. Therefore, a bond separation energy is a "molecular property". Given that a bond separation reaction leads to products, the heats of formation of which are either known experimentally or can be determined from calculations, combining a calculated bond separation energy with experimental (or calculated) heats of formation, gives rise to a unique value for the heat of formation.[2] For example, a heat of formation for methylhydrazine may be obtained from the thermochemical cycle.

$$\Delta H_f(CH_3NHNH_2) = -\Delta E_{rx} - \Delta H_f(NH_3) + \Delta H_f(CH_3NH_2) + \Delta H_f(NH_2NH_2)$$

Here, ΔE_{rx} is the calculated energy of the bond separation reaction of methylhydrazine and $\Delta H_f(NH_3)$, $\Delta H_f(CH_3NH_2)$ and $\Delta H_f(NH_2NH_2)$ are experimental (or calculated) heats of formation.

It has previously been documented (**Chapter 6**) that Hartree-Fock, density functional and MP2 models generally provide excellent descriptions of the energetics of bond separation energies, while semi-empirical models are not successful in this regard (**Tables 6-10** and **A6-36** to **A6-43**). Use of bond separation energies from these models (but not from semi-empirical models) together with

---

[*]   More precisely, a bond separation reaction is unique to a particular valence structure. The bond separation reaction for a molecule which can only be represented by multiple (non-identical) valence structures may not be uniquely defined.

appropriate experimental data, should lead to accurate estimates of heats of formation.

Heats of formation obtained from bond separation reactions from Hartree-Fock, EDF1 and B3LYP density functional and MP2 models all with 6-31G* and 6-311+G** basis sets are compared to experimental values in **Table 13-1**. Errors in calculated quantities are exactly the same as those for the underlying bond separation reactions.[*]

Bond separation reactions and heats of formation obtained from bond separation energies suffer from two serious problems. The first is that "bond types" in reactants and products for some types of processes may not actually be the same or even "similar". The bond separation reaction for benzene is an obvious example. Here to reactant (benzene) incorporates six equivalent "aromatic" carbon-carbon bonds, "midway" between single and double bonds, while the products (three ethanes and three ethylenes) incorporate three distinct carbon-carbon single bonds and three distinct carbon-carbon double bonds.

The second problem is even more serious. The number of product molecules in a bond separation reaction increases with the "size" of the reactant, and (presumably) so too does the overall magnitude of error in the calculated bond separation energy. Whereas errors in bond separation energies (and in heats of formation derived from bond separation reactions) are close to acceptable limits ($\pm$ 2 kcal/mol) for small molecules (see discussion in **Chapter 6**), it is likely that will rapidly move outside of acceptable limits with increasing molecular size.

There is no obvious "best solution" to these problems. One direction is to "redefine" (or generalize) the bond separation reaction such that the products are not restricted to the smallest (two-heavy-atom) molecules, but rather include molecules made up of "larger" components as well (functional groups, rings, etc.). For example, were the phenyl ring and the carboxylic acid functional group included as "fragments", then the "bond separation reaction" for *m*-toluic acid could be written.

---

[*] Note, however, that previous comparisons were with G3 results whereas those presented here are with experimental data.

This process, unlike the "original" bond separation reaction,

comprises many fewer components and does not "force" comparisons between reactants and products with different bond types. There are, however, two potential drawbacks. For one, the increased number of fragments (over those needed only to make two-heavy-atom molecules in the original bond separation reaction) leads directly to an increase in the number of "products" for which experimental (or high-quality calculated) data need to be available. In the above example, data on benzoic acid and toluene (in addition to benzene on the "reactant side") are needed. Second, and potentially even more serious is the difficulty of defining "extended" bond separation reactions in an unambiguous manner. In time these difficulties will be overcome leading to practical schemes for routine and reliable determination of heats of formation.

References

1. Original reference : (a) L.A. Curtiss, K. Raghavachari, P.C. Redfern, V. Rassolov and J.A. Pople, *J. Chem. Phys.*, **109**, 7764 (1998). For an up-to-date, on-line source of G3 data see: (b) L.A. Curtiss, *Computational Thermochemistry,* "chemistry.anl.gov/compmat/comptherm.htm"

2. The idea of using calculated energies of bond separation reactions together with limited experimental thermochemical data to supply "accurate" heats of formation was proposed many years ago, but until recently was not practical. Original reference: R. Ditchfield, W.J. Hehre, J.A. Pople and L. Radom, *Chem. Phys. Ltrs.*, **5**, 13 (1970).

Table 13-1: Heats of Formation from Bond Separation Reactions

molecule	bond separation reaction	Hartree-Fock		EDF1		B3LYP		MP2		expt.
		6-31G*	6-311+G**	6-31G*	6-311+G**	6-31G*	6-311G**	6-31G*	6-311+G**	
propane	$CH_3CH_2CH_3 + CH_4 \rightarrow 2CH_3CH_3$	-23	-23	-23	-22	-23	-23	-24	-24	-25.0
isobutane	$CH(CH_3)_3 + 2CH_4 \rightarrow 3CH_3CH_3$	-26	-26	-25	-25	-27	-27	-30	-31	-32.1
neopentane	$C(CH_3)_4 + 3CH_4 \rightarrow 4CH_3CH_3$	-29	-29	-28	-27	-30	-31	-29	-38	-40.0
dimethylsilane	$SiH_2(CH_3)_2 + 2SiH_4 \rightarrow 3CH_3SiH_3$	2	2	2	2	2	2	2	2	0.8
trimethylsilane	$SiH(CH_3)_3 + 2SiH_4 \rightarrow 3CH_3SiH_3$	4	4	6	4	4	4	4	3	1.6
tetramethylsilane	$Si(CH_3)_4 + 3SiH_4 \rightarrow 4CH_3SiH_3$	6	6	7	7	6	6	6	3	2.8
ethylamine	$CH_3CH_2NH_2 + CH_4 \rightarrow CH_3CH_3 + CH_3NH_2$	-10	-10	-10	-10	-11	-10	-11	-11	-11.4
dimethylamine	$CH_3NHCH_3 + NH_3 \rightarrow 2CH_3NH_2$	-3	-3	-4	-4	-4	-4	-5	-5	-4.5
trimethylamine	$(CH_3)_3N + 2NH_3 \rightarrow 3CH_3NH_2$	0	-1	0	-1	-2	-2	-6	-6	-5.7
ethanol	$CH_3CH_2OH + CH_4 \rightarrow CH_3CH_3 + CH_3OH$	-54	-54	-54	-54	-55	-55	-55	-55	-56.1
dimethyl ether	$CH_3OCH_3 + H_2O \rightarrow 2CH_3OH$	-42	-42	-42	-42	-42	-43	-44	-44	-44.0
ethanethiol	$CH_3CH_2SH + CH_4 \rightarrow CH_3CH_3 + CH_3SH$	-9	-9	-9	-9	-10	-10	-11	-11	-11.1
dimethyl sulfide	$CH_3SCH_3 + H_2S \rightarrow 2CH_3SH$	-7	-7	-7	-8	-8	-8	-9	-9	-9.1
difluoromethane	$CH_2F_2 + CH_4 \rightarrow 2CH_3F$	-107	-104	-109	-104	-108	-104	-109	-105	-106.8
trifluoromethane	$CHF_3 + 2CH_4 \rightarrow 3CH_3F$	-166	-157	-170	-158	-168	-157	-169	-161	-164.5
tetrafluoromethane	$CF_4 + 3CH_4 \rightarrow 4CH_3F$	-226	-209	-230	-209	-228	-210	-232	-214	-221

Table 13-1: Heats of Formation from Bond Separation Reactions (2)

molecule	bond separation reaction	Hartree-Fock		EDF1		B3LYP		MP2		expt.
		6-31G*	6-311+G**	6-31G*	6-311+G**	6-31G*	6-311G**	6-31G*	6-311+G**	
dichloromethane	$CH_3Cl_2 + CH_4 \rightarrow 2CH_3Cl$	-16	-17	-19	-20	-19	-19	-20	-21	**-22.1**
trichloromethane	$CHCl_3 + 2CH_4 \rightarrow 3CH_3Cl$	-11	-12	-19	-20	-17	-18	-20	-24	**-24.7**
tetrachloromethane	$CCl_4 + 3CH_4 \rightarrow 4CH_3Cl$	1	-2	-14	-16	-12	-14	-2	-25	**-25.6**
propene	$CH_3CHCH_2 + CH_4 \rightarrow CH_3CH_3 + CH_2CH_2$	6	6	5	6	5	5	5	5	**4.8**
acetaldehyde	$CH_3CHO + CH_4 \rightarrow CH_3CH_3 + H_2CO$	-39	-39	-40	-40	-40	-40	-40	-39	**-39.6**
propyne	$CH_3CCH + CH_4 \rightarrow CH_3CH + HCCH$	45	45	42	44	43	44	45	46	**44.6**
acetonitrile	$CH_3CN + CH_4 \rightarrow CH_3CH_3 + HCN$	18	18	17	18	17	18	19	20	**15.4**
allene	$CH_2CCH_2 + CH_4 \rightarrow 2CH_2CH_2$	47	49	41	43	42	44	45	46	**45.6**
ketene	$CH_2CO + CH_4 \rightarrow CH_2CH_2 + H_2CO$	-9	-8	-15	-14	-13	-13	-12	-12	**-11.4**
carbon dioxide	$CO_2 + CH_4 \rightarrow 2H_2CO$	-95	-94	-98	-97	-96	-95	-100	-99	**-94.0**
1,3-butadiene	$CH_2CHCHCH_2 + 2CH_4 \rightarrow CH_3CH_3 + 2CH_2CH_2$	29	30	24	27	25	27	26	26	**26.3**
formamide	$NH_2CHO + CH_4 \rightarrow CH_3NH_2 + H_2CO$	-45	-44	-49	-48	-48	-47	-47	-45	**-44.5**
benzene	+ $6CH_4 \rightarrow 3CH_3CH_3 + 3CH_2CH_2$	26	29	13	20	17	22	13	16	**19.8**
pyridine	+ $5CH_4 + NH_3 \rightarrow 2CH_3CH_3 + 2CH_2CH_2 + CH_3NH_3 + CH_2NH$	45	48	35	39	38	41	32	36	**33.8**

389

Table 13-1: Heats of Formation from Bond Separation Reactions (3)

molecule	bond separation reaction	Hartree-Fock 6-31G*	Hartree-Fock 6-311+G**	EDF1 6-31G*	EDF1 6-311+G**	B3LYP 6-31G*	B3LYP 6-311G**	MP2 6-31G*	MP2 6-311+G**	expt.
pyridazine	+ 4CH$_4$ + 2NH$_3$ → 2CH$_3$CH$_3$ + CH$_2$CH$_2$ + 2CH$_2$NH + NH$_2$NH$_2$	89	92	77	79	80	83	73	77	**66.5**
pyrimidine	+ 4CH$_4$ + 2NH$_3$ → CH$_3$CH$_3$ + CH$_2$CH$_2$ + 2CH$_3$NH$_2$ + 2CH$_2$NH	61	63	52	55	55	58	49	53	**47.0**
pyrazine	+ 4CH$_4$ + 2NH$_3$ → CH$_3$CH$_3$ + CH$_2$CH$_2$ + 2CH$_3$NH$_2$ + CH$_2$NH	68	71	57	59	60	63	53	57	**46.8**
cyclopropane	+ 3CH$_4$ → 3CH$_3$CH$_3$	19	18	13	13	18	16	17	16	**12.7**
azacyclopropane	+ 2CH$_3$ + NH$_3$ → CH$_3$CH$_3$ + 2CH$_3$NH$_2$	37	36	30	29	36	34	33	34	**30.2**
oxacyclopropane	+ 2CH$_4$ + H$_2$O → CH$_3$CH$_3$ + 2CH$_3$OH	-2	-5	-13	-15	-7	-9	-11	-10	**-12.6**
thiacyclopropane	+ 2CH$_4$ + H$_2$S → CH$_3$CH$_3$ + 2CH$_3$SH	26	24	18	17	23	22	21	21	**19.6**

Table 13-1: Heats of Formation from Bond Separation Reactions (4)

molecule	bond separation reaction	Hartree-Fock		EDF1		B3LYP		MP2		expt.
		6-31G*	6-311+G**	6-31G*	6-311+G**	6-31G*	6-311G**	6-311G*	6-311+G**	
cyclobutane	$\square$ + $4CH_4 \rightarrow 4CH_3CH_3$	15	14	9	9	13	12	10	10	**6.8**
cyclopropane	$\triangle$ + $3CH_4 \rightarrow 2CH_3CH_3 + CH_2CH_2$	76	77	65	65	71	70	70	70	**66.2**
cyclobutene	$\square$ + $4CH_4 \rightarrow 3CH_3CH_3 + CH_2CH_2$	48	48	39	40	44	44	42	43	**37.5**
cyclopentadiene	+ $5CH_4 \rightarrow 3CH_3CH_3 + 2CH_2CH_2$	42	42	32	34	36	37	32	33	**31.3**
thiophene	+ $4CH_4 + H_2S \rightarrow CH_3CH_3 + 2CH_3SH + 2CH_2CH_2$	28	30	14	16	19	21	11	12	**27.5**
methylenecyclopropane	+ $4CH_4 \rightarrow 3CH_3CH_3 + CH_2CH_2$	54	54	43	44	50	49	50	50	**47.9**
bicyclo[1.1.0]butane	+ $6CH_4 \rightarrow 5CH_3CH_3$	69	66	51	52	65	61	43	43	**51.9**

Chapter 14

Dealing with
Flexible Molecules

This chapter addresses practical issues which arise in dealing with flexible molecules. These include identification of the "important" conformer (or set of conformers) and location of this conformer. The chapter concludes with guidelines for fitting potential energy functions for bond rotation to simple Fourier series.

Introduction

Conformation dictates overall molecular size and shape, and influences molecular properties as well as chemical reactivity. Experimental information about conformation is often scarce, and computational methods may need to stand on their own. There are actually two different problems associated with treatment of conformationally-flexible molecules. The first is to identify the appropriate conformer (or conformers), and the second is to locate it (them). Both of these will be touched on in turn.

Identifying the "Important" Conformer

The equilibrium ("thermodynamic") abundance of conformational forms depends on their relative energies. According to the Boltzmann equation, the lowest-energy conformer (global minimum) will be present in the greatest amount, the second lowest-energy conformer in the next greatest amount, and so forth.[*] This implies that reactions under thermodynamic control and involving conformationally-flexible reagents need to be described in terms of the properties of global

---

* This is not strictly true where certain conformers possess elements of symmetry. Here, the number of occurences of each "unique" conformer also needs to be taken into account.

minima, or more precisely in terms of the properties of all minima weighted by their relative Boltzmann populations.

The situation may be markedly different for reactions under kinetic control. Here, the lowest-energy conformer(s) of the reagent(s) may not be the one(s) involved in the reaction. A simple but obvious example of this is provided by the Diels-Alder cycloaddition of 1,3-butadiene with acrylonitrile.

The diene exists primarily in a *trans* conformation, the *cis* conformer being approximately 2 kcal/mol less stable and separated from the *trans* conformer by a low energy barrier. At room temperature, only about 5% of butadiene molecules will be in a *cis* conformation. Clearly, *trans*-butadiene cannot undergo cycloaddition (as a diene), at least via the concerted pathway which is known to occur, and rotation into a *cis* conformation is required before reaction can proceed.

Diels-Alder cycloaddition of 1,3-butadiene and acrylonitrile is significantly slower than the analogous reaction involving cyclopentadiene. Might this simply be a consequence of the difference in energy between the ground-state *trans* conformer of butadiene and the "*cis* like" conformer which must be adopted for reaction to occur, or does it reflect fundamental differences between the two dienes? That is, are activation energies for Diels-Alder cycloaddition of *cis*-butadiene and of cyclopentadiene actually similar?

According to B3LYP/6-31G* calculations, the activation energy for cycloaddition of *cis*-1,3-butadiene and acrylonitrile is 20 kcal/mol, while the activation energy for the corresponding reaction involving cyclopentadiene is 16 kcal/mol. The two are not the same, and the difference in reactivity is more than the matter of conformation. Interestingly, the difference in activation energies closely matches the difference in the energies of *cis* and *trans* conformers of 1,3-butadiene (4 kcal/mol from B3LYP/6-31G* calculations).

A related example is the observation (from calculations) that Diels-Alder reaction of 1-methoxybutadiene and acrolein gives different regioproducts depending on the conformation of acrolein. Reaction of *trans*-acrolein (the global minimum) gives the *meta* product (not observed experimentally), while reaction of *cis*-acrolein, which is about 2 kcal/mol higher in energy, leads to the observed *ortho* product.

In both of these situations, the reaction actually observed does not occur from the lowest-energy conformation of the reactants. That this need not be the case is a direct consequence the Curtin-Hammett principle[1]. This recognizes that some higher-energy "reactive conformation", will be in rapid equilibrium with the global minimum and, assuming that any barriers which separate these conformations are much smaller than the barrier to reaction, will be replenished throughout the reaction.

In the case of the above-mentioned Diels-Alder reactions, the reactant conformers are separated by energy barriers which are far smaller than the activation required for cycloaddition.

It is clear from the above discussion that the products of kinetically-controlled reactions do not necessarily derive from the lowest-energy conformer. The identity of the "reactive conformer" is, however, not at all apparent. One "reasonable" hypothesis is that this is the conformer which is best "poised to react", or alternatively as the conformer which first results from progression "backward" along the reaction coordinate starting from the transition state. Operationally, such a conformer is easily defined. All that one needs to do is to start at the transition state and, following a "push" along the reaction coordinate in the direction of the reactant, optimize to a stable structure. Given that both the transition state and the reaction coordinate are uniquely defined, the reactive conformer is also uniquely defined. Of course, there is no way to actually prove such an hypothesis (at least in any general context). The best that can be done is to show that it accommodates the available experimental data in specific cases. An example is provided in **Table 14-1**. This compares activation energies calculated using the 3-21G model for Claisen rearrangements of cyano-substituted allyl vinyl ethers, relative to the unsubstituted compound, with experimentally-derived activation energies. Both global and "reactive" conformers of reactant have been considered. Overall, the data based on use of the reactive conformation is in better agreement with the experimental relative activation energies than that based on use of the global minimum, although except for substitution in the 1-position, the noted differences are small.

Locating the Lowest-Energy Conformer

While the discussion in the previous section points out serious ambiguity in kinetically-controlled processes involving flexible molecules, the situation is perfectly clear where thermodynamics is in control. Here, the lowest-energy conformer (or set of low-energy conformers) are important. Identifying the lowest-energy conformation may, however, be difficult, simply because the number of possible conformers can be very large. A systematic search on a molecule with N single bonds and a "step size" of $360°/M$, would need to examine M^N conformers. For a molecule with three single bonds and a step size of $120°$ (M=3), this leads to 27 conformers; for a molecule with eight single bonds, over 6500 conformers would need to be considered. Furthermore, step sizes smaller than $120°$ may be required in order to avoid missing stable conformers. Hence, the problem of systematically searching conformation space is formidable, even for relatively simple molecules. It rapidly becomes

Table 14-1: Activation Energies of Claisen Rearrangements[a]

position of substitution	calculated activation energy[b]		experimental activation energy[c]
	global	reactive	
–	0	0	**0**
1	-1.0	1.3	**1.7**
2	-5.8	-3.4	**-2.6**
4	-2.9	-3.1	**-3.1**
5	-4.9	-5.0	**-2.8**
6	2.6	2.3	**3.4**

a)

b)   W.W. Huang, Ph.D. thesis, University of California, Irvine, 1994.

c)   C.J. Burrows and B.K. Carpenter, *J. Amer. Chem. Soc.*, **103**, 6983 (1981).

insurmountable for larger molecules. Alternative approaches which involve "sampling" as opposed to complete scrutiny of conformation space are needed.

Conformational searching is an active area of research, and it is beyond the scope of the present treatment to elaborate in detail or to assess the available strategies. It is worth pointing out, however, that these generally fall into three categories:

i) systematic methods which "rotate around" bonds and "pucker" ring centers one at a time,

ii) Monte-Carlo and molecular dynamics techniques, which randomly sample conformational space, and

iii) genetic algorithms which randomly "mutate" populations of conformers in search of "survivors".

There are also hybrid methods which combine features from two or all three of the above. Opinions will freely be offered about which technique is "best", but the reality is that different techniques will perform differently depending on the problem at hand. Except for very simple systems with only one or a few degrees of conformational freedom, systematic methods are not practical, and sampling techniques, which do not guarantee location of the lowest-energy structure (because they do not "look" everywhere), are the only viable alternative. By default, Spartan uses systematic searching for systems with only a few degrees of conformational freedom and Monte-Carlo methods for more complicated systems.

A related practical concern is whether a single "energy function" should be used both to locate all "reasonable" conformers and to assign which of these conformers is actually best, or whether two (or more) different energy functions should be employed, i.e.

identification of all minima using "low cost" energy function	$\rightarrow$	assignment of conformer energies using "higher-cost" energy function

In practice, except for very simple molecules, molecular mechanics procedures may be the only choice to survey the full conformational energy surface and to identify low-energy conformers. Even semi-empirical methods are likely to be too costly for extensive conformational searching on systems with more than a few degrees of freedom. Note, however, that even if semi-empirical methods were practical for this task, the data provided in **Chapter 8** (see **Tables 8-1** and **8-2**), indicate that these are not likely to lead to acceptable results. Hartree-Fock and correlated calculations, which do appear to lead to good results, seem out of the question for any but the very simplest systems. Fortunately, the MMFF molecular mechanics model is quite successful in assigning low-energy conformers and in providing quantitative estimates of conformational energy differences. It would appear to be the method of choice for large scale conformational surveys.

Using "Approximate" Equilibrium Geometries to Calculate Conformational Energy Differences

It has previously been shown that equilibrium geometries obtained at one level of calculation more often than not provide a suitable basis for energy evaluation at another (higher) level of calculation (see **Chapter 12**). This applies particularly well to *isodesmic* reactions, in which reactants and products are similar, and where errors resulting from the use of "approximate" geometries might be expected to largely cancel. A closely related issue is whether "approximate" conformational energy differences obtained in this manner would be suitable replacements for "exact" differences. At first glance the answer would appear to be obvious. Conformational energy comparisons are after all *isodesmic* reactions. In fact, bond length and angle changes from one conformer to another would be expected to be very small, and any errors due to the use of "approximate" geometries would therefore be expected to largely cancel. On the other hand, conformational energy differences are likely to be very small (on the order of a few tenths of a kcal/mol to a few kcal/mol) and even small errors due to use of approximate geometries might be intolerable.

Conformational energy differences for a small selection of acyclic and cyclic molecules obtained from 6-31G*, EDF1/6-31G*, B3LYP/6-31G* and MP2/6-31G* models are provided in **Tables 14-2** to **14-5**, respectively. Results from "exact" geometries are compared with those obtained using structures from MMFF, AM1 and 6-31G* calculations.

MMFF geometries appear to be suitable replacements for "exact" structures for obtaining conformational energy differences. For all four calculation methods, the mean absolute error is essentially unchanged, and individual conformational energy differences change by a few tenths of a kcal/mol at most.

AM1 geometries are far less suitable. The mean absolute error in calculated conformational energy differences vs. experiment is significantly increased (relative to use of either MMFF or "exact" geometries), and individual energy differences are in some cases changed by large amounts. In one case (piperidine) the assignment of preferred conformation is reversed (over both experiment and "exact" calculations). Clearly AM1 geometries are not suitable for this purpose.

6-31G* geometries (in EDF1, B3LYP and MP2/6-31G* calculations) provide results comparable to those obtained from full calculations. Their use is strongly recommended.

Although no documentation has been provided here, the same conclusions apply as well to the related problem of barriers to rotation and inversion, where "approximate" geometries from MMFF and small-basis-set Hartree-Fock models can be used with confidence. Again, there are problematic cases (the geometry about nitrogen in amines from small-basis-set Hartree-Fock models), and again caution is urged in the use of geometries from semi-empirical calculations for this purpose.

Table 14-2: **Effect of Choice of Geometry on Conformational Energy Differences. 6-31G* Model**

molecule	low-energy/ high-energy conformer	geometry			
		MMFF	AM1	6-31G*	expt.
n-butane	*trans/gauche*	0.8	1.4	1.0	**0.67**
1-butene	*skew/cis*	0.5	0.6	0.7	**0.22**
1,3-butadiene	*trans/gauche*	3.4	4.3	3.1	**2.89**
acrolein	*trans/cis*	1.4	1.9	1.7	**1.70**
methyl formate	*cis/trans*	6.2	8.3	6.2	**4.75**
methyl ethyl ether	*anti/gauche*	1.7	2.1	1.7	**1.5**
methyl vinyl ether	*cis/skew*	2.0	2.3	2.0	**1.7**
cyclohexane	*chair/twist boat*	7.3	7.8	6.8	**5.5**
methylcyclohexane	*equatorial/axial*	2.3	3.0	2.3	**1.75**
piperidine	*equatorial/axial*	1.0	-0.2	0.8	**0.53**
2-cholorotetrahydropyran	*axial/equatorial*	1.2	1.9	2.5	**1.8**
mean absolute error		0.6	1.1	0.5	–

Table 14-3: **Effect of Choice of Geometry on Conformational Energy Differences. EDF1/6-31G* Model**

molecule	low-energy/ high-energy conformer	geometry			EDF1/ 6-31G*	
		MMFF	AM1	6-31G*		expt.
n-butane	*trans/gauche*	0.9	1.4	1.1	0.8	**0.67**
1-butene	*skew/cis*	0.5	0.6	0.8	0.8	**0.22**
1,3-butadiene	*trans/gauche*	4.2	4.4	3.9	4.0	**2.89**
acrolein	*trans/cis*	1.6	1.9	1.7	1.8	**1.70**
methyl formate	*cis/trans*	5.0	5.9	4.8	4.9	**4.75**
methyl ethyl ether	*anti/gauche*	1.7	2.1	1.5	1.5	**1.5**
methyl vinyl ether	*cis/skew*	2.2	2.2	2.0	1.9	**1.7**
cyclohexane	*chair/twist boat*	6.8	7.3	6.4	6.4	**5.5**
methylcyclohexane	*equatorial/axial*	2.7	3.2	2.8	2.8	**1.75**
piperidine	*equatorial/axial*	0.3	-0.8	0.0	0.1	**0.53**
2-cholorotetrahydropyran	*axial/equatorial*	1.9	2.6	3.4	3.7	**1.8**
mean absolute error		0.5	0.9	0.6	0.6	–

Table 14-4: **Effect of Choice of Geometry on Conformational Energy Differences. B3LYP/6-31G* Model**

molecule	low-energy/ high-energy conformer	geometry				
		MMFF	AM1	6-31G*	B3LYP/ 6-31G*	expt.
n-butane	*trans/gauche*	0.7	1.2	0.8	0.8	**0.67**
1-butene	*skew/cis*	0.3	0.2	0.5	0.4	**0.22**
1,3-butadiene	*trans/gauche*	3.9	4.2	3.6	3.6	**2.89**
acrolein	*trans/cis*	1.4	1.8	1.6	1.7	**1.70**
methyl formate	*cis/trans*	5.4	6.8	5.3	5.3	**4.75**
methyl ethyl ether	*anti/gauche*	1.5	1.8	1.3	1.4	**1.5**
methyl vinyl ether	*cis/skew*	2.5	2.7	2.3	2.3	**1.7**
cyclohexane	*chair/twist boat*	7.0	7.5	6.5	6.5	**5.5**
methylcyclohexane	*equatorial/axial*	2.3	2.9	2.4	2.1	**1.75**
piperidine	*equatorial/axial*	0.5	-0.7	0.3	0.3	**0.53**
2-cholorotetrahydropyran	*axial/equatorial*	2.0	2.7	3.5	3.7	**1.8**
mean absolute error		0.5	0.9	0.5	0.5	–

Table 14-5: **Effect of Choice of Geometry on Conformational Energy Differences. MP2/6-31G* Model**

molecule	low-energy/ high-energy conformer	geometry				
		MMFF	AM1	6-31G*	MP2/ 6-31G*	expt.
n-butane	*trans/gauche*	0.5	1.0	0.7	0.7	**0.67**
1-butene	*skew/cis*	0.5	0.3	0.5	0.5	**0.22**
1,3-butadiene	*trans/gauche*	2.9	3.7	2.7	2.6	**2.89**
acrolein	*trans/cis*	1.3	1.6	1.4	1.5	**1.70**
methyl formate	*cis/trans*	6.3	7.5	6.3	6.4	**4.75**
methyl ethyl ether	*anti/gauche*	1.5	1.8	1.4	1.4	**1.5**
methyl vinyl ether	*cis/skew*	2.7	3.0	2.8	2.8	**1.7**
cyclohexane	*chair/twist boat*	7.3	7.7	6.7	6.4	**5.5**
methylcyclohexane	*equatorial/axial*	1.8	2.3	2.0	1.9	**1.75**
piperidine	*equatorial/axial*	0.3	-0.9	0.5	0.6	**0.53**
2-cholorotetrahydropyran	*axial/equatorial*	2.1	2.5	2.8	2.8	**1.8**
mean absolute error		0.5	0.9	0.5	0.5	–

Using Localized MP2 Models to Calculate Conformational Energy Differences

The MP2/6-311+G** model is among the most accurate and reliable practical method for calculating conformational energy differences (see **Chapter 8**). It has the potential for supplementing or completely replacing experimental data in development of empirical energy functions for use in molecular mechanics/molecular dynamics calculations (see discussion later in this chapter). However, this model is quite costly in terms of overall calculation times and severely restricted in its range of application due to memory and disk requirements. It is, therefore, of considerable practical importance to develop strategies which reduce calculation demands but do not lead to significant degradation in overall quality. As shown in the previous section, one appropriate and highly-effective strategy is to replace MP2 geometries by Hartree-Fock geometries (or even geometries from MMFF molecular mechanics). This eliminates the high cost of geometry optimization with MP2 models but does nothing to extend their range.

Another strategy is to base the MP2 energy correction on Hartree-Fock orbitals which have been localized according to some particular recipe. The resulting method is termed localized MP2 or simply LMP2. While localization results only in modest cost savings (increasing with increasing size of the molecule) the real benefit is significantly reduced memory and disk requirements. Therefore, it leads to a *defacto* increase in the size of system that can be treated.

Table 14-6 compares conformational energy differences from LMP2/6-311+G** and MP2/6-311+G** calculations for a small selection of cyclic and acyclic molecules. 6-31G* geometries have been used throughout. Thus, the data here has been collected to take advantage both of the use of approximate geometries and of localization. The result is clear; energy differences for all systems are identical to within the precision provided (0.1 kcal/mol). LMP2 models may be used with confidence in place of corresponding MP2 models for this purpose.

Table 14-6: **Performance of Localized MP2 Models on Conformational Energy Differences**

molecule	low-energy/ high-energy conformer	LMP2/ 6-311+G**// 6-31G*	MP2/ 6-311+G**// 6-31G*	expt.
n-butane	*trans/gauche*	0.7	0.7	**0.67**
1-butene	*skew/cis*	0.5	0.5	**0.22**
1,3-butadiene	*trans/gauche*	2.7	2.7	**2.89**
acrolein	*trans/cis*	1.4	1.4	**1.70**
methyl formate	*cis/trans*	6.3	6.3	**4.75**
methyl ethyl ether	*anti/gauche*	1.4	1.4	**1.5**
methyl vinyl ether	*cis/skew*	2.8	2.8	**1.7**
cyclohexane	*chair/twist boat*	6.6	6.7	**5.5**
methylcyclohexane	*equatorial/axial*	2.0	2.0	**1.75**
piperidine	*equatorial/axial*	0.5	0.5	**0.53**
2-cholorotetrahydropyran	*axial/equatorial*	2.8	2.8	**1.8**
mean absolute error		0.5	0.5	–

Fitting Energy Functions for Bond Rotation

Empirical force fields used in molecular mechanics/molecular dynamics calculations all share common components, among them components which describe bond-stretching, angle-bending and torsional motions, as well as components which account for non-bonded steric and electrostatic interactions. While much of the information needed to parameterize force fields can be obtained from experiment, quite frequently critical data are missing. Information about torsional potentials, in particular, is often very difficult to obtain from experiment, and here calculations can prove of great value.

The energy of rotation about a single bond is a periodic function of the torsion angle and is, therefore, appropriately described in terms of a truncated Fourier series[2], the simplest acceptable form of which is given by.

$$V(\phi) = \frac{1}{2} V_1(1 - \cos\phi) + \frac{1}{2} V_2(1 - \cos2\phi) + \frac{1}{2} V_3(1 - \cos3\phi)$$
$$= V_1(\phi) + V_2(\phi) + V_3(\phi)$$

$$(1)$$

Here, V_1 is termed the one-fold component (periodic in 360°), V_2 is the two-fold component (periodic in 180°) and V_3 the three-fold component (periodic in 120°). Additional terms are required to account for bond rotations in asymmetric environments. Higher-order components may also be needed, but are not considered here.

A Fourier series is an example of an orthogonal polynomial, meaning that the individual terms which it comprises are independent of each other. It should be possible, therefore, to "dissect" a complex rotational energy profile into a series of N-fold components, and interpret each of these components independent of all others. For example, the one-fold term (the difference between *syn* and *anti* conformers) in *n*-butane probably reflects the crowding of methyl groups,

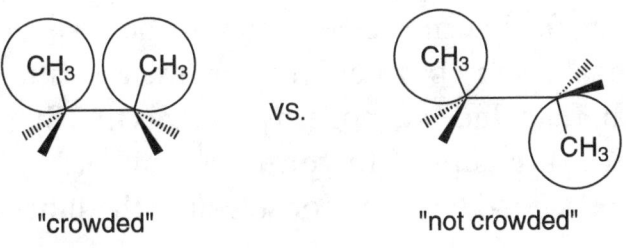

"crowded"   vs.   "not crowded"

while the one-fold term in 1,2-difluoroethane probably reflects differences in electrostatic interactions as represented by bond dipoles.

bond dipoles add   vs.   bond dipoles cancel

The three-fold component is perhaps the most familiar to chemists, as it represents the difference in energy between eclipsed and staggered arrangements about a single bond.

The two-fold component is perhaps the most interesting. It relates to the difference in energy between planar and perpendicular arrangements and often corresponds to turning "on" and "off" of electronic interactions, as for example in benzyl cation.

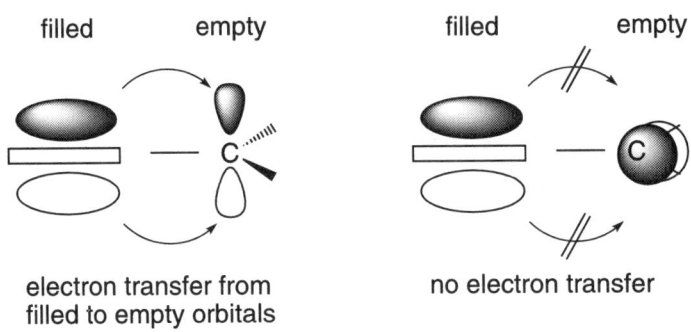

electron transfer from filled to empty orbitals   no electron transfer

In this case, only in the planar arrangement may electrons be transferred from the filled π orbital on benzene to the empty orbital associated with the carbocation center. This leads to delocalization of the positive charge, which in turn contributes to the high stability of the planar cation.

Just as quantum chemical calculations are able to locate and quantify both the stable conformers and the transition states connecting stable conformers for a flexible molecule, so too are they capable of obtaining the full torsional energy profile. This may then be fitted to whatever series is appropriate. Indeed, modern programs like Spartan automate the process. The "quality of fit" of the actual data to the empirical form should be a good criterion for selecting the functional form.

The selection of theoretical model with which to obtain the energy profile should be based on documented performance with regard to calculation of relative conformer energies and barrier heights. Full discussion has already been provided in **Chapter 8**.

References

1. (a) D.Y. Curtin, *Rec. Chem. Prog.*, **15**, 111 (1954); (b) L.P. Hammett, *Physical Organic Chemistry*, 2nd Ed. McGraw Hill, New York, 1970.

2. For a discussion, see: L. Radom, W.J. Hehre and J.A. Pople, *J. Amer. Chem. Soc.*, **94**, 2371 (1972).

Chapter 15

Obtaining and Using Transition-State Geometries

This chapter addresses practical issues associated with establishing, verifying and using transition-state geometries. It outlines a number of practical strategies for finding transition states, and provides criteria for establishing whether or not a particular geometry actually corresponds to the transition state of interest. Most of the remainder of the chapter focuses on choice of transition-state geometry, and in particular, errors introduced by using transition-state (and reactant) geometries from one model for activation energy calculations with another ("better") model. The chapter concludes with a discussion of "reactions without transition states".

Introduction

The usual picture of a chemical reaction is in terms of a one-dimensional potential energy (or reaction coordinate) diagram.

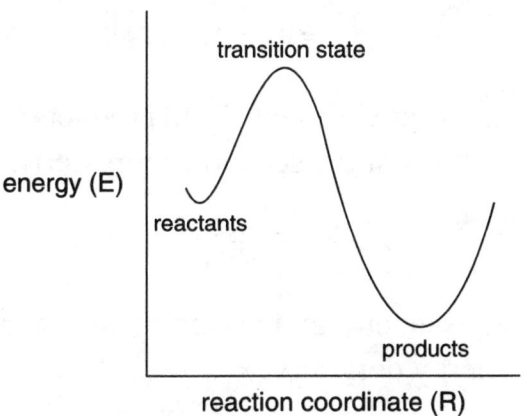

The vertical axis corresponds to the energy of the system and the horizontal axis (the "reaction coordinate") corresponds to the geometry of the system. The starting point on the diagram ("reactants") is an energy minimum, as is the ending point ("products"). In this diagram, the energy of the reactants is higher than that of the products (an "*exothermic* reaction") although this does not need to be the case. The energy of the reactants can be lower than that of the products (an "*endothermic* reaction"), or reactant and product energies may be the same (a "thermoneutral reaction") either by coincidence or because the reactants and products are the same molecule (a "degenerate reaction"). Motion along the reaction coordinate is assumed to be continuous and pass through a single energy maximum (the "transition state"). According to transition-state theory, the height of the transition state above the reactant relates to the overall rate of reaction (see **Chapter 9**).

Reactants, products and transition state are all stationary points on the potential energy diagram. In the one-dimensional case (a "reaction coordinate diagram"), this means that the derivative of the energy with respect to the reaction coordinate is zero.

$$\frac{dE}{dR} = 0 \tag{1}$$

The same must be true in dealing with a many-dimensional potential energy diagram (a "potential energy surface").[*] Here all partial derivatives of the energy with respect to each of the independent geometrical coordinates (R_i) are zero.

$$\frac{\partial E}{\partial R_i} = 0 \quad i = 1,2,...3N\text{-}6 \tag{2}$$

In the one-dimensional case, reactants and products are energy minima and characterized by a positive second energy derivative.

$$\frac{d^2E}{dR^2} > 0 \tag{3}$$

The transition state is an energy maximum and is characterized by a negative second energy derivative.

---

[*] Except for linear molecules, 3N-6 coordinates are required to describe an N atom molecule. 3N-5 coordinates are required to describe a linear N atom molecule. Molecular symmetry may reduce the number of independent coordinates.

$$\frac{d^2E}{dR^2} < 0 \qquad (4)$$

For a molecule with N atoms, each independent coordinate, R_i, gives rise to 3N-6 second derivatives.

$$\frac{\partial^2E}{\partial R_i \partial R_1} \, , \, \frac{\partial^2E}{\partial R_i \partial R_2} \, , \, \frac{\partial^2E}{\partial R_i \partial R_3} \, ,... \, \frac{\partial^2E}{\partial R_i \partial R_{3N-6}} \qquad (5)$$

This leads to a matrix of second derivatives (the "Hessian").

$$\begin{bmatrix} \dfrac{\partial^2E}{\partial R_1^2} & \dfrac{\partial^2E}{\partial R_1 \partial R_2} & \cdots & \\ \dfrac{\partial^2E}{\partial R_2 \partial R_1} & \dfrac{\partial^2E}{\partial R_2^2} & \cdots & \\ \vdots & \vdots & & \dfrac{\partial^2E}{\partial R_{3N-6}^2} \end{bmatrix} \qquad (6)$$

In this form, it is not possible to say whether any given coordinate corresponds to an energy minimum, an energy maximum or neither. In order to see the correspondence, it is necessary to replace the original set of geometrical coordinates (R) by a new set of coordinates (ξ) which leads to a matrix of second derivatives which is diagonal.

$$\begin{bmatrix} \dfrac{\partial^2E}{\partial \xi_1^2} & & & 0 \\ & \dfrac{\partial^2E}{\partial \xi_2^2} & \ddots & \dfrac{\partial^2E}{\partial \xi_{3N-6}^2} \\ 0 & & & \end{bmatrix} \qquad (7)$$

The ξ_i are unique and referred to as "normal coordinates". Stationary points for which all second derivatives (in normal coordinates) are positive are energy minima.

$$\frac{\partial^2E}{\partial \xi_i^2} > 0 \quad i = 1,2,...3N-6 \qquad (8)$$

These correspond to equilibrium forms (reactants and products) Stationary points for which all but one of the second derivatives are

411

positive are so-called (first-order) saddle points, and may correspond to transition states. If they do, the coordinate for which the second derivative is negative is referred to as the reaction coordinate (ξ_p).

$$\frac{\partial^2 E}{\partial \xi_p^2} < 0 \tag{9}$$

In effect, the 3N-6 dimensional system has been "split" into two parts, a one-dimensional system corresponding to motion along the reaction coordinate and a 3N-7 dimensional system accounting for motion along the remaining geometrical coordinates.

An obvious analogy (albeit only in two dimensions) is the crossing of a mountain range, the "goal" being simply to get from one side of the range to the other side with minimal effort.

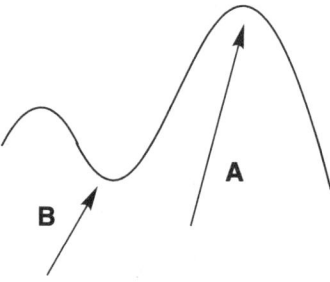

Crossing over the top of a "mountain" (pathway A), which corresponds to crossing through an energy maximum on a (two-dimensional) potential energy surface, accomplishes the goal. However, it is not likely to be the chosen pathway. This is because less effort (energy) will be expended by passing through a valley between two "mountains" (pathway B), a maximum in one dimension but a minimum in the other dimension. This is referred to as a saddle point and corresponds to a transition state.

Note that there are many possible transition states (different coordinates may be singled out as the reaction coordinate). What this means is that merely finding a transition state does not guarantee that this is "the transition state", meaning that it is at the top of the lowest-energy pathway that smoothly connects reactants and products. While it is possible to verify the smooth connection of reactants and products, it will generally not be possible to know with complete certainty that what has been identified as the transition state is in fact the lowest-energy structure over which the reaction might proceed, or whether in fact the actual reaction proceeds over a transition state which is not the lowest energy structure.

It should be clear from the above discussion, that the reactants, products and transition state all correspond to well-defined structures, despite the fact that only the reactants and products (energy minima) can actually be observed experimentally. It should also be clear that the pathway which the reactants actually follow to the products is not well defined. There are many ways to smoothly connect reactants with products which pass through the transition state, just like there are many ways to climb up and over a mountain pass. It is easy to visualize a "reasonable" (but not necessarily the "correct") reaction coordinate for a simple process. For example, the reaction coordinate for isomerization of hydrogen isocyanide to hydrogen cyanide might be thought of in terms of the HNC bond angle which is 180° in the reactant, 0° in the product and perhaps something close to 60° in the transition state.

$$H-N\equiv C \qquad \overset{\displaystyle H}{\underset{}{N\equiv C}} \qquad N\equiv C-H$$

$$< HNC \qquad 180° \qquad \sim 60° \qquad 0°$$

It is obvious, however, that the situation rapidly becomes complex if not completely intractable. Consider, for example, the problem of choosing a reaction coordinate describing as simple a reaction as the thermal elimination of ethylene from ethyl acetate.

$$H_2C\overset{O}{\diagdown}\underset{H_2CH}{\overset{}{C}}\underset{O}{\overset{\parallel}{C}}-CH_3 \longrightarrow \overset{CH_2}{\underset{CH_2}{\parallel}} + \overset{O}{\underset{HO}{C}}-CH_3$$

No single bond distance change or bond angle change provides an adequate description. Some combination of motions is required, the exact nature of which is not at all apparent.

413

What Do Transition States Look Like?

Experiments cannot tell us what transition states look like. The fact is that transition states cannot even be detected experimentally let alone characterized, at least not directly. While measured activation energies relate to the energies of transition states above reactants, and while activation entropies and activation volumes, as well as kinetic isotope effects, may be invoked to imply some aspects of transition-state structure, no experiment can actually provide direct information about the detailed geometries and/or other physical properties of transition states. Quite simply, transition states do not exist in terms of a stable population of molecules on which experimental measurements may be made. Experimental activation parameters provide some guide, but tell us little detail about what actually transpires in going from reactants to products.

On the other hand, quantum chemical calculations, at least non-empirical quantum chemical calculations, do not distinguish between systems which are stable and which may be scrutinized experimentally, and those which are labile (reactive intermediates), or do not even correspond to energy minima (transition states). The generality of the underlying theory, and (hopefully) the lack of intentional bias in formulating practical models, ensures that structures, relative stabilities and other properties calculated for molecules for which experimental data are unavailable will be no poorer (and no better) than the same quantities obtained for stable molecules for which experimental data exist for comparison.

The prognosis is bright. Calculations will uncover systematics in transition-state geometries, just as experiment uncovered systematics in equilibrium structures. These observations will ultimately allow chemists to picture transition states as easily and as realistically as they now view stable molecules.[*]

---

[*]   An effort is underway to provide an extensive library of transition states for organic and organometallic reactions obtained from a variety of theoretical models. The ultimate goal is to produce a transition-state builder inside of Spartan which, much like existing builders for "molecules", will capitalize on systematics and be able to finish accurate structures for reactions of interest.

Finding Transition States

There are several reasons behind the common perception that finding a transition state is more difficult than finding an equilibrium structure:

i)   Relatively little is known about geometries of transition states, at least by comparison with our extensive knowledge about the geometries of stable molecules. "Guessing" transition-state geometries based on prior experience is, therefore, much more difficult than guessing equilibrium geometries. This predicament is obviously due in large part to a complete lack of experimental structural data for transition states. It is also due to a lag in the application of computational methods to the study of transition states (and reaction pathways in general).

ii)   Finding a saddle point is probably (but not necessarily) more difficult than finding a minimum. What is certainly true, is that techniques for locating saddle points are much less well developed than procedures for finding minima (or maxima). After all, minimization is an important chore in many diverse fields of science and technology, whereas saddle point location has few if any "important" applications outside of chemistry.

iii)   The energy surface in the vicinity of a transition state is likely to be more "shallow" than the energy surface in the vicinity of a minimum. This is entirely reasonable; transition states "balance" bond breaking and bond making, whereas bonding is maximized in equilibrium structures. This "shallowness" suggests that the potential energy surface in the vicinity of a transition state is likely to be less well described in terms of a simple quadratic function than the surface in the vicinity of a local minimum. Common optimization algorithms, which assume limiting quadratic behavior, may in the long run be problematic, and new procedures may need to be developed.

iv)   To the extent that transition states incorporate partially (or nearly-completely) broken bonds, it might be anticipated that the simplest quantum-chemical models, including Hartree-Fock models, will not provide satisfactory descriptions, and that models which account explicitly for electron correlation will be required. While this is certainly the case with regard to calculated absolute activation energies, it appears not to be true for comparison of activation energies among closely-related reactions. Nor does it appear to be true for transition-state geometries. Discussion has already been provided in **Chapter 9**.

Key to finding a transition state is providing a "good" guess at its structure. There are several alternatives:

i) Base the guess on the transition structure for a closely-related system which has previously been obtained at the same level of calculation. The idea here is that transition-state geometries, like equilibrium geometries, would be expected to exhibit a high degree of uniformity among closely-related systems. Operationally, what is required is to first perform a transition-state optimization on the model system, and then to modify the model to yield the real system without changing the local geometry around the "reactive centers".*

Figures 15-1 and **15-2** provide evidence for the extent to which transition states for closely-related reactions are very similar. **Figure 15-1** compares the transition state for pyrolysis of ethyl formate (leading to formic acid and ethylene) with that for pyrolysis of cyclohexyl formate (leading to formic acid and cyclohexene). **Figure 15-2** compares the transition state for Diels-Alder cycloaddition of cyclopentadiene and acrylonitrile with both *syn* and *anti* transition states for cycloaddition of 5-methylcyclopentadiene and acrylonitrile. Results for Hartree-Fock 3-21G and 6-31G* models, EDF1/6-31G* and B3LYP/6-31G* density functional models, the MP2/6-31G* model and the AM1 semi-empirical model are provided.

An alternative is to use a transition state for the actual reaction of interest but obtained from a lower-level calculation, for example a semi-empirical or small-basis-set Hartree-Fock

---

* Spartan incorporates a library of transition states and an automated procedure for matching the reaction of interest to a related a reaction in the library.

Figure 15-1: Key Bond Distances in Related Formate Pyrolysis Reactions

model	ethyl formate	cyclohexyl formate
HF/3-21G	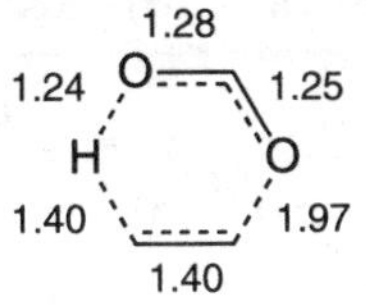	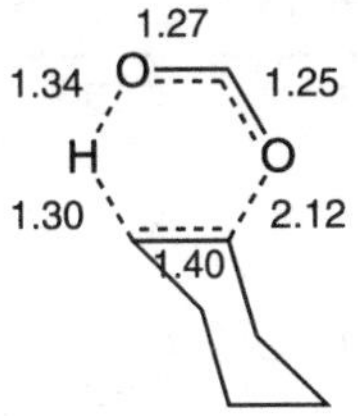
HF/6-31G*	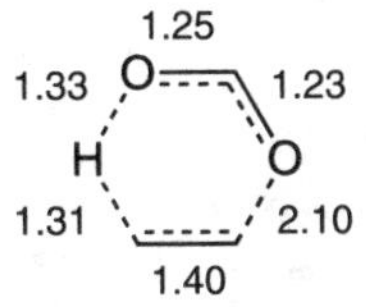	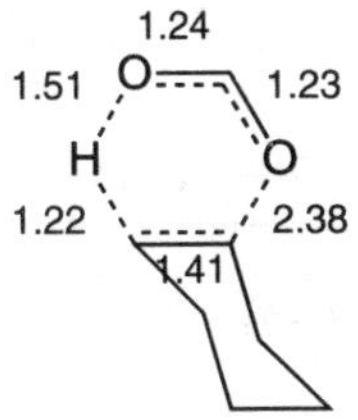
EDF1/6-31G*	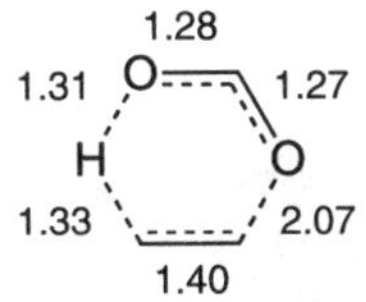	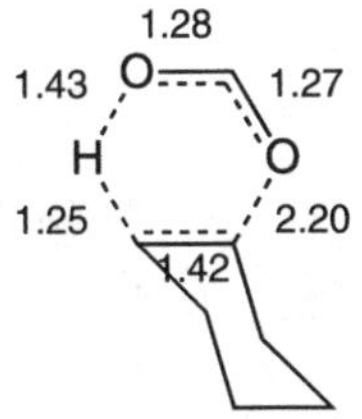
B3LYP/6-31G*	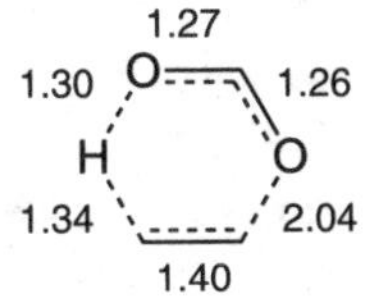	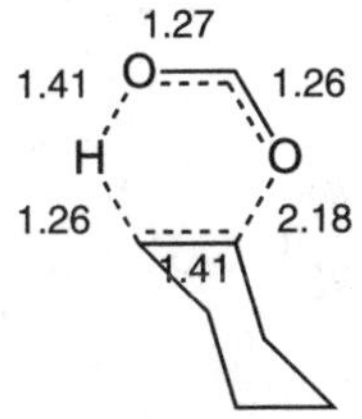
MP2/6-31G*	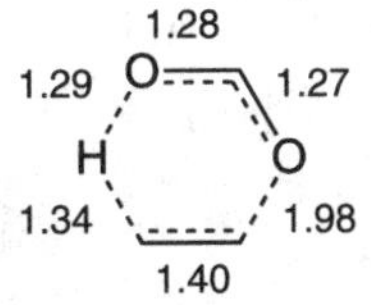	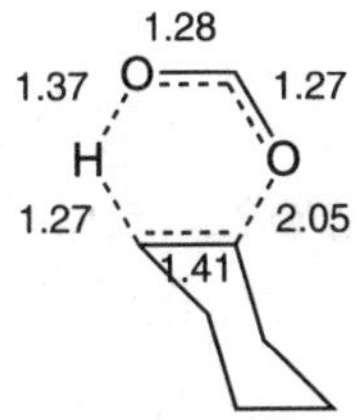
AM1		

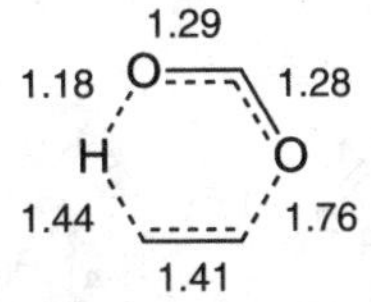

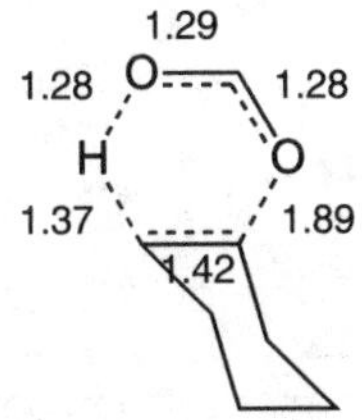

Figure 15-2: Key Bond Distances in Related Diels-Alder Cycloaddition Reactions

model	cyclopentadiene with acrylonitrile	5-methylcyclopentadiene with acrylonitrile (*anti*)	5-methylcyclopentadiene with acrylonitrile (*syn*)
HF/3-21G	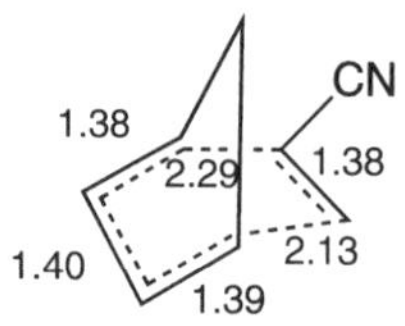 1.38, 2.29, CN 1.38, 1.40, 2.13, 1.39	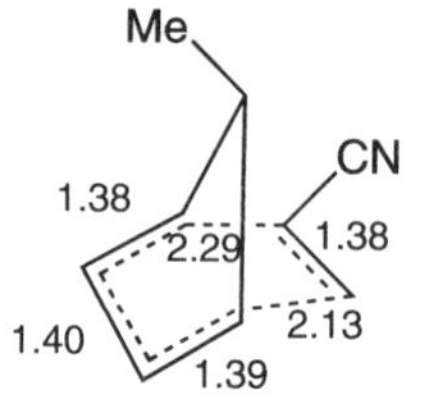 Me, 1.38, 2.29, CN 1.38, 1.40, 2.13, 1.39	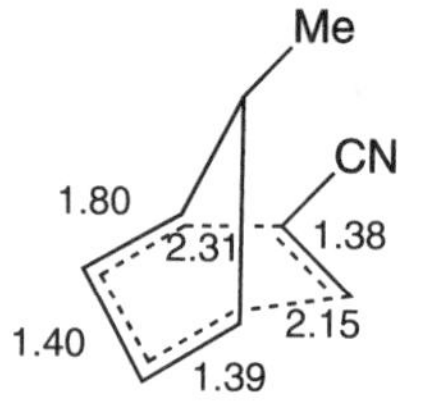 Me, 1.80, 2.31, CN 1.38, 1.40, 2.15, 1.39
HF/6-31G*	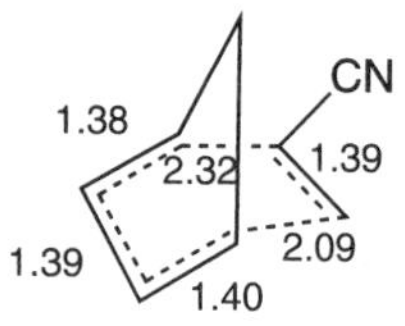 1.38, 2.32, CN 1.39, 1.39, 2.09, 1.40	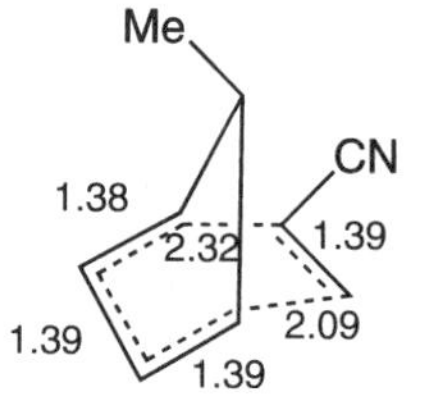 Me, 1.38, 2.32, CN 1.39, 1.39, 2.09, 1.39	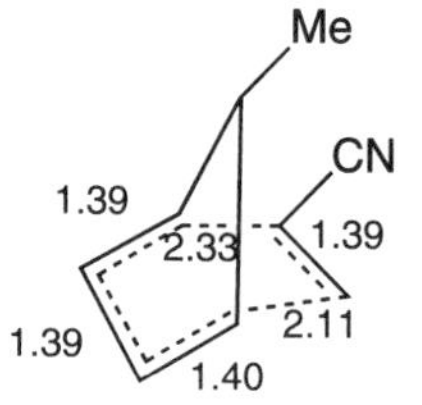 Me, 1.39, 2.33, CN 1.39, 1.39, 2.11, 1.40
EDF1/6-31G*	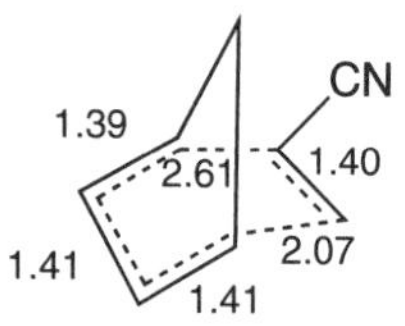 1.39, 2.61, CN 1.40, 1.41, 2.07, 1.41	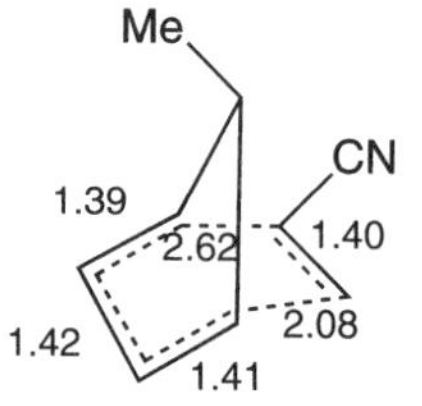 Me, 1.39, 2.62, CN 1.40, 1.42, 2.08, 1.41	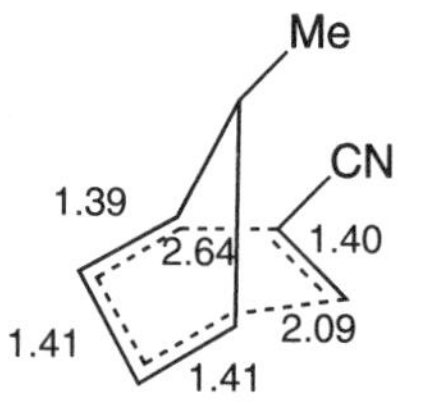 Me, 1.39, 2.64, CN 1.40, 1.41, 2.09, 1.41
B3LYP/6-31G*	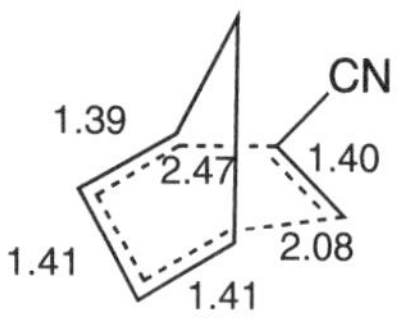 1.39, 2.47, CN 1.40, 1.41, 2.08, 1.41	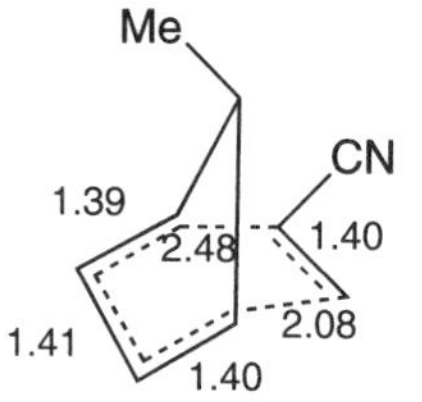 Me, 1.39, 2.48, CN 1.40, 1.41, 2.08, 1.40	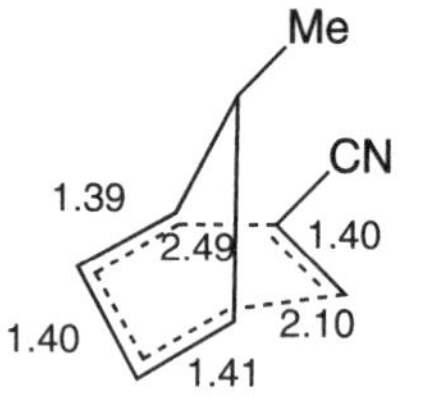 Me, 1.39, 2.49, CN 1.40, 1.40, 2.10, 1.41
MP2/6-31G*	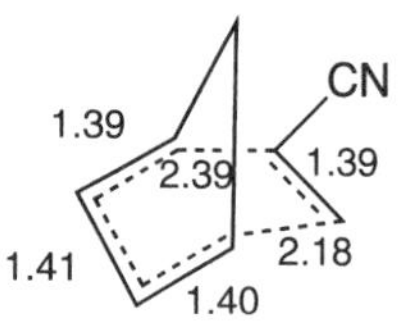 1.39, 2.39, CN 1.39, 1.41, 2.18, 1.40	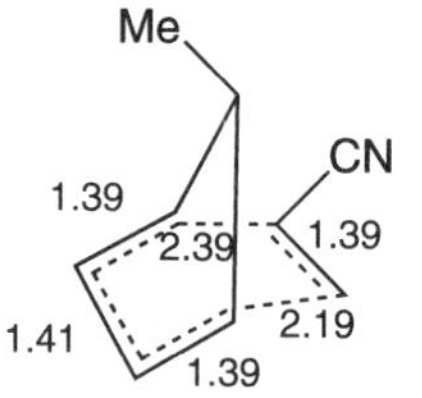 Me, 1.39, 2.39, CN 1.39, 1.41, 2.19, 1.39	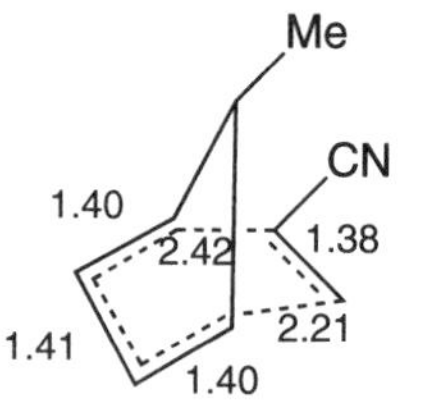 Me, 1.40, 2.42, CN 1.38, 1.41, 2.21, 1.40
AM1	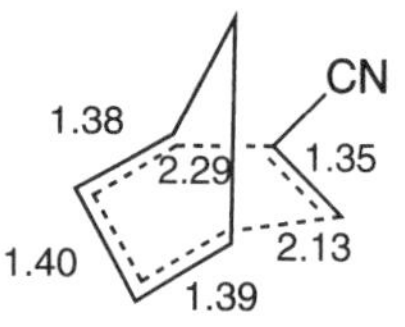 1.38, 2.29, CN 1.35, 1.40, 2.13, 1.39	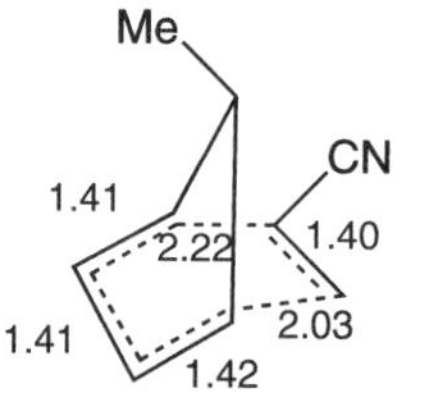 Me, 1.41, 2.22, CN 1.40, 1.41, 2.03, 1.42	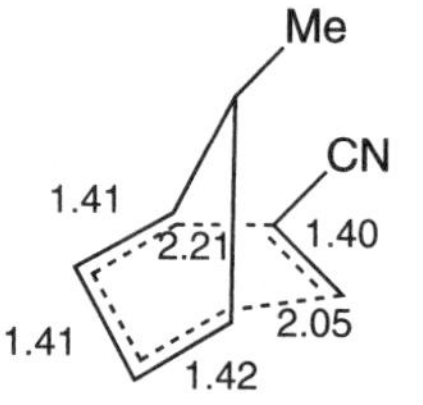 Me, 1.41, 2.21, CN 1.40, 1.41, 2.05, 1.42

calculation. Evidence that such a tactic is likely to be successful also comes from the data provided in **Figures 15-1** and **15-2**. Note the high degree of similarity in bond lengths obtained from different levels of calculation. It is, however, necessary to recognize that low-level methods sometimes lead to very poor transition-state geometries (see discussion in **Chapter 9**).

ii)   Base the guess on an "average" of reactant and product geometries (*Linear Synchronous Transit* method).[*]

iii)   Base the guess on "chemical intuition", specifying critical bond lengths and angles in accord with preconceived notions about mechanism. If possible, do not impose symmetry on the guess, as this may limit its ability to alter the geometry in the event that your "symmetrical" guess was incorrect.

Verifying Calculated Transition-State Geometries

There are two "tests" which need to be performed in order to verify that a particular geometry actually corresponds to a saddle point (transition structure), and further that this saddle point smoothly connects potential energy minima corresponding to reactants and products:[**]

i)   Verify that the Hessian (matrix of second-energy derivatives with respect to coordinates) yields one and only one imaginary frequency. This requires that vibrational frequencies be obtained for the proposed transition structure. Frequency calculation must be carried out using the same model that was employed to obtain the transition state; otherwise the results will be meaningless. The imaginary frequency will typically be in the range of 400-2000 cm^{-1}, quite similar in magnitude to real vibrational frequencies. For molecules with flexible rotors, e.g., methyl groups, or "floppy rings", the analysis may yield one or more additional imaginary frequencies with very small (<200

---

[*]   T.A. Halgren and W.N. Lipscomb, *Chem. Phys. Lett.*, 225 (1977). This is the "fallback" strategy in Spartan, and is automatically invoked when an unknown reaction is encountered.

[**] These "tests" do not guarantee that the "best" (lowest-energy) transition state has been located or, even if it is the lowest-energy transition state, that the reaction actually proceeds over it.

cm^{-1}) values, These typically correspond to torsions or related motions and can usually be ignored. However, identify the motions these small imaginary frequencies actually correspond to before ignoring them. Specifically, make certain they do not correspond to distortion away from any imposed element of symmetry. Also, be wary of structures which yield only very small imaginary frequencies. This suggests a very low energy transition structure, which quite likely will not correspond to the reaction of interest. In this case, it will be necessary to start over with a new guess at the transition structure.

ii) Verify that the normal coordinate corresponding to the imaginary frequency smoothly connects reactants and products. One way to do this is to "animate" the normal coordinate corresponding to the imaginary frequency, that is, to "walk along" this coordinate without any additional optimization. This does not require any further calculation, but will not lead to the precise reactants or to the precise products. The reaction coordinate is "correct" only in the immediate vicinity of the transition state, and becomes less and less "correct" with increased displacement away from the transition state. Even so, experience suggests that this tactic is an inexpensive and effective way to eliminate transition states which do not connect the reactants with the desired products.

An alternative and more costly approach is to actually "follow" the reaction from transition state to both the reactants and (independently) the products. In practice, this involves optimization subject to a fixed position along the reaction coordinate. A number of schemes for doing this have been proposed, and these are collectively termed *Intrinsic Reaction Coordinate* methods.* Note, that no scheme is unique; while the reactants, products and transition state are well defined points on the overall potential energy surface, there are an infinite number of pathways linking them together, just like there are an infinite number of pathways leading over a mountain pass.

---

* C. Gonzalez and H.B. Schlegel, *J. Phys. Chem.*, **90**, 2154 (1989).

Also, note the problem in defining reactants and/or products when they comprise more than a single molecule.

Using "Approximate" Transition-State Geometries to Calculate Activation Energies

Is it always necessary to utilize "exact" transition-state geometries in carrying out activation energy calculations, or will "approximate" geometries suffice?

This question is closely related to that posed previously for thermochemical comparisons (see **Chapter 12**) and may be of even greater practical importance. Finding transition states is more difficult (more costly) than finding equilibrium geometries (see discussion earlier in this chapter). There is reason to be encouraged. As pointed out previously, the potential energy surface in the vicinity of a transition state would be expected to be even more "shallow" than that in the vicinity of an energy minimum. This being the case, it is not unreasonable to expect that even significant differences in transition-state structures should have little effect on calculated activation energies. Small-basis-set Hartree-Fock models or even semi-empirical models might very well provide adequate transition-state geometries, even though their structural descriptions may differ significantly from those of higher-level models.

The question is first addressed with reference to absolute activation energies, with comparisons made using three different models previously shown to produce acceptable results: EDF1/6-31G* and B3LYP/6-31G* density functional models (**Tables 15-1** and **15-2**) and the MP2/6-31G* model (**Table 15-3**). Semi-empirical, Hartree-Fock and local density models have been excluded from the comparisons as these models do not provide good activation energies (see discussion in **Chapter 9** and in particular **Table 9-3**). BP and BLYP density functional models have also been excluded as they provide results broadly comparable to EDF1 and B3LYP models. Transition-state and reactant structures from AM1, 3-21G and 6-31G* calculations have been used for activation energy calculations and compared with activation energies based on the use of "exact"

Table 15-1: Effect of Choice of Geometry on Activation Energies from EDF1/6-31G* Calculations

reaction	geometry of reactant/transition state					
	AM1	3-21G	6-31G*	EDF1/ 6-31G*	MP2/ 6-311+G**	expt.
$CH_3NC \longrightarrow CH_3CN$	40	41	39	**40**	**41**	**38**
$HCO_2CH_2CH_3 \longrightarrow HCO_2H + C_2H_4$	52	48	48	**48**	**56**	**40,44**
(cyclohexadiene → cyclohexadiene)	a	31	32	**31**	**26**	**36**
(O-containing diene ring rearrangement)	34	25	25	**26**	**26**	**31**
(cyclopentadiene + ethylene → norbornene)	22	21	20	**21**	**9**	**20**
(cyclohexene → butadiene + C_2H_4)	53	53	53	**53**	**55**	–
$HCNO + C_2H_2 \longrightarrow$ (isoxazole)	10	16	12	**12**	**9**	–
(hexatriene → cyclohexadiene)	36	35	35	**35**	**34**	–
(cyclobutene → butadiene)	34	34	34	**34**	**34**	–
(lactone → diene + CO_2)	32	35	34	**34**	**41**	–
(sulfolene → diene + SO_2)	7	18	19	**18**	**22**	–
mean absolute error due to use of approximate geometries	3	1	0	–	–	–

a) reasonable transition state cannot be found

422

Table 15-2: Effect of Choice of Geometry on Activation Energies from B3LYP/6-31G* Calculations

reaction	geometry of reactant/transition state					
	AM1	3-21G	6-31G*	B3LYP/ 6-31G*	MP2/ 6-311+G**	expt.
$CH_3NC \longrightarrow CH_3CN$	42	42	40	**41**	**41**	**38**
$HCO_2CH_2CH_3 \longrightarrow HCO_2H + C_2H_4$	57	53	53	**53**	**56**	**40,44**
	a	34	35	**34**	**26**	**36**
	39	27	29	**29**	**26**	**31**
	21	20	19	**20**	**9**	**20**
	58	58	58	**58**	**55**	**–**
$HCNO + C_2H_2 \longrightarrow$	11	15	12	**12**	**9**	**–**
	40	38	39	**39**	**34**	**–**
	36	35	36	**36**	**34**	**–**
$+ CO_2$	37	40	40	**40**	**41**	**–**
$+ SO_2$	11	22	23	**22**	**22**	**–**
mean absolute error due to use of approximate geometries	3	1	0	–	–	–

a) reasonable transition state cannot be found

Table 15-3: Effect of Choice of Geometry on Activation Energies from MP2/6-31G* Calculations

reaction	geometry of reactant/transition state					
	AM1	3-21G	6-31G*	MP2/ 6-31G*	MP2/ 6-311+G**	expt.
$CH_3NC \longrightarrow CH_3CN$	43	44	42	**43**	**41**	**38**
$HCO_2CH_2CH_3 \longrightarrow HCO_2H + C_2H_4$	64	60	61	**60**	**56**	**40,44**
	a	31	31	**28**	**26**	**36**
	34	25	27	**26**	**26**	**31**
	12	11	11	**12**	**9**	**20**
	60	61	61	**60**	**55**	–
$HCNO + C_2H_2 \longrightarrow$	17	15	10	**8**	**9**	–
	39	38	38	**38**	**34**	–
	37	37	37	**37**	**34**	–
	43	43	45	**44**	**41**	–
	16	26	26	**25**	**22**	–
mean absolute error due to use of approximate geometries	3	2	1	–	–	–

a) reasonable transition state cannot be found

geometries. Data from MP2/6-311+G** calculations and (where available) from experiment have been tabulated in order to provide a sense of the magnitudes of errors stemming from use of approximate geometries relative to the magnitude of errors stemming from limitations of the particular model.

All three models show broadly similar behavior. Errors associated with replacement of "exact" reactant and transition-state geometries by AM1 geometries are typically on the order of 2-3 kcal/mol, although there are cases where much larger errors are observed. In addition, AM1 calculations failed to locate a "reasonable" transition state for one of the reactions in the set, the Cope rearrangement of 1,5-hexadiene.

Both 3-21G and 6-31G* Hartree-Fock models provide better and more consistent results in supplying reactant and transition-state geometries than the AM1 calculations. Also the two Hartree-Fock models (unlike the AM1 model) find "reasonable" transition states for all reactions. With only a few exceptions, activation energies calculated using approximate geometries differ from "exact" values by only 1-2 kcal/mol.

The recommendations are clear. While semi-empirical models appear to perform adequately in most cases in the role of supplying reactant and transition-state geometries, some caution needs to be exercised. On the other hand, structures from small-basis-set Hartree-Fock models turn in an overall excellent account. The 3-21G model, in particular, would appear to be an excellent choice for supplying transition-state geometries for organic reactions, at least insofar as initial surveys.

A second set of comparisons assesses the consequences of use of approximate reactant and transition-state geometries for relative activation energy calculations, that is, activation energies for a series of closely related reactions relative to the activation energy of one member of the series. Two different examples have been provided, both of which involve Diels-Alder chemistry. The first involves cycloadditions of cyclopentadiene and a series of electron-deficient dienophiles. Experimental activation energies (relative to Diels-Alder cycloaddition of cyclopentadiene and acrylonitrile) are available. Comparisons are limited to the 6-31G* and MP2/6-31G* models, both of which have previously been shown to correctly reproduce the experimental data. Excluded are density functional models and semi-empirical models, both which did not provide adequate account (discussion has already been provided in **Chapter 9**). AM1 and 3-21G geometries have been considered (in addition to "exact" geometries) for 6-31G* calculations (**Table 15-4**), and AM1, 3-21G and 6-31G* geometries have been considered (in addition to "exact" geometries) for MP2/6-31G* calculations (**Table 15-5**).

In terms of mean absolute error, choice of reactant and transition-state geometry has very little effect on calculated relative activation energies. Nearly perfect agreement between calculated and experimental relative activation energies is found for 6-31G* calculations, irrespective of whether or not "approximate" geometries are employed. Somewhat larger discrepancies are found in the case of MP2/6-31G* calculations, but overall the effects are small.

Comparisons involving reactions of substituted cyclopentadienes and acrylonitrile leading to different regio or stereochemical products are provided in **Tables 15-6** to **15-9** for 6-31G*, EDF1/6-31G*, B3LYP/ 6-31G* and MP2/6-31G* models, respectively. AM1, 3-21G and (except for 6-31G* calculations) 6-31G* geometries have been employed. Here, the experimental data are limited to the identity of the product and some "qualitative insight" about relative directing abilities of different substituents (see previous discussion in **Chapter 9**). The results are again clear and show a modest if not negligible effect of the use of approximate structures.

The overall recommendation following from these types of comparisons is very clear: use approximate geometries for calculations of relative activation energies among closely-related systems. While other examples need to be provided in order to fully generalize such a recommendation (there will no doubt be exceptions), and while calibration studies should be completed before widespread applications, the savings which might be achieved by such a strategy are considerable.

Table 15-4: Effect of Choice of Geometry on Relative Activation Energies of Diels-Alder Cycloadditions of Cyclopentadiene with Electron-Deficient Dienophiles.[a] 6-31G* Model

dienophile	geometry of reactants/transition state			
	AM1	3-21G	6-31G*	expt.
trans-1,2-dicyanoethylene	-4	-3	-3	**-2.6**
cis-1,2-dicyanoethylene	-4	-3	-3	**-3.8**
1,1-dicyanoethylene	-7	-7	-8	**-7.2**
tricyanoethylene	-9	-9	-9	**-9.2**
tetracyanoethylene	-12	-11	-11	**-11.2**
mean absolute error	0	0	0	–

a) energy of reaction

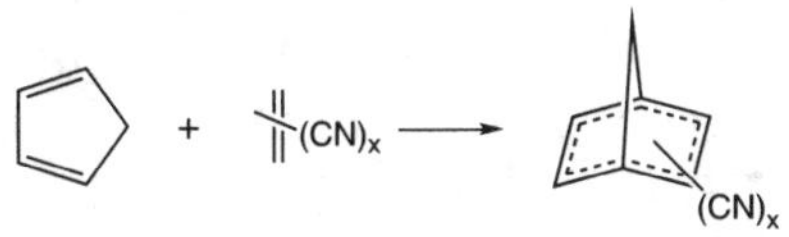

relative to:

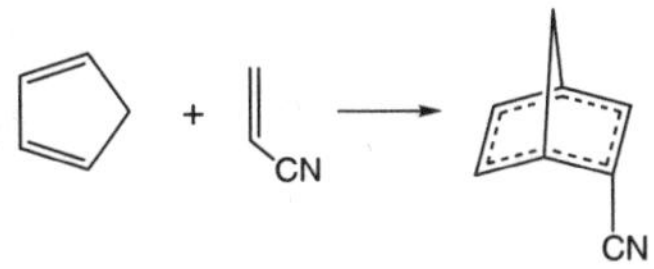

Table 15-5: Effect of Choice of Geometry on Relative Activation Energies of Diels-Alder Cycloadditions of Cyclopentadiene with Electron-Deficient Dienophiles.[a] MP2/6-31G* Model

dienophile	geometry of reactants/transition state				
	AM1	3-21G	6-31G*	MP2/6-31G*	expt.
trans-1,2-dicyanoethylene	-4	-5	-5	-5	**-2.6**
cis-1,2-dicyanoethylene	-5	-5	-4	-4	**-3.8**
1,1-dicyanoethylene	-6	-7	-6	-7	**-7.2**
tricyanoethylene	-10	-11	-9	-10	**-9.2**
tetracyanoethylene	-13	-16	-15	-15	**-11.2**
mean absolute error	1	2	1	1	–

a) energy of reaction

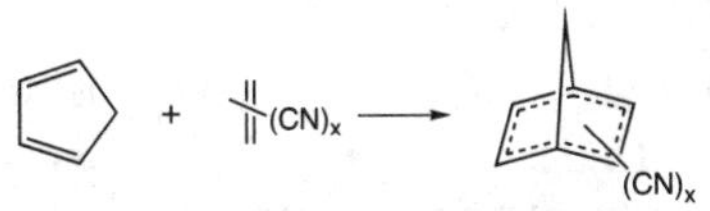

relative to:

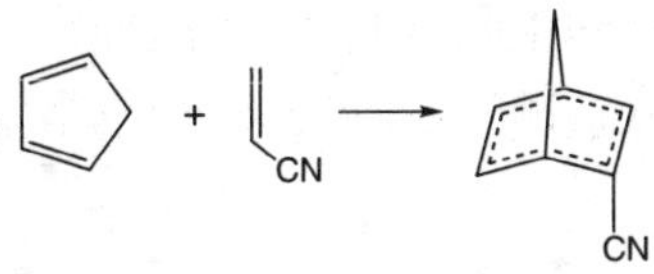

Table 15-6: **Effect of Choice of Geometry on Relative Energies of Regio and Stereochemistry of Diels-Alder Cycloadditions of Substituted Cyclopentadienes with Acrylonitrile.[a] 6-31G* Model**

substituent on cyclopentadiene	transition-state geometry			expt.
	AM1	3-21G	6-31G*	
regioselection				
1-Me	*ortho* (1.1)	*ortho* (1.1)	*ortho* (1.4)	*ortho*
1-OMe	*ortho* (1.7)	*ortho* (3.8)	*ortho* (4.2)	*ortho*
2-Me	*para* (0.5)	*para* (0.7)	*para* (0.6)	*para*
2-OMe	*para* (1.2)	*para* (3.3)	*para* (2.8)	*para*
stereoselection				
5-Me	*anti* (1.8)	*anti* (0.9)	*anti* (1.0)	*anti*
5-OMe	*syn* (6.0)	*syn* (7.5)	*syn* (6.6)	*syn*

a)

Table 15-7: **Effect of Choice of Geometry on Relative Energies of Regio and Stereochemistry of Diels-Alder Cycloadditions of Substituted Cyclopentadienes with Acrylonitrile.[a] EDF1/6-31G* Model**

substituent on cyclopentadiene	transition-state geometry				expt.
	AM1	3-21G	6-31G*	EDF1/6-31G*	
regioselection					
1-Me	*ortho* (1.8)	*ortho* (1.1)	*ortho* (1.5)	*ortho* (2.2)	*ortho*
1-OMe	*ortho* (2.4)	*ortho* (3.6)	*ortho* (4.8)	*ortho* (5.4)	*ortho*
2-Me	*para* (0.6)	*para* (0.3)	*para* (0.5)	*para* (0.3)	*para*
2-OMe	*para* (0.5)	*para* (2.0)	*para* (2.6)	*para* (2.4)	*para*
stereoselection					
5-Me	*anti* (1.7)	*anti* (1.4)	*anti* (1.4)	*anti* (1.6)	*anti*
5-OMe	*syn* (4.4)	*syn* (5.3)	*syn* (4.9)	*syn* (4.9)	*syn*

a)

Table 15-8: **Effect of Choice of Geometry on Relative Energies of Regio and Stereochemistry of Diels-Alder Cycloadditions of Substituted Cyclopentadienes with Acrylonitrile.[a] B3LYP/6-31G* Model**

substituent on cyclopentadiene	transition-state geometry				expt.
	AM1	3-21G	6-31G*	B3LYP/6-31G*	
regioselection					
1-Me	*ortho* (1.7)	*ortho* (1.2)	*ortho* (1.5)	*ortho* (1.6)	***ortho***
1-OMe	*ortho* (2.1)	*ortho* (3.8)	*ortho* (4.5)	*ortho* (4.6)	***ortho***
2-Me	*para* (0.3)	*meta* (0.2)	*none*	*meta* (0.1)	***para***
2-OMe	*para* (0.1)	*para* (1.8)	*para* (2.2)	*para* (2.2)	***para***
stereoselection					
5-Me	*anti* (1.1)	*anti* (0.8)	*anti* (0.9)	*anti* (0.9)	***anti***
5-OMe	*syn* (5.3)	*syn* (5.9)	*syn* (5.6)	*syn* (5.7)	***syn***

a)

Table 15-9: **Effect of Choice of Geometry on Relative Energies of Regio and Stereochemistry of Diels-Alder Cycloadditions of Substituted Cyclopentadienes with Acrylonitrile.[a] MP2/6-31G* Model**

substituent on cyclopentadiene	transition-state geometry				expt.
	AM1	3-21G	6-31G*	MP2/6-31G*	
regioselection					
1-Me	*ortho* (0.7)	*ortho* (0.8)	*ortho* (0.7)	*ortho* (0.7)	***ortho***
1-OMe	*ortho* (0.2)	*ortho* (2.1)	*ortho* (1.3)	*ortho* (1.9)	***ortho***
2-Me	*meta* (0.2)	*meta* (0.7)	*meta* (0.5)	*meta* (0.7)	***para***
2-OMe	*meta* (1.4)	*para* (0.1)	*meta* (0.5)	–	***para***
stereoselection					
5-Me	*anti* (1.1)	*anti* (1.0)	*anti* (1.0)	*anti* (1.0)	***anti***
5-OMe	*syn* (6.3)	*syn* (6.1)	*syn* (6.5)	*syn* (6.4)	***syn***

a)

Using Localized MP2 Models to Calculate Activation Energies

In addition to density functional models, MP2 models provide a good account of activation energies for organic reactions (see discussion in **Chapter 9**). Unfortunately, computer time and even more importantly, memory and disk requirements, seriously limit their application. One potential savings is to base the MP2 calculation on Hartree-Fock orbitals which have been localized. This has a relatively modest effect on overall cost*, but dramatically reduces memory and disk requirements, and allows the range of MP2 models to be extended.

Localized MP2 (LMP2) models have already been shown to provide results which are nearly indistinguishable from MP2 models for both thermochemical calculations (see **Chapter 12**) and for calculation of conformational energy differences (see **Chapter 14**). Activation energy calculations provide an even more stringent test. Transition states necessarily involve delocalized bonding, which may in turn be problematic for localization procedures.

Data presented in **Table 15-10** compare activation energies from LMP2/6-311+G** and MP2/6-311+G** calculations, both sets making use of underlying Hartree-Fock 6-31G* geometries. The results are very clear: localization has an insignificant effect on calculated activation energies. The procedure can be employed with confidence.

---

* Cost savings for localized MP2 models increase with increasing molecular size.

Table 15-10: Performance of Localized MP2 models on Activation Energies for Organic Reactions

reaction	LMP2/6-311+G**//6-31G*	MP2/6-311+G**//6-31G*	expt.
$CH_3NC \longrightarrow CH_3CN$	40	40	38
$HCO_2CH_2CH_3 \longrightarrow HCO_2H + C_2H_4$	58	57	40,44
cyclohexene → 1,3-cyclohexadiene	29	29	36
pyran → dihydropyran	26	26	31
cyclopentadiene + $\|$ → norbornene	9	8	20
(structure, H) → (structure) + C_2H_4	56	56	–
$HCNO + C_2H_2 \longrightarrow$ isoxazole	12	12	–
(diene) → (ring)	34	34	–
cyclobutene → butadiene	34	34	–
lactone $=O$ → diene + CO_2	43	43	–
cyclic SO_2 → diene + SO_2	24	24	–

Reactions Without Transition States

Surprisingly enough, reactions without barriers and discernible transition states are common. Two radicals will typically combine without a barrier, for example, two methyl radicals to form ethane.

$$H_3C^\bullet + {}^\bullet CH_3 \longrightarrow H_3C-CH_3$$

Radicals typically add to multiple bonds with little or no barrier, for example, methyl radical and ethylene to yield 1-propyl radical.

$$H_3C^\bullet + H_2C{=}CH_2 \longrightarrow CH_3CH_2CH_2{}^\bullet$$

In the gas phase, addition of ions to neutral molecules will almost certainly occur without an activation barrier, for example, addition of *tert*-butyl cation to benzene to yield a stable "benzenium" ion.

A more familiar example is S_N2 addition of an anionic nucleophile to an alkyl halide. In the gas phase, this occurs without activation energy, and the known barrier for the process in solution is a solvent effect (see discussion in **Chapter 6**). Finally, reactions of electron-deficient species, including transition-metal complexes, often occur with little or no energy barrier. Processes as hydroboration and β-hydride elimination are likely candidates.

Failure to find a transition state, but instead location of what appears to be a stable intermediate or even the final product, does not necessarily mean failure of the computational model (nor does it rule this out). It may simply mean that there is no transition state! Unfortunately it is very difficult to tell which is the true situation.

An interesting question is why reactions without activation barriers actually occur with different rates. The reason has to do with the pre-exponential term (or "A factor") in the rate expression, which depends both on the frequency of collisions and their overall effectiveness. These factors depend on molecular geometry and accessibility of reagents. Discussion has already been provided in **Chapter 1**.

Chapter 16

Obtaining and Interpreting Atomic Charges

This chapter focuses on the calculation of atomic charges in molecules. It discusses why atomic charges can neither be measured nor calculated unambiguously, and provides two different "recipes" for obtaining atomic charges from quantum chemical calculations. The chapter concludes with a discussion about generating atomic charges for use in molecular mechanics/molecular dynamics calculations.

Introduction

Charges are part of the everyday language of organic chemistry, and aside from geometries and energies, are certainly the most common quantities demanded from quantum chemical calculations. Charge distributions not only assist chemists in assessing overall molecular structure and stability, but also tell them about the "chemistry" which molecules can undergo. Consider, for example, the four resonance structures which a chemist might draw for phenoxy anion.

These not only indicate that all CC and CO bonds are intermediate in length between single and double linkages suggesting a delocalized and hence unusually stable ion, but also reveal that the negative charge resides not only on oxygen, but also on the *ortho* and *para* (but not on the *meta*) ring carbons. This, in turn, suggests that addition of an electrophile will occur only at *ortho* and *para* sites.

Why Can't Atomic Charges be Determined Experimentally or Calculated Uniquely

Despite their obvious utility, atomic charges are, however, not measurable properties, nor may they be determined uniquely from calculation. Overall charge distribution may be inferred from such observables as the dipole moment, but it is not possible to assign discrete atomic charges. The reason that it is not possible either to measure atomic charges or to calculate them, at least not uniquely, is actually quite simple. From the point of view of quantum mechanics, a molecule is made up of nuclei, each of which bears a (positive) charge equal to its atomic number, and electrons, each of which bears unit negative charge. While it is reasonable to assume that the nuclei are point charges, electrons may not be treated in this way. The simplest picture is that they form a distribution of negative charge which, while it extends throughout all space, is primarily concentrated in regions around the individual nuclei and in between nuclei which are close together, i.e., are bonded. The region of space occupied by a conventional space-filling (CPK) model, as defined by atomic van der Waals radii, encloses something on the order of 90-95% of the electrons in the entire distribution. That is to say, the space which molecules occupy in solids and liquids, corresponds to that required to contain 90-95% of the electron distribution.

While the total charge on a molecule (the total nuclear charge and the sum of the charge on all of the electrons) is well defined, it is not possible to uniquely define charges on individual atoms. This would require accounting both for the nuclear charge and for the charge of any electrons uniquely "associated" with the particular atom. As commented above, while it is reasonable to assume that the nuclear contribution to the total charge on an atom is simply the atomic number, it is not at all obvious how to partition the total electron distribution by atoms. Consider, for example, the electron distribution for the heteronuclear diatomic molecule, hydrogen fluoride.

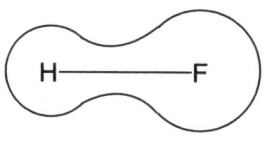

Here, the surrounding "line" is a particular "isodensity surface" (see **Chapter 4**), say that corresponding to a van der Waals surface and enclosing a large fraction of the total electron density. In this picture, the surface has been drawn to suggest that more electrons are associated with fluorine than with hydrogen. This is entirely reasonable, given the known polarity of the molecule, i.e., $^{\delta+}$H-F$^{\delta-}$, as evidenced experimentally by the direction of its dipole moment. It is, however, not at all apparent how to divide this surface between the two nuclei. Are any of the divisions shown below better than the rest?

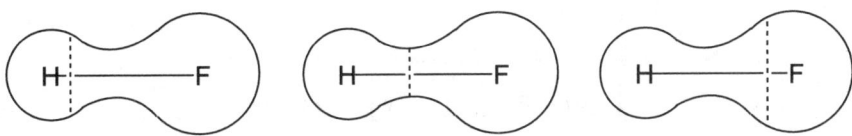

Clearly not! Atomic charges are not molecular properties, and it is not possible to provide a unique definition (or even a definition which will satisfy all). It is possible to calculate (and measure using X-ray diffraction) molecular charge distributions, that is, the number of electrons in a particular volume of space, but it is not possible to uniquely partition them among the atomic centers.

Methods for Calculating Atomic Charges

Several types of methods are now widely employed to assign atomic charges, and two of these will be discussed here. The first is based on partitioning the electron distribution, while the second is based on fitting some property which depends on the electron distribution to a model which replaces this distribution (and the underlying nuclei) by a set of atomic charges. There are many possible variations of each scheme; the criterion on which partitioning is based in the case of the former, and the selection of points and the property to be fit in the case of the latter. We discuss in turn a single variation of each type of scheme.

Population Analyses

In Hartree-Fock theory, the electron density at a point $\mathbf{r}$ may be written.

$$\rho(\mathbf{r}) = \sum_{\mu}^{\text{basis functions}} \sum_{\nu} P_{\mu\nu} \phi_\mu(\mathbf{r}) \phi_\nu(\mathbf{r}) \tag{1}$$

Here, $P_{\mu\nu}$ is an element of the density matrix (see **Chapter 2**), and the summations are carried out over all atom-centered basis functions, ϕ. Summing (integrating) over all space leads to an expression for the total number of electrons, n.

$$\int \rho(\mathbf{r})d\mathbf{r} = \sum_{\mu}^{\text{basis functions}} \sum_{\nu} P_{\mu\nu} \int \phi_\mu(\mathbf{r}) \phi_\nu(\mathbf{r})d\mathbf{r}$$

$$= \sum_{\mu}^{\text{basis functions}} \sum_{\nu} P_{\mu\nu} S_{\mu\nu} = n \tag{2}$$

$S_{\mu\nu}$ are elements of the overlap matrix. Similar types of expressions may be constructed for density functional and correlated models, as well as for semi-empirical models. The important point is that it is possible to equate the total number of electrons in a molecule to a sum of products of density matrix and overlap matrix elements.[*]

$$\sum_{\mu}^{\text{basis functions}} \sum_{\nu} P_{\mu\nu} S_{\mu\nu} = \sum_{\mu}^{\text{basis functions}} \sum_{\nu} P_{\mu\mu} + 2 \sum_{\mu}^{\text{basis functions}} \sum_{\nu} P_{\mu\nu} S_{\mu\nu} = n \tag{3}$$

It is reasonable (but not necessarily "correct") to assign any electrons associated with a particular diagonal element, $P_{\mu\mu}$, to that atom on which the basis function ϕ_μ is located. It is also reasonable to assign electrons associated with off-diagonal elements $P_{\mu\nu}$, where both ϕ_μ and ϕ_ν reside on the same atom, to that atom. However, it is not apparent how to partition electrons from density matrix elements $P_{\mu\nu}$ where ϕ_μ and ϕ_ν reside on different atoms. Mulliken provided a recipe.[1] Give each atom half of the total. Very simple but completely arbitrary!

According to Mulliken's scheme, the gross electron population, q_μ, for basis function ϕ_μ is given by.

---

[*] Note that $S_{\mu\mu} = 1$.

$$q_\mu = P_{\mu\mu} + \overset{\text{basis functions}}{\underset{\mu \neq \nu}{\sum\sum}} P_{\mu\nu}S_{\mu\nu} \tag{4}$$

Atomic electron populations, q_A, and atomic charges, Q_A, follow.

$$q_A = \overset{\substack{\text{basis functions} \\ \text{on atom A}}}{\underset{\mu}{\sum}} q_\mu \tag{5}$$

$$Q_A = Z_A - q_A \tag{6}$$

Z_A is the atomic number of atom A.

The Mulliken procedure for subdivision of the electron density is not unique, and numerous other "recipes" have been proposed. Most of these make use of the overlap between atomic functions to partition the charge, and are identical to the Mulliken method for semi-empirical procedures (where atomic functions do not overlap; see **Chapter 2**). All such procedures contain an element of arbitrariness.

Fitting Schemes

Another approach to providing atomic charges is to fit the value of some property which has been calculated based on the "exact" wavefunction with that obtained from representation of the electronic charge distribution in terms of a collection of atom-centered charges. In practice, the property that has received the most attention is the electrostatic potential, ε_p.[2] This represents the energy of interaction of a unit positive charge at some point in space, p, with the nuclei and the electrons of a molecule (see **Chapter 4**).

$$\varepsilon_p = \overset{\text{nuclei}}{\underset{A}{\sum}} \frac{Z_A}{R_{Ap}} - \overset{\text{basis functions}}{\underset{\mu}{\sum}\underset{\nu}{\sum}} P_{\mu\nu} \int \frac{\phi_\mu(r)\phi_\nu(r)}{r_p} \, dr \tag{7}$$

Here, Z_A are atomic numbers, $P_{\mu\nu}$ are elements of the density matrix and R_{Ap} and r_p are distances separating the point charges from the nuclei and electrons, respectively. The first summation is over nuclei and the second pair of summations is over basis functions. The electrostatic potential may be calculated uniquely from an electronic wavefunction, although it is not clear how it might be measured.

Operationally the fitting scheme is carried out in a series of steps, following calculation of a wavefunction and a density matrix P:

i) Define a grid of points surrounding the molecule. Typically this encloses an area outside the van der Waals surface and extending several Ångstroms beyond this surface. It may comprise several thousand to several tens of thousands of points. It is clear that the detailed selection of a grid introduces arbitrariness into the calculation as the final fit charges depend on it. Note especially, that it is important not to include too many "distant" points in the grid, as the electrostatic potential for a neutral molecule necessarily goes to zero at long distance.

ii) Calculate the electrostatic potential at each grid point.

iii) Determine by least squares, the best fit of the grid points to an "approximate electrostatic potential", ε_p^{approx}, based on replacing the nuclei and electron distribution by a set of atom-centered charges, Q_A, subject to overall charge balance.[*]

$$\varepsilon_p^{approx} = \sum_A^{nuclei} \frac{Q_A}{R_{Ap}} \qquad (8)$$

Electrostatic-fit charges are more "costly" than Mulliken charges, due to the need to evaluate the integrals in equation 7 for many thousands of individual points. Except for semi-empirical models, this cost will be small relative to that for obtaining the wavefunction.

Which Charges are Best?

It is not possible to say which method provides the "better" atomic charges. Each offers distinct advantages and each suffers from disadvantages. The choice ultimately rests with the application and the "level of comfort". Having selected a method, stick with it. As shown from the data in **Table 16-1**, atomic charges calculated from the two different schemes and from different quantum chemical models, may be significantly different.

---

[*] It is also possible to restrict the fit to reproduce the known (or calculated) electric dipole moment, although this is not commonly done.

Table 16-1: Calculated Atomic Charges for Formaldehyde[a]

method	atom	Hartree-Fock				local density		BP	
		STO-3G	3-21G	6-31G*	6-311+G**	6-31G*	6-311+G**	6-31G*	6-311+G**
Mulliken	C	7	21	14	11	0	-9	5	-4
	O	-19	-50	-42	-30	-28	-20	-30	-21
	H	6	14	14	9	14	14	13	12
fits to	C	35	48	44	45	32	38	33	39
electrostatic	O	-29	-46	-46	-48	-35	-40	-35	-40
potentials	H	3	-1	-1	1	1	1	1	0

method	atom	BLYP		EDF1		B3LYP		MP2		semi-empirical		
		6-31G*	6-311+G**	6-31G*	6-311+G**	6-31G*	6-311+G**	6-31G*	6-311+G**	MNDO	AM1	PM3
Mulliken	C	8	1	6	-2	8	2	5	0	29	14	30
	O	-30	-23	-30	-21	-32	-24	-32	-21	-29	-28	-31
	H	11	11	12	11	12	11	13	10	0	7	1
fits to	C	36	42	35	41	37	42	37	43	71	53	68
electrostatic	O	-35	-41	-35	-40	-38	-43	-38	-42	-48	-44	-48
potentials	H	0	-1	0	0	0	0	1	-1	-11	-5	-10

a) units of electrons times 100

We have already commented several times in this guide that molecules like dimethylsulfoxide can either be represented as hypervalent, that is, with more than eight valence electrons surrounding sulfur, or as zwitterions.

Perhaps calculated charges can help to say which representation is more precise.

number of valence electrons	molecule	charge on S	molecule	charge on S
8	dimethylsulfide	-0.2	sulfur difluoride	+0.3
10	dimethylsulfoxide	+0.3	sulfur tetrafluoride	+0.7
12	dimethylsulfone	+1.5	sulfur hexafluoride	+1.1

The results (electrostatic-fit charges based on Hartree-Fock 6-31G* wavefunctions) are ambiguous. Relative to dimethylsulfide as a normal-valent "standard", the sulfur in oxygen "loses" about half an electron, and the sulfur in dimethylsulfone "loses" 1.7 electrons. This would seem to suggest that dimethylsulfoxide is "halfway" to being a zwitterion, but that dimethylsulfone is most of the way. Charges on sulfur in sulfur tetrafluoride and sulfur hexafluoride (relative to sulfur difluoride) show more modest effects, in particular for the latter. Overall, it appears that hypervalent molecules possess significant ionic character.

Hartree-Fock vs. Correlated Charges

Charges from correlated models are typically smaller than those from the corresponding (same basis set) Hartree-Fock model (see **Table 16-1**). One way to rationalize this is to recognize that electron promotion from occupied to unoccupied molecular orbitals (either implicit or explicit in all electron correlation models) takes electrons from "where they are" (negative regions) to "where they are not" (positive regions). In formaldehyde, for example, the lowest-energy promotion is from a non-bonded lone pair localized on oxygen into a π^* orbital principally concentrated on carbon.

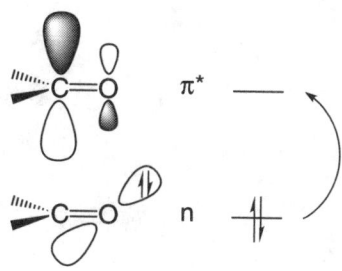

The basic conclusion, that electron correlation acts to reduce overall charge separation, is also supported by the observation that Hartree-Fock dipole moments are typically larger than experimental values, and that these are reduced by inclusion of correlation. Discussion has already been provided in **Chapter 10**.

Using Atomic Charges to Construct Empirical Energy Functions for Molecular Mechanics/Molecular Dynamics Calculations

Quantum chemical calculations may be called on to furnish parameters for use in empirical molecular mechanics/molecular dynamics schemes. Aside from torsional energy contributions (see discussion in **Chapter 14**), the most common quantity is the electrostatic energy, ε^{elect}, given by the following expression (see **Chapter 3**).

$$\varepsilon^{elect} = \sum_{A < B}^{atoms} \sum \frac{q_A q_B}{R_{AB}} \qquad (9)$$

Here, q_A and q_B the charges on atoms A and B, respectively, and R_{AB} is the distance separating the two atoms. Summation is carried out over unique atom pairs B>A. Because it is an energy which is of interest, the obvious procedure to obtain the atomic charges is by way of fits to calculated electrostatic potentials. Commonly, Hartree-Fock models have been employed with the 6-31G* basis set. The known effect of electron correlation in reducing overall charge separation as obtained from Hartree-Fock models suggests that it might be desirable to reduce the 6-31G* charges somewhat, or alternatively, utilize density functional or MP2 correlated models in place of Hartree-Fock models.

Section IV

Case Studies

This section contains a number of "case studies" which are intended both to clarify the relationship between "conventional" and computer-based approaches to molecular modeling, and to illustrate applications of the latter to diverse chemical problems. Coverage focuses exclusively on "organic chemistry". Tactics aimed at *Stabilizing "Unstable" Molecules* are considered first followed by examples of the use of modeling for investigating *Kinetically-Controlled Reactions*. The section concludes with *Applications of Graphical Models*. The choice of problems provided in each of these chapters represents a compromise: sufficiently "complex" to allow the reader to appreciate the essential role of molecular modeling, yet simple enough that the results will not be swamped by details.

The format of each "case study" is intended to follow the manner in which a "research investigation" might actually be carried out. A problem is stated, a "starting move" proposed and calculations performed. The results give rise to new questions, just as they would in an experimental investigation, and new calculations are demanded.

Spartan'02 files associated with each of the case studies have been provided on an accompanying CD-ROM. These are designated by (x-y), x indicating the chapter number and y the number of the file inside the chapter.

Chapter 17

Stabilizing "Unstable" Molecules

This chapter provides examples of the use of molecular modeling to quantify thermochemical stabilities of what might normally be considered "unstable" molecules.

Introduction

One of the major advantages of molecular modeling over experiment is its generality. Thermochemical stability or even existance is not a necessary criterion for investigation by molecular modeling as it is for experiment. This leads to the intriguing and very real possibility that modeling can be used to "explore" how to stabilize "unstable" molecules, and so make them ammenable to scrutiny by experiment. The examples provided in this chapter illustrate some possibilities.

Favoring Dewar Benzene

Among the valence "$(CH)_6$" isomers of benzene, **1**, are Dewar benzene, **2**, prismane, **3**, benzvalene, **4**, and 3,3´-bis (cyclopropene), **5**.

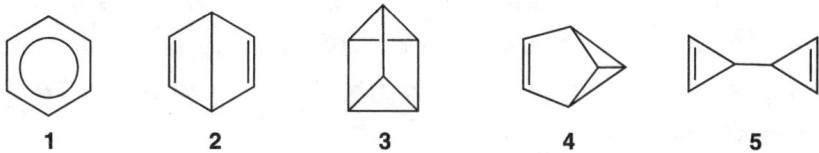

Dewar benzene has actually been isolated[1], and found to revert only slowly to benzene (its half life is approximately 2 days at 25°C). This is remarkable given how similar its geometry is to that of benzene, and what is expected to be a huge thermodynamic driving force for the isomerization. The substituted Dewar benzene, **7**, formed from photolysis of 1,2,4-tri-*tert*-butylbenzene, **6**, is apparently even more stable (relative to the substituted benzene) as it reverts to its precursor only upon heating.[2]

What is the effect of the three bulky *tert*-butyl groups in altering the relative stabilities of benzene and its valence isomer, Dewar benzene? Is it sufficient to overcome what must be the considerable difference in stabilities of the parent compounds? If not, can even more crowded systems be envisioned which would overcome this difference?

i) The first step is to assess the ability of theoretical models to reproduce the experimentally estimated difference in energy between benzene and Dewar benzene. These systems are small, and models which have been shown to provide accurate thermochemistry (see **Chapter 6**) may be easily applied. For the purpose here, the LMP2/6-311+G** model will be applied, using 6-31G* geometries. Calculations on prismane, benzvalene and 3,3´-bis(cyclopropene), in addition to benzene and Dewar benzene, should also be performed.

*At the LMP2/6-311+G**//6-31G* level, Dewar benzene is 80 kcal/mol higher in energy than benzene, in reasonable accord with the experimental estimate of 71 kcal/mol.[1] Interestingly, benzvalene is predicted to be slightly more stable than Dewar benzene. Both prismane and 3,3´-bis(cyclopropene) are much less stable.*

17-1

Energies of benzene valence isomers relative to benzene (kcal/mol)	
benzvalene	74
Dewar benzene	80
prismane	117
3,3´-bis(cyclopropene)	125

ii) The next step is to evaluate the effect which the three *tert*-butyl groups have on the relative stabilities of benzene and Dewar benzene. Calculations could be performed using the LMP2/6-311+G** model, but they would be costly. Alternatively, we can obtain an accurate estimate of the energy difference indirectly by adding the energy of the *isodesmic* reaction obtained from 3-21G calculations,

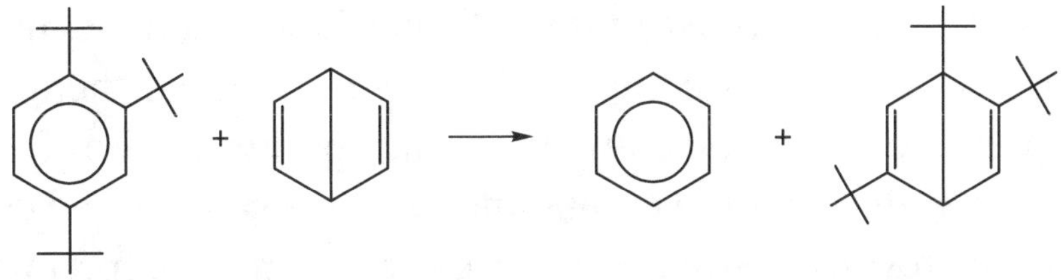

with the previously calculated differences in energies between the two parent compounds. Prior experience (see discussion in **Chapter 6**) suggests that such an approach should be valid.

The above reaction is predicted by the 3-21G calculations to be exothermic by 42 kcal/mol. This reduces the energy difference between benzene and Dewar benzene to approximately 38 kcal/mol (80 kcal/mol - 42 kcal/mol). Thermodynamics still very much favors the crowded benzene isomer over the less-crowded Dewar benzene alternative.

    17-2

iii) Finally, repeat the above process for other bulky groups similarly substituted on benzene. Two reasonable possibilities are the trimethylsilyl and trichloromethyl groups.

The trimethylsilyl group is not as effective as the tert-butyl group in reducing the energy separation between benzene and Dewar benzene (60 kcal/mol according to the above analysis), while the trichloromethyl group is about as effective as (37 kcal/mol). It appears that steric effects alone are not sufficient to overcome the large preference for benzene.

    17-3

While the desired goal, to reverse the thermochemical stabilities of benzene and Dewar benzene, has not been achieved, the calculations have clearly shown their value as a viable alternative to experiment to rapidly explore the limits of what is possible.

Making Stable Carbonyl Hydrates

Carbonyl compounds readily undergo reversible addition of water.[3]

$$K_{eq} = \frac{[\text{hydrate}]}{[H_2O]\,[\text{carbonyl}]} = \frac{[\text{hydrate}]}{55.5\,[\text{carbonyl}]}$$

Carbonyl hydration has been extensively studied primarily because it serves as a model for a number of important reactions, nucleophilic additions to carbonyl compounds foremost around them. While for most common carbonyl compounds the equilibrium lies far to the left (in favor of the carbonyl compound), it is possible to find compounds where the reverse is true. Because there are ample experimental data, it should be possible to identify structural and/or other characteristics which drive the equilibrium one way or the other. Alternatively, quantum chemical models can be employed.

It is straightforward to calculate energies of hydration reactions as a function of the carbonyl compound and, once "calibrated" on the basis of available experimental data, use this as a criterion for selecting systems which might exist primarily as carbonyl compounds, primarily as carbonyl hydrates or anywhere in between. The disadvantage to such an approach (other than it requiring calculations on both the carbonyl compounds and their respective hydrates) is that it provides very little insight into the factors which influence the equilibrium. Another approach is to focus only on the carbonyl compounds (or only on the hydrates) and look for characteristics which correlate with the experimental equilibrium constants. This is the approach illustrated here.

i) To start, obtain structures and other properties for a diverse series of carbonyl compounds for which experimental hydration equilibrium constants are known. "Interesting" properties include the dipole moment, HOMO and LUMO energies and atomic (electrostatic-fit) charges. The Hartree-Fock 6-31G* model is sufficient and 3-21G equilibrium geometries can be used in place of 6-31G* geometries to save time.

Experimental K_{eq} for hydration of carbonyl compounds			
	log $(K_{eq}/55.5)$		log$(K_{eq}/55.5)$
PhCOMe	-6.8	MeCHO	-1.7
Me_2CO	-4.6	$MeCOCO_2H$	-1.4
PhCHO	-3.8	$MeCOCO_2Me$	-1.2
tert-BuCHO	-2.4	CF_3COMe	-0.2
i-Pr CHO	-2.1	$PhCOCF_3$	0.1
n-Pr CHO	-2.1	H_2CO	1.6
EtCHO	-1.9	CF_3CHO	2.7
		CF_3COCF_3	4.3

Experimental data from: J.P. Guthrie, *Can. J.Chem.*, **53**, 898 (1975); **56**, 962 (1978).

Molecular properties from the 6-31G*//3-21G calculations are as follows: (17-4)

molecule	dipole moment	E_{HOMO}	E_{LUMO}	Charge on C(C=O)	Charge on O (C=O)
PhCOMe	3.38	-9.30	2.47	0.65	-0.56
Me_2CO	3.24	-11.12	4.43	0.80	-0.60
PhCHO	3.60	-9.45	2.26	0.46	-0.51
tert-BuCHO	3.08	-11.09	4.23	0.38	-0.51
i-PrCHO	3.25	-11.14	4.21	0.42	-0.52
n-PrCHO	2.90	-11.31	4.32	0.49	-0.51
EtCHO	3.21	-11.31	4.15	0.46	-0.51
MeCHO	3.10	-11.49	4.25	0.59	-0.53
$MeCOCO_2H$	1.37	-11.78	2.13	0.68	-0.48
$MeCOCO_2Me$	1.58	-11.56	2.32	0.49	-0.49
CF_3COMe	2.77	-12.55	3.09	0.54	-0.48
$PhCOCF_3$	4.02	-9.72	1.62	0.41	-0.47
H_2CO	2.79	-11.95	3.77	0.44	-0.47
CF_3CHO	2.02	-13.11	2.63	0.35	-0.41
CF_3COCF_3	0.46	-13.90	1.68	0.36	-0.38

ii) Next, use linear regression analysis to find which single property (from among those tabulated above) correlates best with experimental carbonyl hydration equilibrium constants, which pair of properties, etc.

The best correlation involving a single property is with the charge on the carbonyl oxygen (q_o).

$$log (Keq/55.5) = 47.3q_0 + 22.2 \qquad R^2 = 0.84$$

The best correlation involving a pair of properties is with the charge on the carbonyl oxygen (q_o) and the LUMO energy (E_{LUMO}).

$$log (Keq/55.5) = 56.7q_0 + 0.8 E_{LUMO} + 24.2 \qquad R^2 = 0.91$$

iii) Finally, calculate carbonyl oxygen charges together with LUMO energies for a few "new" compounds and, using the previously established relationship, estimate hydration equilibrium constants. 2-cyclohexenone and δ-valerolactone typify very stable carbonyl compounds and here the equilibrium for hydration should lie far to the left. On the other hand, hydration equilibria for perchloroacetone and for cyclopropenone might be expected to lie to the right. Trichloromethyl groups (like trifluoromethyl groups) should act to withdraw electrons from the (already) electron deficient carbonyl group leading to its destabilization, while hydration of the small-ring ketone should afford some relief of steric strain.

17-5

molecule	E_{LUMO}	Charge on $O (C=O)$	predicted $log (K_{eq}/55.5)$
2-cyclohexenone	2.99	-0.60	-7.4
δ-valerolactone	4.73	-0.60	-5.7
perchloroacetone	1.49	-0.44	0.4
cyclopropanone	3.77	-0.51	-1.7

The results are as expected for 2-cyclohexenone and for δ-valeroactone. Apparently, trichloromethyl is not as effective an electron-withdrawing group as is trifluoromethyl, although the equilibrium for perchloroacetone is predicted to tilt toward the hydrate. Relief of strain in cyclopropanone is not enough to favor the hydrate.

The important point to make is that calculations, like experiments, can be called up to furnish data on diverse molecular species, and attempts can be made to relate these data to a particular "property" of interest. The advantages that calculations hold over experiment is that no molecules need be synthesized and that the "data" need not be restricted to quantities that can actually be measured.

Stabilizing a Carbene: Sterics vs. Aromaticity

Carbenes are only rarely detected, let alone isolated and characterized. Kinetically favorable *exothermic* reactions (among them cycloaddition with alkenes and insertion into CH bonds) generally preclude this. 1,3-Diadamantylimidazol-2-ylidene, **8Ad**, is an exception, in that it forms a stable solid, the crystal structure of which has been determined.[4]

8Ad   8Me   9

Is the stability of **8Ad** due to unfavorable kinetics, i.e., the bulky adamantyl groups blocking reaction, or to unfavorable thermochemistry, i.e., loss of aromaticity of the imidazole ring as a result of reaction, or both? The distinction is potentially important as understanding could assist in designing stable carbenes. To decide, compare the kinetics and thermodynamics of the insertion of **8Ad** into the central CH bond in propane with reactions of **8Me**, which should also be "aromatic" but lacks "shielding groups", and **9**, which is neither aromatic nor crowded.

  i) Use dichlorocarbene to establish a "baseline" for the thermochemistry and kinetics of a carbene insertion reaction. Obtain geometries for **9**, the transition-state for its insertion into the center CH bond of propane, for the product of the insertion, and for propane using the semi-empirical PM3 model. Perform single-point energy calculations (using the PM3 equilibrium and transition-state geometries) with the B3LYP/ 6-31G* model. (Note that better calculations could easily be performed on this reaction, in particular with regard to geometries, and the particular choice of "recipe" is based on the need to investigate the larger systems.) Calculate the overall thermochemistry of the reaction as well as the activation barrier.

The reaction is predicted to be exothermic by 86 kcal/mol with an activation energy of 10 kcal/mol. The former is not unreasonable given that a lone pair (on the carbene) has been "exchanged" for a CC bond. The latter is in line with the observation that (singlet) carbene insertion reactions are known to be very fast.

ii) Obtain geometries for **8Ad** and **8Me**, their corresponding transition states for insertion into propane and the corresponding insertion products using the PM3 model. Following this, perform single-point energy calculations (using the PM3 geometries) with the B3LYP/6-31G* model. Data for propane are already available.

*Calculated activation energies for insertion of both **8Ad** and **8Me** into the central CH bond of propane are 73 and 67 kcal/ mol, respectively, much higher than that seen for insertion of dichlorocarbene (10 kcal/mol). On the other hand, the overall exothermicity of the two reactions (81 kcal/mol for insertion of **8Ad** and 90 kcal/mol for insertion of **8Me**) is quite similar to that obtained for insertion of dichlorocarbene (86 kcal/mol). This suggests that the difference in reactivities (making **8Ad** a "stable" carbene relative to dichlorocarbene which is highly reactive) is due more to kinetics than to thermodynamics.*

*The fact that the activation barriers for **8Ad** and **8Me** are so similar points to an electronic as opposed to a steric origin for the lack of reactivity. While this may at first glance be surprising, inspection of space-filling models for both transition states show that neither transition state is particularly crowded.*

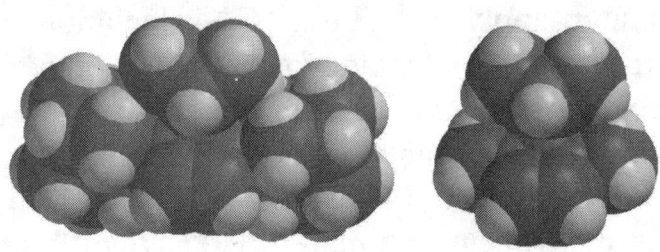

The take home lesson is that calculations can be employed just as can be experiment to pose and answer basic questions as: "is a molecule stable because of thermodynamics or kinetics?" and "what is the origin of the stability?". In so doing calculations provide a powerful means to explore chemistry.

Favoring a Singlet or Triplet Carbene

Singlet and triplet carbenes exhibit different properties and, to great extent, show markedly different "chemistry". For example, a singlet carbene will add to a *cis*-disubstituted alkene to produce only *cis*-disubstituted cyclopropane products (and to a *trans*-disubstituted alkene to produce only *trans*-disubstituted cyclopropane products), while a triplet carbene will add non-stereospecifically to produce a mixture of *cis* and *trans* products.

The origin of the difference lies in the fact that triplet carbenes are biradicals (or diradicals) and exhibit chemistry similar to that exhibited by radicals, while singlet carbenes incorporate both nucleophilic and electrophilic sites, e.g., for singlet and triplet methylene.

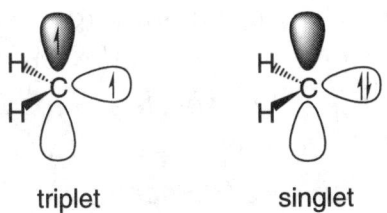

triplet   singlet

It should be possible to take advantage of what is known about stabilizing radical centers vs. stabilizing empty orbitals[5], and to design carbenes which will either be singlets or triplets. Additionally, it should be possible to say with confidence that a specific carbene of interest will either be a singlet or a triplet, and thus to anticipate its chemistry.

453

i) To start, apply B3LYP/6-31G* calculations to a variety of carbenes, X–CH, looking for ways to favor the singlet or triplet electronic state. As detailed in **Chapter 6 (Table 6-4)**, this model successfully accounts for the singlet/triplet energy splitting in parent methylene (14 kcal/mol in favor of the triplet compared to an experimental estimate of 10 kcal/mol). Apply a correction of 4 kcal/mol in favor of the singlet state to bring the B3LYP/6-31G* methylene result in line with the experimental estimate.

The calculations show that it is possible to strongly favor singlet or triplet state depending on the substituent.

	B3LYP/6-31G*		"corrected"	
X= *SiMe₃*	*23*	*triplet*	*19*	*triplet*
CN	*16*	*triplet*	*12*	*triplet*
H	*14*	*triplet*	*10*	*triplet*
C≡CH	*13*	*triplet*	*9*	*triplet*
CH=CH₂	*12*	*triplet*	*8*	*triplet*
Me	*8*	*triplet*	*4*	*triplet*
Ph	*7*	*triplet*	*3*	*triplet*
SO₂Me	*4*	*triplet*	*0*	*–*
Cl	*2*	*singlet*	*6*	*singlet*
F	*12*	*singlet*	*16*	*singlet*
OMe	*26*	*singlet*	*30*	*singlet*
NMe₂	*35*	*singlet*	*39*	*singlet*

These results are easily rationalized. Strong π-donor substituents such as OMe and NMe₂ will preferentially stabilize the singlet by electron donation into the empty p orbital. In a similar way, strong π-acceptor substituents such as SiMe₃ and CN will destabilize the singlet. Substituents which allow "delocalization", such as C≡CH, CH=CH₂ and Ph, will act to stabilize both singlet (delocalization of an empty orbital) and triplet (delocalization of an unpaired electron), and would be expected to have relatively little effect on singlet triplet energy separation.

17-8

ii) Next, assign ground-state electronic configurations and estimate singlet/triplet energy differences for a few additional carbenes.

dibromocarbene   cyclopentadienylidene   fluorenylidene

Use B3LYP/6-31G* calculations (as above), and apply a correction of 4 kcal/mol (in favor or the singlet).

Dibromocarbene prefers a singlet ground state (by 18 kcal/mol following the correction), consistent with the preferences of both fluorocarbene and chlorocarbene. Cyclopentadienylidene shows a strong preference for a triplet ground state (by 23 kcal/ mol following the correction). This can be rationalized by recognizing that the singlet is formally a 4π electron molecule, i.e., it is antiaromatic.

17-9

Fluorenylidene also prefers a triplet ground state, but the corresponding singlet is only 3 kcal/mol higher according to the corrected B3LYP/6-31G calculations. The singlet, like singlet cyclopentadienylidene, is also formally antiaromatic (with 12π electrons), but is much more delocalized than cyclopentadienylidene.*

Like the case study dealing with carbonyl hydration, this exercise places calculation in the role of "gathering data", and following this, seeks to use this data to design systems with specific properties. As before, the calculations offer strong advantages over experimental work.

References

1. M.J. Goldstein and R.S. Leight, *J. Amer. Chem. Soc.*, **99**, 8112 (1977).

2. E.E. van Tamelen, S.P. Pappas and K.L. Kirk, *J. Amer. Chem. Soc.*, **93**, 6092 (1971).

3. For a discussion, see: F.A. Carey and R.J. Sundberg, ***Advanced Organic Chemistry***, 4th Ed., Kluwer, New York, 2000.

4. A.J. Arduengo, III, R.L. Harlow and M. Kline, *J. Amer. Chem. Soc.*, **113**, 361 (1991).

5. For a discussion, see: W.J. Hehre, L. Radom, P.v.R. Schleyer and J.A. Pople, ***Ab Initio Molecular Orbital Theory***, Wiley, New York, 1985, p.346ff.

Chapter 18

Kinetically-Controlled Reactions

This chapter provides examples showing how molecular modeling can be used not only to "rationalize" observed product distributions in kinetically-controlled reactions, but also to anticipate and ultimately control the distribution of products.

Introduction

Quantum chemical models open up a range of possibilities for synthetic organic chemists wishing not only to "rationalize" product distributions in kinetically-controlled reactions, but also to anticipate them. Underlying this is the ability to routinely obtain transition states for organic reactions, and the knowledge based on prior experience, that the models are capable of providing a quantitative account of relative transition-state energies, i.e., those associated with pathways to different products. The examples provided in this chapter are an attempt to illustrate these possibilities.

As elaborated in **Chapter 1**, the "proper" way to anticipate the outcome of a kinetically-controlled reaction is to compare the energies of transition states leading to the different possible products.

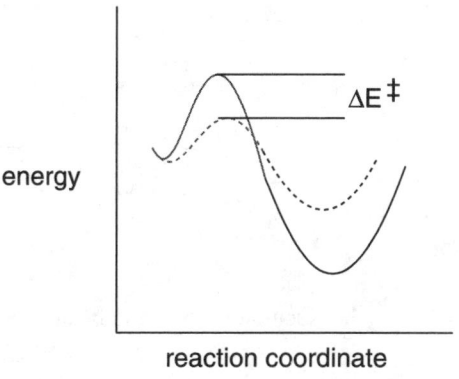

The major ("kinetic") product will pass through the transition state with the lowest energy, and the ratio of major: minor products will increase with increasing difference in transition-state energies:

$\Delta E^{\ddagger}$ (kcal/mol)	major: minor (at room temperature)
1	90:10
2	95:5
3	99:1

Thermodynamic vs. Kinetic Control*

Organic chemists have a keen eye for what is stable and what is not. For example, they will easily recognize that cyclohexyl radical is more stable than methylcyclopentyl radical, because they know that "6-membered rings are better than 5-membered rings", and (more importantly) that "2° radicals are better than 1° radicals". However, much important chemistry is not controlled by what is most stable (thermodynamics) but rather by what forms most readily (kinetics). For example, loss of bromine from 6-bromohexene leading initially to hex-5-enyl radical, results primarily in product from cyclopentylmethyl radical.[1]

---

* The reader may wish to revisit the discussion of thermodynamic vs. kinetic control provided in **Chapter 1**.

There are three possible interpretations for the experimental result: (a) either the rearrangement is not controlled by thermodynamics or, (b) that it is, but our understanding of radical stability is "wrong", or (c) that it is, but entropy rather than enthalpy, exerts the controlling influence. Of course there is always the possibility that the proposed mechanism is at fault.

i) The first objective is to rule out the last two possibilities. First establish that cyclohexyl radical is in fact "better" (thermodynamically) than cyclopentylmethyl radical, and that cyclohexane and not methylcyclopentane is the thermodynamic product. This simply requires calculating energies for the two radicals at their respective equilibrium geometries. To answer the question of whether entropy might play a role, that is, the TΔS component of the free energy overriding the ΔH component, it is necessary to perform a vibrational analysis. Hartree-Fock calculations with the 6-31G* basis set should be adequate.

The calculated energy difference is 7.5 kcal/mol in favor of cyclohexyl radical according to the 6-31G calculations. Including the entropy contribution lowers this number to around 5 kcal/mol. Were the reaction under thermodynamic control, only cyclohexane would be observed, and interpretations (b) and (c) cannot be correct.*   **18-1**

ii) The next objective is to establish which ring closure, to cyclohexyl radical or to cyclopentylmethyl radical, is "easier", that is, which product, cyclohexane or methylcyclopentane is the kinetic product. This requires calculating energies for the two transition states. Again, the 6-31G* model should provide an adequate account.

The calculated difference in transition-state energies in 2.3 kcal/ mol in favor of ring closure to the cyclopentylmethyl radical. Inclusion of entropy increases this difference to around 2.7 kcal/ mol. Methylcyclopentane is in fact the kinetic product and only about 1 - 2% of the total product mixture should be cyclohexane. This is what is observed, suggesting that the radical mechanism is not at fault but that the reaction is under kinetic control.   **18-2**

iii) Finally, see what effect methyl and cyano substituents placed at the 2 position of the hex-5-enyl radical have on the anticipated (kinetic) distribution of products. This requires repeating the transition-state calculations for both substituted systems closing to both 5 and 6-membered rings. Use the 6-31G* model.

Both methyl and cyano substituents preferentially stabilize the transition state for closure to the 6-membered ring, and both lead to reversal of the kinetically-favored product in the parent compound (by 0.7 kcal/mol in the case of methyl substitution and 3.1 kcal/mol in the case of cyano substitution). This result is not unexpected. Ring closure to the 6-membered ring leads to a transition state with the substituent directly attached to the (nascent) radical center, and both methyl and cyano groups are known to stabilize radicals.

The procedure illustrated here, first verifying that kinetics and not thermodynamics exerts product control and then surveying the effect of substituents on product distributions by calculating relative transition-state energies, provides a powerful alternative to experiment for designing selective chemical reactions. Of course, it also raises the possibility (not realized in this example) that calculations can be employed to contest a supposed mechanism.

Rationalizing Product Distributions

Often, just "looking at" the different transition states is sufficient to rationalize the observed product, and ultimately to predict the product of reactions yet to be examined experimentally. Two examples illustrate the point.

i) The observed kinetic product ratios in "anionic" Claisen rearrangements suggest a difference in transition-state energies of around 2 kcal/mol.[2]

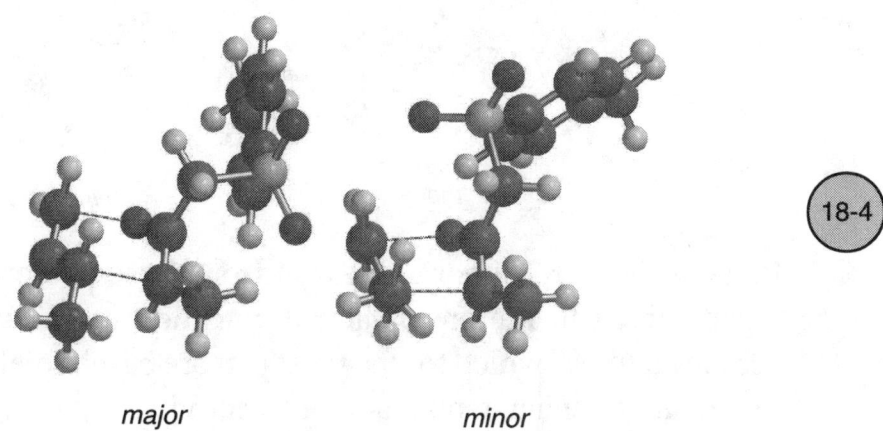

All that is required is to locate transition states leading to the two products, decide which is lower in energy and use the energy difference between the two to obtain the kinetic product distribution. Hartree-Fock 3-21G calculations should be sufficient to perform these tasks. One should also be on the lookout for "clues" in the structures and conformations of the two transition states as to why one is preferred over the other.

The transition state leading to the observed major product is in fact preferred (by 2.1 kcal/mol) according to the 3-21G calculations. Inspection of its structure reveals a "chair-like" geometry, compared to a "boat-like" geometry for the transition state leading to the minor product.

18-4

major   *minor*

ii) Another simple example is provided by the ene reaction.[3]

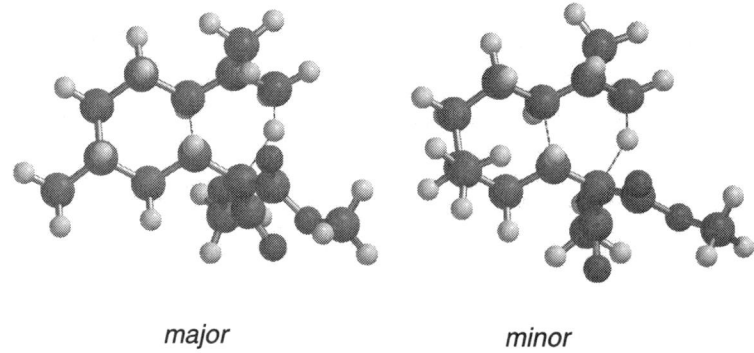

> 99%   < 1%

Here, the observed product ratio suggests that the two transition states are separated by 3 kcal/mol or more. The same "recipe" used in the first example should also be appropriate here.

As before the 3-21G calculations properly account for the observed product, and indicate an energy difference of 1.9 kcal/mol. Inspection reveals that the two transition states are similar, in that both resemble a pair of "fused" six-membered rings. ("One half" of the "pair" leads to the cyclohexane ring in the product, while the "other half" involves the migrating hydrogen.) The only significant difference between the two is the disposition of one methyl group in the cyclohexane ring. In the lower-energy transition state it is equatorial, while in the higher-energy transition state it is axial. It is this axial/equatorial difference which is responsible for the selectivity.

major   minor

Perhaps the most important lesson from these two examples is that quantitative calculations on actual transition states can supplant our "crude pictures" which for the most part are based solely on reactants, in helping to understand reaction selectivity.

Anticipating Product Distributions

The observed product distribution in the Claisen rearrangement[4]

> 99%   < 1%

may easily be rationalized on steric grounds. The minor product derives from a transition state in which all three methyl groups are *axial* leading to severe crowding, while the transition state leading to the major product has one of the methyl groups adjacent to oxygen *equatorial*.

One potentially interesting aspect of this particular system, and Claisen rearrangements in general, is the close structural resemblance of the transition state to the tetrahydropyran ring.

Recall that in the latter, certain types of substituents adjacent to oxygen in the ring actually prefer *axial* arrangements. This observation has been codified in what is commonly referred to as the anomeric effect[5], and is responsible in part for the conformations of carbohydrates. Is it possible that conformational preferences seen in substituted tetrahydropyrans will carry over into preferences in transition-state geometries for Claisen rearrangements?

i) First obtain *equatorial-axial* energy differences for a series of tetrahydropyran analogues of the Claisen transition states eluded to above. Use the Hartree-Fock 3-21G model.

    As expected (and as is known for simpler systems), substitution by "electronegative" groups on the carbon adjacent to the ring oxygen leads to a strong preference for the axial conformer (the anomeric effect), while substitution by methyl and trimethylsilyl groups leads to a favoring of the equatorial conformer.

X=	SiMe$_3$	--> 9.0	
	Me	--> 3.5	
	CN	no preference	
	NMe$_2$	no preference	
	Cl		--> 1.8
	OMe		--> 2.7
	F		--> 5.1

ii) Next, obtain analogous data for the Claisen rearrangement transition states, and see if a correlation between the two in fact exists.[*]

There is a strong correlation between substituent effects on conformational preferences in tetrahydropyrans and Claisen transition states.

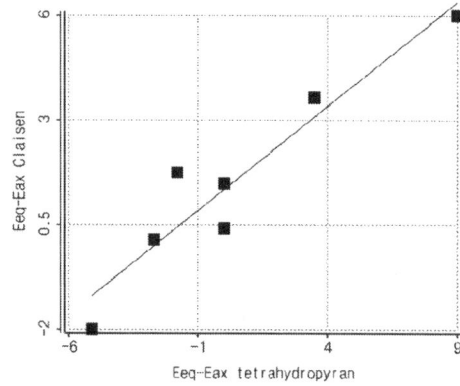

This leads to predictions for product distributions in Claisen rearrangements as a function of substitution.

---

* The idea of a relationship between equilibrium and transition-state energies is not new. See, for example reference 4.

X =		
SiMe$_3$	> 99%	< 1%
Me	99%	1%
Cl	95%	5%
NMe$_2$	95%	5%
CN	80%	20%
OMe	50%	50%
F	5%	95%

This paradigm, anticipating kinetic preferences by examining thermodynamic preferences in analogues, offers a powerful means to explore product distributions in kinetically-controlled reactions.

Altering Product Distributions

Often it is an "unexpected" result that leads to a new way of thinking about how to control product distributions. A good example is the observation that singlet carbenes such as ·CCl$_2$ typically prefer to add to internal (as opposed to external) double bonds.[6]

This is clearly not what one would have expected based on steric considerations. The "secret" lies in the fact that a singlet carbene like

•CCl$_2$ possesses both a high-energy filled molecular orbital in the σ plane and a low-energy unfilled molecular orbital perpendicular to that plane (see discussion in the previous chapter). To interact optimally with the π electrons of an olefin, the carbene must orient itself in the following manner,

and then "twist" by 90° to give the final product.

Quantum chemical calculations show that this is exactly what happens. The bottom line is that singlet carbenes appear to behave as "electrophiles" insofar as their reactivity toward olefins. It should be possible to take advantage of this fact to steer selectivity.

i) The first step is to calibrate the calculations for a reaction for which the product distribution is known experimentally. Use Hartree-Fock 3-21G calculations to obtain transition states for addition to both "internal" and "external" double bonds in the bicyclic for X=H.

The two transition states have approximately the same energy according to 3-21G calculations. It is from this reference that results on substituted systems need to be "measured".

ii) Next, obtain the two analogous transition states (from 3-21G calculations) for X=F. Consider only carbene addition away from the substituent so that steric effects will not be an issue.

The transition state corresponding to addition to the external double bond is favored by 3 kcal/mol over that corresponding to addition to the internal double bond. The "normal" selectivity for addition to the internal double bond has been reversed.

iii) To interpret this result, obtain electrostatic potential maps for both the unsubstituted and the fluorine substituted bicyclics (not for the transition states).

Fluorine substitution has greatly diminished the (negative) electrostatic potential for the internal double bond, but has had little effect on the potential for the external double bond. The change in selectivity (toward favoring addition onto the external double bond) is a direct consequence given that carbene addition is electrophilic addition.

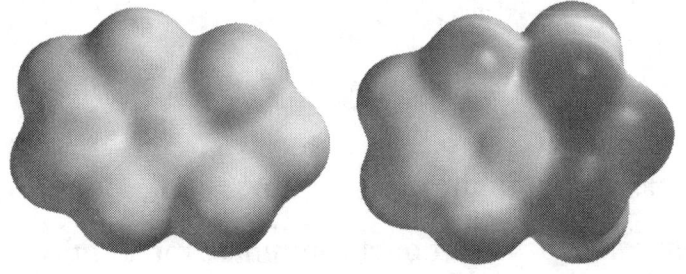

Here, the calculations including graphical models have been used to test a hypothesis, and in so doing provide an avenue for directing product selectivity.

Improving Product Selectivity

Organic chemists have a variety of strategies which they can pursue in order to improve product selectivity. Some of these like temperature and reaction time rest on the balance between thermodynamic and kinetic reaction pathways (see discussion in **Chapter 1**). Others such as solvent and external additives (catalysts) may as well lead to changes in the relative stabilities of competing transition states. Because it has been so widely explored, Diels-Alder chemistry provides a good opportunity to examine these "variables" and, in addition, to survey the use of calculations in anticipating changes in product distributions.

The following general observations have been made about kinetic regioselection in "normal" Diels-Alder reactions, i.e., those involving electron-rich dienes and electron-deficient dienophiles (see also discussion in **Chapter 9**).

EWG = electron-withdrawing group

Kinetic product: *ortho* products dominate for additions to terminally-substituted dienes; *para* products dominate for additions to internally-substituted dienes

Variation with substituent: alkyl groups afford modest regioselectivity; alkoxy groups afford good regioselectivity

Variation with position of substituent: alkyl groups in terminal positions afford greater regioselectivity than alkyl groups in internal positions

i) Start by "assessing" the ability of calculations to reproduce these qualitative observations, insofar as cycloadditions of methyl and methoxy-substituted cyclopentadiene with acrylonitrile. Use the 3-21G model. You only need to worry about the transition states and not reactants or products.

*The 3-21G calculations provide a qualitatively correct account of what is observed (for a fuller account, refer to **Chapter 9**).*

ii) Next, consider the addition of a Lewis Acid as a catalyst. A simple species such as BF_3 would most likely bind to the nitrogen in acrylonitrile.

Before doing any calculations, "take a guess" at what the results might be. Afterall, this is exactly what a chemist would do before carrying out an experiment.

a) The overall rate of cycloaddition should increase, due to an increase in the electron deficiency of the dienophile.

b) The reaction should become less regioselective (not more regioselective as desired), as a consequence of the reactivity-selectivity principle ("haste makes waste").

All in all, this is not very encouraging as it would lead to less selective Diels-Alder chemistry.

The calculations actually yield an entirely different (and much more encouraging) result.

In this instance, not only do the calculations bring into question the validity (or at least the generality) of the reactivity-selectivity principle but, more importantly, they open up a route to actually improving reaction selectivity. Note, in particular, the close parallels between the way this investigation was carried out and the way in which an experimental study would be performed.

References

1. For an excellent review of radical cyclization reactions, see: J. Fossey, D. Lefort and J. Sorba, *Free Radicals in Organic Chemistry*, Wiley, New York, 1995, p. 151.

2. S.E. Denmark, M.A. Harmata and K.S. White, *J. Amer. Chem. Soc.*, **111**, 8878 (1989).

3. (a) L.F. Tietze and U. Biefuss, *Angew. Chem. Int. Ed. Engl.*, **24**, 1042 (1985); (b) L.F. Tietze, U. Biefuss and M. Ruther, *J. Org. Chem.*, **54**, 3120 (1989).

4. D.J. Faulkner and M.R. Peterson, *Tetrahedron Lett.*, 3243 (1969).

5. For a recent series of reviews covering the anomeric effect and related observations, see: G.R.J. Thatcher, ed., *The Anomeric Effect and Associated Stereoelectronic Effects*, ACS Symposium Series, no. 539, American Chemical Society, Washington, DC, 1993.

6. J.J. Sims and V.K. Honwad, *J. Org. Chem.*, **34**, 496 (1969).

Chapter 19

Applications of Graphical Models

This chapter illustrates the way in which graphical models, in particular, electrostatic potential maps and LUMO maps, may be employed to provide insight into molecular structure and chemical reactivity and selectivity.

Introduction

Graphical models, in particular, electrostatic potential maps and LUMO maps have proven of considerable value, not only as a means to "rationalize" trends in molecular structure and stability and chemical reactivity and selectivity, but also as "tools" with which to carry out chemical investigations. A few examples have already been provided in **Chapter 4**. Those which follow have been chosen to further illustrate both "interpretive" and "predictive" aspects of graphical models.

Structure of Benzene in the Solid State

The individual strands which make up DNA are held together by hydrogen bonds between "complementary" nucleotide bases, guanine and cytosine or adenine and thymine. Close inspection of the guanine-cytosine base pair reveals a nearly planar structure with three hydrogen bonds. Guanine contributes two hydrogen-bond donors (electron-poor hydrogens) and one hydrogen-bond acceptor (an electron-rich site) while cytosine contributes one hydrogen-bond donor and two hydrogen-bond acceptors.

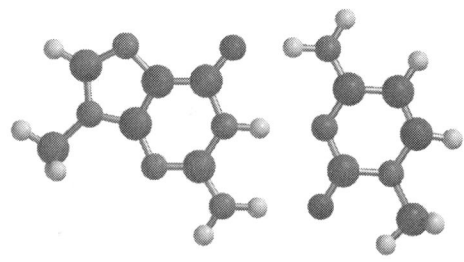

An alternative view is provided by a pair of electrostatic potential maps.* Electron-rich heteroatoms "line up" with the electron-poor (acidic) hydrogens. Attraction between the two bases may be thought of as due to favorable Coulombic interactions.

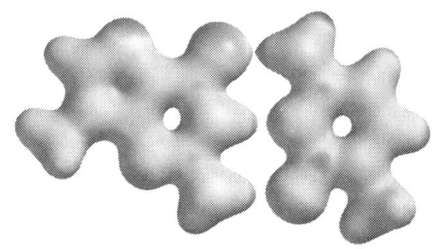

What would be the geometry of the guanine-cytosine complex were there no hydrogen bonds or, more to the point, were chemists not to know about the "benefits" of hydrogen bonding? More than likely, it would be a more closely-packed geometry, such as the following.

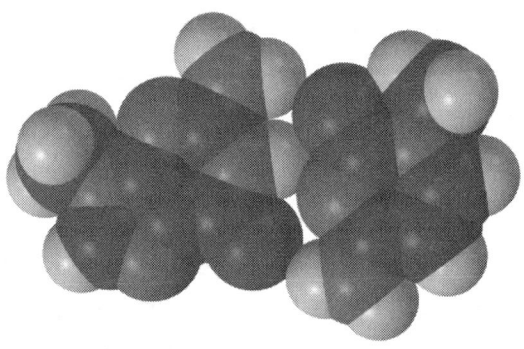

Afterall, this satisfies another "rule", the tendency for molecules to associate as closely as possible.** The question is relevant because

---

* A bond surface has been used in lieu of a size surface to better see the complementary electrostatic interactions.

** This is the so-called closed packing rule.

there are numerous situations where electrostatic and hydrogen-bonding interactions may not be "obvious", but may still be very important in directing intermolecular geometry. Without explicit knowledge, the obvious thing to do will be to associate molecules as closely as steric dictates will permit.

As a case in point, consider the structure of benzene in the solid state. Does it assume a geometry in which the rings are "stacked", or a structure in which the rings are perpendicular (or nearly perpendicular) to one another?

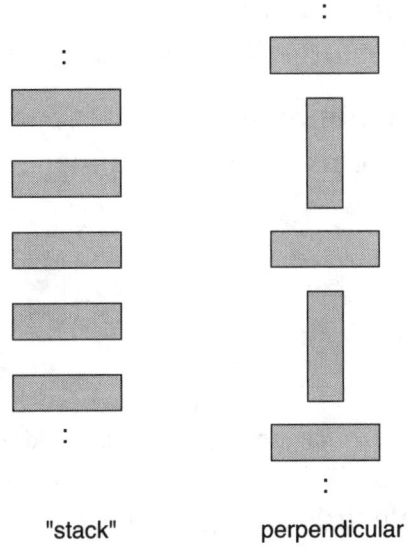

"stack"   perpendicular

Most chemists would probably "guess" the former. Afterall, benzene is "flat" and "flat things stack". However, graphical models lead one to the correct conclusion that a perpendicular arrangement is favored.

i) Start by calculating the electrostatic potential map for benzene. Hartree-Fock 3-21G calculations are sufficient.

The electrostatic potential map for benzene clearly shows that the π face is electron rich (red color) and the periphery is electron poor (blue color), consistent with the observation that electrophilic attack on benzene occurs at the π face.

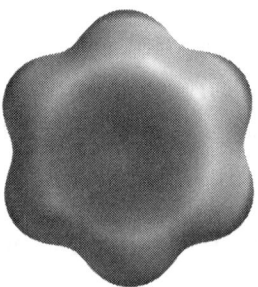

It also suggests that stacking the rings would result in unfavorable electrostatic interactions, while a perpendicular arrangement of benzene rings would result in favorable electrostatic interactions between the (negatively charged) π and (positively charged) σ systems.

ii) Next, obtain equilibrium geometries for both the parallel and perpendicular forms of benzene dimer.

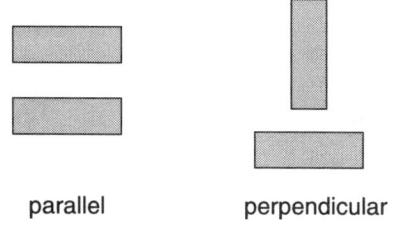

parallel   perpendicular

Also, obtain electrostatic potential maps for the two dimers.

The perpendicular dimer is lower in energy than the parallel dimer. In fact, the parallel arrangement of benzenes essentially "dissociates" at this level. *The electrostatic potential map for the perpendicular dimer shows evidence of charge transfer (charge delocalization) relative to free benzene. The "interacting" π face is less red and the "interacting" hydrogen is less blue. A map for a parallel dimer (artificially held in position) shows that the "interacting" π face has become more red.*

---

* The electrostatic potential map for the parallel dimer has been constructed by artificially holding the two benzenes together.

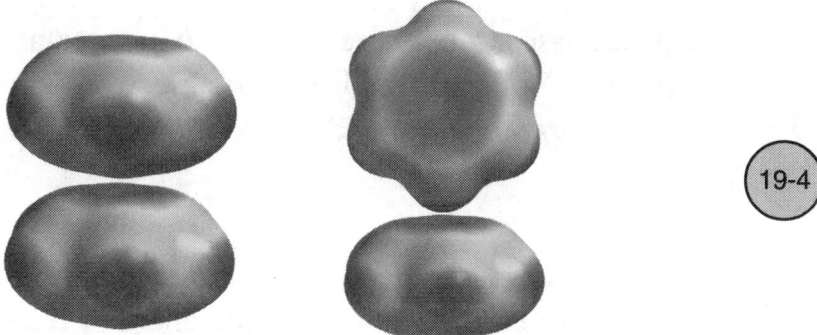

All of this suggests that the crystal structure of benzene is not in the form of a parallel stack but rather is perpendicular. This is exactly what is observed.[1]

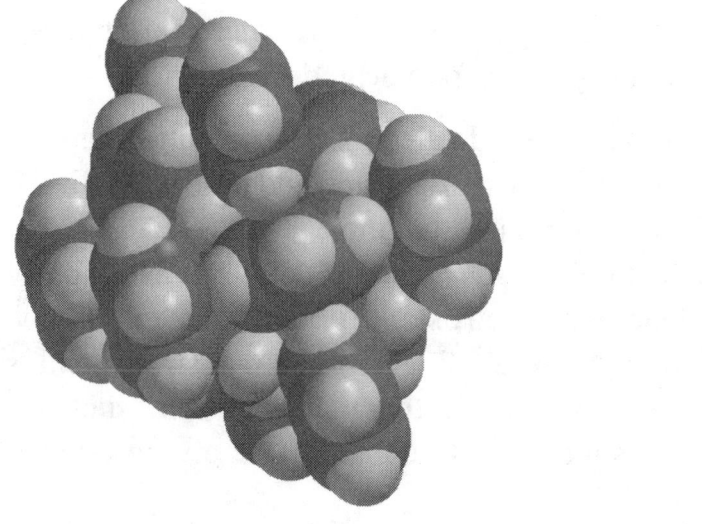

Such a structure does not violate the close-packing rule. In fact, the individual benzene molecules are very closely associated, but not in one dimension as a stack of plates, but rather in three dimensions.

The use of molecular models has not only pushed us beyond the limits of "pencil and paper", but has forced us to think beyond the normal dictates of structure.

Acidities of Carboxylic Acids

Acid and base strength are among the most catalogued of molecular properties. Both are readily available from calculation, either as absolute deprotonation and protonation energies, respectively,

$$AH \longrightarrow A^- + H^+$$

$$B + H^+ \longrightarrow BH^+$$

or as deprotonation and protonation energies relative to a "standard" acid (A°H) and "standard" base (B°), respectively.

$$AH + A^{°-} \longrightarrow A^- + A°H$$

$$B + B°H^+ \longrightarrow BH^+ + B°$$

As seen by the comparisons presented in **Chapter 6**, relative acid and base strengths may be obtained to reasonable accuracy with simple calculation models, whereas obtaining absolute quantities requires more sophisticated treatments.

To what extent do electrostatic potential maps constructed for neutral acids and bases reflect acid and base strengths? If they do, one should be able to replace having to look at a reaction energy by the simpler and more intuitive task of looking at a "property" of a molecule. It is clear that electrostatic potential maps uncover gross trends, for example, the acidic hydrogen in a strong acid, such as nitric acid, is more positive than that in a weak acid, such as acetic acid, which in turn is more positive than that in a very weak acid, such as ethanol.

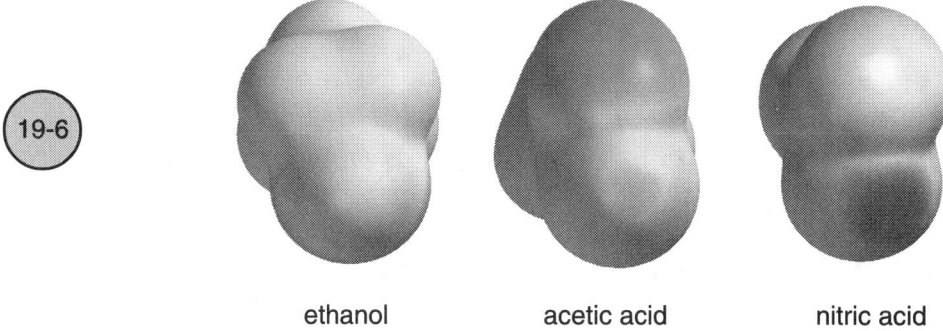

ethanol   acetic acid   nitric acid

Will electrostatic potential maps be able to reveal subtle changes in acid strength among closely-related molecules?

i) Use the 3-21G model to obtain equilibrium structures and follow this with single-point 6-31G* calculations to obtain electrostatic potential maps for the carboxylic acids below.

acid	pK$_a$	acid	pK$_a$
Cl$_3$CCO$_2$H	0.7	HCO$_2$H	3.75
HO$_2$CCO$_2$H	1.23	*trans*-ClCH=CHCO$_2$H	3.79
Cl$_2$CHCO$_2$H	1.48	C$_6$H$_5$CO$_2$H	4.19
NCCH$_2$CO$_2$H	2.45	*p*-ClC$_6$H$_4$CH=CHCO$_2$H	4.41
ClCH$_2$CO$_2$H	2.85	*trans*-CH$_3$CH=CHCO$_2$H	4.70
trans-HO$_2$CCH=CHCO$_2$H	3.10	CH$_3$CO$_2$H	4.75
p-HO$_2$CC$_6$H$_4$CO$_2$H	3.51	(CH$_3$)$_3$CCO$_2$H	5.03

Experimental data from: E.P. Sargeant and B. Dempsey, *Ionization Constants of Organic Acids in Aqueous Solution*, IUPAC no. 23, Permagon Press, 1979.

"Measure" the most positive value of the electrostatic potential associated with the acidic hydrogen in each of these compounds, and plot this against experimental pK$_a$.

The correlation between electrostatic potential on the acidic hydrogen and acid strength is reasonable.

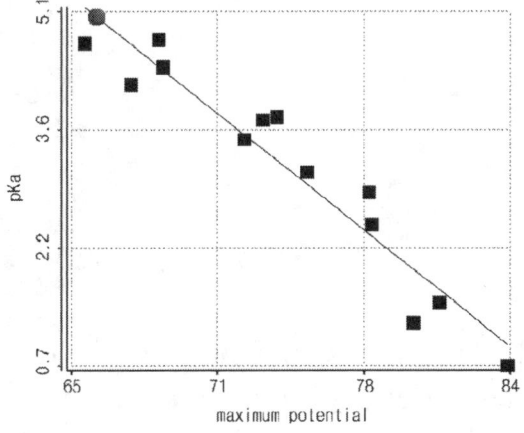

$$pK_a = -0.23 \cdot potential + 20$$

19-7

479

ii) "Predict" pK_a's for a few "similar" carboxylic acids: trifluoroacetic acid, methyl carbonate and glycine. Again, use the 6-31G*//3-21G model.

The maximum value of the electrostatic potential on the acidic hydrogen in trifluoroacetic acid is 88.8 kcal/mol, leading to a predicted pK_a of 0.4. This is an even stronger acid than trichloroacetic acid. The corresponding electrostatic potential for methyl carbonate is 74.9 kcal/mol, leading to a predicted pK_a of 2.8, and for glycine is 70.7 leading to a predicted pK_a of 3.7. Both of these are relatively weak acids.

iii) Obtain electrostatic potential maps for benzoic acid and for benzoic acid attached to chromium tricarbonyl to assess the effect of the $CrCO_3$ "substituent" on acidity. Hartree-Fock models are not suitable for molecules with transition metals (see discussion in **Chapter 5**). Use the semi-empirical PM3 model to establish equilibrium geometries for both molecules and follow with a single-point BP/6-31G* density functional calculation. Visually compare the two electrostatic potential maps.

The maps reveal that the complexed acid is stronger than the free acid, suggesting that $CrCO_3$ acts as an electron-withdrawing "substituent".

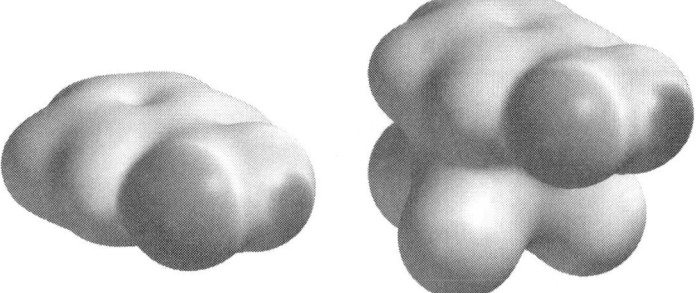

This example, like those in **Chapter 17**, dealing with carbonyl hydration and preferential stabilization of singlet or triplet carbenes, places the calculations in the role of "data gathering". Unlike these previous examples, the data here take the form of images while they can be used to extract "numerical data", can also serve to furnish an overview.

Stereochemistry of Base-Induced Eliminations

Elimination of HX from an alkane or cycloalkane typically follows an E_2 mechanism and occurs with *anti* stereochemistry, e.g., elimination of HCl from chlorocyclohexane.[2]

Labeling studies show, however, that the elimination of HCl from norbornyl chloride in the presence of strong base occurs with *syn* stereochemistry.[3]

The mechanism for elimination in this case probably involves initial deprotonation by alkoxide, followed by loss of Cl⁻.

This suggests that the *syn* proton in norbornyl chloride is more acidic than the *anti* proton.

As shown in the previous example (**Acidities of Carboxylic Acids**), electrostatic potential maps are able to account for the relative acidities of closely-related compounds. They might also be of value here in identifying which hydrogen (*syn* or *anti*) is the more acidic. However, another useful graphical indicator, the LUMO map, will be used instead. As discussed in **Chapter 4**, a LUMO map reveals those regions on the van der Waals contact surface for which the value of the lowest-unoccupied molecular orbital is largest, and hence should be most "appealing" to an incoming base. In this case, the question is which hydrogen, *syn* or *anti*, is likely to be more "attractive".

481

i) Obtain an equilibrium structure for 2-norbornyl chloride using the Hartree-Fock 3-21G model, and calculate a LUMO map.

Inspection of the LUMO map reveals that the syn proton in 2-norbornyl chloride is much more likely to undergo attack by a base than the anti proton.

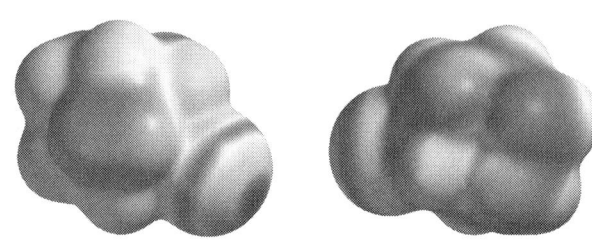

ii) Manipulate the LUMO map to see if any of the other protons in 2-norbornyl chloride might be prone to attack by a base.

The map also shows significant "concentration of the LUMO" on another hydrogen.

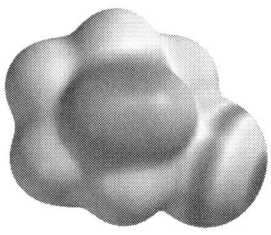

This in turn suggests that nortricyclane might also result, which in fact is actually observed.

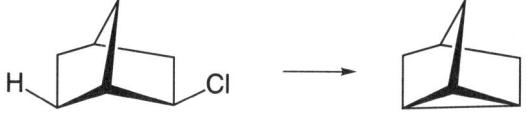

The graphical models have shown us something that we did not ask about. Looking out for the "unusual" is necessary in a modeling experiment, just as it is in a "real" experiment.

Stereochemistry of Carbonyl Additions

Nucleophilic addition to asymmetric carbonyl compounds is stereospecific. For example, most nucleophiles preferentially add to cyclohexanone rings from the more-crowded *axial* face rather than from the less-crowded *equatorial* face.

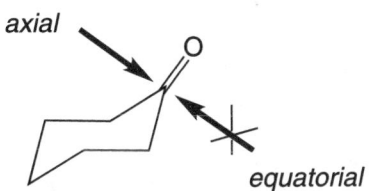

Axial attack is also favored on dioxanone rings, but *equatorial* attack is favored on dithianone rings.

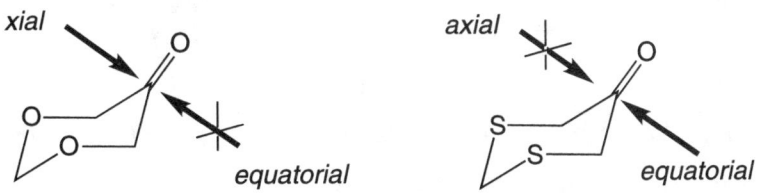

LUMO maps, which reveal the most "electron deficient" sites on a molecule, that is, those which are most susceptible to attack by a nucleophile, should be able to account for differences in direction of nucleophilic attack among closely-related systems. They will be employed here first to verify the above-mentioned preferences and then to explore stereochemical preferences in a number of related systems.

i) Obtain equilibrium geometries for cyclohexanone and its 3,5-dioxanone and 3,5-dithianone derivatives, as well as LUMO maps for the three compounds. Use the Hartree-Fock 3-21G model.

In accord with experimental data, LUMO maps for both cyclohexanone and 1,3-dioxan-5-one clearly anticipate preferential nucleophilic attack onto the axial carbonyl face,

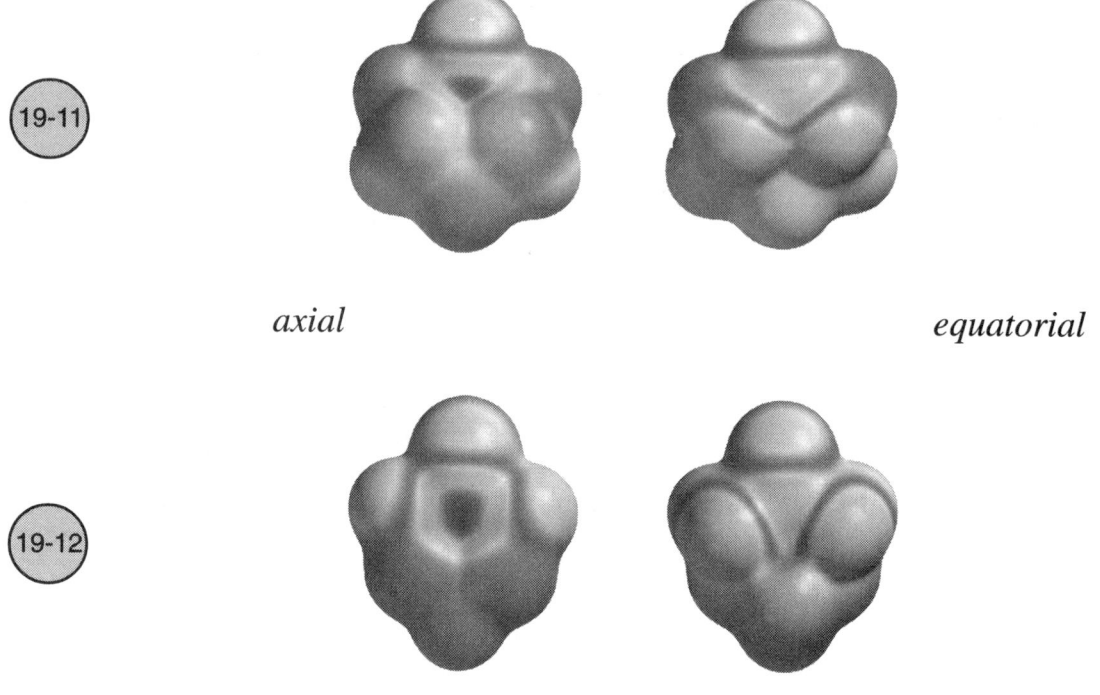

19-11

axial   equatorial

19-12

while the map for 1,3-dithian-5-one anticipates preferential attack onto the equatorial face.

19-13

axial   equatorial

Nucleophilic addition to the spirocyclic ketone also proceeds with high stereoselectivity.[4]

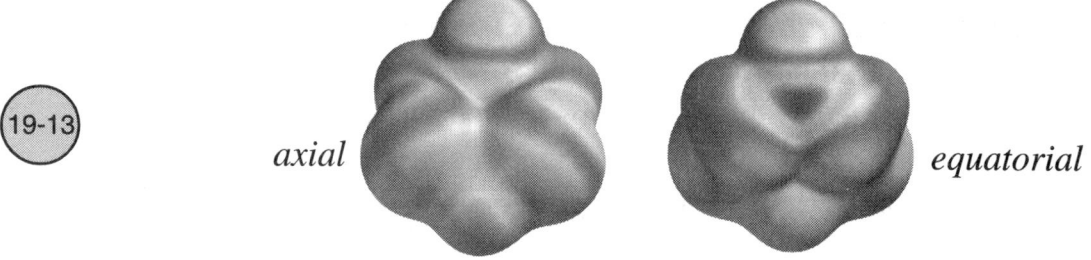

This system, unlike cyclohexanone and its analogues, presents an additional complication in that it can exist in one (or more) of four possible conformers, depending on whether the oxygen in each of the two six-membered rings is an *equatorial* or *axial* substituent of the other ring.

ii) Obtain equilibrium structures and energies for the four possible conformers of the spirocyclic ketone. Identify the lowest-energy conformer and obtain a LUMO map. Use the Hartree-Fock 3-21G model.

The axial-axial arrangement is favored, consistent with expectations based on the anomeric effect. Relative energies (in kcal/mol) are as follows (the first conformation is that relative to the ring incorporating the carbonyl group).

axial-axial	*0*
axial-equatorial	*5*
equatorial-axial	*7*
equatorial-equatorial	*14*

The LUMO map for the axial-axial conformer shows very strong distinction between the axial and equatorial faces. Attack onto the axial face is preferred, consistent with the experimentally observed product.

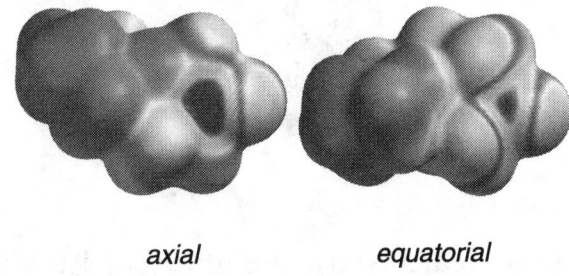

axial   equatorial

19-14

An even more typical situation is exemplified by the following sodium borohydride reduction of a tricyclic diketone.[5]

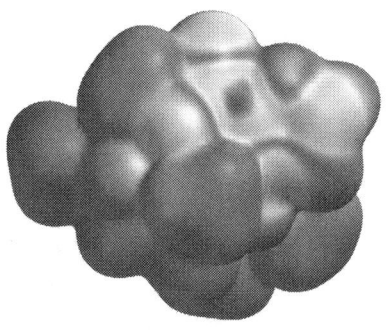

Here, there are two carbonyl groups and the "top" and "bottom" faces of each are different, leading to the possibility of four different reduction products. In addition, there is the question of conformation, especially with regard to the bulky TBS protecting group.

iii) Several conformers involving the TBS group need to be considered. Use the Hartree-Fock 3-21G model to locate the lowest-energy conformer and then obtain a LUMO map for this conformer.

The LUMO map shows a strong preference for nucleophilic attack onto the 7-membered ring carbonyl as opposed to the 5-membered ring carbonyl (in accord with experiment), and a much weaker preference for attack of syn to the adjacent methyl group as opposed to anti to methyl (also in accord with experiment).

These examples clearly show the utility of LUMO maps to assign stereochemistry in nucleophilic additions to complex substrates.

References

1. E.G. Cox, D.W.J. Cruickshank and J.A.S. Smith, *Proc. Roy. Soc.,* **A247**, 1 (1958).

2. For a discussion, see: F.A. Carey and R.J. Sundberg, ***Advanced Organic Chemistry***, 4th Ed., Kluwer Academic/Plenum Publishers, New York, 2000, p. 386.

3. R.A. Bartsch and J.G. Lee, *J. Org. Chem.*, **56**, 212 (1991).

4. P. DeShong and P.J. Rybczynski, *J. Org. Chem.*, **56**, 3207 (1991).

5. J.A. Robinston and S.P. Tanis, poster, American Chemical Society National Meeting, New York, August 1991.

Appendix A

Supplementary Data

This appendix contains data from molecular mechanics and quantum chemical calculations, together with relevant experimental information where available, to supplement that already provided in text and/or to provide evidence for conclusions reached in text. The material is in sections keyed to the corresponding text chapters. **Appendix A5** is keyed to **Chapter 5**, **Appendix A6** to **Chapter 6**, etc., although not all chapters are associated with supplementary materials.

Table A5-1: **Structures of One-Heavy-Atom Hydrides. Molecular Mechanics Models**

molecule	point group	geometrical parameter	SYBYL	MMFF	expt.
H_2	$D_{\infty h}$	r	1.008	0.660	**0.742**
LiH	$C_{\infty v}$	r	1.008	1.595	**1.596**
CH_4	T_d	r	1.100	1.092	**1.092**
NH_3	C_{3v}	r	1.080	1.019	**1.012**
		<	109.5	106.0	**106.7**
H_2O	C_{2v}	r	0.950	0.969	**0.927**
		<	109.5	104.0	**104.5**
HF	$C_{\infty v}$	r	1.008	0.945	**0.917**
NaH	$C_{\infty v}$	r	1.008	1.801	**1.887**
SiH_4	T_d	r	1.500	1.483	**1.481**
PH_3	C_{3v}	r	1.008	1.415	**1.420**
		<	109.5	94.5	**93.3**
H_2S	C_{2v}	r	1.008	1.341	**1.336**
		<	97.0	93.4	**92.1**
HCl	$C_{\infty v}$	r	1.008	1.310	**1.275**

Table A5-2: Structures of One-Heavy-Atom Hydrides. Hartree-Fock Models

molecule	point group	geometrical parameter	STO-3G	3-21G	6-31G*	6-311+G**	expt.
H_2	$D_{\infty h}$	r	0.712	0.735	0.730	0.736	**0.742**
LiH	$C_{\infty v}$	r	1.510	1.640	1.636	1.607	**1.596**
CH_4	T_d	r	1.083	1.083	1.084	1.083	**1.092**
NH_3	C_{3v}	r	1.033	1.003	1.002	1.000	**1.012**
		<	104.2	112.4	107.2	107.9	**106.7**
H_2O	C_{2v}	r	0.990	0.967	0.947	0.943	**0.958**
		<	100.0	107.6	105.5	106.4	**104.5**
HF	$C_{\infty v}$	r	0.956	0.937	0.911	0.900	**0.917**
NaH	$C_{\infty v}$	r	1.926	1.930	1.914	1.915	**1.887**
SiH_4	T_d	r	1.422	1.475	1.475	1.474	**1.481**
PH_3	C_{3v}	r	1.378	1.402	1.403	1.404	**1.420**
		<	95.0	95.2	95.4	95.5	**93.3**
H_2S	C_{2v}	r	1.329	1.327	1.326	1.327	**1.336**
		<	92.5	94.4	94.4	94.1	**92.1**
HCl	$C_{\infty v}$	r	1.313	1.267	1.267	1.266	**1.275**
KH	$C_{\infty v}$	r	2.081	2.353	2.313	-	**2.242**
GeH_4	T_d	r	1.431	1.546	1.520	-	**1.525**
AsH_3	C_{3v}	r	1.456	1.522	1.495	-	**1.511**
		<	93.9	93.3	91.6	-	**92.1**
H_2Se	C_{2v}	r	1.439	1.464	1.452	-	**1.460**
		<	92.4	92.1	91.1	-	**90.6**
HBr	$C_{\infty v}$	r	1.414	1.417	1.413	-	**1.415**
RbH	$C_{\infty v}$	r	2.211	2.490	-	-	**2.367**
SnH_4	T_d	r	1.630	1.744	-	-	**1.711**
SbH_3	C_{3v}	r	1.644	1.727	-	-	**1.704**
		<	94.4	93.6	-	-	**91.6**
TeH_2	C_{2v}	r	1.624	1.675	-	-	**1.658**
		<	92.4	91.8	-	-	**90.3**
HI	$C_{\infty v}$	r	1.599	1.631	-	-	**1.609**

Table A5-3: Structures of One-Heavy-Atom Hydrides. Local Density Models

molecule	point group	geometrical parameter	6-31G*	6-311+G**	expt.
H$_2$	D$_{\infty h}$	r	0.766	0.767	**0.742**
LiH	C$_{\infty v}$	r	1.641	1.608	**1.596**
CH$_4$	T$_d$	r	1.101	1.099	**1.092**
NH$_3$	C$_{3v}$	r	1.027	1.023	**1.012**
		<	106.0	108.4	**106.7**
H$_2$O	C$_{2v}$	r	0.976	1.971	**0.958**
		<	103.6	105.1	**104.5**
HF	C$_{\infty v}$	r	0.940	0.931	**0.917**
NaH	C$_{\infty v}$	r	1.880	1.880	**1.887**
SiH$_4$	T$_d$	r	1.499	1.497	**1.481**
PH$_3$	C$_{3v}$	r	1.437	1.436	**1.420**
		<	92.1	92.0	**93.3**
H$_2$S	C$_{2v}$	r	1.360	1.358	**1.336**
		<	92.2	91.7	**92.1**
HCl	C$_{\infty v}$	r	1.299	1.296	**1.275**
KH	C$_{\infty v}$	r	2.211	-	**2.242**
GeH$_4$	T$_d$	r	1.541	-	**1.525**
AsH$_3$	C$_{3v}$	r	1.536	-	**1.511**
		<	90.5	-	**92.1**
H$_2$Se	C$_{2v}$	r	1.489	-	**1.460**
		<	88.8	-	**90.6**
HBr	C$_{\infty v}$	r	1.442	-	**1.415**

Table A5-4: Structures of One-Heavy-Atom Hydrides. BP Density Functional Models

molecule	point group	geometrical parameter	6-31G*	6-311+G**	expt.
H_2	$D_{\infty h}$	r	0.751	0.752	**0.742**
LiH	$C_{\infty v}$	r	1.636	1.609	**1.596**
CH_4	T_d	r	1.101	1.099	**1.092**
NH_3	C_{3v}	r	1.029	1.024	**1.012**
		<	105.0	107.2	**106.7**
H_2O	C_{2v}	r	0.978	0.971	**0.927**
		<	103.0	104.2	**104.5**
HF	$C_{\infty v}$	r	0.942	0.931	**0.917**
NaH	$C_{\infty v}$	r	1.903	1.903	**1.887**
SiH_4	T_d	r	1.497	1.496	**1.481**
PH_3	C_{3v}	r	1.437	1.435	**1.420**
		<	92.5	92.6	**93.3**
H_2S	C_{2v}	r	1.360	1.357	**1.336**
		<	92.4	92.0	**92.1**
HCl	$C_{\infty v}$	r	1.298	1.250	**1.275**
KH	$C_{\infty v}$	r	2.243	-	**2.242**
GeH_4	T_d	r	1.549	-	**1.525**
AsH_3	C_{3v}	r	1.542	-	**1.511**
		<	91.0	-	**92.1**
H_2Se	C_{2v}	r	1.493	-	**1.466**
		<	89.2	-	**90.6**
HBr	$C_{\infty v}$	r	1.444	-	**1.415**

Table A5-5: **Structures of One-Heavy-Atom Hydrides. BLYP Density Functional Models**

molecule	point group	geometrical parameter	6-31G*	6-311+G**	expt.
H_2	$D_{\infty h}$	r	0.748	0.748	**0.742**
LiH	$C_{\infty v}$	r	1.628	1.599	**1.596**
CH_4	T_d	r	1.100	1.097	**1.092**
NH_3	C_{3v}	r	1.030	1.023	**1.012**
		<	104.8	107.4	**106.7**
H_2O	C_{2v}	r	0.980	0.972	**0.927**
		<	102.7	104.4	**104.5**
HF	$C_{\infty v}$	r	0.945	0.933	**0.917**
NaH	$C_{\infty v}$	r	1.893	1.894	**1.887**
SiH_4	T_d	r	1.495	1.492	**1.481**
PH_3	C_{3v}	r	1.437	1.435	**1.420**
		<	92.8	92.9	**93.3**
H_2S	C_{2v}	r	1.362	1.358	**1.336**
		<	92.5	92.2	**92.1**
HCl	$C_{\infty v}$	r	1.301	1.297	**1.275**
KH	$C_{\infty v}$	r	2.253	-	**2.242**
GeH_4	T_d	r	1.553	-	**1.525**
AsH_3	C_{3v}	r	1.548	-	**1.511**
		<	91.1	-	**92.1**
H_2Se	C_{2v}	r	1.498	-	**1.466**
		<	89.3	-	**90.6**
HBr	$C_{\infty v}$	r	1.450	-	**1.415**

Table A5-6: Structures of One-Heavy-Atom Hydrides. EDF1 Density Functional Models

molecule	point group	geometrical parameter	6-31G*	6-311+G**	expt.
H_2	$D_{\infty h}$	r	0.745	0.746	**0.742**
LiH	$C_{\infty v}$	r	1.635	1.605	**1.596**
CH_4	T_d	r	1.096	1.094	**1.092**
NH_3	C_{3v}	r	1.024	1.018	**1.012**
		<	105.0	107.1	**106.7**
H_2O	C_{2v}	r	0.973	0.966	**0.927**
		<	102.9	104.2	**104.5**
HF	$C_{\infty v}$	r	0.938	0.926	**0.917**
NaH	$C_{\infty v}$	r	1.903	1.905	**1.887**
SiH_4	T_d	r	1.493	1.491	**1.481**
PH_3	C_{3v}	r	1.430	1.429	**1.420**
		<	92.7	92.7	**93.3**
H_2S	C_{2v}	r	1.353	1.351	**1.336**
		<	92.5	92.1	**92.1**
HCl	$C_{\infty v}$	r	1.292	1.289	**1.275**
KH	$C_{\infty v}$	r	2.246	-	**2.242**
GeH_4	T_d	r	1.545	-	**1.525**
AsH_3	C_{3v}	r	1.537	-	**1.511**
		<	91.1	-	**92.1**
H_2Se	C_{2v}	r	1.487	-	**1.466**
		<	89.4	-	**90.6**
HBr	$C_{\infty v}$	r	1.438	-	**1.415**

Table A5-7: Structures of One-Heavy-Atom Hydrides. B3LYP Density Functional Models

molecule	point group	geometrical parameter	6-31G*	6-311+G**	expt.
H_2	$D_{\infty h}$	r	0.743	0.744	**0.742**
LiH	$C_{\infty v}$	r	1.621	1.592	**1.596**
CH_4	T_d	r	1.093	1.091	**1.092**
NH_3	C_{3v}	r	1.020	1.014	**1.012**
		<	105.7	107.9	**106.7**
H_2O	C_{2v}	r	0.969	0.962	**0.958**
		<	103.7	105.0	**104.5**
HF	$C_{\infty v}$	r	0.934	0.922	**0.914**
NaH	$C_{\infty v}$	r	1.883	1.883	**1.887**
SiH_4	T_d	r	1.486	1.484	**1.481**
PH_3	C_{3v}	r	1.424	1.423	**1.420**
		<	93.4	93.5	**93.3**
H_2S	C_{2v}	r	1.349	1.347	**1.336**
		<	92.9	92.6	**92.1**
HCl	$C_{\infty v}$	r	1.290	1.287	**1.275**
KH	$C_{\infty v}$	r	2.245	-	**2.242**
GeH_4	T_d	r	1.542	-	**1.525**
AsH_3	C_{3v}	r	1.533	-	**1.511**
		<	91.6	-	**92.1**
H_2Se	C_{2v}	r	1.483	-	**1.460**
		<	90.1	-	**90.6**
HBr	$C_{\infty v}$	r	1.436	-	**1.415**

Table A5-8: **Structures of One-Heavy-Atom Hydrides. MP2 Models**

molecule	point group	geometrical parameter	6-31G*	6-311+G**	expt.
H_2	$D_{\infty h}$	r	0.738	0.738	**0.742**
LiH	$C_{\infty v}$	r	1.640	1.599	**1.596**
CH_4	T_d	r	1.090	1.090	**1.092**
NH_3	C_{3v}	r	1.017	1.013	**1.012**
		<	106.3	107.4	**106.7**
H_2O	C_{2v}	r	0.969	0.960	**0.958**
		<	104.0	103.5	**104.5**
HF	$C_{\infty v}$	r	0.934	0.917	**0.917**
NaH	$C_{\infty v}$	r	1.918	1.907	**1.887**
SiH_4	T_d	r	1.483	1.475	**1.481**
PH_3	C_{3v}	r	1.414	1.409	**1.420**
		<	94.6	94.3	**93.3**
H_2S	C_{2v}	r	1.340	1.333	**1.336**
		<	93.4	92.1	**92.1**
HCl	$C_{\infty v}$	r	1.280	1.273	**1.275**
KH	$C_{\infty v}$	r	2.320	-	**2.242**
GeH_4	T_d	r	1.545	-	**1.525**
AsH_3	C_{3v}	r	1.525	-	**1.511**
		<	92.8	-	**92.1**
H_2Se	C_{2v}	r	1.473	-	**1.460**
		<	91.5	-	**90.6**
HBr	$C_{\infty v}$	r	1.428	-	**1.415**

497

Table A5-9:   Structures of One-Heavy-Atom Hydrides. Semi-Empirical Models

molecule	point group	geometrical parameter	MNDO	AM1	PM3	expt.
H_2	$D_{\infty h}$	r	0.663	0.677	0.699	**0.742**
LiH	$C_{\infty v}$	r	1.376	-	1.541	**1.596**
CH_4	T_d	r	1.104	1.112	1.087	**1.092**
NH_3	C_{3v}	r	1.007	0.998	0.999	**1.012**
		<	105.2	109.1	108.0	**106.7**
H_2O	C_{2v}	r	0.943	0.961	0.951	**0.958**
		<	106.8	103.6	107.7	**104.5**
HF	$C_{\infty v}$	r	0.956	0.826	0.938	**0.917**
SiH_4	T_d	r	1.410	1.461	1.488	**1.481**
PH_3	C_{3v}	r	1.340	1.364	1.324	**1.420**
		<	96.1	96.4	97.1	**93.3**
H_2S	C_{2v}	r	1.334	1.323	1.290	**1.336**
		<	99.8	95.5	93.5	**92.1**
HCl	$C_{\infty v}$	r	1.320	1.284	1.268	**1.275**
GeH_4	T_d	r	1.482	1.546	1.505	**1.525**
AsH_3	C_{3v}	r	-	-	1.520	**1.511**
		<	-	-	94.2	**92.1**
H_2Se	C_{2v}	r	-	-	1.470	**1.460**
		<	-	-	93.6	**90.6**
HBr	$C_{\infty v}$	r	1.441	1.421	1.470	**1.415**
SnH_4	T_d	r	1.586	1.617	1.701	**1.771**
SbH_3	C_{3v}	r	-	-	1.702	**1.704**
		<	-	-	92.4	**91.6**
TeH_2	C_{2v}	r	-	-	1.675	**1.658**
		<	-	-	88.3	**90.3**
HI	$C_{\infty v}$	r	1.594	1.587	1.677	**1.609**

Table A5-10: Structures of Two-Heavy-Atom Hydrides. Molecular Mechanics Models

molecule	point group	geometrical parameter	SYBYL	MMFF	expt.
Li$_2$	D$_{\infty h}$	r (LiLi)	1.500	2.664	**2.673**
LiOH	C$_{\infty v}$	r (LiO)	1.500	1.624	**1.582**
		r (OH)	0.950	0.957	
LiF	C$_{\infty v}$	r (LiF)	1.500	1.650	**1.564**
LiCl	C$_{\infty v}$	r(LiCl)	1.800	2.135	**2.021**
B$_2$H$_6$	D$_{2h}$	r (BB)	1.459	2.040	**1.763**
		r (BH)	1.023	1.196	**1.201**
		r (BH$_{bridge}$)	1.025	1.232	**1.320**
		< (HBH)	105.1	102.6	**121.0**
C$_2$H$_2$	D$_{\infty h}$	r (CC)	1.204	1.200	**1.203**
		r (CH)	1.056	1.066	**1.061**
C$_2$H$_4$	D$_{2h}$	r (CC)	1.337	1.336	**1.339**
		r (CH)	1.090	1.086	**1.085**
		< (HCH)	118.5	117.9	**117.8**
C$_2$H$_6$	D$_{3d}$	r (CC)	1.549	1.512	**1.531**
		r (CH)	1.102	1.094	**1.096**
		< (HCH)	108.7	108.4	**107.8**
HCN	C$_{\infty v}$	r (CN)	1.158	1.160	**1.153**
		r (CH)	1.056	1.065	**1.065**
HNC	C$_{\infty v}$	r (NC)	1.170	1.170	**1.169**
		r (NH)	1.008	1.018	**0.994**
CH$_2$NH	C$_s$	r (CN)	1.270	1.287	**1.273**
		r (CH$_{syn}$)	1.089	1.102	**1.103**
		r (CH$_{anti}$)	1.089	1.101	**1.081**
		r (NH)	1.008	1.027	**1.023**
		< (H$_{syn}$CN)	120.2	121.9	**123.4**
		< (H$_{anti}$CN)	119.9	120.6	**119.7**
		< (HNC)	120.0	110.1	**110.5**
CH$_3$NH$_2$	C$_s$	r (CN)	1.470	1.452	**1.471**
		r (CH$_{tr}$)	1.100	1.094	**1.099**
		r (CH$_g$)	1.100	1.094	**1.099**
		r (NH)	1.080	1.019	**1.010**
		< (NCH$_{tr}$)	109.4	110.6	**113.9**
		< (NCH$_{gg'}$)	125.2	126.2	**124.4**
		< (H$_g$CH$_{g'}$)	109.5	108.4	**108.0**
		< (CNH$_{gg'}$)	125.3	125.6	**125.7**
		< (HNH)	109.5	106.1	**107.1**
CO	C$_{\infty v}$	r (CO)	1.170	1.114	**1.128**
H$_2$CO	C$_{2v}$	r (CO)	1.220	1.225	**1.208**
		r (CH)	1.089	1.102	**1.116**
		< (HCH)	120.0	115.5	**116.5**

Table A5-10: Structures of Two-Heavy-Atom Hydrides. Molecular Mechanics Models (2)

molecule	point group	geometrical parameter	SYBYL	MMFF	expt.
CH$_3$OH	C$_s$	r (CO)	1.430	1.405	**1.421**
		r (CH$_{tr}$)	1.100	1.101	**1.094**
		r (CH$_g$)	1.100	1.109	**1.094**
		r (OH)	0.950	0.971	**0.963**
		< (OCH$_{tr}$)	109.6	107.1	**107.2**
		< (OCH$_{gg'}$)	125.5	131.7	**129.9**
		< (H$_g$CH$_{g'}$)	109.5	108.5	**108.5**
		< (COH)	109.6	109.0	**108.0**
CH$_3$F	C$_{3v}$	r (CF)	1.360	1.378	**1.383**
		r (CH)	1.100	1.104	**1.100**
		< (HCH)	109.5	109.7	**110.6**
CH$_3$SiH$_3$	C$_{3v}$	r (CSi)	1.880	1.844	**1.867**
		r (CH)	1.100	1.094	**1.093**
		r (SiH)	1.500	1.485	**1.485**
		< (HCH)	109.2	107.7	**107.7**
		< (HSiH)	109.5	108.6	**108.3**
HCP	C$_{\infty v}$	r (CP)	1.404	1.518	**1.540**
		r (CH)	1.056	1.065	**1.069**
CH$_2$PH	C$_s$	r (CP)	1.572	1.701	**1.67**
		r (CH$_{syn}$)	1.095	1.100	–
		r (CH$_{anti}$)	1.089	1.100	–
		r (PH)	1.008	1.424	–
		< (H$_{syn}$CP)	127.0	120.8	–
		< (H$_{anti}$CP)	116.5	120.4	–
		< (HPC)	88.1	96.4	–
CH$_3$PH$_2$	C$_s$	r (CP)	1.832	1.834	**1.862**
		r (CH$_{tr}$)	1.100	1.094	**1.094**
		r (CH$_g$)	1.101	1.094	**1.094**
		r (PH)	1.009	1.415	**1.432**
		< (PCH$_{tr}$)	109.8	110.6	–
		< (PCH$_{gg'}$)	125.2	126.8	–
		< (H$_g$CH$_{g'}$)	109.3	108.4	–
		< (CPH$_{gg'}$)	126.0	101.1	–
		< (HPH)	109.5	95.6	–
CS	C$_{\infty v}$	r (CS)	1.404	1.428	**1.535**
H$_2$CS	C$_{2v}$	r (CS)	1.710	1.670	**1.611**
		r (CH)	1.089	1.102	**1.093**
		< (HCH)	120.0	115.1	**116.9**

Table A5-10: Structures of Two-Heavy-Atom Hydrides. Molecular Mechanics Models (3)

molecule	point group	geometrical parameter	SYBYL	MMFF	expt.
CH₃SH	C$_s$	r (CS)	1.821	1.804	**1.819**
		r (CH$_{tr}$)	1.100	1.093	**1.091**
		r (CH$_g$)	1.101	1.093	**1.091**
		r (SH)	1.009	1.341	**1.336**
		< (SCH$_{tr}$)	109.0	109.9	–
		< (SCH$_{gg'}$)	126.3	125.9	–
		< (H$_g$CH$_{g'}$)	109.4	109.0	**109.8**
		< (CSH)	98.3	96.6	**96.5**
CH₃Cl	C$_{3v}$	r (CCl)	1.767	1.767	**1.781**
		r (CH)	1.100	1.092	**1.096**
		< (HCH)	109.5	109.9	**110.0**
N₂	D$_{∞h}$	r (NN)	1.170	1.118	**1.096**
N₂H₂	C$_{2h}$	r (NN)	1.346	1.245	**1.252**
		r (NH)	1.008	1.031	**1.028**
		< (NNH)	120.0	112.4	**106.9**
N₂H₄	C$_2$	r (NN)	1.500	1.435	**1.449**
		r (NH$_{int}$)	1.080	1.024	**1.021**
		r (NH$_{ext}$)	1.080	1.025	**1.021**
		< (NNH$_{int}$)	109.5	113.2	**106.0**
		< (NNH$_{ext}$)	109.5	111.6	**112.0**
		< (H$_{int}$NH$_{ext}$)	109.5	106.3	
		ω (H$_{int}$NNH$_{ext}$)	60.0	75.6	**112.0**
HNO	C$_s$	r (NO)	1.305	1.235	**1.212**
		r (NH)	1.008	1.018	**1.063**
		< (ONH)	120.0	111.0	**108.6**
NH₂OH	C$_s$	r (NO)	1.500	1.454	**1.453**
		r (NH)	1.080	1.023	**1.016**
		r (OH)	0.950	0.979	**0.962**
		< (ONH$_{gg'}$)	125.4	109.9	**112.6**
		< (HNH)	109.5	105.6	**107.1**
		< (NOH)	109.5	102.3	**101.4**
NP	C$_{∞v}$	r (NP)	1.404	1.394	**1.491**
H₂O₂	C$_2$	r (OO)	1.480	1.454	**1.452**
		r (OH)	0.950	0.976	**0.965**
		< (OOH)	109.4	96.5	**100.0**
		ω (HOOH)	180.0	129.4	**119.1**
HOF	C$_s$	r (OF)	1.500	1.417	**1.442**
		r (OH)	0.950	0.972	**0.966**
		< (HOF)	109.5	110.4	**96.8**
NaOH	C$_{∞v}$	r (NaO)	1.800	1.853	**1.95**
		r (OH)	0.950	0.964	**0.96**

Table A5-10: Structures of Two-Heavy-Atom Hydrides. Molecular Mechanics Models (4)

molecule	point group	geometrical parameter	SYBYL	MMFF	expt.
MgO	$C_{\infty v}$	r (MgO)	1.566	1.922	**1.749**
SiO	$C_{\infty v}$	r (SiO)	1.404	1.342	**1.510**
HPO	C_s	r (PO)	1.566	1.589	**1.512**
		r (PH)	1.008	1.418	-
		< (HPO)	60.0	109.5	**104.7**
HOCl	C_s	r (OCl)	1.800	1.678	**1.690**
		r (OH)	0.950	0.972	**0.975**
		< (HOCl)	109.5	110.4	**102.5**
F_2	$D_{\infty h}$	r (FF)	1.500	1.476	**1.412**
NaF	$C_{\infty v}$	r (NaF)	1.800	1.861	**1.926**
SiH_3F	C_{2v}	r (SiF)	1.800	1.605	**1.596**
		r (SiH)	1.500	1.485	**1.480**
		r (HSiH)	109.5	109.7	**110.6**
ClF	$C_{\infty v}$	r (ClF)	1.800	1.625	**1.628**
Na_2	$D_{\infty h}$	r (NaNa)	2.000	3.070	3.078
NaCl	$C_{\infty v}$	r (NaCl)	2.000	2.344	**2.361**
Si_2H_6	D_{3d}	r (SiSi)	2.000	2.302	**2.327**
		r (SiH)	1.500	1.489	**1.486**
		< (HSiH)	109.5	108.1	**107.8**
SiH_3Cl	C_{3v}	r (SiCl)	2.000	2.047	**2.048**
		r (SiH)	1.500	1.485	**1.481**
		< (HSiH)	109.5	110.9	**110.9**
P_2	$D_{\infty h}$	r (PP)	1.560	1.842	**1.893**
P_2H_4	C_2	r (PP)	2.000	2.279	**2.219**
		r (PH$_{int}$)	1.008	1.415	**1.417**
		r (PH$_{ext}$)	1.008	1.416	**1.414**
		< (PPH$_{int}$)	109.6	98.7	**99.1**
		< (PPH$_{ext}$)	109.6	98.5	**94.3**
		< (H$_{int}$PH$_{ext}$)	109.5	95.3	**92.0**
		ω (H$_{int}$PPH$_{ext}$)	60.4	61.4	**74.0**
H_2S_2	C_2	r (SS)	2.032	2.051	**2.055**
		r (SH)	1.009	1.342	**1.327**
		< (SSH)	97.8	99.6	**91.3**
		ω (HSSH)	61.2	84.2	**90.6**
Cl_2	$D_{\infty v}$	r (ClCl)	2.000	2.012	**1.988**

502

Table A5-11: Structures of Two-Heavy-Atom Hydrides. Hartree-Fock Models

molecule	point group	geometrical parameter	STO-3G	3-21G	6-31G*	6-311+G**	expt.
Li$_2$	D$_{\infty h}$	r (LiLi)	2.698	2.816	2.812	2.784	**2.673**
LiOH	C$_{\infty v}$	r (LiO)	1.432	1.537	1.592	1.591	**1.582**
		r (OH)	0.971	0.955	0.938	0.932	-
LiF	C$_{\infty v}$	r (LiF)	1.407	1.520	1.555	1.565	**1.564**
LiCl	C$_{\infty v}$	r (LiCl)	1.933	2.091	2.072	2.035	**2.021**
B$_2$H$_6$	D$_{2h}$	r (BB)	1.805	1.786	1.778	1.779	**1.763**
		r (BH)	1.154	1.182	1.185	1.184	**1.201**
		r (BH$_{bridge}$)	1.327	1.315	1.316	1.319	**1.320**
		< (HBH)	122.6	112.4	122.1	121.9	**121.0**
C$_2$H$_2$	D$_{\infty h}$	r (CC)	1.168	1.188	1.185	1.185	**1.203**
		r (CH)	1.065	1.051	1.057	1.055	**1.061**
C$_2$H$_4$	D$_{2h}$	r (CC)	1.306	1.315	1.317	1.315	**1.339**
		r (CH)	1.082	1.074	1.076	1.076	**1.085**
		< (HCH)	115.6	116.2	116.4	116.7	**117.8**
C$_2$H$_6$	D$_{3d}$	r (CC)	1.538	1.542	1.527	1.527	**1.531**
		r (CH)	1.086	1.084	1.086	1.086	**1.096**
		< (HCH)	108.2	108.1	107.7	107.7	**107.8**
HCN	C$_{\infty v}$	r (CN)	1.153	1.137	1.133	1.124	**1.153**
		r (CH)	1.070	1.050	1.059	1.058	**1.065**
HNC	C$_{\infty v}$	r (NC)	1.170	1.160	1.154	1.145	**1.169**
		r (NH)	1.011	0.983	0.985	0.984	**0.944**
CH$_2$NH	C$_s$	r (CN)	1.273	1.256	1.250	1.247	**1.273**
		r (CH$_{syn}$)	1.091	1.081	1.084	1.084	**1.103**
		r (CH$_{anti}$)	1.089	1.075	1.080	1.080	**1.081**
		r (NH)	1.048	1.015	1.006	1.005	**1.023**
		< (H$_{syn}$CN)	125.4	125.3	124.7	124.1	**123.4**
		< (H$_{anti}$CN)	119.1	119.2	119.2	119.3	**119.7**
		< (HNC)	109.1	114.9	111.6	111.7	**110.5**
CH$_3$NH$_2$	C$_s$	r (CN)	1.486	1.471	1.453	1.453	**1.471**
		r (CH$_{tr}$)	1.093	1.090	1.091	1.090	**1.099**
		r (CH$_g$)	1.089	1.083	1.084	1.084	**1.099**
		r (NH)	1.033	1.004	1.001	1.000	**1.010**
		< (NCH$_{tr}$)	113.7	114.8	114.8	114.5	**113.9**
		< (NCH$_{gg'}$)	124.0	123.4	123.9	124.0	**124.4**
		< (H$_g$CH$_{g'}$)	108.2	107.6	107.5	107.4	**108.0**
		< (CNH$_{gg'}$)	119.1	135.3	126.3	127.3	**125.7**
		< (HNH)	104.4	111.2	106.9	107.2	**107.1**
CO	C$_{\infty v}$	r (CO)	1.146	1.129	1.114	1.104	**1.128**
H$_2$CO	C$_{2v}$	r (CO)	1.217	1.207	1.181	1.178	**1.208**
		r (CH)	1.101	1.083	1.092	1.094	**1.116**
		< (HCH)	114.5	115.0	115.7	116.3	**116.5**

Table A5-11: Structures of Two-Heavy-Atom Hydrides. Hartree-Fock Models (2)

molecule	point group	geometrical parameter	STO-3G	3-21G	6-31G*	6-311+G**	expt.
CH_3OH	C_s	r (CO)	1.433	1.441	1.400	1.399	**1.421**
		r (CH_{tr})	1.092	1.079	1.081	1.081	**1.094**
		r (CH_g)	1.095	1.085	1.087	1.086	**1.094**
		r (OH)	0.991	0.966	0.946	0.941	**0.963**
		< (OCH_{tr})	107.7	106.3	107.2	107.3	**107.2**
		< ($OCH_{gg'}$)	130.4	130.5	130.1	129.7	**129.9**
		< ($H_gCH_{g'}$)	108.1	108.7	108.7	108.9	**108.5**
		< (COH)	103.8	110.3	109.4	110.1	**108.0**
CH_3F	C_{3v}	r (CF)	1.384	1.404	1.365	1.364	**1.383**
		r (CH)	1.097	1.080	1.082	1.081	**1.110**
		< (HCH)	108.3	109.5	109.8	110.2	**110.6**
CH_3SiH_3	C_{3v}	r (CSi)	1.861	1.883	1.888	1.878	**1.867**
		r (CH)	1.082	1.087	1.086	1.085	**1.093**
		r (SiH)	1.423	1.477	1.478	1.477	**1.485**
		< (HCH)	107.3	107.8	108.3	108.1	**107.7**
		< (HSiH)	108.8	108.3	107.8	108.4	**108.3**
HCP	$C_{\infty v}$	r (CP)	1.472	1.513	1.515	1.511	**1.540**
		r (CH)	1.069	1.059	1.063	1.063	**1.069**
CH_2PH	C_s	r (CP)	1.615	1.645	1.652	1.646	**1.67**
		r (CH_{syn})	1.080	1.074	1.075	1.074	–
		r (CH_{anti})	1.081	1.076	1.076	1.076	–
		r (PH)	1.390	1.408	1.409	1.410	–
		< ($H_{syn}CP$)	126.3	125.1	125.0	124.7	–
		< ($H_{anti}CP$)	120.7	119.9	119.6	119.4	–
		< (HPC)	97.0	99.0	98.9	99.1	–
CH_3PH_2	C_s	r (CP)	1.841	1.855	1.861	1.856	**1.862**
		r (CH_{tr})	1.084	1.082	1.082	1.081	**1.094**
		r (CH_g)	1.083	1.084	1.084	1.084	**1.094**
		r (PH)	1.381	1.404	1.404	1.405	**1.432**
		< (PCH_{tr})	113.1	113.0	113.2	113.2	**109.2**
		< ($PCH_{gg'}$)	125.7	124.1	123.9	123.1	–
		< ($H_gCH_{g'}$)	107.4	107.7	107.6	107.8	–
		< ($CPH_{gg'}$)	98.9	102.3	102.8	103.2	–
		< (HPH)	93.7	94.6	95.1	95.1	–
CS	$C_{\infty v}$	r (CS)	1.519	1.522	1.520	1.513	**1.535**
H_2CS	C_{2v}	r (CS)	1.574	1.594	1.597	1.593	**1.611**
		r (CH)	1.090	1.076	1.078	1.078	**1.093**
		< (HCH)	112.0	115.3	115.5	116.0	**116.9**

Table A5-11: Structures of Two-Heavy-Atom Hydrides. Hartree-Fock Models (3)

molecule	point group	geometrical parameter	STO-3G	3-21G	6-31G*	6-311+G**	expt..
CH₃SH	C_s	r (CS)	1.798	1.823	1.817	1.819	**1.819**
		r (CH_tr)	1.085	1.081	1.082	1.081	**1.091**
		r (CH_g)	1.087	1.080	1.081	1.080	**1.091**
		r (SH)	1.331	1.327	1.327	1.327	**1.336**
		< (SCH_tr)	108.5	106.9	106.7	106.3	–
		< (SCH_gg')	130.1	128.9	129.3	128.9	–
		< (H_gCH_g')	108.1	110.1	110.0	110.5	**109.8**
		< (CSH)	95.4	97.5	97.9	98.0	**96.5**
CH₃Cl	C_3v	r (CCl)	1.802	1.806	1.785	1.792	**1.781**
		r (CH)	1.088	1.076	1.078	1.077	**1.096**
		< (HCH)	110.1	110.8	110.5	110.8	**110.0**
N₂	D_∞h	r (NN)	1.134	1.083	1.078	1.067	**1.098**
N₂H₂	C_2h	r (NN)	1.267	1.239	1.216	1.209	**1.252**
		r (NH)	1.061	1.021	1.014	1.013	**1.028**
		< (NNH)	105.3	109.0	107.5	108.1	**106.9**
N₂H₄	C_2	r (NN)	1.459	1.451	1.413	1.412	**1.449**
		r (NH_int)	1.037	1.003	0.999	1.000	**1.021**
		r (NH_ext)	1.040	1.007	1.003	0.997	**1.021**
		< (NNH_int)	105.4	109.0	107.9	112.4	**106.0**
		< (NNH_ext)	109.0	113.3	112.3	108.5	**112.0**
		< (H_intNH_ext)	104.6	111.8	108.2	108.7	–
		ω (H_intNNH_ext)	91.5	93.8	90.2	89.9	**91.0**
HNO	C_s	r (NO)	1.231	1.217	1.175	1.168	**1.212**
		r (NH)	1.082	1.036	1.032	1.032	**1.063**
		< (ONH)	107.6	109.4	108.8	109.2	**108.6**
NH₂OH	C_s	r (NO)	1.420	1.472	1.403	1.392	**1.453**
		r (NH)	1.043	1.002	1.002	1.001	**1.016**
		r (OH)	1.001	0.959	0.946	0.944	**0.962**
		< (ONH_gg')	119.7	114.7	115.2	123.0	**112.6**
		< (HNH)	104.7	109.6	106.5	108.8	**107.1**
		< (NOH)	105.0	103.6	104.2	109.3	**101.4**
NP	C_∞v	r (NP)	1.459	1.462	1.455	1.449	**1.491**
H₂O₂	C_2	r (OO)	1.396	1.473	1.393	1.388	**1.452**
		r (OH)	1.001	0.971	0.949	0.944	**0.965**
		< (OOH)	101.1	99.4	102.2	102.9	**100.0**
		ω (HOOH)	125.3	180.0	115.2	112.4	**119.1**
HOF	C_s	r (OF)	1.355	1.439	1.376	1.365	**1.442**
		r (OH)	1.006	0.976	0.952	0.948	**0.966**
		< (HOF)	101.4	99.0	99.8	100.7	**96.8**
NaOH	C_∞v	r (NaO)	1.763	1.864	1.922	1.946	**1.95**
		r (OH)	0.988	0.963	0.941	0.935	**0.96**

Table A5-11: Structures of Two-Heavy-Atom Hydrides. Hartree-Fock Models (4)

molecule	point group	geometrical parameter	STO-3G	3-21G	6-31G*	6-311+G**	expt.
MgO	$C_{\infty v}$	r (MgO)		1.731	1.738	1.727	**1.749**
SiO	$C_{\infty v}$	r (SiO)	1.475	1.496	1.487	1.480	**1.510**
HPO	C_s	r (PO)	1.515	1.471	1.460	1.449	**1.512**
		r (PH)	1.410	1.429	1.431	1.437	–
		< (HPO)	99.1	106.0	105.4	104.8	**104.7**
HOCl	C_s	r (OCl)	1.737	1.700	1.670	1.663	**1.690**
		r (OH)	1.004	0.973	0.951	0.946	**0.975**
		< (HOCl)	100.2	106.1	105.1	105.4	**102.5**
F_2	$D_{\infty h}$	r (FF)	1.315	1.402	1.345	1.330	**1.412**
NaF	$C_{\infty v}$	r (NaF)	1.753	1.831	1.885	1.929	**1.926**
SiH_3F	C_{2v}	r (SiF)	1.624	1.593	1.594	1.586	**1.596**
		r (SiH)	1.422	1.469	1.470	1.468	**1.480**
		r (HSiH)	109.6	109.9	110.2	110.8	**110.6**
ClF	$C_{\infty v}$	r (ClF)	1.677	1.636	1.613	1.607	**1.628**
Na_2	$D_{\infty h}$	r (NaNa)	2.359	2.651	3.130	3.198	**3.078**
NaCl	$C_{\infty v}$	r (NaCl)	2.221	2.392	2.397	2.391	**2.361**
Si_2H_6	D_{3d}	r (SiSi)	2.243	2.342	2.353	2.373	**2.327**
		r (SiH)	1.423	1.478	1.478	1.477	**1.486**
		< (HSiH)	108.0	108.6	108.5	108.7	**107.8**
SiH_3Cl	C_{3v}	r (SiCl)	2.089	2.056	2.067	2.072	**2.048**
		r (SiH)	1.423	1.467	1.468	1.466	**1.481**
		< (HSiH)	111.2	110.5	108.3	110.9	**110.9**
P_2	$D_{\infty h}$	r (PP)	1.808	1.853	1.859	1.856	**1.893**
P_2H_4	C_2	r (PP)	2.175	2.205	2.214	2.228	**2.219**
		r (PH_{int})	1.381	1.401	1.401	1.403	**1.417**
		r (PH_{ext})	1.379	1.400	1.402	1.403	**1.414**
		< (PPH_{int})	98.0	100.6	96.8	100.7	**99.1**
		< (PPH_{ext})	95.6	96.3	101.2	96.2	**94.3**
		< ($H_{int}PH_{ext}$)	93.5	95.2	95.6	95.8	**92.0**
		ω ($H_{int}PPH_{ext}$)	79.0	73.0	77.3	77.2	**74.0**
H_2S_2	C_2	r (SS)	2.065	2.057	2.064	2.075	**2.055**
		r (SH)	1.334	1.327	1.327	1.328	**1.327**
		< (SSH)	96.9	99.0	99.1	98.5	**91.3**
		ω (HSSH)	92.6	89.9	87.9	90.2	**90.6**
Cl_2	$D_{\infty v}$	r (ClCl)	2.063	1.996	1.990	2.000	**1.988**

Table A5-12: Structures of Two-Heavy-Atom Hydrides. Local Density Models

molecule	point group	geometrical parameter	6-31G*	6-311+G**	expt.
Li$_2$	D$_{\infty h}$	r (LiLi)	2.735	2.716	**2.673**
LiOH	C$_{\infty v}$	r (LiO)	1.574	1.582	**1.582**
		r (OH)	0.965	0.961	–
LiF	C$_{\infty v}$	r (LiF)	1.539	1.574	**1.564**
LiCl	C$_{\infty v}$	r (LiCl)	2.040	2.010	**2.021**
B$_2$H$_6$	D$_{2h}$	r (BB)	1.735	1.732	**1.763**
		r (BH)	1.201	1.199	**1.201**
		r (BH$_{bridge}$)	1.315	1.316	**1.320**
		< (HBH)	121.4	121.4	**121.0**
C$_2$H$_2$	D$_{\infty h}$	r (CC)	1.211	1.204	**1.203**
		r (CH)	1.077	1.075	**1.061**
C$_2$H$_4$	D$_{2h}$	r (CC)	1.329	1.326	**1.339**
		r (CH)	1.096	1.096	**1.085**
		< (HCH)	116.5	116.7	**117.8**
C$_2$H$_6$	D$_{3d}$	r (CC)	1.511	1.512	**1.531**
		r (CH)	1.103	1.103	**1.096**
		< (HCH)	107.2	107.2	**107.8**
HCN	C$_{\infty v}$	r (CN)	1.162	1.154	**1.153**
		r (CH)	1.081	1.079	**1.065**
HNC	C$_{\infty v}$	r (NC)	1.181	1.172	**1.169**
		r (NH)	1.010	1.011	**0.994**
CH$_2$NH	C$_s$	r (CN)	1.269	1.265	**1.273**
		r (CH$_{syn}$)	1.100	1.109	**1.103**
		r (CH$_{anti}$)	1.104	1.104	**1.081**
		r (NH)	1.035	1.034	**1.023**
		< (H$_{syn}$CN)	126.0	125.2	**123.4**
		< (H$_{anti}$CN)	118.4	118.7	**119.7**
		< (HNC)	110.3	111.0	**110.5**
CH$_3$NH$_2$	C$_s$	r (CN)	1.443	1.444	**1.471**
		r (CH$_{tr}$)	1.115	1.112	**1.099**
		r (CH$_g$)	1.104	1.103	**1.099**
		r (NH)	1.025	1.022	**1.010**
		< (NCH$_{tr}$)	116.9	116.1	**113.9**
		< (NCH$_{gg'}$)	123.7	123.9	**124.4**
		< (H$_g$CH$_{g'}$)	106.3	106.3	**108.0**
		< (CNH$_{gg'}$)	125.5	129.9	**125.7**
		< (HNH)	106.3	107.9	**107.1**
CO	C$_{\infty v}$	r (CO)	1.141	1.130	**1.128**
H$_2$CO	C$_{2v}$	r (CO)	1.205	1.201	**1.208**
		r (CH)	1.124	1.123	**1.116**
		< (HCH)	115.1	115.9	**116.5**

Table A5-12: Structures of Two-Heavy-Atom Hydrides. Local Density Models (2)

molecule	point group	geometrical parameter	6-31G*	6-311+G**	expt.
CH_3OH	C_s	r (CO)	1.399	1.405	**1.421**
		r (CH_{tr})	1.103	1.101	**1.094**
		r (CH_g)	1.112	1.109	**1.094**
		r (OH)	0.975	0.971	**0.963**
		< (OCH_{tr})	107.0	107.1	**107.2**
		< ($OCH_{gg'}$)	132.8	131.7	**129.9**
		< ($H_gCH_{g'}$)	107.8	108.5	**108.5**
		< (COH)	107.7	109.0	**108.0**
CH_3F	C_{3v}	r (CF)	1.364	1.378	**1.383**
		r (CH)	1.107	1.104	**1.100**
		< (HCH)	108.6	109.7	**110.6**
CH_3SiH_3	C_{3v}	r (CSi)	1.867	1.868	**1.867**
		r (CH)	1.102	1.102	**1.093**
		r (SiH)	1.500	1.500	**1.485**
		< (HCH)	108.0	107.9	**107.7**
		< (HSiH)	107.8	108.2	**108.3**
HCP	$C_{\infty v}$	r (CP)	1.546	1.543	**1.540**
		r (CH)	1.086	1.085	**1.069**
CH_2PH	C_s	r (CP)	1.669	1.667	**1.67**
		r (CH_{syn})	1.096	1.095	–
		r (CH_{anti})	1.097	1.097	–
		r (PH)	1.445	1.445	–
		< ($H_{syn}CP$)	125.8	125.4	–
		< ($H_{anti}CP$)	118.8	118.9	–
		< (HPC)	97.0	96.8	–
CH_3PH_2	C_s	r (CP)	1.852	1.853	**1.862**
		r (CH_{tr})	1.100	1.100	**1.094**
		r (CH_g)	1.102	1.102	**1.094**
		r (PH)	1.437	1.437	**1.432**
		< (PCH_{tr})	114.5	114.5	**109.2**
		< ($PCH_{gg'}$)	122.6	122.9	–
		< ($H_gCH_{g'}$)	107.1	107.0	–
		< ($CPH_{gg'}$)	100.5	100.0	–
		< (HPH)	92.2	92.2	–
CS	$C_{\infty v}$	r (CS)	1.550	1.540	**1.535**
H_2CS	C_{2v}	r (CS)	1.612	1.610	**1.611**
		r (CH	1.103	1.102	**1.093**
		< (HCH)	115.6	115.9	**116.9**

Table A5-12: Structures of Two-Heavy-Atom Hydrides. Local Density Models (3)

molecule	point group	geometrical parameter	6-31G*	6-311+G**	expt.
CH$_3$SH	C$_s$	r (CS)	1.808	1.811	**1.819**
		r (CH$_{tr}$)	1.101	1.100	**1.091**
		r (CH$_g$)	1.101	1.100	**1.091**
		r (SH)	1.359	1.360	**1.336**
		< (SCH$_{tr}$)	106.2	106.3	–
		< (SCH$_{gg'}$)	131.1	130.9	–
		< (H$_g$CH$_{g'}$)	110.0	110.1	**109.8**
		< (CSH)	96.9	96.6	**96.5**
CH$_3$Cl	C$_{3v}$	r (CCl)	1.773	1.776	**1.781**
		r (CH)	1.100	1.098	**1.096**
		< (HCH)	109.8	109.9	**110.0**
N$_2$	D$_{\infty h}$	r (NN)	1.110	1.100	**1.098**
N$_2$H$_2$	C$_{2h}$	r (NN)	1.244	1.236	**1.252**
		r (NH)	1.051	1.048	**1.028**
		< (NNH)	106.0	106.9	**106.9**
N$_2$H$_4$	C$_2$	r (NN)	1.404	1.400	**1.449**
		r (NH$_{int}$)	1.029	1.026	**1.021**
		r (NH$_{ext}$)	1.024	1.022	**1.021**
		< (NNH$_{int}$)	113.7	114.8	**106.0**
		< (NNH$_{ext}$)	108.4	110.1	**112.0**
		< (H$_{int}$NH$_{ext}$)	107.8	109.9	–
		ω (H$_{int}$NNH$_{ext}$)	90.3	90.5	**91.0**
HNO	C$_s$	r (NO)	1.204	1.196	**1.212**
		r (NH)	1.092	1.086	**1.063**
		< (ONH)	108.7	109.2	**108.6**
NH$_2$OH	C$_s$	r (NO)	1.403	1.421	**1.453**
		r (NH)	1.031	1.029	**1.016**
		r (OH)	0.984	0.973	**0.962**
		< (ONH$_{gg'}$)	112.4	116.0	**112.6**
		< (HNH)	107.4	106.4	**107.1**
		< (NOH)	108.5	102.9	**101.4**
NP	C$_{\infty v}$	r (NP)	1.500	1.494	**1.491**
H$_2$O$_2$	C$_2$	r (OO)	1.433	1.437	**1.452**
		r (OH)	0.982	0.978	**0.965**
		< (OOH)	99.9	100.6	**100.0**
		ω (HOOH)	117.2	119.4	**119.1**
HOF	C$_s$	r (OF)	1.414	1.423	**1.442**
		r (OH)	0.985	0.984	**0.966**
		< (HOF)	97.9	98.7	**96.8**
NaOH	C$_{\infty v}$	r (NaO)	1.893	1.932	**1.95**
		r (OH)	0.967	0.963	**0.96**

Table A5-12: Structures of Two-Heavy-Atom Hydrides. Local Density Models (4)

molecule	point group	geometrical parameter	6-31G*	6-311+G**	expt.
MgO	$C_{\infty v}$	r (MgO)	1.729	1.756	**1.749**
SiO	$C_{\infty v}$	r (SiO)	1.527	1.527	**1.510**
HPO	C_s	r (PO)	1.502	1.500	**1.512**
		r (PH)	1.491	1.491	–
		< (HPO)	105.4	103.8	**104.7**
HOCl	C_s	r (OCl)	1.702	1.712	**1.690**
		r (OH)	0.983	0.979	**0.975**
		< (HOCl)	102.5	103.3	**102.5**
F_2	$D_{\infty h}$	r (FF)	1.387	1.399	**1.412**
NaF	$C_{\infty v}$	r (NaF)	1.864	1.936	**1.926**
SiH_3F	C_{2v}	r (SiF)	1.606	1.625	**1.596**
		r (SiH)	1.498	1.494	**1.480**
		r (HSiH)	109.6	111.1	**110.6**
ClF	$C_{\infty v}$	r (ClF)	1.642	1.664	**1.628**
Na_2	$D_{\infty h}$	r (NaNa)	2.962	3.005	**3.078**
NaCl	$C_{\infty v}$	r (NaCl)	2.333	2.351	**2.361**
Si_2H_6	D_{3d}	r (SiSi)	2.321	2.328	**2.327**
		r (SiH)	1.500	1.500	**1.486**
		< (HSiH)	108.4	108.6	**107.8**
SiH_3Cl	C_{3v}	r (SiCl)	2.057	2.063	**2.048**
		r (SiH)	1.494	1.494	**1.481**
		< (HSiH)	110.2	110.7	**110.9**
P_2	$D_{\infty h}$	r (PP)	1.906	1.900	**1.893**
P_2H_4	C_2	r (PP)	2.221	2.231	**2.219**
		r(PH_{int})	1.436	1.437	**1.417**
		r (PH_{ext})	1.434	1.435	**1.414**
		< (PPH_{int})	100.3	100.3	**99.1**
		< (PPH_{ext})	94.7	93.8	**94.3**
		< ($H_{int}PH_{ext}$)	92.8	92.6	**92.0**
		ω ($H_{int}PPH_{ext}$)	82.6	79.0	**74.0**
H_2S_2	C_2	r (SS)	2.067	2.083	**2.055**
		r (SH)	1.365	1.366	**1.327**
		< (SSH)	99.4	98.6	**91.3**
		ω (HSSH)	91.2	91.5	**90.6**
Cl_2	$D_{\infty v}$	r (ClCl)	2.012	2.030	**1.988**

Table A5-13: Structures of Two-Heavy-Atom Hydrides. BP Density Functional Models

molecule	point group	geometrical parameter	6-31G*	6-311+G**	expt.
Li$_2$	D$_{\infty h}$	r (LiLi)	2.762	2.738	**2.673**
LiOH	C$_{\infty v}$	r (LiO)	1.598	1.606	**1.582**
		r (OH)	0.967	0.960	–
LiF	C$_{\infty v}$	r (LiF)	1.568	1.601	**1.564**
LiCl	C$_{\infty v}$	r (LiCl)	2.069	2.039	**2.021**
B$_2$H$_6$	D$_{2h}$	r (BB)	1.773	1.768	**1.763**
		r (BH)	1.201	1.197	**1.201**
		r (BH$_{bridge}$)	1.326	1.325	**1.320**
		< (HBH)	122.0	122.0	**121.0**
C$_2$H$_2$	D$_{\infty h}$	r (CC)	1.217	1.210	**1.203**
		r (CH)	1.075	1.071	**1.061**
C$_2$H$_4$	D$_{2h}$	r (CC)	1.340	1.337	**1.339**
		r (CH)	1.096	1.094	**1.085**
		< (HCH)	116.3	116.5	**117.8**
C$_2$H$_6$	D$_{3d}$	r (CC)	1.535	1.532	**1.531**
		r (CH)	1.105	1.102	**1.096**
		< (HCH)	107.5	107.4	**107.8**
HCN	C$_{\infty v}$	r (CN)	1.170	1.162	**1.153**
		r (CH)	1.080	1.076	**1.065**
HNC	C$_{\infty v}$	r (NC)	1.189	1.181	**1.169**
		r (NH)	1.010	1.008	**0.944**
CH$_2$NH	C$_s$	r (CN)	1.281	1.278	**1.273**
		r (CH$_{syn}$)	1.110	1.107	**1.103**
		r (CH$_{anti}$)	1.103	1.101	**1.081**
		r (NH)	1.038	1.034	**1.023**
		< (H$_{syn}$CN)	126.1	125.3	**123.4**
		< (H$_{anti}$CN)	118.4	118.5	**119.7**
		< (HNC)	109.7	110.4	**110.5**
CH$_3$NH$_2$	C$_s$	r (CN)	1.472	1.472	**1.471**
		r (CH$_{tr}$)	1.115	1.110	**1.099**
		r (CH$_g$)	1.105	1.101	**1.099**
		r (NH)	1.028	1.023	**1.010**
		< (NCH$_{tr}$)	116.3	115.6	**113.9**
		< (NCH$_{gg'}$)	123.2	123.4	**124.4**
		< (H$_g$CH$_{g'}$)	107.0	107.0	**108.0**
		< (CNH$_{gg'}$)	122.4	126.2	**125.7**
		< (HNH)	105.2	106.6	**107.1**
CO	C$_{\infty v}$	r (CO)	1.150	1.140	**1.128**
H$_2$CO	C$_{2v}$	r (CO)	1.216	1.212	**1.208**
		r (CH)	1.123	1.120	**1.116**
		< (HCH)	114.9	115.8	**116.5**

Table A5-13: Structures of Two-Heavy-Atom Hydrides. BP Density Functional Models (2)

molecule	point group	geometrical parameter	6-31G*	6-311+G**	expt.
CH_3OH	C_s	r (CO)	1.427	1.433	**1.421**
		r (CH_{tr})	1.102	1.099	**1.094**
		r (CH_g)	1.112	1.107	**1.094**
		r (OH)	0.978	0.971	**0.963**
		< (OCH_{tr})	106.5	106.6	**107.2**
		< ($OCH_{gg'}$)	131.9	130.8	**129.9**
		< ($H_gCH_{g'}$)	108.2	108.9	**108.5**
		< (COH)	106.8	108.0	**108.0**
CH_3F	C_{3v}	r (CF)	1.391	1.405	**1.383**
		r (CH)	1.107	1.101	**1.100**
		< (HCH)	109.1	110.2	**110.6**
CH_3SiH_3	C_{3v}	r (CSi)	1.895	1.891	**1.867**
		r (CH)	1.104	1.101	**1.093**
		r (SiH)	1.501	1.498	**1.485**
		< (HCH)	107.9	108.0	**107.7**
		< (HSiH)	108.0	108.3	**108.3**
HCP	$C_{\infty v}$	r (CP)	1.557	1.554	**1.540**
		r (CH)	1.085	1.083	**1.069**
CH_2PH	C_s	r (CP)	1.687	1.682	**1.67**
		r (CH_{syn})	1.097	1.094	–
		r (CH_{anti})	1.098	1.095	–
		r (PH)	1.446	1.444	–
		< ($H_{syn}CP$)	125.8	125.4	–
		< ($H_{anti}CP$)	119.0	119.0	–
		< (HPC)	97.0	97.1	–
CH_3PH_2	C_s	r (CP)	1.884	1.881	**1.862**
		r (CH_{tr})	1.102	1.099	**1.094**
		r (CH_g)	1.103	1.100	**1.094**
		r (PH)	1.438	1.437	**1.432**
		< (PCH_{tr})	113.9	114.0	**109.2**
		< ($PCH_{gg'}$)	122.9	122.8	–
		< ($H_gCH_{g'}$)	107.5	107.4	–
		< ($CPH_{gg'}$)	100.1	100.1	–
		< (HPH)	95.5	92.5	–
CS	$C_{\infty v}$	r (CS)	1.564	1.554	**1.535**
H_2CS	C_{2v}	r (CS)	1.628	1.625	**1.611**
		r (CH)	1.103	1.100	**1.093**
		< (HCH)	115.3	115.8	**116.9**

Table A5-13: Structures of Two-Heavy-Atom Hydrides. BP Density Functional Models (3)

molecule	point group	geometrical parameter	6-31G*	6-311+G**	expt.
CH_3SH	C_s	r (CS)	1.842	1.840	**1.819**
		r (CH_{tr})	1.101	1.099	**1.091**
		r (CH_g)	1.102	1.099	**1.091**
		r (SH)	1.361	1.359	**1.336**
		< (SCH_{tr})	106.1	106.0	–
		< ($SCH_{gg'}$)	130.0	129.9	–
		< ($H_gCH_{g'}$)	110.2	110.3	**109.8**
		< (CSH)	96.6	96.6	**96.5**
CH_3Cl	C_{3v}	r (CCl)	1.807	1.809	**1.781**
		r (CH)	1.099	1.096	**1.096**
		< (HCH)	110.3	110.5	**110.0**
N_2	$D_{\infty h}$	r (NN)	1.118	1.108	**1.098**
N_2H_2	C_{2h}	r (NN)	1.259	1.252	**1.252**
		r (NH)	1.053	1.047	**1.028**
		< (NNH)	105.4	106.2	**106.9**
N_2H_4	C_2	r (NN)	1.448	1.442	**1.449**
		r (NH_{int})	1.033	1.027	**1.021**
		r (NH_{ext})	1.027	1.023	**1.021**
		< (NNH_{int})	111.5	112.6	**106.0**
		< (NNH_{ext})	105.9	107.5	**112.0**
		< ($H_{int}NH_{ext}$)	106.0	107.8	–
		ω ($H_{int}NNH_{ext}$)	89.7	91.1	**91.0**
HNO	C_s	r (NO)	1.219	1.212	**1.212**
		r (NH)	1.092	1.084	**1.063**
		< (ONH)	108.3	108.7	**108.6**
NH_2OH	C_s	r (NO)	1.465	1.462	**1.453**
		r (NH)	1.034	1.029	**1.016**
		r (OH)	0.980	0.973	**0.962**
		< ($ONH_{gg'}$)	110.6	112.7	**112.6**
		< (HNH)	103.9	105.3	**107.1**
		< (NOH)	100.7	101.7	**101.4**
NP	$C_{\infty v}$	r (NP)	1.512	1.506	**1.491**
H_2O_2	C_2	r (OO)	1.475	1.476	**1.452**
		r (OH)	0.984	0.978	**0.965**
		< (OOH)	98.8	99.6	**100.0**
		ω (HOOH)	118.6	120.8	**119.1**
HOF	C_s	r (OF)	1.452	1.458	**1.442**
		r (OH)	0.987	0.983	**0.966**
		< (HOF)	97.2	97.9	**96.8**
NaOH	$C_{\infty v}$	r (NaO)	1.929	1.968	**1.95**
		r (OH)	0.970	0.963	**0.96**

Table A5-13: Structures of Two-Heavy-Atom Hydrides. BP Density Functional Models (4)

molecule	point group	geometrical parameter	6-31G*	6-311+G**	expt.
MgO	$C_{\infty v}$	r (MgO)	1.757	1.814	**1.749**
SiO	$C_{\infty v}$	r (SiO)	1.542	1.540	**1.510**
HPO	C_s	r (PO)	1.517	1.514	**1.512**
		r (PH)	1.491	1.488	–
		< (HPO)	105.5	104.1	**104.7**
HOCl	C_s	r (OCl)	1.744	1.751	**1.690**
		r (OH)	0.986	0.979	**0.975**
		< (HOCl)	101.7	102.2	**102.5**
F_2	$D_{\infty h}$	r (FF)	1.420	1.432	**1.412**
NaF	$C_{\infty v}$	r (NaF)	1.912	1.974	**1.926**
SiH_3F	C_{2v}	r (SiF)	1.626	1.644	**1.596**
		r (SiH)	1.498	1.492	**1.480**
		r (HSiH)	109.6	111.0	**110.6**
ClF	$C_{\infty v}$	r (ClF)	1.679	1.697	**1.628**
Na_2	$D_{\infty h}$	r (NaNa)	3.072	3.095	**3.078**
NaCl	$C_{\infty v}$	r (NaCl)	2.393	2.399	**2.361**
Si_2H_6	D_{3d}	r (SiSi)	2.354	2.359	**2.327**
		r (SiH)	1.501	1.499	**1.486**
		< (HSiH)	108.2	108.5	**107.8**
SiH_3Cl	C_{3v}	r (SiCl)	2.085	2.089	**2.048**
		r (SiH)	1.494	1.491	**1.481**
		< (HSiH)	110.2	110.6	**110.9**
P_2	$D_{\infty h}$	r (PP)	1.922	1.915	**1.893**
P_2H_4	C_2	r (PP)	2.261	2.268	**2.219**
		r (PH$_{int}$)	1.438	1.436	**1.417**
		r (PH$_{ext}$)	1.436	1.434	**1.414**
		< (PPH$_{int}$)	99.8	99.9	**99.1**
		< (PPH$_{ext}$)	94.4	94.1	**94.3**
		< (H$_{int}$PH$_{ext}$)	92.8	92.8	**92.0**
		ω (H$_{int}$PPH$_{ext}$)	79.3	78.2	**74.0**
H_2S_2	C_2	r (SS)	2.103	2.117	**2.055**
		r (SH)	1.368	1.365	**1.327**
		< (SSH)	99.1	98.4	**91.3**
		ω (HSSH)	90.7	91.3	**90.6**
Cl_2	$D_{\infty v}$	r (ClCl)	2.052	2.063	**1.988**

Table A5-14: Structures of Two-Heavy-Atom Hydrides. BLYP Density Functional Models

molecule	point group	geometrical parameter	6-31G*	6-311+G**	expt.
Li$_2$	D$_{\infty h}$	r (LiLi)	2.727	2.705	**2.673**
LiOH	C$_{\infty v}$	r (LiO)	1.592	1.602	**1.582**
		r (OH)	0.969	0.961	–
LiF	C$_{\infty v}$	r (LiF)	1.559	1.597	**1.564**
LiCl	C$_{\infty v}$	r (LiCl)	2.062	2.036	**2.021**
B$_2$H$_6$	D$_{2h}$	r (BB)	1.783	1.776	**1.763**
		r (BH)	1.197	1.192	**1.201**
		r (BH$_{bridge}$)	1.325	1.322	**1.320**
		< (HBH)	121.9	121.8	**121.0**
C$_2$H$_2$	D$_{\infty h}$	r (CC)	1.215	1.209	**1.203**
		r (CH)	1.073	1.068	**1.061**
C$_2$H$_4$	D$_{2h}$	r (CC)	1.341	1.338	**1.339**
		r (CH)	1.095	1.091	**1.085**
		< (HCH)	116.1	116.5	**117.8**
C$_2$H$_6$	D$_{3d}$	r (CC)	1.542	1.541	**1.531**
		r (CH)	1.103	1.100	**1.096**
		< (HCH)	107.5	107.5	**107.8**
HCN	C$_{\infty v}$	r (CN)	1.169	1.161	**1.153**
		r (CH)	1.078	1.073	**1.065**
HNC	C$_{\infty v}$	r (NC)	1.189	1.180	**1.169**
		r (NH)	1.009	1.007	**0.994**
CH$_2$NH	C$_s$	r (CN)	1.283	1.279	**1.273**
		r (CH$_{syn}$)	1.108	1.104	**1.103**
		r (CH$_{anti}$)	1.102	1.098	**1.081**
		r (NH)	1.038	1.033	**1.023**
		< (H$_{syn}$CN)	126.1	125.2	**123.4**
		< (H$_{anti}$CN)	118.5	118.7	**119.7**
		< (HNC)	109.8	110.7	**110.5**
CH$_3$NH$_2$	C$_s$	r (CN)	1.481	1.480	**1.471**
		r (CH$_{tr}$)	1.113	1.108	**1.099**
		r (CH$_g$)	1.102	1.099	**1.099**
		r (NH)	1.029	1.023	**1.010**
		< (NCH$_{tr}$)	116.2	115.4	**113.9**
		< (NCH$_{gg'}$)	123.1	123.2	**124.4**
		< (H$_g$CH$_{g'}$)	107.3	107.4	**108.0**
		< (CNH$_{gg'}$)	122.0	126.4	**125.7**
		< (HNH)	105.0	106.8	**107.1**
CO	C$_{\infty v}$	r (CO)	1.150	1.140	**1.128**
H$_2$CO	C$_{2v}$	r (CO)	1.218	1.214	**1.208**
		r (CH)	1.121	1.117	**1.116**
		< (HCH)	115.0	115.8	**116.5**

Table A5-14: Structures of Two-Heavy-Atom Hydrides. BLYP Density Functional Models (2)

molecule	point group	geometrical parameter	6-31G*	6-311+G**	expt.
CH_3OH	C_s	r (CO)	1.435	1.443	**1.421**
		r (CH_{tr})	1.101	1.097	**1.094**
		r (CH_g)	1.110	1.104	**1.094**
		r (OH)	0.980	0.972	**0.963**
		< (OCH_{tr})	106.4	106.4	**107.2**
		< ($OCH_{gg'}$)	131.7	130.4	**129.9**
		< ($H_gCH_{g'}$)	108.3	109.1	**108.5**
		< (COH)	106.9	108.2	**108.0**
CH_3F	C_{3v}	r (CF)	1.398	1.416	**1.383**
		r (CH)	1.105	1.098	**1.100**
		< (HCH)	109.2	110.5	**110.6**
CH_3SiH_3	C_{3v}	r (CSi)	1.901	1.899	**1.867**
		r (CH)	1.103	1.100	**1.093**
		r (SiH)	1.500	1.495	**1.485**
		< (HCH)	107.9	107.9	**107.7**
		< (HSiH)	108.0	108.4	**108.3**
HCP	$C_{\infty v}$	r (CP)	1.559	1.554	**1.540**
		r (CH)	1.083	1.079	**1.069**
CH_2PH	C_s	r (CP)	1.691	1.687	**1.67**
		r (CH_{syn})	1.095	1.091	–
		r (CH_{anti})	1.096	1.092	–
		r (PH)	1.447	1.443	–
		< ($H_{syn}CP$)	125.8	125.4	–
		< ($H_{anti}CP$)	119.1	119.1	–
		< (HPC)	97.5	97.4	–
CH_3PH_2	C_s	r (CP)	1.895	1.894	**1.862**
		r (CH_{tr})	1.100	1.097	**1.094**
		r (CH_g)	1.102	1.098	**1.094**
		r (PH)	1.439	1.436	**1.432**
		< (PCH_{tr})	113.8	113.8	**109.2**
		< ($PCH_{gg'}$)	122.8	122.8	–
		< ($H_gCH_{g'}$)	107.6	107.6	–
		< ($CPH_{gg'}$)	100.5	100.3	–
		< (HPH)	92.6	92.8	–
CS	$C_{\infty v}$	r (CS)	1.567	1.556	**1.535**
H_2CS	C_{2v}	r (CS)	1.633	1.631	**1.611**
		r (CH)	1.101	1.097	**1.093**
		< (HCH)	115.1	115.8	**116.9**

Table A5-14: Structures of Two-Heavy-Atom Hydrides. BLYP Density Functional Models (3)

molecule	point group	geometrical parameter	6-31G*	6-311+G**	expt.
CH₃SH	C_s	r (CS)	1.857	1.857	**1.819**
		r (CH_tr)	1.100	1.096	**1.091**
		r (CH_g)	1.100	1.096	**1.091**
		r (SH)	1.362	1.359	**1.336**
		< (SCH_tr)	106.0	105.9	–
		< (SCH_gg')	129.6	129.4	–
		< (H_gCH_g')	110.3	110.5	**109.8**
		< (CSH)	96.7	96.6	**96.5**
CH₃Cl	C_3v	r (CCl)	1.826	1.829	**1.781**
		r (CH)	1.098	1.094	**1.096**
		< (HCH)	110.6	110.8	**110.0**
N₂	D_∞h	r (NN)	1.118	1.108	**1.098**
N₂H₂	C_2h	r (NN)	1.263	1.255	**1.252**
		r (NH)	1.053	1.046	**1.028**
		< (NNH)	105.5	106.4	**106.9**
N₂H₄	C_2	r (NN)	1.463	1.456	**1.449**
		r (NH_int)	1.033	1.026	**1.021**
		r (NH_ext)	1.028	1.022	**1.021**
		< (NNH_int)	111.1	112.3	**106.0**
		< (NNH_ext)	105.5	107.2	**112.0**
		< (H_intNH_ext)	105.7	107.9	–
		ω (H_intNNH_ext)	90.6	91.6	**91.0**
HNO	C_s	r (NO)	1.224	1.217	**1.212**
		r (NH)	1.093	1.082	**1.063**
		< (ONH)	108.3	108.7	**108.6**
NH₂OH	C_s	r (NO)	1.462	1.480	**1.453**
		r (NH)	1.035	1.028	**1.016**
		r (OH)	0.988	0.974	**0.962**
		< (ONH_gg')	117.6	112.0	**112.6**
		< (HNH)	105.6	105.5	**107.1**
		< (NOH)	107.2	101.6	**101.4**
NP	C_∞v	r (NP)	1.515	1.508	**1.491**
H₂O₂	C_2	r (OO)	1.494	1.497	**1.452**
		r (OH)	0.985	0.979	**0.965**
		< (OOH)	98.5	99.2	**100.0**
		ω (HOOH)	120.5	122.9	**119.1**
HOF	C_s	r (OF)	1.468	1.478	**1.442**
		r (OH)	0.989	0.984	**0.966**
		< (HOF)	97.0	97.6	**96.8**
NaOH	C_∞v	r (NaO)	1.930	1.968	**1.95**
		r (OH)	0.972	0.964	**0.96**

Table A5-14: Structures of Two-Heavy-Atom Hydrides. BLYP Density Functional Models (4)

molecule	point group	geometrical parameter	6-31G*	6-311+G**	expt.
MgO	$C_{\infty v}$	r (MgO)	1.739	1.784	**1.749**
SiO	$C_{\infty v}$	r (SiO)	1.544	1.543	**1.510**
HPO	C_s	r (PO)	1.521	1.518	**1.512**
		r (PH)	1.494	1.487	–
		< (HPO)	105.7	104.2	**104.7**
HOCl	C_s	r (OCl)	1.766	1.772	**1.690**
		r (OH)	0.988	0.980	**0.975**
		< (HOCl)	101.2	102.0	**102.5**
F$_2$	$D_{\infty h}$	r (FF)	1.434	1.449	**1.412**
NaF	$C_{\infty v}$	r (NaF)	1.910	1.976	**1.926**
SiH$_3$F	C_{2v}	r (SiF)	1.628	1.650	**1.596**
		r (SiH)	1.496	1.488	**1.480**
		r (HSiH)	109.5	111.0	**110.6**
ClF	$C_{\infty v}$	r (ClF)	1.692	1.716	**1.628**
Na$_2$	$D_{\infty h}$	r (NaNa)	3.051	3.068	**3.078**
NaCl	$C_{\infty v}$	r (NaCl)	2.379	2.403	**2.361**
Si$_2$H$_6$	D_{3d}	r (SiSi)	2.365	2.371	**2.327**
		r (SiH)	1.499	1.495	**1.486**
		< (HSiH)	108.3	108.5	**107.8**
SiH$_3$Cl	C_{3v}	r (SiCl)	2.096	2.103	**2.048**
		r (SiH)	1.492	1.487	**1.481**
		< (HSiH)	110.2	110.7	**110.9**
P$_2$	$D_{\infty h}$	r (PP)	1.930	1.923	**1.893**
P$_2$H$_4$	C_2	r (PP)	2.285	2.293	**2.219**
		r (PH$_{int}$)	1.437	1.435	**1.417**
		r (PH$_{ext}$)	1.436	1.433	**1.414**
		< (PPH$_{int}$)	99.7	99.7	**99.1**
		< (PPH$_{ext}$)	94.5	94.1	**94.3**
		< (H$_{int}$PH$_{ext}$)	92.9	92.9	**92.0**
		ω (H$_{int}$PPH$_{ext}$)	82.6	78.6	**74.0**
H$_2$S$_2$	C_2	r (SS)	2.127	2.145	**2.055**
		r (SH)	1.367	1.364	**1.327**
		< (SSH)	98.9	98.2	**91.3**
		ω (HSSH)	90.8	91.0	**90.6**
Cl$_2$	$D_{\infty v}$	r (ClCl)	2.079	2.091	**1.988**

Table A5-15: Structures of Two-Heavy-Atom Hydrides. EDF1 Density Functional Models

molecule	point group	geometrical parameter	6-31G*	6-311+G**	expt.
Li$_2$	D$_{\infty h}$	r (LiLi)	2.760	2.747	**2.673**
LiOH	C$_{\infty v}$	r (LiO)	1.597	1.603	**1.582**
		r (OH)	0.962	0.955	–
LiF	C$_{\infty v}$	r (LiF)	1.567	1.599	**1.564**
LiCl	C$_{\infty v}$	r (LiCl)	2.065	2.033	**2.021**
B$_2$H$_6$	D$_{2h}$	r (BB)	1.768	1.764	**1.763**
		r (BH)	1.196	1.192	**1.201**
		r (BH$_{bridge}$)	1.321	1.320	**1.320**
		< (HBH)	121.9	121.8	**121.0**
C$_2$H$_2$	D$_{\infty h}$	r (CC)	1.213	1.206	**1.203**
		r (CH)	1.070	1.067	**1.061**
C$_2$H$_4$	D$_{2h}$	r (CC)	1.336	1.334	**1.339**
		r (CH)	1.091	1.089	**1.085**
		< (HCH)	116.2	116.4	**117.8**
C$_2$H$_6$	D$_{3d}$	r (CC)	1.531	1.530	**1.531**
		r (CH)	1.099	1.097	**1.096**
		< (HCH)	107.4	107.4	**107.8**
HCN	C$_{\infty v}$	r (CN)	1.166	1.159	**1.153**
		r (CH)	1.075	1.071	**1.065**
HNC	C$_{\infty v}$	r (NC)	1.185	1.178	**1.169**
		r (NH)	1.005	1.004	**0.994**
CH$_2$NH	C$_s$	r (CN)	1.277	1.274	**1.273**
		r (CH$_{syn}$)	1.104	1.101	**1.103**
		r (CH$_{anti}$)	1.098	1.096	**1.081**
		r (NH)	1.032	1.028	**1.023**
		< (H$_{syn}$CN)	126.1	125.3	**123.4**
		< (H$_{anti}$CN)	118.4	118.7	**119.7**
		< (HNC)	109.8	110.4	**110.5**
CH$_3$NH$_2$	C$_s$	r (CN)	1.466	1.466	**1.471**
		r (CH$_{tr}$)	1.109	1.105	**1.099**
		r (CH$_g$)	1.099	1.096	**1.099**
		r (NH)	1.023	1.018	**1.010**
		< (NCH$_{tr}$)	116.3	115.6	**113.9**
		< (NCH$_{gg'}$)	123.3	123.5	**124.4**
		< (H$_g$CH$_{g'}$)	107.0	107.0	**108.0**
		< (CNH$_{gg'}$)	122.7	126.2	**125.7**
		< (HNH)	105.2	106.5	**107.1**
CO	C$_{\infty v}$	r (CO)	1.147	1.137	**1.128**
H$_2$CO	C$_{2v}$	r (CO)	1.212	1.208	**1.208**
		r (CH)	1.118	1.115	**1.116**
		< (HCH)	114.9	115.6	**116.5**

Table A5-15: Structures of Two-Heavy-Atom Hydrides. EDF1 Density Functional Models (2)

molecule	point group	geometrical parameter	6-31G*	6-311+G**	expt.
CH$_3$OH	C$_s$	r (CO)	1.422	1.427	**1.421**
		r (CH$_{tr}$)	1.097	1.094	**1.094**
		r (CH$_g$)	1.106	1.101	**1.094**
		r (OH)	0.973	0.965	**0.963**
		< (OCH$_{tr}$)	106.6	106.6	**107.2**
		< (OCH$_{gg'}$)	131.8	130.8	**129.9**
		< (H$_g$CH$_{g'}$)	108.2	108.8	**108.5**
		< (COH)	107.0	108.1	**108.0**
CH$_3$F	C$_{3v}$	r (CF)	1.387	1.400	**1.383**
		r (CH)	1.101	1.096	**1.100**
		< (HCH)	109.1	110.2	**110.6**
CH$_3$SiH$_3$	C$_{3v}$	r (CSi)	1.893	1.889	**1.867**
		r (CH)	1.098	1.096	**1.093**
		r (SiH)	1.496	1.494	**1.485**
		< (HCH)	107.9	107.9	**107.7**
		< (HSiH)	108.0	108.2	**108.3**
HCP	C$_{\infty v}$	r (CP)	1.553	1.549	**1.540**
		r (CH)	1.080	1.078	**1.069**
CH$_2$PH	C$_s$	r (CP)	1.682	1.677	**1.67**
		r (CH$_{syn}$)	1.091	1.089	–
		r (CH$_{anti}$)	1.092	1.090	–
		r (PH)	1.439	1.438	–
		< (H$_{syn}$CP)	125.8	125.5	–
		< (H$_{anti}$CP)	119.2	119.1	–
		< (HPC)	97.2	97.2	–
CH$_3$PH$_2$	C$_s$	r (CP)	1.879	1.875	**1.862**
		r (CH$_{tr}$)	1.096	1.093	**1.094**
		r (CH$_g$)	1.098	1.095	**1.094**
		r (PH)	1.431	1.430	**1.432**
		< (PCH$_{tr}$)	113.9	114.0	**109.2**
		< (PCH$_{gg'}$)	123.0	122.9	–
		< (H$_g$CH$_{g'}$)	107.5	107.5	–
		< (CPH$_{gg'}$)	100.4	100.5	–
		< (HPH)	92.7	92.6	–
CS	C$_{\infty v}$	r (CS)	1.559	1.550	**1.535**
H$_2$CS	C$_{2v}$	r (CS)	1.622	1.619	**1.611**
		r (CH)	1.098	1.095	**1.093**
		< (HCH)	115.2	115.6	**116.9**

Table A5-15: Structures of Two-Heavy-Atom Hydrides. EDF1 Density Functional Models (3)

molecule	point group	geometrical parameter	6-31G*	6-311+G**	expt.
CH$_3$SH	C$_s$	r (CS)	1.835	1.834	**1.819**
		r (CH$_{tr}$)	1.096	1.093	**1.091**
		r (CH$_g$)	1.096	1.093	**1.091**
		r (SH)	1.354	1.352	**1.336**
		< (SCH$_{tr}$)	106.3	106.3	–
		< (SCH$_{gg'}$)	130.1	129.9	–
		< (H$_g$CH$_{g'}$)	110.0	110.3	**109.8**
		< (CSH)	96.8	96.7	**96.5**
CH$_3$Cl	C$_{3v}$	r (CCl)	1.800	1.801	**1.781**
		r (CH)	1.094	1.091	**1.096**
		< (HCH)	110.2	110.4	**110.0**
N$_2$	D$_{\infty h}$	r (NN)	1.114	1.105	**1.098**
N$_2$H$_2$	C$_{2h}$	r (NN)	1.254	1.247	**1.252**
		r (NH)	1.047	1.041	**1.028**
		< (NNH)	105.5	106.3	**106.9**
N$_2$H$_4$	C$_2$	r (NN)	1.440	1.434	**1.449**
		r (NH$_{int}$)	1.027	1.021	**1.021**
		r (NH$_{ext}$)	1.022	1.017	**1.021**
		< (NNH$_{int}$)	111.7	112.7	**106.0**
		< (NNH$_{ext}$)	106.1	107.7	**112.0**
		< (H$_{int}$NH$_{ext}$)	106.0	107.8	–
		ω (H$_{int}$NNH$_{ext}$)	90.0	90.8	**91.0**
HNO	C$_s$	r (NO)	1.215	1.207	**1.212**
		r (NH)	1.086	1.078	**1.063**
		< (ONH)	108.4	108.8	**108.6**
NH$_2$OH	C$_s$	r (NO)	1.457	1.454	**1.453**
		r (NH)	1.028	1.024	**1.016**
		r (OH)	0.975	0.967	**0.962**
		< (ONH$_{gg'}$)	110.8	112.9	**112.6**
		< (HNH)	104.0	105.2	**107.1**
		< (NOH)	100.9	101.9	**101.4**
NP	C$_{\infty v}$	r (NP)	1.508	1.502	**1.491**
H$_2$O$_2$	C$_2$	r (OO)	1.468	1.468	**1.452**
		r (OH)	0.979	0.972	**0.965**
		< (OOH)	99.1	99.8	**100.0**
		ω (HOOH)	117.7	120.3	**119.1**
HOF	C$_s$	r (OF)	1.447	1.452	**1.442**
		r (OH)	0.983	0.977	**0.966**
		< (HOF)	97.3	98.1	**96.8**
NaOH	C$_{\infty v}$	r (NaO)	1.933	1.971	**1.95**
		r (OH)	0.965	0.958	**0.96**

Table A5-15: Structures of Two-Heavy-Atom Hydrides. EDF1 Density Functional Models (4)

molecule	point group	geometrical parameter	6-31G*	6-311+G**	expt.
MgO	$C_{\infty v}$	r (MgO)	1.756	1.817	**1.749**
SiO	$C_{\infty v}$	r (SiO)	1.538	1.536	**1.510**
HPO	C_s	r (PO)	1.513	1.509	**1.512**
		r (PH)	1.483	1.481	–
		< (HPO)	105.5	104.2	**104.7**
HOCl	C_s	r (OCl)	1.737	1.742	**1.690**
		r (OH)	0.981	0.973	**0.975**
		< (HOCl)	101.8	102.3	**102.5**
F_2	$D_{\infty h}$	r (FF)	1.417	1.427	**1.412**
NaF	$C_{\infty v}$	r (NaF)	1.916	1.978	**1.926**
SiH_3F	C_{2v}	r (SiF)	1.623	1.642	**1.596**
		r (SiH)	1.493	1.487	**1.480**
		r (HSiH)	109.7	110.9	**110.6**
ClF	$C_{\infty v}$	r (ClF)	1.673	1.691	**1.628**
Na_2	$D_{\infty h}$	r (NaNa)	3.121	3.139	**3.078**
NaCl	$C_{\infty v}$	r (NaCl)	2.396	2.403	**2.361**
Si_2H_6	D_{3d}	r (SiSi)	2.353	2.359	**2.327**
		r (SiH)	1.496	1.494	**1.486**
		< (HSiH)	108.2	108.5	**107.8**
SiH_3Cl	C_{3v}	r (SiCl)	2.080	2.084	**2.048**
		r (SiH)	1.489	1.486	**1.481**
		< (HSiH)	110.2	110.6	**110.9**
P_2	$D_{\infty h}$	r (PP)	1.915	1.909	**1.893**
P_2H_4	C_2	r (PP)	2.253	2.260	**2.219**
		r (PH_{int})	1.431	1.430	**1.417**
		r (PH_{ext})	1.429	1.427	**1.414**
		< (PPH_{int})	100.0	100.1	**99.1**
		< (PPH_{ext})	94.6	94.3	**94.3**
		< ($H_{int}PH_{ext}$)	92.9	92.9	**92.0**
		ω ($H_{int}PPH_{ext}$)	78.6	78.1	**74.0**
H_2S_2	C_2	r (SS)	2.092	2.107	**2.055**
		r (SH)	1.360	1.358	**1.327**
		< (SSH)	99.3	98.6	**91.3**
		ω (HSSH)	90.7	91.0	**90.6**
Cl_2	$D_{\infty v}$	r (ClCl)	2.043	2.053	**1.988**

Table A5-16: Structures of Two-Heavy-Atom Hydrides. B3LYP Density Functional Models

molecule	point group	geometrical parameter	6-31G*	6-311+G**	expt.
Li_2	$D_{\infty h}$	r (LiLi)	2.725	2.705	**2.673**
LiOH	$C_{\infty v}$	r (LiO)	1.584	1.592	**1.582**
		r (OH)	0.958	0.952	–
LiF	$C_{\infty v}$	r (LiF)	1.550	1.584	**1.564**
LiCl	$C_{\infty v}$	r (LiCl)	2.055	2.025	**2.021**
B_2H_6	D_{2h}	r (BB)	1.769	1.765	**1.763**
		r (BH)	1.191	1.187	**1.201**
		r (BH$_{bridge}$)	1.317	1.316	**1.320**
		< (HBH)	121.9	121.8	**121.0**
C_2H_2	$D_{\infty h}$	r (CC)	1.205	1.199	**1.203**
		r (CH)	1.067	1.063	**1.061**
C_2H_4	D_{2h}	r (CC)	1.331	1.328	**1.339**
		r (CH)	1.088	1.085	**1.085**
		< (HCH)	116.3	116.5	**117.8**
C_2H_6	D_{3d}	r (CC)	1.531	1.530	**1.531**
		r (CH)	1.096	1.094	**1.096**
		< (HCH)	107.5	107.5	**107.8**
HCN	$C_{\infty v}$	r (CN)	1.157	1.149	**1.153**
		r (CH)	1.071	1.067	**1.065**
HNC	$C_{\infty v}$	r (NC)	1.177	1.169	**1.169**
		r (NH)	1.001	1.000	**0.994**
CH_2NH	C_s	r (CN)	1.271	1.267	**1.273**
		r (CH$_{syn}$)	1.099	1.097	**1.103**
		r (CH$_{anti}$)	1.094	1.091	**1.081**
		r (NH)	1.027	1.023	**1.023**
		< (H$_{syn}$CN)	125.6	124.9	**123.4**
		< (H$_{anti}$CN)	118.7	118.9	**119.7**
		< (HNC)	110.3	111.0	**110.5**
CH_3NH_2	C_s	r (CN)	1.465	1.465	**1.471**
		r (CH$_{tr}$)	1.105	1.100	**1.099**
		r (CH$_g$)	1.096	1.093	**1.099**
		r (NH)	1.019	1.014	**1.010**
		< (NCH$_{tr}$)	115.8	115.2	**113.9**
		< (NCH$_{gg'}$)	123.4	123.6	**124.4**
		<(H$_g$CH$_{g'}$)	107.2	107.1	**108.0**
		< (CNH$_{gg'}$)	123.8	127.6	**125.7**
		< (HNH)	105.8	107.3	**107.1**
CO	$C_{\infty v}$	r (CO)	1.138	1.128	**1.128**
H_2CO	C_{2v}	r (CO)	1.207	1.202	**1.208**
		r (CH)	1.111	1.109	**1.116**
		< (HCH)	115.2	116.0	**116.5**

Table A5-16: Structures of Two-Heavy-Atom Hydrides. B3LYP Density Functional Models (2)

molecule	point group	geometrical parameter	6-31G*	6-311+G**	expt.
CH₃OH	C_s	r (CO)	1.419	1.424	**1.421**
		r (CH_tr)	1.093	1.090	**1.094**
		r (CH_g)	1.101	1.097	**1.094**
		r (OH)	0.969	0.961	**0.963**
		< (OCH_tr)	106.7	106.7	**107.2**
		< (OCH_gg')	131.3	130.4	**129.9**
		< (H_gCH_g')	108.4	109.0	**108.5**
		< (COH)	107.6	108.8	**108.0**
CH₃F	C_3v	r (CF)	1.383	1.395	**1.383**
		r (CH)	1.097	1.092	**1.100**
		< (HCH)	109.3	110.3	**110.6**
CH₃SiH₃	C_3v	r (CSi)	1.889	1.580	**1.867**
		r (CH)	1.095	1.392	**1.093**
		r (SiH)	1.490	1.784	**1.485**
		< (HCH)	107.9	107.9	**107.7**
		< (HSiH)	108.0	108.3	**108.3**
HCP	C_∞v	r (CP)	1.543	1.540	**1.540**
		r (CH)	1.075	1.073	**1.069**
CH₂PH	C_s	r (CP)	1.674	1.671	**1.67**
		r (CH_syn)	1.087	1.085	–
		r (CH_anti)	1.088	1.086	–
		r (PH)	1.433	1.431	–
		< (H_synCP)	125.6	125.2	–
		< (H_antiCP)	119.2	119.2	–
		< (HPC)	97.7	97.6	–
CH₃PH₂	C_s	r (CP)	1.876	1.873	**1.862**
		r (CH_tr)	1.093	1.090	**1.094**
		r (CH_g)	1.095	1.092	**1.094**
		r (PH)	1.426	1.424	**1.432**
		< (PCH_tr)	113.8	113.7	**109.2**
		< (PCH_gg')	123.1	123.0	–
		< (H_gCH_g')	107.5	108.9	–
		< (CPH_gg')	101.1	101.0	–
		< (HPH)	93.3	93.4	–
CS	C_∞v	r (CS)	1.548	1.539	**1.535**
H₂CS	C_2v	r (CS)	1.618	1.615	**1.611**
		r (CH)	1.093	1.090	**1.093**
		< (HCH)	115.3	115.8	**116.9**

Table A5-16: Structures of Two-Heavy-Atom Hydrides. B3LYP Density Functional Models (3)

molecule	point group	geometrical parameter	6-31G*	6-311+G**	expt.
CH₃SH	C$_s$	r (CS)	1.836	1.836	**1.819**
		r (CH$_{tr}$)	1.093	1.090	**1.091**
		r (CH$_g$)	1.092	1.090	**1.091**
		r (SH)	1.349	1.348	**1.336**
		< (SCH$_{tr}$)	106.1	106.2	–
		< (SCH$_{gg'}$)	129.8	129.5	–
		< (H$_g$CH$_{g'}$)	110.2	110.3	**109.8**
		< (CSH)	97.0	96.9	**96.5**
CH₃Cl	C$_{3v}$	r (CCl)	1.804	1.806	**1.781**
		r (CH)	1.090	1.087	**1.096**
		< (HCH)	110.4	110.6	**110.0**
N₂	D$_{\infty h}$	r (NN)	1.106	1.095	**1.098**
N₂H₂	C$_{2h}$	r (NN)	1.246	1.239	**1.252**
		r (NH)	1.040	1.035	**1.028**
		< (NNH)	106.1	106.9	**106.9**
N₂H₄	C$_2$	r (NN)	1.437	1.432	**1.449**
		r (NH$_{int}$)	1.022	1.017	**1.021**
		r (NH$_{ext}$)	1.017	1.013	**1.021**
		< (NNH$_{int}$)	111.8	112.9	**106.0**
		< (NNH$_{ext}$)	106.7	108.2	**112.0**
		< (H$_{int}$NH$_{ext}$)	106.8	108.6	–
		ω (H$_{int}$NNH$_{ext}$)	90.7	91.2	**91.0**
HNO	C$_s$	r (NO)	1.208	1.200	**1.212**
		r (NH)	1.071	1.064	**1.063**
		< (ONH)	108.4	108.9	**108.6**
NH₂OH	C$_s$	r (NO)	1.432	1.444	**1.453**
		r (NH)	1.023	1.019	**1.016**
		r (OH)	0.975	0.963	**0.962**
		< (ONH$_{gg'}$)	119.6	114.3	**112.6**
		< (HNH)	106.9	106.2	**107.1**
		< (NOH)	107.9	102.7	**101.4**
NP	C$_{\infty v}$	r (NP)	1.495	1.489	**1.491**
H₂O₂	C$_2$	r (OO)	1.456	1.454	**1.452**
		r (OH)	0.973	0.967	**0.965**
		< (OOH)	99.7	100.5	**100.0**
		ω (HOOH)	118.6	121.1	**119.1**
HOF	C$_s$	r (OF)	1.434	1.436	**1.442**
		r (OH)	0.977	0.972	**0.966**
		< (HOF)	97.8	98.7	**96.8**
NaOH	C$_{\infty v}$	r (NaO)	1.916	1.955	**1.95**
		r (OH)	0.961	0.954	**0.96**

Table A5-16: Structures of Two-Heavy-Atom Hydrides. B3LYP Density Functional Models (4)

molecule	point group	geometrical parameter	6-31G*	6-311+G**	expt.
MgO	$C_{\infty v}$	r (MgO)	1.743	1.769	**1.749**
SiO	$C_{\infty v}$	r (SiO)	1.524	1.523	**1.510**
HPO	C_s	r (PO)	1.500	1.497	**1.512**
		r (PH)	1.471	1.469	–
		< (HPO)	105.4	104.1	**104.7**
HOCl	C_s	r (OCl)	1.727	1.732	**1.690**
		r (OH)	0.976	0.968	**0.975**
		< (HOCl)	102.5	103.2	**102.5**
F_2	$D_{\infty h}$	r (FF)	1.403	1.408	**1.412**
NaF	$C_{\infty v}$	r (NaF)	1.890	1.957	**1.926**
SiH_3F	C_{3v}	r (SiF)	1.613	1.631	**1.596**
		r (SiH)	1.485	1.479	**1.480**
		r (HSiH)	109.7	111.0	**110.6**
ClF	$C_{\infty v}$	r (ClF)	1.662	1.679	**1.628**
Na_2	$D_{\infty h}$	r (NaNa)	3.039	3.053	**3.078**
NaCl	$C_{\infty v}$	r (NaCl)	2.368	2.387	**2.361**
Si_2H_6	D_{3d}	r (SiSi)	2.351	2.356	**2.327**
		r (SiH)	1.491	1.487	**1.486**
		< (HSiH)	108.3	108.6	**107.8**
SiH_3Cl	C_{3v}	r (SiCl)	2.079	2.084	**2.048**
		r (SiH)	1.482	1.479	**1.481**
		< (HSiH)	110.3	110.7	**110.9**
P_2	$D_{\infty h}$	r (PP)	1.904	1.898	**1.893**
P_2H_4	C_2	r (PP)	2.251	2.258	**2.219**
		r (PH_{int})	1.424	1.423	**1.417**
		r (PH_{ext})	1.422	1.422	**1.414**
		< (PPH_{int})	100.2	100.2	**99.1**
		< (PPH_{ext})	95.1	94.7	**94.3**
		< ($H_{int}PH_{ext}$)	93.6	93.7	**92.0**
		ω ($H_{int}PPH_{ext}$)	80.4	78.3	**74.0**
H_2S_2	C_2	r (SS)	2.098	2.114	**2.055**
		r (SH)	1.354	1.352	**1.327**
		< (SSH)	98.9	98.3	**91.3**
		ω (HSSH)	90.7	91.2	**90.6**
Cl_2	$D_{\infty h}$	r (ClCl)	2.042	2.054	**1.988**

Table A5-17: Structures of Two-Heavy-Atom Hydrides. MP2 Models

molecule	point group	geometrical parameter	6-31G*	6-311+G**	expt.
Li$_2$	D$_{\infty h}$	r (LiLi)	2.782	2.748	**2.673**
LiOH	C$_{\infty v}$	r (LiO)	1.594	1.606	**1.582**
		r (OH)	0.960	0.952	–
LiF	C$_{\infty v}$	r (LiF)	1.570	1.599	**1.564**
LiCl	C$_{\infty v}$	r (LiCl)	2.069	2.022	**2.021**
B$_2$H$_6$	D$_{2h}$	r (BB)	1.754	1.768	**1.763**
		r (BH)	1.190	1.188	**1.201**
		r (BH$_{bridge}$)	1.311	1.316	**1.320**
		< (HBH)	121.7	122.3	**121.0**
C$_2$H$_2$	D$_{\infty h}$	r (CC)	1.218	1.216	**1.203**
		r (CH)	1.066	1.065	**1.061**
C$_2$H$_4$	D$_{2h}$	r (CC)	1.337	1.339	**1.339**
		r (CH)	1.085	1.086	**1.085**
		< (HCH)	116.6	117.2	**117.8**
C$_2$H$_6$	D$_{3d}$	r (CC)	1.527	1.529	**1.531**
		r (CH)	1.094	1.093	**1.096**
		< (HCH)	107.7	107.8	**107.8**
HCN	C$_{\infty v}$	r (CN)	1.177	1.171	**1.153**
		r (CH)	1.070	1.068	**1.065**
HNC	C$_{\infty v}$	r (NC)	1.187	1.182	**1.169**
		r (NH)	1.002	1.001	**0.994**
CH$_2$NH	C$_s$	r (CN)	1.282	1.281	**1.273**
		r (CH$_{syn}$)	1.096	1.095	**1.103**
		r (CH$_{anti}$)	1.090	1.090	**1.081**
		r (NH)	1.027	1.023	**1.023**
		< (H$_{syn}$CN)	125.4	124.6	**123.4**
		< (H$_{anti}$CN)	116.1	118.6	**119.7**
		< (HNC)	109.7	109.1	**110.5**
CH$_3$NH$_2$	C$_s$	r (CN)	1.465	1.465	**1.471**
		r (CH$_{tr}$)	1.100	1.099	**1.099**
		r (CH$_g$)	1.092	1.092	**1.099**
		r (NH)	1.018	1.014	**1.010**
		< (NCH$_{tr}$)	115.4	114.9	**113.9**
		< (NCH$_{gg'}$)	123.7	123.3	**124.4**
		< (H$_g$CH$_{g'}$)	107.5	107.5	**108.0**
		< (CNH$_{gg'}$)	123.6	125.7	**125.7**
		< (HNH)	105.9	106.7	**107.1**
CO	C$_{\infty v}$	r (CO)	1.151	1.140	**1.128**
H$_2$CO	C$_{2v}$	r (CO)	1.221	1.213	**1.208**
		r (CH)	1.104	1.105	**1.116**
		< (HCH)	115.6	116.1	**116.5**

Table A5-17: Structures of Two-Heavy-Atom Hydrides. MP2 Models (2)

molecule	point group	geometrical parameter	6-31G*	6-311+G**	expt.
CH_3OH	C_s	r (CO)	1.424	1.422	**1.421**
		r (CH_{tr})	1.090	1.090	**1.094**
		r (CH_g)	1.097	1.096	**1.094**
		r (OH)	0.970	0.959	**0.963**
		< (OCH_{tr})	106.4	106.7	**107.2**
		< ($OCH_{gg'}$)	130.7	130.1	**129.9**
		< ($H_gCH_{g'}$)	108.7	109.0	**108.5**
		< (COH)	107.4	107.4	**108.0**
CH_3F	C_{3v}	r (CF)	1.392	1.389	**1.383**
		r (CH)	1.092	1.091	**1.100**
		< (HCH)	109.8	110.2	**110.6**
CH_3SiH_3	C_{3v}	r (CSi)	1.884	1.877	**1.867**
		r (CH)	1.093	1.093	**1.093**
		r (SiH)	1.487	1.478	**1.485**
		< (HCH)	107.9	107.9	**107.7**
		< (HSiH)	108.3	108.5	**108.3**
HCP	$C_{\infty v}$	r (CP)	1.562	1.559	**1.540**
		r (CH)	1.076	1.076	**1.069**
CH_2PH	C_s	r (CP)	1.652	1.673	**1.67**
		r (CH_{syn})	1.075	1.086	–
		r (CH_{anti})	1.076	1.087	–
		r (PH)	1.409	1.416	–
		< ($H_{syn}CP$)	125.0	124.7	–
		< ($H_{anti}CP$)	119.6	119.1	–
		< (HPC)	98.9	96.9	–
CH_3PH_2	C_s	r (CP)	1.860	1.856	**1.862**
		r (CH_{tr})	1.091	1.091	**1.094**
		r (CH_g)	1.092	1.092	**1.094**
		r (PH)	1.417	1.411	**1.432**
		< (PCH_{tr})	113.7	113.8	**109.2**
		< ($PCH_{gg'}$)	123.4	123.4	–
		< ($H_gCH_{g'}$)	107.5	107.5	–
		< ($CPH_{gg'}$)	101.8	101.1	–
		< (HPH)	94.7	94.4	–
CS	$C_{\infty v}$	r (CS)	1.546	1.540	**1.535**
H_2CS	C_{2v}	r (CS)	1.617	1.613	**1.611**
		r (CH)	1.090	1.091	**1.093**
		< (HCH)	116.0	116.1	**116.9**

Table A5-17: Structures of Two-Heavy-Atom Hydrides. MP2 Models (3)

molecule	point group	geometrical parameter	6-31G*	6-311+G**	expt.
CH$_3$SH	C$_s$	r (CS)	1.817	1.813	**1.819**
		r (CH$_{tr}$)	1.091	1.091	**1.091**
		r (CH$_g$)	1.090	1.090	**1.091**
		r (SH)	1.341	1.334	**1.336**
		< (SCH$_{tr}$)	106.7	106.6	–
		< (SCH$_{gg'}$)	129.8	130.1	–
		< (H$_g$CH$_{g'}$)	109.9	109.9	**109.8**
		< (CSH)	96.8	96.3	**96.5**
CH$_3$Cl	C$_{3v}$	r (CCl)	1.778	1.776	**1.781**
		r (CH)	1.088	1.088	**1.096**
		< (HCH)	110.1	109.9	**110.0**
N$_2$	D$_{∞h}$	r (NN)	1.131	1.120	**1.098**
N$_2$H$_2$	C$_{2h}$	r (NN)	1.267	1.258	**1.252**
		r (NH)	1.036	1.032	**1.028**
		< (NNH)	105.4	105.7	**106.9**
N$_2$H$_4$	C$_2$	r (NN)	1.439	1.430	**1.449**
		r (NH$_{int}$)	1.016	1.016	**1.021**
		r (NH$_{ext}$)	1.021	1.012	**1.021**
		< (NNH$_{int}$)	106.3	112.2	**106.0**
		< (NNH$_{ext}$)	111.5	107.5	**112.0**
		< (H$_{int}$NH$_{ext}$)	107.0	108.0	–
		ω (H$_{int}$NNH$_{ext}$)	90.5	90.4	**91.0**
HNO	C$_s$	r (NO)	1.237	1.221	**1.212**
		r (NH)	1.058	1.054	**1.063**
		< (ONH)	107.3	107.9	**108.6**
NH$_2$OH	C$_s$	r (NO)	1.453	1.436	**1.453**
		r (NH)	1.021	1.017	**1.016**
		r (OH)	0.971	0.960	**0.962**
		< (ONH$_{gg'}$)	111.4	114.2	**112.6**
		< (HNH)	105.2	105.9	**107.1**
		< (NOH)	101.2	101.7	**101.4**
NP	C$_{∞v}$	r (NP)	1.537	1.527	**1.491**
H$_2$O$_2$	C$_2$	r (OO)	1.467	1.450	**1.452**
		r (OH)	0.976	0.965	**0.965**
		< (OOH)	98.7	99.6	**100.0**
		ω (HOOH)	121.3	121.2	**119.1**
HOF	C$_s$	r (OF)	1.444	1.431	**1.442**
		r (OH)	0.979	0.968	**0.966**
		< (HOF)	97.2	98.2	**96.8**
NaOH	C$_{∞v}$	r (NaO)	1.934	1.979	**1.95**
		r (OH)	0.962	0.954	**0.96**

Table A5-17: Structures of Two-Heavy-Atom Hydrides. MP2 Models (4)

molecule	point group	geometrical parameter	6-31G*	6-311+G**	expt.
MgO	$C_{\infty v}$	r (MgO)	1.733	1.758	**1.749**
SiO	$C_{\infty v}$	r (SiO)	1.544	1.536	**1.510**
HPO	C_s	r (PO)	1.519	1.509	**1.512**
		r (PH)	1.453	1.447	–
		< (HPO)	105.6	104.3	**104.7**
HOCl	C_s	r (OCl)	1.717	1.715	**1.690**
		r (OH)	0.978	0.966	**0.975**
		< (HOCl)	102.6	102.8	**102.5**
F_2	$D_{\infty h}$	r (FF)	1.421	1.417	**1.412**
NaF	$C_{\infty v}$	r (NaF)	1.920	1.991	**1.926**
SiH_3F	C_{3v}	r (SiF)	1.619	1.625	**1.596**
		r (SiH)	1.481	1.469	**1.480**
		r (HSiH)	110.0	110.9	**110.6**
ClF	$C_{\infty v}$	r (ClF)	1.661	1.673	**1.628**
Na_2	$D_{\infty h}$	r (NaNa)	3.170	3.168	**3.078**
NaCl	$C_{\infty v}$	r (NaCl)	2.393	2.382	**2.361**
Si_2H_6	D_{3d}	r (SiSi)	2.338	2.342	**2.327**
		r (SiH)	1.487	1.478	**1.486**
		< (HSiH)	108.6	108.7	107.8
SiH_3Cl	C_{3v}	r (SiCl)	2.060	2.058	**2.048**
		r (SiH)	1.479	1.469	**1.481**
		< (HSiH)	110.3	110.7	**110.9**
P_2	$D_{\infty h}$	r (PP)	1.936	1.925	**1.893**
P_2H_4	C_2	r (PP)	2.212	2.219	**2.219**
		r (PH$_{int}$)	1.416	1.409	**1.417**
		r (PH$_{ext}$)	1.414	1.407	**1.414**
		< (PPH$_{int}$)	101.5	101.0	**99.1**
		< (PPH$_{ext}$)	96.2	95.4	**94.3**
		< (H$_{int}$PH$_{ext}$)	95.2	94.7	**92.0**
		ω (H$_{int}$PPH$_{ext}$)	77.8	75.7	**74.0**
H_2S_2	C_2	r (SS)	2.069	2.083	**2.055**
		r (SH)	1.344	1.337	**1.327**
		< (SSH)	99.0	98.1	**91.3**
		ω (HSSH)	90.3	90.8	**90.6**
Cl_2	$D_{\infty h}$	r (ClCl)	2.015	2.025	**1.988**

Table A5-18: Structures of Two-Heavy-Atom Hydrides. Semi-Empirical Models

molecule	point group	geometrical parameter	MNDO	AM1	PM3	expt
Li_2	$D_{\infty h}$	r (LiLi)	2.054	-	2.482	**2.673**
LiOH	$C_{\infty v}$	r (LiO)	1.631	-	1.576	**1.582**
		r (OH)	0.921	-	0.933	**–**
LiF	$C_{\infty v}$	r (LiF)	1.614	-	1.586	**1.564**
LiCl	$C_{\infty v}$	r (LiCl)	2.191	-	1.884	**2.021**
B_2H_6	D_{2h}	r (BB)	1.754	1.752	1.773	**1.763**
		r (BH)	1.164	1.192	1.206	**1.201**
		r (BH_{bridge})	1.350	1.329	1.375	**1.320**
		< (HBH)	121.1	123.9	122.0	**121.0**
C_2H_2	$D_{\infty h}$	r (CC)	1.194	1.195	1.190	**1.203**
		r (CH)	1.051	1.061	1.064	**1.061**
C_2H_4	D_{2h}	r (CC)	1.335	1.326	1.322	**1.339**
		r (CH)	1.089	1.098	1.086	**1.085**
		< (HCH)	113.6	114.6	113.8	**117.8**
C_2H_6	D_{3d}	r (CC)	1.521	1.500	1.504	**1.531**
		r (CH)	1.109	1.117	1.098	**1.096**
		< (HCH)	107.7	108.2	107.2	**107.8**
HCN	$C_{\infty v}$	r (CN)	1.160	1.160	1.156	**1.153**
		r (CH)	1.055	1.069	1.071	**1.065**
HNC	$C_{\infty v}$	r (NC)	1.185	1.178	1.178	**1.169**
		r (NH)	0.975	0.967	0.976	**0.994**
CH_2NH	C_s	r (CN)	1.282	1.270	1.276	**1.273**
		r (CH_{syn})	1.098	1.108	1.091	**1.103**
		r (CH_{anti})	1.096	1.107	1.089	**1.081**
		r (NH)	1.006	0.999	0.988	**1.023**
		< ($H_{syn}CN$)	127.0	127.2	126.5	**123.4**
		< ($H_{anti}CN$)	119.2	118.2	117.9	**119.7**
		< (HNC)	114.4	116.6	117.3	**110.5**
CH_3NH_2	C_s	r (CN)	1.461	1.432	1.469	**1.471**
		r (CH_{tr})	1.117	1.126	1.101	**1.099**
		r (CH_g)	1.113	1.122	1.098	**1.099**
		r (NH)	1.008	1.000	0.999	**1.010**
		< (NCH_{tr})	114.1	114.3	114.8	**113.9**
		< ($NCH_{gg'}$)	124.8	123.9	124.0	**124.4**
		<($H_gCH_{g'}$)	108.3	108.5	107.8	**108.0**
		< ($CNH_{gg'}$)	124.0	128.8	125.4	**125.7**
		< (HNH)	105.5	109.0	108.7	**107.1**
CO	$C_{\infty v}$	r (CO)	1.163	1.171	1.135	**1.128**
H_2CO	C_{2v}	r (CO)	1.217	1.227	1.202	**1.208**
		r (CH)	1.106	1.111	1.091	**1.116**
		< (HCH)	113.0	115.5	116.4	**116.5**

Table A5-18: Structures of Two-Heavy-Atom Hydrides. Semi-Empirical Models (2)

molecule	point group	geometrical parameter	MNDO	AM1	PM3	expt
CH_3OH	C_s	r (CO)	1.391	1.411	1.395	**1.421**
		r (CH_{tr})	1.115	1.119	1.097	**1.094**
		r (CH_g)	1.119	1.119	1.094	**1.094**
		r (OH)	0.947	0.964	0.949	**0.963**
		< (OCH_{tr})	108.1	105.1	104.5	**107.2**
		< $(OCH_{gg'})$	129.9	128.3	130.5	**129.9**
		< $(H_gCH_{g'})$	107.3	109.7	109.0	**108.5**
		< (COH)	111.6	107.2	107.5	**108.0**
CH_3F	C_{3v}	r (CF)	1.347	1.375	1.351	**1.383**
		r (CH)	1.118	1.121	1.092	**1.100**
		< (HCH)	108.3	109.4	110.3	**110.6**
CH_3SiH_3	C_{3v}	r (CSi)	1.873	1.807	1.863	**1.867**
		r (CH)	1.116	1.115	1.095	**1.093**
		r (SiH)	1.414	1.463	1.493	**1.485**
		< (HCH)	107.6	107.1	107.2	**107.7**
		< (HSiH)	108.8	108.1	108.3	**108.3**
HCP	$C_{\infty v}$	r (CP)	1.428	1.410	1.409	**1.540**
		r (CH)	1.057	1.063	1.069	**1.069**
CH_2PH	C_s	r (CP)	1.566	1.536	1.570	**1.67**
		r (CH_{syn})	1.087	1.093	1.084	–
		r (CH_{anti})	1.089	1.096	1.087	–
		r (PH)	1.341	1.365	1.352	–
		< $(H_{syn}CP)$	127.1	125.9	128.7	–
		< $(H_{anti}CP)$	120.0	119.0	120.9	–
		< (HPC)	102.8	103.5	105.2	–
CH_3PH_2	C_s	r (CP)	1.749	1.726	1.866	**1.862**
		r (CH_{tr})	1.105	1.109	1.094	**1.094**
		r (CH_g)	1.105	1.111	1.091	**1.094**
		r (PH)	1.343	1.364	1.336	**1.432**
		< (PCH_{tr})	113.9	112.1	113.7	**109.2**
		< $(PCH_{gg'})$	123.4	121.7	123.7	–
		< $(H_gCH_{g'})$	108.6	109.7	108.3	–
		< $(CPH_{gg'})$	106.1	105.0	105.3	–
		< (HPH)	96.1	96.2	97.0	–
CS	$C_{\infty v}$	r (CS)	1.484	1.429	1.447	**1.535**
H_2CS	C_{2v}	r (CS)	1.537	1.512	1.539	**1.611**
		r (CH)	1.093	1.106	1.095	**1.093**
		< (HCH)	112.0	109.3	107.9	**116.9**

Table A5-18: Structures of Two-Heavy-Atom Hydrides. Semi-Empirical Models (3)

molecule	point group	geometrical parameter	MNDO	AM1	PM3	expt
CH₃SH	C_s	r (CS)	1.718	1.754	1.801	**1.819**
		r (CH_tr)	1.107	1.115	1.097	**1.091**
		r (CH_g)	1.107	1.112	1.095	**1.091**
		r (SH)	1.302	1.321	1.306	**1.336**
		< (SCH_tr)	107.9	105.9	107.3	–
		< (SCH_gg')	129.6	130.2	132.0	–
		< (H_gCH_g')	107.9	108.8	107.5	**109.8**
		< (CSH)	102.3	99.6	100.0	**96.5**
CH₃Cl	C_3v	r (CCl)	1.780	1.741	1.764	**1.781**
		r (CH)	1.109	1.111	1.094	**1.096**
		< (HCH)	110.4	110.7	109.0	**110.0**
N₂	D_∞h	r (NN)	1.104	1.106	1.098	**1.098**
N₂H₂	C_2h	r (NN)	1.220	1.212	1.219	**1.252**
		r (NH)	1.025	1.018	0.998	**1.028**
		< (NNH)	111.3	112.4	113.6	**106.9**
N₂H₄	C_2	r (NN)	1.397	1.378	1.440	**1.449**
		r (NH_int)	1.021	1.014	1.001	**1.021**
		r (NH_ext)	1.021	1.014	1.001	**1.021**
		< (NNH_int)	107.2	107.5	106.5	**106.0**
		< (NNH_ext)	107.2	107.5	106.5	**112.0**
		< (H_intNH_ext)	103.1	106.0	108.1	–
		ω (H_intNNH_ext)	180.0	180.0	180.0	**91.0**
HNO	C_s	r (NO)	1.161	1.157	1.176	**1.212**
		r (NH)	1.048	1.043	.0997	**1.063**
		< (ONH)	113.8	115.5	116.2	**108.6**
NH₂OH	C_s	r (NO)	1.320	1.324	1.418	**1.453**
		r (NH)	1.029	1.017	0.994	**1.016**
		r (OH)	0.962	0.980	0.951	**0.962**
		< (ONH_gg')	124.4	126.9	122.9	**112.6**
		< (HNH)	102.8	106.6	110.5	**107.1**
		< (NOH)	114.0	111.1	106.5	**101.4**
NP	C_∞v	r (NP)	1.398	1.382	1.414	**1.491**
H₂O₂	C_2	r (OO)	1.295	1.300	1.482	**1.452**
		r (OH)	0.961	0.983	0.945	**0.965**
		< (OOH)	107.2	106.0	96.5	**100.0**
		ω (HOOH)	159.3	127.2	180.0	**119.1**
HOF	C_s	r (OF)	1.277	1.367	1.396	**1.442**
		r (OH)	0.964	0.972	0.946	**0.966**
		< (HOF)	107.8	103.9	98.4	**96.8**

Table A5-18: Structures of Two-Heavy-Atom Hydrides. Semi-Empirical Models (4)

molecule	point group	geometrical parameter	MNDO	AM1	PM3	expt
MgO	$C_{\infty v}$	r (MgO)	-	-	1.834	**1.749**
SiO	$C_{\infty v}$	r (SiO)	1.518	1.571	1.466	**1.510**
HPO	C_s	r (PO)	1.445	1.461	1.485	**1.512**
		r (PH)	1.361	1.365	1.365	**–**
		< (HPO)	104.7	104.4	106.7	**104.7**
HOCl	C_s	r (OCl)	1.627	1.733	1.708	**1.690**
		r (OH)	0.949	0.961	0.947	**0.975**
		< (HOCl)	110.0	101.1	104.6	**102.5**
F_2	$D_{\infty h}$	r (FF)	1.266	1.427	1.350	**1.412**
SiH_3F	C_{3v}	r (SiF)	1.688	1.621	1.603	**1.596**
		r (SiH)	1.409	1.458	1.499	**1.480**
		r (HSiH)	110.0	108.2	108.6	**110.6**
ClF	$C_{\infty v}$	r (ClF)	1.600	1.647	1.582	**1.628**
Si_2H_6	D_{3d}	r (SiSi)	2.241	2.417	2.396	**2.327**
		r (SiH)	1.421	1.466	1.488	**1.486**
		< (HSiH)	109.7	109.3	109.2	**107.8**
SiH_3Cl	C_{3v}	r (SiCl)	2.069	2.071	2.091	**2.048**
		r (SiH)	1.412	1.462	1.488	**1.481**
		< (HSiH)	110.1	108.9	110.3	**110.9**
P_2	$D_{\infty h}$	r (PP)	1.694	1.623	1.715	**1.893**
P_2H_4	C_2	r (PP)	2.077	1.990	2.088	**2.219**
		r (PH_{int})	1.374	1.363	1.331	**1.417**
		r (PH_{ext})	1.375	1.363	1.331	**1.414**
		< (PPH_{int})	104.3	96.2	109.4	**99.1**
		< (PPH_{ext})	98.8	96.2	101.6	**94.3**
		< ($H_{int}PH_{ext}$)	98.4	97.2	98.0	**92.0**
		ω ($H_{int}PPH_{ext}$)	72.7	180.0	69.3	**74.0**
H_2S_2	C_2	r (SS)	1.924	2.107	2.034	**2.055**
		r (SH)	1.304	1.325	1.311	**1.327**
		< (SSH)	102.6	98.8	103.2	**91.3**
		ω (HSSH)	98.3	99.4	93.2	**90.6**
Cl_2	$D_{\infty h}$	r (ClCl)	1.936	1.918	2.035	**1.988**

Table A5-19: Effect of Polarization Functions on Bond Distances in Hydrocarbons

bond	molecule	EDF1		B3LYP		MP2		expt.
		6-31G	6-31G*	6-31G	6-31G*	6-31G	6-31G*	
C–C	but-1-yne-3-ene	1.423	1.421	1.425	1.424	1.446	1.429	**1.431**
	propyne	1.460	1.458	1.481	1.461	1.481	1.463	**1.459**
	1,3-butadiene	1.459	1.455	1.460	1.458	1.475	1.458	**1.483**
	propene	1.506	1.502	1.499	1.502	1.517	1.499	**1.501**
	cyclopropane	1.521	1.510	1.519	1.509	1.527	1.504	**1.510**
	propane	1.539	1.533	1.537	1.532	1.546	1.526	**1.526**
	cyclobutane	1.565	1.554	1.564	1.553	1.572	1.545	**1.548**
C=C	cyclopropene	1.312	1.302	1.304	1.295	1.321	1.303	**1.300**
	allene	1.316	1.312	1.311	1.307	1.327	1.313	**1.308**
	propene	1.344	1.340	1.337	1.33	1.352	1.338	**1.318**
	cyclobutene	1.355	1.347	1.348	1.341	1.365	1.347	**1.332**
	but-1-yne-3-ene	1.353	1.349	1.345	1.341	1.357	1.344	**1.341**
	1,3-butadiene	1.353	1.348	1.345	1.340	1.358	1.344	**1.345**
	cyclopentadiene	1.362	1.356	1.355	1.349	1.369	1.354	**1.345**
mean absolute error		0.013	0.008	0.010	0.006	0.020	0.007	–

Table A5-20: Effect of Polarization Functions on Bond Distances in Molecules with Heteroatoms

bond	molecule	EDF1		B3LYP		MP2		expt.
		6-31G	6-31G*	6-31G	6-31G*	6-31G	6-31G*	
C-N	formamide	1.370	1.367	1.365	1.362	1.377	1.362	1.376
	methyl isocyanide	1.427	1.419	1.426	1.420	1.449	1.426	1.424
	trimethylamine	1.467	1.456	1.465	1.455	1.479	1.455	1.451
	aziridine	1.504	1.477	1.495	1.473	1.516	1.474	1.475
	nitromethane	1.510	1.507	1.501	1.499	1.506	1.488	1.489
C-O	formic acid	1.383	1.356	1.371	1.347	1.392	1.351	1.343
	furan	1.403	1.369	1.396	1.364	1.416	1.367	1.362
	dimethyl ether	1.453	1.414	1.447	1.410	1.466	1.416	1.410
	oxirane	1.497	1.435	1.491	1.430	1.524	1.438	1.436
	mean absolute error	0.029	0.007	0.024	0.005	0.040	0.005	–

536

Table A5-21: Bond Distances Connecting Heavy Atoms. Molecular Mechanics Models

bond type	molecule	SYBYL	MMFF	expt
Be-C	dimethylberyllium	1.499	1.802	**1.698**
	methyl (cyclopentadienyl) beryllium (Me)	1.505	1.808	**1.706**
	methyl (cyclopentadienyl) beryllium (cyclopentadiene)	1.674	1.954	**1.923**
Be-Cl	beryllium dichloride	1.800	1.945	**1.75**
B-B	pentaborane (base-cap)	1.551	1.725	**1.690**
	(base-base)	1.770	2.034	**1.803**
B-C	trimethylborane	1.507	1.577	**1.578**
B ⋯ N (aromatic)	borazine	1.367	1.304	**1.418**
B-N	difluoroaminoborane	1.500	1.253	**1.402**
B-O	difluorohydroxyborane	1.500	1.277	**1.344**
B-F	trifluoroborane	1.500	1.300	**1.307**
	difluoroborane	1.500	1.300	**1.311**
	difluorohydroxyborane	1.500	1.279	**1.323**
	difluoroaminoborane	1.500	1.264	**1.325**
B-Cl	trichloroborane	1.800	1.750	**1.742**
C≡C	cyanoacetylene	1.204	1.201	**1.205**
	propyne	1.204	1.201	**1.206**
	but-1-yne-3-ene	1.204	1.200	**1.208**
	2-butyne	1.204	1.201	**1.214**
C=C	butatriene (C_2C_3)	1.305	1.200	**1.283**
	cyclopropene	1.317	1.302	**1.300**
	cyclopropenone	1.355	1.323	**1.302**
	allene	1.305	1.297	**1.308**
	ketene	1.305	1.296	**1.314**
	1,1-difluoroethylene	1.337	1.332	**1.315**
	butatriene (C_1C_2)	1.305	1.298	**1.318**
	propene	1.339	1.339	**1.318**
	trans-1,2-difluoroethylene	1.337	1.326	**1.329**
	isobutene	1.341	1.342	**1.330**
	cyclopentenone	1.335	1.337	**1.330**
	cis-1,2-difluoroethylene	1.337	1.327	**1.331**
	methylenecyclopropane	1.335	1.313	**1.332**
	fluoroethylene	1.337	1.330	**1.332**
	cyclobutene	1.327	1.345	**1.332**
	acrylonitrile	1.338	1.339	**1.339**
	benzvalene	1.343	1.345	**1.339**
	but-1-yne-3-ene	1.338	1.337	**1.341**
	cis-methyl vinyl ether	1.343	1.338	**1.341**
	cyclopentadiene	1.335	1.341	**1.345**
	trans-1,3-butadiene	1.338	1.338	**1.345**
	vinylsilane	1.338	1.340	**1.347**
	furan	1.334	1.376	**1.361**

Table A5-21: Bond Distances Connecting Heavy Atoms. Molecular Mechanics Models (2)

bond type	molecule	SYBYL	MMFF	expt
	thiophene	1.335	1.378	**1.369**
	pyrrole	1.335	1.377	**1.370**
C–C (aromatic)	pyridine (C_3C_4)	1.397	1.391	**1.392**
	pyridine (C_1C_2)	1.398	1.386	**1.394**
	benzene	1.398	1.395	**1.397**
C-C	cyanoacetylene	1.380	1.380	**1.378**
	cyclopropenone	1.452	1.410	**1.412**
	pyrrole	1.470	1.416	**1.417**
	acrylonitrile	1.441	1.422	**1.426**
	furan	1.461	1.416	**1.431**
	but-1-yne-3-ene	1.441	1.419	**1.431**
	thiophene	1.469	1.425	**1.433**
	benzvalene (C_4C_6)	1.544	1.489	**1.452**
	methylenecyclopropane (C_2C_3)	1.429	1.407	**1.457**
	acetonitrile	1.458	1.462	**1.458**
	propyne	1.458	1.463	**1.459**
	cyclopentadiene (C_2C_3)	1.466	1.439	**1.468**
	malononitrile	1.458	1.464	**1.468**
	2-butyne	1.458	1.463	**1.468**
	oxirane	1.603	1.503	**1.471**
	aziridine	1.514	1.503	**1.481**
	trans-1,3-butadiene	1.478	1.442	**1.483**
	thiirane	1.549	1.496	**1.484**
	bicyclo[1.1.0]butane (C_1C_3)	1.554	1.529	**1.497**
	bicyclo[1.1.0]butane (C_1C_2)	1.540	1.487	**1.498**
	acetaldehyde	1.505	1.498	**1.501**
	propene	1.509	1.493	**1.501**
	phosphacyclopropane	1.566	1.491	**1.502**
	benzvalene (C_2C_3)	1.506	1.478	**1.503**
	fluoroethane	1.547	1.510	**1.505**
	cyclopentadiene (C_4C_5)	1.505	1.501	**1.506**
	acetone	1.507	1.505	**1.507**
	isobutene	1.512	1.501	**1.507**
	cyclopentenone (C_2C_3)	1.501	1.493	**1.509**
	cyclopropane	1.543	1.502	**1.510**
	cyclopropene	1.517	1.430	**1.515**
	cyclobutene (C_1C_4)	1.501	1.512	**1.517**
	acetic acid	1.501	1.493	**1.517**
	cyclopentenone (C_1C_2)	1.503	1.512	**1.524**
	propane	1.551	1.519	**1.526**
	trans-ethanethiol	1.548	1.517	**1.529**
	benzvalene (C_3C_4)	1.537	1.521	**1.529**
	1,1,1-trifluoroethane	1.545	1.504	**1.530**
	trans-ethanol	1.539	1.514	**1.530**
	ethylsilane	1.548	1.523	**1.540**
	neopentane	1.557	1.535	**1.540**

Table A5-21: Bond Distances Connecting Heavy Atoms. Molecular Mechanics Models (3)

bond type	molecule	SYBYL	MMFF	expt.
	isobutane	1.553	1.527	**1.541**
	methylenecyclopropane (C$_3$C$_4$)	1.666	1.568	**1.542**
	cyclobutane	1.547	1.543	**1.548**
	cyclobutene (C$_3$C$_4$)	1.564	1.542	**1.566**
C≡N	acetonitrile	1.158	1.160	**1.157**
	cyanoacetylene	1.158	1.160	**1.159**
	cyanogen fluoride	1.158	1.160	**1.159**
	acrylonitrile	1.158	1.160	**1.164**
	methyl isocyanide	1.170	1.171	**1.166**
	malononitrile	1.158	1.161	**1.167**
	formonitrile oxide	1.170	1.065	**1.168**
	nitrosyl cyanide	1.158	1.161	**1.170**
C=N	isocyanic acid	1.312	1.170	**1.209**
	isothiocyanic acid	1.312	1.171	**1.216**
	diazomethane	1.305	1.316	**1.300**
C⋯N (aromatic)	pyridine	1.350	1.351	**1.338**
C-N	thioformamide	1.301	1.362	**1.358**
	pyrrole	1.300	1.371	**1.370**
	formamide	1.346	1.359	**1.376**
	nitrosyl cyanide	1.330	1.297	**1.401**
	methyl isocyanide	1.500	1.426	**1.424**
	trimethylamine	1.483	1.462	**1.451**
	dimethylamine	1.471	1.457	**1.462**
	aziridine	1.482	1.459	**1.475**
	diazirine	1.465	1.402	**1.482**
	nitromethane	1.455	1.487	**1.489**
	nitrosomethane	1.443	1.481	**1.49**
C=O	carbonyl sulfide	1.305	1.176	**1.160**
	ketene	1.305	1.175	**1.161**
	carbon dioxide	1.305	1.176	**1.162**
	isocyanic acid	1.308	1.174	**1.166**
	carbonyl fluoride	1.220	1.220	**1.170**
	formyl fluoride	1.220	1.221	**1.181**
	formamide	1.220	1.221	**1.193**
	methyl formate	1.222	1.220	**1.200**
	formic acid	1.221	1.217	**1.202**
	cyclopentenone	1.220	1.228	**1.210**
	cyclopropenone	1.220	1.167	**1.212**
	acetic acid	1.221	1.219	**1.212**
	acetaldehyde	1.221	1.227	**1.216**
	acetone	1.221	1.230	**1.222**
C-O	methyl formate (CH$_3$O-CHO)	1.342	1.356	**1.334**
	formic acid	1.334	1.342	**1.343**
	acetic acid	1.333	1.346	**1.360**
	cis-methyl vinyl ether (CH$_3$O-CHCH$_2$)	1.349	1.365	**1.360**

Table A5-21: Bond Distances Connecting Heavy Atoms. Molecular Mechanics Models (4)

bond type	molecule	SYBYL	MMFF	expt.
	furan	1.337	1.358	**1.362**
	dimethyl ether	1.437	1.421	**1.410**
	trans-ethanol	1.430	1.420	**1.425**
	cis methyl vinyl ether (CH₃-OCHCH₂)	1.441	1.423	**1.428**
	oxirane	1.395	1.433	**1.436**
	methyl formate (CH₃-OCHO)	1.441	1.428	**1.437**
C-F	cyanogen fluoride	1.500	1.267	**1.262**
	fluoroacetylene	1.500	1.267	**1.279**
	tetrafluoromethane	1.360	1.344	**1.317**
	carbonyl fluoride	1.330	1.333	**1.317**
	1,1-difluoroethylene	1.331	1.348	**1.323**
	trifluoromethane	1.360	1.350	**1.332**
	cis-1,2-difluoroethylene	1.330	1.345	**1.335**
	formyl fluoride	1.330	1.339	**1.338**
	trans-1,2-difluoroethylene	1.330	1.345	**1.344**
	fluoroethylene	1.331	1.343	**1.348**
	1,1,1-trifluoroethane	1.361	1.350	**1.348**
	difluoromethane	1.360	1.354	**1.358**
	fluoroethane	1.361	1.359	**1.398**
C–Al	dimethylaluminum chloride dimer	–	2.071	**1.935**
	trimethylaluminum	1.800	2.019	**1.957**
	trimethylaluminum dimer (terminal)	–	2.072	**1.957**
	trimethylaluminum dimer (bridge)	–	2.284	**2.140**
C=Si	methylenedimethylsilane	1.566	1.706	**1.83**
C-Si	vinylsilane	1.802	1.833	**1.853**
	ethylsilane	1.885	1.866	**1.866**
	dimethylsilane	1.878	1.851	**1.867**
	trimethylsilane	1.877	1.861	**1.868**
	tetramethylsilane	1.877	1.875	**1.875**
	methylenedimethylsilane	1.800	1.900	**1.91**
C-P	trimethylphosphine	1.829	1.834	**1.841**
	dimethylphosphine	1.831	1.835	**1.848**
	phosphacyclopropane	1.815	1.860	**1.867**
C=S	carbon disulfide	1.566	1.568	**1.553**
	carbonyl sulfide	1.566	1.568	**1.560**
	isothiocyanic acid	1.570	1.567	**1.561**
	thioformamide	1.711	1.659	**1.626**
C-S	thiophene	1.781	1.712	**1.714**
	dimethyl sulfide	1.820	1.808	**1.802**
	thiirane	1.812	1.725	**1.815**
	trans-ethanethiol	1.824	1.814	**1.820**
C-Cl	trichloromethane	1.767	1.772	**1.758**
	tetrachloromethane	1.767	1.781	**1.767**
	dichloromethane	1.767	1.767	**1.772**

Table A5-21: Bond Distances Connecting Heavy Atoms. Molecular Mechanics Models (5)

bond type	molecule	SYBYL	MMFF	expt
N≡N	diazomethane	1.305	1.140	**1.139**
	diazirine	1.313	1.243	**1.228**
	trans-difluorodiazene	1.346	1.243	**1.214**
N=O	nitrosyl fluoride	1.305	1.235	**1.136**
	nitrosyl chloride	1.305	1.235	**1.139**
	formonitrile oxide	1.305	1.223	**1.199**
	nitrosomethane	1.306	1.236	**1.22**
	nitromethane	1.366	1.237	**1.224**
	nitrosyl cyanide	1.305	1.236	**1.228**
N-F	nitrogen trifluoride	1.500	1.379	**1.365**
	trans-difluorodiazene	1.500	1.379	**1.384**
	nitrosyl fluoride	1.500	1.379	**1.512**
N-Cl	nitrogen trichloride	1.800	1.761	**1.748**
	nitrosyl chloride	1.800	1.724	**1.975**
O-̈-O	ozone	1.365	1.290	**1.278**
O-O	fluorine peroxide	1.480	1.449	**1.217**
O-F	oxygen difluoride	1.500	1.417	**1.405**
	fluorine peroxide	1.500	1.417	**1.575**
O-Cl	oxygen dichloride	1.800	1.678	**1.700**
F-Mg	magnesium difluoride	1.800	2.127	**1.77**
F-Al	aluminum trifluoride	1.800	1.956	**1.63**
F-Si	tetrafluorosilane	1.800	1.603	**1.554**
	trifluorosilane	1.800	1.602	**1.562**
	difluorosilane	1.800	1.603	**1.577**
F-P	trifluorophosphine	1.800	1.575	**1.551**
F-S	difluorosulfide	1.800	1.591	**1.592**
Al-Al	trimethylaluminum dimer	–	2.655	**2.619**
	dimethylaluminum chloride dimer	–	2.574	**3.274**
Al-Cl	trichloroaluminum	2.000	2.247	**2.06**
	dimethylaluminum chloride dimer	–	2.573	**2.303**
Si-Cl	tetrachlorosilane	2.000	2.029	**2.019**
	trichlorosilane	2.000	2.022	**2.021**
	dichlorosilane	2.000	2.032	**2.033**
P-Cl	trichlorophosphine	2.000	2.100	**2.043**
S-Cl	sulfur dichloride	2.000	2.031	**2.014**

Table A5-22: Bond Distances Connecting Heavy Atoms. Hartree-Fock Models

bond type	molecule	STO-3G	3-21G	6-31G*	6-311+G**	expt
Be-C	dimethylberyllium	1.695	1.715	1.704	1.695	**1.698**
	methyl (cyclopentadienyl) beryllium (Me)	1.718	1.730	1.729	1.720	**1.706**
	methyl (cyclopentadienyl) beryllium (cyclopentadiene)	1.904	2.003	1.943	1.939	**1.923**
Be-Cl	beryllium dichloride	1.787	1.811	1.804	1.803	**1.75**
B-B	pentaborane (base-cap)	1.681	1.709	1.702	1.706	**1.690**
	(base-base)	1.809	1.828	1.811	1.816	**1.803**
B-C	trimethylborane	1.581	1.589	1.588	1.583	**1.578**
B ⋯ N (aromatic)	borazine	1.418	1.460	1.427	1.427	**1.418**
B-N	difluoroaminoborane	1.403	1.397	1.392	1.388	**1.402**
B-O	difluorohydroxyborane	1.358	1.354	1.337	1.332	**1.344**
B-F	trifluoroborane	1.309	1.328	1.301	1.298	**1.307**
	difluoroborane	1.302	1.337	1.306	1.305	**1.311**
	difluorohydroxyborane	1.313	1.340	1.317	1.311	**1.323**
	difluoroaminoborane	1.317	1.349	1.320	1.319	**1.325**
B-Cl	trichloroborane	1.768	1.747	1.745	1.748	**1.742**
C≡C	cyanoacetylene	1.175	1.187	1.185	1.182	**1.205**
	propyne	1.170	1.188	1.187	1.185	**1.206**
	but-1-yne-3-ene	1.171	1.190	1.189	1.186	**1.208**
	2-butyne	1.171	1.189	1.188	1.186	**1.214**
C=C	butatriene (C_2C_3)	1.257	1.259	1.265	1.261	**1.283**
	cyclopropene	1.277	1.282	1.276	1.276	**1.300**
	cyclopropenone	1.323	1.333	1.327	1.329	**1.302**
	allene	1.288	1.292	1.296	1.295	**1.308**
	ketene	1.300	1.296	1.306	1.305	**1.314**
	1,1-difluoroethylene	1.316	1.298	1.304	1.303	**1.315**
	butatriene (C_1C_2)	1.296	1.299	1.303	1.302	**1.318**
	propene	1.308	1.316	1.318	1.320	**1.318**
	trans-1,2-difluoroethylene	1.320	1.300	1.307	1.303	**1.329**
	isobutene	1.311	1.318	1.321	1.322	**1.330**
	cyclopentenone	1.310	1.319	1.321	1.322	**1.330**
	cis-1,2-difluoroethylene	1.320	1.301	1.307	1.307	**1.331**
	methylenecyclopropane	1.298	1.301	1.308	1.309	**1.332**
	fluoroethylene	1.312	1.304	1.309	1.308	**1.332**
	cyclobutene	1.314	1.326	1.322	1.323	**1.332**
	acrylonitrile	1.315	1.319	1.320	1.320	**1.339**
	benzvalene	1.313	1.320	1.320	1.322	**1.339**
	but-1-yne-3-ene	1.320	1.320	1.322	1.322	**1.341**
	cis-methyl vinyl ether	1.313	1.316	1.320	1.322	**1.341**
	cyclopentadiene	1.319	1.329	1.329	1.330	**1.345**
	trans-1,3-butadiene	1.313	1.320	1.323	1.324	**1.345**

Table A5-22: Bond Distances Connecting Heavy Atoms. Hartree-Fock Models (2)

bond type	molecule	STO-3G	3-21G	6-31G*	6-311+G**	expt.
	vinylsilane	1.309	1.324	1.325	1.321	**1.347**
	furan	1.340	1.340	1.339	1.339	**1.361**
	thiophene	1.334	1.348	1.345	1.346	**1.369**
	pyrrole	1.351	1.359	1.363	1.359	**1.370**
C--C (aromatic)	pyridine (C_3C_4)	1.355	1.384	1.384	1.384	**1.392**
	pyridine (C_1C_2)	1.388	1.383	1.385	1.385	**1.394**
	benzene	1.387	1.385	1.386	1.386	**1.397**
C-C	cyanoacetylene	1.409	1.370	1.391	1.388	**1.378**
	cyclopropenone	1.440	1.429	1.412	1.411	**1.412**
	pyrrole	1.432	1.432	1.426	1.427	**1.417**
	acrylonitrile	1.461	1.427	1.443	1.441	**1.426**
	furan	1.445	1.450	1.441	1.442	**1.431**
	but-1-yne-3-ene	1.459	1.432	1.439	1.438	**1.431**
	thiophene	1.454	1.438	1.437	1.436	**1.433**
	benzvalene (C_4C_6)	1.445	1.462	1.433	1.436	**1.452**
	methylenecyclopropane (C_2C_3)	1.474	1.472	1.462	1.463	**1.457**
	acetonitrile	1.488	1.457	1.468	1.466	**1.458**
	propyne	1.484	1.466	1.468	1.466	**1.459**
	cyclopentadiene (C_2C_3)	1.490	1.485	1.476	1.476	**1.468**
	malononitrile	1.493	1.461	1.472	1.469	**1.468**
	2-butyne	1.483	1.467	1.469	1.468	**1.468**
	oxirane	1.483	1.474	1.453	1.454	**1.471**
	aziridine	1.491	1.497	1.472	1.473	**1.481**
	trans-1,3-butadiene	1.488	1.479	1.467	1.467	**1.483**
	thiirane	1.507	1.490	1.473	1.472	**1.484**
	bicyclo[1.1.0]butane (C_1C_3)	1.469	1.484	1.466	1.472	**1.497**
	bicyclo[1.1.0]butane (C_1C_2)	1.501	1.513	1.489	1.491	**1.498**
	acetaldehyde	1.537	1.507	1.504	1.502	**1.501**
	propene	1.520	1.510	1.503	1.502	**1.501**
	phosphacyclopropane	1.513	1.516	1.492	1.492	**1.502**
	benzvalene (C_2C_3)	1.519	1.512	1.511	1.510	**1.503**
	fluoroethane	1.547	1.521	1.512	1.510	**1.505**
	cyclopentadiene (C_4C_5)	1.522	1.519	1.507	1.505	**1.506**
	acetone	1.543	1.515	1.514	1.513	**1.507**
	isobutene	1.526	1.516	1.508	1.508	**1.507**
	cyclopentenone (C_2C_3)	1.520	1.512	1.506	1.506	**1.509**
	cyclopropane	1.502	1.513	1.497	1.500	**1.510**
	cyclopropene	1.493	1.523	1.495	1.499	**1.515**
	cyclobutene (C_1C_4)	1.526	1.539	1.514	1.517	**1.517**
	acetic acid	1.537	1.497	1.502	1.500	**1.517**
	cyclopentenone (C_1C_2)	1.553	1.535	1.527	1.526	**1.524**
	propane	1.541	1.541	1.528	1.528	**1.526**
	trans-ethanethiol	1.541	1.541	1.525	1.524	**1.529**
	benzvalene (C_3C_4)	1.520	1.540	1.512	1.514	**1.529**
	1,1,1-trifluoroethane	1.562	1.490	1.500	1.499	**1.530**
	trans-ethanol	1.542	1.525	1.516	1.514	**1.530**

Table A5-22: Bond Distances Connecting Heavy Atoms. Hartree-Fock Models (3)

bond type	molecule	STO-3G	3-21G	6-31G*	6-311+G**	expt.
	ethylsilane	1.543	1.555	1.536	1.536	**1.540**
	neopentane	1.549	1.540	1.535	1.535	**1.540**
	isobutane	1.545	1.541	1.531	1.530	**1.541**
	methylenecyclopropane (C_3C_4)	1.522	1.545	1.527	1.530	**1.542**
	cyclobutane	1.554	1.571	1.548	1.546	**1.548**
	cyclobutene (C_3C_4)	1.565	1.593	1.562	1.564	**1.566**
C≡N	acetonitrile	1.154	1.139	1.135	1.130	**1.157**
	cyanoacetylene	1.159	1.141	1.136	1.130	**1.159**
	cyanogen fluoride	1.160	1.135	1.131	1.126	**1.159**
	acrylonitrile	1.157	1.140	1.136	1.131	**1.164**
	methyl isocyanide	1.170	1.170	1.153	1.149	**1.166**
	malononitrile	1.155	1.138	1.133	1.128	**1.167**
	formonitrile oxide	1.155	1.135	1.130	1.126	**1.168**
	nitrosyl cyanide	1.159	1.139	1.134	1.128	**1.170**
C=N	isocyanic acid	1.245	1.160	1.200	1.197	**1.209**
	isothiocyanic acid	1.226	1.157	1.171	1.172	**1.216**
	diazomethane	1.282	1.281	1.280	1.286	**1.300**
C⁓N (aromatic)	pyridine	1.353	1.331	1.321	1.320	**1.338**
C-N	thioformamide	1.389	1.331	1.324	1.326	**1.358**
	pyrrole	1.389	1.377	1.363	1.363	**1.370**
	formamide	1.436	1.353	1.349	1.349	**1.376**
	nitrosyl cyanide	1.482	1.403	1.416	1.414	**1.401**
	methyl isocyanide	1.447	1.432	1.421	1.425	**1.424**
	trimethylamine	1.486	1.464	1.445	1.448	**1.451**
	dimethylamine	1.484	1.466	1.447	1.448	**1.462**
	aziridine	1.482	1.491	1.449	1.450	**1.475**
	diazirine	1.488	1.522	1.446	1.447	**1.482**
	nitromethane	1.531	1.493	1.479	1.484	**1.489**
	nitrosomethane	1.531	1.499	1.464	1.464	**1.49**
C=O	carbonyl sulfide	1.176	1.147	1.131	1.124	**1.160**
	ketene	1.183	1.162	1.145	1.136	**1.161**
	carbon dioxide	1.188	1.156	1.143	1.136	**1.162**
	isocyanic acid	1.183	1.177	1.148	1.138	**1.166**
	carbonyl fluoride	1.208	1.169	1.157	1.150	**1.170**
	formyl fluoride	1.210	1.180	1.164	1.157	**1.181**
	formamide	1.216	1.212	1.193	1.188	**1.193**
	methyl formate	1.214	1.200	1.184	1.179	**1.200**
	formic acid	1.214	1.198	1.182	1.177	**1.202**
	cyclopentenone	1.215	1.205	1.181	1.183	**1.210**
	cyclopropenone	1.213	1.203	1.190	1.185	**1.212**
	acetic acid	1.216	1.202	1.187	1.182	**1.212**
	acetaldehyde	1.217	1.209	1.188	1.183	**1.216**
	acetone	1.219	1.211	1.192	1.188	**1.222**

Table A5-22: Bond Distances Connecting Heavy Atoms. Hartree-Fock Models (4)

bond type	molecule	STO-3G	3-21G	6-31G*	6-311+G**	expt.
C-O	methyl formate (CH_3O-CHO)	1.388	1.344	1.317	1.314	**1.334**
	formic acid	1.385	1.350	1.323	1.320	**1.343**
	acetic acid	1.391	1.360	1.332	1.331	**1.360**
	cis-methyl vinyl ether (CH_3O-CHCH$_2$)	1.392	1.370	1.341	1.337	**1.360**
	furan	1.376	1.344	1.380	1.342	**1.362**
	dimethyl ether	1.433	1.433	1.391	1.391	**1.410**
	trans-ethanol	1.436	1.444	1.405	1.405	**1.425**
	cis methyl vinyl ether (CH_3-OCHCH$_2$)	1.434	1.437	1.399	1.400	**1.428**
	oxirane	1.433	1.470	1.402	1.400	**1.436**
	methyl formate (CH_3-OCHO)	1.441	1.456	1.419	1.419	**1.437**
C-F	cyanogen fluoride	1.316	1.287	1.253	1.243	**1.262**
	fluoroacetylene	1.318	1.297	1.269	1.258	**1.279**
	tetrafluoromethane	1.366	1.325	1.302	1.298	**1.317**
	carbonyl fluoride	1.347	1.322	1.290	1.285	**1.317**
	1,1-difluoroethylene	1.350	1.334	1.303	1.298	**1.323**
	trifluoromethane	1.371	1.345	1.317	1.313	**1.332**
	cis-1,2-difluoroethylene	1.356	1.358	1.324	1.318	**1.335**
	formyl fluoride	1.351	1.348	1.314	1.312	**1.338**
	trans-1,2-difluoroethylene	1.356	1.360	1.329	1.322	**1.344**
	fluoroethylene	1.354	1.363	1.329	1.326	**1.348**
	1,1,1-trifluoroethane	1.375	1.352	1.325	1.322	**1.348**
	difluoroethylene	1.378	1.372	1.328	1.337	**1.358**
	fluoroethane	1.385	1.410	1.373	1.374	**1.398**
C-Al	dimethylaluminum chloride dimer	1.889	1.960	1.960	1.953	**1.935**
	trimethylaluminum	1.899	1.981	1.980	1.970	**1.957**
	trimethylaluminum dimer (terminal)	1.899	1.978	1.977	1.972	**1.957**
	trimethylaluminum dimer (bridge)	2.081	2.162	2.171	2.166	**2.140**
C=Si	methylenedimethylsilane	1.639	1.690	1.693	1.691	**1.83**
C-Si	vinylsilane	1.852	1.867	1.873	1.869	**1.853**
	ethylsilane	1.869	1.886	1.893	1.889	**1.866**
	dimethylsilane	1.862	1.885	1.890	1.884	**1.867**
	trimethylsilane	1.862	1.887	1.892	1.886	**1.868**
	tetramethylsilane	1.863	1.889	1.894	1.888	**1.875**
	methylenedimethylsilane	1.856	1.876	1.881	1.870	**1.91**
C-P	trimethylphosphine	1.841	1.848	1.853	1.847	**1.841**
	dimethylphosphine	1.841	1.851	1.856	1.851	**1.848**
	phosphacyclopropane	1.812	1.842	1.853	1.852	**1.867**
C=S	carbon disulfide	1.532	1.542	1.545	1.543	**1.553**
	carbonyl sulfide	1.548	1.565	1.572	1.567	**1.560**
	isothiocyanic acid	1.542	1.59	1.584	1.575	**1.561**
	thioformamide	1.591	1.639	1.641	1.637	**1.626**
C-S	thiophene	1.732	1.722	1.726	1.733	**1.714**
	dimethyl sulfide	1.796	1.813	1.809	1.808	**1.802**
	thiirane	1.774	1.817	1.812	1.814	**1.815**
	trans-ethanethiol	1.806	1.829	1.827	1.829	**1.820**

Table A5-22: Bond Distances Connecting Heavy Atoms. Hartree-Fock Models (5)

bond type	molecule	STO-3G	3-21G	6-31G*	6-311+G**	expt.
C-Cl	trichloromethane	1.808	1.776	1.763	1.764	**1.758**
	tetrachloromethane	1.818	1.778	1.767	1.768	**1.767**
	dichloromethane	1.803	1.784	1.768	1.770	**1.772**
N≡N	diazomethane	1.190	1.131	1.116	1.107	**1.139**
N=N	diazirine	1.266	1.217	1.194	1.188	**1.228**
	trans-difluorodiazene	1.373	1.211	1.194	1.188	**1.214**
N=O	nitrosyl fluoride	1.222	1.152	1.128	1.115	**1.136**
	nitrosyl chloride	1.203	1.149	1.116	1.101	**1.139**
	formonitrile oxide	1.294	1.289	1.201	1.194	**1.199**
	nitrosomethane	1.231	1.216	1.177	1.168	**1.22**
	nitromethane	1.275	1.240	1.193	1.186	**1.224**
	nitrosyl cyanide	1.235	1.215	1.171	1.162	**1.228**
N-F	nitrogen trifluoride	1.386	1.402	1.328	1.319	**1.365**
	trans-difluorodiazene	1.277	1.414	1.339	1.326	**1.384**
	nitrosyl fluoride	1.381	1.460	1.384	1.390	**1.512**
N-Cl	nitrogen trichloride	1.803	1.740	1.723	1.721	**1.748**
	nitrosyl chloride	1.862	1.907	1.907	1.931	**1.975**
O=O	ozone	1.285	1.308	1.204	1.194	**1.278**
O-O	fluorine peroxide	1.392	1.440	1.311	1.299	**1.217**
O-F	oxygen difluoride	1.358	1.427	1.384	1.333	**1.405**
	fluorine peroxide	1.358	1.432	1.367	1.353	**1.575**
O-Cl	oxygen dichloride	1.743	1.703	1.671	1.667	**1.700**
F-Mg	magnesium difluoride	1.665	1.701	1.723	1.757	**1.77**
F-Al	aluminum trifluoride	1.600	1.617	1.620	1.632	**1.63**
F-Si	tetrafluorosilane	1.585	1.557	1.557	1.556	**1.554**
	trifluorosilane	1.597	1.569	1.569	1.570	**1.562**
	difluorosilane	1.610	1.581	1.581	1.585	**1.577**
F-P	trifluorophosphine	1.621	1.561	1.564	1.561	**1.551**
F-S	difluorosulfide	1.641	1.592	1.586	1.586	**1.592**
Al-Al	trimethylaluminum dimer	2.519	2.620	2.637	2.642	**2.619**
	dimethylaluminum chloride dimer	3.307	3.373	3.376	3.362	**3.274**
Al-Cl	trichloroaluminum	2.050	2.075	2.077	2.073	**2.06**
	dimethylaluminum chloride dimer	2.260	2.344	2.350	2.344	**2.303**
Si-Cl	tetrachlorosilane	2.071	2.017	2.029	2.029	**2.019**
	trichlorosilane	2.073	2.027	2.038	2.038	**2.021**
	dichlorosilane	2.079	2.039	2.050	2.051	**2.033**
P-Cl	trichlorophosphine	2.104	2.039	2.049	2.054	**2.043**
S-Cl	sulfur dichloride	2.077	2.019	2.020	2.031	**2.014**

Table A5-23: Bond Distances Connecting Heavy Atoms. Local Density Models

bond type	molecule	6-31G*	6-311+G**	expt.
Be-C	dimethylberyllium	1.669	1.664	**1.698**
	methyl (cyclopentadienyl) beryllium (Me)	1.695	1.690	**1.706**
	methyl (cyclopentadienyl) beryllium (cyclopentadiene)	1.900	1.889	**1.923**
Be-Cl	beryllium dichloride	1.791	1.785	**1.75**
B-B	pentaborane (base-cap)	1.682	1.674	**1.690**
	(base-base)	1.771	1.764	**1.803**
B-C	trimethylborane	1.561	1.553	**1.578**
B $\doteq$ N (aromatic)	borazine	1.422	1.422	**1.418**
B-N	difluoroaminoborane	1.394	1.388	**1.402**
B-O	difluorohydroxyborane	1.346	1.341	**1.344**
B-F	trifluoroborane	1.311	1.313	**1.307**
	difluoroborane	1.312	1.316	**1.311**
	difluorohydroxyborane	1.326	1.325	**1.323**
	difluoroaminoborane	1.328	1.332	**1.325**
B-Cl	trichloroborane	1.736	1.736	**1.742**
C≡C	cyanoacetylene	1.217	1.211	**1.205**
	propyne	1.213	1.208	**1.206**
	but-1-yne-3-ene	1.220	1.212	**1.208**
	2-butyne	1.216	1.209	**1.214**
C=C	butatriene (C_2C_3)	1.272	1.266	**1.283**
	cyclopropene	1.298	1.293	**1.300**
	cyclopropenone	1.347	1.345	**1.302**
	allene	1.305	1.301	**1.308**
	ketene	1.311	1.306	**1.314**
	1,1-difluoroethylene	1.323	1.314	**1.315**
	butatriene (C_1C_2)	1.317	1.313	**1.318**
	propene	1.332	1.330	**1.318**
	trans-1,2-difluoroethylene	1.329	1.320	**1.329**
	isobutene	1.336	1.329	**1.330**
	cyclopentenone	1.339	1.337	**1.330**
	cis-1,2-difluoroethylene	1.329	1.321	**1.331**
	methylenecyclopropane	1.321	1.318	**1.332**
	fluoroethylene	1.325	1.320	**1.332**
	cyclobutene	1.343	1.341	**1.332**
	acrylonitrile	1.339	1.335	**1.339**
	benzvalene	1.339	1.338	**1.339**
	but-1-yne-3-ene	1.344	1.338	**1.341**
	cis-methyl vinyl ether	1.337	1.330	**1.341**
	cyclopentadiene	1.351	1.350	**1.345**
	trans-1,3-butadiene	1.340	1.338	**1.345**

Table A5-23: Bond Distances Connecting Heavy Atoms. Local Density Models (2)

bond type	molecule	6-31G*	6-311+G**	expt.
	vinylsilane	1.337	1.329	**1.347**
	furan	1.361	1.359	**1.361**
	thiophene	1.369	1.367	**1.369**
	pyrrole	1.377	1.376	**1.370**
C ⚌ C (aromatic)	pyridine (C_3C_4)	1.389	1.387	**1.392**
	pyridine (C_1C_2)	1.391	1.389	**1.394**
	benzene	1.390	1.389	**1.397**
C-C	cyanoacetylene	1.359	1.356	**1.378**
	cyclopropenone	1.435	1.430	**1.412**
	pyrrole	1.414	1.414	**1.417**
	acrylonitrile	1.414	1.411	**1.426**
	furan	1.423	1.422	**1.431**
	but-1-yne-3-ene	1.408	1.406	**1.431**
	thiophene	1.415	1.414	**1.433**
	benzvalene (C_4C_6)	1.442	1.443	**1.452**
	methylenecyclopropane (C_2C_3)	1.459	1.456	**1.457**
	acetonitrile	1.441	1.436	**1.458**
	propyne	1.441	1.437	**1.459**
	cyclopentadiene (C_2C_3)	1.452	1.451	**1.468**
	malononitrile	1.451	1.447	**1.468**
	2-butyne	1.442	1.437	**1.468**
	oxirane	1.464	1.462	**1.471**
	aziridine	1.478	1.478	**1.481**
	trans-1,3-butadiene	1.439	1.439	**1.483**
	thiirane	1.475	1.474	**1.484**
	bicyclo[1.1.0]butane (C_1C_3)	1.484	1.480	**1.497**
	bicyclo[1.1.0]butane (C_1C_2)	1.488	1.490	**1.498**
	acetaldehyde	1.488	1.483	**1.501**
	propene	1.482	1.481	**1.501**
	phosphacyclopropane	1.481	1.481	**1.502**
	benzvalene (C_2C_3)	1.496	1.496	**1.503**
	fluoroethane	1.498	1.494	**1.505**
	cyclopentadiene (C_4C_5)	1.486	1.485	**1.506**
	acetone	1.500	1.493	**1.507**
	isobutene	1.488	1.482	**1.507**
	cyclopentenone (C_2C_3)	1.487	1.487	**1.509**
	cyclopropane	1.496	1.496	**1.510**
	cyclopropene	1.499	1.501	**1.515**
	cyclobutene (C_1C_4)	1.501	1.502	**1.517**
	acetic acid	1.489	1.482	**1.517**
	cyclopentenone (C_1C_2)	1.521	1.518	**1.524**
	propane	1.511	1.512	**1.526**
	trans-ethanethiol	1.507	1.507	**1.529**
	benzvalene (C_3C_4)	1.517	1.519	**1.529**
	1,1,1-trifluoroethane	1.488	1.485	**1.530**
	trans-ethanol	1.501	1.499	**1.530**

Table A5-23: Bond Distances Connecting Heavy Atoms. Local Density Models (3)

bond type	molecule	6-31G*	6-311+G**	expt.
	ethylsilane	1.517	1.518	**1.540**
	neopentane	1.517	1.518	**1.540**
	isobutane	1.513	1.514	**1.541**
	methylenecyclopropane (C_3C_4)	1.524	1.528	**1.542**
	cyclobutane	1.535	1.537	**1.548**
	cyclobutene (C_3C_4)	1.552	1.555	**1.566**
C≡N	acetonitrile	1.166	1.158	**1.157**
	cyanoacetylene	1.174	1.165	**1.159**
	cyanogen fluoride	1.169	1.160	**1.159**
	acrylonitrile	1.170	1.162	**1.164**
	methyl isocyanide	1.182	1.168	**1.166**
	malononitrile	1.165	1.157	**1.167**
	formonitrile oxide	1.174	1.167	**1.168**
	nitrosyl cyanide	1.176	1.167	**1.170**
C=N	isocyanic acid	1.216	1.208	**1.209**
	isothiocyanic acid	1.210	1.201	**1.216**
	diazomethane	1.286	1.283	**1.300**
C⁻⁻N (aromatic)	pyridine	1.333	1.331	**1.338**
C-N	thioformamide	1.337	1.337	**1.358**
	pyrrole	1.365	1.365	**1.370**
	formamide	1.352	1.351	**1.376**
	nitrosyl cyanide	1.395	1.387	**1.401**
	methyl isocyanide	1.398	1.397	**1.424**
	trimethylamine	1.433	1.434	**1.451**
	dimethylamine	1.435	1.436	**1.462**
	aziridine	1.455	1.456	**1.475**
	diazirine	1.471	1.473	**1.482**
	nitromethane	1.477	1.472	**1.489**
	nitrosomethane	1.462	1.449	**1.49**
C=O	carbonyl sulfide	1.170	1.162	**1.160**
	ketene	1.173	1.164	**1.161**
	carbon dioxide	1.171	1.163	**1.162**
	isocyanic acid	1.177	1.167	**1.166**
	carbonyl fluoride	1.182	1.174	**1.170**
	formyl fluoride	1.188	1.179	**1.181**
	formamide	1.216	1.212	**1.193**
	methyl formate	1.208	1.200	**1.200**
	formic acid	1.206	1.201	**1.202**
	cyclopentenone	1.210	1.206	**1.210**
	cyclopropenone	1.205	1.200	**1.212**
	acetic acid	1.212	1.205	**1.212**
	acetaldehyde	1.211	1.207	**1.216**
	acetone	1.217	1.211	**1.222**

549

Table A5-23: Bond Distances Connecting Heavy Atoms. Local Density Models (4)

bond type	molecule	6-31G*	6-311+G**	expt
C-O	methyl formate (CH₃O-CHO)	1.333	1.329	**1.334**
	formic acid	1.336	1.335	**1.343**
	acetic acid	1.347	1.345	**1.360**
	cis-methyl vinyl ether (CH₃O-CHCH₂)	1.343	1.340	**1.360**
	furan	1.351	1.351	**1.362**
	dimethyl ether	1.390	1.393	**1.410**
	trans-ethanol	1.405	1.413	**1.425**
	cis methyl vinyl ether (CH₃-OCHCH₂)	1.401	1.401	**1.428**
	oxirane	1.415	1.417	**1.436**
	methyl formate (CH₃-OCHO)	1.421	1.422	**1.437**
C-F	cyanogen fluoride	1.263	1.257	**1.262**
	fluoroacetylene	1.273	1.267	**1.279**
	tetrafluoromethane	1.321	1.321	**1.317**
	carbonyl fluoride	1.314	1.315	**1.317**
	1,1-difluoroethylene	1.314	1.312	**1.323**
	trifluoromethane	1.331	1.335	**1.332**
	cis-1,2-difluoroethylene	1.328	1.327	**1.335**
	formyl fluoride	1.334	1.345	**1.338**
	trans-1,2-difluoroethylene	1.330	1.331	**1.344**
	fluoroethylene	1.331	1.339	**1.348**
	1,1,1-trifluoroethane	1.341	1.346	**1.348**
	difluoromethane	1.348	1.354	**1.358**
	fluoroethane	1.375	1.390	**1.398**
C-Al	dimethylaluminum chloride dimer	1.933	1.930	**1.935**
	trimethylaluminum	1.952	1.943	**1.957**
	trimethylaluminum dimer (terminal)	1.949	1.948	**1.957**
	trimethylaluminum dimer (bridge)	2.120	2.118	**2.140**
C=Si	methylenedimethylsilane	1.706	1.698	**1.83**
C-Si	vinylsilane	1.852	1.847	**1.853**
	ethylsilane	1.873	1.873	**1.866**
	dimethylsilane	1.868	1.867	**1.867**
	trimethylsilane	1.869	1.868	**1.868**
	tetramethylsilane	1.873	1.871	**1.875**
	methylenedimethylsilane	1.862	1.849	**1.91**
C-P	trimethylphosphine	1.840	1.829	**1.841**
	dimethylphosphine	1.847	1.837	**1.848**
	phosphacyclopropane	1.867	1.863	**1.867**
C=S	carbon disulfide	1.559	1.558	**1.553**
	carbonyl sulfide	1.566	1.561	**1.560**
	isothiocyanic acid	1.573	1.564	**1.561**
	thioformamide	1.639	1.636	**1.626**
C-S	thiophene	1.714	1.714	**1.714**
	dimethyl sulfide	1.799	1.791	**1.802**
	thiirane	1.813	1.812	**1.815**
	trans-ethanethiol	1.821	1.823	**1.820**

Table A5-23: Bond Distances Connecting Heavy Atoms. Local Density Models (5)

bond type	molecule	6-31G*	6-311+G**	expt.
C-Cl	trichloromethane	1.767	1.766	**1.758**
	tetrachloromethane	1.771	1.770	**1.767**
	dichloromethane	1.767	1.768	**1.772**
N≡N	diazomethane	1.150	1.137	**1.139**
N=N	diazirine	1.231	1.220	**1.228**
	trans-difluorodiazene	1.234	1.222	**1.214**
N=O	nitrosyl fluoride	1.154	1.133	**1.136**
	nitrosyl chloride	1.149	1.137	**1.139**
	formonitrile oxide	1.198	1.191	**1.199**
	nitrosomethane	1.212	1.203	**1.22**
	nitromethane	1.223	1.218	**1.224**
	nitrosyl cyanide	1.209	1.202	**1.228**
N-F	nitrogen trifluoride	1.373	1.375	**1.365**
	trans-difluorodiazene	1.366	1.367	**1.384**
	nitrosyl fluoride	1.470	1.524	**1.512**
N-Cl	nitrogen trichloride	1.781	1.779	**1.748**
	nitrosyl chloride	1.956	1.977	**1.975**
O-∸-O	ozone	1.262	1.255	**1.278**
O-O	fluorine peroxide	1.220	1.181	**1.217**
O-F	oxygen difluoride	1.395	1.400	**1.405**
	fluorine peroxide	1.526	1.594	**1.575**
O-Cl	oxygen dichloride	1.714	1.723	**1.700**
F-Mg	magnesium difluoride	1.722	1.771	**1.770**
F-Al	aluminum trifluoride	1.632	1.650	**1.63**
F-Si	tetrafluorosilane	1.572	1.578	**1.554**
	trifluorosilane	1.582	1.592	**1.562**
	difluorosilane	1.593	1.608	**1.577**
F-P	trifluorophosphine	1.588	1.600	**1.551**
F-S	difluorosulfide	1.613	1.629	**1.592**
Al-Al	trimethylaluminum dimer	2.568	2.580	**2.619**
	dimethylaluminum chloride dimer	3.247	3.255	**3.274**
Al-Cl	trichloroaluminum	2.067	2.067	**2.06**
	dimethylaluminum chloride dimer	2.307	2.305	**2.303**
Si-Cl	tetrachlorosilane	2.027	2.018	**2.019**
	trichlorosilane	2.034	2.028	**2.021**
	dichlorosilane	2.044	2.047	**2.033**
P-Cl	trichlorophosphine	2.067	2.097	**2.043**
S-Cl	sulfur dichloride	2.038	2.052	**2.014**

Table A5-24: Bond Distances Connecting Heavy Atoms. BP Density Functional Models

bond type	molecule	6-31G*	6-311+G**	expt.
Be-C	dimethylberyllium	1.692	1.685	**1.698**
	methyl (cyclopentadienyl) beryllium (Me)	1.720	1.714	**1.706**
	methyl (cyclopentadienyl) beryllium (cyclopentadiene)	1.942	1.934	**1.923**
Be-Cl	beryllium dichloride	1.811	1.807	**1.75**
B-B	pentaborane (base-cap)	1.703	1.701	**1.690**
	(base-base)	1.802	1.802	**1.803**
B-C	trimethylborane	1.585	1.580	**1.578**
B-⁼-N (aromatic)	borazine	1.438	1.438	**1.418**
B-N	difluoroaminoborane	1.408	1.402	**1.402**
B-O	difluorohydroxyborane	1.363	1.357	**1.344**
B-F	trifluoroborane	1.329	1.329	**1.307**
	difluoroborane	1.330	1.333	**1.311**
	difluorohydroxyborane	1.339	1.342	**1.323**
	difluoroaminoborane	1.346	1.350	**1.325**
B-Cl	trichloroborane	1.760	1.757	**1.742**
C≡C	cyanoacetylene	1.223	1.217	**1.205**
	propyne	1.219	1.214	**1.206**
	but-1-yne-3-ene	1.223	1.218	**1.208**
	2-butyne	1.222	1.216	**1.214**
C=C	butatriene (C_2C_3)	1.278	1.273	**1.283**
	cyclopropene	1.306	1.302	**1.300**
	cyclopropenone	1.357	1.355	**1.302**
	allene	1.316	1.312	**1.308**
	ketene	1.323	1.318	**1.314**
	1,1-difluoroethylene	1.334	1.328	**1.315**
	butatriene (C_1C_2)	1.327	1.324	**1.318**
	propene	1.343	1.341	**1.318**
	trans-1,2-difluoroethylene	1.340	1.334	**1.329**
	isobutene	1.347	1.345	**1.330**
	cyclopentenone	1.349	1.346	**1.330**
	cis-1,2-difluoroethylene	1.339	1.335	**1.331**
	methylenecyclopropane	1.332	1.329	**1.332**
	fluoroethylene	1.335	1.330	**1.332**
	cyclobutene	1.352	1.350	**1.332**
	acrylonitrile	1.350	1.346	**1.339**
	benzvalene	1.350	1.348	**1.339**
	but-1-yne-3-ene	1.353	1.350	**1.341**
	cis-methyl vinyl ether	1.348	1.345	**1.341**
	cyclopentadiene	1.361	1.359	**1.345**
	trans-1,3-butadiene	1.352	1.349	**1.345**

Table A5-24: Bond Distances Connecting Heavy Atoms. BP Density Functional Models (2)

bond type	molecule	6-31G*	6-311+G**	expt.
	vinylsilane	1.349	1.345	**1.347**
	furan	1.372	1.369	**1.361**
	thiophene	1.379	1.377	**1.369**
	pyrrole	1.388	1.387	**1.370**
C $\rightleftharpoons$ C (aromatic)	pyridine (C_3C_4)	1.403	1.400	**1.392**
	pyridine (C_1C_2)	1.404	1.402	**1.394**
	benzene	1.405	1.402	**1.397**
C-C	cyanoacetylene	1.371	1.367	**1.378**
	cyclopropenone	1.450	1.444	**1.412**
	pyrrole	1.430	1.429	**1.417**
	acrylonitrile	1.431	1.428	**1.426**
	furan	1.439	1.438	**1.431**
	but-1-yne-3-ene	1.424	1.422	**1.431**
	thiophene	1.433	1.431	**1.433**
	benzvalene (C_4C_6)	1.458	1.459	**1.452**
	methylenecyclopropane (C_2C_3)	1.477	1.473	**1.457**
	acetonitrile	1.463	1.458	**1.458**
	propyne	1.461	1.458	**1.459**
	cyclopentadiene (C_2C_3)	1.472	1.469	**1.468**
	malononitrile	1.474	1.468	**1.468**
	2-butyne	1.463	1.459	**1.468**
	oxirane	1.477	1.475	**1.471**
	aziridine	1.492	1.492	**1.481**
	trans-1,3-butadiene	1.458	1.457	**1.483**
	thiirane	1.491	1.488	**1.484**
	bicyclo[1.1.0]butane (C_1C_3)	1.508	1.506	**1.497**
	bicyclo[1.1.0]butane (C_1C_2)	1.506	1.507	**1.498**
	acetaldehyde	1.513	1.508	**1.501**
	propene	1.505	1.503	**1.501**
	phosphacyclopropane	1.498	1.497	**1.502**
	benzvalene (C_2C_3)	1.515	1.513	**1.503**
	fluoroethane	1.521	1.517	**1.505**
	cyclopentadiene (C_4C_5)	1.510	1.507	**1.506**
	acetone	1.527	1.523	**1.507**
	isobutene	1.512	1.511	**1.507**
	cyclopentenone (C_2C_3)	1.509	1.508	**1.509**
	cyclopropane	1.516	1.516	**1.510**
	cyclopropene	1.517	1.518	**1.515**
	cyclobutene (C_1C_4)	1.523	1.523	**1.517**
	acetic acid	1.514	1.509	**1.517**
	cyclopentenone (C_1C_2)	1.548	1.542	**1.524**
	propane	1.536	1.536	**1.526**
	trans-ethanethiol	1.532	1.530	**1.529**
	benzvalene (C_3C_4)	1.537	1.538	**1.529**
	1,1,1-trifluoroethane	1.513	1.508	**1.530**
	trans-ethanol	1.524	1.521	**1.530**

Table A5-24: Bond Distances Connecting Heavy Atoms. BP Density Functional Models (3)

bond type	molecule	6-31G*	6-311+G**	expt.
	ethylsilane	1.542	1.542	**1.540**
	neopentane	1.545	1.544	**1.540**
	isobutane	1.540	1.539	**1.541**
	methylenecyclopropane (C_3C_4)	1.547	1.549	**1.542**
	cyclobutane	1.559	1.559	**1.548**
	cyclobutene (C_3C_4)	1.578	1.578	**1.566**
C≡N	acetonitrile	1.174	1.166	**1.157**
	cyanoacetylene	1.182	1.173	**1.159**
	cyanogen fluoride	1.177	1.168	**1.159**
	acrylonitrile	1.177	1.170	**1.164**
	methyl isocyanide	1.190	1.183	**1.166**
	malononitrile	1.172	1.164	**1.167**
	formonitrile oxide	1.179	1.173	**1.168**
	nitrosyl cyanide	1.184	1.175	**1.170**
C=N	isocyanic acid	1.231	1.223	**1.209**
	isothiocyanic acid	1.222	1.216	**1.216**
	diazomethane	1.301	1.301	**1.300**
C⁓N (aromatic)	pyridine	1.349	1.347	**1.338**
C-N	thioformamide	1.353	1.353	**1.358**
	pyrrole	1.382	1.382	**1.370**
	formamide	1.370	1.369	**1.376**
	nitrosyl cyanide	1.435	1.425	**1.401**
	methyl isocyanide	1.423	1.424	**1.424**
	trimethylamine	1.461	1.461	**1.451**
	dimethylamine	1.464	1.464	**1.462**
	aziridine	1.484	1.484	**1.475**
	diazirine	1.491	1.492	**1.482**
	nitromethane	1.510	1.513	**1.489**
	nitrosomethane	1.503	1.495	**1.49**
C=O	carbonyl sulfide	1.179	1.171	**1.160**
	ketene	1.183	1.173	**1.161**
	carbon dioxide	1.181	1.173	**1.162**
	isocyanic acid	1.186	1.177	**1.166**
	carbonyl fluoride	1.192	1.182	**1.170**
	formyl fluoride	1.197	1.188	**1.181**
	formamide	1.227	1.223	**1.193**
	methyl formate	1.217	1.212	**1.200**
	formic acid	1.216	1.210	**1.202**
	cyclopentenone	1.220	1.216	**1.210**
	cyclopropenone	1.216	1.212	**1.212**
	acetic acid	1.222	1.216	**1.212**
	acetaldehyde	1.221	1.217	**1.216**
	acetone	1.227	1.223	**1.222**

Table A5-24: Bond Distances Connecting Heavy Atoms. BP Density Functional Models (4)

bond type	molecule	6-31G*	6-311+G**	expt.
C-O	methyl formate (CH$_3$O-CHO)	1.356	1.353	**1.334**
	formic acid	1.360	1.358	**1.343**
	acetic acid	1.372	1.372	**1.360**
	cis-methyl vinyl ether (CH$_3$O-CHCH$_2$)	1.366	1.364	**1.360**
	furan	1.375	1.374	**1.362**
	dimethyl ether	1.418	1.422	**1.410**
	trans-ethanol	1.433	1.441	**1.425**
	cis methyl vinyl ether (CH$_3$-OCHCH$_2$)	1.430	1.433	**1.428**
	oxirane	1.442	1.443	**1.436**
	methyl formate (CH$_3$-OCHO)	1.452	1.455	**1.437**
C-F	cyanogen fluoride	1.283	1.276	**1.262**
	fluoroacetylene	1.293	1.268	**1.279**
	tetrafluoromethane	1.341	1.341	**1.317**
	carbonyl fluoride	1.337	1.337	**1.317**
	1,1-difluoroethylene	1.337	1.337	**1.323**
	trifluoromethane	1.354	1.356	**1.332**
	cis-1,2-difluoroethylene	1.351	1.352	**1.335**
	formyl fluoride	1.361	1.372	**1.338**
	trans-1,2-difluoroethylene	1.355	1.359	**1.344**
	fluoroethylene	1.355	1.362	**1.348**
	1,1,1-trifluoroethane	1.364	1.368	**1.348**
	difluoromethane	1.371	1.378	**1.358**
	fluoroethane	1.402	1.418	**1.398**
C-Al	dimethylaluminum chloride dimer	1.964	1.961	**1.935**
	trimethylaluminum	1.983	1.979	**1.957**
	trimethylaluminum dimer (terminal)	1.981	1.979	**1.957**
	trimethylaluminum dimer (bridge)	2.174	2.170	**2.140**
C=Si	methylenedimethylsilane	1.723	1.721	**1.83**
C-Si	vinylsilane	1.878	1.878	**1.853**
	ethylsilane	1.903	1.899	**1.866**
	dimethylsilane	1.898	1.893	**1.867**
	trimethylsilane	1.900	1.895	**1.868**
	tetramethylsilane	1.904	1.899	**1.875**
	methylenedimethylsilane	1.891	1.883	**1.91**
C-P	trimethylphosphine	1.874	1.869	**1.841**
	dimethylphosphine	1.879	1.874	**1.848**
	phosphacyclopropane	1.897	1.890	**1.867**
C=S	carbon disulfide	1.573	1.570	**1.553**
	carbonyl sulfide	1.581	1.575	**1.560**
	isothiocyanic acid	1.587	1.581	**1.561**
	thioformamide	1.657	1.652	**1.626**
C-S	thiophene	1.742	1.738	**1.714**
	dimethyl sulfide	1.831	1.829	**1.802**
	thiirane	1.843	1.841	**1.815**
	trans-ethanethiol	1.857	1.856	**1.820**

Table A5-24: Bond Distances Connecting Heavy Atoms. BP Density Functional Models (5)

bond type	molecule	6-31G*	6-311+G**	expt.
C-Cl	trichloromethane	1.796	1.794	**1.758**
	tetrachloromethane	1.799	1.797	**1.767**
	dichloromethane	1.767	1.797	**1.772**
N≡N	diazomethane	1.159	1.151	**1.139**
N=N	diazirine	1.245	1.236	**1.228**
	trans-difluorodiazene	1.246	1.235	**1.214**
N=O	nitrosyl fluoride	1.161	1.140	**1.136**
	nitrosyl chloride	1.157	1.145	**1.139**
	formonitrile oxide	1.214	1.208	**1.199**
	nitrosomethane	1.224	1.216	**1.22**
	nitromethane	1.240	1.234	**1.224**
	nitrosyl cyanide	1.221	1.214	**1.228**
N-F	nitrogen trifluoride	1.407	1.408	**1.365**
	trans-difluorodiazene	1.406	1.409	**1.384**
	nitrosyl fluoride	1.524	1.581	**1.512**
N-Cl	nitrogen trichloride	1.825	1.824	**1.748**
	nitrosyl chloride	2.026	2.039	**1.975**
O-∸-O	ozone	1.287	1.280	**1.278**
O-O	fluorine peroxide	1.235	1.195	**1.217**
O-F	oxygen difluoride	1.430	1.432	**1.405**
	fluorine peroxide	1.576	1.632	**1.575**
O-Cl	oxygen dichloride	1.755	1.761	**1.700**
F-Mg	magnesium difluoride	1.751	1.793	**1.77**
F-Al	aluminum trifluoride	1.652	1.670	**1.63**
F-Si	tetrafluorosilane	1.592	1.597	**1.554**
	trifluorosilane	1.602	1.611	**1.562**
	difluorosilane	1.614	1.627	**1.577**
F-P	trifluorophosphine	1.612	1.623	**1.551**
F-S	difluorosulfide	1.642	1.658	**1.592**
Al-Al	trimethylaluminum dimer	2.630	2.641	**2.619**
	dimethylaluminum chloride dimer	3.354	3.356	**3.274**
Al-Cl	trichloroaluminum	2.096	2.092	**2.06**
	dimethylaluminum chloride dimer	2.358	2.355	**2.303**
Si-Cl	tetrachlorosilane	2.052	2.048	**2.019**
	trichlorosilane	2.062	2.059	**2.021**
	dichlorosilane	2.073	2.073	**2.033**
P-Cl	trichlorophosphine	2.100	2.105	**2.043**
S-Cl	sulfur dichloride	2.075	2.085	**2.014**

Table A5-25: Bond Distances Connecting Heavy Atoms. BLYP Density Functional Models

bond type	molecule	6-31G*	6-311+G**	expt.
Be-C	dimethylberyllium	1.690	1.685	**1.698**
	methyl (cyclopentadienyl) beryllium (Me)	1.718	1.714	**1.706**
	methyl (cyclopentadienyl) beryllium (cyclopentadiene)	1.953	1.944	**1.923**
Be-Cl	beryllium dichloride	1.813	1.809	**1.75**
B-B	pentaborane (base-cap)	1.706	1.702	**1.690**
	(base-base)	1.810	1.809	**1.803**
B-C	trimethylborane	1.588	1.581	**1.578**
B ⁻⁻ N (aromatic)	borazine	1.441	1.440	**1.418**
B-N	difluoroaminoborane	1.409	1.402	**1.402**
B-O	difluorohydroxyborane	1.365	1.359	**1.344**
B-F	trifluoroborane	1.330	1.331	**1.307**
	difluoroborane	1.331	1.335	**1.311**
	difluorohydroxyborane	1.341	1.345	**1.323**
	difluoroaminoborane	1.408	1.353	**1.325**
B-Cl	trichloroborane	1.766	1.766	**1.742**
C≡C	cyanoacetylene	1.222	1.215	**1.205**
	propyne	1.218	1.212	**1.206**
	but-1-yne-3-ene	1.222	1.216	**1.208**
	2-butyne	1.220	1.214	**1.214**
C=C	butatriene (C_2C_3)	1.278	1.272	**1.283**
	cyclopropene	1.306	1.301	**1.300**
	cyclopropenone	1.358	1.355	**1.302**
	allene	1.316	1.312	**1.308**
	ketene	1.324	1.319	**1.314**
	1,1-difluoroethylene	1.334	1.328	**1.315**
	butatriene (C_1C_2)	1.328	1.325	**1.318**
	propene	1.344	1.341	**1.318**
	trans-1,2-difluoroethylene	1.339	1.334	**1.329**
	isobutene	1.347	1.345	**1.330**
	cyclopentenone	1.349	1.346	**1.330**
	cis-1,2-difluoroethylene	1.340	1.335	**1.331**
	methylenecyclopropane	1.333	1.329	**1.332**
	fluoroethylene	1.336	1.330	**1.332**
	cyclobutene	1.352	1.350	**1.332**
	acrylonitrile	1.351	1.347	**1.339**
	benzvalene	1.351	1.349	**1.339**
	but-1-yne-3-ene	1.354	1.350	**1.341**
	cis-methyl vinyl ether	1.348	1.346	**1.341**
	cyclopentadiene	1.361	1.359	**1.345**
	trans-1,3-butadiene	1.353	1.350	**1.345**

Table A5-25: Bond Distances Connecting Heavy Atoms. BLYP Density Functional Models (2)

bond type	molecule	6-31G*	6-311+G**	expt.
	vinylsilane	1.349	1.346	**1.347**
	furan	1.372	1.369	**1.361**
	thiophene	1.379	1.377	**1.369**
	pyrrole	1.389	1.388	**1.370**
C--C (aromatic)	pyridine (C_3C_4)	1.405	1.399	**1.392**
	pyridine (C_1C_2)	1.406	1.401	**1.394**
	benzene	1.407	1.405	**1.397**
C-C	cyanoacetylene	1.372	1.368	**1.378**
	cyclopropenone	1.452	1.446	**1.412**
	pyrrole	1.434	1.433	**1.417**
	acrylonitrile	1.434	1.431	**1.426**
	furan	1.443	1.443	**1.431**
	but-1-yne-3-ene	1.426	1.425	**1.431**
	thiophene	1.437	1.435	**1.433**
	benzvalene (C_4C_6)	1.459	1.459	**1.452**
	methylenecyclopropane (C_2C_3)	1.481	1.477	**1.457**
	acetonitrile	1.469	1.463	**1.458**
	propyne	1.466	1.463	**1.459**
	cyclopentadiene (C_2C_3)	1.477	1.475	**1.468**
	malononitrile	1.478	1.474	**1.468**
	2-butyne	1.467	1.465	**1.468**
	oxirane	1.481	1.479	**1.471**
	aziridine	1.496	1.496	**1.481**
	trans-1,3-butadiene	1.462	1.461	**1.483**
	thiirane	1.490	1.489	**1.484**
	bicyclo[1.1.0]butane (C_1C_3)	1.515	1.511	**1.497**
	bicyclo[1.1.0]butane (C_1C_2)	1.510	1.511	**1.498**
	acetaldehyde	1.520	1.515	**1.501**
	propene	1.511	1.510	**1.501**
	phosphacyclopropane	1.499	1.500	**1.502**
	benzvalene (C_2C_3)	1.520	1.518	**1.503**
	fluoroethane	1.527	1.522	**1.505**
	cyclopentadiene (C_4C_5)	1.518	1.515	**1.506**
	acetone	1.534	1.528	**1.507**
	isobutene	1.518	1.517	**1.507**
	cyclopentenone (C_2C_3)	1.523	1.515	**1.509**
	cyclopropane	1.521	1.521	**1.510**
	cyclopropene	1.522	1.524	**1.515**
	cyclobutene (C_1C_4)	1.529	1.530	**1.517**
	acetic acid	1.521	1.516	**1.517**
	cyclopentenone (C_1C_2)	1.555	1.550	**1.524**
	propane	1.543	1.543	**1.526**
	trans-ethanethiol	1.537	1.536	**1.529**
	benzvalene (C_3C_4)	1.544	1.544	**1.529**
	1,1,1-trifluoroethane	1.517	1.513	**1.530**
	trans-ethanol	1.531	1.527	**1.530**

Table A5-25: Bond Distances Connecting Heavy Atoms. BLYP Density Functional Models (3)

bond type	molecule	6-31G*	6-311+G**	expt.
	ethylsilane	1.550	1.550	**1.540**
	neopentane	1.553	1.553	**1.540**
	isobutane	1.547	1.547	**1.541**
	methylenecyclopropane (C₃C₄)	1.554	1.557	**1.542**
	cyclobutane	1.566	1.567	**1.548**
	cyclobutene (C₃C₄)	1.586	1.588	**1.566**
C≡N	acetonitrile	1.173	1.165	**1.157**
	cyanoacetylene	1.181	1.172	**1.159**
	cyanogen fluoride	1.176	1.167	**1.159**
	acrylonitrile	1.177	1.169	**1.164**
	methyl isocyanide	1.190	1.181	**1.166**
	malononitrile	1.172	1.163	**1.167**
	formonitrile oxide	1.179	1.172	**1.168**
	nitrosyl cyanide	1.183	1.174	**1.170**
C=N	isocyanic acid	1.232	1.223	**1.209**
	isothiocyanic acid	1.223	1.215	**1.216**
	diazomethane	1.304	1.305	**1.300**
C-∴-N (aromatic)	pyridine	1.352	1.348	**1.338**
C-N	thioformamide	1.357	1.357	**1.358**
	pyrrole	1.388	1.387	**1.370**
	formamide	1.375	1.373	**1.376**
	nitrosyl cyanide	1.441	1.429	**1.401**
	methyl isocyanide	1.430	1.433	**1.424**
	trimethylamine	1.469	1.469	**1.451**
	dimethylamine	1.471	1.472	**1.462**
	aziridine	1.493	1.493	**1.475**
	diazirine	1.499	1.502	**1.482**
	nitromethane	1.521	1.525	**1.489**
	nitrosomethane	1.513	1.505	**1.49**
C=O	carbonyl sulfide	1.180	1.172	**1.160**
	ketene	1.185	1.175	**1.161**
	carbon dioxide	1.183	1.174	**1.162**
	isocyanic acid	1.188	1.178	**1.166**
	carbonyl fluoride	1.192	1.182	**1.170**
	formyl fluoride	1.198	1.188	**1.181**
	formamide	1.228	1.225	**1.193**
	methyl formate	1.219	1.214	**1.200**
	formic acid	1.217	1.211	**1.202**
	cyclopentenone	1.221	1.217	**1.210**
	cyclopropenone	1.218	1.214	**1.212**
	acetic acid	1.223	1.217	**1.212**
	acetaldehyde	1.223	1.219	**1.216**
	acetone	1.228	1.223	**1.222**

Table A5-25: Bond Distances Connecting Heavy Atoms. BLYP Density Functional Models (4)

bond type	molecule	6-31G*	6-311+G**	expt.
C-O	methyl formate (CH₃O-CHO)	1.361	1.360	**1.334**
	formic acid	1.366	1.366	**1.343**
	acetic acid	1.378	1.380	**1.360**
	cis-methyl vinyl ether (CH₃O-CHCH₂)	1.372	1.371	**1.360**
	furan	1.382	1.381	**1.362**
	dimethyl ether	1.426	1.431	**1.410**
	trans-ethanol	1.442	1.452	**1.425**
	cis methyl vinyl ether (CH₃-OCHCH₂)	1.438	1.443	**1.428**
	oxirane	1.450	1.454	**1.436**
	methyl formate (CH₃-OCHO)	1.460	1.466	**1.437**
C-F	cyanogen fluoride	1.288	1.283	**1.262**
	fluoroacetylene	1.298	1.293	**1.279**
	tetrafluoromethane	1.347	1.347	**1.317**
	carbonyl fluoride	1.342	1.345	**1.317**
	1,1-difluoroethylene	1.342	1.345	**1.323**
	trifluoromethane	1.359	1.363	**1.332**
	cis-1,2-difluoroethylene	1.357	1.361	**1.335**
	formyl fluoride	1.368	1.384	**1.338**
	trans-1,2-difluoroethylene	1.362	1.368	**1.344**
	fluoroethylene	1.363	1.373	**1.348**
	1,1,1-trifluoroethane	1.369	1.375	**1.348**
	difluoromethane	1.378	1.387	**1.358**
	fluoroethane	1.410	1.431	**1.398**
C-Al	dimethylaluminum chloride dimer	1.969	1.967	**1.935**
	trimethylaluminum	1.988	1.980	**1.957**
	trimethylaluminum dimer (terminal)	1.985	1.984	**1.957**
	trimethylaluminum dimer (bridge)	2.170	2.188	**2.140**
C=Si	methylenedimethylsilane	1.724	1.718	**1.83**
C-Si	vinylsilane	1.880	1.884	**1.853**
	ethylsilane	1.909	1.907	**1.866**
	dimethylsilane	1.904	1.900	**1.867**
	trimethylsilane	1.906	1.902	**1.868**
	tetramethylsilane	1.910	1.906	**1.875**
	methylenedimethylsilane	1.899	1.888	**1.91**
C-P	trimethylphosphine	1.888	1.883	**1.841**
	dimethylphosphine	1.892	1.885	**1.848**
	phosphacyclopropane	1.912	1.905	**1.867**
C=S	carbon disulfide	1.578	1.575	**1.553**
	carbonyl sulfide	1.586	1.580	**1.560**
	isothiocyanic acid	1.592	1.585	**1.561**
	thioformamide	1.663	1.658	**1.626**
C-S	thiophene	1.754	1.751	**1.714**
	dimethyl sulfide	1.848	1.844	**1.802**
	thiirane	1.863	1.860	**1.815**
	trans-ethanethiol	1.874	1.874	**1.820**

Table A5-25: Bond Distances Connecting Heavy Atoms. BLYP Density Functional Models (5)

bond type	molecule	6-31G*	6-311+G**	expt.
C-Cl	trichloromethane	1.812	1.809	**1.758**
	tetrachloromethane	1.816	1.812	**1.767**
	dichloromethane	1.814	1.814	**1.772**
N≡N	diazomethane	1.161	1.152	**1.139**
N=N	diazirine	1.246	1.236	**1.228**
	trans-difluorodiazene	1.247	1.234	**1.214**
N=O	nitrosyl fluoride	1.163	1.140	**1.136**
	nitrosyl chloride	1.158	1.146	**1.139**
	formonitrile oxide	1.221	1.216	**1.199**
	nitrosomethane	1.228	1.220	**1.22**
	nitromethane	1.245	1.241	**1.224**
	nitrosyl cyanide	1.225	1.219	**1.228**
N-F	nitrogen trifluoride	1.420	1.423	**1.365**
	trans-difluorodiazene	1.423	1.430	**1.384**
	nitrosyl fluoride	1.543	1.612	**1.512**
N-Cl	nitrogen trichloride	1.850	1.848	**1.748**
	nitrosyl chloride	2.061	2.073	**1.975**
O⁼O	ozone	1.299	1.292	**1.278**
O-O	fluorine peroxide	1.243	1.201	**1.217**
O-F	oxygen difluoride	1.444	1.449	**1.405**
	fluorine peroxide	1.593	1.667	**1.575**
O-Cl	oxygen dichloride	1.775	1.782	**1.700**
F-Mg	magnesium difluoride	1.746	1.796	**1.77**
F-Al	aluminum trifluoride	1.654	1.674	**1.63**
F-Si	tetrafluorosilane	1.595	1.602	**1.554**
	trifluorosilane	1.605	1.617	**1.562**
	difluorosilane	1.616	1.633	**1.577**
F-P	trifluorophosphine	1.618	1.632	**1.551**
F-S	difluorosulfide	1.653	1.672	**1.592**
Al-Al	trimethylaluminum dimer	2.651	2.666	**2.619**
	dimethylaluminum chloride dimer	3.397	3.404	**3.274**
Al-Cl	trichloroaluminum	2.104	2.102	**2.06**
	dimethylaluminum chloride dimer	2.379	2.377	**2.303**
Si-Cl	tetrachlorosilane	2.066	2.062	**2.019**
	trichlorosilane	2.074	2.072	**2.021**
	dichlorosilane	2.083	2.086	**2.033**
P-Cl	trichlorophosphine	2.122	2.120	**2.043**
S-Cl	sulfur dichloride	2.099	2.085	**2.014**

Table A5-26: Bond Distances Connecting Heavy Atoms. EDF1 Density Functional Models

bond type	molecule	6-31G*	6-311+G**	expt.
Be-C	dimethylberyllium	1.690	1.683	**1.698**
	methyl (cyclopentadienyl) beryllium (Me)	1.719	1.713	**1.706**
	methyl (cyclopentadienyl) beryllium (cyclopentadiene)	1.934	1.926	**1.923**
Be-Cl	beryllium dichloride	1.808	1.804	**1.75**
B-B	pentaborane (base-cap)	1.697	1.697	**1.690**
	(base-base)	1.795	1.795	**1.803**
B-C	trimethylborane	1.579	1.578	**1.578**
B ⁻⁻ N (aromatic)	borazine	1.435	1.435	**1.418**
B-N	difluoroaminoborane	1.405	1.399	**1.402**
B-O	difluorohydroxyborane	1.361	1.354	**1.344**
B-F	trifluoroborane	1.327	1.327	**1.307**
	difluoroborane	1.328	1.331	**1.311**
	difluorohydroxyborane	1.336	1.340	**1.323**
	difluoroaminoborane	1.344	1.348	**1.325**
B-Cl	trichloroborane	1.755	1.754	**1.742**
C≡C	cyanoacetylene	1.220	1.214	**1.205**
	propyne	1.215	1.211	**1.206**
	but-1-yne-3-ene	1.220	1.215	**1.208**
	2-butyne	1.218	1.213	**1.214**
C=C	butatriene (C_2C_3)	1.275	1.270	**1.283**
	cyclopropene	1.302	1.299	**1.300**
	cyclopropenone	1.352	1.351	**1.302**
	allene	1.312	1.308	**1.308**
	ketene	1.319	1.314	**1.314**
	1,1-difluoroethylene	1.330	1.325	**1.315**
	butatriene (C_1C_2)	1.324	1.321	**1.318**
	propene	1.340	1.337	**1.318**
	trans-1,2-difluoroethylene	1.336	1.331	**1.329**
	isobutene	1.343	1.341	**1.330**
	cyclopentenone	1.344	1.342	**1.330**
	cis-1,2-difluoroethylene	1.336	1.332	**1.331**
	methylenecyclopropane	1.328	1.326	**1.332**
	fluoroethylene	1.332	1.327	**1.332**
	cyclobutene	1.347	1.345	**1.332**
	acrylonitrile	1.346	1.343	**1.339**
	benzvalene	1.346	1.344	**1.339**
	but-1-yne-3-ene	1.349	1.346	**1.341**
	cis-methyl vinyl ether	1.341	1.341	**1.341**
	cyclopentadiene	1.356	1.355	**1.345**
	trans-1,3-butadiene	1.348	1.346	**1.345**

Table A5-26: Bond Distances Connecting Heavy Atoms. EDF1 Density Functional Models (2)

bond type	molecule	6-31G*	6-311+G**	expt.
	vinylsilane	1.341	1.342	**1.347**
	furan	1.367	1.364	**1.361**
	thiophene	1.374	1.373	**1.369**
	pyrrole	1.384	1.382	**1.370**
C ≈ C (aromatic)	pyridine (C_3C_4)	1.398	1.396	**1.392**
	pyridine (C_1C_2)	1.400	1.398	**1.394**
	benzene	1.400	1.398	**1.397**
C-C	cyanoacetylene	1.368	1.364	**1.378**
	cyclopropenone	1.444	1.438	**1.412**
	pyrrole	1.425	1.424	**1.417**
	acrylonitrile	1.428	1.424	**1.426**
	furan	1.435	1.434	**1.431**
	but-1-yne-3-ene	1.421	1.419	**1.431**
	thiophene	1.428	1.426	**1.433**
	benzvalene (C_4C_6)	1.452	1.452	**1.452**
	methylenecyclopropane (C_2C_3)	1.471	1.468	**1.457**
	acetonitrile	1.460	1.454	**1.458**
	propyne	1.458	1.454	**1.459**
	cyclopentadiene (C_2C_3)	1.467	1.465	**1.468**
	malononitrile	1.470	1.465	**1.468**
	2-butyne	1.459	1.456	**1.468**
	oxirane	1.471	1.469	**1.471**
	aziridine	1.487	1.486	**1.481**
	trans-1,3-butadiene	1.455	1.453	**1.483**
	thiirane	1.486	1.484	**1.484**
	bicyclo[1.1.0]butane (C_1C_3)	1.500	1.499	**1.497**
	bicyclo[1.1.0]butane (C_1C_2)	1.500	1.501	**1.498**
	acetaldehyde	1.511	1.505	**1.501**
	propene	1.502	1.499	**1.501**
	phosphacyclopropane	1.494	1.493	**1.502**
	benzvalene (C_2C_3)	1.510	1.508	**1.503**
	fluoroethane	1.518	1.514	**1.505**
	cyclopentadiene (C_4C_5)	1.505	1.503	**1.506**
	acetone	1.520	1.520	**1.507**
	isobutene	1.510	1.508	**1.507**
	cyclopentenone (C_2C_3)	1.504	1.503	**1.509**
	cyclopropane	1.510	1.510	**1.510**
	cyclopropene	1.510	1.511	**1.515**
	cyclobutene (C_1C_4)	1.517	1.517	**1.517**
	acetic acid	1.511	1.506	**1.517**
	cyclopentenone (C_1C_2)	1.543	1.538	**1.524**
	propane	1.533	1.532	**1.526**
	trans-ethanethiol	1.528	1.526	**1.529**
	benzvalene (C_3C_4)	1.530	1.531	**1.529**
	1,1,1-trifluoroethane	1.510	1.506	**1.530**
	trans-ethanol	1.521	1.518	**1.530**

Table A5-26: Bond Distances Connecting Heavy Atoms. EDF1 Density Functional Models (3)

bond type	molecule	6-31G*	6-311+G**	expt.
	ethylsilane	1.539	1.538	**1.540**
	neopentane	1.544	1.543	**1.540**
	isobutane	1.537	1.536	**1.541**
	methylenecyclopropane (C_3C_4)	1.540	1.542	**1.542**
	cyclobutane	1.554	1.554	**1.548**
	cyclobutene (C_3C_4)	1.572	1.572	**1.566**
C≡N	acetonitrile	1.170	1.163	**1.157**
	cyanoacetylene	1.178	1.170	**1.159**
	cyanogen fluoride	1.173	1.165	**1.159**
	acrylonitrile	1.174	1.166	**1.164**
	methyl isocyanide	1.187	1.180	**1.166**
	malononitrile	1.169	1.161	**1.167**
	formonitrile oxide	1.175	1.169	**1.168**
	nitrosyl cyanide	1.180	1.172	**1.170**
C=N	isocyanic acid	1.227	1.220	**1.209**
	isothiocyanic acid	1.218	1.213	**1.216**
	diazomethane	1.297	1.297	**1.300**
C ⁓ N (aromatic)	pyridine	1.344	1.342	**1.338**
C-N	thioformamide	1.349	1.349	**1.358**
	pyrrole	1.377	1.377	**1.370**
	formamide	1.367	1.365	**1.376**
	nitrosyl cyanide	1.429	1.402	**1.401**
	methyl isocyanide	1.419	1.420	**1.424**
	trimethylamine	1.456	1.457	**1.451**
	dimethylamine	1.459	1.459	**1.462**
	aziridine	1.477	1.477	**1.475**
	diazirine	1.483	1.484	**1.482**
	nitromethane	1.507	1.511	**1.489**
	nitrosomethane	1.499	1.491	**1.49**
C=O	carbonyl sulfide	1.176	1.168	**1.160**
	ketene	1.180	1.170	**1.161**
	carbon dioxide	1.178	1.170	**1.162**
	isocyanic acid	1.183	1.174	**1.166**
	carbonyl fluoride	1.188	1.179	**1.170**
	formyl fluoride	1.193	1.184	**1.181**
	formamide	1.223	1.219	**1.193**
	methyl formate	1.213	1.208	**1.200**
	formic acid	1.212	1.207	**1.202**
	cyclopentenone	1.216	1.212	**1.210**
	cyclopropenone	1.213	1.208	**1.212**
	acetic acid	1.218	1.213	**1.212**
	acetaldehyde	1.217	1.213	**1.216**
	acetone	1.219	1.219	**1.222**

Table A5-26: Bond Distances Connecting Heavy Atoms. EDF1 Density Functional Models (4)

bond type	molecule	6-31G*	6-311+G**	expt.
	cis-methyl vinyl ether (CH$_3$O-CHCH$_2$)	1.360	1.360	**1.360**
	furan	1.369	1.369	**1.362**
	dimethyl ether	1.414	1.417	**1.410**
	trans-ethanol	1.429	1.435	**1.425**
	cis methyl vinyl ether (CH$_3$-OCHCH$_2$)	1.428	1.428	**1.428**
	oxirane	1.435	1.436	**1.436**
	methyl formate (CH$_3$-OCHO)	1.446	1.449	**1.437**
C-F	cyanogen fluoride	1.280	1.273	**1.262**
	fluoroacetylene	1.289	1.283	**1.279**
	tetrafluoromethane	1.338	1.338	**1.317**
	carbonyl fluoride	1.333	1.332	**1.317**
	1,1-difluoroethylene	1.333	1.334	**1.323**
	trifluoromethane	1.351	1.352	**1.332**
	cis-1,2-difluoroethylene	1.347	1.348	**1.335**
	formyl fluoride	1.358	1.367	**1.338**
	trans-1,2-difluoroethylene	1.351	1.355	**1.344**
	fluoroethylene	1.352	1.359	**1.348**
	1,1,1-trifluoroethane	1.361	1.364	**1.348**
	difluoromethane	1.367	1.374	**1.358**
	fluoroethane	1.398	1.414	**1.398**
C-Al	dimethylaluminum chloride dimer	1.964	1.960	**1.935**
	trimethylaluminum	1.983	1.978	**1.957**
	trimethylaluminum dimer (terminal)	1.981	1.978	**1.957**
	trimethylaluminum dimer (bridge)	2.177	2.172	**2.140**
C=Si	methylenedimethylsilane	1.719	1.717	**1.83**
C-Si	vinylsilane	1.876	1.876	**1.853**
	ethylsilane	1.901	1.897	**1.866**
	dimethylsilane	1.895	1.890	**1.867**
	trimethylsilane	1.899	1.893	**1.868**
	tetramethylsilane	1.903	1.897	**1.875**
	methylenedimethylsilane	1.889	1.881	**1.91**
C-P	trimethylphosphine	1.871	1.865	**1.841**
	dimethylphosphine	1.874	1.869	**1.848**
	phosphacyclopropane	1.887	1.881	**1.867**
C=S	carbon disulfide	1.568	1.566	**1.553**
	carbonyl sulfide	1.575	1.570	**1.560**
	isothiocyanic acid	1.581	1.576	**1.561**
	thioformamide	1.651	1.646	**1.626**
C-S	thiophene	1.734	1.731	**1.714**
	dimethyl sulfide	1.825	1.823	**1.802**
	thiirane	1.832	1.830	**1.815**
	trans-ethanethiol	1.850	1.850	**1.820**

Table A5-26: Bond Distances Connecting Heavy Atoms. EDF1 Density Functional Models (5)

bond type	molecule	6-31G*	6-311+G**	expt.
C-Cl	trichloromethane	1.791	1.788	**1.758**
	tetrachloromethane	1.795	1.793	**1.767**
	dichloromethane	1.792	1.790	**1.772**
N≡N	dizaomethane	1.156	1.147	**1.139**
N=N	diazirine	1.241	1.232	**1.228**
	trans-difluorodiazene	1.242	1.231	**1.214**
N=O	nitrosyl fluoride	1.157	1.136	**1.136**
	nitrosyl chloride	1.152	1.141	**1.139**
	formonitrile oxide	1.210	1.204	**1.199**
	nitrosomethane	1.219	1.211	**1.22**
	nitromethane	1.244	1.229	**1.224**
	nitrosyl cyanide	1.216	1.209	**1.228**
N-F	nitrogen trifluoride	1.403	1.402	**1.365**
	trans-difluorodiazene	1.400	1.402	**1.384**
	nitrosyl fluoride	1.522	1.580	**1.512**
N-Cl	nitrogen trichloride	1.818	1.816	**1.748**
	nitrosyl chloride	2.026	2.037	**1.975**
O≔O	ozone	1.281	1.273	**1.278**
O-O	fluorine peroxide	1.226	1.187	**1.217**
O-F	oxygen difluoride	1.425	1.427	**1.405**
	fluorine peroxide	1.579	1.648	**1.575**
O-Cl	oxygen dichloride	1.748	1.754	**1.700**
F-Mg	magnesium difluoride	1.751	1.793	**1.77**
F-Al	aluminum trifluoride	1.651	1.669	**1.63**
F-Si	tetrafluorosilane	1.590	1.595	**1.554**
	trifluorosilane	1.600	1.609	**1.562**
	difluorosilane	1.612	1.625	**1.577**
F-P	trifluorophosphine	1.609	1.620	**1.551**
F-S	difluorosulfide	1.637	1.653	**1.592**
Al-Al	trimethylaluminum dimer	2.639	2.649	**2.619**
	dimethylaluminum chloride dimer	3.373	3.373	**3.274**
Al-Cl	trichloroaluminum	2.094	2.090	**2.06**
	dimethylaluminum chloride dimer	2.362	2.358	**2.303**
Si-Cl	tetrachlorosilane	2.058	2.044	**2.019**
	trichlorosilane	2.058	2.056	**2.021**
	dichlorosilane	2.069	2.069	**2.033**
P-Cl	trichlorophosphine	2.093	2.097	**2.043**
S-Cl	sulfur dichloride	2.066	2.076	**2.014**

Table A5-27: Bond Distances Connecting Heavy Atoms. B3LYP Density Functional Models

bond type	molecule	6-31G*	6-311+G**	expt.
Be-C	dimethylberyllium	1.685	1.679	**1.698**
	methyl (cyclopentadienyl) beryllium (Me)	1.713	1.707	**1.706**
	methyl (cyclopentadienyl) beryllium (cyclopentadiene)	1.937	1.928	**1.923**
Be-Cl	beryllium dichloride	1.804	1.800	**1.75**
B-B	pentaborane (base-cap)	1.696	1.691	**1.690**
	(base-base)	1.799	1.794	**1.803**
B-C	trimethylborane	1.580	1.574	**1.578**
B --- N (aromatic)	borazine	1.431	1.431	**1.418**
B-N	difluoroaminoborane	1.400	1.393	**1.402**
B-O	difluorohydroxyborane	1.352	1.346	**1.344**
B-F	trifluoroborane	1.318	1.318	**1.307**
	difluoroborane	1.320	1.322	**1.311**
	difluorohydroxyborane	1.328	1.331	**1.323**
	difluoroaminoborane	1.336	1.339	**1.325**
B-Cl	trichloroborane	1.753	1.752	**1.742**
C≡C	cyanoacetylene	1.210	1.204	**1.205**
	propyne	1.207	1.202	**1.206**
	but-1-yne-3-ene	1.211	1.206	**1.208**
	2-butyne	1.209	1.204	**1.214**
C=C	butatriene (C_2C_3)	1.271	1.266	**1.283**
	cyclopropene	1.295	1.291	**1.300**
	cyclopropenone	1.346	1.344	**1.302**
	allene	1.307	1.303	**1.308**
	ketene	1.315	1.310	**1.314**
	1,1-difluoroethylene	1.322	1.317	**1.315**
	butatriene (C_1C_2)	1.318	1.315	**1.318**
	propene	1.333	1.331	**1.318**
	trans-1,2-difluoroethylene	1.327	1.320	**1.329**
	isobutene	1.337	1.335	**1.330**
	cyclopentenone	1.338	1.336	**1.330**
	cis-1,2-difluoroethylene	1.327	1.324	**1.331**
	methylenecyclopropane	1.323	1.320	**1.332**
	fluoroethylene	1.325	1.320	**1.332**
	cyclobutene	1.341	1.339	**1.332**
	acrylonitrile	1.339	1.335	**1.339**
	benzvalene	1.339	1.338	**1.339**
	but-1-yne-3-ene	1.341	1.338	**1.341**
	cis-methyl vinyl ether	1.337	1.332	**1.341**
	cyclopentadiene	1.349	1.348	**1.345**
	trans-1,3-butadiene	1.340	1.338	**1.345**

Table A5-27: Bond Distances Connecting Heavy Atoms. B3LYP Density Functional Models (2)

bond type	molecule	6-31G*	6-311+G**	expt.
	vinylsilane	1.339	1.332	**1.347**
	furan	1.361	1.358	**1.361**
	thiophene	1.368	1.366	**1.369**
	pyrrole	1.376	1.377	**1.370**
C ≕ C (aromatic)	pyridine (C_3C_4)	1.394	1.392	**1.392**
	pyridine (C_1C_2)	1.396	1.394	**1.394**
	benzene	1.397	1.395	**1.397**
C-C	cyanoacetylene	1.373	1.369	**1.378**
	cyclopropenone	1.436	1.431	**1.412**
	pyrrole	1.426	1.425	**1.417**
	acrylonitrile	1.431	1.428	**1.426**
	furan	1.436	1.435	**1.431**
	but-1-yne-3-ene	1.424	1.423	**1.431**
	thiophene	1.430	1.428	**1.433**
	benzvalene (C_4C_6)	1.447	1.448	**1.452**
	methylenecyclopropane (C_2C_3)	1.470	1.467	**1.457**
	acetonitrile	1.461	1.456	**1.458**
	propyne	1.461	1.457	**1.459**
	cyclopentadiene (C_2C_3)	1.470	1.468	**1.468**
	malononitrile	1.470	1.465	**1.468**
	2-butyne	1.462	1.459	**1.468**
	oxirane	1.469	1.467	**1.471**
	aziridine	1.485	1.485	**1.481**
	trans-1,3-butadiene	1.458	1.456	**1.483**
	thiirane	1.481	1.480	**1.484**
	bicyclo[1.1.0]butane (C_1C_3)	1.494	1.492	**1.497**
	bicyclo[1.1.0]butane (C_1C_2)	1.499	1.500	**1.498**
	acetaldehyde	1.508	1.504	**1.501**
	propene	1.502	1.500	**1.501**
	phosphacyclopropane	1.492	1.492	**1.502**
	benzvalene (C_2C_3)	1.511	1.510	**1.503**
	fluoroethane	1.517	1.512	**1.505**
	cyclopentadiene (C_4C_5)	1.507	1.505	**1.506**
	acetone	1.521	1.517	**1.507**
	isobutene	1.509	1.507	**1.507**
	cyclopentenone (C_2C_3)	1.506	1.506	**1.509**
	cyclopropane	1.509	1.509	**1.510**
	cyclopropene	1.509	1.510	**1.515**
	cyclobutene (C_1C_4)	1.519	1.519	**1.517**
	acetic acid	1.508	1.504	**1.517**
	cyclopentenone (C_1C_2)	1.540	1.536	**1.524**
	propane	1.532	1.532	**1.526**
	trans-ethanethiol	1.527	1.526	**1.529**
	benzvalene (C_3C_4)	1.529	1.529	**1.529**
	1,1,1-trifluoroethane	1.506	1.503	**1.530**
	trans-ethanol	1.520	1.517	**1.530**

Table A5-27: Bond Distances Connecting Heavy Atoms. B3LYP Density Functional Models (3)

bond type	molecule	6-31G*	6-311+G**	expt.
	ethylsilane	1.540	1.539	**1.540**
	neopentane	1.540	1.540	**1.540**
	isobutane	1.535	1.535	**1.541**
	methylenecyclopropane (C_3C_4)	1.540	1.542	**1.542**
	cyclobutane	1.553	1.554	**1.548**
	cyclobutene (C_3C_4)	1.572	1.573	**1.566**
C≡N	acetonitrile	1.160	1.153	**1.157**
	cyanoacetylene	1.166	1.158	**1.159**
	cyanogen fluoride	1.162	1.153	**1.159**
	acrylonitrile	1.164	1.156	**1.164**
	methyl isocyanide	1.177	1.165	**1.166**
	malononitrile	1.159	1.151	**1.167**
	formonitrile oxide	1.164	1.158	**1.168**
	nitrosyl cyanide	1.165	1.158	**1.170**
C=N	isocyanic acid	1.219	1.212	**1.209**
	isothiocyanic acid	1.208	1.201	**1.216**
	diazomethane	1.294	1.295	**1.300**
C ⁚⁚ N (aromatic)	pyridine	1.339	1.337	**1.338**
C-N	thioformamide	1.344	1.344	**1.358**
	pyrrole	1.376	1.375	**1.370**
	formamide	1.362	1.358	**1.376**
	nitrosyl cyanide	1.420	1.415	**1.401**
	methyl isocyanide	1.420	1.422	**1.424**
	trimethylamine	1.455	1.455	**1.451**
	dimethylamine	1.457	1.458	**1.462**
	aziridine	1.473	1.472	**1.475**
	diazirine	1.479	1.480	**1.482**
	nitromethane	1.499	1.497	**1.489**
	nitrosomethane	1.489	1.481	**1.49**
C=O	carbonyl sulfide	1.165	1.157	**1.160**
	ketene	1.171	1.161	**1.161**
	carbon dioxide	1.169	1.161	**1.162**
	isocyanic acid	1.174	1.165	**1.166**
	carbonyl fluoride	1.180	1.171	**1.170**
	formyl fluoride	1.186	1.177	**1.181**
	formamide	1.216	1.211	**1.193**
	methyl formate	1.206	1.201	**1.200**
	formic acid	1.205	1.199	**1.202**
	cyclopentenone	1.209	1.205	**1.210**
	cyclopropenone	1.207	1.202	**1.212**
	acetic acid	1.210	1.205	**1.212**
	acetaldehyde	1.211	1.206	**1.216**
	acetone	1.216	1.212	**1.222**

Table A5-27: Bond Distances Connecting Heavy Atoms. B3LYP Density Functional Models (4)

bond type	molecule	6-31G*	6-311+G**	expt.
C-O	methyl formate (CH$_3$O-CHO)	1.342	1.340	**1.334**
	formic acid	1.347	1.346	**1.343**
	acetic acid	1.359	1.359	**1.360**
	cis-methyl vinyl ether (CH$_3$O-CHCH$_2$)	1.358	1.355	**1.360**
	furan	1.364	1.363	**1.362**
	dimethyl ether	1.410	1.413	**1.410**
	trans-ethanol	1.425	1.431	**1.425**
	cis methyl vinyl ether (CH$_3$-OCHCH$_2$)	1.420	1.422	**1.428**
	oxirane	1.430	1.431	**1.436**
	methyl formate (CH$_3$-OCHO)	1.441	1.444	**1.437**
C-F	cyanogen fluoride	1.274	1.267	**1.262**
	fluoroacetylene	1.286	1.279	**1.279**
	tetrafluoromethane	1.330	1.328	**1.317**
	carbonyl fluoride	1.322	1.322	**1.317**
	1,1-difluoroethylene	1.326	1.327	**1.323**
	trifluoromethane	1.341	1.344	**1.332**
	cis-1,2-difluoroethylene	1.344	1.343	**1.335**
	formyl fluoride	1.346	1.355	**1.338**
	trans-1,2-difluoroethylene	1.347	1.348	**1.344**
	fluoroethylene	1.348	1.353	**1.348**
	1,1,1-trifluoroethane	1.352	1.354	**1.348**
	difluoromethane	1.361	1.367	**1.358**
	fluoroethane	1.394	1.407	**1.398**
C-Al	dimethylaluminum chloride dimer	1.956	1.953	**1.935**
	trimethylaluminum	1.975	1.967	**1.957**
	trimethylaluminum dimer (terminal)	1.972	1.971	**1.957**
	trimethylaluminum dimer (bridge)	2.167	2.165	**2.140**
C=Si	methylenedimethylsilane	1.710	1.704	**1.83**
C-Si	vinylsilane	1.872	1.868	**1.853**
	ethylsilane	1.895	1.891	**1.866**
	dimethylsilane	1.891	1.886	**1.867**
	trimethylsilane	1.894	1.888	**1.868**
	tetramethylsilane	1.897	1.892	**1.875**
	methylenedimethylsilane	1.884	1.873	**1.91**
C-P	trimethylphosphine	1.868	1.859	**1.841**
	dimethylphosphine	1.872	1.865	**1.848**
	phosphacyclopropane	1.886	1.880	**1.867**
C=S	carbon disulfide	1.563	1.561	**1.553**
	carbonyl sulfide	1.575	1.568	**1.560**
	isothiocyanic acid	1.581	1.574	**1.561**
	thioformamide	1.649	1.645	**1.626**
C-S	thiophene	1.736	1.733	**1.714**
	dimethyl sulfide	1.827	1.824	**1.802**
	thiirane	1.839	1.836	**1.815**
	trans-ethanethiol	1.849	1.850	**1.820**

Table A5-27: Bond Distances Connecting Heavy Atoms. B3LYP Density Functional Models (5)

bond type	molecule	6-31G*	6-311+G**	expt.
C-Cl	trichloromethane	1.749	1.787	**1.758**
	tetrachloromethane	1.793	1.791	**1.767**
	dichloromethane	1.792	1.791	**1.772**
N≡N	diazomethane	1.146	1.137	**1.139**
N=N	diazirine	1.229	1.220	**1.228**
	trans-difluorodiazene	1.229	1.219	**1.214**
N=O	nitrosyl fluoride	1.149	1.127	**1.136**
	nitrosyl chloride	1.143	1.129	**1.139**
	formonitrile oxide	1.209	1.203	**1.199**
	nitrosomethane	1.211	1.203	**1.22**
	nitromethane	1.227	1.222	**1.224**
	nitrosyl cyanide	1.210	1.202	**1.228**
N-F	nitrogen trifluoride	1.384	1.382	**1.365**
	trans-difluorodiazene	1.388	1.387	**1.384**
	nitrosyl fluoride	1.487	1.538	**1.512**
N-Cl	nitrogen trichloride	1.799	1.795	**1.748**
	nitrosyl chloride	2.004	2.020	**1.975**
O⁻O	ozone	1.265	1.256	**1.278**
O-O	fluorine peroxide	1.266	1.216	**1.217**
O-F	oxygen difluoride	1.409	1.407	**1.405**
	fluorine peroxide	1.497	1.550	**1.575**
O-Cl	oxygen dichloride	1.735	1.739	**1.700**
F-Mg	magnesium difluoride	1.733	1.779	**1.77**
F-Al	aluminum trifluoride	1.639	1.657	**1.63**
F-Si	tetrafluorosilane	1.579	1.584	**1.554**
	trifluorosilane	1.590	1.598	**1.562**
	difluorosilane	1.602	1.614	**1.577**
F-P	trifluorophosphine	1.596	1.605	**1.551**
F-S	difluorosulfide	1.626	1.639	**1.592**
Al-Al	trimethylaluminum dimer	2.629	2.641	**2.619**
	dimethylaluminum chloride dimer	3.361	3.363	**3.274**
Al-Cl	trichloroaluminum	2.087	2.084	**2.06**
	dimethylaluminum chloride dimer	2.354	2.351	**2.303**
Si-Cl	tetrachlorosilane	2.047	2.042	**2.019**
	trichlorosilane	2.054	2.052	**2.021**
	dichlorosilane	2.065	2.066	**2.033**
P-Cl	trichlorophosphine	2.089	2.088	**2.043**
S-Cl	sulfur dichloride	2.065	2.076	**2.014**

Table A5-28: Bond Distances Connecting Heavy Atoms. MP2 Models

bond type	molecule	6-31G*	6-311+G**	expt.
Be-C	dimethylberyllium	1.690	1.689	**1.698**
	methyl (cyclopentadienyl) beryllium (Me)	1.715	1.710	**1.706**
	methyl (cyclopentadienyl) beryllium (cyclopentadiene)	1.914	1.924	**1.923**
Be-Cl	beryllium dichloride	1.791	1.793	**1.75**
B-B	pentaborane (base-cap)	1.690	1.702	**1.690**
	(base-base)	1.787	1.801	**1.803**
B-C	trimethylborane	1.575	1.578	**1.578**
B ⁝ N (aromatic)	borazine	1.432	1.434	**1.418**
B-N	difluoroaminoborane	1.398	1.398	**1.402**
B-O	difluorohydroxyborane	1.354	1.351	**1.344**
B-F	trifluoroborane	1.322	1.318	**1.307**
	difluoroborane	1.325	1.323	**1.311**
	difluorohydroxyborane	1.333	1.330	**1.323**
	difluoroaminoborane	1.340	1.338	**1.325**
B-Cl	trichloroborane	1.737	1.740	**1.742**
C≡C	cyanoacetylene	1.223	1.222	**1.205**
	propyne	1.220	1.219	**1.206**
	but-1-yne-3-ene	1.223	1.222	**1.208**
	2-butyne	1.221	1.220	**1.214**
C=C	butatriene (C$_2$C$_3$)	1.274	1.274	**1.283**
	cyclopropene	1.303	1.305	**1.300**
	cyclopropenone	1.354	1.359	**1.302**
	allene	1.313	1.314	**1.308**
	ketene	1.320	1.322	**1.314**
	1,1-difluoroethylene	1.325	1.326	**1.315**
	butatriene (C$_1$C$_2$)	1.322	1.326	**1.318**
	propene	1.338	1.341	**1.318**
	trans-1,2-difluoroethylene	1.330	1.331	**1.329**
	isobutene	1.341	1.344	**1.330**
	cyclopentenone	1.344	1.347	**1.330**
	cis-1,2-difluoroethylene	1.330	1.332	**1.331**
	methylenecyclopropane	1.327	1.329	**1.332**
	fluoroethylene	1.327	1.330	**1.332**
	cyclobutene	1.347	1.352	**1.332**
	acrylonitrile	1.342	1.344	**1.339**
	benzvalene	1.347	1.352	**1.339**
	but-1-yne-3-ene	1.344	1.347	**1.341**
	cis-methyl vinyl ether	1.341	1.345	**1.341**
	cyclopentadiene	1.354	1.359	**1.345**
	trans-1,3-butadiene	1.344	1.347	**1.345**
	vinylsilane	1.345	1.347	**1.347**

Table A5-28: Bond Distances Connecting Heavy Atoms. MP2 Models (2)

bond type	molecule	6-31G*	6-311+G**	expt.
	furan	1.366	1.370	**1.361**
	thiophene	1.376	1.382	**1.369**
	pyrrole	1.383	1.388	**1.370**
C ≕ C (aromatic)	pyridine (C_3C_4)	1.394	1.397	**1.392**
	pyridine (C_1C_2)	1.396	1.399	**1.394**
	benzene	1.397	1.400	**1.397**
C-C	cyanoacetylene	1.377	1.376	**1.378**
	cyclopropenone	1.438	1.442	**1.412**
	pyrrole	1.419	1.423	**1.417**
	acrylonitrile	1.434	1.434	**1.426**
	furan	1.428	1.432	**1.431**
	but-1-yne-3-ene	1.429	1.430	**1.431**
	thiophene	1.420	1.421	**1.433**
	benzvalene (C_4C_6)	1.452	1.464	**1.452**
	methylenecyclopropane (C_2C_3)	1.466	1.470	**1.457**
	acetonitrile	1.463	1.463	**1.458**
	propyne	1.463	1.464	**1.459**
	cyclopentadiene (C_2C_3)	1.463	1.468	**1.468**
	malononitrile	1.469	1.467	**1.468**
	2-butyne	1.464	1.465	**1.468**
	oxirane	1.465	1.468	**1.471**
	aziridine	1.480	1.485	**1.481**
	trans-1,3-butadiene	1.458	1.460	**1.483**
	thiirane	1.481	1.487	**1.484**
	bicyclo[1.1.0]butane (C_1C_3)	1.494	1.515	**1.497**
	bicyclo[1.1.0]butane (C_1C_2)	1.500	1.500	**1.498**
	acetaldehyde	1.503	1.505	**1.501**
	propene	1.499	1.502	**1.501**
	phosphacyclopropane	1.490	1.496	**1.502**
	benzvalene (C_2C_3)	1.503	1.506	**1.503**
	fluoroethane	1.511	1.512	**1.505**
	cyclopentadiene (C_4C_5)	1.501	1.503	**1.506**
	acetone	1.514	1.516	**1.507**
	isobutene	1.504	1.506	**1.507**
	cyclopentenone (C_2C_3)	1.501	1.505	**1.509**
	cyclopropane	1.504	1.511	**1.510**
	cyclopropene	1.507	1.515	**1.515**
	cyclobutene (C_1C_4)	1.514	1.519	**1.517**
	acetic acid	1.526	1.504	**1.517**
	cyclopentenone (C_1C_2)	1.532	1.534	**1.524**
	propane	1.526	1.529	**1.526**
	trans-ethanethiol	1.523	1.525	**1.529**
	benzvalene (C_3C_4)	1.522	1.529	**1.529**
	1,1,1-trifluoroethane	1.497	1.499	**1.530**
	trans-ethanol	1.514	1.515	**1.530**
	ethylsilane	1.534	1.537	**1.540**
	neopentane	1.530	1.532	**1.540**

Table A5-28: Bond Distances Connecting Heavy Atoms. MP2 Models (3)

bond type	molecule	6-31G*	6-311+G**	expt.
	isobutane	1.527	1.530	**1.541**
	methylenecyclopropane (C$_3$C$_4$)	1.538	1.546	**1.542**
	cyclobutane	1.545	1.550	**1.548**
	cyclobutene (C$_3$C$_4$)	1.566	1.571	**1.566**
C≡N	acetonitrile	1.180	1.174	**1.157**
	cyanoacetylene	1.188	1.181	**1.159**
	cyanogen fluoride	1.181	1.175	**1.159**
	acrylonitrile	1.183	1.177	**1.164**
	methyl isocyanide	1.189	1.184	**1.166**
	malononitrile	1.180	1.174	**1.167**
	formonitrile oxide	1.186	1.182	**1.168**
	nitrosyl cyanide	1.186	1.180	**1.170**
C=N	isocyanic acid	1.224	1.221	**1.209**
	isothiocyanic acid	1.215	1.211	**1.216**
	diazomethane	1.312	1.316	**1.300**
C ⁓ N (aromatic)	pyridine	1.345	1.345	**1.338**
C-N	thioformamide	1.346	1.353	**1.358**
	pyrrole	1.373	1.374	**1.370**
	formamide	1.362	1.364	**1.376**
	nitrosyl cyanide	1.425	1.423	**1.401**
	methyl isocyanide	1.426	1.426	**1.424**
	trimethylamine	1.455	1.455	**1.451**
	dimethylamine	1.458	1.458	**1.462**
	aziridine	1.474	1.478	**1.475**
	diazirine	1.482	1.485	**1.482**
	nitromethane	1.488	1.492	**1.489**
	nitrosomethane	1.480	1.479	**1.49**
C=O	carbonyl sulfide	1.179	1.171	**1.160**
	ketene	1.181	1.168	**1.161**
	carbon dioxide	1.180	1.170	**1.162**
	isocyanic acid	1.184	1.173	**1.166**
	carbonyl fluoride	1.186	1.177	**1.170**
	formyl fluoride	1.194	1.185	**1.181**
	formamide	1.225	1.217	**1.193**
	methyl formate	1.216	1.208	**1.200**
	formic acid	1.214	1.205	**1.202**
	cyclopentenone	1.221	1.213	**1.210**
	cyclopropenone	1.214	1.206	**1.212**
	acetic acid	1.218	1.210	**1.212**
	acetaldehyde	1.224	1.215	**1.216**
	acetone	1.228	1.220	**1.222**
C-O	methyl formate (CH$_3$O-CHO)	1.347	1.343	**1.334**
	formic acid	1.351	1.348	**1.343**
	acetic acid	1.362	1.360	**1.360**

Table A5-28: Bond Distances Connecting Heavy Atoms. MP2 Models (4)

bond type	molecule	6-31G*	6-311+G**	expt.
	cis-methyl vinyl ether (CH_3O-$CHCH_2$)	1.363	1.355	**1.360**
	furan	1.367	1.361	**1.362**
	dimethyl ether	1.416	1.411	**1.410**
	trans-ethanol	1.429	1.427	**1.425**
	cis methyl vinyl ether (CH_3-$OCHCH_2$)	1.425	1.420	**1.428**
	oxirane	1.438	1.434	**1.436**
	methyl formate (CH_3-$OCHO$)	1.444	1.439	**1.437**
C-F	cyanogen fluoride	1.281	1.268	**1.262**
	fluoroacetylene	1.294	1.280	**1.279**
	tetrafluoromethane	1.331	1.323	**1.317**
	carbonyl fluoride	1.326	1.319	**1.317**
	1,1-difluoroethylene	1.331	1.322	**1.323**
	trifluoromethane	1.345	1.338	**1.332**
	cis-1,2-difluoroethylene	1.350	1.339	**1.335**
	formyl fluoride	1.352	1.351	**1.338**
	trans-1,2-difluoroethylene	1.354	1.346	**1.344**
	fluoroethylene	1.354	1.348	**1.348**
	1,1,1-trifluoroethane	1.354	1.348	**1.348**
	difluoromethane	1.366	1.360	**1.358**
	fluoroethane	1.401	1.398	**1.398**
C-Al	dimethylaluminum chloride dimer	1.955	1.948	**1.935**
	trimethylaluminum	1.980	1.968	**1.957**
	trimethylaluminum dimer (terminal)	1.971	1.965	**1.957**
	trimethylaluminum dimer (bridge)	2.153	2.148	**2.140**
C=Si	methylenedimethylsilane	1.714	1.710	**1.83**
C-Si	vinylsilane	1.869	1.868	**1.853**
	ethylsilane	1.888	1.883	**1.866**
	dimethylsilane	1.885	1.878	**1.867**
	trimethylsilane	1.886	1.879	**1.868**
	tetramethylsilane	1.889	1.881	**1.875**
	methylenedimethylsilane	1.879	1.870	**1.91**
C-P	trimethylphosphine	1.852	1.846	**1.841**
	dimethylphosphine	1.855	1.850	**1.848**
	phosphacyclopropane	1.869	1.866	**1.867**
C=S	carbon disulfide	1.561	1.561	**1.553**
	carbonyl sulfide	1.565	1.563	**1.560**
	isothiocyanic acid	1.573	1.574	**1.561**
	thioformamide	1.635	1.631	**1.626**
C-S	thiophene	1.718	1.713	**1.714**
	dimethyl sulfide	1.806	1.803	**1.802**
	thiirane	1.817	1.813	**1.815**
	trans-ethanethiol	1.824	1.821	**1.820**
C-Cl	trichloromethane	1.766	1.765	**1.758**
	tetrachloromethane	1.771	1.772	**1.767**
	dichloromethane	1.769	1.767	**1.772**

Table A5-28: Bond Distances Connecting Heavy Atoms. MP2 Models (5)

bond type	molecule	6-31G*	6-311+G**	expt.
N≡N	diazomethane	1.151	1.142	**1.139**
N=N	diazirine	1.257	1.247	**1.228**
	trans-difluorodiazene	1.249	1.236	**1.214**
N=O	nitrosyl fluoride	1.160	1.133	**1.136**
	nitrosyl chloride	1.156	1.139	**1.139**
	formonitrile oxide	1.209	1.197	**1.199**
	nitrosomethane	1.236	1.220	**1.22**
	nitromethane	1.242	1.230	**1.224**
	nitrosyl cyanide	1.241	1.225	**1.228**
N-F	nitrogen trifluoride	1.383	1.369	**1.365**
	trans-difluorodiazene	1.394	1.388	**1.384**
	nitrosyl fluoride	1.516	1.588	**1.512**
N-Cl	nitrogen trichloride	1.775	1.769	**1.748**
	nitrosyl chloride	2.031	2.046	**1.975**
O⁼O	ozone	1.299	1.282	**1.278**
O-O	fluorine peroxide	1.292	1.130	**1.217**
O-F	oxygen difluoride	1.423	1.406	**1.405**
	fluorine peroxide	1.495	1.859	**1.575**
O-Cl	oxygen dichloride	1.729	1.733	**1.700**
F-Mg	magnesium difluoride	1.744	1.785	**1.77**
F-Al	aluminum trifluoride	1.645	1.656	**1.63**
F-Si	tetrafluorosilane	1.583	1.581	**1.554**
	trifluorosilane	1.592	1.582	**1.562**
	difluorosilane	1.604	1.608	**1.577**
F-P	trifluorophosphine	1.595	1.594	**1.551**
F-S	difluorosulfide	1.621	1.628	**1.592**
Al-Al	trimethylaluminum dimer	2.604	2.620	**2.619**
	dimethylaluminum chloride dimer	3.265	3.265	**3.274**
Al-Cl	trichloroaluminum	2.072	2.067	**2.06**
	dimethylaluminum chloride dimer	2.315	2.309	**2.303**
Si-Cl	tetrachlorosilane	2.028	2.044	**2.019**
	trichlorosilane	2.035	2.043	**2.021**
	dichlorosilane	2.043	2.041	**2.033**
P-Cl	trichlorophosphine	2.055	2.058	**2.043**
S-Cl	sulfur dichloride	2.035	2.045	**2.014**

Table A5-29: Bond Distances Connecting Heavy Atoms. Semi-Empirical Models

bond type	molecule	MNDO	AM1	PM3	expt.
Be-C	dimethylberyllium	1.660	-	1.624	**1.698**
	methyl (cyclopentadienyl) beryllium (Me)	1.678	-	1.623	**1.706**
	methyl (cyclopentadienyl) beryllium (cyclopentadiene)	1.998	-	2.100	**1.923**
Be-Cl	beryllium dichloride	1.884	-	1.744	**1.75**
B-B	pentaborane (base-cap)	1.712	1.669	1.695	**1.690**
	(base-base)	1.862	1.816	1.847	**1.803**
B-C	trimethylborane	1.558	1.537	1.554	**1.578**
B ⁝ N (aromatic)	borazine	1.429	1.399	1.439	**1.418**
B-N	difluoroaminoborane	1.408	1.383	1.417	**1.402**
B-O	difluorohydroxyborane	1.362	1.351	1.347	**1.344**
B-F	trifluoroborane	1.316	1.306	1.313	**1.307**
	difluoroborane	1.316	1.308	1.315	**1.311**
	difluorohydroxyborane	1.320	1.306	1.318	**1.323**
	difluoroaminoborane	1.325	1.324	1.325	**1.325**
B-Cl	trichloroborane	1.716	1.707	1.679	**1.742**
C≡C	cyanoacetylene	1.198	1.197	1.192	**1.205**
	propyne	1.197	1.197	1.192	**1.206**
	but-1-yne-3-ene	1.199	1.198	1.193	**1.208**
	2-butyne	1.200	1.198	1.193	**1.214**
C=C	butatriene (C_2C_3)	1.270	1.266	1.267	**1.283**
	cyclopropene	1.328	1.318	1.314	**1.300**
	cyclopropenone	1.359	1.349	1.344	**1.302**
	allene	1.306	1.298	1.297	**1.308**
	ketene	1.319	1.307	1.308	**1.314**
	1,1-difluoroethylene	1.359	1.345	1.337	**1.315**
	butatriene (C_1C_2)	1.311	1.302	1.301	**1.318**
	propene	1.340	1.331	1.328	**1.318**
	trans-1,2-difluoroethylene	1.369	1.357	1.348	**1.329**
	isobutene	1.348	1.336	1.333	**1.330**
	cyclopentenone	1.351	1.346	1.340	**1.330**
	cis-1,2-difluoroethylene	1.367	1.355	1.346	**1.331**
	methylenecyclopropane	1.319	1.309	1.309	**1.332**
	fluoroethylene	1.351	1.340	1.333	**1.332**
	cyclobutene	1.355	1.354	1.349	**1.332**
	acrylonitrile	1.344	1.334	1.331	**1.339**
	benzvalene	1.361	1.360	1.350	**1.339**
	but-1-yne-3-ene	1.345	1.336	1.332	**1.341**
	cis-methyl vinyl ether	1.349	1.335	1.333	**1.341**
	cyclopentadiene	1.317	1.359	1.352	**1.345**
	trans-1,3-butadiene	1.3*f*4	1.335	1.331	**1.345**
	vinylsilane	1.344	1.324	1.322	**1.347**

Table A5-29: Bond Distances Connecting Heavy Atoms. Semi-Empirical Models (2)

bond type	molecule	MNDO	AM1	PM3	expt.
	furan	1.390	1.380	1.373	**1.361**
	thiophene	1.374	1.377	1.366	**1.369**
	pyrrole	1.395	1.402	1.390	**1.370**
C ⁓ C (aromatic)	pyridine (C_3C_4)	1.405	1.396	1.392	**1.392**
	pyridine (C_1C_2)	1.411	1.407	1.395	**1.394**
	benzene	1.407	1.395	1.391	**1.397**
C-C	cyanoacetylene	1.372	1.370	1.376	**1.378**
	cyclopropenone	1.465	1.437	1.435	**1.412**
	pyrrole	1.437	1.435	1.421	**1.417**
	acrylonitrile	1.423	1.419	1.423	**1.426**
	furan	1.444	1.448	1.441	**1.431**
	but-1-yne-3-ene	1.417	1.405	1.414	**1.431**
	thiophene	1.452	1.432	1.437	**1.433**
	benzvalene (C_4C_6)	1.489	1.460	1.456	**1.452**
	methylenecyclopropane (C_2C_3)	1.492	1.467	1.463	**1.457**
	acetonitrile	1.451	1.439	1.441	**1.458**
	propyne	1.445	1.427	1.433	**1.459**
	cyclopentadiene (C_2C_3)	1.478	1.472	1.464	**1.468**
	malononitrile	1.458	1.448	1.450	**1.468**
	2-butyne	1.444	1.425	1.432	**1.468**
	oxirane	1.512	1.484	1.484	**1.471**
	aziridine	1.515	1.496	1.493	**1.481**
	trans-1,3-butadiene	1.466	1.451	1.456	**1.483**
	thiirane	1.512	1.486	1.490	**1.484**
	bicyclo[1.1.0]butane (C_1C_3)	1.537	1.495	1.481	**1.497**
	bicyclo[1.1.0]butane (C_1C_2)	1.527	1.510	1.507	**1.498**
	acetaldehyde	1.517	1.490	1.499	**1.501**
	propene	1.496	1.476	1.480	**1.501**
	phosphacyclopropane	1.511	1.477	1.481	**1.502**
	benzvalene (C_2C_3)	1.506	1.498	1.497	**1.503**
	fluoroethane	1.546	1.519	1.520	**1.505**
	cyclopentadiene (C_4C_5)	1.521	1.509	1.502	**1.506**
	acetone	1.527	1.495	1.505	**1.507**
	isobutene	1.509	1.484	1.487	**1.507**
	cyclopentenone (C_2C_3)	1.512	1.500	1.496	**1.509**
	cyclopropane	1.524	1.522	1.519	**1.510**
	cyclopropene	1.522	1.486	1.497	**1.515**
	cyclobutene (C_1C_4)	1.544	1.520	1.525	**1.517**
	acetic acid	1.522	1.486	1.497	**1.517**
	cyclopentenone (C_1C_2)	1.544	1.520	1.525	**1.524**
	propane	1.561	1.547	1.537	**1.529**
	trans-ethanethiol	1.530	1.507	1.508	**1.526**
	benzvalene (C_3C_4)	1.561	1.547	1.537	**1.529**
	1,1,1-trifluoroethane	1.857	1.540	1.544	**1.530**
	trans-ethanol	1.539	1.572	1.518	**1.530**
	ethylsilane	1.533	1.499	1.500	**1.540**

Table A5-29: Bond Distances Connecting Heavy Atoms. Semi-Empirical Models (3)

bond type	molecule	MNDO	AM1	PM3	expt.
	neopentane	1.555	1.521	1.527	**1.540**
	isobutane	1.541	1.514	1.520	**1.541**
	methylenecyclopropane (C_3C_4)	1.537	1.513	1.514	**1.542**
	cyclobutane	1.549	1.543	1.542	**1.548**
	cyclobutene (C_3C_4)	1.568	1.566	1.568	**1.566**
C≡N	acetonitrile	1.162	1.163	1.159	**1.157**
	cyanoacetylene	1.163	1.165	1.161	**1.159**
	cyanogen fluoride	1.160	1.165	1.159	**1.159**
	acrylonitrile	1.163	1.164	1.160	**1.164**
	methyl isocyanide	1.191	1.181	1.181	**1.166**
	malononitrile	1.161	1.162	1.158	**1.167**
	formonitrile oxide	1.170	1.168	1.167	**1.168**
	nitrosyl cyanide	1.161	1.166	1.160	**1.170**
C=N	isocyanic acid	1.249	1.232	1.251	**1.209**
	isothiocyanic acid	1.231	1.221	1.227	**1.216**
	diazomethane	1.310	1.298	1.295	**1.300**
C⋯N (aromatic)	pyridine	1.353	1.347	1.353	**1.338**
C-N	thioformamide	1.358	1.351	1.364	**1.358**
	pyrrole	1.398	1.392	1.397	**1.370**
	formamide	1.389	1.367	1.391	**1.376**
	nitrosyl cyanide	1.425	1.431	1.430	**1.401**
	methyl isocyanide	1.424	1.395	1.433	**1.424**
	trimethylamine	1.464	1.445	1.480	**1.451**
	dimethylamine	1.462	1.438	1.474	**1.462**
	aziridine	1.479	1.455	1.485	**1.475**
	diazirine	1.488	1.469	1.491	**1.482**
	nitromethane	1.546	1.500	1.514	**1.489**
	nitrosomethane	1.499	1.466	1.475	**1.49**
C=O	carbonyl sulfide	1.182	1.201	1.176	**1.160**
	ketene	1.184	1.193	1.175	**1.161**
	carbon dioxide	1.186	1.189	1.181	**1.162**
	isocyanic acid	1.185	1.202	1.181	**1.166**
	carbonyl fluoride	1.219	1.200	1.199	**1.170**
	formyl fluoride	1.222	1.227	1.202	**1.181**
	formamide	1.227	1.243	1.220	**1.193**
	methyl formate	1.225	1.229	1.208	**1.200**
	ormic acid	1.227	1.230	1.211	**1.202**
	cyclopentenone	1.220	1.228	1.211	**1.210**
	cyclopropenone	1.206	1.214	1.200	**1.212**
	acetic acid	1.231	1.234	1.218	**1.212**
	acetaldehyde	1.221	1.232	1.210	**1.216**
	acetone	1.227	1.235	1.216	**1.222**
C-O	methyl formate (CH_3O-CHO)	1.357	1.362	1.356	**1.334**
	formic acid	1.354	1.357	1.344	**1.343**
	acetic acid	1.359	1.364	1.355	**1.360**

Table A5-29: Bond Distances Connecting Heavy Atoms. Semi-Empirical Models (4)

bond type	molecule	MNDO	AM1	PM3	expt.
	cis-methyl vinyl ether (CH$_3$O-CHCH$_2$)	1.359	1.376	1.366	**1.360**
	furan	1.367	1.395	1.378	**1.362**
	dimethyl ether	1.396	1.417	1.406	**1.410**
	trans-ethanol	1.396	1.420	1.410	**1.425**
	cis methyl vinyl ether (CH$_3$-OCHCH$_2$)	1.400	1.423	1.409	**1.428**
	oxirane	1.417	1.436	1.432	**1.436**
	methyl formate (CH$_3$-OCHO)	1.404	1.428	1.413	**1.437**
C-F	cyanogen fluoride	1.273	1.307	1.297	**1.262**
	fluoroacetylene	1.277	1.297	1.299	**1.279**
	tetrafluoromethane	1.347	1.358	1.337	**1.317**
	carbonyl fluoride	1.316	1.328	1.322	**1.317**
	1,1-difluoroethylene	1.325	1.346	1.333	**1.323**
	trifluoromethane	1.353	1.368	1.346	**1.332**
	cis-1,2-difluoroethylene	1.320	1.346	1.334	**1.335**
	formyl fluoride	1.328	1.342	1.336	**1.338**
	trans-1,2-difluoroethylene	1.321	1.348	1.335	**1.344**
	fluroethylene	1.324	1.351	1.338	**1.348**
	1,1,1-trifluoroethane	1.358	1.373	1.352	**1.348**
	difluoromethane	1.352	1.373	1.349	**1.358**
	fluoroethane	1.353	1.382	1.360	**1.398**
C-Al	dimethylaluminum chloride dimer	2.020	1.772	1.883	**1.935**
	trimethylaluminum	1.827	1.778	1.892	**1.957**
	trimethylaluminum dimer (terminal)	–	1.785	1.928	**1.957**
	trimethylaluminum dimer (bridge)	–	1.791,2.709	1.928,2.480	**2.140**
C=Si	methylenedimethylsilane	1.672	1.608	1.614	**1.83**
C-Si	vinylsilane	1.841	1.762	1.813	**1.853**
	ethylsilane	1.883	1.815	1.877	**1.866**
	dimethylsilane	1.879	1.814	1.873	**1.867**
	trimethylsilane	1.885	1.821	1.882	**1.868**
	tetramethylsilane	1.891	1.829	1.890	**1.875**
	methylenedimethylsilane	1.869	1.801	1.856	**1.91**
C-P	trimethylphosphine	1.762	1.725	1.872	**1.841**
	dimethylphosphine	1.756	1.726	1.869	**1.848**
	phosphacyclopropane	1.766	1.756	1.900	**1.867**
C=S	carbon disulfide	1.493	1.459	1.482	**1.553**
	carbonyl sulfide	1.510	1.458	1.503	**1.560**
	isothiocyanic acid	1.505	1.473	1.500	**1.561**
	thioformamide	1.565	1.571	1.608	**1.626**
C-S	thiophene	1.729	1.672	1.725	**1.714**
	dimethyl sulfide	1.793	1.752	1.801	**1.802**
	thiirane	1.817	1.791	1.821	**1.815**
	trans-ethanethiol	1.801	1.774	1.826	**1.820**
C-Cl	trichloromethane	1.788	1.786	1.779	**1.758**
	tetrachloromethane	1.786	1.760	1.747	**1.767**
	dichloromethane	1.779	1.741	1.758	**1.772**

Table A5-29: Bond Distances Connecting Heavy Atoms. Semi-Empirical Models (5)

bond type	molecule	MNDO	AM1	PM3	expt.
N≡N	diazomethane	1.142	1.139	1.137	**1.139**
N=N	diazirine	1.229	1.225	1.226	**1.228**
	trans-difluorodiazene	1.262	1.244	1.226	**1.214**
N=O	nitrosyl fluoride	1.161	1.148	1.162	**1.136**
	nitrosyl chloride	1.149	1.137	1.156	**1.139**
	formonitrile oxide	1.191	1.180	1.208	**1.199**
	nitrosomethane	1.162	1.158	1.178	**1.22**
	nitromethane	1.210	1.201	1.214	**1.224**
	nitrosyl cyanide	1.159	1.150	1.168	**1.228**
N-F	nitrogen trifluoride	1.315	1.360	1.354	**1.365**
	trans-difluorodiazene	1.278	1.348	1.342	**1.384**
	nitrosyl fluoride	1.305	1.367	1.367	**1.512**
N-Cl	nitrogen trichloride	1.717	1.707	1.718	**1.748**
	nitrosyl chloride	1.740	1.731	1.764	**1.975**
O≡O	ozone	1.191	1.160	1.223	**1.278**
O-O	fluorine peroxide	1.288	1.229	1.359	**1.217**
O-F	oxygen difluoride	1.281	1.355	1.355	**1.405**
	fluorine peroxide	1.286	1.397	1.391	**1.575**
O-Cl	oxygen dichloride	1.644	1.733	1.700	**1.700**
F-Mg	magnesium difluoride	–	–	1.771	**1.77**
F-Al	aluminum trifluoride	1.592	1.580	1.644	**1.63**
F-Si	tetrafluorosilane	1.669	1.604	1.580	**1.554**
	trifluorosilane	1.677	1.609	1.590	**1.562**
	difluorosilane	1.683	1.615	1.597	**1.577**
F-P	trifluorophosphine	1.556	1.543	1.558	**1.551**
F-S	difluorosulfide	1.618	1.556	1.560	**1.592**
Al-Al	trimethylaluminum dimer	–	3.287	2.662	**2.619**
	dimethylaluminum chloride dimer	3.425	3.464	3.774	**3.274**
Al-Cl	trichloroaluminum	2.048	1.874	1.966	**2.06**
	dimethylaluminum chloride dimer	2.290	2.389	2.395	**2.303**
Si-Cl	tetrachlorosilane	2.051	2.039	2.041	**2.019**
	trichlorosilane	2.055	2.050	2.059	**2.021**
	dichlorosilane	2.061	2.061	2.076	**2.033**
P-Cl	trichlorophosphine	1.989	1.919	2.064	**2.043**
S-Cl	sulfur dichloride	1.993	1.958	2.031	**2.014**

Table A5-30: Skeletal Bond Angles. Molecular Mechanics Models

angle	molecule	SYBYL	MMFF	expt.
F–B–F	difluoroaminoborane	120.3	119.6	**117.9**
	difluorohydroxyborane	119.9	120.2	**118.0**
	difluoroborane	120.0	120.0	**118.3**
C=C–C	*trans*-acrolein	122.9	120.8	**119.8**
	isobutene	122.1	122.0	**122.4**
	acrylonitrile	120.7	122.5	**122.6**
	but-1-yne-3-ene	120.7	121.5	**123.1**
	trans-1,3-butadiene	122.9	123.0	**123.1**
	propene	123.3	124.2	**124.3**
	cis-1-butene	125.7	127.4	**126.7**
C–C–C	isobutane	110.5	110.6	**110.8**
	malononitrile	109.4	112.9	**110.9**
	propane	111.3	111.7	**112.4**
	cis-1-butene	116.2	116.6	**114.8**
C=C–O	*cis*-methyl vinyl ether	124.4	129.2	**127.7**
C–C=O	*trans*-glyoxal	120.2	119.9	**121.2**
	acetone	121.2	121.6	**121.4**
	trans-acrolein	120.1	121.3	**123.3**
	acetaldehyde	120.8	123.4	**124.0**
	acetic acid	121.9	126.6	**126.6**
C–C–O	*trans*-ethanol	109.4	109.0	**107.3**
C=C–F	*trans*-1,2-difluoroethylene	120.3	121.7	**119.3**
	fluoroethylene	120.2	121.4	**121.2**
	cis-1,2-difluoroethylene	120.0	121.9	**122.1**
	1,1-difluoroethylene	120.3	119.9	**125.5**
C–C–F	fluoroethane	109.8	108.8	**109.7**
C=C–Si	vinylsilane	121.4	126.6	**122.9**
C–C–Si	ethylsilane	110.6	115.6	**113.2**
C–C–S	*trans*-ethanethiol	108.9	110.6	**108.6**
N–C≡N	nitrosyl cyanide	180.0	179.0	**172.5**
N–C=O	formamide	122.7	123.3	**124.7**
O–C=O	acetic acid	122.7	121.0	**123.0**
	formic acid	121.3	121.8	**124.6**
	methyl formate	123.3	126.7	**125.9**
O=C–F	carbonyl fluoride	120.0	119.8	**126.0**
F–C–F	1,1,1-trifluoroethane	109.2	108.0	**106.7**
	difluoromethane	109.5	107.1	**108.3**
	trifluoromethane	109.5	107.9	**108.8**
Cl–C–Cl	trichloromethane	109.5	110.6	**111.3**
	dichloromethane	109.5	111.1	**111.8**
C–N–C	trimethylamine	113.4	110.5	**110.9**
	dimethylamine	111.0	110.9	**112.0**

Table A5-30: Skeletal Bond Angles. Molecular Mechanics Models (2)

angle	molecule	SYBYL	MMFF	expt.
C–N=O	nitrosomethane	120.9	112.2	**112.6**
	nitrosyl cyanide	120.0	111.7	**114.7**
O-N=O	nitrous acid	120.0	110.8	**110.7**
F–N=O	nitrosyl fluoride	120.0	111.0	**110.1**
Cl–N=O	nitrosyl chloride	120.0	111.0	**113.3**
N=N–F	difluorodiazene	120.0	111.5	**114.5**
O⁼N⁼O	nitromethane	124.9	125.9	**125.3**
	nitric acid	125.1	132.6	**130.3**
F–N–F	nitrogen trifluoride	109.5	110.4	**102.4**
Cl–N–Cl	nitrogen trichloride	109.5	110.4	**107.4**
C–O–C	dimethyl ether	117.1	111.5	**111.7**
	methyl formate	114.2	113.8	**114.8**
	cis-methyl vinyl ether	117.8	117.1	**118.3**
O⁼O⁼O	ozone	60.0	105.6	**116.8**
O–O–F	fluorine peroxide	109.6	110.4	**109.5**
F–O–F	oxygen difluoride	109.5	110.4	**103.1**
Cl–O–Cl	oxygen dichloride	109.5	110.4	**110.9**
C–Si–C	trimethylsilane	109.7	110.7	**110.2**
	dimethylsilane	110.0	111.8	**111.0**
F–Si–F	difluorosilane	109.5	109.2	**107.9**
	trifluorosilane	109.5	109.3	**108.3**
Cl–Si–Cl	trichlorosilane	109.5	107.9	**109.4**
	dichlorosilane	109.5	106.7	**109.4**
C–P–C	trimethylphosphine	109.4	100.0	**98.9**
	dimethylphosphine	109.3	99.6	**99.7**
F–P–F	trifluorophosphine	109.5	94.8	**97.7**
Cl–P–Cl	trichlorophosphine	109.5	98.1	**100.1**
C–S–C	dimethyl sulfide	100.0	98.6	**98.9**
F–S–F	sulfur difluoride	97.0	97.9	**98.2**
Cl–S–Cl	sulfur dichloride	97.0	97.9	**102.7**

Table A5-31: Skeletal Bond Angles. Hartree-Fock Models

angle	molecule	STO-3G	3-21G	6-31G*	6-311+G**	expt.
F–B–F	difluoroaminoborane	118.6	116.4	117.7	117.3	**117.9**
	difluorohydroxyborane	119.6	117.6	118.4	118.0	**118.0**
	difluoroborane	118.2	116.9	118.2	117.8	**118.3**
C=C–C	*trans*-acrolein	122.4	121.1	121.2	121.0	**119.8**
	isobutene	122.4	122.6	122.3	122.2	**122.4**
	acrylonitrile	122.9	122.9	122.2	122.3	**122.6**
	but-1-yne-3-ene	124.0	123.4	123.8	123.8	**123.1**
	trans-1,3-butadiene	124.2	124.0	124.1	124.0	**123.1**
	propene	125.1	124.7	125.2	125.3	**124.3**
	cis-1-butene	127.0	126.6	127.2	127.3	**126.7**
C–C–C	isobutane	110.9	110.4	111.0	111.1	**110.8**
	malononitrile	111.7	111.9	112.2	112.5	**110.9**
	propane	112.4	111.6	112.8	112.9	**112.4**
	cis-1-butene	115.1	114.9	115.9	116.2	**114.8**
C=C–O	*cis*-methyl vinyl ether	129.5	128.1	128.5	128.4	**127.7**
C–C=O	*trans*-glyoxal	122.4	121.4	121.0	121.2	**121.2**
	acetone	122.4	122.5	121.7	121.7	**121.4**
	trans-acrolein	124.0	124.1	123.9	123.9	**123.3**
	acetaldehyde	124.3	124.8	124.4	124.7	**124.0**
	acetic acid	126.8	127.4	125.8	125.7	**126.6**
C–C–O	*trans*-ethanol	107.7	106.2	108.0	108.4	**107.3**
C=C–F	*trans*-1,2-difluoroethylene	122.3	120.4	120.2	120.3	**119.3**
	fluoroethylene	123.4	122.5	122.4	122.2	**121.2**
	cis-1,2-difluoroethylene	123.9	123.3	122.6	122.7	**122.1**
	1,1-difluoroethylene	124.7	125.3	125.2	125.2	**125.5**
C–C–F	fluoroethane	111.2	108.9	109.5	109.9	**109.7**
C=C–Si	vinylsilane	125.0	123.7	123.5	123.7	**122.9**
C–C–Si	ethylsilane	113.8	113.2	114.1	114.4	**113.2**
C–C–S	*trans*-ethanethiol	110.7	109.1	109.7	109.7	**108.6**
N–C≡N	nitrosyl cyanide	175.2	175.2	173.8	173.9	**172.5**
N–C=O	formamide	124.3	125.3	124.9	125.0	**124.7**
O–C=O	acetic acid	121.8	122.1	122.4	122.3	**123.0**
	formic acid	123.7	124.6	124.9	124.9	**124.6**
	methyl formate	124.8	124.6	125.7	125.7	**125.9**
O=C–F	carbonyl fluoride	125.0	122.4	125.9	125.9	**126.0**
F–C–F	1,1,1-trifluoroethane	107.7	107.2	107.2	107.1	**106.7**
	difluoromethane	108.7	108.9	108.6	108.2	**108.3**
	trifluoromethane	108.6	108.3	108.5	108.4	**108.8**
Cl–C–Cl	trichloromethane	111.1	111.0	111.3	111.3	**111.3**
	dichloromethane	112.7	112.0	112.9	112.7	**111.8**
C–N–C	trimethylamine	110.3	113.1	111.9	111.7	**110.9**
	dimethylamine	111.2	114.5	113.4	113.7	**112.0**

Table A5-31: Skeletal Bond Angles. Hartree-Fock Models (2)

angle	molecule	STO-3G	3-21G	6-31G*	6-311+G**	expt.
C–N=O	nitrosomethane	111.5	112.7	113.8	114.5	**112.6**
	nitrosyl cyanide	111.4	113.6	113.4	113.8	**114.7**
O–N=O	nitrous acid	108.3	110.2	111.4	111.9	**110.7**
F–N=O	nitrosyl fluoride	108.2	109.9	110.4	110.6	**110.1**
Cl–N=O	nitrosyl chloride	111.2	112.4	112.9	113.2	**113.3**
N=N–F	difluorodiazene	111.5	105.4	106.7	107.5	**114.5**
O⁚N⁚O	nitromethane	125.2	126.2	125.8	125.6	**125.3**
	nitric acid	129.4	130.1	129.2	129.0	**130.3**
F–N–F	nitrogen trifluoride	102.2	101.6	102.7	103.0	**102.4**
Cl–N–Cl	nitrogen trichloride	107.2	107.6	110.2	110.3	**107.4**
C–O–C	dimethyl ether	108.7	114.0	113.8	114.1	**111.7**
	methyl formate	112.1	118.0	116.9	117.5	**114.8**
	cis-methyl vinyl ether	113.7	119.2	118.2	118.6	**118.3**
O⁚O⁚O	ozone	116.2	117.0	119.0	119.5	**116.8**
O–O–F	fluorine peroxide	104.2	103.4	105.8	106.5	**109.5**
F–O–F	oxygen difluoride	102.4	101.6	103.2	103.7	**103.1**
Cl–O–Cl	oxygen dichloride	109.3	113.2	113.0	112.8	**110.9**
C–Si–C	trimethylsilane	110.1	110.3	110.4	110.4	**110.2**
	dimethylsilane	110.8	111.6	111.5	111.4	**111.0**
F–Si–F	difluorosilane	106.3	106.4	107.6	107.1	**107.9**
	trifluorosilane	107.5	107.6	108.0	107.8	**108.3**
Cl–Si–Cl	trichlorosilane	109.2	109.6	109.6	109.5	**109.4**
	dichlorosilane	109.3	110.4	110.2	110.0	**109.4**
C–P–C	trimethylphosphine	98.3	99.5	100.0	100.4	**98.9**
	dimethylphosphine	98.7	100.5	100.8	101.2	**99.7**
F–P–F	trifluorophosphine	94.7	97.1	97.3	97.3	**97.7**
Cl–P–Cl	trichlorophosphine	98.8	100.5	100.8	100.6	**100.1**
C–S–C	dimethyl sulfide	98.3	99.5	100.0	100.2	**98.9**
F–S–F	sulfur difluoride	94.2	98.3	98.0	97.3	**98.2**
Cl–S–Cl	sulfur dichloride	100.3	102.5	102.8	102.4	**102.7**

Table A5-32: Skeletal Bond Angles. Local Density Models

angle	molecule	6-31G*	6-311+G**	expt.
F–B–F	difluoroaminoborane	117.8	117.5	**117.9**
	difluorohydroxyborane	118.5	118.2	**118.0**
	difluoroborane	118.1	117.6	**118.3**
C=C–C	*trans*-acrolein	119.8	119.8	**119.8**
	isobutene	122.1	122.2	**122.4**
	acrylonitrile	122.7	122.9	**122.6**
	but-1-yne-3-ene	124.4	124.4	**123.1**
	trans-1,3-butadiene	124.0	124.0	**123.1**
	propene	124.9	124.9	**124.3**
	cis-1-butene	126.9	125.5	**126.7**
C–C–C	isobutane	110.6	110.7	**110.8**
	malononitrile	112.9	113.4	**110.9**
	propane	112.3	112.2	**112.4**
	cis-1-butene	116.1	115.0	**114.8**
C=C–O	*cis*-methyl vinyl ether	127.4	127.0	**127.7**
C–C=O	*trans*-glyoxal	121.4	121.4	**121.2**
	acetone	121.9	121.8	**121.4**
	trans-acrolein	124.7	124.6	**123.3**
	acetaldehyde	124.7	124.8	**124.0**
	acetic acid	126.3	126.2	**126.6**
C–C–O	*trans*-ethanol	107.4	107.6	**107.3**
C=C–F	*trans*-1,2-difluoroethylene	120.7	120.2	**119.3**
	fluoroethylene	122.6	122.1	**121.2**
	cis-1,2-difluoroethylene	122.4	121.9	**122.1**
	1,1-difluoroethylene	125.0	125.3	**125.5**
C–C–F	fluoroethane	109.8	109.8	**109.7**
C=C–Si	vinylsilane	122.1	122.2	**122.9**
C–C–Si	ethylsilane	111.9	112.5	**113.2**
C–C–S	ethanethiol	107.9	108.4	**108.6**
N–C≡N	nitrosyl cyanide	161.8	163.1	**172.5**
N–C=O	formamide	124.9	124.8	**124.7**
O–C=O	acetic acid	122.3	122.1	**123.0**
	formic acid	125.0	125.0	**124.6**
	methyl formate	125.1	125.0	**125.9**
O=C–F	carbonyl fluoride	126.2	126.3	**126.0**
F–C–F	1,1,1-trifluoroethane	107.3	106.9	**106.7**
	difluoromethane	109.3	108.5	**108.3**
	trifluoromethane	108.4	108.4	**108.8**
Cl–C–Cl	trichloromethane	111.3	111.3	**111.3**
	dichloromethane	113.3	113.4	**111.8**
C–N–C	trimethylamine	110.7	111.2	**110.9**
	dimethylamine	111.8	112.7	**112.0**

Table A5-32: Skeletal Bond Angles. Local Density Models

angle	molecule	6-31G*	6-311+G**	expt.
C–N=O	nitrosomethane	113.1	114.1	**112.6**
	nitrosyl cyanide	118.4	118.6	**114.7**
O-N=O	nitrous acid	110.7	111.2	**110.7**
F–N=O	nitrosyl fluoride	110.5	110.7	**110.1**
Cl–N=O	nitrosyl chloride	114.1	114.8	**113.3**
N=N–F	difluorodiazene	105.0	105.9	**114.5**
O⚌N⚌O	nitromethane	126.2	125.9	**125.3**
	nitric acid	130.7	130.7	**130.3**
F–N–F	nitrogen trifluoride	101.6	101.9	**102.4**
Cl–N–Cl	nitrogen trichloride	107.8	108.2	**107.4**
C–O–C	dimethyl ether	110.8	111.2	**111.7**
	methyl formate	113.5	113.8	**114.8**
	cis-methyl vinyl ether	115.1	115.2	**118.3**
O⚌O⚌O	ozone	118.0	118.6	**116.8**
O–O–F	fluorine peroxide	109.9	111.3	**109.5**
F–O–F	oxygen difluoride	104.4	104.6	**103.1**
Cl–O–Cl	oxygen dichloride	112.6	114.3	**110.9**
C–Si–C	trimethylsilane	110.7	110.4	**110.2**
	dimethylsilane	111.8	111.0	**111.0**
F–Si–F	difluorosilane	108.0	107.8	**107.9**
	trifluorosilane	108.0	107.9	**108.3**
Cl–Si–Cl	trichlorosilane	109.6	109.3	**109.4**
	dichlorosilane	110.8	110.1	**109.4**
C–P–C	trimethylphosphine	98.7	98.5	**98.9**
	dimethylphosphine	98.7	98.8	**99.7**
F–P–F	trifluorophosphine	97.5	97.2	**97.7**
Cl–P–Cl	trichlorophosphine	101.2	101.6	**100.1**
C–S–C	dimethyl sulfide	98.1	98.1	**98.9**
F–S–F	sulfur difluoride	100.6	99.1	**98.2**
Cl–S–Cl	sulfur dichloride	105.1	104.1	**102.7**

Table A5-33: Skeletal Bond Angles. BP Density Functional Models

angle	molecule	6-31G*	6-311+G**	expt.
F–B–F	difluoroaminoborane	118.0	117.5	**117.9**
	difluorohydroxyborane	118.7	118.2	**118.0**
	difluoroborane	118.3	117.8	**118.3**
C=C–C	*trans*-acrolein	120.9	120.8	**119.8**
	isobutene	122.1	122.1	**122.4**
	acrylonitrile	123.0	123.1	**122.6**
	but-1-yne-3-ene	124.7	124.7	**123.1**
	trans-1,3-butadiene	124.3	124.3	**123.1**
	propene	125.2	125.2	**124.3**
	cis-1-butene	126.6	126.6	**126.7**
C–C–C	isobutane	111.1	111.1	**110.8**
	malononitrile	112.9	113.2	**110.9**
	propane	112.9	112.9	**112.4**
	cis-1-butene	115.7	116.0	**114.8**
C=C–O	*cis*-methyl vinyl ether	128.5	128.2	**127.7**
C–C=O	*trans*-glyoxal	121.7	121.6	**121.2**
	acetone	121.9	121.8	**121.4**
	trans-acrolein	124.4	124.5	**123.3**
	acetaldehyde	124.9	124.9	**124.0**
	acetic acid	126.4	126.4	**126.6**
C–C–O	*trans*-ethanol	107.6	107.7	**107.3**
C=C–F	*trans*-1,2-difluoroethylene	120.5	120.0	**119.3**
	fluoroethylene	122.7	122.1	**121.2**
	cis-1,2-difluoroethylene	123.0	122.8	**122.1**
	1,1-difluoroethylene	125.0	125.3	**125.5**
C–C–F	fluoroethane	109.9	109.9	**109.7**
C=C–Si	vinylsilane	123.5	123.9	**122.9**
C–C–Si	ethylsilane	114.0	114.1	**113.2**
C–C–S	*trans*-ethanethiol	109.3	109.4	**108.6**
N–C≡N	nitrosyl cyanide	161.8	163.0	**172.5**
N–C=O	formamide	125.0	124.9	**124.7**
O–C=O	acetic acid	122.6	122.4	**123.0**
	formic acid	125.4	125.3	**124.6**
	methyl formate	126.1	126.0	**125.9**
O=C–F	carbonyl fluoride	126.2	126.3	**126.0**
F–C–F	1,1,1-trifluoroethane	107.3	106.8	**106.7**
	difluoromethane	109.4	108.6	**108.3**
	trifluoromethane	108.6	108.4	**108.8**
Cl–C–Cl	trichloromethane	111.4	111.4	**111.3**
	dichloromethane	111.1	113.5	**111.8**
C–N–C	trimethylamine	111.1	111.4	**110.9**
	dimethylamine	112.1	112.8	**112.0**

Table A5-33: Skeletal Bond Angles. BP Density Functional Models (2)

angle	molecule	6-31G*	6-311+G**	expt.
C–N=O	nitrosomethane	112.9	113.8	**112.6**
	nitrosyl cyanide	117.0	117.3	**114.7**
O-N=O	nitrous acid	110.5	110.9	**110.7**
F–N=O	nitrosyl fluoride	110.7	110.8	**110.1**
Cl–N=O	nitrosyl chloride	114.4	114.9	**113.3**
N=N–F	difluorodiazene	104.3	104.9	**114.5**
O≃N≃O	nitromethane	126.2	125.8	**125.3**
	nitric acid	131.0	131.1	**130.3**
F–N–F	nitrogen trifluoride	101.7	101.9	**102.4**
Cl–N–Cl	nitrogen trichloride	107.5	107.8	**107.4**
C–O–C	dimethyl ether	111.2	111.7	**111.7**
	methyl formate	114.3	115.1	**114.8**
	cis-methyl vinyl ether	115.7	116.2	**118.3**
O≃O≃O	ozone	117.9	118.4	**116.8**
O–O–F	fluorine peroxide	110.5	111.8	**109.5**
F–O–F	oxygen difluoride	104.8	104.9	**103.1**
Cl–O–Cl	oxygen dichloride	113.0	114.2	**110.9**
C–Si–C	trimethylsilane	110.7	110.6	**110.2**
	dimethylsilane	112.1	111.7	**111.0**
F–Si–F	difluorosilane	108.4	108.1	**107.9**
	trifluorosilane	108.3	108.1	**108.3**
Cl–Si–Cl	trichlorosilane	109.7	109.7	**109.4**
	dichlorosilane	111.0	110.5	**109.4**
C–P–C	trimethylphosphine	99.2	99.2	**98.9**
	dimethylphosphine	100.0	101.1	**99.7**
F–P–F	trifluorophosphine	97.8	97.7	**97.7**
Cl–P–Cl	trichlorophosphine	101.5	101.3	**100.1**
C–S–C	dimethyl sulfide	99.5	99.5	**98.9**
F–S–F	sulfur difluoride	100.9	99.8	**98.2**
Cl–S–Cl	sulfur dichloride	105.0	104.4	**102.7**

Table A5-34: Skeletal Bond Angles. BLYP Density Functional Models

angle	molecule	6-31G*	6-311+G**	expt.
F–B–F	difluoroaminoborane	117.9	117.5	**117.9**
	difluorohydroxyborane	118.6	118.1	**118.0**
	difluoroborane	118.3	117.8	**118.3**
C=C–C	*trans*-acrolein	121.3	121.2	**119.8**
	isobutene	122.0	122.2	**122.4**
	acrylonitrile	123.2	123.4	**122.6**
	but-1-yne-3-ene	124.9	124.9	**123.1**
	trans-1,3-butadiene	124.5	124.5	**123.1**
	propene	125.4	125.4	**124.3**
	cis-1-butene	126.9	127.0	**126.7**
C–C–C	isobutane	111.1	111.2	**110.8**
	malononitrile	112.9	113.4	**110.9**
	propane	113.3	113.2	**112.4**
	cis-1-butene	116.1	116.3	**114.8**
C=C–O	*cis*-methyl vinyl ether	128.7	128.4	**127.7**
C–C=O	*trans*-glyoxal	121.8	121.7	**121.2**
	acetone	121.8	121.6	**121.4**
	trans-acrolein	124.4	124.4	**123.3**
	acetaldehyde	124.9	125.0	**124.0**
	acetic acid	126.3	126.4	**126.6**
C–C–O	*trans*-ethanol	107.7	107.8	**107.3**
C=C–F	*trans*-1,2-difluoroethylene	120.4	119.9	**119.3**
	fluoroethylene	122.5	122.0	**121.2**
	cis-1,2-difluoroethylene	122.8	122.7	122.1
	1,1-difluoroethylene	125.0	125.3	**125.5**
C–C–F	fluoroethane	109.7	109.8	**109.7**
C=C–Si	vinylsilane	123.8	124.5	**122.9**
C–C–Si	ethylsilane	114.3	114.6	**113.2**
C–C–S	*trans*-ethanethiol	109.3	109.7	**108.6**
N–C≡N	nitrosyl cyanide	162.2	163.2	**172.5**
N–C=O	formamide	125.0	124.9	**124.7**
O–C=O	acetic acid	122.5	122.4	**123.0**
	formic acid	125.4	125.3	**124.6**
	methyl formate	126.3	126.1	**125.9**
O=C–F	carbonyl fluoride	126.2	126.4	**126.0**
F–C–F	1,1,1-trifluoroethane	107.3	106.7	**106.7**
	difluoromethane	109.4	108.5	**108.3**
	trifluoromethane	108.6	108.4	**108.8**
Cl–C–Cl	trichloromethane	111.4	111.4	**111.3**
	dichloromethane	113.5	113.5	**111.8**
C–N–C	trimethylamine	111.3	111.6	**110.9**
	dimethylamine	112.5	113.3	**112.0**

Table A5-34: Skeletal Bond Angles. BLYP Density Functional Models (2)

angle	molecule	6-31G*	6-311+G**	expt.
C–N=O	nitrosomethane	112.9	113.8	**112.6**
	nitrosyl cyanide	116.8	117.1	**114.7**
O-N=O	nitrous acid	110.5	110.8	**110.7**
F–N=O	nitrosyl fluoride	110.7	110.8	**110.1**
Cl–N=O	nitrosyl chloride	114.4	115.0	**113.3**
N=N–F	difluorodiazene	104.1	104.7	**114.5**
O⁀N⁀O	nitromethane	126.0	125.7	**125.3**
	nitric acid	131.2	131.3	**130.3**
F–N–F	nitrogen trifluoride	101.8	101.8	**102.4**
Cl–N–Cl	nitrogen trichloride	107.3	107.8	**107.4**
C–O–C	dimethyl ether	111.9	112.3	**111.7**
	methyl formate	114.8	115.8	**114.8**
	cis-methyl vinyl ether	116.3	116.9	**118.3**
O⁀O⁀O	ozone	117.9	118.3	**116.8**
O–O–F	fluorine peroxide	110.5	112.0	**109.5**
F–O–F	oxygen difluoride	104.7	104.9	**103.1**
Cl–O–Cl	oxygen dichloride	112.9	114.3	**110.9**
C–Si–C	trimethylsilane	111.0	110.6	**110.2**
	dimethylsilane	112.4	111.7	**111.0**
F–Si–F	difluorosilane	108.3	108.1	**107.9**
	trifluorosilane	108.3	108.2	**108.3**
Cl–Si–Cl	trichlorosilane	109.9	109.7	**109.4**
	dichlorosilane	111.3	110.6	**109.4**
C–P–C	trimethylphosphine	99.7	99.8	**98.9**
	dimethylphosphine	99.7	100.0	**99.7**
F–P–F	trifluorophosphine	98.1	97.8	**97.7**
Cl–P–Cl	trichlorophosphine	101.7	101.2	**100.1**
C–S–C	dimethyl sulfide	99.5	99.7	**98.9**
F–S–F	sulfur difluoride	101.4	99.8	**98.2**
Cl–S–Cl	sulfur dichloride	105.6	104.4	**102.7**

Table A5-35: Skeletal Bond Angles. EDF1 Density Functional Models

angle	molecule	6-31G*	6-311+G**	expt.
F–B–F	difluoroaminoborane	118.0	117.5	**117.9**
	difluorohydroxyborane	118.6	118.1	**118.0**
	difluoroborane	118.4	117.9	**118.3**
C=C–C	*trans*-acrolein	121.5	121.4	**119.8**
	isobutene	122.1	122.1	**122.4**
	acrylonitrile	123.4	123.4	**122.6**
	but-1-yne-3-ene	125.0	125.0	**123.1**
	trans-1,3-butadiene	124.6	124.6	**123.1**
	propene	125.5	125.6	**124.3**
	cis-1-butene	127.3	127.4	**126.7**
C–C–C	isobutane	111.3	111.4	**110.8**
	malononitrile	113.1	113.5	**110.9**
	propane	113.5	113.6	**112.4**
	cis-1-butene	116.6	116.8	**114.8**
C=C–O	*cis*-methyl vinyl ether	129.0	128.7	**127.7**
C–C=O	*trans*-glyoxal	121.8	121.8	**121.2**
	acetone	121.8	121.7	**121.4**
	trans-acrolein	124.3	124.4	**123.3**
	acetaldehyde	125.0	125.1	**124.0**
	acetic acid	126.2	126.3	**126.6**
C–C–O	*trans*-ethanol	108.0	108.1	**107.3**
C=C–F	*trans*-1,2-difluoroethylene	120.6	120.1	**119.3**
	fluoroethylene	122.7	122.3	**121.2**
	cis-1,2-difluoroethylene	123.2	123.1	**122.1**
	1,1-difluoroethylene	125.0	125.3	**125.5**
C–C–F	fluoroethane	110.1	110.3	**109.7**
C=C–Si	vinylsilane	124.5	124.8	**122.9**
C–C–Si	ethylsilane	115.0	115.2	**113.2**
C–C–S	*trans*-ethanethiol	110.1	110.1	**108.6**
N–C≡N	nitrosyl cyanide	162.0	162.6	**172.5**
N–C=O	formamide	125.1	125.1	**124.7**
O–C=O	acetic acid	122.6	122.4	**123.0**
	formic acid	125.5	125.5	**124.6**
	methyl formate	126.5	126.3	**125.9**
O=C–F	carbonyl fluoride	126.2	126.3	**126.0**
F–C–F	1,1,1-trifluoroethane	107.3	106.8	**106.7**
	difluoromethane	109.5	108.7	**108.3**
	trifluoromethane	108.6	108.5	**108.8**
Cl–C–Cl	trichloromethane	111.6	111.6	**111.3**
	dichloromethane	114.0	114.0	**111.8**
C–N–C	trimethylamine	111.8	112.1	**110.9**
	dimethylamine	113.0	113.7	**112.0**

Table A5-35: Skeletal Bond Angles. EDF1 Density Functional Models (2)

angle	molecule	6-31G*	6-311+G**	expt.
C–N=O	nitrosomethane	113.2	114.0	**112.6**
	nitrosyl cyanide	117.3	117.6	**114.7**
O-N=O	nitrous acid	110.6	111.1	**110.7**
F–N=O	nitrosyl fluoride	110.8	110.9	**110.1**
Cl–N=O	nitrosyl chloride	114.5	115.0	**113.3**
N=N–F	difluorodiazene	104.5	105.1	**114.5**
O≏N≏O	nitromethane	126.1	125.8	**125.3**
	nitric acid	130.9	131.0	**130.3**
F–N–F	nitrogen trifluoride	101.8	102.0	**102.4**
Cl–N–Cl	nitrogen trichloride	107.8	108.1	**107.4**
C–O–C	dimethyl ether	112.3	112.6	**111.7**
	methyl formate	115.3	116.1	**114.8**
	cis-methyl vinyl ether	116.6	117.2	**118.3**
O≏O≏O	ozone	118.1	118.6	**116.8**
O–O–F	fluorine peroxide	110.7	112.2	**109.5**
F–O–F	oxygen difluoride	104.9	105.0	**103.1**
Cl–O–Cl	oxygen dichloride	113.6	114.7	**110.9**
C–Si–C	trimethylsilane	110.9	110.8	**110.2**
	dimethylsilane	112.4	112.1	**111.0**
F–Si–F	difluorosilane	108.4	108.2	**107.9**
	trifluorosilane	108.4	108.2	**108.3**
Cl–Si–Cl	trichlorosilane	109.9	109.8	**109.4**
	dichlorosilane	111.3	110.9	**109.4**
C–P–C	trimethylphosphine	99.9	99.9	**98.9**
	dimethylphosphine	100.7	101.4	**99.7**
F–P–F	trifluorophosphine	98.0	97.8	**97.7**
Cl–P–Cl	trichlorophosphine	101.8	101.6	**100.1**
C–S–C	dimethyl sulfide	100.2	100.3	**98.9**
F–S–F	sulfur difluoride	101.1	100.0	**98.2**
Cl–S–Cl	sulfur dichloride	105.4	104.9	**102.7**

Table A5-36: Skeletal Bond Angles. B3LYP Density Functional Models

angle	molecule	6-31G*	6-311+G**	expt.
F–B–F	difluoroaminoborane	117.9	117.4	**117.9**
	difluorohydroxyborane	118.5	118.1	**118.0**
	difluoroborane	118.3	117.8	**118.3**
C=C–C	*trans*-acrolein	121.1	121.0	**119.8**
	isobutene	122.1	122.2	**122.4**
	acrylonitrile	122.9	123.0	**122.6**
	but-1-yne-3-ene	124.5	124.5	**123.1**
	trans-1,3-butadiene	124.4	124.3	**123.1**
	propene	125.4	125.3	**124.3**
	cis-1-butene	126.7	126.9	**126.7**
C–C–C	isobutane	111.1	111.1	**110.8**
	malononitrile	112.8	113.2	**110.9**
	propane	113.1	113.0	**112.4**
	cis-1-butene	115.9	116.0	**114.8**
C=C–O	*cis*-methyl vinyl ether	128.5	128.2	**127.7**
C–C=O	*trans*-glyoxal	121.5	121.5	**121.2**
	acetone	121.8	121.7	**121.4**
	trans-acrolein	124.2	124.4	**123.3**
	acetaldehyde	124.7	124.8	**124.0**
	acetic acid	126.2	126.2	**126.6**
C–C–O	*trans*-ethanol	107.8	108.0	**107.3**
C=C–F	*trans*-1,2-difluoroethylene	120.4	120.1	**119.3**
	fluoroethylene	122.5	122.0	**121.2**
	cis-1,2-difluoroethylene	122.5	122.6	**122.1**
	1,1-difluoroethylene	125.1	125.3	**125.5**
C–C–F	fluoroethane	109.8	109.9	**109.7**
C=C–Si	vinylsilane	123.6	123.7	**122.9**
C–C–Si	ethylsilane	114.0	114.3	**113.2**
C–C–S	*trans*-ethanethiol	109.3	109.7	**108.6**
N–C≡N	nitrosyl cyanide	167.7	168.0	**172.5**
N–C=O	formamide	125.0	125.0	**124.7**
O–C=O	acetic acid	122.4	122.3	**123.0**
	formic acid	125.2	125.1	**124.6**
	methyl formate	126.0	125.9	125.9
O=C–F	carbonyl fluoride	126.1	126.2	**126.0**
F–C–F	1,1,1-trifluoroethane	107.3	106.8	**106.7**
	difluoromethane	109.2	108.5	**108.3**
	trifluoromethane	108.5	108.4	**108.8**
Cl–C–Cl	trichloromethane	111.4	111.4	**111.3**
	dichloromethane	113.3	113.3	**111.8**
C–N–C	trimethylamine	111.5	111.7	**110.9**
	dimethylamine	112.7	113.4	**112.0**

Table A5-36: Skeletal Bond Angles. B3LYP Density Functional Models (2)

angle	molecule	6-31G*	6-311+G**	expt.
C–N=O	nitrosomethane	113.2	114.0	**112.6**
	nitrosyl cyanide	115.2	115.7	**114.7**
O–N=O	nitrous acid	110.6	111.1	**110.7**
F–N=O	nitrosyl fluoride	110.4	110.4	**110.1**
Cl–N=O	nitrosyl chloride	113.8	114.4	**113.3**
N=N–F	difluorodiazene	105.1	105.7	**114.5**
O∸N∸O	nitromethane	125.9	125.7	**125.3**
	nitric acid	130.4	130.4	**130.3**
F–N–F	nitrogen trifluoride	101.8	102.1	**102.4**
Cl–N–Cl	nitrogen trichloride	107.7	108.2	**107.4**
C–O–C	dimethyl ether	112.3	112.7	**111.7**
	methyl formate	115.3	116.1	**114.8**
	cis-methyl vinyl ether	116.6	117.1	**118.3**
O∸O∸O	ozone	117.9	118.5	**116.8**
O–O–F	fluorine peroxide	108.4	109.8	**109.5**
F–O–F	oxygen difluoride	103.8	104.1	**103.1**
Cl–O–Cl	oxygen dichloride	112.4	113.7	**110.9**
C–Si–C	trimethylsilane	110.7	110.6	**110.2**
	dimethylsilane	112.1	111.6	**111.0**
F–Si–F	difluorosilane	108.1	107.8	**107.9**
	trifluorosilane	108.2	108.0	**108.3**
Cl–Si–Cl	trichlorosilane	109.7	109.6	**109.4**
	dichlorosilane	111.0	110.4	**109.4**
C–P–C	trimethylphosphine	99.4	100.0	**98.9**
	dimethylphosphine	99.9	100.1	**99.7**
F–P–F	trifluorophosphine	97.6	97.5	**97.7**
Cl–P–Cl	trichlorophosphine	101.3	100.8	**100.1**
C–S–C	dimethyl sulfide	99.5	99.7	**98.9**
F–S–F	sulfur difluoride	100.2	98.9	**98.2**
Cl–S–Cl	sulfur dichloride	104.6	103.6	**102.7**

Table A5-37: Skeletal Bond Angles. MP2 Models

angle	molecule	6-31G*	6-311+G**	expt.
F–B–F	difluoroaminoborane	117.9	117.7	**117.9**
	difluorohydroxyborane	118.6	118.2	**118.0**
	difluoroborane	118.4	118.3	**118.3**
C=C–C	*trans*-acrolein	120.6	120.4	**119.8**
	isobutene	122.1	122.1	**122.4**
	acrylonitrile	122.1	122.2	**122.6**
	but-1-yne-3-ene	123.4	123.4	**123.1**
	trans-1,3-butadiene	123.7	123.6	**123.1**
	propene	124.6	124.4	**124.3**
	cis-1-butene	126.2	126.0	**126.7**
C–C–C	isobutane	110.9	110.7	**110.8**
	malononitrile	111.8	112.0	**110.9**
	propane	112.4	112.2	**112.4**
	cis-1-butene	115.1	115.1	**114.8**
C=C–O	*cis*-methyl vinyl ether	128.2	128.0	**127.7**
C–C=O	*trans*-glyoxal	121.2	121.3	**121.2**
	acetone	121.7	121.9	**121.4**
	trans-acrolein	123.9	124.1	**123.3**
	acetaldehyde	124.3	124.6	**124.0**
	acetic acid	126.4	126.3	**126.6**
C–C–O	*trans*-ethanol	107.0	107.6	**107.3**
C=C–F	*trans*-1,2-difluoroethylene	119.7	119.8	**119.3**
	fluoroethylene	122.0	121.8	**121.2**
	cis-1,2-difluoroethylene	121.9	122.5	**122.1**
	1,1-difluoroethylene	125.0	125.1	**125.5**
C–C–F	fluoroethane	109.1	109.7	**109.7**
C=C–Si	vinylsilane	122.7	122.8	**122.9**
C–C–Si	ethylsilane	112.9	113.1	**113.2**
C–C–S	*trans*-ethanethiol	109.0	108.8	**108.6**
N–C≡N	nitrosyl cyanide	171.0	170.3	**172.5**
N–C=O	formamide	124.7	124.8	**124.7**
O–C=O	acetic acid	122.6	122.7	**123.0**
	formic acid	125.1	125.2	**124.6**
	methyl formate	125.7	125.7	**125.9**
O=C–F	carbonyl fluoride	126.3	126.3	**126.0**
F–C–F	1,1,1-trifluoroethane	107.2	107.1	**106.7**
	difluoromethane	108.8	108.5	**108.3**
	trifluoromethane	108.5	108.5	**108.8**
Cl–C–Cl	trichloromethane	111.2	111.3	**111.3**
	dichloromethane	113.0	113.1	**111.8**
C–N–C	trimethylamine	110.4	110.4	**110.9**
	dimethylamine	111.6	111.8	**112.0**

Table A5-37: Skeletal Bond Angles. MP2 Models (2)

angle	molecule	6-31G*	6-311+G**	expt.
C–N=O	nitrosomethane	112.2	113.0	**112.6**
	nitrosyl cyanide	112.6	113.3	**114.7**
O-N=O	nitrous acid	110.4	111.0	**110.7**
F–N=O	nitrosyl fluoride	110.4	110.5	**110.1**
Cl–N=O	nitrosyl chloride	114.0	114.4	**113.3**
N=N–F	difluorodiazene	104.0	104.7	**114.5**
O≈N≈O	nitromethane	125.7	125.6	**125.3**
	nitric acid	130.6	130.5	**130.3**
F–N–F	nitrogen trifluoride	101.6	102.2	**102.4**
Cl–N–Cl	nitrogen trichloride	107.6	108.2	**107.4**
C–O–C	dimethyl ether	111.1	110.8	**111.7**
	methyl formate	113.9	114.2	**114.8**
	cis-methyl vinyl ether	115.3	115.2	**118.3**
O≈O≈O	ozone	116.3	117.1	**116.8**
O–O–F	fluorine peroxide	106.9	114.9	**109.5**
F–O–F	oxygen difluoride	102.6	103.4	**103.1**
Cl–O–Cl	oxygen dichloride	110.7	110.1	**110.9**
C–Si–C	trimethylsilane	110.2	109.9	**110.2**
	dimethylsilane	111.0	110.8	**111.0**
F–Si–F	difluorosilane	108.1	108.0	**107.9**
	trifluorosilane	108.2	108.3	**108.3**
Cl–Si–Cl	trichlorosilane	109.6	109.5	**109.4**
	dichlorosilane	110.4	110.3	**109.4**
C–P–C	trimethylphosphine	99.2	98.5	**98.9**
	dimethylphosphine	99.7	99.3	**99.7**
F–P–F	trifluorophosphine	97.6	97.6	**97.7**
Cl–P–Cl	trichlorophosphine	101.0	100.8	**100.1**
C–S–C	dimethyl sulfide	98.6	98.1	**98.9**
F–S–F	sulfur difluoride	99.3	98.6	**98.2**
Cl–S–Cl	sulfur dichloride	103.4	102.8	**102.7**

Table A5-38: Skeletal Bond Angles. Semi-Empirical Models

angle	molecule	MNDO	AM1	PM3	expt.
F–B–F	difluoroaminoborane	116.7	111.4	112.7	**117.9**
	difluorohydroxyborane	118.5	120.1	118.6	**118.0**
	difluoroborane	114.8	112.9	108.6	**118.3**
C=C–C	*trans*-acrolein	124.9	122.5	121.3	**119.8**
	isobutene	121.9	124.4	122.2	**122.4**
	acrylonitrile	124.9	123.3	122.0	**122.6**
	but-1-yne-3-ene	125.4	124.1	122.5	**123.1**
	trans-1,3-butadiene	125.7	123.4	122.3	**123.1**
	propene	126.9	124.3	123.4	**124.3**
	cis-1-butene	129.9	126.0	125.9	**126.7**
C–C–C	isobutane	112.3	110.6	110.5	**110.8**
	malononitrile	112.4	112.1	111.2	**110.9**
	propane	115.4	111.8	111.7	**112.4**
	cis-1-butene	118.6	114.8	115.6	**114.8**
C=C–O	*cis*-methyl vinyl ether	129.0	126.8	127.9	**127.7**
C–C=O	*trans*-glyoxal	122.0	121.0	120.5	**121.2**
	acetone	121.4	122.3	122.3	**121.4**
	trans-acrolein	123.8	122.9	122.4	**123.3**
	acetaldehyde	124.9	123.5	123.4	**124.0**
	acetic acid	126.8	129.4	129.1	**126.6**
C–C–O	*trans*-ethanol	110.4	107.3	107.8	**107.3**
C=C–F	*trans*-1,2-difluoroethylene	120.9	121.3	120.4	**119.3**
	fluoroethylene	123.3	123.2	122.1	**121.2**
	cis-1,2-difluoroethylene	123.9	124.0	121.8	**122.1**
	1,1-difluoroethylene	125.3	126.7	126.6	**125.5**
C–C–F	fluoroethane	113.1	112.0	112.0	**109.7**
C=C–Si	vinylsilane	125.7	124.3	123.3	**122.9**
C–C–Si	ethylsilane	113.5	110.6	109.6	**113.2**
C–C–S	*trans*-ethanethiol	112.0	107.4	109.0	**108.6**
N–C≡N	nitrosyl cyanide	173.1	173.2	172.7	**172.5**
N–C=O	formamide	120.9	122.0	117.7	**124.7**
O–C=O	acetic acid	118.6	116.6	115.8	**123.0**
	formic acid	120.6	117.7	117.1	**124.6**
	methyl formate	122.0	119.2	120.6	**125.9**
O=C–F	carbonyl fluoride	124.1	124.5	124.6	**126.0**
F–C–F	1,1,1-trifluoroethane	106.1	104.6	104.6	**106.7**
	difluoromethane	106.7	103.9	104.3	**108.3**
	trifluoromethane	107.4	105.7	105.7	**108.8**
Cl–C–Cl	trichloromethane	109.4	111.2	108.6	**111.3**
	dichloromethane	109.5	113.0	107.9	**111.8**
C–N–C	trimethylamine	116.0	113.0	112.3	**110.9**
	dimethylamine	117.6	114.4	113.4	**112.0**

Table A5-38: Skeletal Bond Angles. Semi-Empirical Models (2)

angle	molecule	MNDO	AM1	PM3	expt.
C–N=O	nitrosomethane	118.8	119.0	120.9	**112.6**
	nitrosyl cyanide	117.7	119.6	121.8	**114.7**
O-N=O	nitrous acid	113.3	112.8	109.1	**110.7**
F–N=O	nitrosyl fluoride	113.8	112.3	111.6	**110.1**
Cl–N=O	nitrosyl chloride	115.6	119.3	118.9	**113.3**
N=N–F	difluorodiazene	112.0	113.0	112.0	**114.5**
O=N=O	nitromethane	121.4	122.5	122.3	**125.3**
	nitric acid	126.5	129.0	132.7	**130.3**
F–N–F	nitrogen trifluoride	106.2	102.6	105.0	**102.4**
Cl–N–Cl	nitrogen trichloride	110.0	110.4	110.7	**107.4**
C–O–C	dimethyl ether	120.0	112.9	114.1	**111.7**
	methyl formate	125.7	117.4	119.6	**114.8**
	cis-methyl vinyl ether	123.6	116.3	117.8	**118.3**
O=O=O	ozone	117.6	120.9	114.0	**116.8**
O–O–F	fluorine peroxide	111.0	107.6	101.6	**109.5**
F–O–F	oxygen difluoride	109.1	102.4	102.5	**103.1**
Cl–O–Cl	oxygen dichloride	116.4	111.1	109.2	**110.9**
C–Si–C	trimethylsilane	110.9	111.2	110.5	**110.2**
	dimethylsilane	112.4	113.3	111.7	**111.0**
F–Si–F	difluorosilane	106.3	105.8	104.4	**107.9**
	trifluorosilane	107.6	107.1	106.1	**108.3**
Cl–Si–Cl	trichlorosilane	109.2	109.5	107.8	**109.4**
	dichlorosilane	108.8	109.7	107.0	**109.4**
C–P–C	trimethylphosphine	106.8	102.2	100.6	**98.9**
	dimethylphosphine	107.4	102.7	101.1	**99.7**
F–P–F	trifluorophosphine	98.9	98.0	95.8	**97.7**
Cl–P–Cl	trichlorophosphine	105.2	105.4	99.7	**100.1**
C–S–C	dimethyl sulfide	108.7	102.7	102.6	**98.9**
F–S–F	sulfur difluoride	101.4	99.0	96.5	**98.2**
Cl–S–Cl	sulfur dichloride	105.9	106.3	101.6	**102.7**

Table A5-39: Heavy Atom Bond Distances and Skeletal Bond Angles in Molecules Incorporating Third and Fourth Row, Main-Group Elements. Hartree-Fock Models

molecule	point group	geometrical parameter	STO-3G	3-21G	6-31G*	expt
KOH	$C_{\infty v}$	r (KO)	2.158	2.170	2.174	**2.212**
KF	$C_{\infty v}$	r (KF)	2.286	2.138	2.146	**2.171**
KCl	$C_{\infty v}$	r (KCl)	2.673	2.787	2.744	**2.667**
KBr	$C_{\infty v}$	r (KBr)	2.782	2.885	2.910	**2.821**
KI	$C_{\infty v}$	r (KI)	3.014	3.194	–	**3.048**
GaF	$C_{\infty v}$	r (GaF)	1.754	1.748	1.758	**1.774**
GaCl	$C_{\infty v}$	r (GaCl)	1.930	2.258	2.245	**2.202**
GaBr	$C_{\infty v}$	r (GaBr)	2.054	2.360	2.409	**2.352**
GaI	$C_{\infty v}$	r (GaI)	2.514	2.636	–	**2.575**
GeH_3CH_3	C_{3v}	r (GeC)	1.910	1.979	1.942	**1.945**
GeH_3CN	C_{3v}	r (GeC)	1.842	1.934	1.870	**1.919**
$^-Ge{\equiv}O^+$	$C_{\infty v}$	r (GeO)	1.454	1.619	1.595	**1.625**
GeH_3F	C_{3v}	r (GeF)	1.705	1.725	1.709	**1.732**
GeH_3SiH_3	C_{3v}	r (GeSi)	2.313	2.399	2.363	**2.357**
$^-Ge{\equiv}S^+$	$C_{\infty v}$	r (GeS)	1.872	2.004	2.003	**2.012**
GeH_3Cl	C_{3v}	r (GeCl)	2.158	2.178	2.153	**2.150**
Ge_2H_6	D_{3d}	r (GeGe)	2.360	2.447	2.431	**2.403**
$^-Ge{\equiv}Se^+$	$C_{\infty v}$	r (GeSe)	2.020	2.121	2.130	**2.135**
GeH_3Br	C_{3v}	r (GeBr)	2.277	2.301	2.313	**2.297**
$^-Ge{\equiv}Te^+$	$C_{\infty v}$	r (GeTe)	2.260	2.351	–	**2.340**
AsF_3	C_{3v}	r (AsF)	1.712	1.703	1.689	**1.710**
		< (FAsF)	92.8	95.5	93.4	**96.0**
$^-C{\equiv}Se^+$	$C_{\infty v}$	r (CSe)	1.648	1.674	1.657	**1.676**
CH_3SeH	C_s	r (CSe)	1.930	1.965	1.940	**1.959**
Se=C=O	$C_{\infty v}$	r (SeC)	1.663	1.727	1.718	**1.709**
$^-Si{\equiv}Se^+$	$C_{\infty v}$	r (SiSe)	2.010	2.033	2.048	**2.058**
$^-Sn{\equiv}Se^+$	$C_{\infty v}$	r (SnSe)	2.220	2.320	–	**2.326**
LiBr	$C_{\infty v}$	r (LiBr)	2.047	2.178	2.221	**2.170**
CH_3Br	C_{3v}	r (CBr)	1.906	1.953	1.934	**1.933**
CH_2Br_2	C_{2v}	r (CBr)	1.917	1.937	1.913	**1.927**
		< (BrCBr)	116.0	112.3	113.9	**114.0**
$CHBr_3$	C_{3v}	r (CBr)	1.933	1.932	1.908	**1.930**
		< (BrCBr)	112.6	111.0	112.1	**110.8**
BrF	$C_{\infty v}$	r (BrF)	1.770	1.757	1.740	**1.756**
NaBr	$C_{\infty v}$	r (NaBr)	2.325	2.463	2.545	**2.502**

600

Table A5-39: Heavy Atom Bond Distances and Skeletal Bond Angles in Molecules Incorporating Third and Fourth Row, Main-Group Elements. Hartree-Fock Models (2)

molecule	point group	geometrical parameter	STO-3G	3-21G	6-31G*	expt.
SiH_3Br	C_{3v}	r (SiBr)	2.205	2.201	2.210	**2.210**
BrCl	$C_{\infty v}$	r (BrCl)	2.174	2.138	2.139	**2.136**
Br_2	$D_{\infty h}$	r (BrBr)	2.286	2.273	2.283	**2.281**
RbBr	$C_{\infty v}$	r (RbBr)	2.920	3.038	–	**2.945**
InBr	$C_{\infty v}$	r (InBr)	2.406	2.549	–	**2.543**
SnH_3Br	C_{3v}	r (SnBr)	2.461	2.492	–	**2.469**
$SbBr_3$	C_{3v}	r (SbBr)	2.490	2.483	–	**2.490**
		< (BrSbBr)	98.2	97.8	–	**98.2**
$TeBr_2$	C_{2v}	r (TeBr)	2.512	2.483	–	**2.51**
		< (BrTeBr)	98.0	99.9	–	**98.0**
IBr	$C_{\infty v}$	r (IBr)	2.497	2.486	–	**2.485**
RbOH	$C_{\infty v}$	r (RbO)	2.308	2.270	–	**2.301**
RbF	$C_{\infty v}$	r (RbF)	2.365	2.238	–	**2.270**
RbCl	$C_{\infty v}$	r (RbCl)	2.810	2.927	–	**2.787**
RbI	$C_{\infty v}$	r (RbI)	3.170	3.333	–	**3.177**
$In(CH_3)_3$	C_{3h}	r (InC)	2.092	2.217	–	**2.093**
$In(C_5H_5)$	C_{5v}	r (InC)	2.420	2.646	–	**2.621**
InF	$C_{\infty v}$	r (InF)	1.816	1.973	–	**1.985**
InCl	$C_{\infty v}$	r (InCl)	2.295	2.470	–	**2.401**
InI	$C_{\infty v}$	r (InI)	2.685	2.821	–	**2.754**
$Sn(CH_3)_4$	T_d	r (SnC)	2.110	2.192	–	**2.144**
$^-Sn{\equiv}O^+$	$C_{\infty v}$	r (SnO)	1.712	1.821	–	**1.833**
$^-Sn{\equiv}S^+$	$C_{\infty v}$	r (SnS)	2.078	2.221	–	**2.209**
$SnCl_4$	T_d	r (SnCl)	2.293	2.322	–	**2.281**
$^-Sn{\equiv}Te^+$	$C_{\infty v}$	r (SnTe)	2.360	2.558	–	**2.523**
$SbCl_3$	C_{3v}	r (SbCl)	2.352	2.364	–	**2.333**
		< (ClSbCl)	94.9	96.8	–	**97.2**
Te=C=S	$C_{\infty v}$	r (TeC)	1.858	1.953	–	**1.904**
LiI	$C_{\infty v}$	r (LiI)	2.281	2.453	–	**2.392**
CH_3I	C_{3v}	r (CI)	2.110	2.179	–	**2.132**
ICN	$C_{\infty v}$	r (IC)	1.991	2.028	–	**1.995**
IF	$C_{\infty v}$	r (IF)	1.962	1.911	–	**1.910**
NaI	$C_{\infty v}$	r (NaI)	2.561	2.737	–	**2.711**
SiH_3I	C_{3v}	r (SiI)	2.438	2.467	–	**2.437**
ICl	$C_{\infty v}$	r (ICl)	2.367	2.360	–	**2.321**
I_2	$D_{\infty h}$	r (II)	2.703	2.692	–	**2.666**

Table A5-40: Heavy Atom Bond Distances and Skeletal Bond Angles in Molecules Incorporating Third-Row, Main-Group Elements. Correlated Models with the 6-31G* Basis Set

molecule	point group	geometrical parameter	local density	BP	BLYP	EDF1	B3LYP	MP2	expt.
KOH	$C_{\infty v}$	r (KO)	2.102	2.143	2.158	2.147	2.139	2.186	**2.212**
KF	$C_{\infty v}$	r (KF)	2.051	2.099	2.114	2.105	2.105	2.133	**2.171**
KCl	$C_{\infty v}$	r (KCl)	2.602	2.671	2.691	2.682	2.683	2.849	**2.667**
KBr	$C_{\infty v}$	r (KBr)	2.760	2.828	2.850	2.838	2.841	2.931	**2.821**
GaF	$C_{\infty v}$	r (GaF)	1.762	1.797	1.810	1.800	1.785	1.769	**1.774**
GaCl	$C_{\infty v}$	r (GaCl)	2.210	2.256	2.280	2.261	2.257	2.231	**2.202**
GaBr	$C_{\infty v}$	r (GaBr)	2.362	2.412	2.441	2.418	2.417	2.394	**2.352**
GeH_3CH_3	C_{3v}	r (GeC)	1.937	1.972	1.988	1.971	1.971	1.964	**1.945**
GeH_3CN	C_{3v}	r (GeC)	1.898	1.930	1.941	1.929	1.915	1.932	**1.919**
$^-Ge{\equiv}O^+$	$C_{\infty v}$	r (GeO)	1.632	1.651	1.660	1.649	1.636	1.666	**1.625**
GeH_3F	C_{3v}	r (GeF)	1.732	1.759	1.766	1.757	1.745	1.767	**1.732**
GeH_3SiH_3	C_{3v}	r (GeSi)	2.357	2.399	2.418	2.400	2.397	2.395	**2.357**
$^-Ge{\equiv}S^+$	$C_{\infty v}$	r (GeS)	2.029	2.053	2.068	2.049	2.040	2.035	**2.012**
GeH_3Cl	C_{3v}	r (GeCl)	2.156	2.190	2.212	2.187	2.185	2.167	**2.150**
Ge_2H_6	D_{3d}	r (GeGe)	2.397	2.447	2.473	2.450	2.448	2.450	**2.403**
$^-Ge{\equiv}Se^+$	$C_{\infty v}$	r (GeSe)	2.153	2.181	2.199	2.177	2.170	2.171	**2.135**
GeH_3Br	C_{3v}	r (GeBr)	2.304	2.345	2.371	2.343	2.343	2.330	**2.297**
AsF_3	C_{3v}	r (AsF)	1.724	1.753	1.762	1.751	1.735	1.728	**1.710**
		r (FAsF)	96.4	97.1	97.2	97.1	96.1	96.5	**96.0**
$^-C{\equiv}Se^+$	$C_{\infty v}$	r (CSe)	1.685	1.703	1.710	1.698	1.687	1.689	**1.676**
CH_3SeH	C_s	r (CSe)	1.946	1.983	2.005	1.975	1.980	1.964	**1.959**
Se=C=O	$C_{\infty v}$	r (SeC)	1.703	1.724	1.735	1.719	1.720	1.712	**1.709**
$^-Si{\equiv}Se^+$	$C_{\infty v}$	r (SiSe)	2.078	2.099	2.111	2.093	2.084	2.084	**2.058**
LiBr	$C_{\infty v}$	r (LiBr)	2.183	2.216	2.212	2.212	2.202	2.221	**2.170**
CH_3Br	C_{3v}	r (CBr)	1.930	1.969	1.994	1.962	1.967	1.946	**1.933**
CH_2Br_2	C_{2v}	r (CBr)	1.926	1.963	1.985	1.957	1.955	1.937	**1.927**
		< (BrCBr)	114.1	114.6	114.9	115.2	114.4	113.6	**114.0**
$CHBr_3$	C_{3v}	r (CBr)	1.926	1.963	1.985	1.959	1.950	1.944	**1.930**
		< (BrCBr)	112.0	111.7	112.4	112.5	111.8	112.2	**110.8**
BrF	$C_{\infty v}$	r (BrF)	1.776	1.813	1.830	1.809	1.795	1.788	**1.756**
NaBr	$C_{\infty v}$	r (NaBr)	2.482	2.536	2.538	2.540	2.522	2.546	**2.502**
SiH_3Br	C_{3v}	r (SiBr)	2.224	2.253	2.271	2.249	2.247	2.234	**2.210**
BrCl	$C_{\infty v}$	r (BrCl)	2.157	2.192	2.219	2.184	2.186	2.163	**2.136**
Br_2	$D_{\infty h}$	r (BrBr)	2.289	2.328	2.359	2.322	2.325	2.307	**2.281**

Table A5-41: Heavy-Atom Bond Distances and Skeletal Bond Angles in Molecules Incorporating Third and Fourth Row, Main-Group Elements. Semi Empirical Models

molecule	point group	geometrical parameter	MNDO	AM1	PM3	expt.
GaF	$C_{\infty v}$	r (GaF)	–	–	1.783	**1.774**
GaCl	$C_{\infty v}$	r (GaCl)	–	–	2.306	**2.202**
GaBr	$C_{\infty v}$	r (GaBr)	–	–	2.394	**2.352**
GaI	$C_{\infty v}$	r (GaI)	–	–	2.539	**2.575**
GeH$_3$CH$_3$	C_{3v}	r (GeC)	1.927	1.986	1.956	**1.945**
GeH$_3$CN	C_{3v}	r (GeC)	1.858	1.887	1.834	**1.919**
$^-$Ge≡O$^+$	$C_{\infty v}$	r (GeO)	1.570	1.622	1.637	**1.625**
GeH$_3$F	C_{3v}	r (GeF)	1.738	1.716	1.736	**1.732**
GeH$_3$SiH$_3$	C_{3v}	r (GeSi)	2.339	2.359	2.401	**2.357**
$^-$Ge≡S$^+$	$C_{\infty v}$	r (GeS)	1.925	2.032	1.973	**2.012**
GeH$_3$Cl	C_{3v}	r (GeCl)	2.249	2.132	2.197	**2.150**
Ge$_2$H$_6$	D_{3d}	r (GeGe)	2.523	2.366	2.393	**2.403**
$^-$Ge≡Se$^+$	$C_{\infty v}$	r (GeSe)	–	–	2.255	**2.135**
GeH$_3$Br	C_{3v}	r (GeBr)	2.366	2.256	2.332	**2.297**
$^-$Ge≡Te$^+$	$C_{\infty v}$	r (GeTe)	–	–	2.480	**2.340**
AsF$_3$	C_{3v}	r (AsF)	–	–	1.706	**1.710**
		r (FAsF)	–	–	95.9	**96.0**
$^-$C≡Se$^+$	$C_{\infty v}$	r (CSe)	–	–	1.591	**1.676**
CH$_3$SeH	C_s	r (CSe)	–	–	1.947	**1.959**
Se=C=O	$C_{\infty v}$	r (SeC)	–	–	1.644	**1.709**
$^-$Si≡Se$^+$	$C_{\infty v}$	r (SiSe)	–	–	1.974	**2.058**
$^-$Sn≡Se$^+$	$C_{\infty v}$	r (SnSe)	–	–	2.106	**2.326**
LiBr	$C_{\infty v}$	r (LiBr)	2.242	–	1.805	**2.170**
CH$_3$Br	C_{3v}	r (CBr)	1.917	1.905	1.951	**1.933**
CH$_2$Br$_2$	C_{2v}	r (CBr)	1.916	1.901	1.911	**1.927**
		< (BrCBr)	113.0	113.6	94.4	**114.0**
CHBr$_3$	C_{3v}	r (CBr)	1.922	1.907	1.872	**1.930**
		< (BrCBr)	111.4	111.7	98.4	**110.8**
BrF	$C_{\infty v}$	r (BrF)	1.756	1.777	1.774	**1.756**
SiH$_3$Br	C_{3v}	r (SiBr)	2.243	2.240	1.902	**2.210**
BrCl	$C_{\infty v}$	r (BrCl)	2.112	2.064	2.176	**2.136**
Br$_2$	$D_{\infty h}$	r (BrBr)	2.267	2.184	2.443	**2.281**
InBr	$C_{\infty v}$	r (InBr)	-	-	2.289	**2.543**
SnH$_3$Br	C_{3v}	r (SnBr)	2.400	2.394	2.453	**2.469**

Table A5-41: Heavy-Atom Bond Distances and Skeletal Bond Angles in Molecules Incorporating Third and Fourth Row, Main-Group Elements. Semi Empirical Models (2)

molecule	point group	geometrical parameter	MNDO	AM1	PM3	expt.
SbBr$_3$	C$_{3v}$	r (SbBr)	–	–	2.471	**2.490**
		< (BrSbBr)	–	–	98.4	**98.2**
TeBr$_2$	C$_{2v}$	r (TeBr)	–	–	2.510	**2.51**
		< (BrTeBr)	–	–	99.5	**98.0**
IBr	C$_{\infty v}$	r (IBr)	2.458	2.354	2.561	**2.485**
In(CH$_3$)$_3$	C$_{3h}$	r (InC)	–	–	2.159	**2.093**
In(C$_5$H$_5$)	C$_{5v}$	r (InC)	–	–	2.770	**2.621**
InF	C$_{\infty v}$	r (InF)	–	–	1.986	**1.985**
InCl	C$_{\infty v}$	r (InCl)	–	–	2.400	**2.401**
InI	C$_{\infty v}$	r (InI)	–	–	2.711	**2.754**
Sn(CH$_3$)$_4$	T$_d$	r (SnC)	2.063	2.108	2.148	**2.144**
$^-$Sn$\equiv$O$^+$	C$_{\infty v}$	r (SnO)	1.749	1.882	1.834	**1.833**
$^-$Sn$\equiv$S$^+$	C$_{\infty v}$	r (SnS)	2.001	2.125	2.023	**2.209**
SnCl$_4$	T$_d$	r (SnCl)	2.284	2.260	2.355	**2.281**
$^-$Sn$\equiv$Te$^+$	C$_{\infty v}$	r (SnTe)	–	–	2.573	**2.523**
SbCl$_3$	C$_{3v}$	r (SbCl)	–	–	2.320	**2.333**
		< (ClSbCl)	–	–	97.0	**97.2**
Te=C=S	C$_{\infty v}$	r (TeC)	–	–	2.016	**1.904**
LiI	C$_{\infty v}$	r (LiI)	2.346	–	2.191	**2.392**
CH$_3$I	C$_{3v}$	r (CI)	2.083	2.050	2.028	**2.132**
ICN	C$_{\infty v}$	r (IC)	1.943	1.928	1.909	**1.995**
IF	C$_{\infty v}$	r (IF)	1.949	1.881	1.889	**1.910**
SiH$_3$I	C$_{3v}$	r (SiI)	2.444	2.434	2.012	**2.437**
ICl	C$_{\infty v}$	r (ICl)	2.314	2.218	2.192	**2.321**
I$_2$	D$_{\infty h}$	r (II)	2.635	2.538	2.825	**2.666**

Table A5-42: Structures of Carbenes and Related Compounds. Hartree-Fock Models

molecule	point group	geometrical parameter	STO-3G	3-21G	6-31G*	6-311+G**	expt.
CH_2	C_{2v}	r (CH)	1.123	1.102	1.097	1.098	**1.111**
		< (HCH)	100.5	104.7	103.1	103.6	**102.4**
CH_2[a]	C_{2v}	r (CH)	1.082	1.071	1.071	1.072	**1.078**
		< (HCH)	125.6	131.3	130.7	131.9	**136**
CHF	C_s	r (CF)	1.312	1.339	1.296	1.290	**1.314**
		r (CH)	1.142	1.107	1.104	1.104	–
		< (HCF)	102.3	103.1	102.8	103.2	**101.8**
CHCl	C_s	r (CCl)	1.797	1.737	1.709	1.705	**1.689**
		r (CH)	1.131	1.096	1.092	1.093	**1.12**
		< (HCCl)	100.2	103.9	103.5	103.4	**103.4**
CF_2	C_{2v}	r (CF)	1.323	1.321	1.283	1.276	**1.304**
		< (FCF)	102.7	104.0	104.5	105.0	**104.8**
CCl_2	C_{2v}	r (CCl)	1.803	1.737	1.712	1.710	**1.76**
		< (ClCCl)	106.8	100.0	110.4	110.3	**100±9**
SiH_2	C_{2v}	r (SiH)	1.458	1.506	1.509	1.510	**1.516**
		< (HSiH)	91.5	93.4	93.4	93.7	**92.1**
SiHCl	C_s	r (SiCl)	2.119	2.080	2.093	2.095	**2.064**
		r (SiH)	1.460	1.502	1.505	1.506	**1.561**
		< (HSiCl)	92.6	95.8	95.5	95.6	**102.8**
SiHBr	C_s	r (SiBr)	2.230	2.222	2.264	-	**2.231**
		r (SiH)	1.461	1.502	1.504	-	–
		< (HSiBr)	93.5	96.1	95.1	-	**102.9**
SiF_2	C_{2v}	r (SiF)	1.602	1.587	1.592	1.598	**1.590**
		< (FSiF)	93.2	99.2	99.6	99.2	**100.8**
$SiCl_2$	C_{2v}	r (SiCl)	2.109	2.071	2.093	2.083	**2.083**
		< (ClSiCl)	98.2	101.5	101.5	101.3	**102.8**
$SiBr_2$	C_{2v}	r (SiBr)	2.230	2.219	2.257	-	**2.243**
		< (BrSiBr)	100.0	103.3	102.3	-	**100.1**
GeF_2	C_{2v}	r (GeF)	1.650	1.721	1.722	-	**1.732**
		< (FGeF)	91.9	96.4	96.6	-	**97.2**

a) triplet state

Table A5-43: Structures of Carbenes and Related Compounds. Local Density Models

molecule	point group	geometrical parameter	6-31G*	6-311+G**	expt.
CH_2	C_{2v}	r (CH)	1.136	1.128	**1.111**
		< (HCH)	99.1	100.5	**102.4**
CH_2[a]	C_{2v}	r (CH)	1.093	1.091	**1.078**
		< (HCH)	134.8	137.6	**136**
CHF	C_s	r (CF)	1.306	1.304	**1.314**
		r (CH)	1.152	1.142	–
		< (HCF)	101.6	101.7	**101.8**
CHCl	C_s	r (CCl)	1.709	1.695	**1.689**
		r (CH)	1.136	1.128	**1.12**
		< (HCCl)	101.2	101.6	**103.4**
CF_2	C_{2v}	r (CF)	1.308	1.301	**1.304**
		< (FCF)	103.8	104.4	**104.8**
CCl_2	C_{2v}	r (CCl)	1.741	1.725	**1.76**
		< (ClCCl)	109.0	109.6	**100±9**
SiH_2	C_{2v}	r (SiH)	1.547	1.543	**1.516**
		< (HSiH)	89.7	89.8	**92.1**
SiHCl	C_s	r (SiCl)	2.092	2.091	**2.064**
		r (SiH)	1.552	1.547	**1.561**
		< (HSiCl)	94.0	93.8	**102.8**
SiHBr	C_s	r (SiBr)	2.253	-	**2.231**
		r (SiH)	1.550	-	–
		< (HSiBr)	93.2	-	**102.9**
SiF_2	C_{2v}	r (SiF)	1.610	1.626	**1.590**
		< (FSiF)	100.9	100.0	**100.8**
$SiCl_2$	C_{2v}	r (SiCl)	2.092	2.089	**2.083**
		< (ClSiCl)	101.3	101.1	**102.8**
$SiBr_2$	C_{2v}	r (SiBr)	2.256	-	**2.243**
		< (BrSiBr)	101.5	-	**100.1**
GeF_2	C_{2v}	r (GeF)	1.739	-	**1.732**
		< (FGeF)	98.3	-	**97.2**

a)   triplet state

Table A5-44: Structures of Carbenes and Related Compounds. BP Density Functional Models

molecule	point group	geometrical parameter	6-31G*	6-311+G**	expt.
CH_2	C_{2v}	r (CH)	1.134	1.127	**1.111**
		< (HCH)	99.0	100.4	**102.4**
CH_2[a]	C_{2v}	r (CH)	1.090	1.088	**1.078**
		< (HCH)	133.6	135.8	**136**
CHF	C_s	r (CF)	1.329	1.329	**1.314**
		r (CH)	1.150	1.140	**–**
		< (HCF)	101.2	101.3	**101.8**
CHCl	C_s	r (CCl)	1.740	1.728	**1.689**
		r (CH)	1.135	1.127	**1.12**
		< (HCCl)	100.6	101.0	**103.4**
CF_2	C_{2v}	r (CF)	1.330	1.326	**1.304**
		< (FCF)	103.8	104.2	**104.8**
CCl_2	C_{2v}	r (CCl)	1.773	1.761	**1.76**
		< (ClCCl)	109.1	109.5	**100±9**
SiH_2	C_{2v}	r (SiH)	1.546	1.542	**1.516**
		< (HSiH)	90.4	90.6	**92.1**
SiHCl	C_s	r (SiCl)	2.119	2.120	**2.064**
		r (SiH)	1.550	1.545	**1.561**
		< (HSiCl)	94.6	94.5	**102.8**
SiHBr	C_s	r (SiBr)	2.282	–	**2.231**
		r (SiH)	1.548	–	**–**
		< (HSiBr)	93.8	–	**102.9**
SiF_2	C_{2v}	r (SiF)	1.631	1.648	**1.590**
		< (FSiF)	101.5	100.8	**100.8**
$SiCl_2$	C_{2v}	r (SiCl)	2.119	2.119	**2.083**
		< (ClSiCl)	102.2	102.0	**102.8**
$SiBr_2$	C_{2v}	r (SiBr)	2.288	–	**2.243**
		< (BrSiBr)	102.2	–	**100.1**
GeF_2	C_{2v}	r (GeF)	1.768	–	**1.732**
		< (FGeF)	99.1	–	**97.2**

a) triplet state

Table A5-45: Structures of Carbenes and Related Compounds. BLYP Density Functional Models

molecule	point group	geometrical parameter	6-31G*	6-311+G**	expt.
CH_2	C_{2v}	r (CH)	1.132	1.124	**1.111**
		< (HCH)	99.1	100.8	**102.4**
CH_2[a]	C_{2v}	r (CH)	1.089	1.086	**1.078**
		< (HCH)	133.6	136.0	**136**
CHF	C_s	r (CF)	1.335	1.337	**1.314**
		r (CH)	1.147	1.135	**–**
		< (HCF)	101.3	101.3	**101.8**
CHCl	C_s	r (CCl)	1.756	1.742	**1.689**
		r (CH)	1.131	1.123	**1.12**
		< (HCCl)	100.4	100.9	**103.4**
CF_2	C_{2v}	r (CF)	1.335	1.331	**1.304**
		< (FCF)	103.9	104.3	**104.8**
CCl_2	C_{2v}	r (CCl)	1.788	1.775	**1.76**
		< (ClCCl)	109.1	109.6	**100±9**
SiH_2	C_{2v}	r (SiH)	1.545	1.540	**1.516**
		< (HSiH)	90.7	91.0	**92.1**
SiHCl	C_s	r (SiCl)	2.136	2.138	**2.064**
		r (SiH)	1.548	1.543	**1.561**
		< (HSiCl)	94.8	94.7	**102.8**
SiHBr	C_s	r (SiBr)	2.306	-	**2.231**
		r (SiH)	1.547	-	**–**
		< (HSiBr)	94.0	-	**102.9**
SiF_2	C_{2v}	r (SiF)	1.633	1.640	**1.590**
		< (FSiF)	101.8	101.3	**100.8**
$SiCl_2$	C_{2v}	r (SiCl)	2.135	2.136	**2.083**
		< (ClSiCl)	102.7	102.4	**102.8**
$SiBr_2$	C_{2v}	r (SiBr)	2.308	-	**2.243**
		< (BrSiBr)	102.6	-	**100.1**
GeF_2	C_{2v}	r (GeF)	1.777	-	**1.732**
		< (FGeF)	99.3	-	**97.2**

a) triplet state

608

Table A5-46: Structures of Carbenes and Related Compounds. EDF1 Density Functional Models

molecule	point group	geometrical parameter	6-31G*	6-311+G**	expt.
CH_2	C_{2v}	r (CH)	1.128	1.121	**1.111**
		< (HCH)	99.2	100.5	**102.4**
$CH_2{}^a$	C_{2v}	r (CH)	1.086	1.083	**1.078**
		< (HCH)	133.5	135.6	**136**
CHF	C_s	r (CF)	1.326	1.326	**1.314**
		r (CH)	1.144	1.135	**–**
		< (HCF)	101.3	101.4	**101.8**
CHCl	C_s	r (CCl)	1.732	1.720	**1.689**
		r (CH)	1.128	1.121	**1.12**
		< (HCCl)	100.9	101.2	**103.4**
CF_2	C_{2v}	r (CF)	1.327	1.322	**1.304**
		< (FCF)	104.0	104.4	**104.8**
CCl_2	C_{2v}	r (CCl)	1.765	1.753	**1.76**
		< (ClCCl)	109.5	109.9	**100±9**
SiH_2	C_{2v}	r (SiH)	1.540	1.536	**1.516**
		< (HSiH)	90.7	91.0	**92.1**
SiHCl	C_s	r (SiCl)	2.115	2.115	**2.064**
		r (SiH)	1.543	1.539	**1.561**
		< (HSiCl)	94.8	94.8	**102.8**
SiHBr	C_s	r (SiBr)	2.280	-	**2.231**
		r (SiH)	1.541	-	**–**
		< (HSiBr)	94.1	-	**102.9**
SiF_2	C_{2v}	r (SiF)	1.629	1.646	**1.590**
		< (FSiF)	101.6	101.0	**100.8**
$SiCl_2$	C_{2v}	r (SiCl)	2.115	2.114	**2.083**
		< (ClSiCl)	102.6	102.5	**102.8**
$SiBr_2$	C_{2v}	r (SiBr)	2.284	-	**2.243**
		< (BrSiBr)	102.9	-	**100.1**
GeF_2	C_{2v}	r (GeF)	1.768	-	**1.732**
		< (FGeF)	99.2	-	**97.2**

a) triplet state

Table A5-47: Structures of Carbenes and Related Compounds. B3LYP Density Functional Models

molecule	point group	geometrical parameter	6-31G*	6-311+G**	expt.
CH_2	C_{2v}	r (CH)	1.119	1.114	**1.111**
		< (HCH)	100.3	101.6	**102.4**
CH_2[a]	C_{2v}	r (CH)	1.082	1.080	**1.078**
		< (HCH)	133.2	135.4	**136**
CHF	C_s	r (CF)	1.316	1.316	**1.314**
		r (CH)	1.132	1.124	–
		< (HCF)	101.7	101.9	**101.8**
CHCl	C_s	r (CCl)	1.729	1.718	**1.689**
		r (CH)	1.118	1.112	**1.12**
		< (HCCl)	101.4	101.8	**103.4**
CF_2	C_{2v}	r (CF)	1.313	1.308	**1.304**
		< (FCF)	104.0	104.5	**104.8**
CCl_2	C_{2v}	r (CCl)	1.754	1.742	**1.76**
		< (ClCCl)	109.3	109.8	**100±9**
SiH_2	C_{2v}	r (SiH)	1.531	1.528	**1.516**
		< (HSiH)	91.1	91.5	**92.1**
SiHCl	C_s	r (SiCl)	2.113	2.114	**2.064**
		r (SiH)	1.532	1.529	**1.561**
		< (HSiCl)	94.8	94.7	**102.8**
SiHBr	C_s	r (SiBr)	2.281	–	**2.231**
		r (SiH)	1.531	–	–
		< (HSiBr)	94.1	–	**102.9**
SiF_2	C_{2v}	r (SiF)	1.616	1.633	**1.590**
		< (FSiF)	101.1	100.3	**100.8**
$SiCl_2$	C_{2v}	r (SiCl)	2.110	2.109	**2.083**
		< (ClSiCl)	102.0	101.8	**102.8**
$SiBr_2$	C_{2v}	r (SiBr)	2.282	–	**2.243**
		< (BrSiBr)	102.4	–	**100.1**
GeF_2	C_{2v}	r (GeF)	1.754	–	**1.732**
		< (FGeF)	98.5	–	**97.2**

a) triplet state

Table A5-48: Structures of Carbenes and Related Compounds. MP2 Models

molecule	point group	geometrical parameter	6-31G*	6-311+G**	expt.
CH_2	C_{2v}	r (CH)	1.109	1.110	**1.111**
		< (HCH)	102.0	101.9	**102.4**
CH_2[a]	C_{2v}	r (CH)	1.078	1.079	**1.078**
		< (HCH)	131.6	132.5	**136**
CHF	C_s	r (CF)	1.321	1.311	**1.314**
		r (CH)	1.121	1.118	–
		< (HCF)	101.9	102.1	**101.8**
CHCl	C_s	r (CCl)	1.697	1.688	**1.689**
		r (CH)	1.110	1.109	**1.12**
		< (HCCl)	102.9	102.7	**103.4**
CF_2	C_{2v}	r (CF)	1.315	1.303	**1.304**
		< (FCF)	104.2	104.8	**104.8**
CCl_2	C_{2v}	r (CCl)	1.718	1.709	**1.76**
		< (ClCCl)	109.9	110.3	**100±9**
SiH_2	C_{2v}	r (SiH)	1.519	1.510	**1.516**
		< (HSiH)	92.5	92.5	**92.1**
SiHCl	C_s	r (SiCl)	2.080	2.080	**2.064**
		r (SiH)	1.520	1.510	**1.561**
		< (HSiCl)	95.4	95.2	**102.8**
SiHBr	C_s	r (SiBr)	2.256	–	**2.231**
		r (SiH)	1.518	–	–
		< (HSiBr)	94.7	–	**102.9**
SiF_2	C_{2v}	r (SiF)	1.617	1.625	**1.590**
		< (FSiF)	101.0	100.2	**100.8**
$SiCl_2$	C_{2v}	r (SiCl)	2.075	2.074	**2.083**
		< (ClSiCl)	101.7	101.6	**102.8**
$SiBr_2$	C_{2v}	r (SiBr)	2.253	–	**2.243**
		< (BrSiBr)	102.3	–	**100.1**
GeF_2	C_{2v}	r (GeF)	1.766	–	**1.732**
		< (FGeF)	97.5	–	**97.2**

a) triplet state

Table A5-49: Structures of Carbenes and Related Compounds. Semi-Empirical Models

molecule	point group	geometrical parameter	MNDO	AM1	PM3	expt.
CH_2	C_{2v}	r (CH)	1.091	1.103	1.092	**1.111**
		< (HCH)	111.1	110.6	103.7	**102.4**
$CH_2{}^a$	C_{2v}	r (CH)	1.052	1.062	1.063	**1.078**
		< (HCH)	152.9	151.7	150.0	**136**
CHF	C_s	r (CF)	1.285	1.291	1.284	**1.314**
		r (CH)	1.120	1.127	1.100	**–**
		< (HCF)	111.1	110.7	105.5	**101.8**
CHCl	C_s	r (CCl)	1.706	1.647	1.554	**1.689**
		r (CH)	1.109	1.110	1.100	**1.12**
		< (HCCl)	110.3	111.1	115.5	**103.4**
CF_2	C_{2v}	r (CF)	1.304	1.312	1.298	**1.304**
		< (FCF)	108.3	106.0	106.3	**104.8**
CCl_2	C_{2v}	r (CCl)	1.748	1.672	1.591	**1.76**
		< (ClCCl)	113.9	118.1	120.0	**100±9**
SiH_2	C_{2v}	r (SiH)	1.437	1.457	1.513	**1.516**
		< (HSiH)	99.5	101.0	94.8	**92.1**
SiHCl	C_s	r (SiCl)	2.071	2.021	1.987	**2.064**
		r (SiH)	1.440	1.459	1.513	**1.561**
		< (HSiCl)	102.4	102.5	99.5	**102.8**
SiHBr	C_s	r (SiBr)	2.257	2.198	1.880	**2.231**
		r (SiH)	1.434	1.456	1.521	**–**
		< (HSiBr)	101.9	104.1	101.9	**102.9**
SiF_2	C_{2v}	r (SiF)	1.681	1.612	1.575	**1.590**
		< (FSiF)	102.9	97.1	95.3	**100.8**
$SiCl_2$	C_{2v}	r (SiCl)	2.064	2.027	2.000	**2.083**
		< (ClSiCl)	106.2	104.4	101.9	**102.8**
$SiBr_2$	C_{2v}	r (SiBr)	2.236	2.195	1.855	**2.243**
		< (BrSiBr)	107.9	109.2	98.2	**100.1**
GeF_2	C_{2v}	r (GeF)	1.716	1.672	1.752	**1.732**
		< (FGeF)	94.8	94.1	94.2	**97.2**

a) triplet state

Table A5-50: Structures of Diatomic and Small Polyatomic Radicals. Unrestricted Hartree-Fock Models

radical	point group	electronic state	geometrical parameter	STO-3G	3-21G	6-31G*	6-311+G**	expt.
BeH	$C_{\infty v}$	$^2\Sigma^+$	r (BeH)	1.300	1.365	1.357	1.343	**1.343**
BH$_2$	C_{2v}	$^2B_1$	r (BH)	1.143	1.168	1.168	1.185	**1.181**
			< (HBH)	180.0	180.0	180.0	127.8	**131**
CH	$C_{\infty v}$	$^2\Pi$	r (CH)	1.143	1.119	1.108	1.108	**1.120**
CH$_3$	D_{3h}	$^2A_2$	r (CH)	1.078	1.072	1.073	1.073	**1.079**
NH	$C_{\infty v}$	$^3\Sigma^-$	r (NH)	1.082	1.406	1.024	1.023	**1.036**
NH$_2$	C_{2v}	$^2B_1$	r (NH)	1.015	1.026	1.013	1.012	**1.024**
			< (HNH)	131.3	106.0	104.3	104.6	**103.3**
OH	$C_{\infty v}$	$^2\Pi$	r (OH)	1.014	0.986	0.959	0.952	**0.970**
BO	$C_{\infty v}$	$^2\Sigma^+$	r (BO)	1.190	1.218	1.187	1.182	**1.205**
BN	$C_{\infty v}$	$^3\Pi$	r (BN)	1.305	1.318	1.293	1.289	**1.281**
CN	$C_{\infty v}$	$^2\Sigma^+$	r (CN)	1.235	1.180	1.162	1.154	**1.172**
HCO	C_s	$^2A'$	r (CO)	1.253	1.180	1.159	1.152	**1.175**
			r (CH)	1.101	1.095	1.106	1.108	**1.125**
			< (HCO)	126.3	129.0	126.3	126.9	**124.9**
CF	$C_{\infty v}$	$^2\Pi$	r (CF)	1.293	1.312	1.267	1.258	**1.272**
NO	$C_{\infty v}$	$^2\Pi$	r (NO)	1.186	1.201	1.127	1.118	**1.151**
NF	$C_{\infty v}$	$^3\Sigma^-$	r (NF)	1.342	1.390	1.302	1.293	**1.317**
HOO	C_s	$^2A''$	r (OO)	1.396	1.434	1.369	1.299	**1.335**
			r (OH)	1.002	0.973	0.955	0.948	**0.977**
			< (HOO)	100.1	103.3	102.9	106.5	**104.1**
O$_2$	$D_{\infty h}$	$^3\Sigma^-_g$	r (OO)	1.398	1.240	1.168	1.158	**1.208**

Table A5-51: Structures of Diatomic and Small Polyatomic Radicals. Unrestricted Local Density Models

radical	point group	electronic state	geometrical parameter	6-31G*	6-311+G**	expt.
BeH	$C_{\infty v}$	$^2\Sigma^+$	r (BeH)	1.365	1.359	**1.343**
BH$_2$	C_{2v}	$^2B_1$	r (BH)	1.205	1.199	**1.181**
			< (HBH)	128.0	130.2	**131**
CH	$C_{\infty v}$	$^2\Pi$	r (CH)	1.150	1.141	**1.120**
CH$_3$	D_{3h}	$^2A_2$	r (CH)	1.091	1.089	**1.079**
NH	$C_{\infty v}$	$^3\Sigma^-$	r (NH)	1.060	1.055	**1.036**
NH$_2$	C_{2v}	$^2B_1$	r (NH)	1.044	1.039	**1.024**
			< (HNH)	101.7	102.7	**103.3**
OH	$C_{\infty v}$	$^2\Pi$	r (OH)	0.990	0.985	**0.970**
BO	$C_{\infty v}$	$^2\Sigma^+$	r (BO)	1.212	1.205	**1.205**
BN	$C_{\infty v}$	$^3\Pi$	r (BN)	1.330	1.323	**1.281**
CN	$C_{\infty v}$	$^2\Sigma^+$	r (CN)	1.178	1.169	**1.172**
HCO	C_s	$^2A'$	r (CO)	1.184	1.174	**1.175**
			r (CH)	1.141	1.137	**1.125**
			< (HCO)	123.1	124.0	**124.9**
CF	$C_{\infty v}$	$^2\Pi$	r (CF)	1.277	1.269	**1.272**
NO	$C_{\infty v}$	$^2\Pi$	r (NO)	1.160	1.148	**1.151**
NF	$C_{\infty v}$	$^3\Sigma^-$	r (NF)	1.308	1.305	**1.317**
HOO	C_s	$^2A''$	r (OO)	1.313	1.310	**1.335**
			r (OH)	0.998	0.991	**0.977**
			< (HOO)	105.6	106.7	**104.1**
O$_2$	$D_{\infty h}$	$^3\Sigma^-_g$	r (OO)	1.214	1.204	**1.208**

Table A5-52: Structures of Diatomic and Small Polyatomic Radicals. Unrestricted BP Density Functional Models

radical	point group	electronic state	geometrical parameter	6-31G*	6-311+G**	expt.
BeH	$C_{\infty v}$	$^2\Sigma^+$	r (BeH)	1.364	1.358	**1.343**
BH$_2$	C_{2v}	$^2B_1$	r (BH)	1.204	1.198	**1.181**
			< (HBH)	127.8	129.7	**131**
CH	$C_{\infty v}$	$^2\Pi$	r (CH)	1.148	1.140	**1.120**
CH$_3$	D_{3h}	$^2A_2$	r (CH)	1.091	1.089	**1.079**
NH	$C_{\infty v}$	$^3\Sigma^-$	r (NH)	1.059	1.054	**1.036**
NH$_2$	C_{2v}	$^2B_1$	r (NH)	1.046	1.041	**1.024**
			< (HNH)	101.2	102.1	**103.3**
OH	$C_{\infty v}$	$^2\Pi$	r (OH)	0.993	0.986	**0.970**
BO	$C_{\infty v}$	$^2\Sigma^+$	r (BO)	1.222	1.216	**1.205**
BN	$C_{\infty v}$	$^3\Pi$	r (BN)	1.341	1.336	**1.281**
CN	$C_{\infty v}$	$^2\Sigma^+$	r (CN)	1.178	1.179	**1.172**
HCO	C_s	$^2A'$	r (CO)	1.184	1.186	**1.175**
			r (CH)	1.141	1.137	**1.125**
			< (HCO)	123.1	124.0	**124.9**
CF	$C_{\infty v}$	$^2\Pi$	r (CF)	1.300	1.294	**1.272**
NO	$C_{\infty v}$	$^2\Pi$	r (NO)	1.173	1.162	**1.151**
NF	$C_{\infty v}$	$^3\Sigma^-$	r (NF)	1.335	1.335	**1.317**
HOO	C_s	$^2A''$	r (OO)	1.314	1.343	**1.335**
			r (OH)	0.998	0.990	**0.977**
			< (HOO)	106.9	105.5	**104.1**
O$_2$	$D_{\infty h}$	$^3\Sigma_g^-$	r (OO)	1.231	1.223	**1.208**

Table A5-53: Structures of Diatomic and Small Polyatomic Radicals. Unrestricted BLYP Density Functional Models

radical	point group	electronic state	geometrical parameter	6-31G*	6-311+G**	expt.
BeH	$C_{\infty v}$	$^2\Sigma^+$	r (BeH)	1.356	1.349	**1.343**
BH$_2$	C_{2v}	$^2B_1$	r (BH)	1.200	1.193	**1.181**
			< (HBH)	127.9	129.9	**131**
CH	$C_{\infty v}$	$^2\Pi$	r (CH)	1.146	1.136	**1.120**
CH$_3$	D_{3h}	$^2A_2$	r (CH)	1.090	1.087	**1.079**
NH	$C_{\infty v}$	$^3\Sigma^-$	r (NH)	1.060	1.054	**1.036**
NH$_2$	C_{2v}	$^2B_1$	r (NH)	1.047	1.040	**1.024**
			< (HNH)	101.1	102.3	**103.3**
OH	$C_{\infty v}$	$^2\Pi$	r (OH)	0.998	0.987	**0.970**
BO	$C_{\infty v}$	$^2\Sigma^+$	r (BO)	1.222	1.216	**1.205**
BN	$C_{\infty v}$	$^3\Pi$	r (BN)	1.342	1.336	**1.281**
CN	$C_{\infty v}$	$^2\Sigma^+$	r (CN)	1.187	1.179	**1.172**
HCO	C_s	$^2A'$	r (CO)	1.196	1.187	**1.175**
			r (CH)	1.141	1.135	**1.125**
			< (HCO)	123.0	123.8	**124.9**
CF	$C_{\infty v}$	$^2\Pi$	r (CF)	1.305	1.299	**1.272**
NO	$C_{\infty v}$	$^2\Pi$	r (NO)	1.176	1.165	**1.151**
NF	$C_{\infty v}$	$^3\Sigma^-$	r (NF)	1.346	1.348	**1.317**
HOO	C_s	$^2A''$	r (OO)	1.358	1.358	**1.335**
			r (OH)	0.998	0.990	**0.977**
			< (HOO)	104.6	105.3	**104.1**
O$_2$	$D_{\infty h}$	$^3\Sigma_g^-$	r (OO)	1.239	1.232	**1.208**

Table A5-54: Structures of Diatomic and Small Polyatomic Radicals. Unrestricted EDF1 Density Functional Models

radical	point group	electronic state	geometrical parameter	6-31G*	6-311+G**	expt.
BeH	$C_{\infty v}$	$^2\Sigma^+$	r (BeH)	1.359	1.353	**1.343**
BH$_2$	C_{2v}	$^2B_1$	r (BH)	1.199	1.194	**1.181**
			< (HBH)	127.5	129.2	**131**
CH	$C_{\infty v}$	$^2\Pi$	r (CH)	1.142	1.135	**1.120**
CH$_3$	D_{3h}	$^2A_2$	r (CH)	1.086	1.084	**1.079**
NH	$C_{\infty v}$	$^3\Sigma^-$	r (NH)	1.055	1.050	**1.036**
NH$_2$	C_{2v}	$^2B_1$	r (NH)	1.040	1.035	**1.024**
			< (HNH)	101.2	102.1	**103.3**
OH	$C_{\infty v}$	$^2\Pi$	r (OH)	0.988	0.981	**0.970**
BO	$C_{\infty v}$	$^2\Sigma^+$	r (BO)	1.219	1.213	**1.205**
BN	$C_{\infty v}$	$^3\Pi$	r (BN)	1.339	1.334	**1.281**
CN	$C_{\infty v}$	$^2\Sigma^+$	r (CN)	1.183	1.176	**1.172**
HCO	C_s	$^2A'$	r (CO)	1.191	1.183	**1.175**
			r (CH)	1.137	1.133	**1.125**
			< (HCO)	123.0	123.9	**124.9**
CF	$C_{\infty v}$	$^2\Pi$	r (CF)	1.298	1.292	**1.272**
NO	$C_{\infty v}$	$^2\Pi$	r (NO)	1.169	1.158	**1.151**
NF	$C_{\infty v}$	$^3\Sigma^-$	r (NF)	1.333	1.332	**1.317**
HOO	C_s	$^2A''$	r (OO)	1.338	1.336	**1.335**
			r (OH)	0.992	0.985	**0.977**
			< (HOO)	105.0	105.7	**104.1**
O$_2$	$D_{\infty h}$	$^3\Sigma_g^-$	r (OO)	1.227	1.218	**1.208**

Table A5-55: Structures of Diatomic and Small Polyatomic Radicals. Unrestricted B3LYP Density Functional Models

radical	point group	electronic state	geometrical parameter	6-31G*	6-311+G**	expt.
BeH	$C_{\infty v}$	$^2\Sigma^+$	r (BeH)	1.349	1.343	**1.343**
BH$_2$	C_{2v}	$^2B_1$	r (BH)	1.193	1.187	**1.181**
			< (HBH)	127.9	129.8	**131**
CH	$C_{\infty v}$	$^2\Pi$	r (CH)	1.133	1.126	**1.120**
CH$_3$	D_{3h}	$^2A_2$	r (CH)	1.083	1.080	**1.079**
NH	$C_{\infty v}$	$^3\Sigma^-$	r (NH)	1.048	1.043	**1.036**
NH$_2$	C_{2v}	$^2B_1$	r (NH)	1.034	1.030	**1.024**
			< (HNH)	102.1	103.1	**103.3**
OH	$C_{\infty v}$	$^2\Pi$	r (OH)	0.983	0.976	**0.970**
BO	$C_{\infty v}$	$^2\Sigma^+$	r (BO)	1.209	1.203	**1.205**
BN	$C_{\infty v}$	$^3\Pi$	r (BN)	1.325	1.320	**1.281**
CN	$C_{\infty v}$	$^2\Sigma^+$	r (CN)	1.174	1.166	**1.172**
HCO	C_s	$^2A'$	r (CO)	1.183	1.174	**1.175**
			r (CH)	1.129	1.125	**1.125**
			< (HCO)	123.6	124.4	**124.9**
CF	$C_{\infty v}$	$^2\Pi$	r (CF)	1.287	1.281	**1.272**
NO	$C_{\infty v}$	$^2\Pi$	r (NO)	1.159	1.148	**1.151**
NF	$C_{\infty v}$	$^3\Sigma^-$	r (NF)	1.323	1.322	**1.317**
HOO	C_s	$^2A''$	r (OO)	1.332	1.328	**1.335**
			r (OH)	0.984	0.977	**0.977**
			< (HOO)	105.1	105.9	**104.1**
O$_2$	$D_{\infty h}$	$^3\Sigma_g^-$	r (OO)	1.215	1.206	**1.208**

Table A5-56: Structures of Diatomic and Small Polyatomic Radicals. Unrestricted MP2 Models

radical	point group	electronic state	geometrical parameter	6-31G*	6-311+G**	expt.
BeH	$C_{\infty v}$	$^2\Sigma^+$	r (BeH)	1.349	1.341	1.343
BH$_2$	C_{2v}	$^2B_1$	r (BH)	1.171	1.187	**1.181**
			< (HBH)	180.0	128.6	**131**
CH	$C_{\infty v}$	$^2\Pi$	r (CH)	1.121	1.119	**1.120**
CH$_3$	D_{3h}	$^2A_2$	r (CH)	1.079	1.079	**1.079**
NH	$C_{\infty v}$	$^3\Sigma^-$	r (NH)	1.039	1.035	**1.036**
NH$_2$	C_{2v}	$^2B_1$	r (NH)	1.028	1.025	**1.024**
			< (HNH)	103.3	102.8	**103.3**
OH	$C_{\infty v}$	$^2\Pi$	r (OH)	0.979	0.969	**0.970**
BO	$C_{\infty v}$	$^2\Sigma^+$	r (BO)	1.218	1.212	**1.205**
BN	$C_{\infty v}$	$^3\Pi$	r (BN)	1.252	1.318	**1.281**
CN	$C_{\infty v}$	$^2\Sigma^+$	r (CN)	1.136	1.130	**1.172**
HCO	C_s	$^2A'$	r (CO)	1.192	1.183	**1.175**
			r (CH)	1.123	1.122	**1.125**
			< (HCO)	123.3	124.3	**124.9**
CF	$C_{\infty v}$	$^2\Pi$	r (CF)	1.291	1.278	**1.272**
NO	$C_{\infty v}$	$^2\Pi$	r (NO)	1.143	1.134	**1.151**
NF	$C_{\infty v}$	$^3\Sigma^-$	r (NF)	1.330	1.317	**1.317**
HOO	C_s	$^2A''$	r (OO)	1.396	1.311	**1.335**
			r (OH)	0.981	0.972	**0.977**
			< (HOO)	101.2	105.3	**104.1**
O$_2$	$D_{\infty h}$	$^3\Sigma^-_g$	r (OO)	1.247	1.224	**1.208**

Table A5-57: Structures of Diatomic and Small Polyatomic Radicals. Unrestricted Semi-Empirical Models

radical	point group	electronic state	geometrical parameter	MNDO	AM1	PM3	expt.
BeH	$C_{\infty v}$	$^2\Sigma^+$	r (BeH)	1.293	-	1.315	**1.343**
BH$_2$	C_{2v}	$^2B_1$	r (BH)	1.159	1.198	1.203	**1.181**
			< (HBH)	126.2	127.9	131.3	**131**
CH	$C_{\infty v}$	$^2\Pi$	r (CH)	1.098	1.105	1.089	**1.120**
CH$_3$	D_{3h}	$^2A_2$	r (CH)	1.078	1.086	1.072	**1.079**
NH	$C_{\infty v}$	$^3\Sigma^-$	r (NH)	0.993	0.987	0.974	**1.036**
NH$_2$	C_{2v}	$^2B_1$	r (NH)	1.002	0.996	0.987	**1.024**
			< (HNH)	104.5	107.3	110.8	**103.3**
OH	$C_{\infty v}$	$^2\Pi$	r (OH)	0.937	0.949	0.937	**0.970**
BO	$C_{\infty v}$	$^2\Sigma^+$	r (BO)	1.170	1.168	1.200	**1.205**
BN	$C_{\infty v}$	$^3\Pi$	r (BN)	1.307	1.277	1.333	**1.281**
CN	$C_{\infty v}$	$^2\Sigma^+$	r (CN)	1.155	1.149	1.157	**1.172**
HCO	C_s	$^2A'$	r (CO)	1.185	1.192	1.166	**1.175**
			r (CH)	1.075	1.083	1.089	**1.125**
			< (HCO)	143.1	141.1	136.5	**124.9**
CF	$C_{\infty v}$	$^2\Pi$	r (CF)	1.263	1.259	1.259	**1.272**
NO	$C_{\infty v}$	$^2\Pi$	r (NO)	1.123	1.115	1.127	**1.151**
NF	$C_{\infty v}$	$^3\Sigma^-$	r (NF)	1.223	1.244	1.262	**1.317**
HOO	C_s	$^2A''$	r (OO)	1.208	1.177	1.266	**1.335**
			r (OH)	0.976	1.010	0.957	**0.977**
			< (HOO)	112.3	112.5	107.5	**104.1**
O$_2$	$D_{\infty h}$	$^3\Sigma^-_g$	r (OO)	1.134	1.085	1.169	**1.208**

Table A5-58: Basis Set Effects on Bond Lengths in Hydrogen Bonded Complexes. Hartree-Fock Models

hydrogen-bonded complex	bond	6-31G*	6-31G**	6-31+G*	6-311G*	6-311+G**	expt.
(structure)	OO	2.98	2.98	2.93	2.91	3.00	**2.98**
(structure)	OF	2.72	2.72	2.72	2.67	2.72	**2.69**
(structure)	FF	2.72	2.73	2.79	2.73	2.83	**2.79**
(structure)	ClF	3.36	3.3	3.44	3.35	3.45	**3.37**
(structure)	NF	2.92	2.92	2.90	2.91	2.90	**2.80**
mean absolute error		0.05	0.05	0.05	0.06	0.05	–

Table A5-59: Basis Set Effects on Bond Lengths in Hydrogen Bonded Complexes. EDF1 Density Functional Models

hydrogen-bonded complex	bond	6-31G*	6-31G**	6-31+G*	6-311G*	6-311+G**	expt.
(structure)	OO	2.91	2.94	2.94	2.91	3.01	**2.98**
(structure)	OF	2.67	2.68	2.67	2.63	2.68	**2.69**
(structure)	FF	2.52[a]	2.53[a]	2.80	2.67[a]	2.86	**2.79**
(structure)	ClF	3.20	3.19	3.44	3.25	3.48	**3.37**
(structure)	NF	2.89	2.87	2.84	2.87	2.82	**2.80**
mean absolute error		0.12	0.11	0.04	0.09	0.05	–

a)   gives incorrect structure

Table A5-60: Basis Set Effects on Bond Lengths in Hydrogen Bonded Complexes. B3LYP Density Functional Models

hydrogen-bonded complex	bond	6-31G*	6-31G**	6-31+G*	6-311G*	6-311+G**	expt.
	OO	2.85	2.87	2.88	2.83	2.91	**2.98**
	OF	2.65	2.65	2.65	2.61	2.64	**2.69**
	FF	2.48[a]	2.49[a]	2.72	2.61[a]	2.74	**2.79**
	ClF	3.15	3.15	3.29	3.18	3.31	**3.37**
	NF	2.83	2.82	2.79	2.82	2.77	**2.80**
mean absolute error		0.15	0.14	0.06	0.12	0.05	–

a)   gives incorrect structure

Table A5-61: Basis Set Effects on Bond Lengths in Hydrogen Bonded Complexes. MP2 Models

hydrogen-bonded complex	bond	6-31G*	6-31G**	6-31+G*	6-311G*	6-311+G**	expt.
	OO	2.92	2.91	2.91	2.85	2.91	**2.98**
	OF	2.69	2.68	2.69	2.63	2.66	**2.69**
	FF	2.64	2.54[a]	2.77	2.68	2.79	**2.79**
	ClF	3.24	3.22	3.31	3.27	3.36	**3.37**
	NF	2.88	2.86	2.85	2.88	2.81	**2.80**
mean absolute error		0.08	0.11	0.04	0.10	0.02	–

a)   gives incorrect structure

Table A6-1: Homolytic Bond Dissociation Energies. Hartree-Fock Models

bond dissociation reaction	STO-3G	3-21G*	6-31G*	6-311+G**	G3	expt.
CH_3-$CH_3 \rightarrow 2CH_3$·	95	68	69	66	**96**	**97**
CH_3-$NH_2 \rightarrow CH_3$· + NH_2·	73	59	58	57	**91**	**93**
CH_3-$OH \rightarrow CH_3$· + OH·	67	53	58	58	**97**	**98**
CH_3-$F \rightarrow CH_3$· + F·	66	59	69	69	–	**114**
CH_3-$SiH_3 \rightarrow CH_3$· + SiH_3·	98	72	67	66	**92**	–
CH_3-$PH_2 \rightarrow CH_3$· + PH_2·	74	52	49	47	–	–
CH_3-$SH \rightarrow CH_3$· + SH·	68	48	48	46	**78**	–
CH_3-$Cl \rightarrow CH_3$· + Cl·	65	51	54	53	**86**	–
NH_2-$NH_2 \rightarrow 2NH_2$·	44	37	34	33	**72**	**73**
HO-$OH \rightarrow 2OH$·	22	3	0	-2	**53**	**55**
F-$F \rightarrow 2F$·	5	-29	-33	-39	**38**	**38**
SiH_3-$SiH_3 \rightarrow 2SiH_3$·	96	60	58	57	**79**	–
PH_2-$PH_2 \rightarrow 2PH_2$·	51	35	33	32	–	–
HS-$SH \rightarrow 2SH$·	44	30	29	28	–	–
Cl-$Cl \rightarrow 2Cl$·	19	10	11	12	**57**	–

Table A6-2: Homolytic Bond Dissociation Energies. Local Density Models

bond dissociation reaction	6-31G*	6-311+G**	G3	expt.
CH_3-$CH_3 \rightarrow 2CH_3$·	123	118	**96**	**97**
CH_3-$NH_2 \rightarrow CH_3$· + NH_2·	119	117	**97**	**93**
CH_3-$OH \rightarrow CH_3$· + OH·	128	124	**97**	**98**
CH_3-$F \rightarrow CH_3$· + F·	147	143	–	**114**
CH_3-$SiH_3 \rightarrow CH_3$· + SiH_3·	108	105	**92**	–
CH_3-$PH_2 \rightarrow CH_3$· + PH_2·	96	93	–	–
CH_3-$SH \rightarrow CH_3$· + SH·	101	98	**78**	–
CH_3-$Cl \rightarrow CH_3$· + Cl·	111	108	**86**	–
NH_2-$NH_2 \rightarrow 2NH_2$·	104	103	**72**	**73**
HO-$OH \rightarrow 2OH$·	93	86	**53**	**55**
F-$F \rightarrow 2F$·	85	73	**38**	**38**
SiH_3-$SiH_3 \rightarrow 2SiH_3$·	88	87	**79**	–
PH_2-$PH_2 \rightarrow 2PH_2$·	75	74	–	–
HS-$SH \rightarrow 2SH$·	83	80	–	–
Cl-$Cl \rightarrow 2Cl$·	76	75	**57**	–

Table A6-3: Homolytic Bond Dissociation Energies. BP Density Functional Models

bond dissociation reaction	6-31G*	6-311+G**	G3	expt.
$CH_3\text{-}CH_3 \rightarrow 2CH_3^{\bullet}$	99	95	**96**	**97**
$CH_3\text{-}NH_2 \rightarrow CH_3^{\bullet} + NH_2^{\bullet}$	93	91	**91**	**93**
$CH_3\text{-}OH \rightarrow CH_3^{\bullet} + OH^{\bullet}$	101	98	**97**	**98**
$CH_3\text{-}F \rightarrow CH_3^{\bullet} + F^{\bullet}$	119	116	–	**114**
$CH_3\text{-}SiH_3 \rightarrow CH_3^{\bullet} + SiH_3^{\bullet}$	89	87	**92**	–
$CH_3\text{-}PH_2 \rightarrow CH_3^{\bullet} + PH_2^{\bullet}$	76	74	–	–
$CH_3\text{-}SH \rightarrow CH_3^{\bullet} + SH^{\bullet}$	80	77	**78**	–
$CH_3\text{-}Cl \rightarrow CH_3^{\bullet} + Cl^{\bullet}$	89	87	**86**	–
$NH_2\text{-}NH_2 \rightarrow 2NH_2^{\bullet}$	75	74	**72**	**73**
$HO\text{-}OH \rightarrow 2OH^{\bullet}$	65	59	**53**	**55**
$F\text{-}F \rightarrow 2F^{\bullet}$	58	47	**38**	**38**
$SiH_3\text{-}SiH_3 \rightarrow 2SiH_3^{\bullet}$	75	73	**79**	–
$PH_2\text{-}PH_2 \rightarrow 2PH_2^{\bullet}$	58	57	–	–
$HS\text{-}SH \rightarrow 2SH^{\bullet}$	64	62	–	–
$Cl\text{-}Cl \rightarrow 2Cl^{\bullet}$	57	56	**57**	–

Table A6-4: Homolytic Bond Dissociation Energies. BLYP Density Functional Models

bond dissociation reaction	6-31G*	6-311+G**	G3	expt.
$CH_3\text{-}CH_3 \rightarrow 2CH_3^{\bullet}$	96	91	**96**	**97**
$CH_3\text{-}NH_2 \rightarrow CH_3^{\bullet} + NH_2^{\bullet}$	89	86	**91**	**93**
$CH_3\text{-}OH \rightarrow CH_3^{\bullet} + OH^{\bullet}$	97	93	**97**	**98**
$CH_3\text{-}F \rightarrow CH_3^{\bullet} + F^{\bullet}$	117	113	–	**114**
$CH_3\text{-}SiH_3 \rightarrow CH_3^{\bullet} + SiH_3^{\bullet}$	86	84	**92**	–
$CH_3\text{-}PH_2 \rightarrow CH_3^{\bullet} + PH_2^{\bullet}$	71	69	–	–
$CH_3\text{-}SH \rightarrow CH_3^{\bullet} + SH^{\bullet}$	75	72	**78**	–
$CH_3\text{-}Cl \rightarrow CH_3^{\bullet} + Cl^{\bullet}$	85	82	**86**	–
$NH_2\text{-}NH_2 \rightarrow 2NH_2^{\bullet}$	70	69	**72**	**73**
$HO\text{-}OH \rightarrow 2OH^{\bullet}$	61	55	**53**	**55**
$F\text{-}F \rightarrow 2F^{\bullet}$	56	44	**38**	**38**
$SiH_3\text{-}SiH_3 \rightarrow 2SiH_3^{\bullet}$	72	71	**79**	–
$PH_2\text{-}PH_2 \rightarrow 2PH_2^{\bullet}$	54	52	–	–
$HS\text{-}SH \rightarrow 2SH^{\bullet}$	59	57	–	–
$Cl\text{-}Cl \rightarrow 2Cl^{\bullet}$	52	51	**57**	–

Table A6-5: Homolytic Bond Dissociation Energies. EDF1 Density Functional Models

bond dissociation reaction	6-31G*	6-311+G**	G3	expt.
CH_3-$CH_3 \rightarrow 2CH_3^\bullet$	97	92	**96**	**97**
CH_3-$NH_2 \rightarrow CH_3^\bullet + NH_2^\bullet$	90	88	**91**	**93**
CH_3-$OH \rightarrow CH_3^\bullet + OH^\bullet$	98	95	**97**	**98**
CH_3-$F \rightarrow CH_3^\bullet + F^\bullet$	117	114	–	**114**
CH_3-$SiH_3 \rightarrow CH_3^\bullet + SiH_3^\bullet$	86	84	**92**	–
CH_3-$PH_2 \rightarrow CH_3^\bullet + PH_2^\bullet$	72	70	–	–
CH_3-$SH \rightarrow CH_3^\bullet + SH^\bullet$	77	74	**78**	–
CH_3-$Cl \rightarrow CH_3^\bullet + Cl^\bullet$	86	83	**86**	–
NH_2-$NH_2 \rightarrow 2NH_2^\bullet$	72	71	**72**	**73**
HO-$OH \rightarrow 2OH^\bullet$	61	55	**53**	**55**
F-$F \rightarrow 2F^\bullet$	53	42	**38**	**38**
SiH_3-$SiH_3 \rightarrow 2SiH_3^\bullet$	72	70	**79**	–
PH_2-$PH_2 \rightarrow 2PH_2^\bullet$	54	53	–	–
HS-$SH \rightarrow 2SH^\bullet$	60	58	–	–
Cl-$Cl \rightarrow 2Cl^\bullet$	52	51	**57**	–

Table A6-6: Homolytic Bond Dissociation Energies. B3LYP Density Functional Models

bond dissociation reaction	6-31G*	6-311+G**	G3	expt.
CH_3-$CH_3 \rightarrow 2CH_3^\bullet$	97	92	**96**	**97**
CH_3-$NH_2 \rightarrow CH_3^\bullet + NH_2^\bullet$	89	87	**91**	**93**
CH_3-$OH \rightarrow CH_3^\bullet + OH^\bullet$	96	93	**97**	**98**
CH_3-$F \rightarrow CH_3^\bullet + F^\bullet$	113	110	–	**114**
CH_3-$SiH_3 \rightarrow CH_3^\bullet + SiH_3^\bullet$	88	86	**92**	–
CH_3-$PH_2 \rightarrow CH_3^\bullet + PH_2^\bullet$	72	70	–	–
CH_3-$SH \rightarrow CH_3^\bullet + SH^\bullet$	75	72	**78**	–
CH_3-$Cl \rightarrow CH_3^\bullet + Cl^\bullet$	85	81	**86**	–
NH_2-$NH_2 \rightarrow 2NH_2^\bullet$	70	69	**72**	**73**
HO-$OH \rightarrow 2OH^\bullet$	54	49	**53**	**55**
F-$F \rightarrow 2F^\bullet$	42	32	**38**	**38**
SiH_3-$SiH_3 \rightarrow 2SiH_3^\bullet$	74	73	**79**	–
PH_2-$PH_2 \rightarrow 2PH_2^\bullet$	54	53	–	–
HS-$SH \rightarrow 2SH^\bullet$	58	56	–	–
Cl-$Cl \rightarrow 2Cl^\bullet$	50	47	**57**	–

Table A6-7: Homolytic Bond Dissociation Energies. MP2 Models

bond dissociation reaction	6-31G*	6-311+G**	G3	expt.
$CH_3\text{-}CH_3 \rightarrow 2CH_3^{\bullet}$	99	97	**96**	**97**
$CH_3\text{-}NH_2 \rightarrow CH_3^{\bullet} + NH_2^{\bullet}$	92	93	**91**	**93**
$CH_3\text{-}OH \rightarrow CH_3^{\bullet} + OH^{\bullet}$	98	98	**97**	**98**
$CH_3\text{-}F \rightarrow CH_3^{\bullet} + F^{\bullet}$	113	112	–	**114**
$CH_3\text{-}SiH_3 \rightarrow CH_3^{\bullet} + SiH_3^{\bullet}$	90	90	**92**	–
$CH_3\text{-}PH_2 \rightarrow CH_3^{\bullet} + PH_2^{\bullet}$	74	74	–	–
$CH_3\text{-}SH \rightarrow CH_3^{\bullet} + SH^{\bullet}$	77	77	**78**	–
$CH_3\text{-}Cl \rightarrow CH_3^{\bullet} + Cl^{\bullet}$	84	84	**86**	–
$NH_2\text{-}NH_2 \rightarrow 2NH_2^{\bullet}$	74	74	**72**	**73**
$HO\text{-}OH \rightarrow 2OH^{\bullet}$	56	52	**53**	**55**
$F\text{-}F \rightarrow 2F^{\bullet}$	38	29	**38**	**38**
$SiH_3\text{-}SiH_3 \rightarrow 2SiH_3^{\bullet}$	73	73	**79**	–
$PH_2\text{-}PH_2 \rightarrow 2PH_2^{\bullet}$	53	53	–	–
$HS\text{-}SH \rightarrow 2SH^{\bullet}$	55	55	–	–
$Cl\text{-}Cl \rightarrow 2Cl^{\bullet}$	42	43	**57**	–

Table A6-8: Homolytic Bond Dissociation Energies. Semi-Empirical Models

bond dissociation reaction	MNDO	AM1	PM3	G3	expt.
$CH_3\text{-}CH_3 \rightarrow 2CH_3^{\bullet}$	69	77	74	**96**	**97**
$CH_3\text{-}NH_2 \rightarrow CH_3^{\bullet} + NH_2^{\bullet}$	69	75	68	**91**	**93**
$CH_3\text{-}OH \rightarrow CH_3^{\bullet} + OH^{\bullet}$	82	88	83	**97**	**98**
$CH_3\text{-}F \rightarrow CH_3^{\bullet} + F^{\bullet}$	104	110	101	–	**114**
$CH_3\text{-}SiH_3 \rightarrow CH_3^{\bullet} + SiH_3^{\bullet}$	78	66	74	**92**	–
$CH_3\text{-}PH_2 \rightarrow CH_3^{\bullet} + PH_2^{\bullet}$	64	65	67	–	–
$CH_3\text{-}SH \rightarrow CH_3^{\bullet} + SH^{\bullet}$	63	74	71	**78**	–
$CH_3\text{-}Cl \rightarrow CH_3^{\bullet} + Cl^{\bullet}$	72	78	72	**86**	–
$NH_2\text{-}NH_2 \rightarrow 2NH_2^{\bullet}$	59	62	50	**72**	**73**
$HO\text{-}OH \rightarrow 2OH^{\bullet}$	39	37	46	**53**	**55**
$F\text{-}F \rightarrow 2F^{\bullet}$	30	60	59	**38**	**38**
$SiH_3\text{-}SiH_3 \rightarrow 2SiH_3^{\bullet}$	65	33	66	**79**	–
$PH_2\text{-}PH_2 \rightarrow 2PH_2^{\bullet}$	62	62	61	–	–
$HS\text{-}SH \rightarrow 2SH^{\bullet}$	67	70	66	–	–
$Cl\text{-}Cl \rightarrow 2Cl^{\bullet}$	59	72	70	**57**	–

Table A6-9: **Basis Set Effects on Homolytic Bond Dissociation Energies. EDF1 Density Functional Models**

bond dissociation reaction	6-31G*	6-31G**	6-31+G*	6-311G*	6-311+G**	G3	expt.
$CH_3\text{-}CH_3 \rightarrow 2CH_3^{\cdot}$	97	96	93	95	92	**96**	97
$CH_3\text{-}NH_2 \rightarrow CH_3^{\cdot} + NH_2^{\cdot}$	90	90	89	90	88	**91**	93
$CH_3\text{-}OH \rightarrow CH_3^{\cdot} + OH^{\cdot}$	98	98	96	97	95	**97**	98
$CH_3\text{-}F \rightarrow CH_3^{\cdot} + F^{\cdot}$	117	117	115	114	114	–	114
$NH_2\text{-}NH_2 \rightarrow 2NH_2^{\cdot}$	72	72	72	74	71	**72**	73
$HO\text{-}OH \rightarrow 2OH^{\cdot}$	61	61	57	59	55	**53**	55
$F\text{-}F \rightarrow 2F^{\cdot}$	53	53	46	46	42	**38**	38

Table A6-10: **Basis Set Effects on Homolytic Bond Dissociation Energies. B3LYP Density Functional Models**

bond dissociation reaction	6-31G*	6-31G**	6-31+G*	6-311G*	6-311+G**	G3	expt.
$CH_3\text{-}CH_3 \rightarrow 2CH_3^{\cdot}$	97	96	93	95	92	**96**	97
$CH_3\text{-}NH_2 \rightarrow CH_3^{\cdot} + NH_2^{\cdot}$	89	89	88	89	87	**91**	93
$CH_3\text{-}OH \rightarrow CH_3^{\cdot} + OH^{\cdot}$	96	96	94	95	93	**97**	98
$CH_3\text{-}F \rightarrow CH_3^{\cdot} + F^{\cdot}$	113	113	112	110	110	–	114
$NH_2\text{-}NH_2 \rightarrow 2NH_2^{\cdot}$	70	69	70	72	69	**72**	73
$HO\text{-}OH \rightarrow 2OH^{\cdot}$	54	54	51	53	49	**53**	55
$F\text{-}F \rightarrow 2F^{\cdot}$	42	42	35	35	32	**38**	38

Table A6-11: **Basis Set Effects on Homolytic Bond Dissociation Energies. MP2 Models**

bond dissociation reaction	6-31G*	6-31G**	6-31+G*	6-311G*	6-311+G**	G3	expt.
$CH_3\text{-}CH_3 \rightarrow 2CH_3^{\cdot}$	99	99	96	98	97	**96**	97
$CH_3\text{-}NH_2 \rightarrow CH_3^{\cdot} + NH_2^{\cdot}$	92	93	92	93	93	**91**	93
$CH_3\text{-}OH \rightarrow CH_3^{\cdot} + OH^{\cdot}$	98	99	98	98	98	**97**	98
$CH_3\text{-}F \rightarrow CH_3^{\cdot} + F^{\cdot}$	113	113	114	110	112	–	114
$NH_2\text{-}NH_2 \rightarrow 2NH_2^{\cdot}$	73	73	74	76	74	**72**	73
$HO\text{-}OH \rightarrow 2OH^{\cdot}$	55	55	54	54	52	**53**	55
$F\text{-}F \rightarrow 2F^{\cdot}$	38	38	35	30	29	**35**	38

Table A6-12: Energies of Hydrogenation Reactions. Hartree-Fock Models

hydrogenation reaction	STO-3G	3-21G	6-31G*	6-311+G**	G3	expt.
$Li_2 + H_2 \rightarrow 2LiH$	19	19	20	20	**20**	**20**
$LiOH + H_2 \rightarrow LiH + H_2O$	36	36	24	25	–	**35**
$LiF + H_2 \rightarrow LiH + HF$	31	53	49	43	**49**	**48**
$LiCl + H_2 \rightarrow LiH + HCl$	69	59	60	58	–	**60**
$CH_3CH_3 + H_2 \rightarrow 2CH_4$	-19	-25	-22	-21	**-18**	**-19**
$CH_3NH_2 + H_2 \rightarrow CH_4 + NH_3$	-20	-30	-27	-28	**-26**	**-26**
$CH_3OH + H_2 \rightarrow CH_4 + H_2O$	-16	-28	-27	-31	**-30**	**-30**
$CH_3F + H_2 \rightarrow CH_4 + HF$	-8	-22	-23	-29	–	**-29**
$CH_3SiH_3 + H_2 \rightarrow CH_4 + SiH_4$	-10	-10	-13	-12	**-6**	**-6**
$CH_3SH + H_2 \rightarrow CH_4 + H_2S$	-17	-23	-22	-24	**-19**	**-19**
$CH_3Cl + H_2 \rightarrow CH_4 + HCl$	-14	-25	-22	-25	–	**-22**
$NH_2NH_2 + H_2 \rightarrow 2NH_3$	-28	-47	-46	-50	**-48**	**-48**
$HOOH + H_2 \rightarrow 2H_2O$	-31	-67	-82	-93	**-87**	**-86**
$NaOH + H_2 \rightarrow H_2O + NaH$	15	25	10	14	–	**21**
$HOCl + H_2 \rightarrow H_2O + HCl$	-33	-57	-64	-75	**-64**	**-65**
$F_2 + H_2 \rightarrow 2HF$	-29	-98	-126	-149	**-135**	**-133**
$NaF + H_2 \rightarrow HF + NaH$	37	46	33	33	–	**35**
$SiH_3F + H_2 \rightarrow HF + SiH_4$	12	45	30	18	–	**48**
$ClF + H_2 \rightarrow HF + HCl$	-37	-64	-73	-89	**-77**	**-77**
$Na_2 + H_2 \rightarrow 2NaH$	74	49	40	41	–	**29**
$NaCl + H_2 \rightarrow NaH + HCl$	122	62	60	57	–	**52**
$SiH_3SiH_3 + H_2 \rightarrow 2SiH_4$	-5	-11	-12	-11	**-4**	**-5**
$SiH_3Cl + H_2 \rightarrow SiH_4 + HCl$	23	17	16	12	–	**33**
$PH_2PH_2 + H_2 \rightarrow 2PH_3$	-10	-10	-11	-12	–	**-4**
$HSSH + H_2 \rightarrow 2H_2S$	-9	-21	-21	-25	–	**-14**
$Cl_2 + H_2 \rightarrow 2HCl$	-25	-51	-50	-57	**-46**	**-46**
$CH_2=CH_2 + 2H_2 \rightarrow 2CH_4$	-91	-71	-66	-61	**-57**	**-57**
$CH_2=NH + 2H_2 \rightarrow CH_4 + NH_3$	-78	-68	-61	-62	–	**-64**
$CH_2=O + 2H_2 \rightarrow CH_4 + H_2O$	-65	-64	-54	-59	**-58**	**-59**
$CH_2=S + 2H_2 \rightarrow CH_4 + H_2S$	-83	-65	-64	-66	–	**-54**
$HN=NH + 2H_2 \rightarrow 2NH_3$	-75	-94	-76	-83	–	**-68**
$HN=O + 2H_2 \rightarrow NH_3 + H_2O$	-78	-113	-98	-110	–	**-103**
$O=O + 2H_2 \rightarrow 2H_2O$	-39	-98	-94	-111	**-124**	**-125**
$HP=O + 2H_2 \rightarrow H_2O + PH_3$	-72	-41	-50	-59	–	**-105**
$HC\equiv CH + 3H_2 \rightarrow 2CH_4$	-154	-124	-121	-112	**-105**	**-105**
$HC\equiv N + 3H_2 \rightarrow CH_4 + NH_3$	-97	-85	-78	-78	**-74**	**-76**
$^-C\equiv O+ + 3H_2 \rightarrow CH_4 + H_2O$	-72	-69	-55	-59	**-63**	**-63**
$HC\equiv P + 3H_2 \rightarrow CH_4 + PH_3$	-132	-94	-9	-95	–	**-67**
$^-C\equiv S+ + 3H_2 \rightarrow CH_4 + H_2S$	-134	-114	-111	-110	–	**-91**
$N\equiv N + 3H_2 \rightarrow 2NH_3$	-36	-53	-28	-26	**-38**	**-37**
$P\equiv N + 3H_2 \rightarrow NH_3 + PH_3$	-91	-74	-79	-84	–	**-45**
$^-Si\equiv O+ + 3H_2 \rightarrow H_2O + SiH_4$	-82	-37	-48	-55	**-34**	**-32**
$P\equiv P + 3H_2 \rightarrow 2PH_3$	-98	-56	-57	-59	–	**-44**

Table A6-13: Energies of Hydrogenation Reactions. Local Density Models

hydrogenation reaction	6-31G*	6-311+G**	G3	expt.
$Li_2 + H_2 \rightarrow 2LiH$	17	16	**20**	**20**
$LiOH + H_2 \rightarrow LiH + H_2O$	32	31	–	**35**
$LiF + H_2 \rightarrow LiH + HF$	56	47	**49**	**48**
$LiCl + H_2 \rightarrow LiH + HCl$	57	56	–	**60**
$CH_3CH_3 + H_2 \rightarrow 2CH_4$	-17	-17	**-18**	**-19**
$CH_3NH_2 + H_2 \rightarrow CH_4 + NH_3$	-21	-25	**-26**	**-26**
$CH_3OH + H_2 \rightarrow CH_4 + H_2O$	-22	-29	**-30**	**-30**
$CH_3F + H_2 \rightarrow CH_4 + HF$	-19	-29	–	**-29**
$CH_3SiH_3 + H_2 \rightarrow CH_4 + SiH_4$	-7	-7	**-6**	**-6**
$CH_3SH + H_2 \rightarrow CH_4 + H_2S$	-17	-19	**-19**	**-19**
$CH_3Cl + H_2 \rightarrow CH_4 + HCl$	-18	-23	–	**-22**
$NH_2NH_2 + H_2 \rightarrow 2NH_3$	-37	-45	**-48**	**-48**
$HOOH + H_2 \rightarrow 2H_2O$	-68	-85	**-87**	**-86**
$NaOH + H_2 \rightarrow H_2O + NaH$	12	15	–	**21**
$HOCl + H_2 \rightarrow H_2O + HCl$	-54	-67	**-64**	**-65**
$F_2 + H_2 \rightarrow 2HF$	-106	-137	**-135**	**-133**
$NaF + H_2 \rightarrow HF + NaH$	35	32	–	**35**
$SiH_3F + H_2 \rightarrow HF + SiH_4$	31	16	–	**48**
$ClF + H_2 \rightarrow HF + HCl$	-63	-81	**-77**	**-77**
$Na_2 + H_2 \rightarrow 2NaH$	35	36	–	**29**
$NaCl + H_2 \rightarrow NaH + HCl$	51	49	–	**52**
$SiH_3SiH_3 + H_2 \rightarrow 2SiH_4$	-1	-1	**-4**	**-5**
$SiH_3Cl + H_2 \rightarrow SiH_4 + HCl$	17	13	–	**33**
$PH_2PH_2 + H_2 \rightarrow 2PH_3$	-4	-6	–	**-4**
$HSSH + H_2 \rightarrow 2H_2S$	-14	-19	–	**-14**
$Cl_2 + H_2 \rightarrow 2HCl$	-42	-51	**-46**	**-46**
$CH_2=CH_2 + 2H_2 \rightarrow 2CH_4$	-70	-66	**-57**	**-57**
$CH_2=NH + 2H_2 \rightarrow CH_4 + NH_3$	-65	-69	–	**-64**
$CH_2=O + 2H_2 \rightarrow CH_4 + H_2O$	-58	-68	**-58**	**-59**
$CH_2=S + 2H_2 \rightarrow CH_4 + H_2S$	-65	-67	–	**-54**
$HN=NH + 2H_2 \rightarrow 2NH_3$	-78	-90	–	**-68**
$HN=O + 2H_2 \rightarrow NH_3 + H_2O$	-97	-115	–	**-103**
$O=O + 2H_2 \rightarrow 2H_2O$	-106	-134	**-124**	**-125**
$HP=O + 2H_2 \rightarrow H_2O + PH_3$	-48	-59	–	**-105**
$HC\equiv CH + 3H_2 \rightarrow 2CH_4$	-135	-126	**-105**	**-105**
$HC\equiv N + 3H_2 \rightarrow CH_4 + NH_3$	-96	-98	**-74**	**-76**
$-C\equiv O+ + 3H_2 \rightarrow CH_4 + H_2O$	-83	-90	**-63**	**-63**
$HC\equiv P + 3H_2 \rightarrow CH_4 + PH_3$	-103	-100	–	**-67**
$C\equiv S+ + 3H_2 \rightarrow CH_4 + H_2S$	-128	-126	–	**-91**
$N\equiv N + 3H_2 \rightarrow 2NH_3$	-55	-67	**-38**	**-37**
$P\equiv N + 3H_2 \rightarrow NH_3 + PH_3$	-83	-88	–	**-45**
$-Si\equiv O+ + 3H_2 \rightarrow H_2O + SiH_4$	-44	-53	**-34**	**-32**
$P\equiv P + 3H_2 \rightarrow 2PH_3$	-55	-57	–	**-44**

Table A6-14: Energies of Hydrogenation Reactions. BP Density Functional Models

hydrogenation reaction	6-31G*	6-311+G**	G3	expt.
$Li_2 + H_2 \rightarrow 2LiH$	17	15	**20**	**20**
$LiOH + H_2 \rightarrow LiH + H_2O$	32	32	–	**35**
$LiF + H_2 \rightarrow LiH + HF$	56	48	**49**	**48**
$LiCl + H_2 \rightarrow LiH + HCl$	58	57	–	**60**
$CH_3CH_3 + H_2 \rightarrow 2CH_4$	-18	-19	**-18**	**-19**
$CH_3NH_2 + H_2 \rightarrow CH_4 + NH_3$	-23	-26	**-26**	**-26**
$CH_3OH + H_2 \rightarrow CH_4 + H_2O$	-23	-29	**-30**	**-30**
$CH_3F + H_2 \rightarrow CH_4 + HF$	-18	-28	–	**-29**
$CH_3SiH_3 + H_2 \rightarrow CH_4 + SiH_4$	-8	-8	**-6**	**-6**
$CH_3SH + H_2 \rightarrow CH_4 + H_2S$	-17	-20	**-19**	**-19**
$CH_3Cl + H_2 \rightarrow CH_4 + HCl$	-18	-22	–	**-22**
$NH_2NH_2 + H_2 \rightarrow 2NH_3$	-38	-46	**-48**	**-48**
$HOOH + H_2 \rightarrow 2H_2O$	-65	-81	**-87**	**-86**
$NaOH + H_2 \rightarrow H_2O + NaH$	15	17	–	**21**
$HOCl + H_2 \rightarrow H_2O + HCl$	-51	-64	**-64**	**-65**
$F_2 + H_2 \rightarrow 2HF$	-99	-129	**-135**	**-133**
$NaF + H_2 \rightarrow HF + NaH$	37	36	–	**35**
$SiH_3F + H_2 \rightarrow HF + SiH_4$	31	17	–	**48**
$ClF + H_2 \rightarrow HF + HCl$	-59	-76	**-77**	**-77**
$Na_2 + H_2 \rightarrow 2NaH$	35	36	–	**29**
$NaCl + H_2 \rightarrow NaH + HCl$	54	53	–	**52**
$SiH_3SiH_3 + H_2 \rightarrow 2SiH_4$	-3	-3	**-4**	**-5**
$SiH_3Cl + H_2 \rightarrow SiH_4 + HCl$	17	13	–	**33**
$PH_2PH_2 + H_2 \rightarrow 2PH_3$	-5	-6	–	**-4**
$HSSH + H_2 \rightarrow 2H_2S$	-13	-18	–	**-14**
$Cl_2 + H_2 \rightarrow 2HCl$	-40	-48	**-46**	**-46**
$CH_2=CH_2 + 2H_2 \rightarrow 2CH_4$	-61	-58	**-57**	**-57**
$CH_2=NH + 2H_2 \rightarrow CH_4 + NH_3$	-54	-59	**-58**	**-64**
$CH_2=O + 2H_2 \rightarrow CH_4 + H_2O$	-46	-56	–	**-59**
$CH_2=S + 2H_2 \rightarrow CH_4 + H_2S$	-55	-58	–	**-54**
$HN=NH + 2H_2 \rightarrow 2NH_3$	-65	-77	–	**-68**
$HN=O + 2H_2 \rightarrow NH_3 + H_2O$	-82	-101	–	**-103**
$O=O + 2H_2 \rightarrow 2H_2O$	-88	-114	**-124**	**-125**
$HP=O + 2H_2 \rightarrow H_2O + PH_3$	-36	-47	–	**-105**
$HC\equiv CH + 3H_2 \rightarrow 2CH_4$	-115	-108	**-105**	**-105**
$HC\equiv N + 3H_2 \rightarrow CH_4 + NH_3$	-73	-78	**-74**	**-76**
$-C\equiv O+ + 3H_2 \rightarrow CH_4 + H_2O$	-57	-67	**-63**	**-63**
$HC\equiv P + 3H_2 \rightarrow CH_4 + PH_3$	-84	-83	–	**-67**
$-C\equiv S+ + 3H_2 \rightarrow CH_4 + H_2S$	-103	-104	–	**-91**
$N\equiv N + 3H_2 \rightarrow 2NH_3$	-28	-43	**-38**	**-37**
$P\equiv N + 3H_2 \rightarrow NH_3 + PH_3$	-60	-67	–	**-45**
$-Si\equiv O+ + 3H_2 \rightarrow H_2O + SiH_4$	-26	-37	**-34**	**-32**
$P\equiv P + 3H_2 \rightarrow 2PH_3$	-38	-41	–	**-44**

Table A6-15: Energies of Hydrogenation Reactions. BLYP Density Functional Models

hydrogenation reaction	6-31G*	6-311+G**	G3	expt.
$Li_2 + H_2 \rightarrow 2LiH$	16	14	20	20
$LiOH + H_2 \rightarrow LiH + H_2O$	36	34	–	35
$LiF + H_2 \rightarrow LiH + HF$	60	51	49	48
$LiCl + H_2 \rightarrow LiH + HCl$	58	58	–	60
$CH_3CH_3 + H_2 \rightarrow 2CH_4$	-19	-20	-18	-19
$CH_3NH_2 + H_2 \rightarrow CH_4 + NH_3$	-22	-27	-26	-26
$CH_3OH + H_2 \rightarrow CH_4 + H_2O$	-22	-29	-30	-30
$CH_3F + H_2 \rightarrow CH_4 + HF$	-16	-27	–	-29
$CH_3SiH_3 + H_2 \rightarrow CH_4 + SiH_4$	-9	-9	-6	-6
$CH_3SH + H_2 \rightarrow CH_4 + H_2S$	-18	-20	-19	-19
$CH_3Cl + H_2 \rightarrow CH_4 + HCl$	-18	-22	–	-22
$NH_2NH_2 + H_2 \rightarrow 2NH_3$	-38	-46	-48	-48
$HOOH + H_2 \rightarrow 2H_2O$	-62	-80	-87	-86
$NaOH + H_2 \rightarrow H_2O + NaH$	16	18	–	21
$HOCl + H_2 \rightarrow H_2O + HCl$	-49	-62	-64	-65
$F_2 + H_2 \rightarrow 2HF$	-94	-126	-135	-133
$NaF + H_2 \rightarrow HF + NaH$	39	37	–	35
$SiH_3F + H_2 \rightarrow HF + SiH_4$	35	19	–	48
$ClF + H_2 \rightarrow HF + HCl$	-56	-74	-77	-77
$Na_2 + H_2 \rightarrow 2NaH$	35	36	–	29
$NaCl + H_2 \rightarrow NaH + HCl$	53	52	–	52
$SiH_3SiH_3 + H_2 \rightarrow 2SiH_4$	-5	-6	-4	-5
$SiH_3Cl + H_2 \rightarrow SiH_4 + HCl$	17	13	–	33
$PH_2PH_2 + H_2 \rightarrow 2PH_3$	-5	-7	–	-4
$HSSH + H_2 \rightarrow 2H_2S$	-13	-18	–	-14
$Cl_2 + H_2 \rightarrow 2HCl$	-38	-45	-46	-46
$CH_2=CH_2 + 2H_2 \rightarrow 2CH_4$	-57	-55	-57	-57
$CH_2=NH + 2H_2 \rightarrow CH_4 + NH_3$	-49	-55	–	-64
$CH_2=O + 2H_2 \rightarrow CH_4 + H_2O$	-41	-52	-58	-59
$CH_2=S + 2H_2 \rightarrow CH_4 + H_2S$	-51	-54	–	-54
$HN=NH + 2H_2 \rightarrow 2NH_3$	-60	-75	–	-68
$HN=O + 2H_2 \rightarrow NH_3 + H_2O$	-77	-97	–	-103
$O=O + 2H_2 \rightarrow 2H_2O$	-82	-110	-124	-125
$HP=O + 2H_2 \rightarrow H_2O + PH_3$	-29	-42	–	-105
$HC\equiv CH + 3H_2 \rightarrow 2CH_4$	-108	-101	-105	-105
$HC\equiv N + 3H_2 \rightarrow CH_4 + NH_3$	-64	-69	-74	-76
$-C\equiv O+ + 3H_2 \rightarrow CH_4 + H_2O$	-49	-59	-63	-63
$HC\equiv P + 3H_2 \rightarrow CH_4 + PH_3$	-77	-76	–	-67
$-C\equiv S+ + 3H_2 \rightarrow CH_4 + H_2S$	-95	-96	–	-91
$N\equiv N + 3H_2 \rightarrow 2NH_3$	-19	-35	-38	-37
$P\equiv N + 3H_2 \rightarrow NH_3 + PH_3$	-51	-59	–	-45
$-Si\equiv O+ + 3H_2 \rightarrow H_2O + SiH_4$	-17	-29	-34	-32
$P\equiv P + 3H_2 \rightarrow 2PH_3$	-32	-35	–	-44

Table A6-16: Energies of Hydrogenation Reactions. EDF1 Density Functional Models

hydrogenation reaction	6-31G*	6-311+G**	G3	expt.
$Li_2 + H_2 \rightarrow 2LiH$	18	17	**20**	**20**
$LiOH + H_2 \rightarrow LiH + H_2O$	33	32	–	**35**
$LiF + H_2 \rightarrow LiH + HF$	58	49	**49**	**48**
$LiCl + H_2 \rightarrow LiH + HCl$	58	57	–	**60**
$CH_3CH_3 + H_2 \rightarrow 2CH_4$	-19	-20	**-18**	**-19**
$CH_3NH_2 + H_2 \rightarrow CH_4 + NH_3$	-23	-27	**-26**	**-26**
$CH_3OH + H_2 \rightarrow CH_4 + H_2O$	-23	-30	**-30**	**-30**
$CH_3F + H_2 \rightarrow CH_4 + HF$	-18	-28	–	**-29**
$CH_3SiH_3 + H_2 \rightarrow CH_4 + SiH_4$	-9	-8	**-6**	**-6**
$CH_3SH + H_2 \rightarrow CH_4 + H_2S$	-18	-20	**-19**	**-19**
$CH_3Cl + H_2 \rightarrow CH_4 + HCl$	-19	-22	–	**-22**
$NH_2NH_2 + H_2 \rightarrow 2NH_3$	-39	-46	**-48**	**-48**
$HOOH + H_2 \rightarrow 2H_2O$	-66	-83	**-87**	**-86**
$NaOH + H_2 \rightarrow H_2O + NaH$	15	17	–	**21**
$HOCl + H_2 \rightarrow H_2O + HCl$	-52	-65	**-64**	**-65**
$F_2 + H_2 \rightarrow 2HF$	-100	-130	**-135**	**-133**
$NaF + H_2 \rightarrow HF + NaH$	39	37	–	**35**
$SiH_3F + H_2 \rightarrow HF + SiH_4$	32	18	–	**48**
$ClF + H_2 \rightarrow HF + HCl$	-59	-77	**-77**	**-77**
$Na_2 + H_2 \rightarrow 2NaH$	35	36	–	**29**
$NaCl + H_2 \rightarrow NaH + HCl$	54	52	–	**52**
$SiH_3SiH_3 + H_2 \rightarrow 2SiH_4$	-4	-4	**-4**	**-5**
$SiH_3Cl + H_2 \rightarrow SiH_4 + HCl$	17	13	–	**33**
$PH_2PH_2 + H_2 \rightarrow 2PH_3$	-5	-7	–	**-4**
$HSSH + H_2 \rightarrow 2H_2S$	-13	-18	–	**-14**
$Cl_2 + H_2 \rightarrow 2HCl$	-41	-49	**-46**	**-46**
$CH_2{=}CH_2 + 2H_2 \rightarrow 2CH_4$	-58	-56	**-57**	**-57**
$CH_2{=}NH + 2H_2 \rightarrow CH_4 + NH_3$	-51	-56	–	**-64**
$CH_2{=}O + 2H_2 \rightarrow CH_4 + H_2O$	-43	-53	**-58**	**-59**
$CH_2{=}S + 2H_2 \rightarrow CH_4 + H_2S$	-52	-55	–	**-54**
$HN{=}NH + 2H_2 \rightarrow 2NH_3$	-62	-74	–	**-68**
$HN{=}O + 2H_2 \rightarrow NH_3 + H_2O$	-79	-98	–	**-103**
$O{=}O + 2H_2 \rightarrow 2H_2O$	-83	-110	**-124**	**-125**
$HP{=}O + 2H_2 \rightarrow H_2O + PH_3$	-32	-44	–	**-105**
$HC{\equiv}CH + 3H_2 \rightarrow 2CH_4$	-109	-103	**-105**	**-105**
$HC{\equiv}N + 3H_2 \rightarrow CH_4 + NH_3$	-67	-72	**-74**	**-76**
$-C{\equiv}O+ + 3H_2 \rightarrow CH_4 + H_2O$	-51	-61	**-63**	**-63**
$HC{\equiv}P + 3H_2 \rightarrow CH_4 + PH_3$	-78	-77	–	**-67**
$-C{\equiv}S+ + 3H_2 \rightarrow CH_4 + H_2S$	-97	-98	–	**-91**
$N{\equiv}N + 3H_2 \rightarrow 2NH_3$	-22	-36	**-38**	**-37**
$P{\equiv}N + 3H_2 \rightarrow NH_3 + PH_3$	-54	-62	–	**-45**
$-Si{\equiv}O+ + 3H_2 \rightarrow H_2O + SiH_4$	-20	-31	**-34**	**-32**
$P{\equiv}P + 3H_2 \rightarrow 2PH_3$	-32	-35	–	**-44**

Table A6-17: Energies of Hydrogenation Reactions. B3LYP Density Functional Models

hydrogenation reaction	6-31G*	6-311+G**	G3	expt.
$Li_2 + H_2 \rightarrow 2LiH$	16	14	20	20
$LiOH + H_2 \rightarrow LiH + H_2O$	53	33	–	35
$LiF + H_2 \rightarrow LiH + HF$	57	50	49	48
$LiCl + H_2 \rightarrow LiH + HCl$	59	58	–	60
$CH_3CH_3 + H_2 \rightarrow 2CH_4$	-19	-20	-18	-19
$CH_3NH_2 + H_2 \rightarrow CH_4 + NH_3$	-24	-27	-26	-26
$CH_3OH + H_2 \rightarrow CH_4 + H_2O$	-24	-30	-30	-30
$CH_3F + H_2 \rightarrow CH_4 + HF$	-18	-29	–	-29
$CH_3SiH_3 + H_2 \rightarrow CH_4 + SiH_4$	-10	-9	-6	-6
$CH_3SH + H_2 \rightarrow CH_4 + H_2S$	-19	-21	-19	-19
$CH_3Cl + H_2 \rightarrow CH_4 + HCl$	-19	-23	–	-22
$NH_2NH_2 + H_2 \rightarrow 2NH_3$	-40	-47	-48	-48
$HOOH + H_2 \rightarrow 2H_2O$	-69	-85	-87	-86
$NaOH + H_2 \rightarrow H_2O + NaH$	14	18	–	21
$HOCl + H_2 \rightarrow H_2O + HCl$	-54	-66	-64	-65
$F_2 + H_2 \rightarrow 2HF$	-105	-134	-135	-133
$NaF + H_2 \rightarrow HF + NaH$	37	36	–	35
$SiH_3F + H_2 \rightarrow HF + SiH_4$	33	19	–	48
$ClF + H_2 \rightarrow HF + HCl$	-62	-79	-77	-77
$Na_2 + H_2 \rightarrow 2NaH$	34	37	–	29
$NaCl + H_2 \rightarrow NaH + HCl$	55	54	–	52
$SiH_3SiH_3 + H_2 \rightarrow 2SiH_4$	-6	-6	-4	-5
$SiH_3Cl + H_2 \rightarrow SiH_4 + HCl$	17	13	–	33
$PH_2PH_2 + H_2 \rightarrow 2PH_3$	-6	-8	–	-4
$HSSH + H_2 \rightarrow 2H_2S$	-15	-19	–	-14
$Cl_2 + H_2 \rightarrow 2HCl$	-42	-49	-46	-46
$CH_2{=}CH_2 + 2H_2 \rightarrow 2CH_4$	-62	-59	-57	-57
$CH_2{=}NH + 2H_2 \rightarrow CH_4 + NH_3$	-55	-60	–	-64
$CH_2{=}O + 2H_2 \rightarrow CH_4 + H_2O$	-48	-57	-58	-59
$CH_2{=}S + 2H_2 \rightarrow CH_4 + H_2S$	-57	-59	–	-54
$HN{=}NH + 2H_2 \rightarrow 2NH_3$	-67	-79	–	-68
$HN{=}O + 2H_2 \rightarrow NH_3 + H_2O$	-86	-104	–	-103
$O{=}O + 2H_2 \rightarrow 2H_2O$	-92	-118	-124	-125
$HP{=}O + 2H_2 \rightarrow H_2O + PH_3$	-38	-49	–	-105
$HC{\equiv}CH + 3H_2 \rightarrow 2CH_4$	-116	-108	-105	-105
$HC{\equiv}N + 3H_2 \rightarrow CH_4 + NH_3$	-74	-77	-74	-76
$-C{\equiv}O+ + 3H_2 \rightarrow CH_4 + H_2O$	-57	-66	-63	-63
$HC{\equiv}P + 3H_2 \rightarrow CH_4 + PH_3$	-86	-85	–	-67
$-C{\equiv}S+ + 3H_2 \rightarrow CH_4 + H_2S$	-106	-106	–	-91
$N{\equiv}N + 3H_2 \rightarrow 2NH_3$	-28	-42	-38	-37
$P{\equiv}N + 3H_2 \rightarrow NH_3 + PH_3$	-63	-70	–	-45
$-Si{\equiv}O+ + 3H_2 \rightarrow H_2O + SiH_4$	-29	-39	-34	-32
$P{\equiv}P + 3H_2 \rightarrow 2PH_3$	-41	-44	–	-44

Table A6-18: Energies of Hydrogenation Reactions. MP2 Models

hydrogenation reaction	6-31G*	6-311+G**	G3	expt.
$Li_2 + H_2 \rightarrow 2LiH$	24	20	**20**	**20**
$LiOH + H_2 \rightarrow LiH + H_2O$	35	33	–	**35**
$LiF + H_2 \rightarrow LiH + HF$	58	49	**49**	**48**
$LiCl + H_2 \rightarrow LiH + HCl$	64	60	–	**60**
$CH_3CH_3 + H_2 \rightarrow 2CH_4$	-16	-17	**-18**	**-19**
$CH_3NH_2 + H_2 \rightarrow CH_4 + NH_3$	-23	-26	**-26**	**-26**
$CH_3OH + H_2 \rightarrow CH_4 + H_2O$	-25	-31	**-30**	**-30**
$CH_3F + H_2 \rightarrow CH_4 + HF$	-21	-32	–	**-29**
$CH_3SiH_3 + H_2 \rightarrow CH_4 + SiH_4$	-6	-7	**-6**	**-6**
$CH_3SH + H_2 \rightarrow CH_4 + H_2S$	-15	-19	**-19**	**-19**
$CH_3Cl + H_2 \rightarrow CH_4 + HCl$	-26	-22	–	**-22**
$NH_2NH_2 + H_2 \rightarrow 2NH_3$	-42	-48	**-48**	**-48**
$HOOH + H_2 \rightarrow 2H_2O$	-75	-92	**-87**	**-86**
$NaOH + H_2 \rightarrow H_2O + NaH$	19	22	–	**21**
$HOCl + H_2 \rightarrow H_2O + HCl$	-55	-72	**-64**	**-65**
$F_2 + H_2 \rightarrow 2HF$	-116	-145	**-135**	**-133**
$NaF + H_2 \rightarrow HF + NaH$	66	39	–	**35**
$SiH_3F + H_2 \rightarrow HF + SiH_4$	56	19	–	**48**
$ClF + H_2 \rightarrow HF + HCl$	-83	-86	**-77**	**-77**
$Na_2 + H_2 \rightarrow 2NaH$	42	41	–	**29**
$NaCl + H_2 \rightarrow NaH + HCl$	62	57	–	**52**
$SiH_3SiH_3 + H_2 \rightarrow 2SiH_4$	-4	-7	**-4**	**-5**
$SiH_3Cl + H_2 \rightarrow SiH_4 + HCl$	21	13	–	**33**
$PH_2PH_2 + H_2 \rightarrow 2PH_3$	-3	-8	–	**-4**
$HSSH + H_2 \rightarrow 2H_2S$	-13	-22	–	**-14**
$Cl_2 + H_2 \rightarrow 2HCl$	-43	-55	**-46**	**-46**
$CH_2=CH_2 + 2H_2 \rightarrow 2CH_4$	-58	-58	**-57**	**-57**
$CH_2=NH + 2H_2 \rightarrow CH_4 + NH_3$	-52	-59	–	**-64**
$CH_2=O + 2H_2 \rightarrow CH_4 + H_2O$	-46	-58	**-58**	**-59**
$CH_2=S + 2H_2 \rightarrow CH_4 + H_2S$	-49	-57	–	**-54**
$HN=NH + 2H_2 \rightarrow 2NH_3$	-66	-79	–	**-68**
$HN=O + 2H_2 \rightarrow NH_3 + H_2O$	-87	-104	–	**-103**
$O=O + 2H_2 \rightarrow 2H_2O$	-98	-125	**-124**	**-125**
$HP=O + 2H_2 \rightarrow H_2O + PH_3$	-47	-47	–	**-105**
$HC\equiv CH + 3H_2 \rightarrow 2CH_4$	-104	-104	**-105**	**-105**
$HC\equiv N + 3H_2 \rightarrow CH_4 + NH_3$	-60	-60	**-74**	**-76**
$-C\equiv O+ + 3H_2 \rightarrow CH_4 + H_2O$	-48	-60	**-63**	**-63**
$HC\equiv P + 3H_2 \rightarrow CH_4 + PH_3$	-67	-74	–	**-67**
$-C\equiv S+ + 3H_2 \rightarrow CH_4 + H_2S$	-92	-98	–	**-91**
$N\equiv N + 3H_2 \rightarrow 2NH_3$	-13	-30	**-38**	**-37**
$P\equiv N + 3H_2 \rightarrow NH_3 + PH_3$	-36	-53	–	**-45**
$-Si\equiv O+ + 3H_2 \rightarrow H_2O + SiH_4$	-16	-35	**-34**	**-32**
$P\equiv P + 3H_2 \rightarrow 2PH_3$	-18	-35	–	**-44**

Table A6-19: Energies of Hydrogenation Reactions. Semi-Empirical Models

hydrogenation reaction	MNDO	AM1	PM3	G3	expt.
$Li_2 + H_2 \rightarrow 2LiH$	-10	-	30	**20**	**20**
$LiOH + H_2 \rightarrow LiH + H_2O$	16	-	50	–	**35**
$LiF + H_2 \rightarrow LiH + HF$	47	-	60	**49**	**48**
$LiCl + H_2 \rightarrow LiH + HCl$	47	-	82	–	**60**
$CH_3CH_3 + H_2 \rightarrow 2CH_4$	-5	5	5	**-18**	**-19**
$CH_3NH_2 + H_2 \rightarrow CH_4 + NH_3$	-11	-3	2	**-26**	**-26**
$CH_3OH + H_2 \rightarrow CH_4 + H_2O$	-16	-6	-1	**-30**	**-30**
$CH_3F + H_2 \rightarrow CH_4 + HF$	-11	-17	-9	–	**-29**
$CH_3SiH_3 + H_2 \rightarrow CH_4 + SiH_4$	6	12	17	**-6**	**-6**
$CH_3SH + H_2 \rightarrow CH_4 + H_2S$	-8	2	5	**-19**	**-19**
$CH_3Cl + H_2 \rightarrow CH_4 + HCl$	-5	-9	-5	–	**-22**
$NH_2NH_2 + H_2 \rightarrow 2NH_3$	-28	-23	-13	**-48**	**-48**
$HOOH + H_2 \rightarrow 2H_2O$	-84	-78	-53	**-87**	**-86**
$NaOH + H_2 \rightarrow H_2O + NaH$	-	-	22	–	**21**
$HOCl + H_2 \rightarrow H_2O + HCl$	-50	-57	-26	**-64**	**-65**
$F_2 + H_2 \rightarrow 2HF$	-128	-121	-90	**-135**	**-133**
$NaF + H_2 \rightarrow HF + NaH$	-	-	33	–	**35**
$SiH_3F + H_2 \rightarrow HF + SiH_4$	41	21	41	–	**48**
$ClF + H_2 \rightarrow HF + HCl$	-47	-83	-48	**-77**	**-77**
$Na_2 + H_2 \rightarrow 2NaH$	-	-	60	–	**29**
$NaCl + H_2 \rightarrow NaH + HCl$	-	-	79	–	**52**
$SiH_3SiH_3 + H_2 \rightarrow 2SiH_4$	-4	-3	20	**-4**	**-5**
$SiH_3Cl + H_2 \rightarrow SiH_4 + HCl$	30	22	33	–	**33**
$PH_2PH_2 + H_2 \rightarrow 2PH_3$	7	19	6	–	**-4**
$HSSH + H_2 \rightarrow 2H_2S$	-2	-1	3	–	**-14**
$Cl_2 + H_2 \rightarrow 2HCl$	-21	-30	-16	**-46**	**-46**
$CH_2=CH_2 + 2H_2 \rightarrow 2CH_4$	-41	-24	-16	**-57**	**-57**
$CH_2=NH + 2H_2 \rightarrow CH_4 + NH_3$	-40	-24	-10	–	**-64**
$CH_2=O + 2H_2 \rightarrow CH_4 + H_2O$	-41	-26	-6	**-58**	**-59**
$CH_2=S + 2H_2 \rightarrow CH_4 + H_2S$	-48	-32	-25	–	**-54**
$HN=NH + 2H_2 \rightarrow 2NH_3$	-46	-36	-17	–	**-68**
$HN=O + 2H_2 \rightarrow NH_3 + H_2O$	-72	-58	-23	–	**-103**
$O=O + 2H_2 \rightarrow 2H_2O$	-107	-80	-76	**-124**	**-125**
$HP=O + 2H_2 \rightarrow H_2O + PH_3$	-23	-17	4	–	**-105**
$HC\equiv CH + 3H_2 \rightarrow 2CH_4$	-84	-57	-37	**-105**	**-105**
$HC\equiv N + 3H_2 \rightarrow CH_4 + NH_3$	-56	-32	-9	**-74**	**-76**
$-C\equiv O+ + 3H_2 \rightarrow CH_4 + H_2O$	-69	-47	-7	**-63**	**-63**
$HC\equiv P + 3H_2 \rightarrow CH_4 + PH_3$	-71	-30	-19	–	**-67**
$-C\equiv S+ + 3H_2 \rightarrow CH_4 + H_2S$	-115	-87	-71	–	**-91**
$N\equiv N + 3H_2 \rightarrow 2NH_3$	-23	-10	16	**-38**	**-37**
$P\equiv N + 3H_2 \rightarrow NH_3 + PH_3$	-55	-14	4	–	**-45**
$-Si\equiv O+ + 3H_2 \rightarrow H_2O + SiH_4$	-32	-37	25	**-34**	**-32**
$P\equiv P + 3H_2 \rightarrow 2PH_3$	-28	11	9	–	**-44**

Table A6-20: Basis Set Effects on Energies of Hydrogenation Reactions. Hartree-Fock Models

hydrogenation reaction	6-31G*	6-31G**	6-31+G*	6-311G*	6-311+G**	expt.
$CH_3\text{-}CH_3 + H_2 \rightarrow 2CH_4$	-22	-21	-22	-22	-21	**-19**
$CH_3\text{-}NH_2 + H_2 \rightarrow CH_4 + NH_3$	-27	-28	-28	-27	-28	**-26**
$CH_3\text{-}OH + H_2 \rightarrow CH_4 + H_2O$	-27	-30	-29	-27	-31	**-30**
$CH_3\text{-}F + H_2 \rightarrow CH_4 + HF$	-23	-27	-25	-23	-29	**-29**
$NH_2NH_2 + H_2 \rightarrow 2NH_3$	-46	-48	-47	-46	-50	**-48**
$HOOH + H_2 \rightarrow 2H_2O$	-82	-87	-86	-82	-93	**-86**
$F\text{-}F + H_2 \rightarrow 2HF$	-126	-134	-137	-131	-149	**-133**
$H_2C{=}CH_2 + 2H_2 \rightarrow 2CH_4$	-66	-64	-64	-64	-61	**-57**
$HC{\equiv}CH + 3H_2 \rightarrow 2CH_4$	-121	-118	-118	-117	-112	**-105**

Table A6-21: Basis Set Effects on Energies of Hydrogenation Reactions. EDF1 Density Functional Models

hydrogenation reaction	6-31G*	6-31G**	6-31+G*	6-311G*	6-311+G**	expt.
$CH_3\text{-}CH_3 + H_2 \rightarrow 2CH_4$	-19	-19	-20	-19	-20	**-19**
$CH_3\text{-}NH_2 + H_2 \rightarrow CH_4 + NH_3$	-23	-25	-25	-24	-27	**-26**
$CH_3\text{-}OH + H_2 \rightarrow CH_4 + H_2O$	-23	-26	-26	-23	-30	**-30**
$CH_3\text{-}F + H_2 \rightarrow CH_4 + HF$	-18	-21	-24	-19	-28	**-29**
$NH_2NH_2 + H_2 \rightarrow 2NH_3$	-39	-42	-43	-41	-46	**-48**
$HOOH + H_2 \rightarrow 2H_2O$	-66	-72	-74	-67	-83	**-86**
$F\text{-}F + H_2 \rightarrow 2HF$	-100	-108	-118	-106	-130	**-133**
$H_2C{=}CH_2 + 2H_2 \rightarrow 2CH_4$	-58	-58	-58	-56	-56	**-57**
$HC{\equiv}CH + 3H_2 \rightarrow 2CH_4$	-109	-109	-108	-105	-103	**-105**

Table A6-22: Basis Set Effects on Energies of Hydrogenation Reactions. B3LYP Density Functional Models

hydrogenation reaction	6-31G*	6-31G**	6-31+G*	6-311G*	6-311+G**	expt.
$CH_3\text{-}CH_3 + H_2 \rightarrow 2CH_4$	-19	-19	-20	-20	-20	**-19**
$CH_3\text{-}NH_2 + H_2 \rightarrow CH_4 + NH_3$	-24	-25	-26	-25	-27	**-26**
$CH_3\text{-}OH + H_2 \rightarrow CH_4 + H_2O$	-24	-26	-27	-24	-30	**-30**
$CH_3\text{-}F + H_2 \rightarrow CH_4 + HF$	-18	-22	-24	-20	-29	**-29**
$NH_2NH_2 + H_2 \rightarrow 2NH_3$	-40	-43	-44	-42	-47	**-48**
$HOOH + H_2 \rightarrow 2H_2O$	-69	-74	-77	-70	-85	**-86**
$F\text{-}F + H_2 \rightarrow 2HF$	-105	-112	-122	-111	-134	**-133**
$H_2C=CH_2 + 2H_2 \rightarrow 2CH_4$	-62	-61	-61	-60	-59	**-57**
$HC\equiv CH + 3H_2 \rightarrow 2CH_4$	-116	-115	-114	-111	-108	**-105**

Table A6-23: Basis Set Effects on Energies of Hydrogenation Reactions. MP2 Models

hydrogenation reaction	6-31G*	6-31G**	6-31+G*	6-311G*	6-311+G**	expt.
$CH_3\text{-}CH_3 + H_2 \rightarrow 2CH_4$	-16	-18	-17	-17	-17	**-19**
$CH_3\text{-}NH_2 + H_2 \rightarrow CH_4 + NH_3$	-23	-25	-23	-24	-26	**-26**
$CH_3\text{-}OH + H_2 \rightarrow CH_4 + H_2O$	-25	-28	-26	-25	-31	**-30**
$CH_3\text{-}F + H_2 \rightarrow CH_4 + HF$	-21	-26	-25	-23	-32	**-29**
$NH_2NH_2 + H_2 \rightarrow 2NH_3$	-42	-46	-44	-43	-48	**-48**
$HOOH + H_2 \rightarrow 2H_2O$	-75	-81	-82	-76	-92	**-86**
$F\text{-}F + H_2 \rightarrow 2HF$	-116	-123	-131	-121	-145	**-133**
$H_2C=CH_2 + 2H_2 \rightarrow 2CH_4$	-58	-61	-56	-58	-58	**-57**
$HC\equiv CH + 3H_2 \rightarrow 2CH_4$	-104	-110	-102	-104	-104	**-105**

Table A6-24: Relative Energies of Structural Isomers. Hartree-Fock Models

formula (ref. compound)	molecule	STO-3G	3-21G	6-31G*	6-311+G**	G3	expt.
CHN (hydrogen cyanide)	hydrogen isocyanide	19	9	13	11	–	15.1
CH_2N_2 (diazomethane)	diazirine	-16	30	5	7	–	8
CH_2O (formaldehyde)	hydroxymethylene	48	47	52	47	–	54.2
CH_3NO (formamide)	nitrosomethane	24	55	64	68	–	60.5
CH_3NO_2 (nitromethane)	methyl nitrite	-46	-14	-4	-1	2.7	2.0
$C_2H_2Cl_2$ (1,1-dichloroethylene)	cis-1,2-dichloroethylene	-1	-3	-3	-2	–	0.5
	trans-1,2-dichloroethylene	-2	-3	-3	-2	–	0.6
C_2H_3N (acetonitrile)	methyl isocyanide	24	21	24	21	–	20.5
C_2H_4O (acetaldehyde)	oxacyclopropane	11	34	31	32	26.4	27.1
$C_2H_4O_2$ (acetic acid)	methyl formate	7	13	13	17	16.3	18.3
$C_2H_5NO_2$ (glycine)	nitroethane	69	104	82	87	–	69.7
C_2H_6O (ethanol)	dimethyl ether	-1	6	7	11	11.7	12.0
$C_2H_6O_2$ (1,2-ethanediol)	ethanehydroperoxide	14	41	56	62	–	45.1
	dimethylperoxide	11	46	61	71	–	62.6
C_2H_6Si (vinylsilane)	1-methylsilaethylene	26	12	17	16	–	11
C_2H_6S (ethanethiol)	dimethylsulfide	-5	0	0	2	1.2	2.1
C_2H_7N (ethylamine)	dimethylamine	2	6	6	7	7.8	6.9
$C_2H_7PO_3$ (dimethylphosphonate)	dimethylphosphorous acid	-75	11	11	7	–	6.5
C_2H_8Si (dimethylsilane)	ethylsilane	12	15	10	11	–	13
C_3H_4 (propyne)	allene	17	3	2	2	1.1	1.0
	cyclopropene	30	40	26	28	24.2	21.6
$C_3H_4N_2$ (imidazole)	pyrazole	7	15	13	13	–	12.7
C_3H_6 (propene)	cyclopropane	-4	14	8	10	8.1	6.7

Table A6-24: Relative Energies of Structural Isomers. Hartree-Fock Models (2)

formula (ref. compound)	molecule	STO-3G	3-21G	6-31G*	6-311+G**	G3	expt.
C₄H₆O (acetone)	propanal	6	6	7	8	—	7.1
	2-hydroxypropene	20	11	19	14	—	7.7
	allyl alcohol	32	22	28	23	—	22.2
	methyl vinyl ether	25	20	29	28	—	25.6
	propylene oxide	13	37	33	35	—	29.5
	oxetane	2	31	33	35	—	32.7
C₄H₄N₂ (pyrazine)	pyrimidine	-3	-6	-7	-7	—	0.1
	1,2-ethanedinitrile	-7	-16	-5	-7	—	3.3
	pyridazine	13	24	21	21	—	19.7
C₄H₆ (1,3-butadiene)	2-butyne	-13	4	7	7	8.5	8.5
	cyclobutene	-13	18	13	15	12.5	11.2
	1,2-butadiene	9	11	13	15	—	12.5
	1-butyne	-5	9	13	13	—	13.2
	methylenecyclopropane	6	26	20	24	19.9	21.7
	bicyclo[1.1.0]butane	12	46	30	33	27.6	25.6
	1-methylcyclopropene	17	44	32	35	—	32.0
C₄H₈ (isobutene)	trans-2-butene	0	1	0	1	—	1.1
	cis-2-butene	1	2	2	2	—	2.2
	skew-1-butene	4	2	3	3	—	3.9
	methylcyclopropane	-2	15	9	13	—	9.7
	cyclobutane	-18	10	9	12	—	10.8
C₄H₈O₂ (1,3-dioxane)	1,4-dioxane	4	7	6	5	—	7.6
C₄H₁₀ (isobutane)	n-butane	0	1	0	0	1.7	2.1
C₆H₄F₂ (meta-difluorobenzene)	para-difluorobenzene	1	1	1	1	—	0.7
	ortho-difluorobenzene	3	5	5	4	—	3.7
C₈H₁₄ (cis-cyclooctene)	trans-cyclooctene	16	13	14	14	—	9.3
C₉H₁₆ (cis-cyclononene)	trans-cyclononene	7	6	5	8	—	2.9
C₁₀H₁₈ (trans-decalin)	cis-decalin	3	3	3	4	—	3.2

Table A6-25: Relative Energies of Structural Isomers. Local Density Models

formula (ref. compound)	molecule	6-31G*	6-311+G**	G3	expt.
CHN (hydrogen cyanide)	hydrogen isocyanide	17	15	–	15.1
CH₂N₂ (diazomethane)	diazirine	13	17	–	8
CH₂O (formaldehyde)	hydroxymethylene	59	54	–	54.2
CH₃NO (formamide)	nitrosomethane	63	69	–	60.5
CH₃NO₂ (nitromethane)	methyl nitrite	5	10	2.7	2.0
C₂H₂Cl₂ (1,1-dichloroethylene)	cis-1,2-dichloroethylene	-2	-1	–	0.5
	trans-dichloroethylene	-1	0	–	0.6
C₂H₃N (acetonitrile)	methyl isocyanide	25	25	–	20.5
C₂H₄O (acetaldehyde)	oxacyclopropane	22	23	26.4	27.1
C₂H₄O₂ (acetic acid)	methyl formate	13	18	16.3	18.3
C₂H₅NO₂ (glycine)	nitroethane	54	67	–	69.7
C₂H₆O (ethanol)	dimethyl ether	6	12	11.7	12.0
C₂H₆O₂ (1,2-ethanediol)	ethanehydroperoxide	49	57	–	45.1
	dimethylperoxide	53	66	–	62.6
C₂H₆Si (vinylsilane)	1-methylsilaethylene	4	3	–	11
C₂H₆S (ethanethiol)	dimethylsulfide	0	2	1.2	2.1
C₂H₇N (ethylamine)	dimethylamine	5	8	7.8	6.9
C₂H₇PO₃ (dimethylphosphonate)	dimethylphosphorous acid	7	2	–	6.5
C₂H₈Si (dimethylsilane)	ethylsilane	10	12	–	13
C₃H₄ (propyne)	allene	-4	-3	1.1	1.0
	cyclopropene	15	18	24.2	21.6
C₃H₄N₂ (imidazole)	pyrazole	9	10	–	12.7
C₃H₆ (propene)	cyclopropane	1	3	8.1	6.7

Table A6-25: Relative Energies of Structural Isomers. Local Density Models (2)

formula (ref. compound)	molecule	6-31G*	6-311+G**	G3	expt.
C₃H₆O (acetone)	propanal	9	7	–	7.1
	2-hydroxypropene	14	9	–	7.7
	allyl alcohol	28	22	–	22.2
	methyl vinyl ether	24	24	–	25.6
	propylene oxide	24	26	–	29.5
	oxetane	25	28	–	32.7
C₄H₄N₂ (pyrazine)	pyrimidine	–4	–4	–	0.1
	1,2-ethanedinitrile	30	24	–	3.3
	pyridazine	19	18	–	19.7
C₄H₆ (1,3-butadiene)	2-butyne	7	9	8.5	8.5
	cyclobutene	3	6	12.5	11.2
	1,2-butadiene	9	11	–	12.5
	1-butyne	17	15	–	13.2
	methylenecyclopropane	11	13	19.9	21.7
	bicyclo[1.1.0]butane	13	16	27.6	25.6
	1-methylcyclopropene	23	25	–	32.0
C₄H₈ (isobutene)	trans-2-butene	1	1	–	1.1
	cis-2-butene	2	2	–	2.2
	skew-1-butene	5	5	–	3.9
	methylcyclopropane	3	5	–	9.7
	cyclobutane	2	4	–	10.8
C₄H₈O₂ (1,3-dioxane)	1,4-dioxane	6	6	–	7.6
C₄H₁₀ (isobutane)	n-butane	2	2	1.7	2.1
C₆H₄F₂ (meta-difluorobenzene)	para-difluorobenzene	1	1	–	0.7
	ortho-difluorobenzene	4	4	–	3.7
C₈H₁₄ (cis-cyclooctene)	trans-cyclooctene	10	10	–	9.3
C₉H₁₆ (cis-cyclononene)	trans-cyclononene	6	6	–	2.9
C₁₀H₁₈ (trans-decalin)	cis-decalin	3	2	–	3.2

Table A6-26: Relative Energies of Structural Isomers. BP Density Functional Models

formula (ref. compound)	molecule	6-31G*	6-311+G**	G3	expt.
CHN (hydrogen cyanide)	hydrogen isocyanide	17	15	–	**15.1**
CH₂N₂ (diazomethane)	diazirine	12	15	–	**8**
CH₂O (formaldehyde)	hydroxymethylene	58	53	–	**54.2**
CH₃NO (formamide)	nitrosomethane	58	64	–	**60.5**
CH₃NO₂ (nitromethane)	methyl nitrite	4	5	2.7	**2.0**
C₂H₂Cl₂ (1,1-dichloroethylene)	cis-1,2-dichloroethylene	-2	-1	–	**0.5**
	trans-1,2-dichloroethylene	-2	-1	–	**0.6**
C₂H₃N (acetonitrile)	methyl isocyanide	25	25	–	**20.5**
C₂H₄O (acetaldehyde)	oxacyclopropane	25	27	26.4	**27.1**
C₂H₄O₂ (acetic acid)	methyl formate	12	17	16.3	**18.3**
C₂H₅NO₂ (glycine)	nitroethane	53	62	–	**69.7**
C₂H₆O (ethanol)	dimethyl ether	6	12	11.7	**12.0**
C₂H₆O₂ (1,2-ethanediol)	ethanehydroperoxide	44	52	–	**45.1**
	dimethylperoxide	47	60	–	**62.6**
C₂H₆Si (vinylsilane)	1-methylsilaethylene	7	5	–	**11**
C₂H₆S (ethanethiol)	dimethylsulfide	-1	1	1.2	**2.1**
C₂H₇N (ethylamine)	dimethylamine	5	8	7.8	**6.9**
C₂H₇PO₃ (dimethylphosphonate)	dimethylphosphorous acid	8	3	–	**6.5**
C₂H₈Si (dimethylsilane)	ethylsilane	11	12	–	**13**
C₃H₄ (propyne)	allene	-4	-3	1.1	**1.0**
	cyclopropene	19	21	24.2	**21.6**
C₃H₄N₂ (imidazole)	pyrazole	9	10	–	**12.7**
C₃H₆ (propene)	cyclopropane	6	7	8.1	**6.7**

Table A6-26: Relative Energies of Structural Isomers. BP Density Functional Models (2)

formula (ref. compound)	molecule	6-31G*	6-311+G**	G3	expt.
C_3H_6O (acetone)	propanal	7	7	–	7.1
	2-hydroxypropene	17	12	–	7.7
	allyl alcohol	27	23	–	22.2
	methyl vinyl ether	26	26	–	25.6
	propylene oxide	28	29	–	29.5
	oxetane	31	32	–	32.7
$C_4H_4N_2$ (pyrazine)	pyrimidine	-4	-4	–	0.1
	1,2-ethanedinitrile	14	10	–	3.3
	pyridazine	17	18	–	19.7
C_4H_6 (1,3-butadiene)	2-butyne	9	9	8.5	8.5
	cyclobutene	9	12	12.5	11.2
	1,2-butadiene	11	12	–	12.5
	1-butyne	17	16	–	13.2
	methylenecyclopropane	16	18	19.9	21.7
	bicyclo[1.1.0]butane	23	26	27.6	25.6
	1-methylcyclopropene	27	29	–	32.0
C_4H_8 (isobutene)	trans-2-butene	0	0	–	1.1
	cis-2-butene	1	1	–	2.2
	skew-1-butene	4	4	–	3.9
	methylcyclopropane	8	9	–	9.7
	cyclobutane	7	9	–	10.8
$C_4H_8O_2$ (1,3-dioxane)	1,4-dioxane	5	5	–	7.6
C_4H_{10} (isobutane)	n-butane	1	1	1.7	2.1
$C_6H_4F_2$ (meta-difluorobenzene)	para-difluorobenzene	1	1	–	0.7
	ortho-difluorobenzene	4	4	–	3.7
C_8H_{14} (cis-cyclooctene)	trans-cyclooctene	11	11	–	9.3
C_9H_{16} (cis-cyclononene)	trans-cyclononene	6	6	–	2.9
$C_{10}H_{18}$ (trans-decalin)	cis-decalin	3	3	–	3.2

Table A6-27: Relative Energies of Structural Isomers. BLYP Density Functional Models

formula (ref. compound)	molecule	6-31G*	6-311+G**	G3	expt.
CHN (hydrogen cyanide)	hydrogen isocyanide	18	15	–	15.1
CH_2N_2 (diazomethane)	diazirine	14	18	–	8
CH_2O (formaldehyde)	hydroxymethylene	58	53	–	54.2
CH_3NO (formamide)	nitrosomethane	56	63	–	60.5
CH_3NO_2 (nitromethane)	methyl nitrite	-1	3	2.7	2.0
$C_2H_2Cl_2$ (1,1-dichloroethylene)	cis-1,2-dichloroethylene	-2	-1	–	0.5
	trans-1,2-dichloroethylene	-2	-1	–	0.6
C_2H_3N (acetonitrile)	methyl isocyanide	25	25	–	20.5
C_2H_4O (acetaldehyde)	oxacyclopropane	28	30	26.4	27.1
$C_2H_4O_2$ (acetic acid)	methyl formate	10	15	16.3	18.3
$C_2H_5NO_2$ (glycine)	nitroethane	53	62	–	69.7
C_2H_6O (ethanol)	dimethyl ether	4	11	11.7	12.0
$C_2H_6O_2$ (1,2-ethanediol)	ethanehydroperoxide	42	50	–	45.1
	dimethylperoxide	44	57	–	62.6
C_2H_6Si (vinylsilane)	1-methylsilaethylene	8	7	–	11
C_2H_6S (ethanethiol)	dimethylsulfide	-1	1	1.2	2.1
C_2H_7N (ethylamine)	dimethylamine	5	7	7.8	6.9
$C_2H_7PO_3$ (dimethylphosphonate)	dimethylphosphorous acid	7	2	–	6.5
C_2H_8Si (dimethylsilane)	ethylsilane	11	12	–	13
C_3H_4 (propyne)	allene	-4	-3	1.1	1.0
	cyclopropene	30	25	24.2	21.6
$C_3H_4N_2$ (imidazole)	pyrazole	9	10	–	12.7
C_3H_6 (propene)	cyclopropane	10	11	8.1	6.7

Table A6-27: Relative Energies of Structural Isomers. BLYP Density Functional Models (2)

formula (ref. compound)	molecule	6-31G*	6-311+G**	G3	expt.
C_3H_6O (acetone)	propanal	8	7	—	7.1
	2-hydroxypropene	17	13	—	7.7
	allyl alcohol	28	22	—	22.2
	methyl vinyl ether	25	26	—	25.6
	propylene oxide	30	32	—	29.5
	oxetane	33	35	—	32.7
$C_4H_4N_2$ (pyrazine)	pyrimidine	-4	-4	—	0.1
	1,2-ethanedinitrile	9	4	—	3.3
	pyridazine	18	18	—	19.7
C_4H_6 (1,3-butadiene)	2-butyne	9	10	8.5	8.5
	cyclobutene	14	17	12.5	11.2
	1,2-butadiene	10	12	—	12.5
	1-butyne	17	16	—	13.2
	methylenecyclopropane	20	23	19.9	21.7
	bicyclo[1.1.0]butane	31	35	27.6	25.6
	1-methylcyclopropene	31	34	—	32.0
C_4H_8 (isobutene)	trans-2-butene	0	0	—	1.1
	cis-2-butene	2	1	—	2.2
	skew-1-butene	4	3	—	3.9
	methylcyclopropane	18	13	—	9.7
	cyclobutane	11	14	—	10.8
$C_4H_8O_2$ (1,3-dioxane)	1,4-dioxane	5	5	—	7.6
C_4H_{10} (isobutane)	n-butane	0	1	1.7	2.1
$C_6H_4F_2$ (meta-difluorobenzene)	para-difluorobenzene	1	0	—	0.7
	ortho-difluorobenzene	4	4	—	3.7
C_8H_{14} (cis-cyclooctene)	trans-cyclooctene	11	11	—	9.3
C_9H_{16} (cis-cyclononene)	trans-cyclononene	6	6	—	2.9
$C_{10}H_{18}$ (trans-decalin)	cis-decalin	4	4	—	3.2

Table A6-28: Relative Energies of Structural Isomers. EDF1 Density Functional Models

formula (ref. compound)	molecule	6-31G*	6-311+G**	G3	expt.
CHN (hydrogen cyanide)	hydrogen isocyanide	18	15	–	**15.1**
CH₂N₂ (diazomethane)	diazirine	12	15	–	**8**
CH₂O (formaldehyde)	hydroxymethylene	59	54	–	**54.2**
CH₃NO (formamide)	nitrosomethane	58	64	–	**60.5**
CH₃NO₂ (nitromethane)	methyl nitrite	4	6	2.7	**2.0**
C₂H₂Cl₂ (1,1-dichloroethylene)	*cis*-1,2-dichloroethylene	-2	-1	–	**0.5**
	trans-1,2-dichloroethylene	-2	-1	–	**0.6**
C₂H₃N (acetonitrile)	methyl isocyanide	25	25	–	**20.5**
C₂H₄O (acetaldehyde)	oxacyclopropane	26	27	26.4	**27.1**
C₂H₄O₂ (acetic acid)	methyl formate	12	17	16.3	**18.3**
C₂H₅NO₂ (glycine)	nitroethane	52	61	–	**69.7**
C₂H₆O (ethanol)	dimethyl ether	6	12	11.7	**12.0**
C₂H₆O₂ (1,2-ethanediol)	ethanehydroperoxide	44	52	–	**45.1**
	dimethylperoxide	48	60	–	**62.6**
C₂H₆Si (vinylsilane)	1-methylsilaethylene	7	5	–	**11**
C₂H₆S (ethanethiol)	dimethylsulfide	-1	1	1.2	**2.1**
C₂H₇N (ethylamine)	dimethylamine	5	8	7.8	**6.9**
C₂H₇PO₃ (dimethylphosphonate)	dimethylphosphorous acid	8	4	–	**6.5**
C₂H₈Si (dimethylsilane)	ethylsilane	11	12	–	**13**
C₃H₄ (propyne)	allene	-4	-3	1.1	**1.0**
	cyclopropene	18	21	24.2	**21.6**
C₃H₄N₂ (imidazole)	pyrazole	9	10	–	**12.7**
C₃H₆ (propene)	cyclopropane	5	7	8.1	**6.7**

Table A6-28: Relative Energies of Structural Isomers. EDF1 Density Functional Models (2)

formula (ref. compound)	molecule	6-31G*	6-311+G**	G3	expt.
C_3H_6O (acetone)	propanal	7	7	—	7.1
	2-hydroxypropene	17	13	—	7.7
	allyl alcohol	29	23	—	22.2
	methyl vinyl ether	27	27	—	25.6
	propylene oxide	28	29	—	29.5
	oxetane	31	33	—	32.7
$C_4H_4N_2$ (pyrazine)	pyrimidine	-4	-4	—	0.1
	1,2-ethanedinitrile	11	7	—	3.3
	pyridazine	18	18	—	19.7
C_4H_6 (1,3-butadiene)	2-butyne	8	8	8.5	8.5
	cyclobutene	10	12	12.5	11.2
	1,2-butadiene	11	11	—	12.5
	1-butyne	16	15	—	13.2
	methylenecyclopropane	16	17	19.9	21.7
	bicyclo[1.1.0]butane	23	26	27.6	25.6
	1-methylcyclopropene	27	29	—	32.0
C_4H_8 (isobutene)	trans-2-butene	0	0	—	1.1
	cis-2-butene	1	1	—	2.2
	skew-1-butene	4	3	—	3.9
	methylcyclopropane	8	8	—	9.7
	cyclobutane	7	9	—	10.8
$C_4H_8O_2$ (1,3-dioxane)	1,4-dioxane	5	5	—	7.6
C_4H_{10} (isobutane)	n-butane	-1	0	1.7	2.1
$C_6H_4F_2$ (meta-difluorobenzene)	para-difluorobenzene	0	1	—	0.7
	ortho-difluorobenzene	4	4	—	3.7
C_8H_{14} (cis-cyclooctene)	trans-cyclooctene	11	11	—	9.3
C_9H_{16} (cis-cyclononene)	trans-cyclononene	6	6	—	2.9
$C_{10}H_{18}$ (trans-decalin)	cis-decalin	4	4	—	3.2

647

Table A6-29: Relative Energies of Structural Isomers. B3LYP Density Functional Models

formula (ref. compound)	molecule	6-31G*	6-311+G**	G3	expt.
CHN (hydrogen cyanide)	hydrogen isocyanide	17	14	–	**15.1**
CH₂N₂ (diazomethane)	diazirine	12	15	–	**8**
CH₂O (formaldehyde)	hydroxymethylene	57	53	–	**54.2**
CH₃NO (formamide)	nitrosomethane	68	66	–	**60.5**
CH₃NO₂ (nitromethane)	methyl nitrite	1	5	2.7	**2.0**
C₂H₂Cl₂ (1,1-dichloroethylene)	cis-1,2-dichloroethylene	-2	-2	–	**0.5**
	trans-1,2-dichloroethylene	-2	-1	–	**0.6**
C₂H₃N (acetonitrile)	methyl isocyanide	27	24	–	**20.5**
C₂H₄O (acetaldehyde)	oxacyclopropane	28	29	26.4	**27.1**
C₂H₄O₂ (acetic acid)	methyl formate	12	16	16.3	**18.3**
C₂H₅NO₂ (glycine)	nitroethane	60	68	–	**69.7**
C₂H₆O (ethanol)	dimethyl ether	5	11	11.7	**12.0**
C₂H₆O₂ (1,2-ethanediol)	ethanehydroperoxide	47	54	–	**45.1**
	dimethylperoxide	50	62	–	**62.6**
C₂H₆Si (vinylsilane)	1-methylsilaethylene	9	8	–	**11**
C₂H₆S (ethanethiol)	dimethylsulfide	-1	1	1.2	**2.1**
C₂H₇N (ethylamine)	dimethylamine	5	7	7.8	**6.9**
C₂H₇PO₃ (dimethylphosphonate)	dimethylphosphorous acid	8	3	–	**6.5**
C₂H₈Si (dimethylsilane)	ethylsilane	11	12	–	**13**
C₃H₄ (propyne)	allene	-3	-2	1.1	**1.0**
	cyclopropene	22	24	24.2	**21.6**
C₃H₄N₂ (imidazole)	pyrazole	10	11	–	**12.7**
C₃H₆ (propene)	cyclopropane	8	9	8.1	**6.7**

Table A6-29: Relative Energies of Structural Isomers. B3LYP Density Functional Models (2)

formula (ref. compound)	molecule	6-31G*	6-311+G**	G3	expt.
C_3H_6O (acetone)	propanal	8	7	—	7.1
	2-hydroxypropene	17	12	—	7.7
	allyl alcohol	28	22	—	22.2
	methyl vinyl ether	26	26	—	25.6
	propylene oxide	30	31	—	29.5
	oxetane	32	34	—	32.7
$C_4H_4N_2$ (pyrazine)	pyrimidine	-4	-4	—	0.1
	1,2-ethanedinitrile	10	6	—	3.3
	pyridazine	19	19	—	19.7
C_4H_6 (1,3-butadiene)	2-butyne	8	9	8.5	8.5
	cyclobutene	12	15	12.5	11.2
	1,2-butadiene	10	12	—	12.5
	1-butyne	16	15	—	13.2
	methylenecyclopropane	19	20	19.9	21.7
	bicyclo[1.1.0]butane	28	31	27.6	25.6
	1-methylcyclopropene	30	32	—	32.0
C_4H_8 (isobutene)	trans-2-butene	0	0	—	1.1
	cis-2-butene	2	2	—	2.2
	skew-1-butene	4	3	—	3.9
	methylcyclopropane	10	11	—	9.7
	cyclobutane	9	11	—	10.8
$C_4H_8O_2$ (1,3-dioxane)	1,4-dioxane	5	5	—	7.6
C_4H_{10} (isobutane)	n-butane	5	1	1.7	2.1
$C_6H_4F_2$ (meta-difluorobenzene)	para-difluorobenzene	1	1	—	0.7
	ortho-difluorobenzene	4	4	—	3.7
C_8H_{14} (cis-cyclooctene)	trans-cyclooctene	12	12	—	9.3
C_9H_{16} (cis-cyclononene)	trans-cyclononene	7	7	—	2.9
$C_{10}H_{18}$ (trans-decalin)	cis-decalin	3	3	—	3.2

649

Table A6-30: Relative Energies of Structural Isomers. MP2 Models

formula (ref. compound)	molecule	6-31G*	6-311+G**	G3	expt.
CHN (hydrogen cyanide)	hydrogen isocyanide	21	19	–	**15.1**
CH$_2$N$_2$ (diazomethane)	diazirine	6	8	–	**8**
CH$_2$O (formaldehyde)	hydroxymethylene	61	55	–	**54.2**
CH$_3$NO (formamide)	nitrosomethane	61	65	–	**60.5**
CH$_3$NO$_2$ (nitromethane)	methyl nitrite	4	8	2.7	**2.0**
C$_2$H$_2$Cl$_2$ (1,1-dichloroethylene)	cis-1,2-dichloroethylene	-2	-1	–	**0.5**
	trans-1,2-dichloroethylene	-1	0	–	**0.6**
C$_2$H$_3$N (acetonitrile)	methyl isocyanide	29	27	–	**20.5**
C$_2$H$_4$O (acetaldehyde)	oxacyclopropane	27	28	26.4	**27.1**
C$_2$H$_4$O$_2$ (acetic acid)	methyl formate	14	18	16.3	**18.3**
C$_2$H$_5$NO$_2$ (glycine)	nitroethane	61	69	–	**69.7**
C$_2$H$_6$O (ethanol)	dimethyl ether	9	14	11.7	**12.0**
C$_2$H$_6$O$_2$ (1,2-ethanediol)	ethanehydroperoxide	52	60	–	**45.1**
	dimethylperoxide	58	71	–	**62.6**
C$_2$H$_6$Si (vinylsilane)	1-methylsilaethylene	9	8	–	**11**
C$_2$H$_6$S (ethanethiol)	dimethylsulfide	-1	2	1.2	**2.1**
C$_2$H$_7$N (ethylamine)	dimethylamine	6	8	7.8	**6.9**
C$_2$H$_7$PO$_3$ (dimethylphosphonate)	dimethylphosphorous acid	11	7	–	**6.5**
C$_2$H$_8$Si (dimethylsilane)	ethylsilane	12	11	–	**13**
C$_3$H$_4$ (propyne)	allene	5	5	1.1	**1.0**
	cyclopropene	23	24	24.2	**21.6**
C$_3$H$_4$N$_2$ (imidazole)	pyrazole	10	10	–	**12.7**
C$_3$H$_6$ (propene)	cyclopropane	4	5	8.1	**6.7**

Table A6-30: Relative Energies of Structural Isomers. MP2 Models (2)

formula (ref. compound)	molecule	6-31G*	6-311+G**	G3	expt.
C_3H_6O (acetone)	propanal	8	6	—	7.1
	2-hydroxypropene	18	13	—	7.7
	allyl alcohol	29	23	—	22.2
	methyl vinyl ether	30	29	—	25.6
	propylene oxide	29	29	—	29.5
	oxetane	33	33	—	32.7
$C_4H_4N_2$ (pyrazine)	pyrimidine	-4	-3	—	0.1
	1,2-ethanedinitrile	-2	-3	—	3.3
	pyridazine	18	19	—	19.7
C_4H_6 (1,3-butadiene)	2-butyne	4	5	8.5	8.5
	cyclobutene	8	9	12.5	11.2
	1,2-butadiene	12	14	—	12.5
	1-butyne	10	9	—	13.2
	methylenecyclopropane	16	17	19.9	21.7
	bicyclo[1.1.0]butane	21	22	27.6	25.6
	1-methylcyclopropene	26	27	—	32.0
C_4H_8 (isobutene)	trans-2-butene	1	1	—	1.1
	cis-2-butene	2	3	—	2.2
	skew-1-butene	4	3	—	3.9
	methylcyclopropane	6	7	—	9.7
	cyclobutane	6	7	—	10.8
$C_4H_8O_2$ (1,3-dioxane)	1,4-dioxane	6	6	—	7.6
C_4H_{10} (isobutane)	n-butane	2	2	1.7	2.1
$C_6H_4F_2$ (meta-difluorobenzene)	para-difluorobenzene	1	1	—	0.7
	ortho-difluorobenzene	4	4	—	3.7
C_8H_{14} (cis-cyclooctene)	trans-cyclooctene	11	10	—	9.3
C_9H_{16} (cis-cyclononene)	trans-cyclononene	6	6	—	2.9

Table A6-31: Relative Energies of Structural Isomers. Semi-Empirical Models

formula (ref. compound)	molecule	MNDO	AM1	PM3	G3	expt.
CHN (hydrogen cyanide)	hydrogen isocyanide	33	18	24	–	**15.1**
CH_2N_2 (diazomethane)	diazirine	5	24	31	–	**8**
CH_2O (formaldehyde)	hydroxymethylene	46	48	51	–	**54.2**
CH_3NO (formamide)	nitrosomethane	38	43	48	–	**60.5**
CH_3NO_2 (nitromethane)	methyl nitrite	-40	-22	7	2.7	**2.0**
$C_2H_2Cl_2$ (1,1-dichloroethylene)	cis-1,2-dichloroethylene	0	-3	1	–	**0.5**
	trans-1,2-dichloroethylene	-1	-3	1	–	**0.6**
C_2H_3N (acetonitrile)	methyl isocyanide	41	31	31	–	**20.5**
C_2H_4O (acetaldehyde)	oxacyclopropane	27	33	36	26.4	**27.1**
$C_2H_4O_2$ (acetic acid)	methyl formate	16	12	15	16.3	**18.3**
$C_2H_5NO_2$ (glycine)	nitroethane	89	84	71	–	**69.7**
C_2H_6O (ethanol)	dimethyl ether	12	10	9	11.7	**12.0**
$C_2H_6O_2$ (1,2-ethanediol)	ethanehydroperoxide	69	73	57	–	**45.1**
	dimethylperoxide	79	84	64	–	**62.6**
C_2H_6Si (vinylsilane)	1-methylsilaethylene	14	-6	2	–	**11**
C_2H_6S (ethanethiol)	dimethylsulfide	2	1	-2	1.2	**2.1**
C_2H_7N (ethylamine)	dimethylamine	7	10	5	7.8	**6.9**
$C_2H_7PO_3$ (dimethylphosphonate)	dimethylphosphorous acid	12	-14	-10	–	**6.5**
C_2H_8Si (dimethylsilane)	ethylsilane	10	9	11	–	**13**
C_3H_4 (propyne)	allene	3	3	7	1.1	**1.0**
	cyclopropene	27	31	28	24.2	**21.6**
$C_3H_4N_2$ (imidazole)	pyrazole	12	15	18	–	**12.7**
C_3H_6 (propene)	cyclopropane	6	11	10	8.1	**6.7**

Table A6-31: Relative Energies of Structural Isomers. Semi-Empirical Models (2)

formula (ref. compound)	molecule	MNDO	AM1	PM3	G3	expt.
C_3H_6O (acetone)	propanal	2	2	5	—	7.1
	2-hydroxypropene	7	10	13	—	7.7
	allyl alcohol	12	10	20	—	22.2
	methyl vinyl ether	23	22	28	—	25.6
	propylene oxide	26	34	37	—	29.5
	oxetane	12	24	27	—	32.7
$C_4H_4N_2$ (pyrazine)	pyrimidine	-3	0	-1	—	0.1
	1,2-ethanedinitrile	11	1	18	—	3.3
	pyridazine	6	11	17	—	19.7
C_4H_6 (1,3-butadiene)	2-butyne	-4	2	-1	8.5	8.5
	cyclobutene	2	16	7	12.5	11.2
	1,2-butadiene	5	8	8	—	12.5
	1-butyne	7	8	5	—	13.2
	methylenecyclopropane	9	18	14	19.9	21.7
	bicyclo[1.1.0]butane	35	48	38	27.6	25.6
	1-methylcyclopropene	25	35	26	—	32.0
C_4H_8 (isobutene)	trans-2-butene	-3	-2	0	—	1.1
	cis-2-butene	-2	-1	0	—	2.2
	skew-1-butene	2	1	5	—	3.9
	methylcyclopropane	4	12	12	—	9.7
	cyclobutane	-9	0	-1	—	10.8
$C_4H_8O_2$ (1,3-dioxane)	1,4-dioxane	5	7	5	—	7.6
C_4H_{10} (isobutane)	n-butane	-3	-2	1	1.7	2.1
$C_6H_4F_2$ (meta-difluorobenzene)	para-difluorobenzene	0	0	0	—	0.7
	ortho-difluorobenzene	0	2	0	—	3.7
C_8H_{14} (cis-cyclooctene)	trans-cyclooctene	16	12	13	—	9.3
C_9H_{16} (cis-cyclononene)	trans-cyclononene	6	4	6	—	2.9
$C_{10}H_{18}$ (trans-decalin)	cis-decalin	1	2	2	—	3.2

Table A6-32: Basis Set Effects on Relative Energies of Structural Isomers. Hartree-Fock Models

formula (ref. compound)	isomer	6-31G*	6-31G**	6-31+G*	6-311G*	6-311+G**	expt.
CH_2O (acetonitrile)	hydroxymethylene	52	49	50	51	47	**54.2**
CH_3NO (formamide)	nitrosomethane	64	67	65	65	68	**60.5**
$C_2H_4O_2$ (acetic acid)	methyl formate	13	17	14	14	17	**18.3**
C_2H_6O (ethanol)	dimethyl ether	7	10	8	7	11	**12.0**
C_3H_6 (1,2-ethanediol)	ethanehydroperoxide	56	59	58	57	62	**45.1**
	dimethylperoxide	61	67	63	62	71	**62.6**
C_3H_6O (acetone)	propanal	7	6	7	6	8	**7.1**
	2-hydroxypropene	19	15	18	18	14	**7.7**
	allyl alcohol	28	25	26	27	23	**22.2**

Table A6-33: Basis Set Effects on Relative Energies of Structural Isomers. EDF1 Density Functional Models

formula (ref. compound)	isomer	6-31G*	6-31G**	6-31+G*	6-311G*	6-311+G**	expt.
CH_2O (acetonitrile)	hydroxymethylene	59	56	56	57	54	**54.2**
CH_3NO (formamide)	nitrosomethane	58	61	59	61	64	**60.5**
$C_2H_4O_2$ (acetic acid)	methyl formate	12	15	13	13	17	**18.3**
C_2H_6O (ethanol)	dimethyl ether	6	9	8	7	12	**12.0**
C_3H_6 (1,2-ethanediol)	ethanehydroperoxide	44	47	52	46	52	**45.1**
	dimethylperoxide	48	53	47	50	60	**62.6**
C_3H_6O (acetone)	propanal	7	7	7	6	7	**7.1**
	2-hydroxypropene	17	14	16	17	13	**7.7**
	allyl alcohol	29	26	27	28	23	**22.2**

Table A6-34: Basis Set Effects on Relative Energies of Structural Isomers. B3LYP Density Functional Models

formula (ref. compound)	isomer	6-31G*	6-31G**	6-31+G*	6-311G*	6-311+G**	expt.
CH_2O (acetonitrile)	hydroxymethylene	57	55	55	56	53	**54.2**
CH_3NO (formamide)	nitrosomethane	68	62	61	62	66	**60.5**
$C_2H_4O_2$ (acetic acid)	methyl formate	12	15	13	13	16	**18.3**
C_2H_6O (ethanol)	dimethyl ether	5	8	7	7	11	**12.0**
C_3H_6 (1,2-ethanediol)	ethanehydroperoxide	47	50	50	50	54	**45.1**
	dimethylperoxide	50	55	54	53	62	**62.6**
C_3H_6O (acetone)	propanal	8	7	7	6	7	**7.1**
	2-hydroxypropene	17	14	16	16	12	**7.7**
	allyl alcohol	28	25	26	27	22	**22.2**

Table A6-35: Basis Set Effects on Relative Energies of Structural Isomers. MP2 Models

formula (ref. compound)	isomer	6-31G*	6-31G**	6-31+G*	6-311G*	6-311+G**	expt.
CH_2O (acetonitrile)	hydroxymethylene	61	59	59	60	55	**54.2**
CH_3NO (formamide)	nitrosomethane	61	63	63	62	65	**60.5**
$C_2H_4O_2$ (acetic acid)	methyl formate	14	16	15	15	18	**18.3**
C_2H_6O (ethanol)	dimethyl ether	9	11	10	9	14	**12.0**
C_3H_6 (1,2-ethanediol)	ethanehydroperoxide	52	54	55	53	60	**45.1**
	dimethylperoxide	58	63	61	59	71	**62.6**
C_3H_6O (acetone)	propanal	8	6	7	6	6	**7.1**
	2-hydroxypropene	18	16	17	18	13	**7.7**
	allyl alcohol	29	27	26	30	23	**22.2**

Table A6-36: Energies of Bond Separation Reactions. Hartree-Fock Models

molecule	bond separation reaction	STO-3G	3-21G	6-31G*	6-311+G**	G3	expt.
propane	$CH_3CH_2CH_3 + CH_4 \rightarrow 2CH_3CH_3$	1	1	1	1	2.3	2.6
isobutane	$CH(CH_3)_3 + 2CH_4 \rightarrow 3CH_3CH_3$	1	4	2	2	6.4	7.5
neopentane	$C(CH_3)_4 + 3CH_4 \rightarrow 4CH_3CH_3$	2	7	3	3	–	13.5
ethylsilane	$CH_3CH_2SiH_3 + CH_4 \rightarrow CH_3CH_3 + CH_3SiH_3$	-2	-1	-2	-2	–	–
dimethylsilane	$SiH_2(CH_3)_2 + SiH_4 \rightarrow 2CH_3SiH_3$	0	0	0	0	–	0.8
trimethylsilane	$SiH(CH_3)_3 + 2SiH_4 \rightarrow 3CH_3SiH_3$	1	0	0	0	–	1.6
tetramethylsilane	$Si(CH_3)_4 + 3SiH_4 \rightarrow 4CH_3SiH_3$	2	0	0	0	–	2.8
ethylamine	$CH_3CH_2NH_2 + CH_4 \rightarrow CH_3CH_3 + CH_3NH_2$	2	3	3	3	4.0	3.6
dimethylamine	$CH_3NHCH_3 + NH_3 \rightarrow 2CH_3NH_2$	1	2	2	2	4.2	4.4
trimethylamine	$(CH_3)_3N + 2NH_3 \rightarrow 3CH_3NH_2$	3	6	5	6	11.6	11.1
trans-ethanol	$CH_3CH_2OH + CH_4 \rightarrow CH_3CH_3 + CH_3OH$	3	5	4	4	5.4	5.7
dimethyl ether	$CH_3OCH_3 + H_2O \rightarrow 2CH_3OH$	1	2	3	3	5.1	5.4
trans-ethanethiol	$CH_3CH_2SH + CH_4 \rightarrow CH_3CH_3 + CH_3SH$	0	2	1	1	3.1	3.3
dimethyl sulfide	$CH_3SCH_3 + H_2S \rightarrow 2CH_3SH$	1	1	1	1	2.8	2.9
difluoromethane	$CH_2F_2 + CH_4 \rightarrow 2CH_3F$	10	14	14	11	–	13.9
trifluoromethane	$CHF_3 + 2CH_4 \rightarrow 3CH_3F$	29	38	36	27	–	34.6
tetrafluoromethane	$CF_4 + 3CH_4 \rightarrow 4CH_3F$	54	62	58	41	–	52.8
dichloromethane	$CH_2Cl_2 + CH_4 \rightarrow 2CH_3Cl$	-4	-4	-4	-3	0.9	1.5
trichloromethane	$CHCl_3 + 2CH_4 \rightarrow 3CH_3Cl$	-11	-13	-12	-11	0.6	1.9
tetrachloromethane	$CCl_4 + 3CH_4 \rightarrow 4CH_3Cl$	-21	-26	-29	-26	-2.4	-1.8
propene	$CH_3CHCH_2 + CH_4 \rightarrow CH_3CH_3 + CH_2CH_2$	5	4	4	4	4.9	5.4
acetaldehyde	$CH_3CHO + CH_4 \rightarrow CH_3CH_3 + H_2CO$	8	10	10	10	10.6	11.4
propyne	$CH_3CCH + CH_4 \rightarrow CH_3CH_3 + HCCH$	8	8	8	8	7.8	7.6
acetonitrile	$CH_3CN + CH_4 \rightarrow CH_3CH_3 + HCN$	11	13	12	12	11.0	14.7

Table A6-36: Energies of Bond Separation Reactions. Hartree-Fock Models (2)

molecule	bond separation reaction	STO-3G	3-21G	6-31G*	6-311+G**	G3	expt.
allene	$CH_2CCH_2 + CH_4 \rightarrow 2CH_2CH_2$	0	-3	-4	-6	-3.2	-2.8
ketene	$CH_2CO + CH_4 \rightarrow CH_2CH_2 + H_2CO$	16	19	14	13	15.2	15.7
carbon dioxide	$CO_2 + CH_4 \rightarrow 2H_2CO$	54	59	61	60	61.3	59.9
trans-1,3-butadiene	$CH_2CHCHCH_2 + 2CH_4 \rightarrow CH_3CH_3 + 2CH_2CH_2$	13	11	11	10	13.2	14.2
formamide	$NH_2CHO + CH_4 \rightarrow CH_3NH_2 + H_2CO$	20	37	31	30	–	30.8
benzene	+ $6CH_4 \rightarrow 3CH_3CH_3 + 3CH_2CH_2$	70	60	58	55	61.0	64.2
pyridine	+ $5CH_4 + NH_3 \rightarrow 2CH_3CH_3 +$ $2CH_2CH_2 + CH_3NH_2 + CH_2NH$	70	71	61	58	–	71.9
pyridazine	+ $4CH_4 + 2NH_3 \rightarrow 2CH_3CH_3 +$ $CH_2CH_2 + 2CH_2NH + NH_2NH_2$	62	52	53	50	–	74.6
pyrimidine	+ $4CH_4 + 2NH_3 \rightarrow CH_3CH_3 +$ $CH_2CH_2 + 2CH_3NH_2 + 2CH_2NH$	71	69	66	64	–	80.4
pyrazine	+ $4CH_4 + 2NH_3 \rightarrow CH_3CH_3 +$ $CH_2CH_2 + 2CH_3NH_2 + 2CH_2NH$	69	64	60	57	–	80.5
cyclopropane	+ $3CH_4 \rightarrow 3CH_3CH_3$	-45	-31	-26	-25	-23.5	-19.6
azacyclopropane	+ $2CH_4 + NH_3 \rightarrow CH_3CH_3 + 2CH_3NH_2$	-40	-33	-22	-21	-17.9	-14.7

Table A6-36: Energies of Bond Separation Reactions. Hartree-Fock Models (3)

molecule	bond separation reaction	STO-3G	3-21G	6-31G*	6-311+G**	G3	expt.
oxacyclopropane	$\triangle$O + 2CH$_4$ + H$_2$O → CH$_3$CH$_3$ + 2CH$_3$OH	-35	-31	-22	-19	-14.1	-10.5
thiacyclopropane	$\triangle$S + 2CH$_4$ + H$_2$S → CH$_3$CH$_3$ + 2CH$_3$SH	-35	-30	-16	-14	-11.5	-10.1
cyclobutane	$\square$ + 4CH$_4$ → 4CH$_3$CH$_3$	-27	-24	-24	-23	-19.3	-15.9
cyclopropene	$\triangle$ + 3CH$_4$ → 2CH$_3$CH$_3$ + CH$_2$CH$_2$	-66	-60	-51	-50	-46.6	-40.5
cyclobutene	$\square$ + 4CH$_4$ → 3CH$_3$CH$_3$ + CH$_2$CH$_2$	-28	-28	-24	-24	-19.5	-14.0
cyclopentadiene	+ 5CH$_4$ → 3CH$_3$CH$_3$ + 2CH$_2$CH$_2$	16	11	11	11	–	22.4
thiophene	S + 4CH$_4$ + H$_2$S → CH$_3$CH$_3$ + 2CH$_3$SH + 2CH$_2$CH$_2$	41	30	29	27	40.1	29.2
methylenecyclopropane	+ 4CH$_4$ → 3CH$_3$CH$_3$ + CH$_2$CH$_2$	-46	-36	-31	-31	-27.0	-24.5
bicyclo[1.1.0]butane	+ 6CH$_4$ → 5CH$_3$CH$_3$	-106	-76	-63	-60	-55.0	-45.5

Table A6-37: Energies of Bond Separation Reactions. Local Density Models

molecule	bond separation reaction	6-31G*	6-311+G**	G3	expt.
propane	$CH_3CH_2CH_3 + CH_4 \rightarrow 2CH_3CH_3$	2	3	2.3	2.6
isobutane	$CH(CH_3)_3 + 2CH_4 \rightarrow 3CH_3CH_3$	6	7	6.4	7.5
neopentane	$C(CH_3)_4 + 3CH_4 \rightarrow 4CH_3CH_3$	11	12	–	13.5
ethylsilane	$CH_3CH_2SiH_3 + CH_4 \rightarrow CH_3CH_3 + CH_3SiH_3$	0	0	–	–
–dimethylsilane	$SiH_2(CH_3)_2 + SiH_4 \rightarrow 2CH_3SiH_3$	0	0	–	0.8
trimethylsilane	$SiH(CH_3)_3 + 2SiH_4 \rightarrow 3CH_3SiH_3$	1	1	–	1.6
tetramethylsilane	$Si(CH_3)_4 + 3SiH_4 \rightarrow 4CH_3SiH_3$	1	3	–	2.8
ethylamine	$CH_3CH_2NH_2 + CH_4 \rightarrow CH_3CH_3 + CH_3NH_2$	5	5	4.0	3.6
dimethylamine	$CH_3NHCH_3 + NH_3 \rightarrow 2CH_3NH_2$	4	5	4.2	4.4
trimethylamine	$(CH_3)_3N + 2NH_3 \rightarrow 3CH_3NH_2$	11	12	11.6	11.1
trans-ethanol	$CH_3CH_2OH + CH_4 \rightarrow CH_3CH_3 + CH_3OH$	6	6	5.4	5.7
dimethyl ether	$CH_3OCH_3 + H_2O \rightarrow 2CH_3OH$	5	6	5.1	5.4
trans-ethanethiol	$CH_3CH_2SH + CH_4 \rightarrow CH_3CH_3 + CH_3SH$	3	3	3.1	3.3
dimethyl sulfide	$CH_3SCH_3 + H_2S \rightarrow 2CH_3SH$	3	3	2.8	2.9
difluoromethane	$CH_2F_2 + CH_4 \rightarrow 2CH_3F$	18	14	–	13.9
trifluoromethane	$CHF_3 + 2CH_4 \rightarrow 3CH_3F$	47	36	–	34.6
tetrafluoromethane	$CF_4 + 3CH_4 \rightarrow 4CH_3F$	75	54	–	52.8
dichloromethane	$CH_2Cl_2 + CH_4 \rightarrow 2CH_3Cl$	1	2	0.9	1.5
trichloromethane	$CHCl_3 + 2CH_4 \rightarrow 3CH_3Cl$	0	2	0.6	1.9
tetrachloromethane	$CCl_4 + 3CH_4 \rightarrow 4CH_3Cl$	-5	-2	-2.4	-1.8
propene	$CH_3CHCH_2 + CH_4 \rightarrow CH_3CH_3 + CH_2CH_2$	7	6	4.9	5.4
acetaldehyde	$CH_3CHO + CH_4 \rightarrow CH_3CH_3 + H_2CO$	13	13	10.6	11.4
propyne	$CH_3CCH + CH_4 \rightarrow CH_3CH_3 + HCCH$	12	10	7.8	7.6
acetonitrile	$CH_3CN + CH_4 \rightarrow CH_3CH_3 + HCN$	14	13	11.0	14.7

Table A6-37: Energies of Bond Separation Reactions. Local Density Models (2)

molecule	bond separation reaction	6-31G*	6-311+G**	G3	expt.
allene	$CH_2CCH_2 + CH_4 \rightarrow 2CH_2CH_2$	4	3	-3.2	-2.8
ketene	$CH_2CO + CH_4 \rightarrow CH_2CH_2 + H_2CO$	24	24	15.2	15.7
carbon dioxide	$CO_2 + CH_4 \rightarrow 2H_2CO$	68	67	61.3	59.9
trans-1,3-butadiene	$CH_2CHCHCH_2 + 2CH_4 \rightarrow CH_3CH_3 + 2CH_2CH_2$	18	17	13.2	14.2
formamide	$NH_2CHO + CH_4 \rightarrow CH_3NH_2 + H_2CO$	39	39	–	30.8
benzene	$+ 6CH_4 \rightarrow 3CH_3CH_3 + 3CH_2CH_2$	73	68	61.0	64.2
pyridine	$+ 5CH_4 + NH_3 \rightarrow 2CH_3CH_3 + 2CH_2CH_2 + CH_3NH_2 + CH_2NH$	75	71	–	71.9
pyridazine	$+ 4CH_4 + 2NH_3 \rightarrow 2CH_3CH_3 + CH_2CH_2 + 2CH_2NH + NH_2NH_2$	68	65	–	74.6
pyrimidine	$+ 4CH_4 + 2NH_3 \rightarrow CH_3CH_3 + CH_2CH_2 + 2CH_3NH_2 + 2CH_2NH$	78	76	–	80.4
pyrazine	$+ 4CH_4 + 2NH_3 \rightarrow CH_3CH_3 + CH_2CH_2 + 2CH_3NH_2 + 2CH_2NH$	75	72	–	80.5
cyclopropane	$+ 3CH_4 \rightarrow 3CH_3CH_3$	-30	-28	-23.5	-19.6
azacyclopropane	$+ 2CH_4 + NH_3 \rightarrow CH_3CH_3 + 2CH_3NH_2$	-26	-24	-17.9	-14.7
oxacyclopropane	$+ 2CH_4 + H_2O \rightarrow CH_3CH_3 + 2CH_3OH$	-22	-20	-14.1	-10.5

Table A6-37: Energies of Bond Separation Reactions. Local Density Models (3)

molecule	bond separation reaction	6-31G*	6-311+G**	G3	expt.
thiacyclopropane	$\triangle\!\!S$ + 2CH$_4$ + H$_2$S $\rightarrow$ CH$_3$CH$_3$ + 2CH$_3$SH	-18	-16	-11.5	-10.1
cyclobutane	$\square$ + 4CH$_4$ $\rightarrow$ 4CH$_3$CH$_3$	-24	-23	-19.3	-15.9
cyclopropene	$\triangle$ + 3CH$_4$ $\rightarrow$ 2CH$_3$CH$_3$ + CH$_2$CH$_2$	-50	-49	-46.6	-40.5
cyclobutene	$\square$ + 4CH$_4$ $\rightarrow$ 3CH$_3$CH$_3$ + CH$_2$CH$_2$	-20	-20	-19.5	-14.0
cyclopentadiene	$\pentagon$ + 5CH$_4$ $\rightarrow$ 3CH$_3$CH$_3$ + 2CH$_2$CH$_2$	22	21	–	22.4
thiophene	$\pentagon\!\!S$ + 4CH$_4$ + H$_2$S $\rightarrow$ CH$_3$CH$_3$ + 2CH$_3$SH + 2CH$_2$CH$_2$	45	44	40.1	29.2
methylenecyclopropane	$\triangle\!\!=$ + 4CH$_4$ $\rightarrow$ 3CH$_3$CH$_3$ + CH$_2$CH$_2$	-29	-28	-27.0	-24.5
bicyclo[1.1.0]butane	$\bowtie$ + 6CH$_4$ $\rightarrow$ 5CH$_3$CH$_3$	-67	-63	-55.0	-45.5

Table A6-38: Energies of Bond Separation Reactions. BP Density Functional Models

molecule	bond separation reaction	6-31G*	6-311+G**	G3	expt.
propane	$CH_3CH_2CH_3 + CH_4 \rightarrow 2CH_3CH_3$	2	1	2.3	2.6
isobutane	$CH(CH_3)_3 + 2CH_4 \rightarrow 3CH_3CH_3$	4	3	6.4	7.5
neopentane	$C(CH_3)_4 + 3CH_4 \rightarrow 4CH_3CH_3$	7	5	–	13.5
ethylsilane	$CH_3CH_2SiH_3 + CH_4 \rightarrow CH_3CH_3 + CH_3SiH_3$	–1	–1	–	–
dimethylsilane	$SiH_2(CH_3)_2 + SiH_4 \rightarrow 2CH_3SiH_3$	–1	0	–	0.8
trimethylsilane	$SiH(CH_3)_3 + 2SiH_4 \rightarrow 3CH_3SiH_3$	–2	0	–	1.6
tetramethylsilane	$Si(CH_3)_4 + 3SiH_4 \rightarrow 4CH_3SiH_3$	–1	0	–	2.8
ethylamine	$CH_3CH_2NH_2 + CH_4 \rightarrow CH_3CH_3 + CH_3NH_2$	4	3	4.0	3.6
dimethylamine	$CH_3NHCH_3 + NH_3 \rightarrow 2CH_3NH_2$	3	3	4.2	4.4
trimethylamine	$(CH_3)_3N + 2NH_3 \rightarrow 3CH_3NH_2$	6	7	11.6	11.1
trans-ethanol	$CH_3CH_2OH + CH_4 \rightarrow CH_3CH_3 + CH_3OH$	5	5	5.4	5.7
dimethyl ether	$CH_3OCH_3 + H_2O \rightarrow 2CH_3OH$	3	4	5.1	5.4
trans-ethanethiol	$CH_3CH_2SH + CH_4 \rightarrow CH_3CH_3 + CH_3SH$	2	2	3.1	3.3
dimethyl sulfide	$CH_3SCH_3 + H_2S \rightarrow 2CH_3SH$	2	2	2.8	2.9
difluoromethane	$CH_2F_2 + CH_4 \rightarrow 2CH_3F$	16	11	–	13.9
trifluoromethane	$CHF_3 + 2CH_4 \rightarrow 3CH_3F$	41	28	–	34.6
tetrafluoromethane	$CF_4 + 3CH_4 \rightarrow 4CH_3F$	64	41	–	52.8
dichloromethane	$CH_2Cl_2 + CH_4 \rightarrow 2CH_3Cl$	0	0	0.9	1.5
trichloromethane	$CHCl_3 + 2CH_4 \rightarrow 3CH_3Cl$	–2	–2	0.6	1.9
tetrachloromethane	$CCl_4 + 3CH_4 \rightarrow 4CH_3Cl$	–11	–9	–2.4	–1.8
propene	$CH_3CHCH_2 + CH_4 \rightarrow CH_3CH_3 + CH_2CH_2$	6	5	4.9	5.4
acetaldehyde	$CH_3CHO + CH_4 \rightarrow CH_3CH_3 + H_2CO$	12	12	10.6	11.4
propyne	$CH_3CCH + CH_4 \rightarrow CH_3CH_3 + HCCH$	11	9	7.8	7.6
acetonitrile	$CH_3CN + CH_4 \rightarrow CH_3CH_3 + HCN$	14	12	11.0	14.7

662

Table A6-38: Energies of Bond Separation Reactions. BP Density Functional Models (2)

molecule	bond separation reaction	6-31G*	6-311+G**	G3	expt.
allene	$CH_2CCH_2 + CH_4 \rightarrow 2CH_2CH_2$	2	1	-3.2	-2.8
ketene	$CH_2CO + CH_4 \rightarrow CH_2CH_2 + H_2CO$	21	20	15.2	15.7
carbon dioxide	$CO_2 + CH_4 \rightarrow 2H_2CO$	64	63	61.3	59.9
trans-1,3-butadiene	$CH_2CHCHCH_2 + 2CH_4 \rightarrow CH_3CH_3 + 2CH_2CH_2$	16	14	13.2	14.2
formamide	$NH_2CHO + CH_4 \rightarrow CH_3NH_2 + H_2CO$	35	34	–	30.8
benzene	$+ 6CH_4 \rightarrow 3CH_3CH_3 + 3CH_2CH_2$	71	65	61.0	64.2
pyridine	$+ 5CH_4 + NH_3 \rightarrow 2CH_3CH_3 + 2CH_2CH_2 + CH_3NH_2 + CH_2NH$	71	67	–	71.9
pyridazine	$+ 4CH_4 + 2NH_3 \rightarrow 2CH_3CH_3 + CH_2CH_2 + 2CH_2NH + NH_2NH_2$	65	62	–	74.6
pyrimidine	$+ 4CH_4 + 2NH_3 \rightarrow CH_3CH_3 + CH_2CH_2 + 2CH_3NH_2 + 2CH_2NH$	74	71	–	80.4
pyrazine	$+ 4CH_4 + 2NH_3 \rightarrow CH_3CH_3 + CH_2CH_2 + 2CH_3NH_2 + 2CH_2NH$	71	68	–	80.5
cyclopropane	$+ 3CH_4 \rightarrow 3CH_3CH_3$	-23	-22	-23.5	-19.6
azacyclopropane	$+ 2CH_4 + NH_3 \rightarrow CH_3CH_3 + 2CH_3NH_2$	-18	-17	-17.9	-14.7

Table A6-38: Energies of Bond Separation Reactions. BP Density Functional Models (3)

molecule	bond separation reaction	6-31G*	6-311+G**	G3	expt.
oxacyclopropane	[O-triangle] + $2CH_4$ + H_2O → CH_3CH_3 + $2CH_3OH$	-14	-13	-14.1	-10.5
thiacyclopropane	[S-triangle] + $2CH_4$ + H_2S → CH_3CH_3 + $2CH_3SH$	-11	-11	-11.5	-10.1
cyclobutane	[square] + $4CH_4$ → $4CH_3CH_3$	-18	-19	-19.3	-15.9
cyclopropene	[triangle] + $3CH_4$ → $2CH_3CH_3$ + CH_2CH_2	-43	-43	-46.6	-40.5
cyclobutene	[square] + $4CH_4$ → $3CH_3CH_3$ + CH_2CH_2	-16	-17	-19.5	-14.0
cyclopentadiene	[ring] + $5CH_4$ → $3CH_3CH_3$ + $2CH_2CH_2$	21	18	–	22.4
thiophene	[S-ring] + $4CH_4$ + H_2S → CH_3CH_3 + $2CH_3SH$ + $2CH_2CH_2$	42	40	40.1	29.2
methylenecyclopropane	[structure] + $4CH_4$ → $3CH_3CH_3$ + CH_2CH_2	-23	-24	-27.0	-24.5
bicyclo[1.1.0]butane	[structure] + $6CH_4$ → $5CH_3CH_3$	-53	-52	-55.0	-45.5

Table A6-39: Energies of Bond Separation Reactions. BLYP Density Functional Models

molecule	bond separation reaction	6-31G*	6-311+G**	G3	expt.
propane	$CH_3CH_2CH_3 + CH_4 \rightarrow 2CH_3CH_3$	1	1	2.3	2.6
isobutane	$CH(CH_3)_3 + 2CH_4 \rightarrow 3CH_3CH_3$	3	3	6.4	7.5
neopentane	$C(CH_3)_4 + 3CH_4 \rightarrow 4CH_3CH_3$	4	4	–	13.5
ethylsilane	$CH_3CH_2SiH_3 + CH_4 \rightarrow CH_3CH_3 + CH_3SiH_3$	-1	-1	–	–
dimethylsilane	$SiH_2(CH_3)_2 + SiH_4 \rightarrow 2CH_3SiH_3$	0	0	–	0.8
trimethylsilane	$SiH(CH_3)_3 + 2SiH_4 \rightarrow 3CH_3SiH_3$	0	0	–	1.6
tetramethylsilane	$Si(CH_3)_4 + 3SiH_4 \rightarrow 4CH_3SiH_3$	-1	0	–	2.8
ethylamine	$CH_3CH_2NH_2 + CH_4 \rightarrow CH_3CH_3 + CH_3NH_2$	4	3	4.0	3.6
dimethylamine	$CH_3NHCH_3 + NH_3 \rightarrow 2CH_3NH_2$	3	3	4.2	4.4
trimethylamine	$(CH_3)_3N + 2NH_3 \rightarrow 3CH_3NH_2$	7	7	11.6	11.1
trans-ethanol	$CH_3CH_2OH + CH_4 \rightarrow CH_3CH_3 + CH_3OH$	5	5	5.4	5.7
dimethyl ether	$CH_3OCH_3 + H_2O \rightarrow 2CH_3OH$	3	3	5.1	5.4
trans-ethanethiol	$CH_3CH_2SH + CH_4 \rightarrow CH_3CH_3 + CH_3SH$	2	2	3.1	3.3
dimethyl sulfide	$CH_3SCH_3 + H_2S \rightarrow 2CH_3SH$	2	2	2.8	2.9
difluoromethane	$CH_2F_2 + CH_4 \rightarrow 2CH_3F$	15	11	–	13.9
trifluoromethane	$CHF_3 + 2CH_4 \rightarrow 3CH_3F$	38	26	–	34.6
tetrafluoromethane	$CF_4 + 3CH_4 \rightarrow 4CH_3F$	60	38	–	52.8
dichloromethane	$CH_2Cl_2 + CH_4 \rightarrow 2CH_3Cl$	-1	0	0.9	1.5
trichloromethane	$CHCl_3 + 2CH_4 \rightarrow 3CH_3Cl$	-4	-3	0.6	1.9
tetrachloromethane	$CCl_4 + 3CH_4 \rightarrow 4CH_3Cl$	-13	-11	-2.4	-1.8
propene	$CH_3CHCH_2 + CH_4 \rightarrow CH_3CH_3 + CH_2CH_2$	5	5	4.9	5.4
acetaldehyde	$CH_3CHO + CH_4 \rightarrow CH_3CH_3 + H_2CO$	11	11	10.6	11.4
propyne	$CH_3CCH + CH_4 \rightarrow CH_3CH_3 + HCCH$	10	9	7.8	7.6
acetonitrile	$CH_3CN + CH_4 \rightarrow CH_3CH_3 + HCN$	13	12	11.0	14.7

Table A6-39: Energies of Bond Separation Reactions. BLYP Density Functional Models (2)

molecule	bond separation reaction	6-31G*	6-311+G**	G3	expt.
allene	$CH_2CCH_2 + CH_4 \rightarrow 2CH_2CH_2$	2	0	-3.2	-2.8
ketene	$CH_2CO + CH_4 \rightarrow CH_2CH_2 + H_2CO$	19	18	15.2	15.7
carbon dioxide	$CO_2 + CH_4 \rightarrow 2H_2CO$	61	61	61.3	59.9
trans-1,3-butadiene	$CH_2CHCHCH_2 + 2CH_4 \rightarrow CH_3CH_3 + 2CH_2CH_2$	16	14	13.2	14.2
formamide	$NH_2CHO + CH_4 \rightarrow CH_3NH_2 + H_2CO$	34	34	–	30.8
benzene	$+ 6CH_4 \rightarrow 3CH_3CH_3 + 3CH_2CH_2$	67	62	61.0	64.2
pyridine	$+ 5CH_4 + NH_3 \rightarrow 2CH_3CH_3 + 2CH_2CH_2 + CH_3NH_2 + CH_2NH$	69	65	–	71.9
pyridazine	$+ 4CH_4 + 2NH_3 \rightarrow 2CH_3CH_3 + CH_2CH_2 + 2CH_2NH + NH_2NH_2$	62	60	–	74.6
pyrimidine	$+ 4CH_4 + 2NH_3 \rightarrow CH_3CH_3 + CH_2CH_2 + 2CH_3NH_2 + 2CH_2NH$	72	69	–	80.4
pyrazine	$+ 4CH_4 + 2NH_3 \rightarrow CH_3CH_3 + CH_2CH_2 + 2CH_3NH_2 + 2CH_2NH$	68	65	–	80.5
cyclopropane	$+ 3CH_4 \rightarrow 3CH_3CH_3$	-24	-21	-23.5	-19.6
azacyclopropane	$+ 2CH_4 + NH_3 \rightarrow CH_3CH_3 + 2CH_3NH_2$	-18	-16	-17.9	-14.7

Table A6-39: Energies of Bond Separation Reactions. BLYP Density Functional Models (3)

molecule	bond separation reaction	6-31G*	6-311+G**	G3	expt.
oxacyclopropane	$+ 2CH_4 + H_2O \rightarrow CH_3CH_3 + 2CH_3OH$	-15	-12	**-14.1**	**-10.5**
thiacyclopropane	$+ 2CH_4 + H_2S \rightarrow CH_3CH_3 + 2CH_3SH$	-11	-10	**-11.5**	**-10.1**
cyclobutane	$+ 4CH_4 \rightarrow 4CH_3CH_3$	-20	-19	**-19.3**	**-15.9**
cyclopropene	$+ 3CH_4 \rightarrow 2CH_3CH_3 + CH_2CH_2$	-44	-42	**-46.6**	**-40.5**
cyclobutene	$+ 4CH_4 \rightarrow 3CH_3CH_3 + CH_2CH_2$	-18	-18	**-19.5**	**-14.0**
cyclopentadiene	$+ 5CH_4 \rightarrow 3CH_3CH_3 + 2CH_2CH_2$	6	17	–	**22.4**
thiophene	$+ 4CH_4 + H_2S \rightarrow CH_3CH_3 + 2CH_3SH + 2CH_2CH_2$	38	37	**40.1**	**29.2**
methylenecyclopropane	$+ 4CH_4 \rightarrow 3CH_3CH_3 + CH_2CH_2$	-24	-23	**-27.0**	**-24.5**
bicyclo[1.1.0]butane	$+ 6CH_4 \rightarrow 5CH_3CH_3$	-54	-50	**-55.0**	**-45.5**

Table A6-40: Energies of Bond Separation Reactions. EDF1 Density Functional Models

molecule	bond separation reaction	6-31G*	6-311+G**	G3	expt.
propane	$CH_3CH_2CH_3 + CH_4 \rightarrow 2CH_3CH_3$	1	0	2.3	2.6
isobutane	$CH(CH_3)_3 + 2CH_4 \rightarrow 3CH_3CH_3$	1	1	6.4	7.5
neopentane	$C(CH_3)_4 + 3CH_4 \rightarrow 4CH_3CH_3$	2	1	–	13.5
ethylsilane	$CH_3CH_2SiH_3 + CH_4 \rightarrow CH_3CH_3 + CH_3SiH_3$	-2	-2	–	–
dimethylsilane	$SiH_2(CH_3)_2 + SiH_4 \rightarrow 2CH_3SiH_3$	0	0	–	0.8
trimethylsilane	$SiH(CH_3)_3 + 2SiH_4 \rightarrow 3CH_3SiH_3$	-2	0	–	1.6
tetramethylsilane	$Si(CH_3)_4 + 3SiH_4 \rightarrow 4CH_3SiH_3$	-1	-1	–	2.8
ethylamine	$CH_3CH_2NH_2 + CH_4 \rightarrow CH_3CH_3 + CH_3NH_2$	3	3	4.0	3.6
dimethylamine	$CH_3NHCH_3 + NH_3 \rightarrow 2CH_3NH_2$	2	3	4.2	4.4
trimethylamine	$(CH_3)_3N + 2NH_3 \rightarrow 3CH_3NH_2$	5	6	11.6	11.1
trans-ethanol	$CH_3CH_2OH + CH_4 \rightarrow CH_3CH_3 + CH_3OH$	4	4	5.4	5.7
dimethyl ether	$CH_3OCH_3 + H_2O \rightarrow 2CH_3OH$	3	3	5.1	5.4
trans-ethanethiol	$CH_3CH_2SH + CH_4 \rightarrow CH_3CH_3 + CH_3SH$	1	1	3.1	3.3
dimethyl sulfide	$CH_3SCH_3 + H_2S \rightarrow 2CH_3SH$	1	2	2.8	2.9
difluoromethane	$CH_2F_2 + CH_4 \rightarrow 2CH_3F$	16	11	–	13.9
trifluoromethane	$CHF_3 + 2CH_4 \rightarrow 3CH_3F$	40	28	–	34.6
tetrafluoromethane	$CF_4 + 3CH_4 \rightarrow 4CH_3F$	62	41	–	52.8
dichloromethane	$CH_2Cl_2 + CH_4 \rightarrow 2CH_3Cl$	-1	0	0.9	1.5
trichloromethane	$CHCl_3 + 2CH_4 \rightarrow 3CH_3Cl$	-4	-3	0.6	1.9
tetrachloromethane	$CCl_4 + 3CH_4 \rightarrow 4CH_3Cl$	-14	-12	-2.4	-1.8
propene	$CH_3CHCH_2 + CH_4 \rightarrow CH_3CH_3 + CH_2CH_2$	5	4	4.9	5.4
acetaldehyde	$CH_3CHO + CH_4 \rightarrow CH_3CH_3 + H_2CO$	11	11	10.6	11.4
propyne	$CH_3CCH + CH_4 \rightarrow CH_3CH_3 + HCCH$	11	9	7.8	7.6
acetonitrile	$CH_3CN + CH_4 \rightarrow CH_3CH_3 + HCN$	13	12	11.0	14.7

Table A6-40: Energies of Bond Separation Reactions. EDF1 Density Functional Models (2)

molecule	bond separation reaction	6-31G*	6-311+G**	G3	expt.
allene	$CH_2CCH_2 + CH_4 \rightarrow 2CH_2CH_2$	2	0	-3.2	-2.8
ketene	$CH_2CO + CH_4 \rightarrow CH_2CH_2 + H_2CO$	20	19	15.2	15.7
carbon dioxide	$CO_2 + CH_4 \rightarrow 2H_2CO$	64	63	61.3	59.9
trans-1,3-butadiene	$CH_2CHCHCH_2 + 2CH_4 \rightarrow CH_3CH_3 + 2CH_2CH_2$	16	13	13.2	14.2
formamide	$NH_2CHO + CH_4 \rightarrow CH_3NH_2 + H_2CO$	35	34	–	30.8
benzene	$+ 6CH_4 \rightarrow 3CH_3CH_3 + 3CH_2CH_2$	71	64	61.0	64.2
pyridine	$+ 5CH_4 + NH_3 \rightarrow 2CH_3CH_3 + 2CH_2CH_2 + CH_3NH_2 + CH_2NH$	71	67	–	71.9
pyridazine	$+ 4CH_4 + 2NH_3 \rightarrow 2CH_3CH_3 + CH_2CH_2 + 2CH_2NH + NH_2NH_2$	65	63	–	74.6
pyrimidine	$+ 4CH_4 + 2NH_3 \rightarrow CH_3CH_3 + CH_2CH_2 + 2CH_3NH_2 + 2CH_2NH$	75	72	–	80.4
pyrazine	$+ 4CH_4 + 2NH_3 \rightarrow CH_3CH_3 + CH_2CH_2 + 2CH_3NH_2 + 2CH_2NH$	71	69	–	80.5
cyclopropane	$+ 3CH_4 \rightarrow 3CH_3CH_3$	-20	-20	-23.5	-19.6
azacyclopropane	$+ 2CH_4 + NH_3 \rightarrow CH_3CH_3 + 2CH_3NH_2$	-15	-14	-17.9	-14.7

Table A6-40: Energies of Bond Separation Reactions. EDF1 Density Functional Models (2)

molecule	bond separation reaction	6-31G*	6-311+G**	G3	expt.
oxacyclopropane	$\triangle$O + $2CH_4$ + H_2O → CH_3CH_3 + $2CH_3OH$	-11	-9	-14.1	-10.5
thiacyclopropane	$\triangle$S + $2CH_4$ + H_2S → CH_3CH_3 + $2CH_3SH$	-8	-7	-11.5	-10.1
cyclobutane	$\square$ + $4CH_4$ → $4CH_3CH_3$	-18	-18	-19.3	-15.9
cyclopropene	$\triangle$ + $3CH_4$ → $2CH_3CH_3$ + CH_2CH_2	-40	-40	-46.6	-40.5
cyclobutene	$\square$ + $4CH_4$ → $3CH_3CH_3$ + CH_2CH_2	-15	-16	-19.5	-14.0
cyclopentadiene	+ $5CH_4$ → $3CH_3CH_3$ + $2CH_2CH_2$	21	19	–	22.4
thiophene	S + $4CH_4$ + H_2S → CH_3CH_3 + $2CH_3SH$ + $2CH_2CH_2$	43	41	40.1	29.2
methylenecyclopropane	$\triangle$ + $4CH_4$ → $3CH_3CH_3$ + CH_2CH_2	-20	-21	-27.0	-24.5
bicyclo[1.1.0]butane	+ $6CH_4$ → $5CH_3CH_3$	-48	-46	-55.0	-45.5

Table A6-41: Energies of Bond Separation Reactions. B3LYP Density Functional Models

molecule	bond separation reaction	6-31G*	6-311+G**	G3	expt.
propane	$CH_3CH_2CH_3 + CH_4 \rightarrow 2CH_3CH_3$	1	1	2.3	2.6
isobutane	$CH(CH_3)_3 + 2CH_4 \rightarrow 3CH_3CH_3$	3	3	6.4	7.5
neopentane	$C(CH_3)_4 + 3CH_4 \rightarrow 4CH_3CH_3$	4	5	–	13.5
ethylsilane	$CH_3CH_2SiH_3 + CH_4 \rightarrow CH_3CH_3 + CH_3SiH_3$	-1	-1	–	–
dimethylsilane	$SiH_2(CH_3)_2 + SiH_4 \rightarrow 2CH_3SiH_3$	0	0	–	0.8
trimethylsilane	$SiH(CH_3)_3 + 2SiH_4 \rightarrow 3CH_3SiH_3$	0	0	–	1.6
tetramethylsilane	$Si(CH_3)_4 + 3SiH_4 \rightarrow 4CH_3SiH_3$	0	0	–	2.8
ethylamine	$CH_3CH_2NH_2 + CH_4 \quad CH_3CH_3 + CH_3NH_2$	4	3	4.0	3.6
dimethylamine	$CH_3NHCH_3 + NH_3 \rightarrow 2CH_3NH_2$	3	3	4.2	4.4
trimethylamine	$(CH_3)_3N + 2NH_3 \quad 3CH_3NH_2$	7	7	11.6	11.1
trans-ethanol	$CH_3CH_2OH + CH_4 \rightarrow CH_3CH_3 + CH_3OH$	5	5	5.4	5.7
dimethyl ether	$CH_3OCH_3 + H_2O \rightarrow 2CH_3OH$	3	4	5.1	5.4
trans-ethanethiol	$CH_3CH_2SH + CH_4 \rightarrow CH_3CH_3 + CH_3SH$	2	2	3.1	3.3
dimethyl sulfide	$CH_3SCH_3 + H_2S \rightarrow 2CH_3SH$	2	2	2.8	2.9
difluoromethane	$CH_2F_2 + CH_4 \rightarrow 2CH_3F$	15	11	–	13.9
trifluoromethane	$CHF_3 + 2CH_4 \rightarrow 3CH_3F$	38	27	–	34.6
tetrafluoromethane	$CF_4 + 3CH_4 \rightarrow 4CH_3F$	60	40	–	52.8
dichloromethane	$CH_2Cl_2 + CH_4 \rightarrow 2CH_3Cl$	-1	-1	0.9	1.5
trichloromethane	$CHCl_3 + 2CH_4 \rightarrow 3CH_3Cl$	-6	-5	0.6	1.9
tetrachloromethane	$CCl_4 + 3CH_4 \rightarrow 4CH_3Cl$	-16	-14	-2.4	-1.8
propene	$CH_3CHCH_2 + CH_4 \rightarrow CH_3CH_3 + CH_2CH_2$	5	5	4.9	5.4
acetaldehyde	$CH_3CHO + CH_4 \rightarrow CH_3CH_3 + H_2CO$	11	11	10.6	11.4
propyne	$CH_3CCH + CH_4 \rightarrow CH_3CH_3 + HCCH$	10	9	7.8	7.6
acetonitrile	$CH_3CN + CH_4 \rightarrow CH_3CH_3 + HCN$	13	12	11.0	14.7

Table A6-41: Energies of Bond Separation Reactions. B3LYP Density Functional Models (2)

molecule	bond separation reaction	6-31G*	6-311+G**	G3	expt.
allene	$CH_2CCH_2 + CH_4 \rightarrow 2CH_2CH_2$	1	-1	-3.2	-2.8
ketene	$CH_2CO + CH_4 \rightarrow CH_2CH_2 + H_2CO$	18	18	15.2	15.7
carbon dioxide	$CO_2 + CH_4 \rightarrow 2H_2CO$	62	61	61.3	59.9
trans-1,3-butadiene	$CH_2CHCHCH_2 + 2CH_4 \rightarrow CH_3CH_3 + 2CH_2CH_2$	15	13	13.2	14.2
formamide	$NH_2CHO + CH_4 \quad CH_3NH_2 + H_2CO$	34	33	–	30.8
benzene	$+ 6CH_4 \rightarrow 3CH_3CH_3 + 3CH_2CH_2$	67	62	61.0	64.2
pyridine	$+ 5CH_4 + NH_3 \rightarrow 2CH_3CH_3 + 2CH_2CH_2 + CH_3NH_2 + CH_2NH$	68	65	–	71.9
pyridazine	$+ 4CH_4 + 2NH_3 \rightarrow 2CH_3CH_3 + CH_2CH_2 + 2CH_2NH + NH_2NH_2$	62	59	–	74.6
pyrimidine	$+ 4CH_4 + 2NH_3 \rightarrow CH_3CH_3 + CH_2CH_2 + 2CH_3NH_2 + 2CH_2NH$	72	69	–	80.4
pyrazine	$+ 4CH_4 + 2NH_3 \rightarrow CH_3CH_3 + CH_2CH_2 + 2CH_3NH_2 + 2CH_2NH$	68	65	–	80.5
cyclopropane	$+ 3CH_4 \rightarrow 3CH_3CH_3$	-25	-23	-23.5	-19.6
azacyclopropane	$+ 2CH_4 + NH_3 \rightarrow CH_3CH_3 + 2CH_3NH_2$	-21	-19	-17.9	-14.7

Table A6-41: Energies of Bond Separation Reactions. B3LYP Density Functional Models (3)

molecule	bond separation reaction	6-31G*	6-311+G**	G3	expt.
oxacyclopropane	$\triangle$O + 2CH$_4$ + H$_2$O → CH$_3$CH$_3$ + 2CH$_3$OH	-17	-15	-14.1	-10.5
thiacyclopropane	$\triangle$S + 2CH$_4$ + H$_2$S → CH$_3$CH$_3$ + 2CH$_3$SH	-13	-12	-11.5	-10.1
cyclobutane	$\square$ + 4CH$_4$ → 4CH$_3$CH$_3$	-22	-21	-19.3	-15.9
cyclopropene	$\triangle$ + 3CH$_4$ → 2CH$_3$CH$_3$ + CH$_2$CH$_2$	-46	-45	-46.6	-40.5
cyclobutene	$\square$ + 4CH$_4$ → 3CH$_3$CH$_3$ + CH$_2$CH$_2$	-20	-20	-19.5	-14.0
cyclopentadiene	+ 5CH$_4$ → 3CH$_3$CH$_3$ + 2CH$_2$CH$_2$	17	16	–	22.4
thiophene	+ 4CH$_4$ + H$_2$S → CH$_3$CH$_3$ + 2CH$_3$SH + 2CH$_2$CH$_2$	38	36	40.1	29.2
methylenecyclopropane	+ 4CH$_4$ → 3CH$_3$CH$_3$ + CH$_2$CH$_2$	-27	-26	-27.0	-24.5
bicyclo[1.1.0]butane	+ 6CH$_4$ → 5CH$_3$CH$_3$	-59	-55	-55.0	-45.5

Table A6-42: Energies of Bond Separation Reactions. MP2 Models

molecule	bond separation reaction	6-31G*	6-311+G**	G3	expt.
propane	$CH_3CH_2CH_3 + CH_4 \rightarrow 2CH_3CH_3$	2	2	2.3	2.6
isobutane	$CH(CH_3)_3 + 2CH_4 \rightarrow 3CH_3CH_3$	6	7	6.4	7.5
neopentane	$C(CH_3)_4 + 3CH_4 \rightarrow 4CH_3CH_3$	3	12	–	13.5
ethylsilane	$CH_3CH_2SiH_3 + CH_4 \rightarrow CH_3CH_3 + CH_3SiH_3$	0	0	–	–
dimethylsilane	$SiH_2(CH_3)_2 + SiH_4 \rightarrow 2CH_3SiH_3$	0	0	–	0.8
trimethylsilane	$SiH(CH_3)_3 + 2SiH_4 \rightarrow 3CH_3SiH_3$	0	1	–	1.6
tetramethylsilane	$Si(CH_3)_4 + 3SiH_4 \rightarrow 4CH_3SiH_3$	0	3	–	2.8
ethylamine	$CH_3CH_2NH_2 + CH_4 \rightarrow CH_3CH_3 + CH_3NH_2$	4	4	4.0	3.6
dimethylamine	$CH_3NHCH_3 + NH_3 \rightarrow 2CH_3NH_2$	4	4	4.2	4.4
trimethylamine	$(CH_3)_3N + 2NH_3 \rightarrow 3CH_3NH_2$	11	11	11.6	11.1
trans-ethanol	$CH_3CH_2OH + CH_4 \rightarrow CH_3CH_3 + CH_3OH$	5	5	5.4	5.7
dimethyl ether	$CH_3OCH_3 + H_2O \rightarrow 2CH_3OH$	5	5	5.1	5.4
trans-ethanethiol	$CH_3CH_2SH + CH_4 \rightarrow CH_3CH_3 + CH_3SH$	3	3	3.1	3.3
dimethyl sulfide	$CH_3SCH_3 + H_2S \rightarrow 2CH_3SH$	3	3	2.8	2.9
difluoromethane	$CH_2F_2 + CH_4 \rightarrow 2CH_3F$	16	12	–	13.9
trifluoromethane	$CHF_3 + 2CH_4 \rightarrow 3CH_3F$	41	31	–	34.6
tetrafluoromethane	$CF_4 + 3CH_4 \rightarrow 4CH_3F$	64	46	–	52.8
dichloromethane	$CH_2Cl_2 + CH_4 \rightarrow 2CH_3Cl$	0	1	0.9	1.5
trichloromethane	$CHCl_3 + 2CH_4 \rightarrow 3CH_3Cl$	-3	1	0.6	1.9
tetrachloromethane	$CCl_4 + 3CH_4 \rightarrow 4CH_3Cl$	-26	-3	-2.4	-1.8
propene	$CH_3CHCH_2 + CH_4 \rightarrow CH_3CH_3 + CH_2CH_2$	5	5	4.9	5.4
acetaldehyde	$CH_3CHO + CH_4 \rightarrow CH_3CH_3 + H_2CO$	11	10	10.6	11.4
propyne	$CH_3CCH + CH_4 \rightarrow CH_3CH_3 + HCCH$	8	7	7.8	7.6
acetonitrile	$CH_3CN + CH_4 \rightarrow CH_3CH_3 + HCN$	11	10	11.0	14.7

674

Table A6-42: Energies of Bond Separation Reactions. MP2 Models (2)

molecule	bond separation reaction	6-31G*	6-311+G**	G3	expt.
allene	$CH_2CCH_2 + CH_4 \rightarrow 2CH_2CH_2$	-2	-3	-3.2	-2.8
ketene	$CH_2CO + CH_4 \rightarrow CH_2CH_2 + H_2CO$	17	17	15.2	15.7
carbon dioxide	$CO_2 + CH_4 \rightarrow 2H_2CO$	66	65	61.3	59.9
trans-1,3-butadiene	$CH_2CHCHCH_2 + 2CH_4 \rightarrow CH_3CH_3 + 2CH_2CH_2$	14	14	13.2	14.2
formamide	$NH_2CHO + CH_4 \rightarrow CH_3NH_2 + H_2CO$	33	31	–	30.8
benzene	$+ 6CH_4 \rightarrow 3CH_3CH_3 + 3CH_2CH_2$	71	68	61.0	64.2
pyridine	$+ 5CH_4 + NH_3 \rightarrow 2CH_3CH_3 + 2CH_2CH_2 + CH_3NH_2 + CH_2NH$	74	70	–	71.9
pyridazine	$+ 4CH_4 + 2NH_3 \rightarrow 2CH_3CH_3 + CH_2CH_2 + 2CH_2NH + NH_2NH_2$	69	65	–	74.6
pyrimidine	$+ 4CH_4 + 2NH_3 \rightarrow CH_3CH_3 + CH_2CH_2 + 2CH_3NH_2 + 2CH_2NH$	78	74	–	80.4
pyrazine	$+ 4CH_4 + 2NH_3 \rightarrow CH_3CH_3 + CH_2CH_2 + 2CH_3NH_2 + 2CH_2NH$	75	71	–	80.5
cyclopropane	$+ 3CH_4 \rightarrow 3CH_3CH_3$	-24	-23	-23.5	-19.6
azacyclopropane	$+ 2CH_4 + NH_3 \rightarrow CH_3CH_3 + 2CH_3NH_2$	-18	-19	-17.9	-14.7

Table A6-42: Energies of Bond Separation Reactions. MP2 Models (3)

molecule	bond separation reaction	6-31G*	6-311+G**	G3	expt.
oxacyclopropane	(△O) + 2CH$_4$ + H$_2$O → CH$_3$CH$_3$ + 2CH$_3$OH	-13	-14	-14.1	-10.5
thiacyclopropane	(△S) + 2CH$_4$ + H$_2$S → CH$_3$CH$_3$ + 2CH$_3$SH	-11	-11	-11.5	-10.1
cyclobutane	(□) + 4CH$_4$ → 4CH$_3$CH$_3$	-19	-19	-19.3	-15.9
cyclopropene	(△) + 3CH$_4$ → 2CH$_3$CH$_3$ + CH$_2$CH$_2$	-45	-45	-46.6	-40.5
cyclobutene	(□) + 4CH$_4$ → 3CH$_3$CH$_3$ + CH$_2$CH$_2$	-18	-19	-19.5	-14.0
cyclopentadiene	+ 5CH$_4$ → 3CH$_3$CH$_3$ + 2CH$_2$CH$_2$	21	20	–	22.4
thiophene	(S) + 4CH$_4$ + H$_2$S → CH$_3$CH$_3$ + 2CH$_3$SH + 2CH$_2$CH$_2$	46	45	40.1	29.2
methylenecyclopropane	+ 4CH$_4$ → 3CH$_3$CH$_3$ + CH$_2$CH$_2$	-27	-27	-27.0	-24.5
bicyclo[1.1.0]butane	+ 6CH$_4$ → 5CH$_3$CH$_3$	-55	-55	-55.0	-45.5

Table A6-43: Energies of Bond Separation Reactions. Semi-Empirical Models

molecule	bond separation reaction	MNDO	AM1	PM3	G3	expt..
propane	$CH_3CH_2CH_3 + CH_4 \rightarrow 2CH_3CH_3$	-3	-2	0	2.3	2.6
isobutane	$CH(CH_3)_3 + 2CH_4 \rightarrow 3CH_3CH_3$	-9	-5	1	6.4	7.5
neopentane	$C(CH_3)_4 + 3CH_4 \rightarrow 4CH_3CH_3$	-18	-11	2	–	13.5
ethylsilane	$CH_3CH_2SiH_3 + CH_4 \rightarrow CH_3CH_3 + CH_3SiH_3$	0	-3	1	–	–
dimethylsilane	$SiH_2(CH_3)_2 + SiH_4 \rightarrow 2CH_3SiH_3$	-1	-1	0	–	0.8
trimethylsilane	$SiH(CH_3)_3 + 2SiH_4 \rightarrow 3CH_3SiH_3$	-2	-2	0	–	1.6
tetramethylsilane	$Si(CH_3)_4 + 3SiH_4 \rightarrow 4CH_3SiH_3$	-4	-5	-1	–	2.8
ethylamine	$CH_3CH_2NH_2 + CH_4 \rightarrow CH_3CH_3 + CH_3NH_2$	-2	-1	2	4.0	3.6
dimethylamine	$CH_3NHCH_3 + NH_3 \rightarrow 2CH_3NH_2$	-2	-2	1	4.2	4.4
trimethylamine	$(CH_3)_3N + 2NH_3 \rightarrow 3CH_3NH_2$	-7	-6	1	11.6	11.1
trans-ethanol	$CH_3CH_2OH + CH_4 \rightarrow CH_3CH_3 + CH_3OH$	-2	-3	0	5.4	5.7
dimethyl ether	$CH_3OCH_3 + H_2O \rightarrow 2CH_3OH$	-3	-2	-2	5.1	5.4
trans-ethanethiol	$CH_3CH_2SH + CH_4 \rightarrow CH_3CH_3 + CH_3SH$	-2	-2	-2	3.1	3.3
dimethyl sulfide	$CH_3SCH_3 + H_2S \rightarrow 2CH_3SH$	-1	-1	1	2.8	2.9
difluoromethane	$CH_2F_2 + CH_4 \rightarrow 2CH_3F$	2	3	9	–	13.9
trifluoromethane	$CHF_3 + 2CH_4 \rightarrow 3CH_3F$	5	7	27	–	34.6
tetrafluoromethane	$CF_4 + 3CH_4 \rightarrow 4CH_3F$	6	8	49	–	52.8
dichloromethane	$CH_2Cl_2 + CH_4 \rightarrow 2CH_3Cl$	-5	-3	1	0.9	1.5
trichloromethane	$CHCl_3 + 2CH_4 \rightarrow 3CH_3Cl$	-15	-10	3	0.6	1.9
tetrachloromethane	$CCl_4 + 3CH_4 \rightarrow 4CH_3Cl$	-29	-21	6	-2.4	-1.8
propene	$CH_3CHCH_2 + CH_4 \rightarrow CH_3CH_3 + CH_2CH_2$	3	1	5	4.9	5.4
acetaldehyde	$CH_3CHO + CH_4 \rightarrow CH_3CH_3 + H_2CO$	2	0	5	10.6	11.4
propyne	$CH_3CCH + CH_4 \rightarrow CH_3CH_3 + HCCH$	9	3	5	7.8	7.6
acetonitrile	$CH_3CN + CH_4 \rightarrow CH_3CH_3 + HCN$	8	3	5	11.0	14.7

Table A6-43: Energies of Bond Separation Reactions. Semi-Empirical Models (2)

molecule	bond separation reaction	MNDO	AM1	PM3	G3	expt.
allene	$CH_2CCH_2 + CH_4 \rightarrow 2CH_2CH_2$	-1	-4	-1	**-3.2**	**-2.8**
ketene	$CH_2CO + CH_4 \rightarrow CH_2CH_2 + H_2CO$	1	-1	5	**15.2**	**15.7**
carbon dioxide	$CO_2 + CH_4 \rightarrow 2H_2CO$	21	26	30	**61.3**	**59.9**
trans-1,3-butadiene	$CH_2CHCHCH_2 + 2CH_4 \rightarrow CH_3CH_3 + 2CH_2CH_2$	6	3	10	**13.2**	**14.2**
formamide	$NH_2CHO + CH_4 \rightarrow CH_3NH_2 + H_2CO$	11	15	14		**30.8**
benzene	[benzene ring] $+\ 6CH_4 \rightarrow 3CH_3CH_3 + 3CH_2CH_2$	37	28	50	**61.0**	**64.2**
pyridine	[pyridine ring] $+\ 5CH_4 + NH_3 \rightarrow 2CH_3CH_3 + 2CH_2CH_2 + CH_3NH_2 + CH_2NH$	41	29	51	–	71.9
pyridazine	[pyridazine ring] $+\ 4CH_4 + 2NH_3 \rightarrow 2CH_3CH_3 + CH_2CH_2 + 2CH_2NH + NH_2NH_2$	46	27	45	–	74.6
pyrimidine	[pyrimidine ring] $+\ 4CH_4 + 2NH_3 \rightarrow CH_3CH_3 + CH_2CH_2 + 2CH_3NH_2 + 2CH_2NH$	46	27	51	–	**80.4**
pyrazine	[pyrazine ring] $+\ 4CH_4 + 2NH_3 \rightarrow CH_3CH_3 + CH_2CH_2 + 2CH_3NH_2 + 2CH_2NH$	43	27	49	–	**80.5**
cyclopropane	[triangle] $+\ 3CH_4 \rightarrow 3CH_3CH_3$	-35	-44	-32	**-23.5**	**-19.6**
azacyclopropane	[triangle NH] $+\ 2CH_4 + NH_3 \rightarrow CH_3CH_3 + 2CH_3NH_2$	-30	-40	-31	**-17.9**	**-14.7**

Table A6-43: Energies of Bond Separation Reactions. Semi-Empirical Models (3)

molecule	bond separation reaction	MNDO	AM1	PM3	G3	expt..
oxacyclopropane	$\triangle$O + $2CH_4$ + H_2O → CH_3CH_3 + $2CH_3OH$	-34	-46	-34	**-14.1**	**-10.5**
thiacyclopropane	$\triangle$S + $2CH_4$ + H_2S → CH_3CH_3 + $2CH_3SH$	-24	-40	-31	**-11.5**	**-10.1**
cyclobutane	$\square$ + $4CH_4$ → $4CH_3CH_3$	1	-16	1	**-19.3**	**-15.9**
cyclopropene	$\triangle$ + $3CH_4$ → $2CH_3CH_3$ + CH_2CH_2	-57	-67	-49	**-46.6**	**-40.5**
cyclobutene	$\square$ + $4CH_4$ → $3CH_3CH_3$ + CH_2CH_2	-27	-45	-23	**-19.5**	**-14.0**
cyclopentadiene	+ $5CH_4$ → $3CH_3CH_3$ + $2CH_2CH_2$	-1	-12	12	–	**22.4**
thiophene	S + $4CH_4$ + H_2S → CH_3CH_3 + $2CH_3SH$ + $2CH_2CH_2$	22	13	18	**40.1**	**29.2**
methylenecyclopropane	+ $4CH_4$ → $3CH_3CH_3$ + CH_2CH_2	-34	-48	-30	**-27.0**	**-24.5**
bicyclo[1.1.0]butane	+ $6CH_4$ → $5CH_3CH_3$	-91	-113	-81	**-55.0**	**-45.5**

Table A6-44: Basis Set Effects on Energies of Bond Separation Reactions. Hartree-Fock Models

molecule	bond separation reaction	6-31G*	6-31G**	6-31+G*	6-311G*	6-311+G**	expt.
neopentane	$C(CH_3)_4 + 3CH_4 \rightarrow 4CH_3CH_3$	3	3	3	3	3	**13.5**
tetramethylsilane	$Si(CH_3)_4 + 3SiH_4 \rightarrow 4CH_3SiH_3$	0	0	0	0	0	**2.8**
difluoromethane	$CH_2F_2 + CH_4 \rightarrow 2CH_3F$	14	14	11	13	11	**13.9**
trifluoromethane	$CHF_3 + 2CH_4 \rightarrow 3CH_3F$	36	36	28	33	27	**34.6**
tetrafluoromethane	$CF_4 + 3CH_4 \rightarrow 4CH_3F$	58	57	43	51	41	**52.8**
trichloromethane	$CHCl_3 + 2CH_4 \rightarrow 3CH_3Cl$	-12	-12	-12	-13	-11	**1.9**
tetrachloromethane	$CCl_4 + 3CH_4 \rightarrow 4CH_3Cl$	-29	-26	-27	-27	-26	**-1.8**
benzene	$+ 6CH_4 \rightarrow 3CH_3CH_3 + 3CH_2CH_2$	58	58	56	55	55	**64.2**
bicyclo[1.1.0]butane	$+ 6CH_4 \rightarrow 5CH_3CH_3$	-63	-62	-62	-63	-60	**-45.5**

Table A6-45: Basis Set Effects on Energies of Bond Separation Reactions. EDF1 Density Functional Models

molecule	bond separation reaction	6-31G*	6-31G**	6-31+G*	6-311G*	6-311+G**	expt.
neopentane	$C(CH_3)_4 + 3CH_4 \rightarrow 4CH_3CH_3$	2	2	2	1	1	**13.5**
tetramethylsilane	$Si(CH_3)_4 + 3SiH_4 \rightarrow 4CH_3SiH_3$	-1	-1	-1	-1	-1	**2.8**
difluoromethane	$CH_2F_2 + CH_4 \rightarrow 2CH_3F$	16	15	12	15	11	**13.9**
trifluoromethane	$CHF_3 + 2CH_4 \rightarrow 3CH_3F$	40	39	29	36	28	**34.6**
tetrafluoromethane	$CF_4 + 3CH_4 \rightarrow 4CH_3F$	62	61	42	55	41	**52.8**
trichloromethane	$CHCl_3 + 2CH_4 \rightarrow 3CH_3Cl$	-4	-4	-4	-5	-3	**1.9**
tetrachloromethane	$CCl_4 + 3CH_4 \rightarrow 4CH_3Cl$	-14	-13	-14	-14	-12	**-1.8**
benzene	$+ 6CH_4 \rightarrow 3CH_3CH_3 + 3CH_2CH_2$	71	69	67	66	64	**64.2**
bicyclo[1.1.0]butane	$+ 6CH_4 \rightarrow 5CH_3CH_3$	-48	-47	-46	-48	-46	**-45.5**

681

Table A6-46: Basis Set Effects on Energies of Bond Separation Reactions. B3LYP Density Functional Models

molecule	bond separation reaction	6-31G*	6-31G**	6-31+G*	6-311G*	6-311+G**	expt.
neopentane	$C(CH_3)_4 + 3CH_4 \rightarrow 4CH_3CH_3$	4	5	5	4	5	**13.5**
tetramethylsilane	$Si(CH_3)_4 + 3SiH_4 \rightarrow 4CH_3SiH_3$	0	0	0	0	0	**2.8**
difluoromethane	$CH_2F_2 + CH_4 \rightarrow 2CH_3F$	15	15	11	14	11	**13.9**
trifluoromethane	$CHF_3 + 2CH_4 \rightarrow 3CH_3F$	38	38	27	35	27	**34.6**
tetrafluoromethane	$CF_4 + 3CH_4 \rightarrow 4CH_3F$	60	60	40	53	40	**52.8**
trichloromethane	$CHCl_3 + 2CH_4 \rightarrow 3CH_3Cl$	-6	-6	-6	-6	-5	**1.9**
tetrachloromethane	$CCl_4 + 3CH_4 \rightarrow 4CH_3Cl$	-16	-15	-15	-16	-14	**-1.8**
benzene	$+ 6CH_4 \rightarrow 3CH_3CH_3 + 3CH_2CH_2$	67	66	64	63	62	**64.2**
bicyclo[1.1.0]butane	$+ 6CH_4 \rightarrow 5CH_3CH_3$	-59	-58	-56	-58	-55	**-45.5**

Table A6-47: Basis Set Effects on Energies of Bond Separation Reactions. MP2 Models

molecule	bond separation reaction	6-31G*	6-31G**	6-31+G*	6-311G*	6-311+G**	expt.
neopentane	$C(CH_3)_4 + 3CH_4 \rightarrow 4CH_3CH_3$	3	11	12	13	12	13.5
tetramethylsilane	$Si(CH_3)_4 + 3SiH_4 \rightarrow 4CH_3SiH_3$	0	2	4	3	3	2.8
difluoromethane	$CH_2F_2 + CH_4 \rightarrow 2CH_3F$	16	15	11	15	12	13.9
trifluoromethane	$CHF_3 + 2CH_4 \rightarrow 3CH_3F$	41	40	28	39	31	34.6
tetrafluoromethane	$CF_4 + 3CH_4 \rightarrow 4CH_3F$	64	63	42	60	46	52.8
trichloromethane	$CHCl_3 + 2CH_4 \rightarrow 3CH_3Cl$	-3	-2	-1	-2	1	1.9
tetrachloromethane	$CCl_4 + 3CH_4 \rightarrow 4CH_3Cl$	-26	-7	-6	-7	-3	-1.8
benzene	$+ 6CH_4 \rightarrow 3CH_3CH_3 + 3CH_2CH_2$	71	70	69	68	68	64.2
bicyclo[1.1.0]butane	$+ 6CH_4 \rightarrow 5CH_3CH_3$	-55	-57	-53	-56	-55	-45.5

Table A6-48: Proton Affinites of Oxygen Bases Relative to Water[a]

base, B	Hartree-Fock			local density	BP	BLYP	EDF1	B3LYP	MP2	semi-empirical			expt.
	STO-3G	3-21G	6-31G*	6-31G*	6-31G*	6-31G*	6-31G*	6-31G*	6-31G*	MNDO	AM1	PM3	
methanol	11	14	17	11	14	16	15	15	16	4	15	11	15.1
oxirane	5	12	21	10	16	18	17	17	17	7	20	17	19.8
ethanol	17	18	22	17	21	22	22	21	21	6	21	18	20.4
dimethyl ether	16	22	26	15	21	23	22	23	23	6	20	14	24.1
isopropyl alcohol	22	22	27	21	26	27	27	26	25	8	25	23	24.3
1,4-dioxane	21	21	28	18	24	26	25	25	24	6	21	16	25.4
oxetane	20	29	31	18	23	25	24	26	24	10	19	11	26.1
tert-butyl alcohol	26	25	31	25	30	32	31	30	28	11	30	28	26.6
tetrahydrofuran	26	28	33	23	29	31	30	30	30	11	29	23	31.3
tetrahydropyran	29	29	35	25	31	33	32	32	31	11	28	22	31.5
1,3-dioxane	23	22	29	22	28	31	30	28	27	8	29	23	32.1
oxepane	31	31	37	28	34	36	35	34	33	11	32	27	34.2
diisopropyl ether	37	35	40	33	38	40	40	39	37	15	38	34	39.3
di (tert-butyl) ether	46	40	49	40	47	49	49	47	44	18	44	40	46.9
mean absolute error	4	4	3	6	2	2	2	1	2	19	3	7	–

a) energy of reaction: $BH^+ + H_2O \rightarrow B + H_3O^+$

Table A6-49: Proton Affinites of Carbonyl Compounds Relative to Formaldehyde[a]

base, B	Hartree-Fock			local density	BP	BLYP	EDF1	B3LYP	MP2	semi-empirical			expt.
	STO-3G	3-21G	6-31G*	6-31G*	6-31G*	6-31G*	6-31G*	6-31G*	6-31G*	MNDO	AM1	PM3	
acetaldehyde	15	13	13	18	16	16	16	15	13	7	10	9	13.3
propanal	18	14	15	21	19	19	20	18	15	7	11	11	17.5
cyclobutanone	25	18	20	25	24	24	24	23	19	10	12	13	21.4
acetone	28	22	22	29	27	27	27	26	23	12	18	18	23.7
1,4-cyclohexanedione	27	18	18	25	25	25	25	23	19	8	13	13	23.8
cyclopentanone	31	25	26	31	30	30	30	29	26	12	18	18	26.5
7-norbornanone	32	25	26	30	29	30	30	29	26	12	15	15	28.5
benzaldehyde	36	28	29	34	35	36	36	35	29	24	23	23	28.9
cyclohexanone	34	29	28	35	35	35	35	33	29	14	21	20	30.6
cycloheptanone	37	32	31	37	36	36	36	35	32	15	23	22	31.7
2-norbornanone	34	30	29	36	35	35	35	33	30	14	19	19	32.1
camphor	37	32	32	39	38	38	39	36	33	16	21	21	35.0
acetophenone	44	33	34	40	40	41	41	40	35	23	29	29	35.4
benzophenone	51	40	41	48	49	50	49	48	43	30	35	34	40.5
mean absolute error	4	2	2	4	3	4	4	2	2	12	9	9	–

a) energy of reaction: $BH^+ + H_2CO \rightarrow B + H_2COH^+$

Table A6-50: Acidities of *p*-Substituted Phenols Relative to Phenol[a]

p substituent, X	Hartree-Fock STO-3G	Hartree-Fock 3-21G	Hartree-Fock 6-31G*	local density 6-31G*	BP 6-31G*	BLYP 6-31G*	EDF1 6-31G*	B3LYP 6-31G*	MP2 6-31G*	semi-empirical MNDO	semi-empirical AM1	semi-empirical PM3	expt.
NH_2	-2	-7	-4	-5	-4	-4	-4	-4	-3	3	0	0	**-3.3**
CH_3	0	-1	-1	0	0	0	0	0	0	3	1	0	**-1.1**
F	2	5	3	2	3	3	3	3	4	7	6	6	**2.4**
Cl	15	10	9	8	9	9	9	9	8	9	7	6	**6.1**
SOMe	15	15	16	14	14	14	14	14	14	19	15	15	**11.7**
CF_3	15	18	14	13	14	14	14	14	14	19	16	16	**12.2**
CN	24	21	21	21	21	21	21	21	20	16	16	16	**17.0**
SO_2Me	24	24	23	21	20	20	20	21	21	32	26	22	**18.1**
NO_2	31	31	28	26	27	27	27	27	24	28	27	27	**21.4**
mean absolute error	4	4	3	2	2	2	2	2	2	6	4	3	–

a) energy of reaction: $X-C_6H_4O^- + C_6H_5OH \rightarrow X-C_6H_4OH + C_6H_5O^-$

Table A7-1: Vibrational Frequencies for One-Heavy-Atom Hydrides. Hartree-Fock Models

molecule	symmetry	description	STO-3G	3-21G	6-31G*	6-311+G**	expt. meas.	harmonic
LiH			1865	1426	1416	1429	**1360**	**1406**
CH$_4$	a$_1$	s-stretch	3527	3187	3197	3149	**2917**	**3137**
	e	deform	1904	1740	1703	1666	**1534**	**1567**
	t$_2$	stretch	3788	3281	3301	3251	**3019**	**3158**
		deform	1675	1520	1488	1453	**1306**	**1357**
NH$_3$	a$_1$	s-stretch	3833	3644	3688	3697	**3337**	**3506**
		deform	1413	857	1208	1091	**950**	**1022**
	e	stretch	4108	3800	3821	3827	**3444**	**3577**
		deform	2077	1858	1850	1791	**1627**	**1691**
H$_2$O	a$_1$	s-stretch	4141	3813	4070	4142	**3657**	**3832**
		bend	2170	1799	1827	1726	**1595**	**1648**
	b$_1$	a-stretch	4392	3946	4188	4244	**3756**	**3943**
HF			4475	4065	4357	4490	**3962**	**4139**
NaH			2175	1168	1198	1180	**1172**	
SiH$_4$	a$_1$	s-stretch	2862	2406	2393	2351	**2187**	**2337**
	e	deform	1267	1057	1051	1051	**975**	**975**
	t$_2$	stretch	2929	2397	2385	2337	**2191**	**2319**
		deform	1170	1021	1016	1019	**914**	**945**
PH$_3$	a$_1$	s-stretch	3040	2613	2610	2555	**2323**	**2452**
		deform	1427	1148	1139	1112	**992**	**1041**
	e	stretch	3090	2597	2605	2550	**2328**	**2457**
		deform	1543	1290	1271	1245	**1118**	**1154**
H$_2$S	a$_1$	s-stretch	3275	2903	2918	2869	**2615**	**2722**
		bend	1610	1381	1368	1320	**1183**	**1215**
	b$_1$	a-stretch	3323	2906	2929	2878	**2626**	**2733**
HCl			3373	3152	3186	3146	**2890**	–
KH			1633	987	973	–	**985**	–
GeH$_4$	a$_1$	s-stretch	2504	2231	2248	–	**2106**	–
	e	deform	1198	991	992	–	**931**	–
	t$_2$	stretch	2431	2221	2234	–	**2114**	–
		deform	1054	905	904	–	**819**	–
AsH$_3$	a$_1$	s-stretch	2724	2350	2353	–	**2116**	–
		deform	1237	1016	1025	–	**906**	–
	e	stretch	2703	2351	2358	–	**2123**	–
		deform	1362	1118	1125	–	**1003**	–
H$_2$Se	a$_1$	s-stretch	2915	2591	2601	–	**2345**	–
		bend	1374	1192	1187	–	**1034**	–
	b$_1$	a-stretch	2917	2599	2613	–	**2358**	–
HBr			3120	2804	2817	–	**2649**	–

Table A7-2: Vibrational Frequencies for One-Heavy-Atom Hydrides. Local Density Models

molecule	symmetry	description	6-31G*	6-311+G**	expt. meas.	expt. harmonic
H_2			4027	4185	**4160**	**4401**
LiH			1353	1366	**1360**	**1406**
CH_4	a_1	s-stretch	2988	2959	**2917**	**3137**
	e	deform	1526	1482	**1534**	**1567**
	t_2	stretch	3121	3089	**3019**	**3158**
		deform	1295	1256	**1306**	**1357**
NH_3	a_1	s-stretch	3372	3411	**3337**	**3506**
		deform	1047	914	**950**	**1022**
	e	stretch	3519	3554	**3444**	**3577**
		deform	1655	1582	**1627**	**1697**
H_2O	a_1	s-stretch	3656	3728	**3657**	**3832**
		bend	1646	1514	**1595**	**1648**
	b_1	a-stretch	3788	3843	**3756**	**3943**
HF			3911	3996	**3962**	**4139**
NaH			1171	1156	**1172**	–
SiH_4	a_1	s-stretch	2182	2165	**2187**	**2337**
	e	deform	928	931	**975**	**975**
	t_2	stretch	2207	2189	**2191**	**2319**
		deform	840	841	**914**	**945**
PH_3	a_1	s-stretch	2330	2321	**2323**	**2452**
		deform	991	949	**992**	**1041**
	e	stretch	2357	2340	**2328**	**2457**
		deform	1115	1083	**1118**	**1154**
H_2S	a_1	s-stretch	2622	2617	**2615**	**2722**
		bend	1188	1134	**1183**	**1215**
	b_1	a-stretch	2649	2638	**2626**	**2733**
HCl			2873	2879	**2890**	–
KH			984	–	**985**	–
GeH_4	a_1	s-stretch	2093	–	**2106**	–
	e	deform	892	–	**931**	–
	t_2	stretch	2123	–	**2114**	–
		deform	761	–	**819**	–
AsH_3	a_1	s-stretch	2143	–	**2116**	–
		deform	907	–	**906**	–
	e	stretch	2176	–	**2123**	–
		deform	994	–	**1003**	–
H_2Se	a_1	s-stretch	2349	–	**2345**	–
		bend	1045	–	**1034**	–
	b_1	a-stretch	2369	–	**2358**	–
HBr			2566	–	**2649**	–

Table A7-3: Vibrational Frequencies for One-Heavy-Atom Hydrides. BP Density Functional Models

molecule	symmetry	description	6-31G*	6-311+G**	expt. meas.	expt. harmonic
H_2			4339	4317	**4160**	**4401**
LiH			1356	1368	**1360**	**1406**
CH_4	a_1	s-stretch	2981	2958	**2917**	**3137**
	e	deform	1549	1571	**1534**	**1567**
	t_2	stretch	3097	3071	**3019**	**3158**
		deform	1326	1292	**1306**	**1357**
NH_3	a_1	s-stretch	3329	3381	**3337**	**3506**
		deform	1115	996	**950**	**1022**
	e	stretch	3467	3511	**3444**	**3577**
		deform	1685	1623	**1627**	**1697**
H_2O	a_1	s-stretch	3612	3700	**3657**	**3832**
		bend	1679	1568	**1595**	**1648**
	b_1	a-stretch	3739	3608	**3756**	**3943**
HF			3870	3976	**3962**	**4139**
NaH			1144	1132	**1172**	–
SiH_4	a_1	s-stretch	2190	2173	**2187**	**2337**
	e	deform	943	946	**975**	**975**
	t_2	stretch	2208	2190	**2191**	**2319**
		deform	873	875	**914**	**945**
PH_3	a_1	s-stretch	2328	2316	**2323**	**2452**
		deform	1017	984	**992**	**1041**
	e	stretch	2349	2330	**2328**	**2457**
		deform	1130	1104	**1118**	**1154**
H_2S	a_1	s-stretch	2616	2610	**2615**	**2722**
		bend	1210	1164	**1183**	**1215**
	b_1	a-stretch	2641	2629	**2626**	**2733**
HCl			2868	2872	**2890**	–
KH			957	–	**985**	–
GeH_4	a_1	s-stretch	2070	–	**2106**	–
	e	deform	893	–	**931**	–
	t_2	stretch	2096	–	**2114**	–
		deform	780	–	**819**	–
AsH_3	a_1	s-stretch	2116	–	**2116**	–
		deform	924	–	**906**	–
	e	stretch	2144	–	**2123**	–
		deform	999	–	**1003**	–
H_2Se	a_1	s-stretch	2328	–	**2345**	–
		bend	1056	–	**1034**	–
	b_1	a-stretch	2346	–	**2358**	–
HBr			2548	–	**2649**	–

Table A7-4: **Vibrational Frequencies for One-Heavy-Atom Hydrides. BLYP Density Functional Models**

molecule	symmetry	description	6-31G*	6-311+G**	expt. meas.	expt. harmonic
H_2			4373	4344	**4160**	**4401**
LiH			1372	1384	**1360**	**1406**
CH_4	a_1	s-stretch	2980	2956	**2917**	**3137**
	e	deform	1557	1522	**1534**	**1567**
	t_2	stretch	3083	3054	**3019**	**3158**
		deform	1340	1310	**1306**	**1357**
NH_3	a_1	s-stretch	3305	3369	**3337**	**3506**
		deform	1129	995	**950**	**1022**
	e	stretch	3433	3492	**3444**	**3577**
		deform	1690	1631	**1627**	**1697**
H_2O	a_1	s-stretch	3567	3673	**3657**	**3832**
		bend	1682	1571	**1595**	**1648**
	b_1	a-stretch	3689	3778	**3756**	**3943**
HF			3809	3938	**3962**	**4139**
NaH			1153	1142	**1172**	–
SiH_4	a_1	s-stretch	2189	2178	**2187**	**2337**
	e	deform	950	956	**975**	**975**
	t_2	stretch	2206	2191	**2191**	**2319**
		deform	887	892	**914**	**945**
PH_3	a_1	s-stretch	2311	2303	**2323**	**2452**
		deform	1031	1000	**992**	**1041**
	e	stretch	2330	1112	**2328**	**2457**
		deform	1135	2313	**1118**	**1154**
H_2S	a_1	s-stretch	2587	2584	**2615**	**2722**
		bend	1214	1175	**1183**	**1215**
	b_1	a-stretch	2610	2601	**2626**	**2733**
HCl			2832	2840	**2890**	–
KH			947	–	**985**	–
GeH_4	a_1	s-stretch	2048	–	**2106**	–
	e	deform	891	–	**931**	–
	t_2	stretch	2073	–	**2114**	–
		deform	785	–	**819**	–
AsH_3	a_1	s-stretch	2086	–	**2116**	–
		deform	931	–	**906**	–
	e	stretch	2113	–	**2123**	–
		deform	997	–	**1003**	–
H_2Se	a_1	s-stretch	2295	–	**2345**	–
		bend	1055	–	**1034**	–
	b_1	a-stretch	2312	–	**2358**	–
HBr			2507	–	**2649**	–

Table A7-5: Vibrational Frequencies for One-Heavy-Atom Hydrides. EDF1 Density Functional Models

molecule	symmetry	description	6-31G*	6-311+G**	expt. meas.	expt. harmonic
H_2			4409	4376	**4160**	**4401**
LiH			1361	1372	**1360**	**1406**
CH_4	a_1	s-stretch	3014	2993	**2917**	**3137**
	e	deform	1565	1529	**1534**	**1567**
	t_2	stretch	3134	3106	**3019**	**3158**
		deform	1343	1310	**1306**	**1357**
NH_3	a_1	s-stretch	3372	3427	**3337**	**3506**
		deform	1134	1011	**950**	**1022**
	e	stretch	3512	3558	**3444**	**3577**
		deform	1704	1643	**1627**	**1697**
H_2O	a_1	s-stretch	3654	3751	**3657**	**3832**
		bend	1699	1591	**1595**	**1648**
	b_1	a-stretch	3780	3859	**3756**	**3943**
HF			3899	4025	**3962**	**4139**
NaH			1139	1126	**1172**	–
SiH_4	a_1	s-stretch	2210	2191	**2187**	**2337**
	e	deform	950	955	**975**	**975**
	t_2	stretch	2227	2206	**2191**	**2319**
		deform	885	889	**914**	**945**
PH_3	a_1	s-stretch	2405	2346	**2323**	**2452**
		deform	1054	997	**992**	**1041**
	e	stretch	2420	2360	**2328**	**2457**
		deform	1170	1117	**1118**	**1154**
H_2S	a_1	s-stretch	2657	2645	**2615**	**2722**
		bend	1226	1180	**1183**	**1215**
	b_1	a-stretch	2681	2664	**2626**	**2733**
HCl			2909	2915	**2890**	–
KH			977	–	**985**	–
GeH_4	a_1	s-stretch	2080	–	**2106**	–
	e	deform	899	–	**931**	–
	t_2	stretch	2105	–	**2114**	–
		deform	790	–	**819**	–
AsH_3	a_1	s-stretch	2135	–	**2116**	–
		deform	936	–	**906**	–
	e	stretch	2163	–	**2123**	–
		deform	1009	–	**1003**	–
H_2Se	a_1	s-stretch	2355	–	**2345**	–
		bend	1068	–	**1034**	–
	b_1	a-stretch	2374	–	**2358**	–
HBr			2579	–	**2649**	–

Table A7-6: Vibrational Frequencies for One-Heavy-Atom Hydrides. B3LYP Density Functional Models

molecule	symmetry	description	6-31G*	6-311+G**	expt. meas.	expt. harmonic
H_2			4453	4419	**4160**	**4401**
LiH			1400	1414	**1360**	**1406**
CH_4	a_1	s-stretch	3053	3024	**2917**	**3137**
	e	deform	1593	1557	**1534**	**1567**
	t_2	stretch	3162	3129	**3019**	**3158**
		deform	1374	1341	**1306**	**1357**
NH_3	a_1	s-stretch	3435	3480	**3337**	**3506**
		deform	1133	1001	**950**	**1022**
	e	stretch	3566	3608	**3444**	**3577**
		deform	1727	1667	**1627**	**1697**
H_2O	a_1	s-stretch	3727	3817	**3657**	**3832**
		bend	1713	1603	**1595**	**1648**
	b_1	a-stretch	3848	3922	**3756**	**3943**
HF			3977	4098	**3962**	**4139**
NaH			1188	1179	**1172**	–
SiH_4	a_1	s-stretch	2255	2236	**2187**	**2337**
	e	deform	975	980	**975**	**975**
	t_2	stretch	2268	2244	**2191**	**2319**
		deform	916	921	**914**	**945**
PH_3	a_1	s-stretch	2405	2387	**2323**	**2452**
		deform	1054	1021	**992**	**1041**
	e	stretch	2420	2395	**2328**	**2457**
		deform	1170	1145	**1118**	**1154**
H_2S	a_1	s-stretch	2691	2676	**2615**	**2722**
		bend	1250	1208	**1183**	**1215**
	b_1	a-stretch	2712	2691	**2626**	**2733**
HCl			2937	2934	**2890**	–
KH			977	–	**985**	–
GeH_4	a_1	s-stretch	2122	–	**2106**	–
	e	deform	919	–	**931**	–
	t_2	stretch	2138	–	**2114**	–
		deform	813	–	**819**	–
AsH_3	a_1	s-stretch	2175	–	**2116**	–
		deform	954	–	**906**	–
	e	stretch	2198	–	**2123**	–
		deform	1032	–	**1003**	–
H_2Se	a_1	s-stretch	2394	–	**2345**	–
		bend	1088	–	**1034**	–
	b_1	a-stretch	2411	–	**2358**	–
HBr			2610	–	**2649**	–

Table A7-7: Vibrational Frequencies for One-Heavy-Atom Hydrides. MP2 Models

molecule	symmetry	description	6-31G*	6-311+G**	expt. meas.	expt. harmonic
H_2			4534	4533	**4160**	**4401**
LiH			1395	1437	**1360**	**1406**
CH_4	a_1	s-stretch	3111	3076	**2917**	**3137**
	e	deform	1625	1575	**1534**	**1567**
	t_2	stretch	3248	3213	**3019**	**3158**
		deform	1414	1362	**1306**	**1357**
NH_3	a_1	s-stretch	3503	3530	**3337**	**3506**
		deform	1163	1062	**950**	**1022**
	e	stretch	3658	3682	**3444**	**3577**
		deform	1755	1661	**1627**	**1697**
H_2O	a_1	s-stretch	3775	3883	**3657**	**3832**
		bend	1735	1629	**1595**	**1648**
	b_1	a-stretch	3916	4002	**3756**	**3943**
HF			4043	4199	**3962**	**4139**
NaH			1182	1187	**1172**	**–**
SiH_4	a_1	s-stretch	2323	2329	**2187**	**2337**
	e	deform	1005	1013	**975**	**975**
	t_2	stretch	2336	2333	**2191**	**2319**
		deform	957	973	**914**	**945**
PH_3	a_1	s-stretch	2510	2514	**2323**	**2452**
		deform	1080	1049	**992**	**1041**
	e	stretch	2526	2522	**2328**	**2457**
		deform	1182	1166	**1118**	**1154**
H_2S	a_1	s-stretch	2797	2818	**2615**	**2722**
		bend	1279	1232	**1183**	**1215**
	b_1	a-stretch	2825	2837	**2626**	**2733**
HCl			3048	3087	**2890**	**–**
KH			951	–	**985**	**–**
GeH_4	a_1	s-stretch	2177	–	**2106**	**–**
	e	deform	942	–	**931**	**–**
	t_2	stretch	2183	–	**2114**	**–**
		deform	943	–	**819**	**–**
AsH_3	a_1	s-stretch	2260	–	**2116**	**–**
		deform	968	–	**906**	**–**
	e	stretch	2281	–	**2123**	**–**
		deform	1060	–	**1003**	**–**
H_2Se	a_1	s-stretch	2486	–	**2345**	**–**
		bend	1111	–	**1034**	**–**
	b_1	a-stretch	2510	–	**2358**	**–**
HBr			2694	–	**2649**	**–**

Table A7-8: **Vibrational Frequencies for One-Heavy-Atom Hydrides. Semi-Empirical Models**

molecule	symmetry	description	MNDO	AM1	PM3	expt. meas.	harmonic
H_2			4293	4341	4478	**4160**	**4401**
LiH			1466	-	1476	**1360**	**1406**
CH_4	a_1	s-stretch	2556	3215	3311	**2917**	**3137**
	e	deform	1819	1412	1451	**1534**	**1567**
	t_2	stretch	2586	3103	3208	**3019**	**3158**
		deform	1958	1380	1363	**1306**	**1357**
NH_3	a_1	s-stretch	3635	3536	3662	**3337**	**3506**
		deform	1479	1151	1401	**950**	**1022**
	e	stretch	3572	3464	3460	**3444**	**3577**
		deform	1848	1765	1759	**1627**	**1697**
H_2O	a_1	s-stretch	4084	3585	3990	**3657**	**3832**
		bend	1957	1882	1739	**1595**	**1648**
	b_1	a-stretch	4048	3505	3870	**3756**	**3943**
HF			4594	4460	4352	**3962**	**4139**
NaH			-	-	1004	**1172**	**–**
SiH_4	a_1	s-stretch	1862	2261	1937	**2187**	**2337**
	e	deform	1320	854	930	**975**	**975**
	t_2	stretch	1881	2255	1910	**2191**	**2319**
		deform	1420	770	801	**914**	**945**
PH_3	a_1	s-stretch	2734	2496	1549	**2323**	**2452**
		deform	1217	1012	923	**992**	**1041**
	e	stretch	2692	2460	1606	**2328**	**2457**
		deform	1210	1035	979	**1118**	**1154**
H_2S	a_1	s-stretch	3005	2074	1813	**2615**	**2722**
		bend	1219	1199	1192	**1183**	**1215**
	b_1	a-stretch	2989	2038	1809	**2626**	**2733**
HCl			3162	2658	2704	**2890**	**–**
GeH_4	a_1	s-stretch	1620	1941	2009	**2106**	**–**
	e	deform	1127	675	745	**931**	**–**
	t_2	stretch	1639	1955	1898	**2114**	**–**
		deform	1211	614	679	**819**	**–**
AsH_3	a_1	s-stretch	–	–	2200	**2116**	**–**
		deform	–	–	862	**906**	**–**
	e	stretch	–	–	2213	**2123**	**–**
		deform	–	–	878	**1003**	**–**
H_2Se	a_1	s-stretch	–	–	1607	**2345**	**–**
		bend	–	–	885	**1034**	**–**
	b_1	a-stretch	–	–	1612	**2358**	**–**
HBr			2776	2439	2122	**2649**	**–**

Table A7-9: Vibrational Frequencies for CH₃X Molecules. Hartree-Fock Models

molecule	symmetry of vibration	description of mode	STO-3G	3-21G	6-31G*	6-311+G**	expt.
CH_3CH_3	a_{1g}	CH_3 s-stretch	3570	3198	3203	3162	**2954**
		CH_3 s-deform	1763	1571	1580	1550	**1388**
		CC stretch	1194	1002	1060	1052	**995**
	a_{1u}	torsion	318	318	330	331	**289**
	a_{2u}	CH_3 s-stretch	3569	3194	3197	3154	**2986**
		CH_3 s-deform	1716	1580	1548	1517	**1379**
	e_g	CH_3 d-stretch	3746	3239	3247	3201	**2969**
		CH_3 d-deform	1848	1678	1650	1618	**1468**
		CH_3 rock	1460	1352	1338	1320	**1190**
	e_u	CH_3 d-stretch	3758	3265	3272	3227	**2985**
		CH_3 d-deform	1832	1677	1644	1615	**1469**
		CH_3 rock	981	922	891	881	**822**
CH_3NH_2	a'	NH_2 s-stretch	3901	3677	3730	3737	**3361**
		CH_3 d-stretch	3710	3230	3246	3201	**2961**
		CH_3 s-stretch	3540	3135	3158	3122	**2820**
		NH_2 scis.	2023	1852	1841	1801	**1623**
		CH_3 d-deform	1822	1680	1648	1618	**1473**
		CH_3 s-deform	1749	1615	1608	1582	**1430**
		CH_3 rock	1387	1257	1290	1265	**1130**
		CN stretch	1260	1096	1149	1138	**1044**
		NH_2 wag	1056	759	947	900	**780**
	a"	NH_2 a-stretch	4085	3773	3812	3816	**3427**
		CH_3 d-stretch	3743	3264	3282	3235	**2985**
		CH_3 d-deform	1834	1702	1665	1635	**1485**
		NH_2 twist	1611	1457	1480	1454	**1419**
		CH_3 rock	1118	1047	1052	1041	**1195**
		torsion	355	317	342	323	**268**
CH_3OH	a'	OH stretch	4229	3868	4117	4192	**3681**
		CH_3 d-stretch	3715	3294	3306	3261	**3000**
		CH_3 s-stretch	3517	3179	3186	3147	**2844**
		CH_3 d-deform	1830	1698	1663	1629	**1477**
		CH_3 s-deform	1769	1638	1638	1609	**1455**
		OH bend	1723	1480	1508	1472	**1345**
		CH_3 rock	1316	1153	1189	1173	**1060**
		CO stretch	1208	1090	1164	1148	**1033**
	a"	CH_3 d-stretch	3681	3218	3232	3194	**2960**
		CH_3 d-deform	1815	1686	1652	1620	**1477**
		CH_3 rock	1308	1254	1290	1278	**1165**
		torsion	398	360	348	318	**295**

Table A7-9: Vibrational Frequencies for CH₃X Molecules. Hartree-Fock Models (2)

molecule	symmetry of vibration	description of mode	STO-3G	3-21G	6-31G*	6-311+G**	expt.
CH_3F	a_1	CH_3 s-stretch	3511	3229	3233	3193	**2930**
		CH_3 s-deform	1784	1663	1652	1611	**1464**
		CF stretch	1355	1141	1187	1156	**1049**
	e	CH_3 d-stretch	3684	3294	3314	3277	**3006**
		CH_3 d-deform	1808	1686	1653	1615	**1467**
		CH_3 rock	1329	1278	1312	1296	**1182**
CH_3SiH_3	a_1	CH_3 s-stretch	3573	3174	3196	3156	**2898**
		SiH_3 s-stretch	2871	2387	2376	2332	**2169**
		CH_3 s-deform	1683	1486	1455	1413	**1260**
		SiH_3 s-deform	1194	1049	1040	1038	**940**
		CSi stretch	870	738	725	723	**700**
	a_2	torsion	182	195	194	197	**187**
	e	CH_3 d-stretch	3755	3239	3268	3228	**2982**
		SiH_3 d-stretch	2920	2372	2363	2314	**2166**
		CH_3 d-deform	1853	1634	1607	1575	**1403**
		SiH_3 d-deform	1241	1053	1045	1044	**980**
		CH_3 rock	1122	992	971	956	**869**
		SiH_3 rock	696	579	562	559	**540**
CH_3SH	a'	CH_3 d-stretch	3731	3306	3325	3282	**3000**
		CH_3 s-stretch	3554	3224	3238	3196	**2931**
		SH stretch	3275	2892	2910	2863	**2572**
		CH_3 s-deform	1841	1662	1635	1606	**1475**
		CH_3 d-deform	1711	1540	1521	1491	**1319**
		CH_3 rock	1355	1232	1220	1199	**976**
		SH bend	1002	890	872	860	**803**
		CS stretch	953	744	776	764	**708**
	a"	CH_3 d-stretch	3748	3299	3325	3281	**3000**
		CH_3 d-deform	1830	1653	1623	1592	**1430**
		CH_3 rock	1169	1089	1080	1062	**1074**
		torsion	263	257	261	256	**–**
CH_3Cl	a_1	CH_3 s-stretch	3569	3257	3264	3224	**2937**
		CH_3 s-deform	1651	1545	1539	1511	**1355**
		CCl stretch	936	716	782	773	**732**
	e	CH_3 d-stretch	3769	3354	3367	3326	**3039**
		CH_3 d-deform	1772	1655	1629	1599	**1452**
		CH_3 rock	1188	1139	1139	1121	**1017**

Table A7-10: Vibrational Frequencies for CH₃X Molecules. Local Density Models

molecule	symmetry of vibration	description of mode	6-31G*	6-311+G**	expt.
CH₃CH₃	a₁g	CH₃ s-stretch	2990	2969	**2954**
		CH₃ s-deform	1395	1360	**1388**
		CC stretch	1042	1028	**995**
	a₁u	torsion	317	311	**289**
	a₂u	CH₃ s-stretch	2995	2972	**2986**
		CH₃ s-deform	1364	1337	**1379**
	eg	CH₃ d-stretch	3059	3032	**2969**
		CH₃ d-deform	1467	1432	**1468**
		CH₃ rock	1182	1164	**1190**
	eu	CH₃ d-stretch	3083	3056	**2985**
		CH₃ d-deform	1470	1433	**1469**
		CH₃ rock	804	796	**822**
CH₃NH₂	a'	NH₂ s-stretch	3420	3457	**3361**
		CH₃ d-stretch	3015	2997	**2961**
		CH₃ s-stretch	2889	2888	**2820**
		NH₂ scis.	1638	1596	**1623**
		CH₃ d-deform	1463	1426	**1473**
		CH₃ s-deform	1423	1391	**1430**
		CH₃ rock	1155	1123	**1130**
		CN stretch	1101	1090	**1044**
		NH₂ wag	820	751	**780**
	a"	NH₂ a-stretch	3511	3542	**3427**
		CH₃ d-stretch	3062	3040	**2985**
		CH₃ d-deform	1482	1447	**1485**
		NH₂ twist	1315	1283	**1419**
		CH₃ rock	958	939	**1195**
		torsion	334	303	**268**
CH₃OH	a'	OH stretch	3693	3767	**3681**
		CH₃ d-stretch	3065	3049	**3000**
		CH₃ s-stretch	2913	2913	**2844**
		CH₃ d-deform	1480	1437	**1477**
		CH₃ s-deform	1450	1413	**1455**
		OH bend	1346	1298	**1345**
		CH₃ rock	1129	1096	**1060**
		CO stretch	1047	1022	**1033**
	a"	CH₃ d-stretch	2958	2965	**2960**
		CH₃ d-deform	1456	1422	**1477**
		CH₃ rock	1148	1129	**1165**
		torsion	355	305	**295**

Table A7-10: Vibrational Frequencies for CH₃X Molecules. Local Density Models (2)

molecule	symmetry of vibration	description of mode	6-31G*	6-311+G**	expt.
CH₃F	a₁	CH₃ s-stretch	2953	2954	**2930**
		CH₃ s-deform	1476	1414	**1464**
		CF stretch	1138	1075	**1049**
	e	CH₃ d-stretch	3030	3045	**3006**
		CH₃ d-deform	1456	1417	**1467**
		CH₃ rock	1169	1144	**1182**
CH₃SiH₃	a₁	CH₃ s-stretch	3006	2983	**2898**
		SiH₃ s-stretch	2179	2166	**2169**
		CH₃ s-deform	1261	1229	**1260**
		SiH₃ s-deform	886	881	**940**
		CSi stretch	704	705	**700**
	a₂	torsion	198	195	**187**
	e	CH₃ d-stretch	3101	3073	**2982**
		SiH₃ d-stretch	2192	2177	**2166**
		CH₃ d-deform	1426	1388	**1403**
		SiH₃ d-deform	909	910	**980**
		CH₃ rock	864	852	**869**
		SiH₃ rock	492	495	**540**
CH₃SH	a'	CH₃ d-stretch	3110	3083	**3000**
		CH₃ s-stretch	3009	2987	**2931**
		SH stretch	2621	2614	**2572**
		CH₃ s-deform	1449	1409	**1475**
		CH₃ d-deform	1337	1308	**1319**
		CH₃ rock	1075	1055	**976**
		SH bend	790	783	**803**
		CS stretch	721	714	**708**
	a"	CH₃ d-stretch	3102	3076	**3000**
		CH₃ d-deform	1439	1396	**1430**
		CH₃ rock	956	943	**1074**
		torsion	253	241	**–**
CH₃Cl	a₁	CH₃ s-stretch	3022	3006	**2937**
		CH₃ s-deform	1359	1330	**1355**
		CCl stretch	732	751	**732**
	e	CH₃ d-stretch	3126	3102	**3039**
		CH₃ d-deform	1444	1404	**1452**
		CH₃ rock	1011	994	**1017**

698

Table A7-11: Vibrational Frequencies for CH$_3$X Molecules. BP Density Functional Models

molecule	symmetry of vibration	description of mode	6-31G*	6-311+G**	expt.
CH$_3$CH$_3$	a$_{1g}$	CH$_3$ s-stretch	2971	2956	2954
		CH$_3$ s-deform	1406	1374	1388
		CC stretch	995	980	995
	a$_{1u}$	torsion	307	300	289
	a$_{2u}$	CH$_3$ s-stretch	2973	2957	2986
		CH$_3$ s-deform	1384	1359	1379
	e$_g$	CH$_3$ d-stretch	3027	3007	2969
		CH$_3$ d-deform	1486	1456	1468
		CH$_3$ rock	1195	1177	1190
	e$_u$	CH$_3$ d-stretch	3052	3032	2985
		CH$_3$ d-deform	1491	1457	1469
		CH$_3$ rock	809	802	822
CH$_3$NH$_2$	a'	NH$_2$ s-stretch	3359	3409	3361
		CH$_3$ d-stretch	2999	2985	2961
		CH$_3$ s-stretch	2878	2881	2820
		NH$_2$ scis.	1657	1620	1623
		CH$_3$ d-deform	1480	1448	1473
		CH$_3$ s-deform	1436	1409	1430
		CH$_3$ rock	1159	1127	1130
		CN stretch	1045	1031	1044
		NH$_2$ wag	856	807	780
	a"	NH$_2$ a-stretch	3446	3489	3427
		CH$_3$ d-stretch	3042	3024	2985
		CH$_3$ d-deform	1500	1469	1485
		NH$_2$ twist	1332	1302	1419
		CH$_3$ rock	962	946	1195
		torsion	333	305	268
CH$_3$OH	a'	OH stretch	3632	3723	3681
		CH$_3$ d-stretch	3051	3041	3000
		CH$_3$ s-stretch	2904	2907	2844
		CH$_3$ d-deform	1494	1456	1477
		CH$_3$ s-deform	1462	1428	1455
		OH bend	1365	1320	1345
		CH$_3$ rock	1068	1042	1060
		CO stretch	1031	1004	1033
	a"	CH$_3$ d-stretch	2944	2955	2960
		CH$_3$ d-deform	1475	1444	1477
		CH$_3$ rock	1143	1127	1165
		torsion	347	298	295

Table A7-11: Vibrational Frequencies for CH₃X Molecules. BP Density Functional Models (2)

molecule	symmetry of vibration	description of mode	6-31G*	6-311+G**	expt.
CH₃F	a₁	CH₃ *s*-stretch	2946	2951	**2930**
		CH₃ *s*-deform	1485	1439	**1464**
		CF stretch	1063	1000	**1049**
	e	CH₃ *d*-stretch	3018	3037	**3006**
		CH₃ *d*-deform	1474	1429	**1467**
		CH₃ rock	1167	1144	**1182**
CH₃SiH₃	a₁	CH₃ *s*-stretch	3064	2966	**2898**
		SiH₃ *s*-stretch	2173	2162	**2169**
		CH₃ *s*-deform	1282	1251	**1260**
		SiH₃ *s*-deform	908	905	**940**
		CSi stretch	675	676	**700**
	a₂	torsion	188	185	**187**
	e	CH₃ *d*-stretch	2983	3044	**2982**
		SiH₃ *d*-stretch	2181	2168	**2166**
		CH₃ *d*-deform	1450	1418	**1403**
		SiH₃ *d*-deform	925	927	**980**
		CH₃ rock	876	864	**869**
		SiH₃ rock	502	506	**540**
CH₃SH	a'	CH₃ *d*-stretch	3090	3071	**3000**
		CH₃ *s*-stretch	2995	2979	**2931**
		SH stretch	2598	2595	**2572**
		CH₃ *s*-deform	1469	1435	**1475**
		CH₃ *d*-deform	1347	1319	**1319**
		CH₃ rock	1080	1061	**976**
		SH bend	783	776	**803**
		CS stretch	685	677	**708**
	a"	CH₃ *d*-stretch	3081	3082	**3000**
		CH₃ *d*-deform	1460	1423	**1430**
		CH₃ rock	960	944	**1074**
		torsion	244	230	**–**
CH₃Cl	a₁	CH₃ *s*-stretch	3013	2996	**2937**
		CH₃ *s*-deform	1368	1341	**1355**
		CCl stretch	710	699	**732**
	e	CH₃ *d*-stretch	3113	3096	**3039**
		CH₃ *d*-deform	1463	1429	**1452**
		CH₃ rock	1014	998	**1017**

Table A7-12: Vibrational Frequencies for CH₃X Molecules. BLYP Density Functional Models

molecule	symmetry of vibration	description of mode	6-31G*	6-311+G**	expt.
CH₃CH₃	a_{1g}	CH₃ s-stretch	3047	2950	**2954**
		CH₃ s-deform	1456	1385	**1388**
		CC stretch	1011	959	**995**
	a_{1u}	torsion	314	300	**289**
	a_{2u}	CH₃ s-stretch	3048	2951	**2986**
		CH₃ s-deform	1434	1376	**1379**
	e_g	CH₃ d-stretch	3098	2990	**2969**
		CH₃ d-deform	1532	1468	**1468**
		CH₃ rock	1237	1187	**1190**
	e_u	CH₃ d-stretch	3123	3016	**2985**
		CH₃ d-deform	1538	1471	**1469**
		CH₃ rock	835	811	**822**
CH₃NH₂	a'	NH₂ s-stretch	3333	3393	**3361**
		CH₃ d-stretch	2994	2980	**2961**
		CH₃ s-stretch	2876	2878	**2820**
		NH₂ scis.	1660	1628	**1623**
		CH₃ d-deform	1489	1459	**1473**
		CH₃ s-deform	1446	1421	**1430**
		CH₃ rock	1164	1131	**1130**
		CN stretch	1023	1005	**1044**
		NH₂ wag	859	809	**780**
	a"	NH₂ a-stretch	3415	3468	**3427**
		CH₃ d-stretch	3034	3015	**2985**
		CH₃ d-deform	1509	1479	**1485**
		NH₂ twist	1337	1309	**1419**
		CH₃ rock	965	951	**1195**
		torsion	330	299	**268**
CH₃OH	a'	OH stretch	3589	3696	**3681**
		CH₃ d-stretch	3046	3036	**3000**
		CH₃ s-stretch	2905	2911	**2844**
		CH₃ d-deform	1502	1466	**1477**
		CH₃ s-deform	1470	1438	**1455**
		OH bend	1368	1320	**1345**
		CH₃ rock	1061	1037	**1060**
		CO stretch	1011	977	**1033**
	a"	CH₃ d-stretch	2940	2953	**2960**
		CH₃ d-deform	1483	1455	**1477**
		CH₃ rock	1145	1130	**1165**
		torsion	346	295	**295**

Table A7-12: Vibrational Frequencies for CH₃X Molecules. BLYP Density Functional Models (2)

molecule	symmetry of vibration	description of mode	6-31G*	6-311+G**	expt.
CH₃F	a₁	CH₃ s-stretch	2947	2955	**2930**
		CH₃ s-deform	1495	1436	**1464**
		CF stretch	1037	965	**1049**
	e	CH₃ d-stretch	3012	3036	**3006**
		CH₃ d-deform	1481	1450	**1467**
		CH₃ rock	1169	1144	**1182**
CH₃SiH₃	a₁	CH₃ s-stretch	2976	2958	**2898**
		SiH₃ s-stretch	2172	2163	**2169**
		CH₃ s-deform	1295	1270	**1260**
		SiH₃ s-deform	922	921	**940**
		CSi stretch	664	665	**700**
	a₂	torsion	190	190	**187**
	e	CH₃ d-stretch	3048	3026	**2982**
		SiH₃ d-stretch	2180	2168	**2166**
		CH₃ d-deform	1463	1434	**1403**
		SiH₃ d-deform	934	939	**980**
		CH₃ rock	883	875	**869**
		SiH₃ rock	508	515	**540**
CH₃SH	a'	CH₃ d-stretch	3083	3061	**3000**
		CH₃ s-stretch	2995	2977	**2931**
		SH stretch	2571	2571	**2572**
		CH₃ s-deform	1480	1449	**1475**
		CH₃ d-deform	1352	1329	**1319**
		CH₃ rock	1080	1064	**976**
		SH bend	780	776	**803**
		CS stretch	660	651	**708**
	a"	CH₃ d-stretch	3076	3055	**3000**
		CH₃ d-deform	1470	1438	**1430**
		CH₃ rock	959	950	**1074**
		torsion	241	229	**–**
CH₃Cl	a₁	CH₃ s-stretch	3016	2997	**2937**
		CH₃ s-deform	1370	1347	**1355**
		CCl stretch	678	665	**732**
	e	CH₃ d-stretch	3111	3091	**3039**
		CH₃ d-deform	1472	1442	**1452**
		CH₃ rock	1011	998	**1017**

Table A7-13: Vibrational Frequencies for CH₃X Molecules. EDF1 Density Functional Models

molecule	symmetry of vibration	description of mode	6-31G*	6-311+G**	expt.
CH_3CH_3	a_{1g}	CH_3 s-stretch	3008	2991	**2954**
		CH_3 s-deform	1424	1394	**1388**
		CC stretch	1001	989	**995**
	a_{1u}	torsion	309	306	**289**
	a_{2u}	CH_3 s-stretch	3009	2991	**2986**
		CH_3 s-deform	1402	1378	**1379**
	e_g	CH_3 d-stretch	3066	3043	**2969**
		CH_3 d-deform	1504	1475	**1468**
		CH_3 rock	1209	1191	**1190**
	e_u	CH_3 d-stretch	3091	3068	**2985**
		CH_3 d-deform	1509	1476	**1469**
		CH_3 rock	822	816	**822**
CH_3NH_2	a'	NH_2 s-stretch	3408	3456	**3361**
		CH_3 d-stretch	3037	3022	**2961**
		CH_3 s-stretch	2917	2917	**2820**
		NH_2 scis.	1676	1641	**1623**
		CH_3 d-deform	1498	1467	**1473**
		CH_3 s-deform	1455	1428	**1430**
		CH_3 rock	1173	1142	**1130**
		CN stretch	1056	1041	**1044**
		NH_2 wag	866	820	**780**
	a"	NH_2 a-stretch	3497	3536	**3427**
		CH_3 d-stretch	3081	3062	**2985**
		CH_3 d-deform	1518	1487	**1485**
		NH_2 twist	1347	1319	**1419**
		CH_3 rock	975	960	**1195**
		torsion	334	306	**268**
CH_3OH	a'	OH stretch	3682	3777	**3681**
		CH_3 d-stretch	3090	3077	**3000**
		CH_3 s-stretch	2942	2944	**2844**
		CH_3 d-deform	1512	1475	**1477**
		CH_3 s-deform	1480	1448	**1455**
		OH bend	1379	1344	**1345**
		CH_3 rock	1080	1054	**1060**
		CO stretch	1043	1018	**1033**
	a"	CH_3 d-stretch	2985	2993	**2960**
		CH_3 d-deform	1493	1462	**1477**
		CH_3 rock	1158	1144	**1165**
		torsion	347	301	**295**

Table A7-13: Vibrational Frequencies for CH₃X Molecules. EDF1 Density Functional Models (2)

molecule	symmetry of vibration	description of mode	6-31G*	6-311+G**	expt.
CH_3F	a_1	CH_3 s-stretch	2983	2986	**2930**
		CH_3 s-deform	1502	1447	**1464**
		CF stretch	1069	1007	**1049**
	e	CH_3 d-stretch	3059	3075	**3006**
		CH_3 d-deform	1492	1457	**1467**
		CH_3 rock	1181	1158	**1182**
CH_3SiH_3	a_1	CH_3 s-stretch	3019	3001	**2898**
		SiH_3 s-stretch	2195	2177	**2169**
		CH_3 s-deform	1300	1269	**1260**
		SiH_3 s-deform	920	920	**940**
		CSi stretch	677	678	**700**
	a_2	torsion	189	190	**187**
	e	CH_3 d-stretch	3103	3082	**2982**
		SiH_3 d-stretch	2203	2183	**2166**
		CH_3 d-deform	1469	1437	**1403**
		SiH_3 d-deform	935	938	**980**
		CH_3 rock	885	875	**869**
		SiH_3 rock	512	518	**540**
CH_3SH	a'	CH_3 d-stretch	3129	3107	**3000**
		CH_3 s-stretch	3031	3014	**2931**
		SH stretch	2640	2630	**2572**
		CH_3 s-deform	1487	1455	**1475**
		CH_3 d-deform	1366	1339	**1319**
		CH_3 rock	1093	1075	**976**
		SH bend	795	788	**803**
		CS stretch	693	685	**708**
	a"	CH_3 d-stretch	3119	3099	**3000**
		CH_3 d-deform	1478	1443	**1430**
		CH_3 rock	973	960	**1074**
		torsion	245	236	**–**
CH_3Cl	a_1	CH_3 s-stretch	3049	3030	**2937**
		CH_3 s-deform	1387	1360	**1355**
		CCl stretch	716	707	**732**
	e	CH_3 d-stretch	3151	3131	**3039**
		CH_3 d-deform	1481	1448	**1452**
		CH_3 rock	1028	1012	**1017**

Table A7-14: Vibrational Frequencies for CH₃X Molecules. B3LYP Density Functional Models

molecule	symmetry of vibration	description of mode	6-31G*	6-311+G**	expt.
CH₃CH₃	a₁g	CH₃ s-stretch	3047	3023	**2954**
		CH₃ s-deform	1456	1425	**1388**
		CC stretch	1011	996	**995**
	a₁u	torsion	314	310	**289**
	a₂u	CH₃ s-stretch	3048	3023	**2986**
		CH₃ s-deform	1434	1410	**1379**
	eg	CH₃ d-stretch	3098	3069	**2969**
		CH₃ d-deform	1532	1504	**1468**
		CH₃ rock	1237	1219	**1190**
	eu	CH₃ d-stretch	3123	3094	**2985**
		CH₃ d-deform	1538	1506	**1469**
		CH₃ rock	835	830	**822**
CH₃NH₂	a'	NH₂ s-stretch	3466	3508	**3361**
		CH₃ d-stretch	3079	3056	**2961**
		CH₃ s-stretch	2971	2962	**2820**
		NH₂ scis.	1704	1669	**1623**
		CH₃ d-deform	1528	1498	**1473**
		CH₃ s-deform	1486	1461	**1430**
		CH₃ rock	1193	1164	**1130**
		CN stretch	1071	1056	**1044**
		NH₂ wag	878	823	**780**
	a"	NH₂ a-stretch	3550	3586	**3427**
		CH₃ d-stretch	3119	3093	**2985**
		CH₃ d-deform	1547	1518	**1485**
		NH₂ twist	1371	1342	**1419**
		CH₃ rock	988	974	**1195**
		torsion	333	305	**268**
CH₃OH	a'	OH stretch	3754	3849	**3681**
		CH₃ d-stretch	3133	3111	**3000**
		CH₃ s-stretch	2997	2988	**2844**
		CH₃ d-deform	1541	1506	**1477**
		CH₃ s-deform	1511	1480	**1455**
		OH bend	1401	1358	**1345**
		CH₃ rock	1097	1072	**1060**
		CO stretch	1067	1042	**1033**
	a"	CH₃ d-stretch	3040	3035	**2960**
		CH₃ d-deform	1525	1496	**1477**
		CH₃ rock	1183	1168	**1165**
		torsion	347	299	**295**

Table A7-14: Vibrational Frequencies for CH₃X Molecules. B3LYP Density Functional Models (2)

molecule	symmetry of vibration	description of mode	6-31G*	6-311+G**	expt.
CH₃F	a₁	CH₃ s-stretch	3037	3032	**2930**
		CH₃ s-deform	1532	1480	**1464**
		CF stretch	1091	1033	**1049**
	e	CH₃ d-stretch	3111	3117	**3006**
		CH₃ d-deform	1524	1490	**1467**
		CH₃ rock	1205	1185	**1182**
CH₃SiH₃	a₁	CH₃ s-stretch	3054	3029	**2898**
		SiH₃ s-stretch	2238	2222	**2169**
		CH₃ s-deform	1333	1303	**1260**
		SiH₃ s-deform	950	949	**940**
		CSi stretch	688	689	**700**
	a₂	torsion	192	194	**187**
	e	CH₃ d-stretch	3130	3102	**2982**
		SiH₃ d-stretch	2242	2222	**2166**
		CH₃ d-deform	1497	1466	**1403**
		SiH₃ d-deform	961	964	**980**
		CH₃ rock	905	895	**869**
		SiH₃ rock	521	526	**540**
CH₃SH	a'	CH₃ d-stretch	3166	3137	**3000**
		CH₃ s-stretch	3075	3050	**2931**
		SH stretch	2677	2644	**2572**
		CH₃ s-deform	1517	1486	**1475**
		CH₃ d-deform	1395	1369	**1319**
		CH₃ rock	1116	1099	**976**
		SH bend	806	801	**803**
		CS stretch	702	693	**708**
	a"	CH₃ d-stretch	3160	3132	**3000**
		CH₃ d-deform	1508	1474	**1430**
		CH₃ rock	990	977	**1074**
		torsion	247	237	**–**
CH₃Cl	a₁	CH₃ s-stretch	3094	3070	**2937**
		CH₃ s-deform	1415	1389	**1355**
		CCl stretch	721	710	**732**
	e	CH₃ d-stretch	3193	3168	**3039**
		CH₃ d-deform	1511	1479	**1452**
		CH₃ rock	1046	1031	**1017**

Table A7-15: Vibrational Frequencies for CH₃X Molecules. MP2 Models

molecule	symmetry of vibration	description of mode	6-31G*	6-311+G**	expt.
CH₃CH₃	a₁g	CH₃ s-stretch	3112	3079	**2954**
		CH₃ s-deform	1492	1445	**1388**
		CC stretch	1050	1033	**995**
	a₁u	torsion	331	329	**289**
	a₂u	CH₃ s-stretch	3112	3079	**2986**
		CH₃ s-deform	1465	1419	**1379**
	eg	CH₃ d-stretch	3188	3153	**2969**
		CH₃ d-deform	1569	1522	**1468**
		CH₃ rock	1271	1238	**1190**
	eu	CH₃ d-stretch	3201	3175	**2985**
		CH₃ d-deform	1573	1523	**1469**
		CH₃ rock	849	833	**822**
CH₃NH₂	a'	NH₂ s-stretch	3526	3557	**3361**
		CH₃ d-stretch	3163	3134	**2961**
		CH₃ s-stretch	3060	3041	**2820**
		NH₂ scis.	1732	1667	**1623**
		CH₃ d-deform	1565	1520	**1473**
		CH₃ s-deform	1516	1480	**1430**
		CH₃ rock	1224	1194	**1130**
		CN stretch	1101	1089	**1044**
		NH₂ wag	911	873	**780**
	a"	NH₂ a-stretch	3626	3654	**3427**
		CH₃ d-stretch	3208	3176	**2985**
		CH₃ d-deform	1584	1539	**1485**
		NH₂ twist	1397	1363	**1419**
		CH₃ rock	1002	985	**1195**
		torsion	343	323	**268**
CH₃OH	a'	OH stretch	3797	3914	**3681**
		CH₃ d-stretch	3222	3191	**3000**
		CH₃ s-stretch	3076	3055	**2844**
		CH₃ d-deform	1578	1534	**1477**
		CH₃ s-deform	1539	1508	**1455**
		OH bend	1417	1379	**1345**
		CH₃ rock	1113	1097	**1060**
		CO stretch	1083	1076	**1033**
	a"	CH₃ d-stretch	3143	3125	**2960**
		CH₃ d-deform	1565	1519	**1477**
		CH₃ rock	1204	1196	**1165**
		torsion	348	309	**295**

Table A7-15: Vibrational Frequencies for CH_3X Molecules. MP2 Models (2)

molecule	symmetry of vibration	description of mode	6-31G*	6-311+G**	expt.
CH_3F	a_1	CH_3 s-stretch	3115	3093	**2930**
		CH_3 s-deform	1556	1517	**1464**
		CF stretch	1106	1076	**1049**
	e	CH_3 d-stretch	3215	3197	**3006**
		CH_3 d-deform	1566	1522	**1467**
		CH_3 rock	1225	1217	**1182**
CH_3SiH_3	a_1	CH_3 s-stretch	3114	3081	**2898**
		SiH_3 s-stretch	2305	2308	**2169**
		CH_3 s-deform	1376	1319	**1260**
		SiH_3 s-deform	981	988	**940**
		CSi stretch	716	717	**700**
	a_2	torsion	202	202	**187**
	e	CH_3 d-stretch	3209	3176	**2982**
		SiH_3 d-stretch	2310	2309	**2166**
		CH_3 d-deform	1528	1473	**1403**
		SiH_3 d-deform	995	1004	**980**
		CH_3 rock	934	906	**869**
		SiH_3 rock	536	527	**540**
CH_3SH	a'	CH_3 d-stretch	3237	3198	**3000**
		CH_3 s-stretch	3133	3098	**2931**
		SH stretch	2791	2809	**2572**
		CH_3 s-deform	1554	1499	**1475**
		CH_3 d-deform	1448	1411	**1319**
		CH_3 rock	1152	1127	**976**
		SH bend	937	832	**803**
		CS stretch	759	754	**708**
	a"	CH_3 d-stretch	3233	3195	**3000**
		CH_3 d-deform	1540	1472	**1430**
		CH_3 rock	1033	1008	**1074**
		torsion	261	247	–
CH_3Cl	a_1	CH_3 s-stretch	3150	3116	**2937**
		CH_3 s-deform	1468	1442	**1355**
		CCl stretch	785	785	**732**
	e	CH_3 d-stretch	3262	3223	**3039**
		CH_3 d-deform	1546	1494	**1452**
		CH_3 rock	1086	1066	**1017**

Table A7-16: Vibrational Frequencies for CH₃X Molecules. Semi-Empirical Models

molecule	symmetry of vibration	description of mode	MNDO	AM1	PM3	expt.
CH_3CH_3	a_{1g}	CH_3 s-stretch	3339	3147	3186	**2954**
		CH_3 s-deform	1499	1446	1444	**1388**
		CC stretch	1217	1248	1138	**995**
	a_{1u}	torsion	277	243	266	**289**
	a_{2u}	CH_3 s-stretch	3347	3164	3185	**2986**
		CH_3 s-deform	1479	1432	1360	**1379**
	e_g	CH_3 d-stretch	3256	3048	3076	**2969**
		CH_3 d-deform	1444	1394	1414	**1468**
		CH_3 rock	1185	1184	1121	**1190**
	e_u	CH_3 d-stretch	3278	3072	3089	**2985**
		CH_3 d-deform	1440	1396	1409	**1469**
		CH_3 rock	910	893	881	**822**
CH_3NH_2	a'	NH_2 s-stretch	3586	3470	3522	**3361**
		CH_3 d-stretch	3214	2981	3034	**2961**
		CH_3 s-stretch	3316	3104	3140	**2820**
		NH_2 scis.	1815	1735	1675	**1623**
		CH_3 d-deform	1426	1376	1390	**1473**
		CH_3 s-deform	1520	1478	1381	**1430**
		CH_3 rock	1236	1138	1173	**1130**
		CN stretch	1331	1350	1103	**1044**
		NH_2 wag	1042	940	958	**780**
	a"	NH_2 a-stretch	3548	3425	3386	**3427**
		CH_3 d-stretch	3251	3021	3060	**2985**
		CH_3 d-deform	1432	1381	1384	**1485**
		NH_2 twist	1383	1348	1184	**1419**
		CH_3 rock	1064	1016	922	**1195**
		torsion	258	282	287	**268**
CH_3OH	a'	OH stretch	4009	3501	3896	**3681**
		CH_3 d-stretch	3229	3074	3069	**3000**
		CH_3 s-stretch	3300	3149	3141	**2844**
		CH_3 d-deform	1171	1132	992	**1477**
		CH_3 s-deform	1565	1471	1408	**1455**
		OH bend	1506	1522	1367	**1345**
		CH_3 rock	1417	1382	1362	**1060**
		CO stretch	1439	1362	1164	**1033**
	a"	CH_3 d-stretch	3200	3056	3036	**2960**
		CH_3 d-deform	1416	1370	1366	**1477**
		CH_3 rock	1238	1191	1027	**1165**
		torsion	307	327	336	**295**

Table A7-16: Vibrational Frequencies for CH₃X Molecules. Semi-Empirical Models (2)

molecule	symmetry of vibration	description of mode	MNDO	AM1	PM3	expt.
CH₃F	a₁	CH₃ s-stretch	3298	3122	3141	**2930**
		CH₃ s-deform	1612	1502	1534	**1464**
		CF stretch	1480	1402	1203	**1049**
	e	CH₃ d-stretch	3224	3036	3087	**3006**
		CH₃ d-deform	1416	1367	1357	**1467**
		CH₃ rock	1281	1199	1019	**1182**
CH₃SiH₃	a₁	CH₃ s-stretch	3299	3171	3231	**2898**
		SiH₃ s-stretch	2409	2252	1937	**2169**
		CH₃ s-deform	1421	1373	1323	**1260**
		SiH₃ s-deform	1083	804	833	**940**
		CSi stretch	815	677	647	**700**
	a₂	torsion	167	99	184	**187**
	e	CH₃ d-stretch	3232	3066	3125	**2982**
		SiH₃ d-stretch	2389	2243	1915	**2166**
		CH₃ d-deform	1429	1391	1412	**1403**
		SiH₃ d-deform	1050	863	896	**980**
		CH₃ rock	914	826	880	**869**
		SiH₃ rock	619	512	530	**540**
CH₃SH	a'	CH₃ d-stretch	3260	3073	3105	**3000**
		CH₃ s-stretch	3320	3167	3205	**2931**
		SH stretch	2980	2046	1715	**2572**
		CH₃ s-deform	1429	1381	1396	**1475**
		CH₃ d-deform	1452	1355	1335	**1319**
		CH₃ rock	1089	1064	1061	**976**
		SH bend	842	817	827	**803**
		CS stretch	934	754	753	**708**
	a"	CH₃ d-stretch	3248	3075	3110	**3000**
		CH₃ d-deform	1422	1367	1388	**1430**
		CH₃ rock	1031	929	930	**1074**
		torsion	207	219	214	**–**
CH₃Cl	a₁	CH₃ s-stretch	3322	3150	3194	**2937**
		CH₃ s-deform	1448	1346	1346	**1355**
		CCl stretch	942	837	678	**732**
	e	CH₃ d-stretch	3276	3675	3123	**3039**
		CH₃ d-deform	1415	1368	1384	**1452**
		CH₃ rock	1086	988	1012	**1017**

Table A7-17: CC Stretching Frequencies. Hartree-Fock Models

molecule	STO-3G	3-21G	6-31G*	6-311+G**	expt.
cyclobutene	1997	1761	1804	1766	**1570**
tetrachloroethylene	1971	1809	1824	1809	**1571**
trans-1,2-dichloroethylene	1996	1821	1837	1808	**1578**
cis-1,2-dichloroethylene	2007	1822	1842	1811	**1587**
cyclopentene	2058	1830	1852	1814	**1614**
ethylene	2041	1842	1856	1814	**1623**
1,1-dichloroethylene	2007	1835	1849	1818	**1627**
cyclopropene	2085	1833	1886	1848	**1641**
propene	2076	1861	1881	1841	**1656**
isobutene	2092	1875	1888	1849	**1661**
cis-2-butene	2112	1883	1907	1870	**1669**
trans-2-butene	2115	1890	1915	1880	**1680**
tetramethylethylene	2123	1901	1911	1882	**1683**
cis-1,2-difluoroethylene	2082	1964	1971	1936	**1715**
tetrafluoroethylene	2106	2131	2145	2129	**1872**

Table A7-18: CC Stretching Frequencies. Local Density Models

molecule	6-31G*	6-311+G**	expt.
cyclobutene	1631	1600	**1570**
tetrachloroethylene	1612	1599	**1571**
trans-1,2-dichloroethylene	1646	1622	**1578**
cis-1,2-dichloroethylene	1650	1625	**1587**
cyclopentene	1673	1642	**1614**
ethylene	1685	1650	**1623**
1,1-dichloroethylene	1659	1634	**1627**
cyclopropene	1730	1705	**1641**
propene	1713	1680	**1656**
isobutene	1723	1690	**1661**
cis-2-butene	1734	1703	**1669**
trans-2-butene	1746	1716	**1680**
tetramethylethylene	1737	1714	**1683**
cis-1,2-difluoroethylene	1773	1735	**1715**
tetrafluoroethylene	1906	1888	**1872**

Table A7-19: CC Stretching Frequencies. BP Density Functional Models

molecule	6-31G*	6-311+G**	expt.
cyclobutene	1606	1576	**1570**
tetrachloroethylene	1564	1552	**1571**
trans-1,2-dichloroethylene	1605	1583	**1578**
cis-1,2-dichloroethylene	1612	1589	**1587**
cyclopentene	1645	1617	**1614**
ethylene	1666	1632	**1623**
1,1-dichloroethylene	1625	1601	**1627**
cyclopropene	1701	1675	**1641**
propene	1684	1652	**1656**
isobutene	1689	1657	**1661**
cis-2-butene	1699	1670	**1669**
trans-2-butene	1708	1679	**1680**
tetramethylethylene	1693	1670	**1683**
cis-1,2-difluoroethylene	1726	1693	**1715**
tetrafluoroethylene	1849	1832	**1872**

Table A7-20: CC Stretching Frequencies. BLYP Density Functional Models

molecule	6-31G*	6-311+G**	expt.
cyclobutene	1598	1567	**1570**
tetrachloroethylene	1547	1534	**1571**
trans-1,2-dichloroethylene	1594	1570	**1578**
cis-1,2-dichloroethylene	1600	1574	**1587**
cyclopentene	1638	1608	**1614**
ethylene	1664	1628	**1623**
1,1-dichloroethylene	1616	1590	**1627**
cyclopropene	1695	1669	**1641**
propene	1677	1643	**1656**
isobutene	1680	1647	**1661**
cis-2-butene	1689	1657	**1669**
trans-2-butene	1698	1667	**1680**
tetramethylethylene	1680	1655	**1683**
cis-1,2-difluoroethylene	1713	1675	**1715**
tetrafluoroethylene	1832	1811	**1872**

Table A7-21: CC Stretching Frequencies. EDF1 Density Functional Models

molecule	6-31G*	6-311+G**	expt.
cyclobutene	1623	1596	**1570**
tetrachloroethylene	1570	1562	**1571**
trans-1,2-dichloroethylene	1619	1599	**1578**
cis-1,2-dichloroethylene	1626	1605	**1587**
cyclopentene	1663	1637	**1614**
ethylene	1682	1649	**1623**
1,1-dichloroethylene	1637	1614	**1627**
cyclopropene	1719	1695	**1641**
propene	1699	1670	**1656**
isobutene	1702	1673	**1661**
cis-2-butene	1714	1688	**1669**
trans-2-butene	1723	1697	**1680**
tetramethylethylene	1704	1684	**1683**
cis-1,2-difluoroethylene	1741	1709	**1715**
tetrafluoroethylene	1863	1847	**1872**

Table A7-22: CC Stretching Frequencies. B3LYP Density Functional Models

molecule	6-31G*	6-311+G**	expt.
cyclobutene	1660	1628	**1570**
tetrachloroethylene	1629	1614	**1571**
trans-1,2-dichloroethylene	1666	1643	**1578**
cis-1,2-dichloroethylene	1671	1647	**1587**
cyclopentene	1702	1671	**1614**
ethylene	1720	1683	**1623**
1,1-dichloroethylene	1685	1659	**1627**
cyclopropene	1753	1726	**1641**
propene	1739	1704	**1656**
isobutene	1744	1710	**1661**
cis-2-butene	1755	1724	**1669**
trans-2-butene	1765	1733	**1680**
tetramethylethylene	1751	1726	**1683**
cis-1,2-difluoroethylene	1791	1756	**1715**
tetrafluoroethylene	1928	1909	**1872**

Table A7-23: CC Stretching Frequencies. MP2 Models

molecule	6-31G*	6-311+G**	expt.
cyclobutene	1645	1598	**1570**
tetrachloroethylene	1635	1607	**1571**
trans-1,2-dichloroethylene	1675	1641	**1578**
cis-1,2-dichloroethylene	1679	1643	**1587**
cyclopentene	1692	1646	**1614**
ethylene	1719	1675	**1623**
1,1-dichloroethylene	1686	1650	**1627**
cyclopropene	1730	1683	**1641**
propene	1741	1698	**1656**
isobutene	1752	1712	**1661**
cis-2-butene	1761	1721	**1669**
trans-2-butene	1769	1728	**1680**
tetramethylethylene	1765	1732	**1683**
cis-1,2-difluoroethylene	1801	1764	**1715**
tetrafluoroethylene	1944	1921	**1872**

Table A7-24: CC Stretching Frequencies. Semi-Empirical Models

molecule	MNDO	AM1	PM3	expt.
cyclobutene	1720	1742	1772	**1570**
tetrachloroethylene	1739	1765	1759	**1571**
trans-1,2-dichloroethylene	1744	1797	1783	**1578**
cis-1,2-dichloroethylene	1761	1807	1781	**1587**
cyclopentene	1772	1810	1829	**1614**
ethylene	1783	1827	1829	**1623**
1,1-dichloroethylene	1759	1799	1798	**1627**
cyclopropene	1875	1926	1924	**1641**
propene	1826	1867	1862	**1656**
isobutene	1819	1884	1880	**1661**
cis-2-butene	1869	1910	1888	**1669**
trans-2-butene	1852	1896	1840	**1680**
tetramethylethylene	1839	1923	1910	**1683**
cis-1,2-difluoroethylene	1929	1866	1913	**1715**
tetrafluoroethylene	1980	1930	2018	**1872**

Table A7-25: CO Stretching Frequencies. Hartree-Fock Models

molecule	STO-3G	3-21G	6-31G*	6-311+G**	expt.
trans-acrolein	2092	1916	2014	1980	**1724**
acetone	2130	1940	2022	1983	**1731**
acetaldehyde	2121	1926	2033	1999	**1743**
formaldehyde	2100	1916	2028	1996	**1746**
methyl formate	2102	1915	2012	1971	**1754**
formic acid	2117	1939	2035	1996	**1770**
acetic acid	2142	1971	2042	2001	**1788**

Table A7-26: CO Stretching Frequencies. Local Density Models

molecule	6-31G*	6-311+G**	expt.
trans-acrolein	1792	1760	**1724**
acetone	1809	1771	**1731**
acetaldehyde	1830	1796	**1743**
formaldehyde	1838	1807	**1746**
methyl formate	1822	1785	**1754**
formic acid	1843	1806	**1770**
acetic acid	1845	1805	**1788**

Table A7-27: CO Stretching Frequencies. BP Density Functional Models

molecule	6-31G*	6-311+G**	expt.
trans-acrolein	1733	1700	**1724**
acetone	1755	1718	**1731**
acetaldehyde	1776	1742	**1743**
formaldehyde	1788	1754	**1746**
methyl formate	1770	1741	**1754**
formic acid	1790	1753	**1770**
acetic acid	1792	1754	**1788**

Table A7-28: CO Stretching Frequencies. BLYP Density Functional Models

molecule	6-31G*	6-311+G**	expt.
trans-acrolein	1714	1676	**1724**
acetone	1737	1697	**1731**
acetaldehyde	1761	1722	**1743**
formaldehyde	1775	1734	**1746**
methyl formate	1750	1708	**1754**
formic acid	1773	1732	**1770**
acetic acid	1774	1733	**1788**

Table A7-29: CO Stretching Frequencies. EDF1 Density Functional Models

molecule	6-31G*	6-311+G**	expt.
trans-acrolein	1749	1717	**1724**
acetone	1771	1736	**1731**
acetaldehyde	1795	1762	**1743**
formaldehyde	1808	1773	**1746**
methyl formate	1788	1752	**1754**
formic acid	1810	1773	**1770**
acetic acid	1808	1773	**1788**

Table A7-30: CO Stretching Frequencies. B3LYP Density Functional Models

molecule	6-31G*	6-311+G**	expt.
trans-acrolein	1804	1770	**1724**
acetone	1823	1785	**1731**
acetaldehyde	1842	1807	**1743**
formaldehyde	1849	1814	**1746**
methyl formate	1831	1791	**1754**
formic acid	1854	1814	**1770**
acetic acid	1842	1818	**1788**

Table A7-31: CO Stretching Frequencies. MP2 Models

molecule	6-31G*	6-311+G**	expt.
trans-acrolein	1770	1745	**1724**
acetone	1792	1767	**1731**
acetaldehyde	1796	1774	**1743**
formaldehyde	1788	1762	**1746**
methyl formate	1815	1783	**1754**
formic acid	1838	1808	**1770**
acetic acid	1858	1825	**1788**

Table A7-32: CO Stretching Frequencies. Semi-Empirical Models

molecule	MNDO	AM1	PM3	expt.
trans-acrolein	2112	2052	1979	**1724**
acetone	2120	2069	1979	**1731**
acetaldehyde	2128	2060	1984	**1743**
formaldehyde	2115	2053	1988	**1746**
methyl formate	2088	2044	1945	**1754**
formic acid	2089	2051	1944	**1770**
acetic acid	2111	2089	1982	**1788**

Table A9-1: **Key Bond Distances in Transition States for Organic Reactions. Hartree-Fock Models**

reaction	transition state	bond length	STO-3G	3-21G	6-31G*	6-311+G**	MP2/ 6-311+G**
CH₃NC ⟶ CH₃CN		a	1.85	1.98	1.90	1.91	**1.87**
		b	1.73	1.87	1.74	1.75	**1.76**
		c	1.22	1.19	1.17	1.17	**1.20**
HCO₂CH₂CH₃ ⟶ HCO₂H + C₂H₄		a	1.85	1.97	2.10	2.12	**1.94**
		b	1.26	1.25	1.23	1.23	**1.26**
		c	1.32	1.28	1.25	1.24	**1.27**
		d	1.08	1.24	1.33	1.36	**1.28**
		e	1.50	1.40	1.31	1.29	**1.33**
		f	1.39	1.40	1.40	1.40	**1.40**
		a	1.79	2.02	2.05	2.07	**1.80**
		b	1.42	1.39	1.39	1.39	**1.43**
		a	1.63	1.88	1.92	1.96	**1.80**
		b	1.32	1.29	1.26	1.26	**1.30**
		c	1.40	1.37	1.37	1.37	**1.39**
		d	1.90	2.14	2.27	2.32	**2.22**
		e	1.40	1.38	1.38	1.38	**1.39**
		f	1.43	1.39	1.39	1.39	**1.41**
		a	1.42	1.40	1.39	1.39	**1.41**
		b	1.37	1.38	1.39	1.39	**1.40**
		c	2.22	2.20	2.19	2.18	**2.28**
		d	1.36	1.37	1.38	1.39	**1.38**
		a	1.43	1.40	1.40	1.40	**1.43**
		b	1.37	1.37	1.38	1.38	**1.39**
		c	2.00	2.11	2.12	2.13	**2.07**
		d	1.39	1.40	1.40	1.40	**1.41**
		e	1.53	1.45	1.45	1.43	**1.53**
		f	1.22	1.35	1.36	1.37	**1.25**
HCNO + C₂H₂ ⟶		a	2.37	2.27	2.15	2.13	**2.04**
		b	1.19	1.16	1.17	1.17	**1.24**
		c	1.29	1.31	1.22	1.22	**1.21**
		d	2.14	2.13	2.21	2.20	**2.22**
		e	1.19	1.21	1.21	1.21	**1.25**

Table A9-1: **Key Bond Distances in Transition States for Organic Reactions. Hartree-Fock Models (2)**

reaction	transition state	bond length	STO-3G	3-21G	6-31G*	6-311+G**	MP2/ 6-311+G**
		a	1.39	1.39	1.39	1.39	**1.40**
		b	1.42	1.41	1.41	1.41	**1.42**
		c	1.36	1.45	1.44	1.44	**1.41**
		a	1.40	1.42	1.41	1.42	**1.43**
		b	2.10	2.14	2.13	2.13	**2.13**
		c	1.39	1.37	1.37	1.37	**1.38**
		a	1.41	1.39	1.38	1.38	**1.40**
		b	1.37	1.37	1.37	1.37	**1.38**
		c	1.99	2.12	2.26	2.28	**2.06**
		d	1.24	1.23	1.22	1.22	**1.24**
		e	2.04	1.88	1.74	1.72	**1.83**
		f	1.38	1.40	1.43	1.43	**1.41**
		a	1.39	1.39	1.39	1.38	**1.40**
		b	1.40	1.39	1.39	1.39	**1.41**
		c	2.26	2.33	2.33	2.32	**2.29**

718

Table A9-2: Key Bond Distances in Transition States for Organic Reactions. Local Density Models

reaction	transition state	bond length	6-31G*	6-311+G**	MP2/ 6-311+G**
CH₃NC ⟶ CH₃CN		a	1.74	1.75	**1.87**
		b	1.74	1.75	**1.76**
		c	1.21	1.20	**1.20**
HCO₂CH₂CH₃ ⟶ HCO₂H + C₂H₄		a	1.97	1.94	**1.94**
		b	1.26	1.25	**1.26**
		c	1.27	1.27	**1.27**
		d	1.24	1.26	**1.28**
		e	1.38	1.35	**1.33**
		f	1.39	1.39	**1.40**
		a	1.77	1.79	**1.80**
		b	1.42	1.41	**1.43**
		a	1.71	1.77	**1.80**
		b	1.30	1.29	**1.30**
		c	1.38	1.38	**1.39**
		d	2.10	2.17	**2.22**
		e	1.39	1.38	**1.39**
		f	1.41	1.40	**1.41**
		a	1.41	1.41	**1.41**
		b	1.38	1.38	**1.40**
		c	2.37	2.31	**2.28**
		d	1.36	1.37	**1.38**
		a	1.43	1.42	**1.43**
		b	1.37	1.37	**1.39**
		c	2.12	2.12	**2.07**
		d	1.39	1.39	**1.41**
		e	1.59	1.55	**1.53**
		f	1.24	1.26	**1.25**
HCNO + C₂H₂ ⟶		a	2.38	2.31	**2.04**
		b	1.20	1.20	**1.24**
		c	1.20	1.20	**1.21**
		d	2.48	2.43	**2.22**
		e	1.23	1.22	**1.25**

reaction	transition state	bond length	6-31G*	6-311+G**	MP2/ 6-311+G**
		a	1.39	1.39	**1.40**
		b	1.41	1.41	**1.42**
		c	1.41	1.41	**1.41**
		a	1.42	1.42	**1.43**
		b	2.14	2.13	**2.13**
		c	1.38	1.37	**1.38**
		a	1.40	1.40	**1.40**
		b	1.37	1.37	**1.38**
		c	2.24	2.21	**2.06**
		d	1.23	1.23	**1.24**
		e	1.79	1.77	**1.83**
		f	1.40	1.40	**1.41**
			a	a	

a) reasonable transition state cannot be found

Table A9-3: Key Bond Distances in Transition States for Organic Reactions. BP Density Functional Models

reaction	transition state	bond length	6-31G*	6-311+G**	MP2/ 6-311+G**
$CH_3NC \longrightarrow CH_3CN$		a	1.80	1.82	**1.87**
		b	1.79	1.80	**1.76**
		c	1.21	1.21	**1.20**
$HCO_2CH_2CH_3 \longrightarrow HCO_2H + C_2H_4$		a	2.06	2.04	**1.94**
		b	1.27	1.27	**1.26**
		c	1.29	1.28	**1.27**
		d	1.31	1.33	**1.28**
		e	1.35	1.32	**1.33**
		f	1.41	1.41	**1.40**
		a	1.88	1.93	**1.80**
		b	1.42	1.42	**1.43**
		a	1.87	1.93	**1.80**
		b	1.30	1.30	**1.30**
		c	1.39	1.39	**1.39**
		d	2.31	2.39	**2.22**
		e	1.39	1.39	**1.39**
		f	1.41	1.41	**1.41**
		a	1.42	1.41	**1.41**
		b	1.40	1.41	**1.40**
		c	2.30	2.27	**2.28**
		d	1.39	1.39	**1.38**
		a	1.43	1.42	**1.43**
		b	1.39	1.39	**1.39**
		c	2.12	2.14	**2.07**
		d	1.41	1.41	**1.41**
		e	1.53	1.49	**1.53**
		f	1.29	1.31	**1.25**
$HCNO + C_2H_2 \longrightarrow$		a	2.29	2.25	**2.04**
		b	1.22	1.22	**1.24**
		c	1.22	1.22	**1.21**
		d	2.57	2.51	**2.22**
		e	1.24	1.23	**1.25**

Table A9-3: Key Bond Distances in Transition States for Organic Reactions. BP Density Functional Models (2)

reaction	transition state	bond length	6-31G*	6-311+G**	MP2/ 6-311+G**
		a	1.40	1.40	**1.40**
		b	1.43	1.42	**1.42**
		c	1.43	1.42	**1.41**
		a	1.43	1.43	**1.43**
		b	2.16	2.15	**2.13**
		c	1.39	1.38	**1.38**
		a	1.41	1.40	**1.40**
		b	1.39	1.39	**1.38**
		c	2.20	2.19	**2.06**
		d	1.26	1.25	**1.24**
		e	1.78	1.76	**1.83**
		f	1.43	1.43	**1.41**
		a	1.41	1.40	**1.40**
		b	1.41	1.41	**1.41**
		c	2.39	2.35	**2.29**

722

Table A9-4: Key Bond Distances in Transition States for Organic Reactions. BLYP Density Functional Models

reaction	transition state	bond length	6-31G*	6-311+G**	MP2/ 6-311+G**
CH$_3$NC → CH$_3$CN		a	1.84	1.85	**1.87**
		b	1.83	1.85	**1.76**
		c	1.21	1.20	**1.20**
HCO$_2$CH$_2$CH$_3$ → HCO$_2$H + C$_2$H$_4$		a	2.11	2.10	**1.94**
		b	1.27	1.27	**1.26**
		c	1.29	1.28	**1.27**
		d	1.34	1.37	**1.28**
		e	1.33	1.30	**1.33**
		f	1.41	1.41	**1.40**
		a	2.02	2.14	**1.80**
		b	1.42	1.41	**1.43**
		a	1.96	2.02	**1.80**
		b	1.30	1.30	**1.30**
		c	1.39	1.39	**1.39**
		d	2.42	2.51	**2.22**
		e	1.39	1.39	**1.39**
		f	1.41	1.41	**1.41**
		a	1.41	1.41	**1.41**
		b	1.41	1.41	**1.40**
		c	2.26	2.23	**2.28**
		d	1.40	1.41	**1.38**
		a	1.43	1.42	**1.43**
		b	1.40	1.39	**1.39**
	→ + C$_2$H$_4$	c	2.15	2.21	**2.07**
		d	1.42	1.42	**1.41**
		e	1.48	1.42	**1.53**
		f	1.34	1.37	**1.25**
HCNO + C$_2$H$_2$ →		a	2.25	2.21	**2.04**
		b	1.22	1.22	**1.24**
		c	1.23	1.23	**1.21**
		d	2.57	2.50	**2.22**
		e	1.24	1.23	**1.25**

Table A9-4: Key Bond Distances in Transition States for Organic Reactions. BLYP Density Functional Models (2)

reaction	transition state	bond length	6-31G*	6-311+G**	MP2/ 6-311+G**
		a	1.41	1.41	**1.40**
		b	1.43	1.43	**1.42**
		c	1.44	1.43	**1.41**
		a	1.44	1.44	**1.43**
		b	2.16	2.15	**2.13**
		c	1.38	1.38	**1.38**
		a	1.40	1.40	**1.40**
		b	1.39	1.39	**1.38**
		c	2.18	2.17	**2.06**
		d	1.26	1.26	**1.24**
		e	1.78	1.76	**1.83**
		f	1.44	1.44	**1.41**
		a	1.40	1.39	**1.40**
		b	1.42	1.43	**1.41**
		c	2.32	2.28	**2.29**

Table A9-5: Key Bond Distances in Transition States for Organic Reactions. EDF1 Density Functional Models

reaction	transition state	bond length	6-31G*	6-311+G**	MP2/ 6-311+G**
CH₃NC ⟶ CH₃CN		a	1.79	1.81	**1.87**
		b	1.78	1.79	**1.76**
		c	1.21	1.20	**1.20**
HCO₂CH₂CH₃ ⟶ HCO₂H + C₂H₄		a	2.07	2.06	**1.94**
		b	1.27	1.27	**1.26**
		c	1.28	1.27	**1.27**
		d	1.31	1.34	**1.28**
		e	1.33	1.31	**1.33**
		f	1.40	1.40	**1.40**
		a	1.87	1.92	**1.80**
		b	1.42	1.41	**1.43**
		a	1.89	1.95	**1.80**
		b	1.29	1.29	**1.30**
		c	1.39	1.38	**1.39**
		d	2.34	2.42	**2.22**
		e	1.39	1.38	**1.39**
		f	1.41	1.40	**1.41**
		a	1.41	1.41	**1.41**
		b	1.40	1.40	**1.40**
		c	2.29	2.26	**2.28**
		d	1.39	1.39	**1.38**
		a	1.43	1.42	**1.43**
		b	1.39	1.39	**1.39**
⟶ + C₂H₄		c	2.11	2.14	**2.07**
		d	1.41	1.41	**1.41**
		e	1.51	1.47	**1.53**
		f	1.29	1.30	**1.25**
		a	2.21	2.18	**2.04**
		b	1.22	1.21	**1.24**
HCNO + C₂H₂ ⟶		c	1.22	1.21	**1.21**
		d	2.65	2.58	**2.22**
		e	1.23	1.23	**1.25**

Table A9-5: **Key Bond Distances in Transition States for Organic Reactions. EDF1 Density Functional Models (2)**

reaction	transition state	bond length	6-31G*	6-311+G**	MP2/ 6-311+G**
		a	1.40	1.40	**1.40**
		b	1.42	1.42	**1.42**
		c	1.42	1.41	**1.41**
		a	1.43	1.43	**1.43**
		b	2.15	2.15	**2.13**
		c	1.38	1.38	**1.38**
$+ CO_2$		a	1.40	1.40	**1.40**
		b	1.39	1.39	**1.38**
		c	2.20	2.19	**2.06**
		d	1.25	1.25	**1.24**
		e	1.76	1.74	**1.83**
		f	1.43	1.43	**1.41**
SO_2 $+ SO_2$		a	1.40	1.39	**1.40**
		b	1.41	1.41	**1.41**
		c	2.38	2.34	**2.29**

Table A9-6: Key Bond Distances in Transition States for Organic Reactions. B3LYP Density Functional Models

reaction	transition state	bond length	6-31G*	6-311+G**	MP2/ 6-311+G**
CH$_3$NC $\longrightarrow$ CH$_3$CN		a	1.83	1.85	**1.87**
		b	1.79	1.80	**1.76**
		c	1.20	1.19	**1.20**
HCO$_2$CH$_2$CH$_3$ $\longrightarrow$ HCO$_2$H + C$_2$H$_4$		a	2.04	2.05	**1.94**
		b	1.26	1.26	**1.26**
		c	1.27	1.27	**1.27**
		d	1.30	1.33	**1.28**
		e	1.34	1.31	**1.33**
		f	1.40	1.40	**1.40**
		a	1.96	2.04	**1.80**
		b	1.41	1.40	**1.43**
		a	1.90	1.96	**1.80**
		b	1.29	1.28	**1.30**
		c	1.38	1.38	**1.39**
		d	2.31	2.38	**2.22**
		e	1.38	1.38	**1.39**
		f	1.40	1.40	**1.41**
		a	1.41	1.40	**1.41**
		b	1.40	1.40	**1.40**
		c	2.25	2.22	**2.28**
		d	1.39	1.39	**1.38**
		a	1.42	1.41	**1.43**
		b	1.39	1.39	**1.39**
		c	2.11	2.14	**2.07**
		d	1.41	1.41	**1.41**
		e	1.48	1.44	**1.53**
		f	1.32	1.34	**1.25**
HCNO + C$_2$H$_2$ $\longrightarrow$		a	2.22	2.18	**2.04**
		b	1.21	1.20	**1.24**
		c	1.22	1.22	**1.21**
		d	2.43	2.39	**2.22**
		e	1.23	1.22	**1.25**

Table A9-6: **Key Bond Distances in Transition States for Organic Reactions. B3LYP Density Functional Models (2)**

reaction	transition state	bond length	6-31G*	6-311+G**	MP2/ 6-311+G**
		a	1.39	1.40	**1.40**
		b	1.42	1.42	**1.42**
		c	1.43	1.43	**1.41**
		a	1.43	1.43	**1.43**
		b	2.14	2.14	**2.13**
		c	1.38	1.37	**1.38**
		a	1.40	1.39	**1.40**
		b	1.38	1.38	**1.38**
		c	2.18	2.18	**2.06**
		d	1.24	1.24	**1.24**
		e	1.78	1.76	**1.83**
		f	1.42	1.42	**1.41**
		a	1.39	1.39	**1.40**
		b	1.40	1.41	**1.41**
		c	2.35	2.32	**2.29**

728

Table A9-7: **Key Bond Distances in Transition States for Organic Reactions. MP2 Models**

reaction	transition state	bond length	6-31G*	6-311+G**
$CH_3NC \longrightarrow CH_3CN$		a	1.86	**1.87**
		b	1.75	**1.76**
		c	1.21	**1.20**
$HCO_2CH_2CH_3 \longrightarrow HCO_2H + C_2H_4$		a	1.98	**1.94**
		b	1.27	**1.26**
		c	1.28	**1.27**
		d	1.29	**1.27**
		e	1.34	**1.33**
		f	1.40	**1.40**
		a	1.78	**1.80**
		b	1.43	**1.43**
		a	1.80	**1.80**
		b	1.31	**1.30**
		c	1.38	**1.39**
		d	2.20	**2.22**
		e	1.39	**1.39**
		f	1.41	**1.41**
		a	1.41	**1.41**
		b	1.39	**1.40**
		c	2.27	**2.28**
		d	1.38	**1.38**
		a	1.43	**1.43**
		b	1.39	**1.39**
		c	2.02	**2.07**
		d	1.41	**1.41**
		e	1.55	**1.53**
		f	1.25	**1.25**
$HCNO + C_2H_2 \longrightarrow$		a	2.08	**2.04**
		b	1.24	**1.24**
		c	1.22	**1.21**
		d	2.21	**2.22**
		e	1.24	**1.25**

Table A9-7: Key Bond Distances in Transition States for Organic Reactions. MP2 Models (2)

reaction	transition state	bond length	6-31G*	6-311+G**
		a	1.40	**1.40**
		b	1.42	**1.42**
		c	1.41	**1.41**
		a	1.43	**1.43**
		b	2.13	**2.13**
		c	1.38	**1.38**
		a	1.40	**1.40**
		b	1.38	**1.38**
		c	2.08	**2.06**
		d	1.25	**1.24**
		e	1.83	**1.83**
		f	1.41	**1.41**
		a	1.40	**1.40**
		b	1.40	**1.41**
		c	2.30	**2.29**

Table A9-8: **Key Bond Distances in Transition States for Organic Reactions. Semi-Empirical Models**

reaction	transition state	bond length	MNDO	AM1	PM3	MP2/ 6-311+G**
$CH_3NC \longrightarrow CH_3CN$		a	1.80	1.80	1.83	**1.87**
		b	1.70	1.70	1.78	**1.76**
		c	1.22	1.23	1.22	**1.20**
$HCO_2CH_2CH_3 \longrightarrow HCO_2H + C_2H_4$		a	1.53	1.76	1.68	**1.94**
		b	1.29	1.28	1.28	**1.26**
		c	1.30	1.29	1.28	**1.27**
		d	1.03	1.18	1.08	**1.28**
		e	1.72	1.44	1.51	**1.33**
		f	1.44	1.41	1.42	**1.40**
		a	a	a	a	**1.80**
		b				**1.43**
		a	1.46	1.58	1.68	**1.80**
		b	1.34	1.32	1.30	**1.30**
		c	1.41	1.41	1.40	**1.39**
		d	1.88	1.84	1.94	**2.22**
		e	1.40	1.40	1.39	**1.39**
		f	1.48	1.43	1.42	**1.41**
		a	1.42	1.41	1.41	**1.41**
		b	1.41	1.42	1.40	**1.40**
		c	2.16	2.11	2.14	**2.28**
		d	1.40	1.39	1.38	**1.38**
		a	1.45	1.41	1.41	**1.43**
		b	1.44	1.39	1.39	**1.39**
		c	1.65	2.02	1.97	**2.07**
		d	1.44	1.40	1.40	**1.41**
		e	1.85	1.44	1.51	**1.53**
		f	1.18	1.33	1.29	**1.25**
$HCNO + C_2H_2 \longrightarrow$		a	1.90	1.95	2.04	**2.04**
		b	1.24	1.25	1.23	**1.24**
		c	1.19	1.19	1.23	**1.21**
		d	3.17	2.31	2.09	**2.22**
		e	1.22	1.23	1.23	**1.25**

Table A9-8: Key Bond Distances in Transition States for Organic Reactions. Semi-Empirical Models (2)

reaction	transition state	bond length	MNDO	AM1	PM3	MP2/ 6-311+G**
		a	1.40	1.39	a	**1.40**
		b	1.43	1.41		**1.42**
		c	1.40	1.42		**1.41**
		a	1.42	1.43	1.41	**1.43**
		b	2.13	2.12	2.11	**2.13**
		c	1.40	1.39	1.39	**1.38**
		a	1.40	1.40	1.39	**1.40**
		b	1.39	1.37	1.38	**1.38**
		c	2.37	2.25	2.02	**2.06**
		d	1.26	1.25	1.24	**1.24**
		e	1.65	1.71	1.93	**1.83**
		f	1.46	1.41	1.40	**1.41**
		a	1.43	1.40	1.42	**1.40**
		b	1.39	1.39	1.37	**1.41**
		c	2.46	2.22	2.37	**2.29**

a) reasonable transition state cannot be found

Table A10-1: Dipole Moments in Diatomic and Small Polyatomic Molecules. Hartree-Fock Models

molecule	STO-3G	3-21G	6-31G*	6-311+G**	expt.
CO	0.1	0.4	0.3	0.2	**0.11**
HCP	0.0	0.5	0.7	0.4	**0.39**
PH_3	0.6	0.9	0.9	0.8	**0.58**
ClF	0.6	1.3	1.2	1.4	**0.88**
H_2S	1.0	1.4	1.4	1.4	**0.97**
HCl	1.8	1.5	1.5	1.4	**1.08**
SiH_3F	1.0	1.4	1.5	1.8	**1.27**
SiH_3Cl	3.1	2.0	2.1	2.1	**1.30**
NH_3	1.9	1.8	1.9	1.7	**1.47**
HF	1.3	2.2	2.0	2.0	**1.82**
H_2O	1.7	2.4	2.2	2.2	**1.85**
CH_3F	1.2	2.3	2.0	2.2	**1.85**
CH_3Cl	2.3	2.3	2.3	2.3	**1.87**
CS	1.0	1.4	1.3	1.6	**1.98**
H_2CO	1.5	2.7	2.7	2.8	**2.34**
HCN	2.5	3.0	3.2	3.3	**2.99**
LiH	4.8	6.0	6.0	6.0	**5.83**
LiF	3.1	5.8	6.2	6.7	**6.28**
NaH	5.7	6.9	6.9	7.1	**6.96**
LiCl	5.4	7.8	7.7	7.5	**7.12**
NaF	6.2	7.3	7.8	8.7	**8.16**
NaCl	9.1	9.5	9.5	9.7	**9.00**

Table A10-2: Dipole Moments in Diatomic and Small Polyatomic Molecules. Local Density Models

molecule	6-31G*	6-311+G**	expt.
CO	0.2	0.2	**0.11**
HCP	0.8	0.4	**0.39**
PH_3	1.2	1.0	**0.58**
ClF	0.9	1.2	**0.88**
H_2S	1.6	1.5	**0.97**
HCl	1.6	1.5	**1.08**
SiH_3F	1.2	1.7	**1.27**
SiH_3Cl	1.6	1.7	**1.30**
NH_3	2.0	1.7	**1.47**
HF	1.9	2.1	**1.82**
H_2O	2.1	2.2	**1.85**
CH_3F	1.6	1.9	**1.85**
CH_3Cl	2.0	2.0	**1.87**
CS	1.6	2.0	**1.98**
H_2CO	2.0	2.3	**2.34**
HCN	2.9	3.0	**2.99**
LiH	5.5	5.6	**5.83**
LiF	5.4	6.2	**6.28**
NaH	5.6	5.6	**6.96**
LiCl	6.8	6.9	**7.12**
NaF	6.5	8.0	**8.16**
NaCl	8.3	8.7	**9.00**

Table A10-3: Dipole Moments in Diatomic and Small Polyatomic Molecules. BP Density Functional Models

molecule	6-31G*	6-311+G**	expt.
CO	0.1	0.2	**0.11**
HCP	0.8	0.4	**0.39**
PH_3	1.0	0.9	**0.58**
ClF	0.9	1.2	**0.88**
H_2S	1.5	1.4	**0.97**
HCl	1.5	1.4	**1.08**
SiH_3F	1.2	1.8	**1.27**
SiH_3Cl	1.7	1.8	**1.30**
NH_3	1.9	1.7	**1.47**
HF	1.8	2.0	**1.82**
H_2O	2.1	2.1	**1.85**
CH_3F	1.6	2.0	**1.85**
CH_3Cl	2.0	2.0	**1.87**
CS	1.5	1.9	**1.98**
H_2CO	2.1	2.3	**2.34**
HCN	2.8	3.0	**2.99**
LiH	5.6	5.7	**5.83**
LiF	5.5	6.4	**6.28**
NaH	5.9	6.0	**6.96**
LiCl	7.0	7.1	**7.12**
NaF	6.7	8.2	**8.16**
NaCl	8.7	9.0	**9.00**

Table A10-4: Dipole Moments in Diatomic and Small Polyatomic Molecules. BLYP Density Functional Models

molecule	6-31G*	6-311+G**	expt.
CO	0.1	0.1	**0.11**
HCP	0.7	0.4	**0.39**
PH$_3$	1.0	0.8	**0.58**
ClF	0.9	1.3	**0.88**
H$_2$S	1.4	1.3	**0.97**
HCl	1.4	1.4	**1.08**
SiH$_3$F	1.2	1.8	**1.27**
SiH$_3$Cl	1.7	1.8	**1.30**
NH$_3$	1.9	1.7	**1.47**
HF	1.8	2.0	**1.82**
H$_2$O	2.0	2.1	**1.85**
CH$_3$F	1.6	2.1	**1.85**
CH$_3$Cl	2.1	2.1	**1.87**
CS	1.5	1.9	**1.98**
H$_2$CO	2.0	2.4	**2.34**
HCN	2.8	3.0	**2.99**
LiH	5.5	5.6	**5.83**
LiF	5.4	6.3	**6.28**
NaH	5.6	5.6	**6.96**
LiCl	6.8	7.0	**7.12**
NaF	6.4	8.0	**8.16**
NaCl	8.3	8.8	**9.00**

Table A10-5: Dipole Moments in Diatomic and Small Polyatomic Molecules. EDF1 Density Functional Models

molecule	6-31G*	6-311+G**	expt.
CO	0.1	0.1	**0.11**
HCP	0.7	0.4	**0.39**
PH$_3$	1.0	0.9	**0.58**
ClF	0.9	1.2	**0.88**
H$_2$S	1.5	1.4	**0.97**
HCl	1.5	1.4	**1.08**
SiH$_3$F	1.2	1.8	**1.27**
SiH$_3$Cl	1.7	1.8	**1.30**
NH$_3$	2.0	1.7	**1.47**
HF	1.8	1.9	**1.82**
H$_2$O	2.1	2.1	**1.85**
CH$_3$F	1.6	2.0	**1.85**
CH$_3$Cl	2.0	2.0	**1.87**
CS	1.5	1.8	**1.98**
H$_2$CO	2.1	2.3	**2.34**
HCN	2.8	3.0	**2.99**
LiH	5.6	5.7	**5.83**
LiF	5.5	6.3	**6.28**
NaH	5.9	5.9	**6.96**
LiCl	7.0	7.1	**7.12**
NaF	6.7	8.1	**8.16**
NaCl	8.6	9.0	**9.00**

Table A10-6: Dipole Moments in Diatomic and Small Polyatomic Molecules. B3LYP Density Functional Models

molecule	6-31G*	6-311+G**	expt.
CO	0.1	0.1	**0.11**
HCP	0.7	0.4	**0.39**
PH_3	1.0	0.8	**0.58**
ClF	1.0	1.3	**0.88**
H_2S	1.4	1.4	**0.97**
HCl	1.5	1.4	**1.08**
SiH_3F	1.3	1.8	**1.27**
SiH_3Cl	1.8	1.9	**1.30**
NH_3	1.9	1.7	**1.47**
HF	1.9	2.0	**1.82**
H_2O	2.1	2.2	**1.85**
CH_3F	1.7	2.1	**1.85**
CH_3Cl	2.1	2.1	**1.87**
CS	1.5	1.9	**1.98**
H_2CO	2.2	2.5	**2.34**
HCN	2.9	3.1	**2.99**
LiH	5.6	5.7	**5.83**
LiF	5.6	6.4	**6.28**
NaH	6.0	6.0	**6.96**
LiCl	7.1	7.1	**7.12**
NaF	7.0	8.3	**8.16**
NaCl	8.7	9.0	**9.00**

Table A10-7: Dipole Moments in Diatomic and Small Polyatomic Molecules. MP2 Models

molecule	6-31G*	6-311+G**	expt.
CO	0.2	0.3	**0.11**
HCP	0.8	0.4	**0.39**
PH$_3$	1.0	0.8	**0.58**
ClF	1.1	1.3	**0.88**
H$_2$S	1.5	1.3	**0.97**
HCl	1.5	1.4	**1.08**
SiH$_3$F	1.4	1.8	**1.27**
SiH$_3$Cl	1.8	1.8	**1.30**
NH$_3$	2.0	1.7	**1.47**
HF	1.9	2.0	**1.82**
H$_2$O	2.2	2.2	**1.85**
CH$_3$F	1.9	2.1	**1.85**
CH$_3$Cl	2.0	1.9	**1.87**
CS	2.0	2.4	**1.98**
H$_2$CO	2.3	2.4	**2.34**
HCN	3.0	3.0	**2.99**
LiH	5.8	5.9	**5.83**
LiF	5.9	6.5	**6.28**
NaH	6.6	6.8	**6.96**
LiCl	7.3	7.2	**7.12**
NaF	7.5	8.6	**8.16**
NaCl	9.2	9.4	**9.00**

Table A10-8: Dipole Moments in Diatomic and Small Polyatomic Molecules. Semi-Empirical Models

molecule	MNDO	AM1	PM3	expt.
CO	0.2	0.1	0.2	**0.11**
HCP	1.6	1.8	0.4	**0.39**
PH_3	1.4	2.3	1.2	**0.58**
ClF	0.7	0.9	1.4	**0.88**
H_2S	1.5	1.9	1.8	**0.97**
HCl	1.1	1.4	1.4	**1.08**
SiH_3F	0.7	1.3	1.0	**1.27**
SiH_3Cl	0.6	1.6	2.3	**1.30**
NH_3	1.8	1.9	1.6	**1.47**
HF	2.0	1.7	1.4	**1.82**
H_2O	1.8	1.9	1.7	**1.85**
CH_3F	1.8	1.6	1.4	**1.85**
CH_3Cl	1.9	1.5	1.4	**1.87**
CS	0.4	2.2	1.4	**1.98**
H_2CO	2.2	2.3	2.2	**2.34**
HCN	2.5	2.4	2.7	**2.99**
LiH	5.8	-	5.7	**5.83**
LiF	5.7	-	5.3	**6.28**
LiCl	7.8	-	6.5	**7.12**

Table A10-9: Dipole Moments in Molecules Containing Heteroatoms. Hartree-Fock Models

heteratom	molecule	STO-3G	3-21G	6-31G*	6-311+G**	expt.
Nitrogen	trimethylamine	1.1	0.9	0.7	0.8	**0.61**
	dimethylamine	1.4	1.2	1.1	1.1	**1.03**
	ethylamine	1.6	1.4	1.5	1.4	**1.22**
	methylamine	1.6	1.4	1.5	1.5	**1.31**
	ammonia	1.9	1.8	1.9	1.7	**1.47**
	aniline	1.5	1.6	1.5	1.4	**1.53**
	aziridine	1.8	2.2	1.9	1.9	**1.90**
	pyridine	2.0	2.4	2.3	2.4	**2.19**
Oxygen	furan	0.5	1.1	0.8	0.8	**0.66**
	dimethyl ether	1.3	1.8	1.5	1.5	**1.30**
	ethanol	1.4	1.9	1.7	1.8	**1.69**
	methanol	1.5	2.1	1.9	1.9	**1.70**
	water	1.7	2.4	2.2	2.2	**1.85**
	oxirane	1.5	2.8	2.3	2.3	**1.89**
Silicon	trimethylsilane	0.0	0.5	0.5	0.5	**0.53**
	methylsilane	0.1	0.7	0.7	0.6	**0.74**
	dimethylsilane	0.1	0.7	0.7	0.6	**0.75**
	ethylsilane	0.1	0.8	0.7	0.8	**0.81**
Phosphorus	phosphine	0.6	0.9	0.9	0.8	**0.58**
	methylphosphine	0.5	1.2	1.3	1.3	**1.10**
	ethylphosphine	0.5	1.3	1.4	1.4	**1.17**
	trimethylphosphine	0.3	1.3	1.4	1.4	**1.19**
	dimethylphosphine	0.5	1.3	1.4	1.4	**1.23**
Sulfur	thiophene	0.1	0.8	0.9	0.8	**0.55**
	hydrogen sulfide	1.0	1.4	1.4	1.4	**0.97**
	dimethyl sulfide	0.9	1.7	1.8	1.8	**1.50**
	methane thiol	1.0	1.7	1.8	1.8	**1.52**
	ethane thiol	1.0	1.8	1.9	1.9	**1.58**
	thiirane	0.8	2.1	2.3	2.3	**1.85**

Table A10-10: Dipole Moments in Molecules Containing Heteroatoms. Local Density Models

heteratom	molecule	6-31G*	6-311+G**	expt.
Nitrogen	trimethylamine	0.5	0.4	**0.61**
	dimethylamine	1.0	1.0	**1.03**
	ethylamine	1.5	1.3	**1.22**
	methylamine	1.4	1.3	**1.31**
	ammonia	2.0	1.7	**1.47**
	aniline	2.0	1.9	**1.53**
	aziridine	1.7	1.8	**1.90**
	pyridine	2.2	2.4	**2.19**
Oxygen	furan	0.5	0.6	**0.66**
	dimethyl ether	1.2	1.3	**1.30**
	ethanol	1.5	1.7	**1.69**
	methanol	1.7	1.8	**1.70**
	water	2.2	2.2	**1.85**
	oxirane	1.8	2.0	**1.89**
Silicon	trimethylsilane	0.7	0.6	**0.53**
	methylsilane	1.0	0.8	**0.74**
	dimethylsilane	1.0	0.8	**0.75**
	ethylsilane	1.0	0.9	**0.81**
Phosphorus	phosphine	1.2	1.0	**0.58**
	methylphosphine	1.5	1.3	**1.10**
	ethylphosphine	1.5	1.4	**1.17**
	trimethylphosphine	1.5	1.3	**1.19**
	dimethylphosphine	1.6	1.4	**1.23**
Sulfur	thiophene	0.6	0.4	**0.55**
	hydrogen sulfide	1.6	1.5	**0.97**
	dimethyl sulfide	1.8	1.7	**1.50**
	methane thiol	1.8	1.7	**1.52**
	ethane thiol	1.8	1.8	**1.58**
	thiirane	2.1	2.0	**1.85**

Table A10-11: Dipole Moments in Molecules Containing Heteroatoms. BP Density Functional Models

heteratom	molecule	6-31G*	6-311+G**	expt.
Nitrogen	trimethylamine	0.5	0.5	**0.61**
	dimethylamine	1.0	1.0	**1.03**
	ethylamine	1.5	1.4	**1.22**
	methylamine	1.5	1.4	**1.31**
	ammonia	2.0	1.7	**1.47**
	aniline	1.8	1.7	**1.53**
	aziridine	1.7	1.8	**1.90**
	pyridine	2.2	2.3	**2.19**
Oxygen	furan	0.6	0.7	**0.66**
	dimethyl ether	1.2	1.4	**1.30**
	ethanol	1.5	1.7	**1.69**
	methanol	1.6	1.8	**1.70**
	water	2.1	2.1	**1.85**
	oxirane	1.9	2.0	**1.89**
Silicon	trimethylsilane	0.7	0.6	**0.53**
	methylsilane	0.9	0.7	**0.74**
	dimethylsilane	0.9	0.8	**0.75**
	ethylsilane	1.0	0.9	**0.81**
Phosphorus	phosphine	1.0	0.9	**0.58**
	methylphosphine	1.4	1.2	**1.10**
	ethylphosphine	1.4	1.3	**1.17**
	trimethylphosphine	1.4	1.2	**1.19**
	dimethylphosphine	1.5	1.3	**1.23**
Sulfur	thiophene	0.6	0.5	**0.55**
	hydrogen sulfide	1.5	1.3	**0.97**
	dimethyl sulfide	1.7	1.4	**1.50**
	methane thiol	1.8	1.7	**1.52**
	ethane thiol	1.8	1.8	**1.58**
	thiirane	2.1	2.0	**1.85**

Table A10-12: Dipole Moments in Molecules Containing Heteroatoms. BLYP Density Functional Models

heteratom	molecule	6-31G*	6-311+G**	expt.
Nitrogen	trimethylamine	0.5	0.5	**0.61**
	dimethylamine	1.0	1.0	**1.03**
	ethylamine	1.4	1.4	**1.22**
	methylamine	1.5	1.4	**1.31**
	ammonia	1.9	1.7	**1.47**
	aniline	1.8	1.6	**1.53**
	aziridine	1.7	1.9	**1.90**
	pyridine	2.1	2.3	**2.19**
Oxygen	furan	0.6	0.8	**0.66**
	dimethyl ether	1.2	1.4	**1.30**
	ethanol	1.5	1.7	**1.69**
	methanol	1.6	1.9	**1.70**
	water	2.0	2.1	**1.85**
	oxirane	1.9	1.1	**1.89**
Silicon	trimethylsilane	0.6	0.5	**0.53**
	methylsilane	0.9	0.7	**0.74**
	dimethylsilane	0.9	0.7	**0.75**
	ethylsilane	0.9	0.8	**0.81**
Phosphorus	phosphine	1.0	0.8	**0.58**
	methylphosphine	1.3	1.2	**1.10**
	ethylphosphine	1.4	1.2	**1.17**
	trimethylphosphine	1.3	1.2	**1.19**
	dimethylphosphine	1.4	1.2	**1.23**
Sulfur	thiophene	0.6	0.5	**0.55**
	hydrogen sulfide	1.4	1.3	**0.97**
	dimethyl sulfide	1.7	1.6	**1.50**
	methane thiol	1.7	1.7	**1.52**
	ethane thiol	1.8	1.8	**1.58**
	thiirane	2.1	2.1	**1.85**

Table A10-13: Dipole Moments in Molecules Containing Heteroatoms. EDF1 Density Functional Models

heteratom	molecule	6-31G*	6-311+G**	expt.
Nitrogen	trimethylamine	0.5	0.5	**0.61**
	dimethylamine	1.0	1.0	**1.03**
	ethylamine	1.4	1.4	**1.22**
	methylamine	1.1	1.4	**1.31**
	ammonia	1.9	1.7	**1.47**
	aniline	1.8	1.7	**1.53**
	aziridine	1.7	1.8	**1.90**
	pyridine	2.2	2.3	**2.19**
Oxygen	furan	0.6	0.7	**0.66**
	dimethyl ether	1.2	1.4	**1.30**
	ethanol	1.5	1.7	**1.69**
	methanol	1.5	1.8	**1.70**
	water	2.1	2.1	**1.85**
	oxirane	1.8	2.0	**1.89**
Silicon	trimethylsilane	0.7	0.6	**0.53**
	methylsilane	0.8	0.7	**0.74**
	dimethylsilane	0.9	0.7	**0.75**
	ethylsilane	0.9	0.9	**0.81**
Phosphorus	phosphine	1.0	0.9	**0.58**
	methylphosphine	1.4	1.2	**1.10**
	ethylphosphine	1.4	1.3	**1.17**
	trimethylphosphine	1.4	1.3	**1.19**
	dimethylphosphine	1.4	1.3	**1.23**
Sulfur	thiophene	0.6	0.5	**0.55**
	hydrogen sulfide	1.5	1.4	**0.97**
	dimethyl sulfide	1.7	1.7	**1.50**
	methane thiol	1.8	1.7	**1.52**
	ethane thiol	1.8	1.8	**1.58**
	thiirane	2.1	2.0	**1.85**

745

Table A10-14: Dipole Moments in Molecules Containing Heteroatoms. B3LYP Density Functional Models

heteratom	molecule	6-31G*	6-311+G**	expt.
Nitrogen	trimethylamine	0.6	0.6	**0.61**
	dimethylamine	1.0	1.1	**1.03**
	ethylamine	1.4	1.4	**1.22**
	methylamine	1.5	1.4	**1.31**
	ammonia	1.9	1.7	**1.47**
	aniline	1.7	1.6	**1.53**
	aziridine	1.8	1.9	**1.90**
	pyridine	2.2	2.4	**2.19**
Oxygen	furan	0.6	0.8	**0.66**
	dimethyl ether	1.3	1.5	**1.30**
	ethanol	1.6	1.8	**1.69**
	methanol	1.7	1.9	**1.70**
	water	2.1	2.2	**1.85**
	oxirane	2.0	2.1	**1.89**
Silicon	trimethylsilane	0.6	0.5	**0.53**
	methylsilane	0.8	0.7	**0.74**
	dimethylsilane	0.8	0.7	**0.75**
	ethylsilane	0.9	0.8	**0.81**
Phosphorus	phosphine	1.0	0.8	**0.58**
	methylphosphine	1.3	1.2	**1.10**
	ethylphosphine	1.4	1.3	**1.17**
	trimethylphosphine	1.4	1.2	**1.19**
	dimethylphosphine	1.4	1.3	**1.23**
Sulfur	thiophene	0.6	0.5	**0.55**
	hydrogen sulfide	1.4	1.4	**0.97**
	dimethyl sulfide	1.7	1.7	**1.50**
	methane thiol	1.7	1.7	**1.52**
	ethane thiol	1.8	1.8	**1.58**
	thiirane	2.2	2.1	**1.85**

Table A10-15: Dipole Moments in Molecules Containing Heteroatoms. MP2 Models

heteratom	molecule	6-31G*	6-311+G**	expt.
Nitrogen	trimethylamine	0.7	0.7	**0.61**
	dimethylamine	1.2	1.2	**1.03**
	ethylamine	1.5	1.5	**1.22**
	methylamine	1.6	1.5	**1.31**
	ammonia	2.0	1.7	**1.47**
	aniline	1.6	1.6	**1.53**
	aziridine	1.9	1.9	**1.90**
	pyridine	2.3	2.4	**2.19**
Oxygen	furan	0.7	0.7	**0.66**
	dimethyl ether	1.4	1.5	**1.30**
	ethanol	1.7	1.8	**1.69**
	methanol	1.8	1.9	**1.70**
	water	2.2	2.2	**1.85**
	oxirane	2.1	2.1	**1.89**
Silicon	trimethylsilane	0.6	0.5	**0.53**
	methylsilane	0.7	0.6	**0.74**
	dimethylsilane	0.7	0.6	**0.75**
	ethylsilane	0.8	0.7	**0.81**
Phosphorus	phosphine	1.0	0.8	**0.58**
	methylphosphine	1.3	1.1	**1.10**
	ethylphosphine	1.4	1.2	**1.17**
	trimethylphosphine	1.4	1.2	**1.19**
	dimethylphosphine	1.4	1.2	**1.23**
Sulfur	thiophene	0.5	0.3	**0.55**
	hydrogen sulfide	1.5	1.3	**0.97**
	dimethyl sulfide	1.8	1.6	**1.50**
	methane thiol	1.8	1.6	**1.52**
	ethane thiol	1.8	1.7	**1.58**
	thiirane	2.1	1.9	**1.85**

Table A10-16: **Dipole Moments in Molecules Containing Heteroatoms. Semi-Empirical Models**

heteratom	molecule	MNDO	AM1	PM3	expt.
Nitrogen	trimethylamine	0.8	1.0	1.2	**0.61**
	dimethylamine	1.2	1.2	1.3	**1.03**
	ethylamine	1.5	1.6	1.4	**1.22**
	methylamine	1.5	1.5	1.4	**1.31**
	ammonia	1.8	1.9	1.6	**1.47**
	aniline	1.5	1.5	1.3	**1.53**
	aziridine	1.8	1.8	1.7	**1.90**
	pyridine	2.0	2.0	1.9	**2.19**
Oxygen	furan	0.4	0.5	0.2	**0.66**
	dimethyl ether	1.3	1.4	1.3	**1.30**
	ethanol	1.4	1.6	1.5	**1.69**
	methanol	1.5	1.6	1.5	**1.70**
	water	1.8	1.9	1.7	**1.85**
	oxirane	1.9	1.9	1.8	**1.89**
Silicon	trimethylsilane	0.9	0.4	0.4	**0.53**
	methylsilane	1.0	0.4	0.4	**0.74**
	dimethylsilane	1.0	0.5	0.5	**0.75**
	ethylsilane	0.1	0.4	0.4	**0.81**
Phosphorus	phosphine	1.4	2.3	1.2	**0.58**
	methylphosphine	1.6	2.0	1.2	**1.10**
	ethylphosphine	1.6	2.1	1.2	**1.17**
	trimethylphosphine	1.9	1.5	1.1	**1.19**
	dimethylphosphine	1.8	1.8	1.1	**1.23**
Sulfur	thiophene	0.2	0.3	0.7	**0.55**
	hydrogen sulfide	1.1	1.9	1.8	**0.97**
	dimethyl sulfide	1.6	1.6	2.0	**1.50**
	methane thiol	1.4	1.8	2.0	**1.52**
	ethane thiol	1.4	1.8	2.0	**1.58**
	thiirane	1.7	2.1	2.4	**1.85**

Table A10-17: Dipole Moments in Hypervalent Molecules. Hartree-Fock Models

molecule	STO-3G	3-21G	6-31G*	6-311+G**	expt.
F_3PO	1.6	1.7	2.0	1.9	**1.76**
F_3PS	1.8	1.2	1.4	0.8	**0.64**
SO_2	2.2	2.3	2.2	2.3	**1.63**
$(CH_3)_2SO$	3.7	4.3	4.5	5.0	**3.96**
SF_4	1.3	1.4	1.0	1.1	**0.63**
F_2SO	2.6	2.4	2.2	2.3	**1.63**
NSF	1.4	2.2	2.1	2.3	**1.90**
$(CH_3)_2SO_2$	2.6	5.0	5.1	5.4	**4.49**
ClF_3	0.5	1.2	0.9	1.0	**0.55**
ClF_5	1.7	1.3	0.8	1.0	**0.54**
$FClO_2$	1.3	2.6	2.3	2.5	**1.72**
$FClO_3$	1.2	0.5	0.5	0.3	**0.02**

Table A10-18: Dipole Moments in Hypervalent Molecules. Local Density Models

molecule	6-31G*	6-311+G**	expt.
F_3PO	1.7	1.5	**1.76**
F_3PS	0.9	0.0	**0.64**
SO_2	1.6	1.9	**1.63**
$(CH_3)_2SO$	3.8	4.3	**3.96**
SF_4	0.8	1.1	**0.63**
F_2SO	1.6	1.9	**1.63**
NSF	1.8	2.1	**1.90**
$(CH_3)_2SO_2$	4.5	4.8	**4.49**
ClF_3	0.8	1.0	**0.55**
ClF_5	0.8	1.0	**0.54**
$FClO_2$	1.7	2.3	**1.72**
$FClO_3$	0.0	0.7	**0.02**

Table A10-19: **Dipole Moments in Hypervalent Molecules. BP Density Functional Models**

molecule	6-31G*	6-311+G**	expt.
F_3PO	1.6	1.4	**1.76**
F_3PS	0.9	0.0	**0.64**
SO_2	1.6	1.9	**1.63**
$(CH_3)_2SO$	3.7	4.2	**3.96**
SF_4	0.8	1.1	**0.63**
F_2SO	1.6	2.0	**1.63**
NSF	1.7	2.2	**1.90**
$(CH_3)_2SO_2$	4.4	4.7	**4.49**
ClF_3	0.8	1.0	**0.55**
ClF_5	0.8	0.9	**0.54**
$FClO_2$	1.8	2.4	**1.72**
$FClO_3$	0.2	0.9	**0.02**

Table A10-20: **Dipole Moments in Hypervalent Molecules. BLYP Density Functional Models**

molecule	6-31G*	6-311+G**	expt.
F_3PO	1.6	1.3	**1.76**
F_3PS	0.9	0.1	**0.64**
SO_2	1.6	1.9	**1.63**
$(CH_3)_2SO$	3.7	4.2	**3.96**
SF_4	0.8	1.2	**0.63**
F_2SO	1.6	2.1	**1.63**
NSF	1.7	2.3	**1.90**
$(CH_3)_2SO_2$	4.4	4.8	**4.49**
ClF_3	0.8	1.0	**0.55**
ClF_5	0.8	0.9	**0.54**
$FClO_2$	1.8	2.5	**1.72**
$FClO_3$	0.3	1.0	**0.02**

Table A10-21: Dipole Moments in Hypervalent Molecules. EDF1 Density Functional Models

molecule	6-31G*	6-311+G**	expt.
F_3PO	1.6	1.4	**1.76**
F_3PS	0.9	0.0	**0.64**
SO_2	1.6	1.9	**1.63**
$(CH_3)_2SO$	3.7	4.2	**3.96**
SF_4	0.8	1.1	**0.63**
F_2SO	1.6	2.0	**1.63**
NSF	1.7	2.2	**1.90**
$(CH_3)_2SO_2$	4.4	4.7	**4.49**
ClF_3	0.8	0.9	**0.55**
ClF_5	0.7	0.9	**0.54**
$FClO_2$	1.8	2.4	**1.72**
$FClO_3$	0.2	0.9	**0.02**

Table A10-22: Dipole Moments in Hypervalent Molecules. B3LYP Density Functional Models

molecule	6-31G*	6-311+G**	expt.
F_3PO	1.7	1.5	**1.76**
F_3PS	0.9	0.1	**0.64**
SO_2	1.8	2.0	**1.63**
$(CH_3)_2SO$	3.9	4.5	**3.96**
SF_4	0.9	1.2	**0.63**
F_2SO	1.8	2.1	**1.63**
NSF	1.9	2.3	**1.90**
$(CH_3)_2SO_2$	4.6	5.0	**4.49**
ClF_3	0.9	1.0	**0.55**
ClF_5	0.8	1.0	**0.54**
$FClO_2$	1.9	2.5	**1.72**
$FClO_3$	0.0	0.7	**0.02**

Table A10-23: Dipole Moments in Hypervalent Molecules. MP2 Models

molecule	6-31G*	6-311+G**	expt.
F_3PO	1.6	1.5	**1.76**
F_3PS	0.8	0.1	**0.64**
SO_2	1.7	1.9	**1.63**
$(CH_3)_2SO$	4.1	4.4	**3.96**
SF_4	1.0	1.1	**0.63**
F_2SO	1.8	2.1	**1.63**
NSF	2.1	2.8	**1.90**
$(CH_3)_2SO_2$	4.6	4.8	**4.49**
ClF_3	1.0	1.1	**0.55**
ClF_5	0.9	1.0	**0.54**
$FClO_2$	2.1	3.2	**1.72**
$FClO_3$	0.3	1.1	**0.02**

Table A10-24: Dipole Moments in Hypervalent Molecules. Semi-Empirical Models

molecule	MNDO	AM1	PM3	expt.
F_3PO	1.2	2.6	2.0	**1.76**
F_3PS	0.4	4.7	2.7	**0.64**
SO_2	3.6	4.3	3.6	**1.63**
$(CH_3)_2SO$	4.2	4.0	4.5	**3.96**
SF_4	2.4	2.7	2.4	**0.63**
F_2SO	3.2	4.2	3.5	**1.63**
NSF	2.7	3.0	2.7	**1.90**
$(CH_3)_2SO_2$	4.8	4.2	5.0	**4.49**
ClF_3	0.9	0.0	0.0	**0.55**
ClF_5	1.9	1.0	1.1	**0.54**
$FClO_2$	4.0	4.7	7.0	**1.72**
$FClO_3$	0.2	0.3	0.3	**0.02**

Appendix B
Common Terms and Acronyms

3-21G. A **Split-Valence Basis Set** in which each **Core Basis Function** is written in terms of three **Gaussians**, and each **Valence Basis Function** is split into two parts, written in terms of two and one **Gaussians**, respectively. 3-21G basis sets have been determined to yield the lowest total **Hartree-Fock Energies** for atoms.

3-21G$^{(*)}$. The **3-21G Basis Set** supplemented by d-type **Gaussians** for each second-row and heavier main-group elements only. 3-21G$^{(*)}$ is a supplemented **Split-Valence Basis Set**.

6-31G. A **Split-Valence Basis Set** in which each **Core Basis Function** is written in terms of six **Gaussians**, and each **Valence Basis Function** is split into two parts, written in terms of three and one **Gaussians**, respectively. 6-31G basis sets have been determined to yield the lowest total **Hartree-Fock Energies** for atoms.

6-31G*, **6-31G****. The **6-31G Basis Set** in which non-hydrogen atoms are supplemented by d-type **Gaussians** and (for 6-31G**) hydrogen atoms are supplemented by p-type **Gaussians** (**Polarization Functions**). 6-31G* and 6-31G** are **Polarization Basis Sets**.

6-31+G*, **6-31+G****. **Basis Sets** that are identical to **6-31G*** and **6-31G**** except that all non-hydrogen atoms are supplemented by diffuse s and p-type **Gaussians** (**Diffuse Functions**). 6-31+G* and 6-31+G** are supplemented **Polarization Basis Sets**.

6-311G. A **Split-Valence Basis Set** in which each **Core Basis Function** is written in terms of six **Gaussians**, and each **Valence Basis Function** is split into three parts, written in terms of three, one and one **Gaussians**, respectively. 6-311G basis sets have been determined to yield the lowest total **MP2 Energies** for atoms.

6-311G*, 6-311G**. The **6-311G Basis Set** in which non-hydrogen atoms are supplemented by d-type **Gaussians** and (for 6-311G**) hydrogen atoms are supplemented by p-type **Gaussians** (**Polarization Functions**). 6-311G* and 6-311G** are **Polarization Basis Sets**.

6-311+G*, 6-311+G**. Basis Sets that are identical to **6-311G*** and **6-311G**** except that non-hydrogen atoms are supplemented by diffuse s and p-type **Gaussians** (**Diffuse Functions**). 6-311+G* and 6-311+G** are supplemented **Polarization Basis Sets**.

***Ab Initio* Models.** The general term used to describe methods seeking approximate solutions to the many-electron **Schrödinger Equation**, but which do not involve empirical parameters. *Ab initio* models include **Hartree-Fock Models**, **Møller-Plesset Models** and **Density Functional Models**.

Activation Energy. The energy of a **Transition State** above that of reactants. Activation energy is related to reaction rate by way of the **Arrhenius Equation**.

AM1. Austin Method **1**. A **Semi-Empirical Model**.

Antibonding Molecular Orbital. A **Molecular Orbital** which is antibonding between particular atomic centers. The opposite is a **Bonding Molecular Orbital**.

Arrhenius Equation. An equation governing the rate of a chemical reaction as a function of the **Activation Energy** and the temperature.

Atomic Orbital. A **Basis Function** centered on an atom. Atomic orbitals typically take on the form of the solutions to the hydrogen atom (s, p, d, f... type orbitals).

Atomic Units. The set of units which remove all of the constants from inside the **Schrödinger Equation**. The **Bohr** is the atomic unit of length and the **Hartree** is the atomic unit of energy.

Atomization Energy. The energy of dissociation of an atom or molecule into separated nuclei and electrons. Atomization energy is the quantity which is calculated in **G2** and **G3 Models**.

B3LYP Model. A **Hybrid Density Functional Model** which improves on the **Local Density Model** by accounting explicitly for non-uniformity in electron distributions, and which also incorporates the **Exchange Energy** from the **Hartree-Fock Model**. The B3LYP model involves three adjustable parameters.

Basis Functions. Functions usually centered on atoms which are linearly combined to make up the set of **Molecular Orbitals**. Except for **Semi-Empirical Models** where basis functions are **Slater** type, basis functions are **Gaussian** type.

Basis Set. The entire collection of **Basis Functions**.

Becke-Lee-Yang-Parr Model. *See* **BLYP Model**, **B3LYP Model**.

Becke-Perdew Model. *See* **BP Model**.

BLYP Model. A **Density Functional Model** which improves on the **Local Density Model** by accounting explicitly for non-uniformity in electron distributions.

Bohr. The **Atomic Unit** of length. 1 bohr = 0.529167Å.

Boltzmann Equation. The equation governing the distribution of products in **Thermodynamically-Controlled Reaction**.

Bonding Molecular Orbital. A **Molecular Orbital** which is bonding between particular atom centers. The opposite is an **Antibonding Molecular Orbital**.

Bond Separation Reaction. An *Isodesmic* **Reaction** in which a molecule described in terms of a conventional valence structure is broken down into the simplest (two-heavy-atom) molecules containing the same component bonds.

Bond Surface. An **Isodensity Surface** used to elucidate the bonding in molecules. The value of the density is typically taken as 0.1 electrons/**bohr**.[3]

Born-Oppenheimer Approximation. An approximation based on the assumption that nuclei are stationary. Applied to the **Schrödinger Equation**, it leads to the **Electronic Schrödinger Equation**.

BP Model; A **Density Functional Model** which improves on the **Local Density Model** by accounting explicitly for non-uniformity in electron distributions.

Cambridge Structural Database. A collection of >270K experimental structures for organic and organometallic compounds from X-ray crystallography and neutron diffraction.

Cartesian Coordinates. x, y, z spatial coordinates.

cc-pVDZ. A **Polarization Basis Set** which has been developed specifically for use with **Correlated Models**. The **Valence** has been split into two parts. May be supplemented with **Diffuse Functions**.

cc-pVQZ. A **Polarization Basis Set** which has been developed specifically for use with **Correlated Models**. The **Valence** has been split into four parts, and functions though g have been included. May be supplemented with **Diffuse Functions**.

cc-pVTZ. A **Polarization Basis Set** which has been developed specifically for use with **Correlated Models**. The **Valence** has been split into three parts, and functions though f have been included. May be supplemented with **Diffuse Functions**.

CCSD Model. Coupled Cluster Singles and Doubles. A **Correlated Model**.

CCSD (T) Model. Coupled Cluster Singles and Doubles with Triples correction. A **Correlated Model**.

CID Model. Configuration Interaction Doubles. A limited **Configuration Interaction** scheme in which only double excitations from occupied to unoccupied molecular orbitals are considered.

CIS Model. Configuration Interactions Singles. A limited **Configuration Interaction** scheme in which only single excitations from occupied to unoccupied molecular orbitals are considered. This is perhaps the simplest method available to the description of **Excited States** of molecules.

CISD Model. Configuration Interaction, Singles and Doubles. A limited **Configuration Interaction** scheme in which only single and double excitations from occupied to unoccupied molecular orbitals are considered.

Configuration Interaction. Provides an account of **Electron Correlation** by way of explicit promotion (excitation) of electrons from occupied molecular orbitals into unoccupied molecular orbitals. Full configuration interaction (all possible promotions) is not a practical method and limited schemes, for example, **CIS**, **CID** and **CISD Models**, need to be employed.

Closed Shell. An atom or molecule in which all electrons are paired.

Conformation. The arrangement about single bonds and of flexible rings.

Correlated Models. Models which take implicit or explicit account of the **Correlation** of electron motions. **Møller-Plesset Models**, **Configuration Interaction Models** and **Density Functional Models** are correlated models.

Correlation. The coupling of electron motions not explicitly taken into account in **Hartree-Fock Models**.

Correlation Energy. The difference in energy between the **Hartree-Fock Energy** and the experimental energy.

Coulomb Energy. The electron-electron repulsion energy according to Coulomb's law.

Coulombic Interactions. Charge-charge interactions which follow Coulomb's law. Stabilizing when charges are of opposite sign and destabilizing when they are of the same sign.

Core. Electrons which are primarily associated with individual atoms and do not participate significantly in chemical bonding (1s electrons for first-row elements, 1s, 2s, $2p_x$, $2p_y$, $2p_z$ electrons for second-row elements, etc.).

CPK Model. A molecular model in which atoms are represented by spheres, the radii of which correspond to **van der Waals Radii**. Intended to portray molecular size and shape.

CSD; *See* **Cambridge Structural Database**.

Curtin-Hammett Principle. The idea that reactive conformer in a **Kinetically-Controlled Reaction** is not necessarily the lowest-energy conformer. The rationale is that the energy barriers separating conformers are typically much smaller than barriers to chemical reaction, and that any reactive conformers will be replenished.

Density; *See* **Electron Density**.

Density Functional Models. Methods in which the energy is evaluated as a function of the **Electron Density**. **Electron Correlation** is taken into account explicitly by incorporating into the **Hamiltonian** terms which derive from exact solutions of idealized many-electron systems.

Diffuse Functions. Functions added to a **Basis Set** to allow description of electron distributions far away from atomic positions. Important for descriptions of anions.

Diffusion-Controlled Reactions. Chemical reactions without **Transition States** (or energy barriers), the rates of which are determined by the speed in which molecules encounter each other and how likely these encounters are to lead to reaction.

EDF1 Model. A **Density Functional Model** which improves on the **Local Density Model** by accounting explicitly for non-uniformity in electron distributions.

Electron Correlation. *See* **Correlation**.

Electron Density. The number of electrons per unit volume at a point in space. This is the quantity which is measured in an X-ray diffraction experiment.

Electronic Schrödinger Equation. The equation which results from incorporation of the **Born-Oppenheimer Approximation** to the **Schrödinger Equation**.

Electrostatic Charges. Atomic charges chosen to best match the **Electrostatic Potential** at points surrounding a molecule, subject to overall charge balance.

Electrostatic Potential. A function describing the energy of interaction of a positive point charge with the nuclei and fixed electron distribution of a molecule.

Electrostatic Potential Map. A graph that shows the value of **Electrostatic Potential** on an **Electron Density Isosurface** corresponding to a **van der Waals Surface**.

Equilibrium Geometry. A **Local Minimum** on a **Potential Energy Surface**.

Exchange/Correlation Functional. A function of the **Electron Density** and perhaps as well the gradient of the **Electron Density**. The functional form derives from "exact" solution of the **Schrödinger Equation** for an idealized many-electron problem. Used in **Density Functional Models**.

Exchange Energy. An "attractive" (negative) component of the electron-electron interaction energy. Arises due to an overestimation of the "repulsive" (positive) component or **Coulomb energy**.

Excited State. An electronic state for an atom or molecule which is not the lowest-energy or **Ground State**.

Force Field. The set of rules underlying **Molecular Mechanics Models**. Comprises terms which account for distortions from ideal bond distances and angles and for **Non-Bonded van der Waals** and **Coulombic Interactions**.

Frontier Molecular Orbitals. The **HOMO** and **LUMO**.

G2, G3 Models. "Recipes" which combine a series of **Hartree-Fock** and **Correlated Models** to properly account for **Atomization Energies** of atoms and molecules. Useful for providing accurate thermochemical data. G3 is a more recent implementation of G2.

Gaussian. A function of the form $x^l y^m z^n \exp(\alpha r^2)$ where l, m, n are integers $(0, 1, 2 \ldots)$ and α is a constant. Used in the construction of **Basis Sets** for **Hartree-Fock**, **Density Functional**, **MP2** and other **Correlated Models**.

Gaussian Basis Set. A **Basis Set** made up of **Gaussian Basis Functions**.

Gaussian Function; *See* **Gaussian**.

Global Minimum. The lowest energy **Local Minimum** on a **Potential Energy Surface**.

Gradient Corrected Density Functional Models; *See* **Non-Local Density Functional Models**.

Ground State. The lowest energy electronic state for an atom or molecule.

Hamiltonian. An operator which accounts for the kinetic and potential energy of an atom or a molecule.

Hammond Postulate. The idea that the **Transition State** for an *exothermic* reaction will more closely resemble reactants than products. This provides the basis for "modeling" properties of **Transition States** in terms of the properties of reactants.

Harmonic Frequency. A **Vibrational Frequency** which has been corrected to remove all non-quadratic (non-harmonic) components. Calculated **Vibrational Frequencies** correspond to harmonic frequencies. The corrections require data on isotopically-substituted systems and are typically available only for small molecules.

Hartree. The **Atomic Unit** of energy. 1 hartree = 627.47 kcal/mol.

Hartree-Fock Approximation. Separation of electron motions in many-electron systems into a product form of the motions of the individual electrons.

Hartree-Fock Energy. The energy resulting from **Hartree-Fock Models**.

Hartree-Fock Equations. The set of differential equations resulting from application of the **Born-Oppenheimer** and **Hartree-Fock Approximations** to the many-electron **Schrödinger Equation**.

Hartree-Fock Models. Methods in which the many-electron wavefunction in written terms of a product of one-electron wavefunctions. Electrons are assigned in pairs to functions called **Molecular Orbitals**.

Hartree-Fock Wavefunction. The simplest quantum-mechanically correct representation of the many-electron wavefunction. Electrons are treated as independent particles and are assigned in pairs to functions termed **Molecular Orbitals**. Also known as **Single-Determinant Wavefunction**.

Hessian. The matrix of second energy derivatives with respect to geometrical coordinates. The Hessian together with the atomic masses lead to the **Vibrational Frequencies** of molecular systems.

Heterolytic Bond Dissociation. A process in which a bond is broken and a cation and anion result. The number of electron pairs is conserved, but a non-bonding electron pair has been substituted for a bond.

HOMO. Highest Occupied Molecular Orbital.

Homolytic Bond Dissociation. A process in which a bond is broken and two radicals result. The number of electron pairs is not conserved.

HOMO Map. A graph of the absolute value of the **HOMO** on an **Isodensity Surface** corresponding to a **van der Waals Surface**.

Hybrid Density Functional Models. **Density Functional Models** which incorporate the **Exchange Energy** from the **Hartree-Fock Model**. The **B3LYP Model** is a hybrid density functional model.

Hypervalent Molecule. A molecule containing one or more main-group elements in which the normal valence of eight electrons has been exceeded. Hypervalent molecules are common for second-row and heavier main-group elements but are uncommon for first-row elements.

Imaginary Frequency. A frequency which results from a negative element in the diagonal form of the **Hessian**. **Equilibrium Geometries** are characterized by all real frequencies while **Transition States** are characterized by one imaginary frequency.

Internal Coordinates. A set of bond lengths, bond angles and dihedral angles (among other possible variables) describing molecular geometry. The **Z Matrix** is a set of internal coordinates.

Intrinsic Reaction Coordinate. The name given to procedures with seek to "walk" in a smooth pathway connecting reactant and product.

Isodensity Surface. An **Electron Density Isosurface**. **Bond Surfaces** and **Size Surfaces** may be used to elucidate bonding or to characterize overall molecular size and shape, respectively.

Isodesmic **Reaction**. A chemical reaction in which the number of formal chemical bonds of each type is conserved.

Isopotential Surface. An **Electrostatic Potential Isosurface**. It may be used to elucidate regions in a molecule which are particularly electron rich and subject to electrophilic attack and those which are particularly electron poor, subject to nucleophilic attack.

Isosurface. A three-dimensional surface defined by the set of points in space where the value of the function is constant.

Isotope Effect. Dependence of molecular properties and chemical behavior on atomic masses.

Isovalue Surface; *See* **Isosurface**.

Kinetically-Controlled Reaction. Refers to a chemical reaction which has not gone all the way to completion, and the ratio of products is not related to their thermochemical stabilities, but rather inversely to the heights of the energy barriers separating reactants to products.

Kinetic Product. The product of a **Kinetically-Controlled Reaction**.

Kohn-Sham Equations. The set of equations obtained by applying the **Local Density Approximation** to a general multi-electron system. An **Exchange/Correlation Functional** which depends on the electron density has replaced the **Exchange Energy** expression used in the **Hartree-Fock Equations**. The Kohn-Sham equations become the **Roothaan-Hall Equations** if this functional is set equal to the **Hartree-Fock Exchange Energy** expression.

LCAO Approximation. Linear Combination of Atomic Orbitals approximation. Approximates the unknown **Hartree-Fock Wavefunctions** (**Molecular Orbitals**) by linear combinations of atom-centered functions (**Atomic Orbitals**) and leads to the **Roothaan-Hall Equations**.

LACVP*, LACVP**. **Pseudopotentials** which include not only **Valence** s and p-type orbitals (s, p and d-type for transition metals) but also the highest set of **Core** orbitals, and in which non-hydrogen atoms are supplemented by d-type **Gaussians** and (for LACVP**) hydrogen atoms are supplemented by p-type **Gaussians**.

Linear Synchronous Transit. The name given to procedures which estimate the geometries of transition states based on "averages" of the geometries of reactants and products, sometimes weighted by the overall thermodynamics of reaction.

LMP2 Model. An **MP2 Model** in which the **Hartree-Fock** orbitals are first localized.

Local Density Models. **Density Functional Models** which are based on the assumption that the **Electron Density** is constant (or slowly varying) throughout all space.

Local Ionization Potential. A function of the relative ease of electron removal (ionization) from a molecule.

Local Ionization Potential Map. A graph of the **Local Ionization Potential** on an **Isodensity Surface** corresponding to a **van der Waals Surface**.

Local Minimum. Any **Stationary Point** on a **Potential Energy Surface** for which all elements in the diagonal representation of the **Hessian** are positive.

Local Spin Density Models; *See* **Local Density Models**.

Lone Pair. A **Non-Bonded Molecular Orbital** which is typically associated with a single atom.

LUMO. Lowest Unoccupied Molecular Orbital.

LUMO Map. A graph of the absolute value of the **LUMO** on an **Isodensity Surface** corresponding to a **van der Waals Surface**.

Mechanics Models; *See* **Molecular Mechanics Models**.

Mechanism. The sequence of steps connecting reactants and products in an overall chemical reaction. Each step starts from an equilibrium form (reactant or intermediate) and ends in an equilibrium form (intermediate or product).

Merck Molecular Force Field; *See* **MMFF94**.

Minimal Basis Set. A **Basis Set** which contains the fewest functions needed to hold all the electrons on an atom and still maintain spherical symmetry. **STO-3G** is a minimal basis set.

MM2, MM3, MM4. A series of **Molecular Mechanics Force Fields** developed by the Allinger group at the University of Georgia.

MMFF94. Merck Molecular Force Field. A **Molecular Mechanics Force Field** for organic molecules and biopolymers developed by Merck Pharmaceuticals.

MNDO. Modified Neglect of Differential Overlap. A **Semi-Empirical Model.**

MNDO/d. An extension of the **MNDO Semi-Empirical Model** in which second-row (and heavier) main-group elements are provided a set of d-type functions.

Molecular Mechanics; *See* **Molecular Mechanics Models**.

Molecular Mechanics Models. Methods for structure, conformation and strain energy calculation based on bond stretching, angle bending and torsional distortions, together with **Non-Bonded Interactions**, and parameterized to fit experimental data.

Molecular Orbital. A one-electron function made of contributions of **Basis Functions** on individual atoms (**Atomic Orbitals**) and delocalized throughout the entire molecule.

Molecular Orbital Models. Methods based on writing the many-electron solution of the **Electronic Schrödinger Equation** in terms of a product of one-electron solutions (**Molecular Orbitals**).

Møller-Plesset Energy. The energy resulting from **Møller-Plesset Models** terminated to a given order, e.g., the **MP2** energy is the energy of the second-order **Møller-Plesset Model** (or **MP2**).

Møller-Plesset Models. Methods which partially account for **Electron Correlation** by way of the perturbation theory of Møller and Plesset.

MP2 Energy. The energy resulting from **MP2 Models**.

MP2 Model. A **Møller-Plesset Model** terminated to be second order in the energy.

MP3 Model. A **Møller-Plesset Model** terminated to be third order in the energy.

MP4 Model. A **Møller-Plesset Model** terminated to be fourth order in the energy.

Mulliken Charges. Atomic charges obtained from a **Mulliken Population Analysis**.

Mulliken Population Analysis. A partitioning scheme in which electrons are shared equally between different **Basis Functions**.

Multiplicity. The number of unpaired electrons (number of electrons with "down" spin) +1. 1=singlet; 2=doublet; 3=triplet, etc.

NDDO Approximation. **N**eglect of **D**iatomic **D**ifferential **O**verlap approximation. The approximation underlying all present generation **Semi-Empirical Models**. It says that two **Atomic Orbitals** on different atoms do not overlap.

Non-Bonded Interactions. Interactions between atoms which are not directly bonded. **van der Waals Interactions** and **Coulombic Interactions** are non-bonded interactions.

Non-Bonded Molecular Orbital. A molecular orbital which does not show any significant **Bonding** or **Antibonding** characteristics. A **Lone Pair** is a non-bonded molecular orbital.

Non-Local Density Functional Models. **Density Functional Models** such as the **BLYP, BP, EDF1** and **B3LYP Models** which explicitly take into account electron inhomogeneities.

Normal Coordinates. Coordinates that lead to a diagonal **Hessian**, and which correspond to vibrational motions in molecules.

Normal Mode Analysis. The process for calculating **Normal Coordinates** leading to **Vibrational Frequencies**. This involves diagonalizing the **Hessian** and accounting for atomic masses.

Octet Rule. The notion that main-group elements prefer to be "surrounded" by eight electrons (going into s, p_x, p_y, p_z orbitals).

Orbital Symmetry Rules; *See* **Woodward-Hoffmann Rules**.

Open Shell. An atom or molecule in which one or more electrons are unpaired.

PM3. **P**arameterization **M**ethod **3**. A **Semi-Empirical Model**.

Polarization Basis Set. A **Basis Set** which contains functions of higher angular quantum number (**Polarization Functions**) than required for the **Ground State** of the atom, e.g., p-type functions for hydrogen and d-type functions for main-group elements. **6-31G***, **6-31G****, **6-311G***, **6-311G****, **cc-pVDZ**, **cc-pVTZ** and **cc-pVQZ** are polarization basis sets.

Polarization Functions. Functions of higher angular quantum than required for the **Ground State** atomic description. Added to a **Basis Set** to allow displacement of **Valence Basis Functions** away from atomic positions.

Polarization Potential. A function describing the energy of electronic relaxation of a molecular charge distribution following interaction with a point positive charge. The polarization potential may be added to the **Electrostatic Potential** to provide a more accurate account of the interaction of a point-positive charge and a molecule.

Pople-Nesbet Equations. The set of equations describing the best **Unrestricted Single Determinant Wavefunction** within the **LCAO Approximation**. These reduce the **Roothaan-Hall Equations** for **Closed Shell** (paired electron) systems.

Potential Energy Surface. A function of the energy of a molecule in terms of the geometrical coordinates of the atoms.

Property Map. A representation or "map" of a "property" on top of an **Isosurface**, typically an **Isodensity Surface**. **Electrostatic Potential Maps**, and **HOMO** and **LUMO Maps** and **Spin Density Maps** are useful property maps.

Pseudopotential. A **Basis Set** which treats only **Valence** electrons in an explicit manner, all other electrons being considered as a part of a "**Core**". **LAVCP** (and extensions including **Polarization** and/or **Diffuse Functions**) are pseudopotentials.

Pseudorotation. A mechanism for interconversion of *equatorial* and *axial* sites around trigonal bipyramidal centers, e.g., fluorines in phosphorous pentafluoride.

QCISD Model. Quadratic **CISD Model**. A **Correlated Model**.

QSAR. **Q**uantitative **S**tructure **A**ctivity **R**elationships. The name given to attempts to relate measured or calculated properties to molecular structure.

Quantum Mechanics. Methods based on approximate solution of the **Schrödinger Equation**.

Rate Limiting Step. The step in an overall chemical reaction (**Mechanism**) which proceeds via the highest-energy **Transition State**.

Reaction Coordinate. The coordinate that connects the **Local Minima** corresponding to the reactant and product, and which passes through a **Transition State**.

Reaction Coordinate Diagram. A plot of energy vs. **Reaction Coordinate**.

Restricted SCF. An **SCF** procedure which restricts electrons to be paired in orbitals or permits orbitals to be singly occupied.

Roothaan-Hall Equations. The set of equations describing the best **Hartree-Fock** or **Single-Determinant Wavefunction** within the **LCAO Approximation**.

SCF. Self Consistent Field. An iterative procedure whereby a one-electron orbital is determined under the influence of a potential made up of all the other electrons. Iteration continues until self consistency. **Hartree-Fock, Density Functional** and **MP2 Models** all employ SCF procedures.

Schrödinger Equation. The quantum mechanical equation which accounts for the motions of nuclei and electrons in atomic and molecular systems.

Self Consistent Field; *See* **SCF**.

Semi-Empirical Models. **Quantum Mechanics** methods that seek approximate solutions to the many electron **Schrödinger Equation**, but which involve empirical parameters.

Single-Determinant Wavefunction; *See* **Hartree-Fock Wavefunction**.

Size Consistent. Methods for which the total error in the calculated energy is more or less proportional to the (molecular) size. **Hartree-Fock** and **Møller-Plesset** models are size consistent, while **Density Functional Models**, (limited) **Configuration Interaction Models** and **Semi-Empirical Models** are not size consistent.

Size Surface. An **Isodensity Surface** used to establish overall molecular size and shape. The value of the density is typically taken as 0.002 electrons/**bohr^3**.

Slater. A function of the form $x^l y^m z^n \exp(-\zeta r)$ where l, m, n are integers (0, 1, 2 . . .) and ζ is a constant. Related to the exact solutions to the **Schrödinger Equation** for the hydrogen atom. Used as **Basis Functions** in **Semi-Empirical Models**.

Slater Determinant; *See* **Hartree-Fock Wavefunction**

Slater Function. *See* **Slater**.

Space-Filling Model; *See* **CPK Model**.

Spin Density. The difference in the number of electrons per unit volume of "up" spin and "down" spin at a point in space.

Spin Density Map. A graph that shows the value of the **Spin Density** on an **Isodensity Surface** corresponding to a **van der Waals Surface**.

Spin Multiplicity; *See* **Multiplicity**.

Spin Orbital. The form of **Wavefunction** resulting from application of the **Hartree-Fock Approximation** to the **Electronic Schrödinger Equation**. Comprises a space part (**Molecular Orbital**) and one of two possible spin parts ("spin-up" and "spin-down").

Split-Valence Basis Set. A **Basis Set** in which the **Core** is represented by a single set of **Basis Functions** (a **Minimal Basis Set**) and the **Valence** is represented by two or more sets of **Basis Functions**. This allows for description of aspherical atomic environments in molecules. **3-21G**, **6-31G** and **6-311G** are split-valence basis sets.

Stationary Point. A point on a **Potential Energy Surface** for which all energy first derivatives with respect to the coordinates are zero. **Local Minima** and **Transition States** are stationary points.

STO-3G. A **Minimal Basis Set**. Each atomic orbital is written in terms of a sum of three **Gaussian** functions taken as best fits to **Slater**-type (exponential) functions.

SVWN Model. (Slater, Vosko, Wilk, Nusair) A **Density Functional Model** which involves the **Local Density Approximation**.

SYBYL. A **Molecular Mechanics Force Field** developed by Tripos, Inc.

Theoretical Model. A "recipe" leading from the **Schrödinger Equation** to a general computational scheme. A theoretical models needs to be unique and well defined and, to the maximum extent possible, be unbiased by preconceived ideas. It should lead to **Potential Energy Surfaces** which are continuous. It is also desirable (but not required) that a theoretical model be **Size Consistent** and

Variational. Finally, a theoretical model will be applicable for the problems of interest.

Theoretical Model Chemistry. The set of results following from application of a particular **Theoretical Model**.

Thermodynamically-Controlled Reaction. A chemical reaction which has gone all the way to completion, and the ratio of different possible products is related to their thermochemical stabilities according to the **Boltzmann Equation**.

Thermodynamic Product. The product of a reaction which is under **Thermodynamic Control**.

Total Electron Density; *See* **Electron Density**.

Transition State. A **Stationary Point** on a **Potential Energy Surface** in which all but one of the elements in the diagonal representation of the **Hessian** are positive, and one element is negative. Corresponds to the highest-energy point on the **Reaction Coordinate**.

Transition-State Geometry. The geometry (bond lengths and angles) of a **Transition State**.

Transition State Theory. The notion that all molecules react through a single well-defined **Transition State**.

Transition Structure; *See* **Transition-State Geometry**.

Unrestricted SCF. An **SCF** procedure which does not restrict electrons to be paired on orbitals.

Valence. Electrons which are delocalized throughout the molecule and participate in chemical bonding ($2s$, $2p_x$, $2p_y$, $2p_z$ for first-row elements, $3s$, $3p_x$, $3p_y$, $3p_z$ for second-row elements, etc.).

van der Waals Interactions. Interactions which account for short-range repulsion of non-bonded atoms as well as for weak long-range attraction.

van der Waals Radius. The radius of an atom (in a molecule), which is intended to reflect its overall size.

van der Waals Surface. A surface formed by a set of interpreting spheres (atoms) with specific **van der Waals radii**, and which is intended to represent overall molecular size and shape.

Variational. Methods for which the calculated energy represents an upper bound to the exact (experimental) energy. **Hartree-Fock** and **Configuration Interaction Models** are variational while **Møller-Plesset Models**, **Density Functional Models** and **Semi-Empirical Models** are not variational.

Vibrational Analysis; *See* **Normal Mode Analysis.**

VWN; *See* **Local Density Models, SVWN Models.**

Vibrational Frequencies. The energies at which molecules vibrate. Vibrational frequencies correspond to the peaks in an infrared and Raman spectrum.

Wavefunction. The solution of the **Schrödinger Equation**. In the case of the hydrogen atom, a function of the coordinates which describes the motion of the electron as fully as possible. In the case of a many-electron system a function which describes the motion of the individual electrons.

Woodward-Hoffmann Rules. A set of rules based on the symmetry of **Frontier Molecular Orbitals** on interacting molecules which indicate whether a particular reaction will be favorable or unfavorable.

Zero Point Energy. The energy of molecular vibration at 0K, given as half the sum of the **Vibrational Frequencies** times Planck's constant.

Z Matrix. A set of internal coordinates (bond lengths, bond angles and dihedral angles only) describing molecular geometry. Of historical interest only.

Zwitterion. A neutral valence structure which incorporates both a formal positive charge and a formal negative charge.

Index

Page numbers in **bold type** refer to page numbers of tables or figures.

Q

R

Index of Tables

Index of Figures

Energies of structural isomers